HUMAN BODY
from
A TO Z

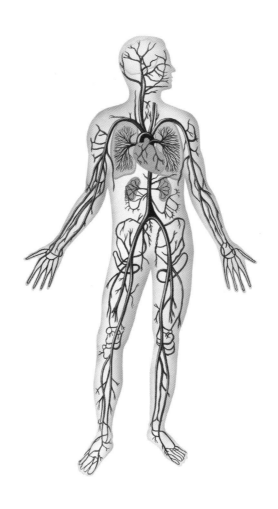

 **Marshall Cavendish**
Reference
New York

Other Marshall Cavendish Offices:
Marshall Cavendish International (Asia) Private Limited, 1 New Industrial Road, Singapore 536196 • Marshall Cavendish International (Thailand) Co Ltd. 253 Asoke, 12th Flr, Sukhumvit 21 Road, Klongtoey Nua, Wattana, Bangkok 10110, Thailand • Marshall Cavendish (Malaysia) Sdn Bhd, Times Subang, Lot 46, Subang Hi-Tech Industrial Park, Batu Tiga, 40000 Shah Alam, Selangor Darul Ehsan, Malaysia.

Marshall Cavendish is a trademark of Times Publishing Limited.

All websites were available and accurate when this book was sent to press.

Library of Congress Cataloging-in-Publication Data

Human body from A to Z.
       p. cm.
Includes bibliographical references and index.
  ISBN 978-0-7614-7946-8 (alk. paper)
1. Human physiology--Popular works. 2.   Human anatomy--Popular works.
I. Marshall Cavendish Reference.
QP38.H835 2012
612--dc22   2011006783

Printed in Malaysia

15 14 13 12 11 1 2 3 4 5

Marshall Cavendish

Publisher: Paul Bernabeo
Project Editor: Brian Kinsey
Production Manager: Michael Esposito
Indexer: Cynthia Crippen, AEIOU, Inc.

Medical Consultant

Rita Washko, MD MPH
Westat Research Corporation
Rockville, Maryland

CONTENTS

INTRODUCTION

The human body is an amazing organism. When all the parts are working well, the person barely notices how the body functions. However, when some element stops working, there is immediate cause for concern. For example, a person rarely worries about the shoulders until they are injured or suddenly become a source of pain. When the body

malfunctions, the individual will obviously have questions. *Human Body from A to Z* is a comprehensive source of answers to questions about the human organism.

The 168 articles in this volume provide authoritative information on all the elements, large and small, that make up the human body. Subjects range from major body systems, such as the circulatory system and the digestive system, to the components that make up those systems, such as arteries and the stomach, to physical features of the human body, such as birthmarks and blushing.

This volume provides students and the general public with valuable information that is concise and accessible. It offers clear presentations of medical concerns, allowing for immediate comprehension of general health issues related to the human body. The volume's easy and attractive style will help readers understand medical terms and issues and thereby assist in maintaining an optimal body condition and preventing the occurrence of problems. The substantial level of medical knowledge contained here will allow users to dispel misunderstandings and apprehensions that are associated with health care and the human body.

The content in these articles can be accessed in a variety of ways because of their structured organization, cross-referencing, and the simple A-Z format. Therefore, it is easy for the reader to find reliable information from a trusted source. Valuable information is also conveyed through numerous photographs, charts, graphs, and artworks with clear descriptive captions. The immediately accessible questions-and-answers feature that appears in every article addresses issues that are most likely to be of concern to readers. At the end of the volume is a list of resources for further study, as well as an index.

While *Human Body from A to Z* is not a substitute for obtaining advice and treatment from a licensed medical practitioner, the knowledge offered in this reference work will help promote good health and the proper maintenance of the human body. The more a person knows about human anatomy and physiology, the better prepared that person will be to prevent illness and maintain a healthy body and mind.

Additional health care information is available in the single-volume *Family Health from A to Z*, the 18-volume set *Encyclopedia of Family Health*, and the online *Family Health* database at www.marshallcavendishdigital.com.

Abdomen

When I auditioned for the school play my stomach rumbled the whole time. How can I stop this from happening again?

Your stomach rumbles all the time as the intestine churns and moves food along, but usually it can be heard only through a stethoscope. Louder rumblings happen when the intestine is empty or contains too much air. When people are nervous, they often swallow air without realizing it. Next time, have a good meal beforehand, try to relax, and chew a candy to help keep you from swallowing air.

My little sister often says she has a stomachache on Monday mornings before school. Is she just faking?

She probably does have a real pain in her abdomen, caused not by some sort of physical problem, but by nerves. On Monday mornings your sister has a whole week of school to worry about; after a day in school, things may not seem so bad. Alternatively, she may be being bullied, or she may be afraid of one of her teachers. Once you get to the root of the problem, talk through her worries, and the trouble should disappear.

My baby son has a huge abdomen, like one of those starving children seen on television. Is this normal?

Yes, it is. A baby's abdomen looks swollen because the liver is very large—it has to be, to do many of the jobs essential for growth. Also, the abdominal muscles are not very strong yet. By the time your son is five, he will lose this potbellied look. Starving children have swollen abdomens whatever their age. This is a part of kwashiorkor, a disease in which the child gets no dietary fats or proteins and has to eat a lot of low-calorie cereal carbohydrates. These fill up the abdomen and also produce gas.

The abdomen contains most of the digestive system, the urinary system, and, in women, the reproductive organs. Pains in this area have a variety of causes, but any sudden, severe abdominal pain requires immediate medical attention.

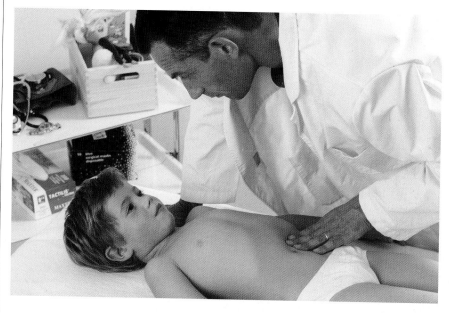

▲ *Abdominal pain in a child may be serious—if in doubt, take the child to a doctor.*

The abdomen is the largest cavity in the body, extending from underneath the diaphragm—the sheet of muscle that forms the lower boundary of the chest—down to the groin. At the back of the abdomen is the spine; around its upper sides are the ribs, and the front is covered by a thick sheet of muscle. This muscle is extremely elastic; for example, it must stretch enormously in order to accommodate a growing baby when a woman is pregnant.

The abdomen contains many organs. Almost all of the alimentary canal lies inside the abdomen, starting with the stomach—sited just under the diaphragm—and ending with the rectum, which empties via the anus. The alimentary canal is the body's food-processing system; it breaks food down into substances that can be absorbed into the blood, and it ejects indigestible wastes. Connected to the alimentary canal are glands such as the liver, pancreas, and gallbladder. A huge network of blood vessels serves all the abdominal organs and nerves.

Behind the alimentary canal lie the two kidneys, each joined to the bladder by a tube called a ureter. The bladder, which stores urine before it is released, is in the lower part of the abdomen. Closely connected to the urinary system is the reproductive system. In females, nearly all the sex organs are inside the abdomen, but in males, part of the sex organs descend to a position outside the abdomen while the fetus is developing before birth.

Around 33 feet (10 m) or so of gut are coiled and twisted to fit neatly inside the abdomen. To keep it all in place, the abdomen is lined with a tissue sac called the peritoneum. The organs, often called collectively the viscera, are attached to the rear abdominal wall by sheets or strings of tissue known as mesenteries.

When something goes wrong

Because people cannot feel the abdominal organs at work, it is hard to know when something is wrong, except when another part of the body is affected. For example, the pain of a stomach ulcer is felt not in the stomach itself but in nerve signals reaching the upper part of the abdomen.

Common causes of abdominal pain

POSITION	TYPE OF PAIN/SYMPTOMS	POSSIBLE CAUSES	WHAT TO DO
Top center, behind breastbone	Severe discomfort or burning sensation. Nausea and headache. Heavy feeling in abdomen, which may be swollen. Gas.	Commonly, acid reflux (heartburn) due to an excess of rich or spicy food, alcohol, or nicotine. Less likely: peptic ulcer or chronic gastritis.	Take antacids, plenty of fluids, and eat a bland diet. Cut down on drinking and smoking. Avoid stress. See a doctor if pain continues.
Top center, moving through to back	Colicky, persistent, or intermittent. Upper abdomen may feel full and be swollen. May be relieved by food.	Peptic ulcer or inflammation of the pancreas.	Take antacids for temporary relief, plus small meals with milk, but see your doctor within a week.
Top center, behind ribs, moving to right	Constant, may be made worse by fatty foods. Heartburn. Abdomen feels very full. Possible jaundice.	Inflamed gallbladder, inflammation of liver (hepatitis).	See your doctor right away.
Top center, shooting to right shoulder	Agonizing, with sweating, nausea, and vomiting.	Trapped gallstones (biliary colic).	See your doctor immediately. Take painkillers for temporary relief.
Center, around or above navel	Persistent, may be burning, cramping, or gripping, often with vomiting and/or diarrhea. Abdomen swollen.	Food poisoning, gastritis, or another infection of the intestine.	Take plenty of fluid and a kaolin mixture. No food for 24 hours. See doctor if symptoms continue.
Center or in either loin, may shoot down to groin; burning on urination	Colicky or persistent, may be worse when the person is moving; may be accompanied by vomiting.	Kidney infection or stone (renal colic).	See your doctor immediately if pain is very severe.
Center, around navel, moving to lower right	Persistent, may be accompanied by flatulence, nausea, and vomiting. Constipation.	Appendicitis.	See your doctor immediately if pain is severe. Do not take laxatives or antacids.
Center, above, around, or just below navel, right or left	Gripping. Abdomen may be swollen. Hard stools may alternate with diarrhea.	Constipation, irritable colon, or diverticulitis; early appendicitis.	Add more bran and roughage to your diet. See your doctor if symptoms continue.
Center, around, and below navel	Cramping, with frequent vomiting and constipation. Abdomen swollen.	Intestinal obstruction.	See your doctor immediately.
Center and below navel	Gripping, recurrent. Abdomen may be swollen.	Inflammation of the colon, miscarriage.	See your doctor right away.
Center and below navel (women only)	Continual, occurring during or just before menstruation. Abdomen may be swollen.	Premenstrual tension, period pain.	Take painkillers, but see your doctor if symptoms continue or become severe.
Below navel, right or left (women only)	Continuous dull ache, worse during periods. Lower abdomen may be swollen.	Congestion in area of the uterus, inflammation of pelvic or sex organs, pelvic inflammatory disease (PID).	See your doctor within a week.
Below navel, right or left (women only)	Colicky, recurrent, with scanty or missed periods.	Ectopic pregnancy (fetus lodged in the fallopian tubes).	See your doctor immediately.
Very low, center, may move down into groin	Constant, may be worse as bladder fills. Burning sensation during urination. Urine may be cloudy, darker in color, or contain blood.	Cystitis or urinary infection.	See your doctor right away. Drink plenty of fluids.
All over abdomen	Agonizing, very sudden, very tender all over abdomen. Person is generally very ill, possibly collapsed.	Peritonitis from perforated peptic ulcer, perforated bowel, or burst appendix. Ruptured aorta.	Get to hospital urgently.

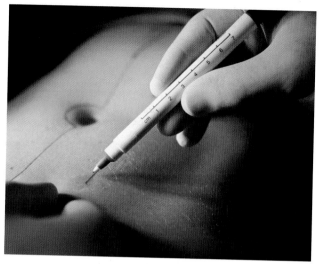

▲ *A surgeon marks a woman's abdomen before surgery.*

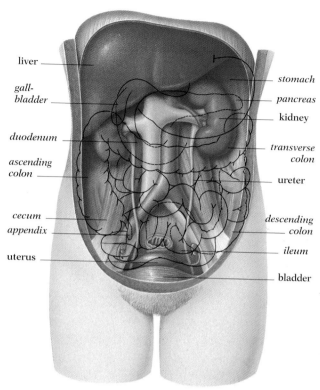

INTESTINAL ORGANS (IN A WOMAN)

liver

gall-
bladder

duodenum

ascending
colon

cecum

appendix

uterus

stomach

pancreas

kidney

transverse
colon

ureter

descending
colon

ileum

bladder

▲ *In the diagram above, the annotations in italics refer to the organs drawn in outline; these organs lie in front of the organs shown in color.*

Pains in the abdomen vary in position, type, and intensity according to the cause of the trouble and are in themselves good clues for a doctor's diagnosis. The pains are usually accompanied by other symptoms—nausea, vomiting, or diarrhea, for example, in the case of food poisoning. Diseases of the abdominal organs can also be detected by aches and pains in other parts of the body, such as in the back and shoulders.

Symptoms linked with reproduction

Many girls and women are worried about an uncomfortable low abdominal pain that they get midway between periods. This is completely normal—it is not worth calling a doctor unless the pain lasts for more than 24 hours or is accompanied by bleeding.

This type of pain is caused by an increase in tension in the ovary as the egg is released halfway between periods. Many women do not suffer this pain at all, and for others it is mild. As it is a sign that an egg is being released, this pain is an accurate indication of the time a woman is most likely to get pregnant.

Also connected with reproduction, but more unusual, is the set of symptoms known as sympathetic pregnancy. Sometimes it is possible for a man whose partner is pregnant to feel some of the same symptoms as a pregnant woman. The condition is caused by extreme anxiety; the man may have weight gain, and sometimes even morning sickness and food cravings. After the baby's birth, his symptoms usually go away.

When to get help

Abdominal pains are something that everyone has experienced, but it is important to remember that although they often have a minor cause—such as overeating, constipation, or a mild "stomach bug"—such symptoms can be a sign that something is very wrong. Abdominal pain should always be taken seriously and should not just be covered up with painkillers; these may hide important symptoms that require a doctor's immediate attention.

For everyday complaints, people should wait 48 hours before seeking medical advice. However, with severe, sudden abdominal pain, they should never just put up with it or take painkillers or laxatives. If an intense pain lasts more than a few hours, especially if it is accompanied by a swollen abdomen that feels tender to the touch or by blood or tarry substances in the stool, the sufferer should see a doctor as soon as possible. This is very important, because it could be the sign of serious illness. It is always better to catch and treat an illness or disease in the early stages.

Treating children's pains

The same rules apply if children have pains in the abdominal area; however, if parents are in doubt, they should always ask a doctor. In children, abdominal pain may also be the result of problems that are uncommon in adults; for example, tonsillitis, middle-ear infections, and lead poisoning can all result in symptoms of stomach upset.

Anxiety is another cause of stomachache in both children and adults. The pain results from the tendency of the body to intensify the squeezing action of the intestine during times of stress. It is best for people not to eat too much during stressful times, because food is hard to digest if the abdomen is tightened into "knots." A hot drink may bring some relief; a hot water bottle placed against the stomach may also help. During stressful events, it is best to eat a light diet.

See also: Alimentary canal; Colon; Digestive system; Menstruation; Stomach

Acid-base balance

Acid-base balance is essential for the normal functioning of the body so that its internal environment does not become too acidic or too alkaline. The body has effective mechanisms to maintain this balance.

▲ *Blood is spun in a centrifuge to separate it into red cells and plasma.*

Blood and other bodily fluids are slightly alkaline. The degree of alkalinity of fluids is expressed on the pH scale (a logarithmic scale of the hydrogen ion concentration); strongly acidic fluids have a low pH of about 1, whereas strong bases or alkalis have a high pH of around 14. A neutral pH is 7, above 7 is basic or alkaline, and below 7 is acidic. The stability of the acid-base balance in the body is important because blood has a pH range of 7.35 to 7.45, which must be maintained, since any deviation can affect the acid-base balance and severely affect the body's organs.

There are three ways that acid-base balance or blood pH is maintained. The first two are metabolic; the third is respiratory. First, any excess acid is removed via the kidneys. Second, buffers are used that are able to balance each other. These are bicarbonate, a base buffer; and carbon dioxide, an acidic buffer. If more acid enters the bloodstream, the body produces more bicarbonate and less carbon dioxide to balance the pH. Conversely, if more base enters the bloodstream, the body produces more carbon dioxide and less bicarbonate. Third, the respiratory method is by excretion of carbon dioxide, which is produced by oxygen metabolism in the cells. Carbon dioxide is expelled from the bloodstream via the lungs, where it is exhaled. If breathing is deep and fast, too much carbon dioxide is exhaled and the blood becomes basic or alkaline. If breathing is inadequate, carbon dioxide builds up in the blood, making the pH more acidic. The blood pH can be regulated by adjusting the depth and rate of breathing.

Acidosis and alkalosis

These conditions can be metabolic or respiratory, depending on the cause. Generally metabolic acidosis is too much acidity in the blood with a correspondingly low level of bicarbonate

blood oxygen levels, fever, or aspirin overdose. The symptoms are obvious anxiety and tingling around the face, followed by muscle spasm and a feeling of disorientation. Diagnosis is made by measuring blood pH from arterial blood. The pH will be elevated.

Effective treatment is to slow the rate of breathing. If it is caused by anxiety, calming patients and getting them to make an effort to slow their breathing may relieve the condition; alternatively, they can be made to hold the breath as long as they can, then take a shallow breath and again hold it as long as possible six to 10 times.

For anxiety, a simple remedy is for affected persons to breathe into a paper bag and thus breathe back in the carbon dioxide that they have breathed out. If pain is the problem, relieving the pain usually corrects the condition. When the levels of carbon dioxide in the blood rise, the symptoms disappear, and the attack stops.

Metabolic acidosis

As the blood pH is lowered, breathing becomes faster and deeper in an attempt to remove excess acid by reducing the amount of carbon dioxide in the blood. The kidneys also try to excrete more acid in the urine. If the body continues to produce more acid, the compensatory systems can be overwhelmed, resulting in severe acidosis and eventually coma.

Metabolic acidosis can be caused by ingesting acid; by the production of excess acid through metabolism, for example in poorly controlled diabetes (when the body breaks down fats and produces acids called ketones); by kidney malfunction; or by kidney failure, which reduces the kidneys' ability to excrete acid.

The first mild symptoms of metabolic acidosis are nausea, vomiting, and fatigue. If the condition does not improve through the efforts of the compensatory mechanisms or through treatment, the patient may become weak, sleepy, confused, and more nauseated. Eventually blood pressure falls; shock, coma, and death follow. The condition is diagnosed by measuring blood pH from a sample of arterial blood (rather than venous blood). The levels of carbon dioxide and bicarbonate are also measured, and if appropriate, blood sugar levels and ketones are determined.

Treatment depends on the reasons for the acidosis. Uncontrolled diabetes is treated with insulin and intravenous fluids, whereas if the cause is poisoning, the substance may be removed from the blood in some cases, or treated with an antidote.

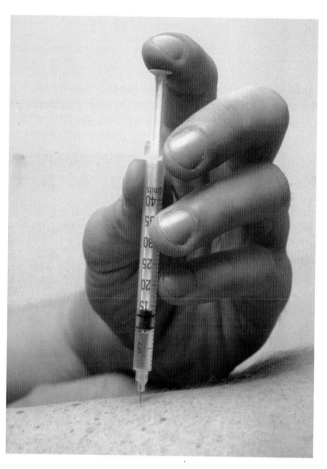

▲ *A diabetic injects insulin to prevent high levels of blood sugar, which cause acidosis.*

whereas in the condition of metabolic alkalosis the blood is alkaline with a high level of bicarbonate. Respiratory acidosis is caused by a buildup of carbon dioxide in the blood due to inadequate breathing or because the lungs are not working properly. Respiratory alkalosis means that the blood is too alkaline because fast or very deep breathing causes a low carbon dioxide level.

Respiratory acidosis

If the blood is acidic with high levels of carbon dioxide, the brain is stimulated to produce faster and deeper breathing. If someone is suffering from a disease such as emphysema, or the muscles and nerves of the lungs are impaired, the lungs cannot expel carbon dioxide efficiently. An overdose with certain drugs can also cause this problem. The symptoms of respiratory acidosis can be drowsiness and headache followed by coma. The kidneys attempt to compensate by retaining bicarbonate, but this process can be lengthy. A diagnosis is made by measuring arterial blood pH and carbon dioxide. Treatment is to improve lung function, by either drugs to aid breathing or, in serious cases, a ventilator.

Respiratory alkalosis

Hyperventilation or rapid, deep breathing can cause respiratory alkalosis. This condition can be caused by acute anxiety, pain, low

Metabolic alkalosis

The body can lose stomach acid during severe vomiting; or, rarely, the blood becomes alkaline owing to ingestion of too much alkali, such as bicarbonate of soda. Alkalosis can also develop if sodium and potassium are lost; this loss affects the kidney's function of controlling acid-base balance.

The symptoms are muscle cramps, twitching, and irritability. If the condition is severe, muscle spasm and contraction, known as tetany, can develop. Diagnosis is confirmed by measuring blood pH from arterial blood or measuring bicarbonate in venous blood.

Treatment depends on the cause but may involve the replacement of water, as well as sodium and potassium. If alkalosis is very severe, a dilute acid, such as ammonium chloride, may be given intravenously.

> *See also:* **Breathing;**
> **Homeostasis; Ketones**

Adenoids

My sister breathes through her mouth most of the time. Does this mean she has adenoid trouble?

If she also has a nasal discharge and a nighttime cough, adenoids are probably the cause. Otherwise, it may be a habit; this is best cured by regular nose-blowing. Also, to make sure there is nothing lodged in her nose, check her ability to breathe through each nostril in turn. A few children are born with bone abnormalities that prevent normal breathing through the nose.

Can you diagnose adenoid trouble just by looking at a child's face?

No. "Adenoidal faces"—a snub nose, a high arched palate, and protuberant upper teeth—have also been found in connection with many other conditions.

If my brother has his adenoids operated on, can they regrow?

Yes. They can regrow because it is surgically impossible to remove all of the tissue. The operation leaves a small stump, which can grow large again. However, as the stimulus for adenoids to grow disappears around the age of six, regrowth seldom causes a problem.

How long is it safe to use decongestant nose drops?

Decongestant drops will dry up a runny nose. However, if the cause of the problem is not dealt with, the discharge will return when you stop using the drops.
  A runny nose caused by adenoids should improve after a week or so; stop using the drops then. A slight discharge may return for a day or two, but do not resort to the drops. If drops are used for more than three weeks, there is a risk that the nose will counterbalance their effect by running all the time.

The adenoids have been called "the watchdogs of the throat." In common with the tonsils, they guard against respiratory infection in young children; however, they sometimes become infected and swollen themselves.

Adenoids are small masses of lymph tissue located at the back of the nose, where the air passages join those of the back of the mouth. The lymph system is the body's defense against infection, and the lymph nodes, including the adenoids, are full of infection-fighting cells, the lymphocytes. The adenoids are placed so that any germs breathed in through the nose may be trapped by them and—it is hoped—killed.
  Adenoids are present from birth but usually disappear before puberty. They are most obvious between the ages of one and four; it is during these years that a child is continually exposed to new types of infection and viruses.
  Germs produce toxins that can kill cells, even those whose function it is to attack the germs. Inflammation then attracts more immune system cells to the site of the infection. If the adenoids become damaged, chronic infection may

Some common symptoms of swollen adenoids
Hearing difficulties.
Dry gums, which result in tooth decay and halitosis.
Loss of smell, taste, and appetite.
Listless expression, because breathing is such hard work.

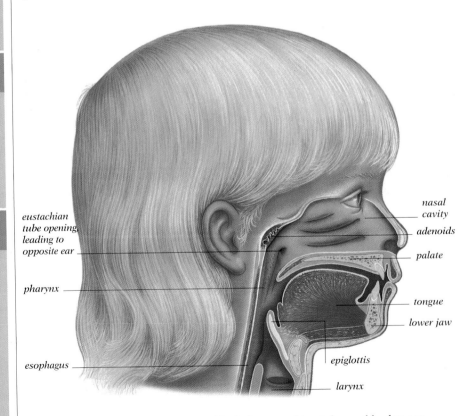

▲ *Inflamed and swollen adenoids can block the eustachian tubes and lead to ear infections and even deafness.*

Questions and Answers

Are sneezing and a runny nose always due to adenoids?

Usually not. These symptoms suggest an allergic nose problem. Research has shown that many childhood illnesses are due to allergies, often caused by dust and pollen. Treatment with decongestants and antihistamines is the same as for adenoids; an adenoidectomy would not help.

Is there a link between adenoids and mental deficiency?

Not at all. Though a child with an adenoid problem may look dull and be slow to respond, studies have not shown any such link. However, if swollen adenoids are left untreated, the complication of deafness can prevent a child from living up to his or her potential.

My sister snores heavily. Could enlarged adenoids be the cause?

Snoring can certainly be due to enlarged adenoids; however, by itself, this symptom does not need treating. Otherwise, snoring could be due to an abnormality of the facial structure, present since birth. Many children breathe noisily at night; this is not true snoring, but a result of a catarrhal complaint such as a cold.

Can antihistamines and antibiotics do anything to help treat a child with adenoid problems?

Antihistamines reduce the swelling of the adenoids when the problem has been caused by an allergic reaction rather than by an infection. They also tend to reduce the discharge. A further benefit is that antihistamines usually cause drowsiness and combat nausea, so a bedtime dose will reduce any symptoms of coughing and morning vomiting.
 Antibiotics are beneficial for enlarged adenoids only when there is also a bacterial infection present. Adenoidal symptoms may remain in such cases, but painful complications, such as ear infections, can be prevented.

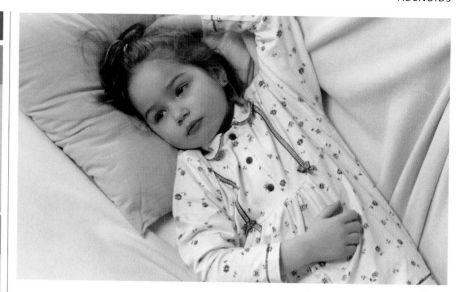

▲ *Although the adenoids defend against disease, they can also cause a child to feel unwell.*

set in. Also, if they become inflamed over and over again, the adenoids tend to swell; this can also result in the child becoming ill.

Symptoms

If the adenoids become swollen by an infection, they interfere with the flow of air through the nose so that the child has to breathe through his or her mouth. This may cause heavy snoring at night. Swollen adenoids also create a nasal tone in speech. For normal articulation of sounds, the cavities of the nose and mouth must be normal and the passage of air unobstructed. The effect of swollen adenoids is similar to that caused by swelling of the nose lining in the course of a cold. Breathing through the mouth also makes it very dry, and the child may feel continually thirsty. As the adenoids fight off infection, lymphocytes—both dead and alive—are released in the form of pus. The pus will be seen as a discharge from the nose; this is very different from the clear, watery discharge seen with a cold. The child sniffs to try to clear the discharge, but it then runs down the back of his or her throat, leading to coughing. The cough is particularly obvious at night and is a typical sign of infected adenoids. In the morning, the swallowed pus may cause vomiting.

Dangers

Swollen adenoids can block the eustachian tubes, which run through the skull bones from the middle ear to the pharynx. Their function is to equalize the pressure inside the middle ear with the pressure outside. If the tubes are blocked by enlarged adenoids, the pressure cannot be balanced. The main hazard is that secretions in the ear cannot drain from the inside. This causes "glue ear," in which the hearing apparatus becomes stuck together by secretions. The secretions in turn may become infected, causing a condition called otitis media. This is painful and can affect a child's hearing permanently. If it is left untreated, the eardrums usually burst to release the fluid.

Treatment and outlook

Decongestants, antihistamines, and antibiotics can all help in treating adenoid problems. The first two can be bought over the counter; antibiotics are available by prescription.
 As a last resort, the adenoids can be removed by an operation called an adenoidectomy. The procedure is fairly simple and is carried out under a general anesthetic in a hospital. The tonsils may also be removed in the same operation.
 The decline in infectious diseases means that adenoids are not the problem they once were. Provided that serious complications do not occur, symptoms often go away on their own when the child is about six years old.

> *See also:* Hearing; Lymphatic system; Lymphocytes; Tonsils

Adrenal glands

Questions and Answers

My father suffers from high blood pressure when he gets excited, and he says it is due to his adrenals. Could this be true?

Raised blood pressure is one of the symptoms of certain rare adrenal diseases, but such a disease is unlikely to be the cause of your father's problem. Other factors, such as heavy smoking, excess weight, and stress, are much more likely to be the cause of high blood pressure. If your father is under constant pressure at work, or if he has an excitable temperament, it could be that his adrenals are continually being required to produce adrenaline—the hormone that helps us to cope in emergencies—and thus his body is not being allowed to return to normal, so the excessive adrenaline in his system could, in time, have led to high blood pressure. This does not mean his adrenals are at fault.

My mother has to have her adrenals removed. Will she still be able to lead a normal life?

Yes. Patients who have their adrenals removed for medical reasons are able to lead a perfectly normal life by taking the drug cortisone regularly.

I am overweight and have tried dieting and exercise without success. Is there a problem with my adrenal glands?

If your excess weight is distributed evenly, the answer is no. There is only one disease of the adrenals which gives rise to obesity—Cushing's syndrome—and this is extremely rare. It is easy to spot because there is an uneven distribution of fat on the body; the arms and legs remain thin, and fat is concentrated on the chest and abdomen. If you really want to find the cause of the problem, talk to your doctor.

The adrenal glands produce hormones that affect vital body functions such as breathing, metabolism, and circulation. The adrenals are also responsible for releasing a hormone that increases the heart rate in response to stress.

The adrenal glands—known as the adrenals—are located immediately above the kidneys, where they sit on cushions of fat above each kidney. Each adrenal gland consists of two distinct parts: the inner medulla and the outer covering, called the cortex. These parts secrete different hormones, each of which has an entirely separate function.

The adrenal medulla

The medulla, or core, of the adrenals is the part of the gland that secretes adrenaline and its close relative, noradrenaline. Together these are known as the "fight or flight" hormones, because they

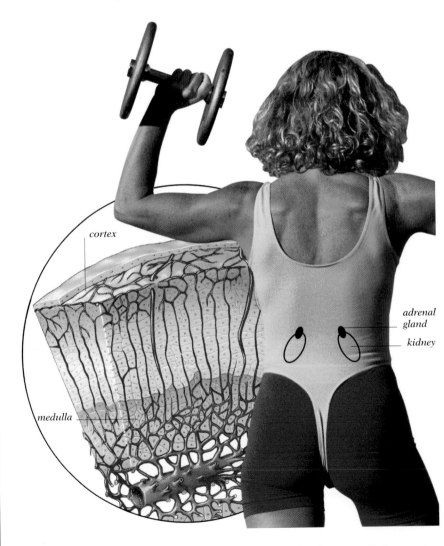

cortex

medulla

adrenal gland

kidney

▲ *The adrenal glands are located just above the kidneys. They have two distinct parts (see cross section), which perform different functions.*

Hormones and their uses

SOURCE	HORMONE	FUNCTIONS	WHEN SYNTHETIC HORMONES ARE USED AS DRUGS
Adrenal medulla	Adrenaline	Prepares the body for physical action	To treat very severe asthma In severe allergic collapse During some surgical procedures (such as suturing the skin), mixed with local anesthetic
	Noradrenaline	Maintains even blood pressure	
Adrenal cortex	Aldosterone	Regulates excretion of salt by kidney Keeps balance of salt (sodium) and potassium Plays a part in the body's use of carbohydrates	As replacement therapy
	Cortisol	Stimulates manufacture and storage of energy-giving glucose Reduces inflammation Regulates distribution of fat in the body	As replacement therapy when adrenals are missing or defective As replacement therapy when pituitary is defective For shock after severe injuries or burns For severe allergic reactions For rheumatoid arthritis (either as pills or injected into painful joints) and related diseases For skin diseases such as eczema (as ointment) As anticancer treatment, especially when the lymphatic system is affected To prevent rejection after transplants
	Sex hormones	Supplement sex hormones secreted by gonads	To correct deficiencies For contraception To promote muscle and bone growth

prepare the body for the extra effort that is required to meet danger, cope with stress, or carry out a difficult task.

The adrenal medulla is unique among the hormone-producing glands known as the endocrine glands in that it influences the autonomic nervous system by releasing adrenaline and noradrenaline; these hormones increase heart rate and blood flow.

The dangers and stresses modern people face are as likely to be psychological as physical, but either way, the body has the same physical reaction. There is a surge in the production of adrenaline, which makes the heart beat faster and more strongly. This raises the blood pressure, while at the same time it constricts the blood vessels near the surface of the body and in the gut and redirects the flow of blood toward the heart, causing the skin to turn pale. It also turns glycogen stored in the liver and muscles into glucose, which is used for extra energy. When the danger is over or the stress is removed, adrenaline production is reduced, and the body returns to normal. However, if the danger or stress is constant, or if a person is continually overexcited or under constant pressure, the body remains primed for action—and in time this can lead to stress-related conditions, such as high blood pressure.

The adrenal cortex

Wrapped around the adrenal core, the adrenal cortex secretes three types of hormones known as steroids, each one performing a very different function. Two of the most important steroid hormones are aldosterone and cortisol.

Aldosterone: This hormone affects the kidneys by helping them to regulate salt excretion and to reduce the amount of salt being lost in the urine, and in turn helps to maintain blood pressure. Salt determines the volume of blood in circulation, which then affects the heart's efficiency as a pump. Every molecule of salt in the body is accompanied by a large number of water molecules. Therefore, in losing salt, the body loses even more water, and this reduces the volume and pressure of the circulating blood. As a result, the heart has difficulty pumping enough blood around the body. The secretion of aldosterone is controlled by the hormone renin, which is produced by the kidneys. The system works like a seesaw: when aldosterone is low, the kidneys produce renin and the hormone level rises; when it is too high, the kidneys reduce their level of activity, and the amount of hormone present in the blood returns to a normal level.

Cortisol: The hormone cortisol is released from the adrenals as a response to physical or emotional stress. Cortisol is also responsible

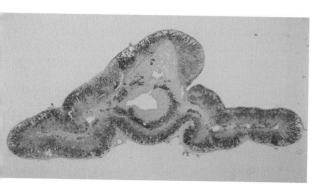

▲ *A cross section of an adrenal gland, showing the outer cortex and the inner medulla.*

EFFECTS OF ADRENALINE ON THE BODY

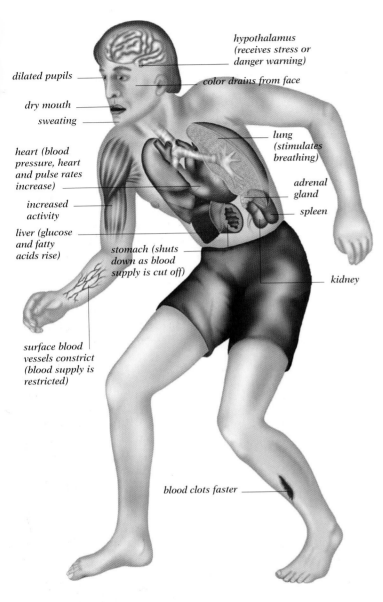

hypothalamus (receives stress or danger warning)

dilated pupils

color drains from face

dry mouth

sweating

lung (stimulates breathing)

heart (blood pressure, heart and pulse rates increase)

adrenal gland

increased activity

spleen

liver (glucose and fatty acids rise)

stomach (shuts down as blood supply is cut off)

kidney

surface blood vessels constrict (blood supply is restricted)

blood clots faster

medical treatment (for example, to prevent rejection after transplant surgery), the resistance to infection is reduced. Insufficient cortisol production also impairs immune function.

Sex hormones: The final group of hormones produced by the adrenals are male sex hormones, called androgens. These are secreted in small amounts by the adrenal cortex, and they complement the sex hormones produced in much larger quantities by the gonads, or sex glands. The principal male sex hormone—also present in women to a smaller degree—is testosterone, which is responsible for increasing the size of muscles. Steroid drugs are synthetic derivatives of the male sex hormones.

Control of cortisol

Cortisol is so crucial to body function that its secretion has to be strictly controlled. The gland that regulates its production—and that of the other steroids—is the pituitary, situated at the base of the brain. The pituitary is also known as the master gland. The pituitary secretes adrenocorticotropic hormone (ACTH), which stimulates cortisol production, and, as with the hormones renin and aldosterone, the two substances work by a feedback mechanism. When the cortisol level is too low, the pituitary secretes ACTH, and the level rises; and when the level is too high, the gland slows production, and the level of cortisol falls.

Cortisol as a drug

Cortisol is used as a replacement treatment in the debilitating hormonal disorder known as Addison's disease, from which President John F. Kennedy suffered. When this condition occurs, either the adrenal cortex does not produce sufficient cortisol of its own or the pituitary gland is defective. In Addison's disease, the adrenal glands may be destroyed by an autoimmune reaction, by cancer, or by infection.

Cortisol derivatives are also valuable for a number of other complaints, although they are best used as short-term remedies. Because they reduce inflammation, they are sometimes used to treat a painful form of arthritis called rheumatoid arthritis, and related rheumatoid diseases; for skin complaints such as eczema; and in certain allergic reactions, such as asthma or drug allergies. They are also used to overcome the body's natural immune reaction, so that the body will not reject foreign tissue in organ transplant operations, and they are used in combination with other drugs to treat cancers. Steroid therapy is not without its dangers, however. The most serious drawback is that the adrenal cortex is likely to stop producing its own cortisol when synthetic cortisol, or a related drug, is prescribed for any length of time.

Diseases of the adrenal glands are rare, but some occasionally do occur. However, with the availability of hormone replacement treatments, the more serious consequences can be avoided.

for increasing the flow of glucose, fat, and protein out of the tissues and into the bloodstream. Glucose is the body's principal fuel, and when extra energy is needed, for example in response to a threat or some type of stress, cortisol triggers the conversion of protein to glucose.

Many hormones act to boost the level of sugar in the blood, but cortisol is the most important. Insulin is a hormone produced by the pancreas that reduces the level of glucose in the blood by stimulating body cells to take up glucose. Insulin also stimulates the liver to store glucose in the form of glycogen, and when this storage capacity is exceeded, insulin converts excess glucose into fatty acids.

In addition to playing a key part in metabolism (the life-maintaining processes of the body), cortisol is also vital to the functioning of the immune system, the body's defense against illness and injury. If the normal level of cortisol is raised through

> **See also:** Hormones; Insulin; Pituitary gland

Aging

Questions and Answers

I am already a fairly heavy smoker at age 20. Will this affect my longevity?

Yes, it will. Many people underestimate the risk they are taking by smoking 25 cigarettes a day (even low-tar brands). Smoking will almost always speed the aging process; it can cause death from heart disease and diseases of the lungs and arteries before the age of 60. It affects the skin; people who smoke heavily often wrinkle prematurely.

Smoking is not just a minor hazard; it is a major cause of premature death.

I have an active and satisfactory sex life, and it worries me that I may lose interest as I grow older. Is this likely to happen?

Not necessarily. Libido does lessen with age, but there is no reason for sexual activity to disappear from your life. A decrease in sexual desire may be related to the physical changes that occur in the sexual organs with age. After menopause, the vagina becomes drier, and a woman may need to use lubricating jelly to allow easier penetration.

An older man may find that he is slower to become erect and ejaculate, but there is no reason that this should inhibit lovemaking in later life.

My grandmother has shrunk over the past two years. Why?

Height loss is mainly due to thinning of the spinal bones and shrinkage of the disks between the bones. Old people's backs tend to bow slightly, making them bend forward. Maintaining a correct posture in earlier life will help vertebrae to retain some strength, and this helps to reduce later height loss. A doctor can prescribe treatments to prevent osteoporosis.

Medical advances, together with improved living conditions, have increased life expectancy over the past two centuries, for both men and women, and helped to prepare people for a healthy and happy old age.

In the United States in the 19th century, a male child who survived the first year could expect to live for 48 years, and a female child for 50 years. In the 21st century, many people in the developed world live for longer than 70 or 80 years. Life expectancy has now reached 75 for men and 80 for women. The vast improvement in life expectancy is largely the result of public health measures. Better sanitation and an improved standard of living have eradicated many of the conditions that encourage disease, and mass immunization programs protect young children from diseases that were often killers in the past. New methods of treating some diseases, such as antibiotics, radiotherapy, and transplants, also help save lives. The increase in the number of older people has led to expansion in the field of geriatric medicine, the aim of which is to alleviate the diseases and disabilities of old age.

The causes of aging

The reasons for the variation of longevity in different people lie in a combination of factors; the genes people inherit from their parents (those vital parts of the cells that determine inherited characteristics) and their mother's health during pregnancy, for example, can both have an effect

▲ *Taking an interest in a grandchild's activities can help an elderly person keep an active mind and remain connected to the modern world.*

▲ It is important that older people who may otherwise feel isolated have directions in which they can channel their affection. A grandchild, or even a pet, can fulfill that need.

The facts about aging

Exposed skin loses elasticity, causing wrinkling, especially around the eyes and mouth.
Hair turns gray, then white, becoming sparse—baldness in men starts at between 20 and 30 years; this is a progressive condition.
Eyesight deteriorates and eyes are more prone to cataracts.
A loss of calcium makes bones brittle and liable to fracture. Spinal bones become thinner, resulting in height loss.
The loss of teeth, frequently caused by gum disease, alters facial structure.
The body's balance and gait alter with reduced height and stiffening joints, making the body's carriage less stable.
Smoking causes heart, circulatory, and lung problems.
Alcohol can damage the liver and nervous system.

on longevity. Cigarette smoking and drinking to excess can undermine health, as can living in poor conditions.

The peak of physical, mental, sexual, and social well-being thus varies from person to person and may be greatly influenced by a whole set of elements. For example, a successful businessman may be in his prime at 50 years of age and enjoy good health; but if he loses his job, his situation may change overnight—and the psychological effects of being out of work can be far-reaching, causing him to age faster than before.

The importance of heredity

A person who comes from a long-lived family will generally have a long life, despite any adverse conditions. Just as genetic inheritance has a part to play in determining longevity, it also affects the health and strength of individual organs. A good inheritance means that a person is more likely to have a strong heart, good circulation, a healthy brain, and sharp eyesight and hearing.

There is nothing anyone can do about the genes that people inherit, but individuals who have inherited "bad" genes might be able to live a reasonably long life if they take care of themselves and try to minimize the risk factors that cause, or at least contribute to, poor physical and mental health and aging.

Environment

Genetic inheritance can explain only some of the traits that appear to run in families. Many physical and mental patterns can be explained only by the shared environment of parents and their

children. Overweight mothers are more likely to have fat children—though this result is more likely due to the children's imitating their mother's eating habits rather than inheriting her genes. Similarly, those who suffer from bronchitis may have children who grow into bronchitic adults because their parents smoked, and there was enough smoke in the home environment to affect the children's lungs. Also, they are more likely to imitate their parents and smoke themselves, becoming more likely to develop cancer and emphysema.

The possibility of a short life—whether it is due to inherited factors or environmental factors—can be reduced if people ensure that they lead a healthy life, especially in later years. Eating the right balance of healthy foods, exercising regularly, drinking in moderation, cutting out smoking, and practicing routine safety precautions at home and at work can all help to increase the potential life span.

Cell aging

Common health problems that may increase with age are not, in fact, caused by the aging process itself—for example, arthritis is due to the same causes no matter what the individual's age. Certain conditions, however, are more likely to occur in later years, simply because of a general decline in the body's strength and resistance to infection. This happens as a result of cell aging. Most of the body cells are continually renewing themselves as they wear out. Exact replicas of the old cells are made by a process of cell division in which cells simply divide into two. The main exception to this rule is the cells of the brain; these cells stop dividing after birth.

Cell deterioration

As the body ages, the cells begin to deteriorate and function less efficiently. The explanation is partly general wear and tear. For example, the skin can be compared to a piece of elastic—when it stretched, it returns to its original shape. Over time, it loses its elasticity and remains permanently stretched. Various theories have been put forward to explain the loss of the body's cellular efficiency. One theory is that some of the body cells have a built-in life cycle. A "program" switches them on and off—an example can be seen in declining muscle strength. In old age, certain cells do switch off, causing muscle fibers to decrease in size, but recent research suggests that this decrease can be counteracted somewhat by regular exercise throughout middle age, and even later into old age.

Another theory is the Hayflick limit. According to this theory, human cells seem to have a predetermined number of cell divisions; once they have divided a certain number of times, they simply stop dividing. Even a small difference in the time between cell divisions could radically alter a person's life span.

A third theory is that cells gradually fail to reproduce themselves as accurately as before. This can lead either to the death of the cell or to the production of a whole series of cells that are not as efficient as the original. Another possibility is that cells in the elderly are less resistant to oxidant injury, which can lead to cell damage and cell death.

The reason for this limit on cell division has recently been discovered. At the end of each chromosome are lengths of spare DNA, called telomeres. With each cell division, the telomeres are shortened because some DNA is lost at each replication of the chromosomes. Chromosomal division can occur only if the telomere ends are present, and after 50 to 100 cell divisions some telomeres will have been completely lost. At that point, no further cell replication is possible.

Women outlive men by an average of five to six years. Various explanations have been given for this difference, but many of them

▲ *Regular exercise for older people, as long as it is not too strenuous, can help combat aging and enhance longevity.*

have little supporting evidence. Some people believe that women are under less stress than men, because women usually do not hold such demanding or highly responsible positions at work or in society. Others claim that women have to manage far more stress than men, because they deal with dual roles at home and at work, but that they learn to cope with it better.

The fact that until recently women smoked far less than men could certainly be a factor contributing to female's longevity. However, young women smoke far more today, and this could cause a radical change in their life expectancy.

It has been suggested that female hormones play a significant part in women's longer life span by protecting them in some way, but research in this area is still inconclusive. Contrary to this view, there is evidence that the drop in the level of female sex hormones after menopause can have the opposite effect. The lack of the hormone estrogen can hasten wrinkling and cause the vagina to become less lubricated. Also, the body's calcium balance can be affected, and this makes women more susceptible to osteoporosis (thinning of the bones). This condition makes bones more likely to fracture and can cause curvature of the spine. Hormone replacement therapy should be used with caution, as it carries long-term risks.

Acceptance of aging

In the past, older people occupied a respected place in society. People remained in their jobs until retirement, family members lived close to one another, and grandparents played a pivotal role in family life. However, this respect has gradually been undermined, with young people being considered more capable of dealing with newer, more scientific work methods, as well as being considered more creative. Families move apart more, and roles are no longer clear. One of the effects of these attitudes, together with the effects of recession, has been an increase in the number of people taking early retirement. This may make older people feel useless and unwanted, but it can also give them a positive chance to start life afresh and take up new interests.

How to stay healthy
Avoid exposure to the sun and wind. Moisturize and handle the skin gently.
Keep off certain drugs, like marijuana, in youth—research suggests that this could cause premature hair loss.
See your doctor regularly.
Osteoporosis in women following menopause can be largely prevented by calcium supplements, exercise, and treatment with drugs that slow down and reverse this process. A family doctor can advise about any risks that may be involved.
Have regular dental checkups, and maintain an effective oral hygiene routine.
Exercise regularly from youth to middle age. Older people should avoid strenuous sports, since such exertions can cause bones to crack.
Stop smoking.
Drink moderately.

See also: Cells and chromosomes; Hormones

Alimentary canal

The alimentary canal is a digestive tube extending from the mouth to the anus. Its task and that of certain other organs in the body is the physical and chemical breakdown of food to release energy for growth and repair.

Food is the fuel that powers the activities of the body. Before it can fulfill this role, it must be properly processed. The body's food-processing plant is the alimentary canal, a muscular tube about 30 feet (9 m) long that starts at the mouth and ends at the anus, from where undigested waste material is expelled as feces.

Chewing and swallowing

When food is put into the mouth, it is tested for taste and temperature by the tongue. Solid food is bitten off by the front teeth (incisors), then chewed by the back teeth, or molars. Even before the food is tasted, and during chewing, saliva pours into the mouth from salivary glands situated near the lower jaw. Saliva moistens food, and the enzymes it contains start the process of digestion. By the time food is ready to be swallowed, the original mouthful has been transformed into a soft ball, called a bolus, and warmed or cooled to the right temperature.

Swallowing is quick but fairly complex. First the tongue pushes the bolus of food up against the roof of the mouth and into the muscle-lined cavity at the back of the mouth, the pharynx. Once food is in the pharynx, several activities take place within the space of a couple of seconds to prevent swallowing from interfering with breathing. The soft palate, the nonbony part of the roof of the mouth, is pushed upward by the tongue to shut off the inner entrance to the nose. The vocal cords are quickly drawn together, and a flap of tissue called the epiglottis closes over the entrance to the tubes that lead to the lungs.

The esophagus

From the pharynx, the bolus passes into the esophagus, or gullet, the tube joining the mouth to the stomach. The bolus does not just fall down the esophagus because of the force of gravity, but is pushed along by waves of muscle action called peristalsis. Except during eating, the esophagus is kept closed by a ring of muscle called the cardiac sphincter just above where it enters the stomach. This prevents the highly acidic contents of the stomach from being regurgitated into the esophagus. As a bolus passes down through the esophagus, the sphincter relaxes enough to open the pathway into the stomach, which prepares for the arrival of food and liquids.

Digestion

The stomach is a collapsible muscular bag designed to store food that has been eaten (so that it is not usually necessary to eat small meals all day long), to mix food with various digestive juices, and to release the food slowly into the intestine.

Food is mixed as the stomach wall contracts and relaxes and is moved along by waves of peristalsis. By the time it has spent two to six hours being processed in the stomach, the partially digested food has been converted by various chemicals into a liquid called chyme.

HOW LONG FOOD IS IN THE BODY

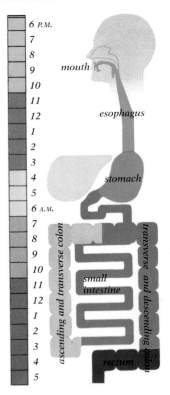

The stomach exit is guarded by a muscle known as the pyloric sphincter, which is similar to the sphincter at the stomach entrance except that it is never completely closed. As the waves of peristalsis push chyme through the stomach, the sphincter lets small amounts of it flow into the small intestine.

The small intestine is the longest section of the alimentary canal, measuring 20 feet (6 m). Its name comes from its width—about 1.5 inches (4 cm). The first 10 inches (25 cm) of the small intestine are the duodenum, the next 8 feet (2.4 m) are the jejunum, and the final 11 feet (3.3 m) are the ileum.

Aids to digestion

The alimentary canal processes around 35 tons of food during an average lifetime of 70 years, so it is little wonder that things can go wrong. Some problems are so common that they have become household words: ulcers, appendicitis, constipation, diarrhea, heartburn. Some disorders are unavoidable, but there are ways to keep the digestive system healthy.

Do not eat (or feed children) too much. This puts a strain on the digestion and can create a weight problem. In infants, it can lead to vomiting and regurgitation.

Chew your food thoroughly before swallowing. The digestion of carbohydrates begins in the mouth with ptyalin, an enzyme in saliva.

Include sufficient fiber in your diet: fruit with the skin on, lightly cooked vegetables, and bran with your breakfast cereal. Fiber cannot be digested, but it stimulates the passage of food through the large intestine, helping to prevent constipation and perhaps some intestinal diseases. Fiber-rich foods are also useful for dieting because they are filling but not fattening.

Drink sparingly with meals, because liquids of any kind dilute digestive juices.

Avoid any foods to which you know that you or your family react badly.

Minimize stress, which increases the acid secretion of the stomach and the muscular action of the whole system, causing food to be pushed along so fast that it is not properly digested.

Stop smoking, because, like stress, smoking stimulates acid secretion.

ALIMENTARY CANAL

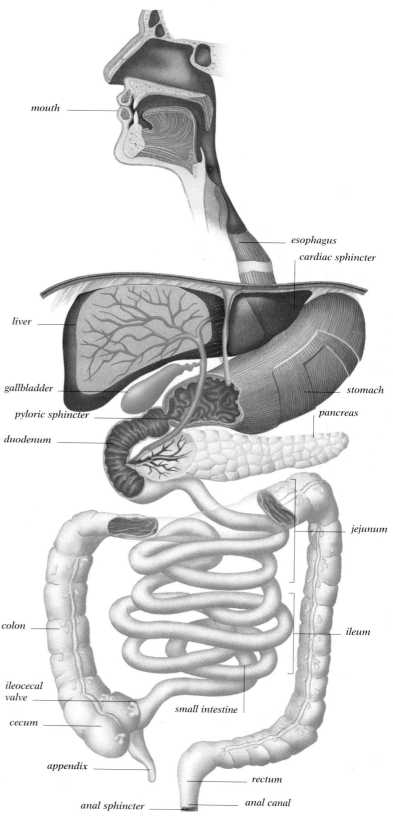

mouth

esophagus

cardiac sphincter

liver

gallbladder

stomach

pyloric sphincter

pancreas

duodenum

jejunum

colon

ileum

ileocecal valve

cecum

small intestine

appendix

rectum

anal sphincter

anal canal

Common alimentary problems in babies and children

SYMPTOM	RELATED SYMPTOMS	POSSIBLE CAUSES	ACTION
Diarrhea in babies	Fever, loss of appetite, vomiting	Tonsillitis, infection of ear canal or middle ear, common cold	Boiled water only. Call doctor if symptoms persist for more than 24 hours.
		Food allergy	Boiled water only. Reintroduce foods one by one to find cause. See doctor if it persists.
Diarrhea in children	Fever, loss of appetite, vomiting, abdominal pain	Infection of ear canal	Fluids for 24 hours. See doctor if symptoms persist.
		Effects of antibiotic drug	Bland diet; fluids. See doctor if symptoms persist.
Vomiting in babies	Diarrhea, loss of appetite	Too much air taken in with food	Use nipple with smaller hole. Give small, frequent feedings.
	Constipation	Bowel obstruction	See doctor at once
Projectile vomiting in babies under two months	Constipation, no loss of appetite	Pyloric stenosis	See doctor at once
Vomiting in children	Diarrhea, loss of appetite, pain	Appendicitis	See doctor at once
Constipation in babies	Vomiting	Bowel obstruction	See doctor at once
		Anxiety about potty training	Stop training temporarily
Constipation in children	Poor appetite	Lack of fluids, lack of exercise, too little roughage	Plenty of fluids; increase roughage

PERISTALSIS—HOW PARTIALLY DIGESTED FOOD (CHYME) IS MOVED THROUGH THE INTESTINE

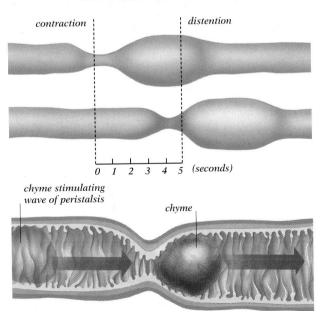

contraction | *distention*

0 1 2 3 4 5 *(seconds)*

chyme stimulating wave of peristalsis

chyme

▲ *Food mixed with gastric juice is known as chyme; it is propelled through the intestine by peristaltic waves.*

Most of the digestive process occurs in the small intestine, through the action of digestive juices from the intestine, the liver, and the pancreas, which are closely linked to the alimentary canal. The liver produces bile, a substance stored by the gallbladder and needed to digest fats; and the pancreas produces secretions, which, like bile, pass into the duodenum. As waves of peristalsis move chyme along the small intestine, more mixing occurs. In the ileum, digested foodstuffs are absorbed into the blood via thousands of villi, tiny fingerlike projections in the gut wall, then are carried first to the liver and next to all body cells. At the end of the ileum, a sphincter called the ileocecal valve keeps chyme trapped in the small intestine until the next meal. When more food enters the stomach, the valve opens, and the chyme passes into the large intestine, a tube about 13 feet (3.9 m) long and about 3 inches (7.5 cm) in diameter.

Excretion

The large intestine has three parts: the cecum, colon, and rectum. The cecum and the appendix have no known functions. In the colon—the largest section, at 4.5 feet (1.3 m)—water is absorbed into the blood from the liquid portion of digestion. When this liquid reaches the rectum, it is solid feces, moved along by huge propulsions that take place a few times a day. Finally, feces enter the anal canal, which is kept closed by the last sphincter in the alimentary canal. In infancy, it opens automatically when the anus is full; as the nervous system matures, humans override the automatic signals.

See also: Anus; Digestive system; Mouth; Pharynx

Anus

Questions and Answers

Is it true that eating spicy food can make the anus itch?

Eating highly spiced food with cayenne or hot chili powder can irritate the anus because the spices in the feces pass through the anal canal. If this is a problem, then it is better to avoid spicy foods.

The last couple of times that I have been to the toilet, I have noticed blood on the toilet paper. Could this be serious?

It is always wise to consult your doctor if you notice blood in your feces or on the toilet paper, or experience any bleeding from the anus. The underlying problem may well be a minor one, but it is not worth taking any risks.

My baby has diarrhea, and a rash has developed around his anus. What should I do?

Diarrhea can cause a rash because of the continual irritation. To clear it up quickly, change your baby's diapers as soon as they become dirty, clean the area with mild soap and water, pat it dry, and dress the area with antiseptic cream or zinc ointment. Consider allowing your baby to play for a while without a diaper on, since fresh air is an effective healer.

I am sometimes woken in the night by an intense pain in my rectum. What causes this?

This type of fleeting pain is called proctalgia fugax and is thought to be caused by muscle spasms in the lower part of the alimentary canal. It seems to be brought on by anxiety. The pain should gradually subside, and the best solution is to get up and have a glass of water or a hot drink. If it persists, occurs frequently, or is adding to your worries, talk to your doctor and get his or her advice.

Once all the goodness has been extracted from the food people eat, the unwanted residue is excreted as feces via the anus or anal canal. This is a sensitive area where irritating and painful disorders can occur.

The anus is about 1.5 inches (4 cm) long and is situated at the end of the rectum, forming the last part of the digestive system. As the waste products of digestion near the end of their journey down the intestines, they slowly harden as the body absorbs liquids; the solid waste, or feces, is then pushed into the rectum. Two rings of muscle called the internal and external sphincters normally keep the ends of the anus closed, but during defecation they relax to allow the feces to escape. The internal sphincter (which is under the control of the nervous system) senses the presence of the feces and relaxes, allowing them to enter the anal canal. The external sphincter is deliberately kept closed (a skill people learn as babies) until a convenient moment occurs when the feces can be passed. To ease the passage of the feces, the lining of the anus secretes a lubricating mucus.

Most people will suffer from problems in the anal area at one time or another, and these can be irritating and uncomfortable. Problems in this area are most common in childhood, during pregnancy, and from middle age onward, although washing the anus regularly with warm water and mild, unscented soap and then drying it with a soft towel can help prevent some of them.

Anal fissures

Difficulty in passing hard feces often results in other trouble. The lining of the anal canal is delicate, and passing hard feces can tear it. Not only can infection enter the wound—called an anal fissure—but it can become painful to defecate. The problem can be difficult to deal with because the person may try to hold back the feces to avoid the pain; this has the effect of hardening the feces even further, worsening the situation. Usually the problem is treated with antibiotics for the infection and a laxative to soften the feces.

In the long term, the best way that both adults and children can prevent this problem is to eat a diet rich in bran and roughage. This speeds the passage of feces through the last part of the alimentary canal, reducing the time available for water to be absorbed and making the feces softer. A diet high in roughage also helps prevent other anal problems such as piles (hemorrhoids), and some disorders of the large intestine.

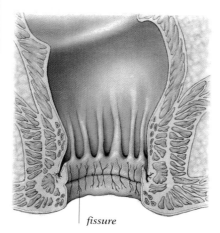

▲ *Cracks in the anus are called fissures. They are usually caused by passing a hard or very large stool. Because of repeated irritation (the passing of further feces), fissures do not heal easily.*

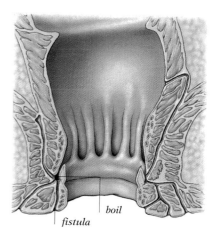

▲ *A boil in the anus can lead to a fistula. The boil bursts but does not drain into the anus. Instead, drainage channels form through the skin, opening to the outside of the body at sites beside the anus.*

Common causes of pain in the anal area

SYMPTOMS	POSSIBLE CAUSE	ACTION
Sharp pain during bowel movement and pain or aching for up to an hour afterward; bleeding from anus	Anal fissure	Apply soothing ointment; have a warm salt bath for temporary relief; see a doctor. A lubricant may be prescribed and surgery recommended to remove the affected tissue and stretch the anal opening.
Soreness not made better or worse by defecation; discharge from anal region	Anal fistula	See a doctor. Antibiotics will be prescribed; an operation may be needed.
Severe, throbbing pain	Abscess or boil	See a doctor. The pus will need to be surgically drained.
Intense pain, not worse on passing feces; anus painful to touch, may itch; painful, deep purple protrusion may appear on one side of anus	External hemorrhoids	Apply ice-cold water on a tissue or have a hot bath for temporary relief; see a doctor—surgery may be needed. Long-term, get more exercise; eat more roughage.
No pain but itching, soreness, and the passage of much mucus; bright red blood may be lost on defecation; feeling of fullness in anus	Internal hemorrhoids	See a doctor. An injection will provide temporary relief, but surgical treatment may be needed.
Pain may be worse on defecation; protrusion of pink tissue from anus	Prolapse (dropping) of rectum	See a doctor. An operation will probably be necessary, but in the case of a child, the doctor may be able to push the collapsed part back into place.

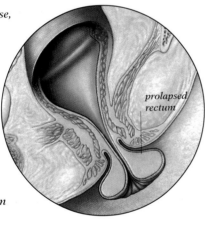

► *In a rectal prolapse, the muscles surrounding the rectum weaken so that during defecation, part of the rectum collapses and protrudes from the anus. An operation will probably be necessary to stitch the prolapsed rectum back into place.*

prolapsed rectum

Piles

Piles or hemorrhoids are a common anal problem. They are caused by abnormalities of the blood vessels surrounding the anus, which form clumps of swollen, contorted varicose veins both inside and outside the anus. Hemorrhoids cause irritation and itching, often accompanied by bleeding. If suppositories fail, hemorrhoids can be treated with injections or by surgery if the condition is severe.

In adults

In adults, constant anal itching—called pruritus ani—can be caused by infection of the skin around the anus or by severe anxiety. Scratching can result in considerable skin damage, causing the infection to spread and the area to bleed. The best treatments are antiseptic ointments that kill the bacteria and calamine lotion to cool the inflammation. A doctor should be consulted, however, as itching and irritation may be symptoms of other problems, such as diseases of the spinal cord, which can lead to an inability to control excretion.

Boils and abscesses are the most common anal infections, and they can be very uncomfortable. They may form drainage channels called fistulas, which open outside the body next to the anus. A fistula looks like a small red spot from which fluid leaks. This complaint is normally treated with antibiotics and, in severe cases, surgery.

In children

Children are often susceptible to irritations around the anus, and in younger children this is usually caused by diaper rash. Another anal irritation that is more common in children than adults is caused by threadworms (pinworms)—a type of parasite that lives in the lower bowel. Threadworm infestation can cause persistent itching around the anus, but the problem can be treated with drugs.

Other complications

Another complication can arise through the weakening of the internal muscles that surround the rectum. This causes the lining of the rectum to drop so that it protrudes from the anal opening, a complaint called a prolapse. The weakness may be present at birth, may develop with old age, or can be brought on by pushing too hard during defecation, by diarrhea, by whooping cough, or during labor. The complaint should be reported to a doctor, who may push the lining back into position or secure it with a few stitches.

Sometimes unwanted growths protrude from the anus. Usually these are benign lumps called polyps, which are common in young children. Warts and wartlike tumors called papillomata may also occur, especially in middle and old age. For warts, a doctor may recommend diathermy, during which the growths are burned off electrically. Papillomata require surgery to ensure complete removal, as they can lead to cancer.

See also: Alimentary canal; Digestive system; Mucus; Rectum

Appetite

Questions and Answers

My young son is just getting over a bad cold, and he doesn't want to eat anything. What can I do?

Try to tempt him with foods he really likes—it is possible to try any food. There is not much that can be done aside from this. If he really does not want to eat anything, food-substitute drinks that contain added vitamins and minerals should provide more than enough nourishment until his normal appetite returns.

How can I know when I've eaten just enough—and stop overeating?

It is possible to become more aware of your body's signals, and with practice you can gauge more accurately how much food you really need. One way of finding out is to chew more slowly; this signals the part of the brain that controls appetite to stop you from overeating.

I gain and lose weight during my menstrual cycle. Is the cause food or fluid retention?

Some theories attribute weight gain to an excess of body fluid caused by hormonal changes during the menstrual cycle. However, a new theory suggests that weight is gained because the hormones alter the appetite, causing the woman to overeat.

My father says there is no need to eat both lunch and dinner—dinner is enough for him. Is he right?

No. People are controlled by their hormones, which influence their sleeping and waking patterns and their eating habits and appetite, and help the brain to regulate the times at which they start and stop eating. Therefore, what is right for your father may not be right for everyone. Each person's eating habits and needs are different.

If appetite and hunger were the same, no one would be overweight or too thin—everyone would eat just the right amount to satisfy hunger and nourish himself or herself. What causes appetite, and what is its purpose?

When people want to eat something because it looks good, smells nice, and tastes delicious, the appetite is working, but when they want to eat something because the stomach is rumbling and they feel the need for food, then the hunger drive is working. This is the basic difference between appetite and hunger.

This is an important difference, because it is not hunger that causes overeating, but appetite. In the same way, appetite can stop people from eating even though they may feel hunger and the need for food. When this happens, the body's needs are not satisfied, as occurs in some diseases and in a condition such as anorexia nervosa, in which the person becomes obsessed with dieting.

It seems that hunger is the body's way of letting people know their personal food requirements. However, it seems that appetite sometimes interferes, preventing the body from getting the right amount of nourishment. Therefore, it is important to learn how appetite develops and if there are times when it has a useful role to play.

Developing an appetite

When babies are born, one of the first things they feel is the need for food. In most cases, this need is satisfied with milk, either from the mother's breast or from a bottle. Babies show no real preference for any type of food at this stage in life; they cry when hungry and are calm and smiling when fed. It is only later that they learn which foods they really prefer. With this learning, a growing child's appetite gradually develops. Because humans are so complex, they have different appetites and likes and dislikes for different foods.

Interestingly, some foods that Westerners consider to be inedible are thought to be delicious delicacies in some parts of the world. It is also revealing that in countries where there is starvation, people's appetites may change to the point when they are hungry enough to eat almost anything edible to keep themselves alive and functioning.

How appetite works

Appetite is the regulator of daily food intake and therefore the eventual regulator of weight. Therefore, many scientists are interested in finding out precisely what factors control it—and so

▲ *Color helps to stimulate the appetite. The food on the left looks appealing, but the same food colored blue (the most unappetizing color) is less likely to stimulate the appetite.*

far they have found that appetite is a complicated process. In most people, when the appetite is satisfied, eating stops soon after. It might seem obvious that when the stomach is full, eating should stop, but experiments have shown that it is not just a full stomach that tells the brain this. For example, it has been found that a hormone produced by the intestines is another signal that eating should stop. Other signals come from the amount of food passing through the mouth, the concentration of nutrients in the blood, and the degree of fullness of the stomach.

All these signals are picked up by an area of the brain called the hypothalamus, and scientists have discovered two separate areas in the hypothalamus that are in charge of eating. One of the areas controls the actual eating, and the other controls satiety, or satisfaction of the appetite. The name given to these two areas together is "appestat."

The various signals go to the appestat, and when there are enough signals from the areas that are concerned with eating, the appestat tells the brain that the body has had enough. Theoretically, eating should stop at this time, but it does not always. If it did, no one would be overweight or too thin.

Fat and thin

One theory explaining the differences in body weight of heavy people, or those who eat more than they need, is that they have an appestat that is set too high. In other words, the appestat does not tell the brain to stop eating soon enough, so that overeating occurs. The opposite is true for thin people, who stop eating too soon. However, this theory does not take into account any of the other reasons for being over- or underweight.

▲ *The amount that people tend to eat and their preference for certain foods varies from person to person.*

It has also been found that heavy people are more likely to eat when they see food in front of them, while people with a normal weight will eat only when hungry. Also, heavier people tend to pay more attention to the taste of food; people of normal weight seem to care less about the food than the fact that they are taking in fuel to keep the body running. Consequently, heavy people will eat out of habit at mealtimes and will eat more than they need if the food tastes good, while normal-weight people will eat only enough to satisfy their hunger, without really caring too much about the taste of the food.

This still does not take into account eating binges and the fact that some people indulge in food for reasons that have nothing to do with taste or the regularity of mealtimes. For example, many psychological studies have shown that people sometimes eat to compensate for feelings of frustration and inadequacy or because they are bored and lonely. People who raid the refrigerator in the middle of the night usually realize that they do so because they want to gain some sort of temporary emotional comfort from the act of eating, rather than because they want to relieve any feelings of hunger.

See also: Hormones; Hypothalamus; Taste

FACTS ABOUT APPETITE

During illness, it is common for people to lose their appetite, and doctors think that subtle changes in the body chemistry reduce both hunger and appetite at this time. Sick people who do not want to eat should be tempted with interesting dishes so that they take at least a little nourishment.

Another type of unusual appetite can happen during pregnancy, and there are many stories of pregnant women eating all sorts of strange foods or combinations of foods. Generally speaking, fads during pregnancy need not be a cause of anxiety. It is likely that hormonal changes in early pregnancy change the sense of taste, making previously pleasant foods or drinks taste strange and unpalatable. Pregnant women have been known to consume unusual things such as coal and earth; doctors think that this behavior was caused by a need to make up a mineral deficiency. In this case the body appears to be more clever than many people think, as it can tell what food to eat without the person knowing why.

A central factor in appetite control is a hormone called leptin, which is synthesized by fat cells and released in proportion to the amount of fat present. Leptin acts on the hypothalamus by inhibiting a substance that stimulates eating. Leptin also increases the metabolic rate so that fat is burned up more rapidly. To date, no satisfactory antiobesity drug has emerged from leptin studies.

Arteries

Questions and Answers

I have been told that smoking causes artery disease. Is this true?

Yes. Atherosclerosis, the most common and dangerous arterial disease, is always aggravated by smoking. How this happens is unclear, but there is no doubt that nicotine in the bloodstream does cause arteries to narrow and eventually become permanently rigid and less able to carry blood to all parts of the body, especially the heart, resulting in heart attacks; the brain, resulting in strokes; and the extremities (hands and feet). If blood cannot reach the extremities, tissues degenerate, gangrene sets in, and a limb may have to be amputated. However, if smoking is given up in time, damage can be avoided.

I jog regularly. Will this reduce the risk of my having a heart attack?

Yes. It is now established that taking regular exercise can reduce the risk of a heart attack. One study, which compared the chances of heart attacks among bus drivers and bus conductors in England, showed that conductors were less prone to heart attacks because their job is more active than the drivers'. So exercise does appear to be a preventive measure.

Three of my male relatives, including my father, have died of heart attacks. Are men more prone to atherosclerosis than women?

Yes. This seems to be true for several reasons. First, female hormones seem to protect women from atheroma, the buildup of fatty deposits that cause blockages in the arteries. After menopause, when hormone levels fall, atheroma increases. Second, until recently more men smoked than women. The pattern is now changing, however, and the number of women suffering from atheroma is therefore rising.

The arteries carry blood containing nutrients and oxygen to all parts of the body. Healthy arteries are therefore very important. Taking measures to prevent arterial disease is vital, as is early treatment.

The arteries and veins are the two types of large blood vessels in the body. The arteries are like pipes, carrying blood outward from the heart to the tissues, while the veins carry the blood on the return journey. The entire body depends on blood for its supply of oxygen and other vital substances without which life could not go on. The heart is a pump that propels blood around the body through the arteries. The main pumping chamber on the left side of the heart, which is called the left ventricle, ejects blood into the main artery of the body—the aorta. The aorta is a tube about 1 inch (2.5 cm) across on the inside. Three arteries lead from the aorta: the right artery and the left, which divides into two main branches. These are the coronary arteries that supply blood to the heart itself. The coronary arteries are particularly likely to be affected by disease. A blocked coronary artery, called a coronary thrombosis, causes a heart attack. After leading to the coronary arteries, the aorta passes upward before doubling back on itself in an arch. Originating from this arch are the two main arteries supplying the head—the left and right carotid arteries—and one artery leading to each arm. The aorta descends down the chest and into the abdomen.

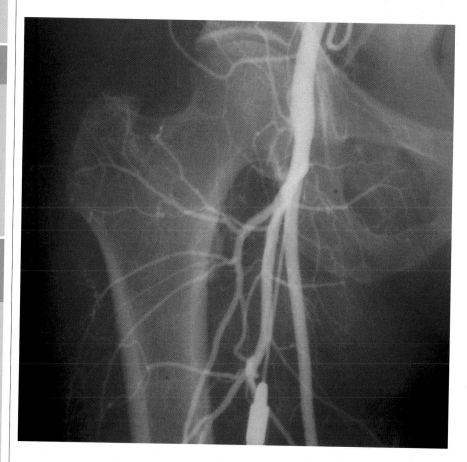

▲ *An arteriogram is an X ray in which dye is injected into the bloodstream to detect blockages. This arteriogram of an upper leg is normal and shows no blockage.*

ARTERIAL SYSTEM

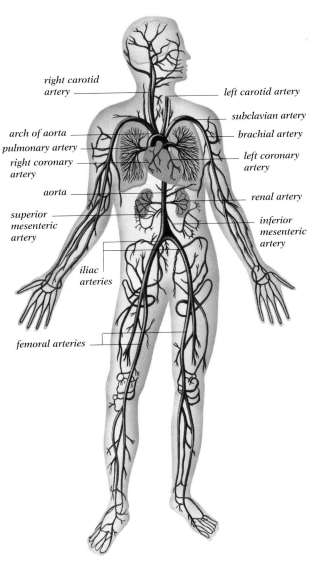

right carotid artery

left carotid artery

subclavian artery

arch of aorta

brachial artery

pulmonary artery

right coronary artery

left coronary artery

aorta

renal artery

superior mesenteric artery

inferior mesenteric artery

iliac arteries

femoral arteries

In the abdomen three main arteries lead to the intestines and the liver, and one to each kidney, before the aorta divides into the left and right iliac arteries, which supply blood to the pelvis and the legs. After passing through the capillaries (a network of tiny blood vessels that link the smallest arteries and veins from which oxygen and nourishment enter the tissues), blood returns toward the heart in the veins. In general, the artery and vein supplying an area run side by side. The veins empty into the right side of the heart; from there, blood is pumped to the lungs and recharged with oxygen. From the lungs, oxygen-rich blood is drained by the pulmonary veins into the left side of the heart. Here, blood circulates around the body by being pumped into the aorta by the left ventricle of the heart. The left ventricle generates a large amount of pressure to force the blood through the arterial network. When blood pressure is measured, the tightness of the inflatable cuff strapped around the arm is equal to the amount of pressure of the maximum squeeze in the left ventricle that occurs with each heartbeat.

The structure of the arteries

Since the arteries are subjected to this force with each heartbeat, they have to be thick-walled to be able to cope with the pressure. The outer wall of an artery is a loose, fibrous tissue sheath, inside of which is a thick elastic sheath that gives the artery its strength. Amid the elastic tissue, there are rings of muscle fiber encircling the artery. The inner layer of the artery is made of a smooth layer of cells that allows the blood to flow freely (the endothelium). The thick elastic walls are important to the system, since they encounter much of the force of each heartbeat. The elastic walls continue to push the blood forward in the pause between heartbeats.

Artery disease

Arterial disease in any part of the body is very dangerous, because if an artery is blocked or narrowed, the part it supplies may die from oxygen starvation. There are two basic ways in which a blockage can happen.

The term "arteriosclerosis," which means hardening and loss of elasticity in the walls of the arteries, and which was formerly extensively used, has virtually disappeared from medical textbooks and journals. The reason for this is that doctors now know that pure arteriosclerosis is actually a rare disorder and that in the majority of cases these changes in the arterial wall are associated with, and related to, other, more serious changes involving damage to the inner lining of the arteries with the deposition of plaques of material called atheroma. These plaques consist of fatty material containing cholesterol, degenerate muscle cells, collagen, phagocytes of the immune system, and other materials.

The atheromatous plaques can grow to the point where they significantly obstruct the flow of blood through the artery. Even more seriously, because they cause local roughening of the arterial lining they may form the site of blood clotting, which can block off the lumen of the artery altogether. This is called thrombosis. It is for these reasons that the term "arteriosclerosis" has been virtually replaced by the term "atherosclerosis."

Atherosclerosis is the principal cause of death in the Western world. It is the underlying cause of angina pectoris, heart attacks, strokes, kidney failure, aortic ballooning (aneurysm), and gangrene of the limbs. It can cause all these disasters not because the arteries have hardened, but because they are prone to become obstructed by atheroma and thrombosis so that the vital blood supply to heart, brain, other organs, or limbs is cut off.

Cholesterol

The atheromatous process first starts with a deposit of cholesterol—a normal ingredient of the blood and one of the building blocks of normal cells—in the wall of the artery. However, it seems that cholesterol leaks into the inner surface (intima) of the artery, and a fatty streak forms within the arterial wall.

As the fatty streak grows in size, two other things happen. First, the surface of the streak may break down and expose the middle portions of the arterial wall to blood. When this happens it triggers the mechanism for clotting the blood. A clot normally forms a plug of fibrous tissue to help stop bleeding from a wound. When the process occurs around a fatty streak, a mixture of fibrous and fatty tissue is formed in the arterial wall, and this is called an atheromatous plaque. As the plaque grows, it starts to encroach upon the central blood-filled space—the lumen of the artery. Finally, the development of the

I know that smoking is bad for me, but will alcohol also increase my chances of getting a bad heart?

There is no evidence that moderate amounts of alcohol make atheroma worse. Indeed, there are pointers in the opposite direction. However, a moderate approach to alcohol should still be observed. An excess is harmful in other ways.

I have arterial disease. Should I try to keep my legs warm?

If you are suffering from peripheral vascular disease, which reduces—and can prevent—blood flow to the extremities, you may be feeling the cold. However, warming up your legs will only increase their demand for oxygen, which is supplied by the blood. In the hospital, legs badly afflicted by arterial disease are deliberately kept cool in the hope of preserving them. Take further advice from your own doctor, who knows the history of your condition.

My husband and I eat lots of meat and dairy products. Is our cholesterol intake too high to be healthy?

There are several important points here. First, if you or your husband smokes, that habit should be eliminated, because smoking counteracts the good effects of any other measures. Second, you should be aware that there is more to a balanced diet than just cutting down on cholesterol. Most cholesterol comes from the body's own chemical processes anyway, and it is these that the prudent diet should aim to alter. It seems sensible to eat more unrefined carbohydrates—foods containing a lot of fiber, such as whole wheat bread and bran. Next, reduce all fats, especially animal fats and dairy products, called "saturated fats," replacing them with "unsaturated" vegetable fats. This diet will reduce cholesterol intake, while also having a dramatic effect on the cholesterol already present in the blood.

plaque involves changes occurring deep in the arterial wall. Fibrous tissue forms on the inner surface of the original fatty streak, but there is also a growth of this tissue on the wall side of the plaque, growing from the outside of the artery toward its center. The end result is a mixture of fibrous and fatty tissue blocking a proportion of the arterial lumen. The disease extends to a considerable depth in the wall of the vessel and encroaches on a large proportion of its circumference. Once a large atheromatous plaque has formed, it may have a number of consequences. It may steadily enlarge to block the artery. Because the artery is partially blocked, the flow of blood past the obstruction is reduced. This may activate the clotting system at the site of narrowing. The clot may well produce a total obstruction, known as a thrombosis.

Atheromatous plaques that are partly blocking an artery may become displaced and swing across the lumen of the artery, like a locked gate, to block it completely. Parts of an atheromatous plaque may break off and travel toward a smaller artery, which then becomes blocked. This is known as embolism.

Atheroma can affect arteries with as small a diameter as 0.08 inch (2 mm). However, the process is most likely to occur in areas of arterial wall that are subjected to movement and the most stress. For this reason, atheroma is most common at sites where arteries branch into smaller arteries. There is a greater stretching of the lining of the arteries at these points, allowing more cholesterol to get into the wall.

The results of blockage

Since arteries are necessary to supply oxygen to every part of the body, there is no organ that is completely immune to the effects of arterial disease. If an organ or a limb has its blood supply cut off by atheroma, then it will eventually die.

An area of tissue that has lost its oxygen supply is called an area of infarction. When this process occurs in an arm or a leg, the term "gangrene" is usually applied. There are some areas where the effects of atheroma cause especially severe problems. The most important are the heart, the brain, the legs, and the aorta itself.

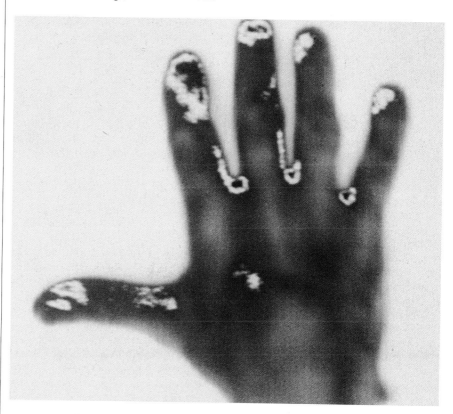

▲ *This thermograph, or heat-sensitive picture, shows the effects of smoking on the circulation. Note the decreased blood flow (the white areas) to the fingertips.*

DEVELOPMENT OF ATHEROSCLEROSIS

tunica externa (loose, fibrous tissue)
tunica media (elastic, muscular sheath)
tunica intima (endothelium)

lumen

fatty streaks (atherosclerotic lesions)

lumen

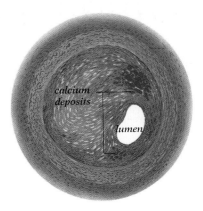

calcium deposits

lumen

▲ *The normal artery is thick-walled; three layers surround the lumen, through which the blood flows.*

▲ *In a moderate case of atherosclerosis, fatty deposits begin to build up in the inner layer of the artery.*

▲ *Here atherosclerosis is almost total: fatty deposits have severely decreased the lumen, and calcium is forming.*

Heart attacks

Atheroma particularly affects the heart, because the coronary arteries, which supply blood to the heart, are under more mechanical stress than practically any other artery in the body. The heart is continuously contracting and relaxing with each heartbeat, and in so doing, the coronary arteries—which lie on the outer surface of the heart—are alternately stretched and relaxed. The conditions are ideal for atheromatous plaques to form. When a coronary artery becomes blocked as a result of atheroma, a heart attack results. This is also known as a coronary thrombosis, or a myocardial infarction (the myocardium is the heart muscle, and infarction is the formation of a dead area of the muscle when it is deprived of blood).

Heart attacks are not the only problem that atheroma causes in the heart. Where there is a fixed obstruction that is not totally blocking the artery, the supply of blood to the heart may be sufficient to meet the needs of the body only when it is at rest. Exercise increases the need for blood in the heart, and it becomes starved of oxygen. This causes pain in the heart, which is known as angina pectoris, or simply angina. The two problems—angina and myocardial infarction—are often lumped together under the heading ischemic heart disease; "ischemia" means a lack of oxygen without total deprivation or infarction.

Strokes

In the brain, atherosclerosis may result in a stroke. Strokes may vary from trivial to fatal and may result when an artery becomes blocked through atheroma or embolism, or when an artery leaks blood into the brain as a result of a weakened wall.

When the legs are severely affected by atheroma, they become painful. This pain is worse during exercise, just as with angina. If the disease is severe, then gangrene results and the affected leg may have to be amputated.

Finally, the aorta itself is a very important area of atheromatous disease. Two different things can happen. The wall of the aorta may start to balloon out as a result of the weakening effect of the disease. This produces a saclike swelling called an aneurysm, in place of the regular tubular structure of the normal aorta. Aneurysms are usually found in the abdomen but may occur in the chest. An aneurysm may continue to expand and eventually start to leak. Surgery is the only treatment, and it is necessary to strengthen the aorta with a mesh tube. The results of this sort of surgery, when undertaken in an emergency, are often good, although there are some failures. Another form of aneurysm, which tends to occur in the chest rather than the abdomen, is called dissection of the aorta; the layers of the aortic wall become split by escaping blood, and the end result is much the same. Occasionally, patients survive aortic dissection without surgery, but surgery is usually necessary.

Those affected

There is now a well-established list of risk factors indicating people who are more likely to suffer from "accelerated" or "early" atheroma. For instance, some diseases put people at greater risk. The two most important ones are high blood pressure and diabetes. People from a family in which atherosclerosis has occurred are at greater risk of developing problems themselves. Finally, a raised cholesterol level in the blood is a risk factor, although the value of cholesterol measurements in individuals has perhaps been overemphasized. Generally, it is sensible to reduce meat and dairy products in the diet.

Prevention

Certain measures can be taken in order to prevent or postpone the development of atheromatous disease. Both diabetes and high blood pressure must be treated. If there are no predisposing illnesses, the risk factor of greatest importance is family history. A doctor can prescribe a range of drugs to lower cholesterol and help prevent atheromatous disease.

The most effective thing a person can do to prevent or postpone atherosclerosis is to stop smoking. Aside from reducing the chances of developing heart disease, giving up cigarettes will improve overall health. Other simple measures are to engage in vigorous exercise several times a week, keep body weight within normal limits, and follow a diet that is high in fruit and vegetables and low in animal fats and dairy products.

See also: **Blood pressure; Coronary arteries; Immune system; Kidneys**

Autonomic nervous system

Involuntary body processes such as glandular action, digestion, heartbeat, and many other vital functions are managed with no consciousness or effort on our part by the hidden workings of the autonomic nervous system.

As people go through the day, many processes within the body, such as control of the heartbeat and digestion, keep them functioning smoothly. People usually take these for granted because they happen automatically. All processes need some kind of controlling mechanism, however, and in the human body two different systems provide this control.

One of these, the endocrine system, affects much of the body's chemistry through the production of hormones, which regulate growth and reproduction, among other functions. On the whole, hormone-controlled systems do not have to work as quickly or as spontaneously as those controlled by the second system, the autonomic nervous system.

This system, which is part of the whole nervous system, is mainly concerned with keeping up the automatic functions, without deliberate mental or conscious effort on our part, of the blood vessels and organs such as the heart, lungs, stomach, intestines, bladder, and sex organs.

Why is it that I often feel sleepy after I have eaten?

Many digestive processes depend on the autonomic nervous system. It has two subsystems: the sympathetic, which is the system of action and activity; and the parasympathetic, which is mostly concerned with relaxation. Digestion is mainly controlled by the parasympathetic system, so when it is at work the sympathetic system is less active than usual. This minor degree of parasympathetic overactivity and sympathetic underactivity combines to produce sleepiness.

Why does our hair literally "stand on end" when we are frightened?

Fear influences the part of the autonomic nervous system known as the sympathetic, which controls the small muscles that actually lift the hair off the skin. The pupils will also widen, and sweating may start, particularly in the palms of the hands, where the sweat glands are controlled more directly by the sympathetic nervous system than those in other parts of the body.

Are drugs that affect the autonomic nervous system dangerous?

Like all drugs, those that affect the autonomic system must be supervised. Most of those in common use have a wide margin of safety, but since this system controls basic life functions, such as the heart and lungs, an overdose can be very serious.

Do some people have an overactive autonomic nervous system?

Yes. Those who suffer from this problem complain of palpitations and churning of the stomach, but these symptoms are harmless and do not need treatment.

HOW THE AUTONOMIC NERVOUS SYSTEM CONTROLS THE BODY

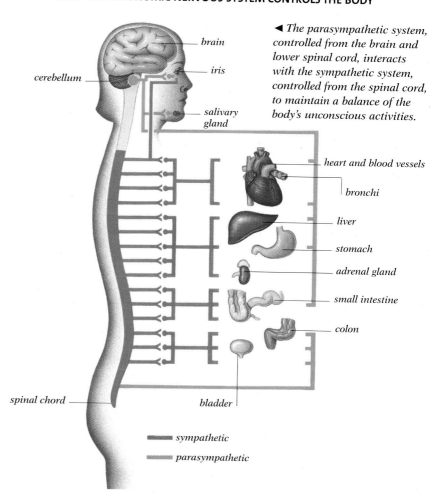

◄ *The parasympathetic system, controlled from the brain and lower spinal cord, interacts with the sympathetic system, controlled from the spinal cord, to maintain a balance of the body's unconscious activities.*

brain
iris
cerebellum
salivary gland
heart and blood vessels
bronchi
liver
stomach
adrenal gland
small intestine
colon
spinal chord
bladder

sympathetic
parasympathetic

▶ *Asthma causes the bronchial tubes to narrow, making breathing difficult. An inhaler mimics nature by delivering rapid relief in the form of a drug based on noradrenaline, the chemical transmitter for the central nervous system.*

How it works

An understanding of how the autonomic nervous system operates is vital to medicine, mainly because its workings—part chemical, part electrical—can be controlled or blocked by drugs. Therefore, the organs it influences can also be controlled with drugs.

The nervous system can be thought of as the control system of the body. Thought and other conscious or deliberate activities go on in the brain hemispheres called the "highest" part of the system. The activities dealt with in the autonomic nervous system go on in the "lower" parts of the brain and spinal cord.

The actual process by which the nervous system works is complicated, but in simple terms it can be thought of as electrical signals transmitted down nerve fibers, each of which ends in nerve cells and has nerve cells along its path. The complexity and quantity of these cells and fibers within the body are astonishing: there are more than 10 billion in the brain and spinal cord.

Cells and axons

Nerve cells are the tiny bodies that either transmit or receive messages or sensations. The fibers, or processes—known in medical science as axons—are the "wires" along which the impulses, or stimuli, travel to and from the control centers of the brain and spinal cord.

Nerve cells are not physically connected to each other. There is a gap, called a synapse, between the ending of an axon and the cell itself; the "message" is carried across the gap by means of a chemical. It is this gap, with the chemical bridge, that enables doctors to control the nervous system; the action of the chemical transmitters can be imitated with similar, artificial chemicals.

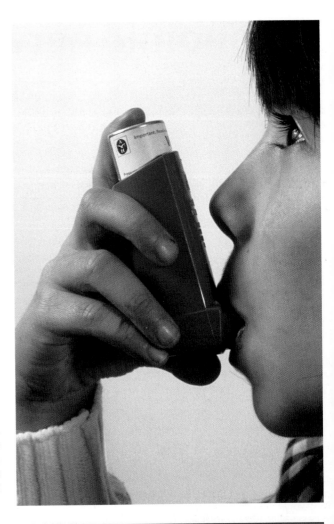

Treatment of the autonomic nervous system

DISEASE	TREATMENT	DISEASE	TREATMENT
Abnormal heart rhythms (palpitations)	Some abnormalities of the heart rhythms are treated with drugs that block the sympathetic system.	Glaucoma (high pressure within the eye)	Parasympathetic-stimulating and sympathetic-blocking drugs constrict the pupil of the eye, which helps lower the pressure.
Angina	Sympathetic-blocking drugs are the nonsurgical treatment.	Heart attack	Blocking the sympathetics is a mainstay of treatment. A very slow heart rate may be improved with a pacemaker or atropine—a parasympathetic-blocking drug.
Asthma	Stimulant drugs are the mainstay of treatment. They mimic the sympathetic nerves of the autonomic system.		
		Hypertension	Sympathetic-blocking drugs are a major form of therapy.
Common cold	Blocking drugs may dry up a runny nose or other mucus secretions. They act on the parasympathetic nerves.	Ulcers (stomach and duodenal)	In rare cases, when all other therapies fail, selective branches of the vagus nerve are cut to reduce stomach acid.

Two types of control

The autonomic nervous system is divided into two parts, the sympathetic and parasympathetic. Each uses a different transmitter where the nerve fiber reaches its target organ, each is built differently, and each has a different effect on the organ it serves.

Parasympathetic nerves serving, for example, the bronchial airway leading to and from the lungs, make them constrict, or grow narrow. The sympathetic nerves leading to the same area cause widening, that is, dilating of the bronchial passages.

The chemical transmitter for parasympathetic nerves is acetylcholine, and the transmitter for sympathetic nerves is noradrenaline—a close relative of adrenaline, which is the main hormone released to get the body's processes moving fast in "fight or flight" situations.

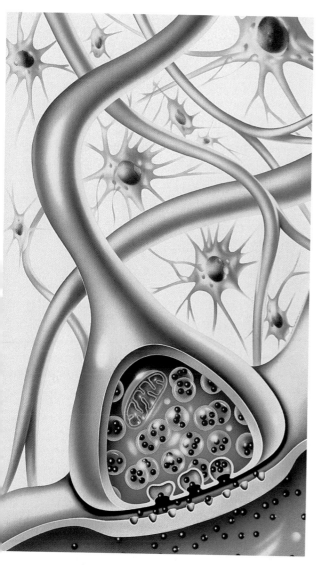

▲ *A synapse. Synapses—the gaps between the ending of an axon and the adjoining cell—play an important part in the autonomic nervous system and are the means by which doctors can control the system.*

Armed with this knowledge, doctors have learned how to manipulate the autonomic nervous system—for example, to treat asthma successfully. If a patient inhales a drug similar to noradrenaline, the sympathetic nervous system is assisted, the bronchial airways dilate, and the asthma attack subsides and stops. When a person is congested—producing too much mucus—a drug that blocks the parasympathetic system may help.

A drug commonly used to block the parasympathetic system is atropine. This is also called belladonna—from the Italian *bella*, "beautiful"; and *donna*, "lady"—because women used to put drops of atropine in their eyes to widen the pupils and increase their allure. The drug was extracted from plants such as deadly nightshade, which is also sometimes called belladonna.

If a patient is suffering from angina, a drug to block the sympathetic nervous system and its noradrenaline activity may be used as treatment. This sort of drug, called a beta-blocker, may cause asthma as a side effect if the "nonselective" form is used.

Surgery

Surgery can also be used on the autonomic nervous system. One such operation is a vagotomy, or cutting of the vagus nerve—a large bundle of parasympathetic nerves in the chest. This reduces acid secretions in the stomach, which in turn relieves gastric or duodenal ulcers. Nowadays it is more commonly an emergency procedure.

Disorders of the autonomic system

The autonomic system itself rarely causes serious trouble, but one minor problem is relatively common—fainting.

Fainting attacks are technically known as vasovagal attacks. Vasa are blood vessels, and the vagus is the major nerve of the parasympathetic system. Overactivity of the parasympathetic system causes dilation, or widening, of the blood vessels, particularly the smallest arteries (the arterioles). As a result, blood pressure falls, and there is a reduction in the flow of blood to the brain, causing loss of consciousness.

The treatment is simple. Once the patient is laid flat, a position in which he or she is likely to have fallen anyway, blood no longer pools in the lower part of the body, and blood flow and consciousness return to the brain.

More serious problems

Any disease process that can have a widespread effect on nerve tissue can affect the autonomic nervous system. Such diseases include infections of the nervous system, tumors of the brain or spinal cord, spinal cord injuries, diabetes, AIDS, and poisoning of nerves through excessive use of alcohol or misuse of drugs. Primary failure of the autonomic nervous system can also occur independently of damage to the central nervous system, most often in elderly men. Treatment depends on which part of the autonomic nervous system is involved. Overall autonomic failure is characterized by dizziness, fainting, fatigue, an inability to sweat, loss of sexual interest, and, in men, impotence. Some of the worst effects often result from a lowering of blood pressure and may be worse after eating.

See also: **Brain; Digestive system; Endocrine system; Heart; Hormones; Nervous system; Spinal cord**

Back

Questions and Answers

I want to use a bicycle to get to work. Will cycling be bad for my back?

No, because on a bicycle your body weight is supported and balanced by the shoulder girdle and the pelvic girdle, with the spine between them. The continual movement of cycling also keeps the back supple and well exercised. You will need to adjust the handlebars and saddle, however, to suit your height and the proportion of your limbs for comfort and to reduce possible strain on the back. If you use a sports bike, lower the handlebars slightly for long journeys.

I am nearing the end of my first pregnancy and am getting bad backaches. Why is this, and what can I do to reduce the pain?

Backache is a problem in late pregnancy because the weight of the fetus pulls the lumbar region forward, forcing the natural curve of the spine toward the waist to become exaggerated. This puts a strain on the lumbosacral joints in the back, causing pain. You can ease this problem by watching your posture and resting every afternoon in bed.

If pain persists after the first postnatal visit at six weeks, tell your doctor. The common suggestion that back pain is caused by a tipped womb is untrue.

My sister often wears shoes with heels. I have warned her that they may hurt her feet, but will they also be bad for her back?

Yes. High-heeled shoes change the body's balance and weight-bearing axis, tilting the lower spine and pushing the weight forward. This puts excessive pressure on the lower part of the lumbar region, causing pain and faster aging of the joints, including the knee joints.

Most of us suffer from backache or back pain at some time in our lives. Exercise and treatment can help, but prevention is always better than cure, and there is much we can do to protect and help the back.

Ever since human beings have stood upright, they have been having trouble with their backs. This is why people gain such relief from getting down on all fours—the upright posture has brought some disadvantages with which evolution has not yet caught up. The natural all-fours position exaggerates the curvature of the spine at the waist and reduces pressure on the back. Standing, especially for long periods of time, provides little relief from back pain. Poor posture and uncoordinated or erratic movements can cause strain on the back and the main structure that supports it, the spine.

The back

Anatomically, the back is the area of the trunk that runs from the base of the neck down to the base of the spine—a column of bones known collectively as vertebrae. Aside from the neck, the spine is made up of 12 thoracic vertebrae at the back of the chest; five much larger bones in the lumbar region (in and below the curvature of the waist); and five fused vertebrae, called the sacrum, that form a triangular bone at the back of the pelvis. At the base of the spine is the coccyx, three to five small vertebrae: these are all that remains of the tail, a structure from humanity's evolutionary past.

BONES IN THE BACK

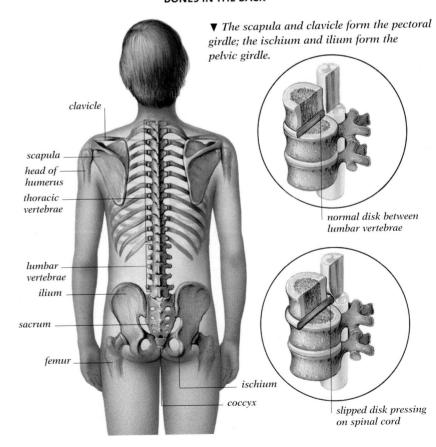

▼ *The scapula and clavicle form the pectoral girdle; the ischium and ilium form the pelvic girdle.*

clavicle

scapula

head of humerus

thoracic vertebrae

lumbar vertebrae

ilium

sacrum

femur

ischium

coccyx

normal disk between lumbar vertebrae

slipped disk pressing on spinal cord

Questions and Answers

I have chronic back pain, and my doctor says I should wear a corset for relief. I am afraid my friends will laugh at me if they find out. Do I have to do this?

If the pain is bad, it may be better to face some teasing from your friends than to continue to suffer. They will soon forget, and anyway, they should be glad that you are getting some relief at last.

This kind of corset, measured and fitted by a qualified practitioner, will provide support for the spine while holding it in the correct position. This will immobilize the spine and prevent you from bending. It should relieve any nagging back pain that you have.

A corset is also used to support the spine when a degenerative disk disease, in which the spine has developed curvatures, is present. In both cases it is important that the spine, though immobilized, does not lose its strength. Carrying out a regular program of stretching exercises will prevent the muscles from weakening and wasting away.

My grandmother says her back trouble is due to a weak spine, which she thinks is hereditary. Could this be true, and could she have done anything to prevent the problem from getting worse as she grew older?

No, back pain is not hereditary. Spinal disorders that run in families are rare. There is no way of avoiding the effects of aging, during which the cartilage of the disks becomes brittle and dry. Your grandmother may have damaged her spine and made it worse by lifting heavy items or by having poor posture. Also, she could have developed osteoporosis, which made her back more prone to injury.

Can a chill cause backache?

Yes. The cold may make your muscles stiff and inflamed, especially after exercise has made you sweat. Take a hot bath and get a good night's sleep.

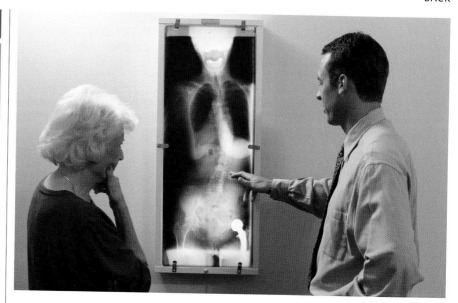

▲ *A chiropractor shows a female patient an X ray of her spine, pointing out vertebrae that have moved out of alignment and the area of the spine that is in need of treatment.*

At each end of the spinal column is a ring, or girdle, of bones that provides support for the limbs. At the top, the pectoral girdle, to which the arms are attached, consists of the collarbone and shoulder blades, together with their various muscular attachments to the spine. At the base of the vertebrae is the pelvic girdle—the pubic bones, the iliac bones, and the sacrum (part of the spinal column). The vertebrae in the lumbar region take the greatest strain during physical activity; the thoracic vertebrae can also be under strain.

Causes of back pain

Backache and back pain can be due to any number of causes; they may be related to a physical defect in the spinal column, for instance, or they may be the result of another disease or condition in some other part of the body. Backache and back pain can even be psychological.

The most common form of backache is lower back pain. It can occur suddenly or develop over hours or days and is caused by lifting or twisting following injury or overuse; or it can happen for no apparent reason. The result is a tearing of the ligaments, inflammation of the joints between the vertebrae, or a combination of both. Back strain can also result from spending long hours at a computer, particularly in a chair of the wrong height.

Slipped disk

Backache is a problem that most people can live with; a slipped disk is not. Each thoracic and lumbar vertebra is cushioned against friction by a disk that has a tough outer layer and a soft jellylike center. This provides easy, comfortable movement, the disks expanding and contracting as the spine moves. If the ligaments are in a relaxed position when the vertebrae to which they are attached suffer a sudden jolt, one or more disks can be damaged. The phrase "slipped disk" is misleading and wrong. Intervertebral disks are firmly attached to the vertebrae above and below and cannot slip. What happens is that degeneration of part of the outer fibrocartilaginous ring of the disk allows some of the disk's pulpy center to be squeezed through a split in the ring. If this happens at the rear of the disk, the pulp may press on the emerging roots of the spinal nerves, causing severe pain that may be aggravated by coughing, sneezing, or straining.

Wear and tear

Disks are subject to a degenerative disease in later life. They can literally wear away, causing great pain and disability. This condition is due to the general aging process and rarely occurs in anyone under 50 years of age. Women are the main sufferers, usually as a result of postmenopausal osteoporosis; men in physically demanding occupations are also potential victims.

As people age, the disks become gradually thinner, starting with those in the lumbar region. Disk and bone shrinkage can reduce a person's height by 4 or 5 inches (10 or 13 cm). Increasing strain is placed on the vertebrae and their supporting ligaments, which often develop bony growths called osteophytes. These can cause a lot of pain, and when the patient undergoes manipulation by an osteopath, sharp, cracking sounds can be heard as the growths break up. Problems with bones can also cause backache and back pain. The most common of these problems is osteoporosis, in which the bones suffer demineralization and substantial loss of protein. The main structure of bone is protein, which is stiffened and strengthened by calcium and phosphate salts. Loss of protein and minerals severely damages bones, which become increasingly weak and brittle. The condition usually occurs to a great extent as part of the aging process, and it can suddenly accelerate. This leads to a continual backache. The spine becomes increasingly bent, and in rare instances the vertebrae can even crumble. A fall, compression of the spine in an accident, and bone tumors will make bones affected by osteoporosis more likely to fracture.

Disease

Cancer of the vertebrae is rare, but cancer can spread to bones from other body sites. An X ray or bone scan may raise suspicion for cancer involving bone; confirmation must be made by a biopsy of the site. Tumors respond to radiotherapy, which will relieve the pain and may delay or stop further growth.

Tuberculosis of the spine, while rare, can be accompanied by abscesses that extend into the adjacent muscles. X rays can raise suspicion for this disease, but diagnosis must be made by appropriate laboratory testing.

The pain from a gallstone can lead to intense back pain. These stones cause trouble when they block the tubes that empty the gallbladder. When the blockage occurs, the patient suffers pain that can last from two or three hours to several days.

Shingles is an infection of the nerve endings that can cause pain along the course of the nerve. If the nerve supplies the back, the patient may confuse this pain with pain caused by a slipped disk.

Other causes and symptoms

One of the most common causes of persistent, nagging, lower back pain in women is pelvic inflammatory disease (PID). This is a chronic infection affecting any or all of the pelvic organs. It is usually sexually transmitted and caused initially by chlamydia or gonorrhea. Later, other germs infect the damaged tissues. The infection usually starts in the fallopian tubes, then spreads to infect the uterus, the ovaries, and the lining membrane of the abdomen (the peritoneum). Apart from pelvic and lower back pain there may be pain during sexual intercourse.

"Psychosomatic back pain" is the term used when no physical cause can be found. Such pain is usually due to stressful situations—for example, because of personal problems at home or conflict at work. It may be a short-term condition or more persistent, depending on the underlying cause and the type of treatment given by the doctor.

Lower back pain is often difficult to locate specifically. It is usually located in the lumbar region but can spread to the buttocks and thighs. Sometimes there is a history of injury (from an automobile crash, for example) or excessive use (as among laborers and gardeners) preceding the pain. Bending, straightening, or any side-to-side movement can make the pain worse. These motions can also make it difficult to stand up straight without acute discomfort.

A slipped disk differs in that the onset of pain is usually the result of a particularly awkward movement. Patients may tell a doctor that they "felt their back go." If the disk pulp presses on the nerve roots emerging and entering the spinal cord, pain may extend down one of the legs as far as the foot. This is called sciatica. It can be aggravated by bending forward, and coughing or sneezing may

Types of back pain and their causes	
CONDITION **Related to the spine**	**CAUSES**
Lower back pain—strained muscles or ligaments	Repetitive bending and lifting; bad posture—standing or sitting; twisting of spine while lifting or moving; obesity; fatigue; lack of fitness; history of injury
Slipped disk (or disks)	Same as above; accidents (car, falling, etc.); excessive physical activity
Degenerative disk disease	Childbearing; housework; aging; heavy physical work over a long period of time
Ankylosing spondylitis (form of arthritis)	Hereditary; sometimes unknown
Osteoporosis (bone disease)	Loss of the anabolic (building up) effect of estrogens after menopause
Fracture	Accident
Bone cancer	Abnormal cell division
Relating to condition in another part of the body	
Kidneys (pain in)	Injury; stones
Gallstones	Blocking of tubes that empty gallbladder
Shingles	Virus infection that travels along nerves
Muscle abscess	Tuberculosis of spine
Gynecological disorders – low back pain – pelvic pain	Prolapse of womb (protrudes into vagina) Pelvic inflammatory disease (PID)
Psychosomatic illness	Imaginary; no physical cause; due to stress and anxiety

DAILY EXERCISES TO AVOID BACKACHE

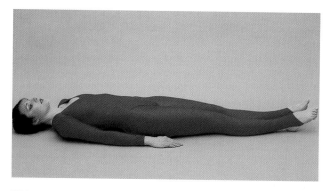

▲ *Lie flat on your back, legs straight, arms at your sides. Slowly raise one knee at a time toward your chest. Repeat the movement 10 times.*

▲ *Lie with your head supported on a folded towel, knees bent, feet flat on the ground. Gently raise and lower your buttocks. Repeat 10 times.*

▲ *Stand upright with your arms loosely by your sides and shoulders back.*

▲ *Bend forward slightly from the waist, letting your head and arms fall forward.*

▲ *Continue the movement by reaching gently toward your toes 10 times.*

send a sharp pain through the back. In the most severe cases, pressure on the lower nerve roots may cause weakness in the legs and an inability to pass urine. Emergent evaluation is required.

Identifying the problem

The backache of wear-and-tear arthritis and worn disks can mimic the symptoms of a slipped disk, with intermittent bouts occurring every one or two weeks over several months.

Osteoporosis can be confirmed by X ray or a bone scan. The bones are sensitive to pressure, and patients with osteoporosis may complain of a constant backache, increased by movement. Night pain is common, and resting makes little difference. Morning stiffness is also a symptom, as it also is in ankylosing spondylitis, a form of arthritis in which the vertebrae fuse to form a "poker back." The pain starts low on both sides, and the spine becomes more rigid as the condition worsens.

Any fracture of the vertebrae results in a localized, relentless pain that is intensified by movement. Bone cancer, fortunately very rare, is accompanied by feeling generally sick, tired, and run-down. A low backache in the last three months of pregancy is normal, as the weight of the fetus places an increasing strain on the lumbar region.

Pain from gallstones is usually felt below the right front ribs but can also be felt in the back, sometimes accompanied by nausea. With shingles, back pain coincides with a blistering rash.

Treatment

The main treatment for moderate to severe back pain is a limited period of bed rest and painkilling drugs. The back needs total relief, with a consequent easing of strain on torn muscles and strained disks or ligaments, as the case may be.

The sufferer should lie on a firm surface; if the mattress is not firm enough, the patient should lie on the floor. In cases of lower back pain, relief can be gained by bending the knees or lying on the side, with a cushion placed in the small of the back. After the back is made comfortable, the patient should keep still.

Bed rest is ineffectual, however, for the intermittent pain caused by degenerative disease. Shorter periods of rest are then best, with activity helping to keep muscle tone and bones healthy.

Persistent pain

If pain persists, acetaminophen or something stronger will help. If a muscle goes into spasm, resulting in a throbbing pain, muscle relaxants can help.

A doctor diagnoses the causes of back pain by a process of elimination. Strained muscles will improve within several days; torn muscles and ligaments take a bit longer for the pain to subside; and a slipped disk takes even longer, sometimes weeks or months. If no improvement occurs after two weeks, a doctor's help must be sought. If there is a concern for a serious underlying condition, such as a tumor or abscess of the spine, the doctor may advise hospitalization. If testing reveals a fracture in the spine, the doctor

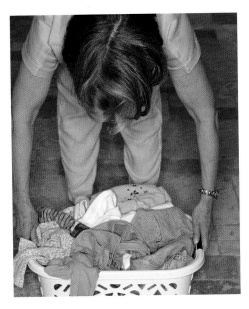

◀ *Many backaches are caused by lifting objects incorrectly; always bend your knees before lifting things.*

may ask the patient to wear a surgical corset or plaster cast once he or she is no longer bedridden.

For severe pain that reaches down to the leg, a spinal injection may reduce the swollen tissues and relieve the pain. If it does not, the hospital may perform a myelogram, in which a dye is injected into the spinal column while the patient is under a local anesthetic; the back is then X-rayed to reveal the damage and to indicate whether surgery is necessary.

Prevention

The likelihood that a slipped disk will recur increases each time it happens, so even a single occurrence should signal a warning that the patient's lifestyle should be changed to avoid recurrence. For example, a teenager who has slipped a disk playing sports or dancing should minimize those activities in the future; in the same way, a gardener with lower back pain should find a more sedentary pursuit. It is important not to resume the habits that led to the problem.

Backache can be avoided by following some simple preventive measures. If a person must bend repeatedly, it is best to make sure the knees are bent to reduce strain on the spine. Slouching should be avoided. All work surfaces should be at a comfortable height, and only chairs that have a good, supportive back should be used. A firm bed is an essential support for the spine at night.

If it is necessary to twist the body while lifting, the movement should be made in two separate stages. In shoveling sand, some should be lifted onto the spade, then turned and dumped elsewhere: the body should not be twisted to dump the sand. In getting out of a vehicle, the body should be twisted first, then the feet should be placed on the curb before one stands up. Maintaining an appropriate body weight reduces strain on the back and aids posture. If osteoporosis is likely to become a problem, the doctor may prescribe the drug alendronate to prevent it. If back pain still occurs, special care must be taken. Meals should be eaten at a table. Stairs should be climbed with a straight back, and lifting or bending avoided. A heavy workout program is not advisable; the doctor will recommend exercises to recover muscle strength. Gentle walking is good—as is swimming.

Improvements in furniture design are being researched, planned, and undertaken to improve posture at home and at work. The field of wear-and-tear arthritis is giving doctors clues to advance the treatment of back pain.

Trained physiotherapists, chiropractors, and osteopaths can all be consulted on the advice of the family doctor.

See also: Aging; Bones; Gallbladder; Neck; Pelvis; Posture; Pregnancy; Spinal cord

Balance

Questions and Answers

Why does my grandfather find it so difficult to keep his balance?

Like all senses, the sense of balance begins to deteriorate in old age, so it is just as common to have difficulty in balancing as it is to have failing eyesight or hearing. Poor balance in the elderly can be dangerous, so your grandfather should be particularly careful, for example, when climbing up or down a staircase, where a fall could be serious.

Why don't ballerinas get dizzy when they keep twirling around?

If you watch a dancer carefully you will notice that her head does not move around with her body—she turns her head in one swift movement. This prevents the fluid in the balancing mechanism in the inner ear from swirling back and forth, making her feel dizzy.

My baby girl can balance well but prefers to crawl rather than walk. Is this normal?`

Your baby may need more practice at balancing before walking. At the moment, she is sure that she can crawl, so this is the most obvious way for her get around. Don't worry about her walking—it will come when she is confident that her body is ready.

Why do some soldiers collapse while on guard?

To keep the body in a stiff, upright position, the muscles work hard and use some of the blood that would normally go to the brain. The heart has to work harder than usual to pump blood up to the brain. The soldier on guard keels over, not because of a fault with the balancing mechanism in the ear, but because the brain has been suddenly deprived of blood.

A sense of balance is essential for human beings. It is something that normally is taken for granted, and it is really noticed only when something goes wrong. So how do humans manage to keep upright?

Humans are unique among mammals, being the only ones who always walk upright on two legs. To keep to this unusual stance, humans rely on a highly developed sense of balance. The body's balancing mechanism is intricate and is mostly contained inside the ears. As well as being an organ of hearing, they are also responsible for moment-by-moment monitoring of the position and movements of the head. If the exact position of the head is correctly monitored, then the body can adjust itself to stay balanced.

Well-protected by the bones of the skull, the delicate organs of balance lie in the innermost part of the ear, the inner ear. Inside lies a maze of tubes filled with fluid, at various levels and at differing angles. Of these tubes, the ones directly involved in balance are the utricle, the saccule, and the semicircular canals.

Detecting position

The utricle and the saccule are concerned with detecting the position of the head. Each of these two cavities contains a pad of cells overlaid with a jellylike substance, which is embedded with small granules of chalk. When the body is upright and the head still, these remain unbent and a resting signal is sent by the hair cells along the auditory nerve to the brain. If the head makes a rotary movement, the bodies of the hair cells move but the inertia of the fluid causes the hairs to bend in the same direction as the head. Bending the hairs prompts nerve impulses that pass to the brain. When the head leans forward, backward, or sidewise, the hairs bend in a different

HOW THE BODY BALANCES

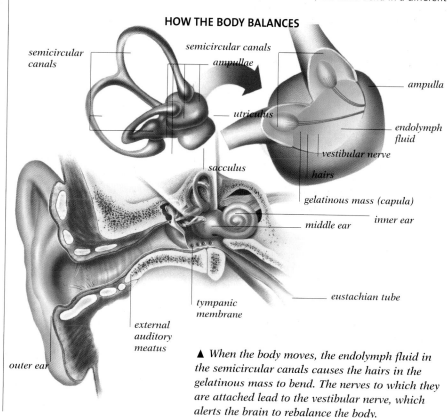

semicircular canals

semicircular canals

ampullae

ampulla

utriculus

endolymph fluid

vestibular nerve

sacculus

hairs

gelatinous mass (capula)

inner ear

middle ear

tympanic membrane

eustachian tube

external auditory meatus

outer ear

▲ *When the body moves, the endolymph fluid in the semicircular canals causes the hairs in the gelatinous mass to bend. The nerves to which they are attached lead to the vestibular nerve, which alerts the brain to rebalance the body.*

Causes of dizziness associated with ear problems

CAUSE	EFFECT	OTHER SYMPTOMS	ACTION
Bacterial infection	Middle ear infection	Pain, hearing loss	Never try to "unblock" ears. See a doctor.
Infection, careless use of ear cleaners	Perforated eardrum	Nausea, vomiting, blood in ear canal, pain, hearing loss	See a doctor. Put cotton ball outside the entrance to the ear canal for temporary relief.
Colds, infections, allergies, ascending to heights, travel in poorly pressurized aircraft	Blockage of eustachian tube	Impaired hearing, "full" feeling, and noises in ears	See a doctor. Use nose drops or nasal spray decongestants for temporary relief.
Ménière's disease	Severe loss of balance	Loss of hearing, loud noises in ears	See a doctor.
Car, sea, or airplane travel	Motion sickness	Nausea, vomiting	Prevention with medication.

way, firing off new messages to the brain, which instructs the muscles to adjust the position of the body. The utricle is also in action when the body starts to move forward or backward. If a child starts to run, the hairs are pushed back as though the child were falling backward. When the brain receives this information, it sends signals to the muscles, which make the body lean forward, restoring its balance. These reactions are reversed if the child falls backward off a chair.

Starting and stopping
Jutting out just above the utricle of the ear are three fluid-filled semicircular canals. At the base of each is an oval mass of jelly containing the tips of sensitive hairs, which become bent by movements of fluid in the canals as the head moves. The canals pick up information about when the head starts and stops moving, which is important during quick movements. As the head begins to move one way, the fluid in the canals tends to stay still, pushing against the sensitive hairs. The hairs send messages to the brain, which can take action. When the head stops moving, particularly when it stops turning around and around, the fluid goes on moving inside the canals for up to a minute or more, causing dizziness.

The part of the brain most responsible for directing the action of the muscles in keeping the body balanced is the cerebellum. As well as taking in messages from the balancing organs in the ears, this section of the brain receives quantities of other information. It coordinates messages from the eyes, the neck, the spine, and the limbs. The eyes play a part in balance, because they provide vital information about the body's relationship to its surroundings. The eyes also have an important linkup with the semicircular canals. When the head moves to the left, the movement of fluid in the

◀ *This gymnast's skill on the beam demonstrates the highest development of the balancing process.*

semicircular canals makes the eyes move to the right, but the balance mechanism then makes them move to the left to adjust to the same position as the head. This eye movement partly explains why people are likely to be sick if they try to read while traveling in a moving vehicle. The reading tends to counteract these natural eye movements, triggering an episode of travel sickness.

Learning to balance
Learning to balance is a long process, taking almost the first two years of life. It takes another year for a child to master the art of standing on one leg, and even longer to balance on a narrow beam. Before a child can achieve perfect balance, the brain and muscles must both be mature enough to provide essential coordination and strength.

What can go wrong
The most serious disorder of balance is Ménière's disease. The underlying cause is not known, but the possibilities include allergies, infections, tumors, and the side effects of some drugs. Because the hearing and balancing mechanisms are so close together in the ear, hearing problems can affect the balancing mechanism; for this reason, dizziness can accompany ear infections and bad colds. Tests can determine if the balancing mechanism is working. These include tests of how movement affects balance and other specialized tests.

See also: Earwax; Hearing

Baldness

Most men worry about losing their hair and may in fact do so at some point in their lives. Baldness can also affect women. What causes the condition, and can anything be done to prevent it?

Washing the hair does not cause baldness, but always use a mild shampoo, and condition your hair. Otherwise it can become dry and brittle, with the result that hairs break off, making the hair coverage look thin. However, the roots will be undamaged, and the hair will grow back to its normal thickness if treated with care.

My 17-year-old sister wears her hair in cornrows. Recently I noticed that it has been thinning. Could she be getting bald patches?

If the hair is pulled for any length of time, there is a danger that it break or fall out. Your sister may have been braiding her hair too tightly. If she wears it loose for a few weeks, the lost hair should grow back.

My bald patch is growing. Should I try a hair restorer?

It is worth a try. Minoxidil does cause some hair growth in many people, but don't expect it to work miracles.

I like wearing hats, but my father says that this is why he became bald. Is this true?

No. This is a common myth. A man with a receding hairline sometimes blames it on the brim of a hat rubbing against his hair. Or, if he is thin on top, he may say that the crown of the hat restricted the growth of the hair there. The only effect a hat can have on baldness is to conceal it.

Is baldness inherited?

You can inherit the tendency to go bald from either your mother's or your father's side.

The biggest single factor in baldness—either partial or total—is heredity. But men who have not inherited the trait may find themselves going bald for many other reasons. Occasionally, baldness can happen to women too.

Male-pattern baldness

This hereditary condition is the most common cause of baldness in men. It can be inherited from the mother's or the father's side of the family and involves the male sex hormone androgen. It is thought that androgen shortens the growing time of the hair, making all the hairs increasingly thin and short until the roots are producing nothing but a fine down. The drug minoxidil may increase hair growth.

Other causes of hair loss

Mistreating hair can cause temporary baldness. Wearing a ponytail, braids, or other styles that pull hair tightly, or even just frequent tugging, can cause hair loss in men, women, and children. The medical term for this is "traction alopecia." Sometimes temporary baldness results from overbleaching the hair or getting a permanent wave. Protein and vitamin deficiencies can also damage the hair. These types of hair loss are generally temporary; the hair grows back to normal strength and thickness once the cause is removed.

Hormonal changes in the body frequently cause thinning of the hair in women—for example, after a pregnancy or on stopping the Pill. These hormonal swings generally adjust themselves naturally, and thinning or damaged hair will be replaced by a good, strong growth. Some women find that they lose quite a lot of hair after menopause; they may even develop a receding hairline or thinness of the hair on top of the head. Disorders of thyroid hormone production may also cause hair loss.

Certain illnesses bring with them the possibility of temporary baldness. Alopecia areata results in hair loss in patches from the scalp and sometimes from the rest of the body. Its cause is not yet known, and there is no satisfactory cure, but often the hair does grow back of its own accord. Finally, acute physical or mental stress can result in hair loss, but permanent damage is unlikely, and the hair usually grows back.

Outlook for the future

For those people who wish to treat or disguise their baldness, there are a number of options. In addition to the drug minoxidil, various types of hairpieces and hair transplants are available.

See also: **Hair; Hormones; Menopause**

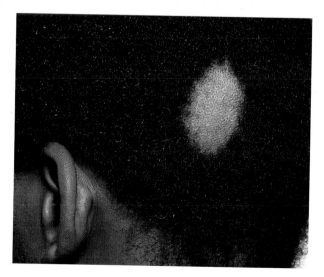

▲ *Alopecia areata can affect both men and women.*

Basal metabolic rate

The energy a person expends when he or she is completely at rest is known as the basal or resting metabolic rate and consists of the energy required to maintain basic functions such as breathing.

Yes. Men generally tend to have higher BMRs than women. This is because more calories are required to fuel muscle, and men usually have a higher proportion of muscle to fat than women. On average, women have about 5 to 10 percent more body fat than men of a similar size.

BMR is the amount of energy used for essential functions, such as maintaining breathing, body heat, and heart rate. It is raised in the short term by some illnesses, fever, stress, cold or hot temperatures, and pregnancy.
BMR decreases with age, when there tends to be a lower proportion of muscle to fat. It also decreases in people who are malnourished or who reduce their food intake by dieting or fasting.

Unfortunately, less food causes the body to lower its BMR in an attempt to to slow down and conserve energy. This slowing down can cause what is known as a plateau effect when people are dieting—no further weight loss can be achieved. A regular routine of cardiovascular exercise will help to counteract this effect by raising the BMR and thus the rate at which calories are used.
Even if you don't have the time or money to go to a gym, you can burn calories and help raise your BMR by walking to work or walking up the escalator when leaving the subway or climbing any stairs at work instead of catching the elevator.

The basal metabolic rate (BMR) is the rate at which a body uses up energy when completely at rest just to maintain vital functions such as breathing, circulation, heartbeat, digestion, and nervous activity. A key regulator of BMR is thyroxine, a hormone produced by the thyroid gland.

BMR varies from person to person, depending on how much energy a person uses when resting. As a general guide, BMR is about half a calorie per pound of body weight per hour. So if someone weighs 120 pounds (54 kg) he or she will use about 60 calories an hour or 1,440 calories each day when resting. People who weigh more therefore have higher BMRs; very heavy people have a BMR about 25 percent higher than that of lighter people. Although metabolism affects a person's weight, it seldom causes obesity except in rare cases of serious metabolic disease.

▲ *Physical exercise such as jogging burns up calories and changes the BMR.*

How to calculate BMR

"Burning" fuel in the form of calories to create energy produces heat, so BMR can be estimated by recording the amount of heat released from a resting person's body surface over a period of time. Energy production also requires oxygen, so BMR can also be determined by measuring the amount of oxygen consumed over a period of rest. It is also possible to estimate BMR using variations of the Harris-Benedict equation, derived by Harris and Benedict in 1919 for estimating BMR according to gender, age, height, and weight. For men, BMR = 67 + (6.24 x weight in pounds) + (12.7 x height in inches) − (6.9 x age). For women, BMR = 661 + (4.38 x weight in pounds) + (4.33 x height in inches) − (4.7 x age). In addition to life-sustaining calories, the body requires more energy for daily activities like eating, sitting, or more strenuous exercise. Once the BMR is known, the extra calories required can be estimated for these activities. To work out the approximate number of calories needed each day the BMR is multiplied by the following activity factors. For minimal physical activity, multiply BMR by 1.3; for moderate physical activity, multiply BMR by 1.4; for high physical activity (exercising four times a week), multiply BMR by 1.7; for extremely high physical activity (exercising daily), multiply BMR by 1.8. After the approximate number of calories required is calculated in relation to the amount of activity, calorie intake can be adjusted to the optimum level.

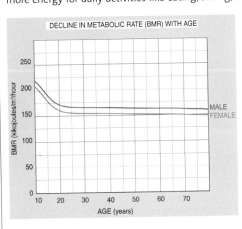

DECLINE IN METABOLIC RATE (BMR) WITH AGE

MALE
FEMALE

BMR (kilkojoules/m²/hour)

AGE (years)

▲ *After about age 10, BMR decreases with age and tends to be lower in females (red line) than males (blue line).*

See also: **Metabolism**

Bile

Bile, a bitter-tasting substance, is manufactured by the liver and stored in the gallbladder. It plays an important role in the digestive process, and its importance has been recognized since the early days of medicine.

Questions and Answers

My father has to have his gallbladder removed. Will he be able to eat normally afterward?

The digestive system can manage remarkably well without a gallbladder. Normally, bile is stored in the gallbladder before being sent to the duodenum. When the gallbladder is removed, the bile simply passes straight into the duodenum. This bile is less concentrated than normal, which means that it may be less efficient in its job, so the doctor will probably advise your father to go easy on fatty foods and to eat smaller, but more frequent, meals. In a short time, he will probably find that his body readjusts to the new circumstances, and he will be able to eat perfectly normally.

Are there really two different kinds of bile?

This idea is a holdover from the early days of medicine. The ancient Greeks believed that bile could be black or yellow. Too much black bile was supposed to cause depression; too much yellow bile made a person irritable.

We now know that there is only one sort of bile, which is greenish yellow in color. The word "bilious," however, still has its meaning of peevish or bad-tempered.

Does feeling bilious indicate a bile problem?

A bilious attack often occurs if the duodenum becomes irritated and distended by infection—or by too much food and drink. The bitter, bile-containing fluid in the duodenum gets flushed back into the stomach, resulting in nausea and vomiting. The bitter taste in the mouth and the greenish color of the vomit are due to the presence of bile. So a bilious attack means bile is in the wrong place, not that there is something wrong with its production.

Bile is a thick, bitter, greenish yellow fluid that is made in the liver and stored in the gallbladder. Released from the gallbladder into the small intestine in response to the presence of food, it is essential to the digestion of fats. Bile also contains the remnants of worn-out blood cells and is, therefore, also part of the body's excretory, or waste-disposal, system. Every day, the liver produces about 2.1 pints (1 l) of bile. Although bile consists of over 95 percent water, it contains a great variety of chemicals, including bile salts, mineral salts, cholesterol, and the pigments that give the bile its characteristic greenish yellow color.

PASSAGE OF BILE

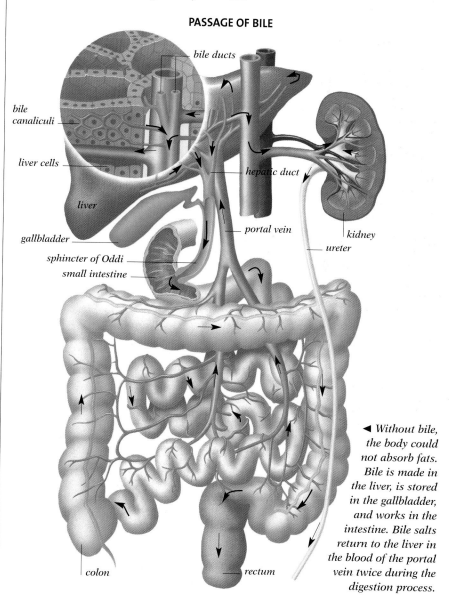

◀ *Without bile, the body could not absorb fats. Bile is made in the liver, is stored in the gallbladder, and works in the intestine. Bile salts return to the liver in the blood of the portal vein twice during the digestion process.*

▶ *Gallstones are hard lumps of cholesterol, bile pigments, calcium salts, and protein, which form in the gallbladder. They cause the gallbladder to become inflamed and may make its removal necessary.*

Bile is made continuously, in small quantities, by every cell in the liver. As it flows from the cells, it collects between groups of liver cells in minute channels called bile canaliculi. These channels empty into bile ducts, or tubes, placed between the lobes, or projecting parts, of the liver. From the bile ducts, bile drains into exit tubes known as the hepatic ducts. Unless bile is needed immediately for digestion, it flows into the gallbladder, a storage sac located under the liver.

Bile and digestion
The bile remains in the gallbladder until it is needed to play its part in the digestive process. As food—particularly fatty food—enters the duodenum (the first portion of the intestine) from the stomach, the duodenum makes a hormone called cholecystokinin. This hormone travels in the bloodstream to the gallbladder and makes its walls contract so that bile is squeezed out. The bile then flows down another tube, the common bile duct, and through a narrow gap, the sphincter of Oddi, which lets bile into the small intestine.

Bile's mineral salts, which include bicarbonate, then neutralize the acidity of partly digested food in the stomach. The bile salts, chemicals called sodium glycocholate and sodium taurocholate, break down fats so that the digestive chemicals (enzymes) can do their work.

Ferrying action
In addition to this detergent-like action, the bile salts are also believed to act as "ferries" further down the intestine, enabling digested fats to travel through the intestinal wall. They are also carriers of the vitamins A, D, E, and K.

The body is very conservative in its use of bile salts. They are not destroyed after use; instead 80 to 90 percent of them are carried by the blood back to the liver, where they stimulate the secretion of more bile and are used again.

Many colors
Bile gets its color from the presence of a pigment called bilirubin. One of the many jobs of the liver is to break down worn-out red blood cells. As this happens, the red pigment hemoglobin in the cells is chemically split and forms the green pigment biliverdin, which is quickly converted to the yellow-brown bilirubin.

The greenish tinge of bile comes from unconverted biliverdin. In addition to pigmenting the bile, bilirubin colors and partially deodorizes the feces and also encourages the intestine to work as effectively as possible.

Bile pigment is also partly responsible for the yellow color of urine. In the intestine, bilirubin is attacked by bacteria (minute living creatures) permanently stationed there and converted to a chemical known as urobilinogen, which is carried to the kidneys before being released in the urine.

Problems in bile production
When something is wrong with the liver or gallbladder, bilirubin tends to accumulate in the blood and the skin, and the whites of the eyes look yellow. Because too little bile is reaching the intestines, the feces may be a pale grayish color.

Gallstones
Even if the liver's bile production system is working normally, things can go wrong in the gallbladder. Most notorious of gallbladder problems are gallstones. These hard lumps, mostly of a chemical substance called cholesterol, form in the gallbladder and may travel down the bile duct into the duodenum. The gallbladder becomes inflamed, and the movement of the stones down the bile duct can be very painful.

Treatment
The treatment of the different bile problems and related diseases depends on their cause. Liver infections and cirrhosis of the liver are difficult to treat with drugs, and the usual cure is rest, no alcohol, and a nourishing diet.

Surgery may be the only way of treating gallbladder problems. In the case of gallstones, this means removing not just the stones but the whole gallbladder. The operation is usually carried out using minimally invasive techniques, or keyhole surgery. Unless the gallstones are so severe that they are a medical emergency, the doctor may first advise a low-fat diet to reduce stress on the digestive system and lower the amount of cholesterol in the blood. When surgery is the only permanent answer, the body can manage without a gallbladder. People who have chronic gallstones find remarkable relief from their previous symptoms of pain and discomfort, and the body's tendency to make gallstones vanishes.

See also: **Digestive system; Gallbladder; Liver**

▲ *Bile is a greenish fluid, simulated here to show how its emulsifying salts act like detergents. They physically break down globules of fat during digestion.*

Biological clocks

Like plants and animals, we possess internal clocks that govern our eating, sleeping, and waking times. However, unlike plants and animals, we are able to reset these clocks to fit in more usefully with everyday life.

Many people wake up early on weekends even when they have the chance to sleep late. Others suffer from jet lag when they fly long distances across time zones. Some women get irritable at certain times of the month or find that their hair sometimes becomes oilier than usual. These people may have one thing in common: they are feeling the effects of one or more of the biological clocks within their bodies.

Rhythmic cycles in the behavior of animals and plants as well as in humans have been observed for hundreds of years. Yet little is known about how they work, except that they are probably genetically coded—that is, built in as a result of instructions from the genes acquired at the moment of conception—and that they act by releasing hormones into the body.

The three clocks

The brain has an accurate interval timer based on the firing of nerve cells in the outer layer (cortex). This timer can measure intervals ranging from a few seconds to hours. It operates to time conscious events so that their duration can be compared with that of past experience of the same events. This allows appropriate behavior modification. The interval timer is essential for rhythmic activity such as dancing or beating time to music.

A second clock, the circadian clock, times the many physiological processes that work on 24-hour cycles of maximum and minimum activity. It is synchronized by light signals entering the eyes during daytime. Cells in the retina send signals to a small nerve nucleus, the suprachiasmic nucleus, in the middle of the underside of the brain, and this nucleus sends out messages to other parts of the brain that control circadian functions. In dark conditions, some of these messages cause the pineal gland in the brain to secrete the hormone melatonin, which has been

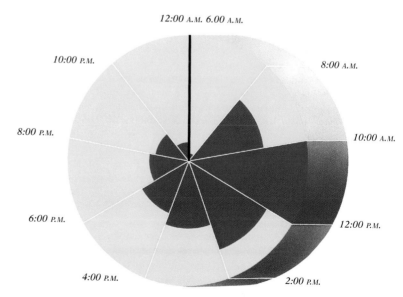

WHEN ARE YOU BEST AT MATH?

▲ *On average, people are most accurate between 10:00 A.M. and noon. This level of accuracy falls gradually until midnight. The cycle begins again, provided you have had a normal night's sleep, at 6:00 A.M.—the time of least accuracy.*

used to treat jet lag. A third clock determines the total number of times many cells can divide. During life this is between 60 and 100 times. This clock is driven by shortening which occurs in the ends of the chromosomes each time a cell divides and which eventually stops cell division. The ultimate life span is thought to be limited by this mechanism.

The effects

All humans show evidence of daily and monthly cycles in their behavior. The regularity of the sleep cycle is an obvious example, but oxygen intake rate, urine excretion rate, the amount of sugar in the blood, and the levels of adrenaline and many other chemicals in the body all vary over 24 hours, as do both the body's resting pulse rate and temperature.

Tolerance to pain is highest at about midnight. Alcohol is broken down by the body much more slowly between 2:00 A.M. and 2:00 P.M., and the effect is therefore felt much longer between these hours. This may partially explain why lunchtime drinking can affect working ability so noticeably.

The monthly biological clock is most clearly shown in a woman's menstrual cycle. It affects both the physical state—changes in weight, body-fluid retention, skin and hair condition—and the psychological state, so that moods of depression, sudden irritability, or nervousness may occur. Evidence that these periodic changes are controlled by internal mechanisms—rather than by the cycles of darkness and light or changes in external temperature—is fairly strong, although it is known that humans are better able than animals or plants to adapt their biological clock messages to suit their environment. For example, at first babies fall asleep and wake up with complete disregard for either the day or night cycle or their parents' behavior, but beginning at about six weeks of age they establish a regular pattern of sleeping and waking that is more or less 24 hours long.

Two experiences that demand a drastic altering of the biological clock are doing shift work and flying to a country in a different time zone. A change in the activity-rest cycle is inevitable.

Fast and slow clocks

Since not everyone's daily cycle covers exactly 24 hours, many people need to adjust their biological clocks regularly to synchronize with the external clock by which they have to live. In extreme cases, an individual's daily clock may run fast or slow by as much as half an hour a day. This condition helps to account for some very common, if puzzling, minor difficulties in sustaining regular sleeping patterns.

If a person's daily biological clock runs a little fast, he or she will tend to become sleepy early in the evening but will be able to wake at a reasonable hour in the morning with little difficulty. This type of person is often known as an "early bird." Normal exposure to the day-night cycle of the outside world will help "early birds" reset their clock each day; otherwise they would end up going to bed and getting up earlier and earlier. Because of this need to readjust the internal clock, the person is continually under pressure from the external everyday environment. "Early birds" therefore have to try to compensate for the sleep deficit that gradually builds up by taking an occasional afternoon nap or sometimes going to bed especially early. Studies have also shown that "early to bed, early to rise" people are generally shyer, more reserved, and more anxious than "late to bed, late to rise" people.

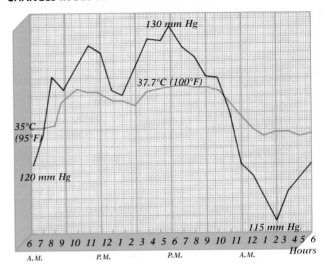

130 mm Hg

37.7°C (100°F)

35°C (95°F)

120 mm Hg

115 mm Hg

6 7 8 9 10 11 12 1 2 3 4 5 6 7 8 9 10 11 12 1 2 3 4 5 6 Hours
A.M.   P.M.   P.M.   A.M.

A person whose clock is running slow, on the other hand, can easily stay up until the early hours, because the internal clock signals bedtime later and later each night. However, getting up in the morning is far more of an ordeal than it is for the "early bird." These people are often known as "night owls." Since there is no message from the biological clock signaling sleep in the evening, "night owls" often lie awake in bed. Getting up a little later occasionally may help reduce the built-up sleep deficit.

In both cases, if problems persist with sleep-wake cycles or with jet lag, a doctor can prescribe a drug to help reset the biological clock.

The reasons for biological clocks

Animals are thought to have developed biological clocks to help them survive. The clocks of some species, for example, discourage activity during darkness, when the animal is more vulnerable to predators because of reduced vision. The clocks also order fertility cycles to maximize breeding efficiency and regularize migration or hibernation, thereby offsetting the effect of adverse weather or reduced food supplies and enhancing the animal's chance of survival.

Humans have retained daily and monthly clocks for probably much the same set of reasons; however, the clocks now help with our regular habits. For example, people eat at certain times, and with a signal from the biological clock, the body produces gastric juices at the right time to digest that food. At night, the clock reduces the speed at which urine accumulates in the bladder so that we are less likely to have our sleep disturbed by having to get up and go to the bathroom.

Some people can wake up from normal sleep at any given time with punctuality, even when cues such as light, birdsong, or traffic noises are absent. Experiments using hypnotic sleep show that, with practice, people can increase this accuracy so that the clock alerts them to perform an action at a certain time. It is possible to develop this ability simply by being aware and practicing it.

See also: **Blood pressure; Brain; Cells and chromosomes; Eyes; Hormones; Melatonin; Menstruation; Retina; Temperature**

Biorhythms

There are some people who believe that biorhythms underlie the patterns of human behavior and that understanding these rhythms can help people to plan their lives for the optimum timing of important events.

Questions and Answers

Can knowing my biorhythms help my relationship with my boyfriend?

Yes, a knowledge of your own biorhythms and those of your boyfriend could help your relationship run more smoothly. The most important biorhythm in all personal relationships is the emotional cycle, so compare yours with your boyfriend's. You might choose to postpone meeting if it is going to coincide with a critical phase, and make allowances for each other in the negative phases.

I have a big exam coming up. Can my biorhythms affect my grade?

Although you cannot alter the date of an exam, you can take some precautions to ensure that you perform as well as possible, whatever the state of your biorhythms. Get a good night's sleep beforehand and try to avoid any emotional upsets. Have everything ready in advance so that there is no last-minute panic. For exams, the intellectual cycle is the most important of the three biorhythms. The more you exercise your mind, the better it will be able to resist the effects of a negative intellectual phase, should this coincide with your exam.

Can I change my biorhythms?

No. The three cycles begin on the day you are born and continue in the same automatic rhythm throughout life. However, you can be aware of your rhythms and plan activities accordingly.

How can I tell the difference between my menstrual cycle and my emotional cycle?

You can track your menstrual cycle by noting the dates of your periods. You can calculate your emotional cycle by dividing the total number of days since you were born by 28.

▲ *At times when the physical cycle is at its peak, optimum performance in sports and other activities that require physical energy can be achieved.*

Most people have certain days when they feel good and perform well and other days when, for no apparent reason, they lack energy, cannot concentrate, and find themselves making mistakes that they would not normally make. Biorhythms are a way of explaining such variations in mood and behavior.

According to the biorhythm theory, human beings have three different cycles—physical, emotional, and intellectual—which rise and fall independently of each other. These three cycles work like three internal clocks, controlling people's physical well-being and behavior, their feelings, and their ability to think and reason.

The idea of cycles of growth and decline is, of course, nothing new. Cycles are seen in nature in the four seasons, and also in a woman's menstruation. Most people now accept that the menstrual cycle can have a great influence on a woman's behavior and physical and emotional states, which vary depending on where she is in the cycle. If she is in the first half, a woman will often feel physically well, cheerful, and energetic. If she is in the latter half, particularly the few days before the onset of her menstrual period, she may seem like a different person; she may be depressed, irritable, weepy, and accident-prone, among other symptoms.

There are also daily cycles that vary from one individual to another. Some people feel most dynamic and productive during the daytime hours, while others do not reach their peak until late at night. However, a few general rules have been observed. According to recent scientific

45

research on human behavior patterns, it seems that the best time for most people to make decisions is around the middle of the day, and the best time to study or rehearse is just before going to bed.

It takes three to four weeks to complete all three biorhythmic cycles. By calculating the timing of the future highs and lows in each cycle, a person can predict how they are likely to feel and behave on any one day.

The discovery of biorhythms

Three people were separately responsible for the discovery of biorhythms: a psychologist, a doctor, and an engineer. The psychologist was Hermann Swoboda of the University of Vienna. Shortly before the end of the 19th century, Swoboda became aware that there was a repetitive pattern in the behavior of human beings. He continued his research into this pattern and was able to chart a physical cycle lasting 23 days that affected a person's physical abilities and reactions.

Swoboda also noticed a 28-day cycle governing the emotions. This sometimes coincided with the menstrual cycle in women, making it easy to confuse the two. However, he was able to establish that this emotional cycle was separate from the menstrual cycle because it also occurred in men. When he was convinced that his findings were accurate, Swoboda published his first book, *Periodicity in Man's Life*.

By coincidence, a nose-and-throat specialist in Berlin, Wilhelm Fliess, was making similar discoveries at the same time as Swoboda. He, too, noticed repeating patterns over 23 and 28 days—the physical and emotional cycles.

About 20 years after Swoboda and Fliess made their observations, Alfred Teltscher, an engineer and student of mathematics, became intrigued by the possibility of an intellectual cycle. Teltscher was aware that individuals showed varying intellectual abilities at different times. On one day, a person would grasp new ideas easily and would be able to think clearly and reason well. On another day, the same person would have difficulty with intellectual tasks. Teltscher investigated further and discovered that these fluctuations

▲ *For optimum performance in business activities, such as important meetings, it is wise to choose days when the intellectual and emotional cycles are at their peak.*

> ### Planning your own biogram
>
> Calculate the total number of days in your life so far, starting from the day of your birth. Remember to include an extra day for each leap year (do not just add an extra day for every four years; you will need to check when the last leap year was and count backward from there to get the total number to date).
>
> Divide the total number of days in your life by the number of days in each cycle. If there are any days left over, this will show you which day you are on in that cycle now. If there are no days left over, you are at the beginning of a new cycle.
>
> Draw up a chart (see chart, page 47). Divide it into narrow columns for the days, with a horizontal line across the center. Using a different color for each cycle, plot all three, showing where they peak, where they cross the line, and where they are at their lowest point. By plotting your biorhythms in this way, you will be able to see what your energies and capabilities will be on any day.

in ability formed a pattern spread over a period of 33 days—the intellectual cycle.

Two doctors at the University of Pennsylvania, Rexford Hersey and Michael John Bennett, who were working independently of Teltscher, confirmed his findings.

How biorhythms work

Biorhythms have been compared to astrology. However, astrology is based on a belief that human beings are influenced by the position of the stars and planets in relation to each other. In biorhythms, the influence comes from a person's own internal rhythms, not from external forces.

The three biorhythms are said to begin at birth and continue rising and falling throughout a lifetime. They can be plotted on a graph representing a particular stretch of time, say a month. Such a graph is called a biogram. Looking at it allows a person to anticipate favorable and less favorable days in the near future.

If a biogram covers one month, it would be divided up vertically into the number of days in that month. It would also be divided horizontally by a line across the center. The "waves" of each cycle rise above, cross, and fall below the "horizon line" in an endlessly repeating pattern.

The first half of each cycle begins with a rising wave: this is the active, energy-packed phase. A person will reach the peak of his or her powers on the crest of the wave, the highest point, halfway through the first half of the cycle. After this, the person's energies and abilities will gradually diminish as the wave falls below the horizon and passes into the second half of the cycle. This is the passive phase, when energy is being restored in preparation for the next active period.

Critical days

Surprisingly, the most unstable time in each cycle is not the lowest point in the passive phase but the day on which the cycle crosses the dividing line between its active and passive halves—like an engine changing from one gear to another. It is at such sensitive turning

SAMPLE BIORHYTHM CHART FOR 33-DAY PERIOD

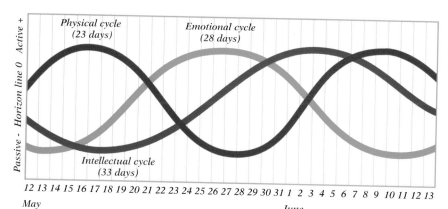

Physical cycle (23 days)

Emotional cycle (28 days)

Intellectual cycle (33 days)

Active + · Horizon line 0 · Passive –

12 13 14 15 16 17 18 19 20 21 22 23 24 25 26 27 28 29 30 31 1 2 3 4 5 6 7 8 9 10 11 12 13

May · *June*

around day 18. The emotional cycle peaks around day 8 and has its lowest point around day 22. The intellectual cycle peaks at about day 9 and is at its lowest around day 26.

In any one month, there are at least six critical days in everyone's biorhythms. Out of an average month of 30 days, these six days represent one-fifth or 20 percent; therefore, people are in a critical phase of some kind for at least 20 percent of their lives. A 30-year-old woman, for example, will have had a minimum of at least six whole years in her life of being particularly off-kilter physically, mentally, and emotionally.

points that people are more likely to make mistakes, have accidents, or behave in an uncharacteristic way because they are physically, emotionally, or mentally off-kilter. The days when these gear changes occur are called critical days. The effects of a critical day can last between 24 and 48 hours.

Each cycle has two such days—one at the beginning and one in the middle. There is also a third critical day as one cycle ends and another one begins. In the 23-day physical cycle, the critical days fall on the first and 12th days, with the next cycle beginning on day 24. In the 28-day emotional cycle, the first and 15th are the critical days, and the next cycle starts on day 29. In the 33-day intellectual cycle, the first and 17th days are critical days, and the next cycle begins on day 34.

In addition to these critical days, there are other times that are less troublesome but still need to be guarded against. These are the days when a particular cycle is either at its peak or at its lowest point. On peak days, people may have such an excess of energy in one area that they become overconfident and go overboard. On days when the passive phase is at its lowest, the opposite is true and people may lack confidence and will tend to be least effective physically or mentally, or at their most negative emotionally. The physical cycle peaks around day 7 and is at its lowest point

Double and triple critical days

Because there are three separate biorhythms operating in any one person at the same time, two or even three critical days from different cycles can sometimes coincide. On a double critical day in,

▶ *The three biorhythm cycles all begin simultaneously at birth, but quickly fall out of sync with each other. Like the phases of the moon, the patterns are unchangeable.*

say, the emotional and physical cycles, an individual who is normally calm may flare up in an angry show of emotion and may express feelings physically by throwing or breaking things or even lashing out at another person. On a triple critical day, when the intellectual cycle is also at one of its turning points, the potential for uncharacteristic behavior is multiplied three times.

Although the three biorhythms all begin at the same time at the moment a person is born, they soon fall out of step with each other because they last for different lengths of time. The very first physical cycle will end when a baby is 23 days old, the first emotional cycle at 28 days of age, and the first intellectual cycle at 33 days.

The three rhythms continue rising and falling throughout life, beginning and ending at different times. There is one point, however, when they do coincide once more, as they did when the person was born. This "grand triple critical day" when all three cycles start together again has been calculated to fall 21,252 days after a person's birth, or at 58.2 years of age. Up to this moment, a person will have experienced 924 physical cycles, 759 emotional cycles, and 644 intellectual cycles.

Biorhythms in action

Biorhythms have no effect on the external conditions and events in a person's life, as the planets and stars of astrology are said to do. What they do affect, however, is the way in which someone reacts to those conditions and events. Of course, different reactions produce different outcomes, so in one sense, biorhythms can be said to have an indirect effect on a person's fortunes. For example, if people have to go for job interviews, they should choose appointment times when they will be in a positive frame of mind, both emotionally and intellectually. If a date suggested for an interview coincides with a critical day in either of these cycles, another date could be chosen during a more favorable phase.

Knowing in advance when active and passive phases and critical days are going to fall enables people to make allowances for their behavior. For example, a student may use exactly the same route and method of transportation to get to college every day. During the active physical phase, the journey will seem easy and effortless, and the student will arrive at the destination full of energy for the day to come. However, as the cycle passes into its negative phase, the journey to college may become unbearable, and just getting through the day may seem a struggle. Here, there has been no change in external circumstances—what has changed is the person's reactions, under the influence of biorhythms.

Critical days in the physical cycle might lead to accidents, such as pulling a muscle while playing tennis, owing to misjudgment of one's physical abilities on that day.

The active phase of the emotional cycle brings a sense of friendliness and optimism, so it is a good time for socializing and mixing with other people. However, as the cycle moves into its negative phase, friends, family members, and coworkers may suddenly become a source of irritation. Under the biorhythm's negative influence, a person may feel depressed and be hurt or angered by words or actions that would not normally bother him or her.

The positive phase of the intellectual cycle is a good time for mental work of all kinds, such as writing essays or reports, studying, having discussions, working out a budget, and starting a new job. During the negative phase of this cycle, however, a person's ability to learn, understand, and concentrate will be lowered, and even the

▲ *Emotional relationships can be put under some strain when one or both partner's emotional cycles are in a critical phase. Understanding and planning can help.*

easiest mental task may seem difficult. This is the time to review work already done, not to begin something new.

Statistical evidence

In 1939 Dr. Hans Schwing of the Swiss Federal Institute of Technology published a study on biorhythms and their effects on accidents and accidental deaths. For his study, Dr. Schwing used two sets of figures. The first were a number of cases kept on record in the city of Zurich: of these, 700 were accidental injuries and 300 were accidental deaths. The second figure he used was the total biorhythmic span of 21,252 days (the time it takes for all three cycles to coincide again). Out of this figure, 16,925 days, or 79.6 percent of the total, were noncritical days, and 4,327 days, or 20.4 percent, were critical days.

When Schwing compared the two sets of figures, he discovered an interesting ratio: nearly 60 percent of the accidents had occurred in the 20.4 percent of critical days, while only 40 percent had occurred in the remaining 79.6 percent of the time. So, over half the accidents had taken place in only a fifth of the time. Equally astonishing findings were published in 1954 by Rheinhold Bochow at Humboldt University in Berlin. This report was based on a study of 497 accidents that had involved agricultural machinery. Only 2.2 percent of the accidents fell on noncritical, mixed-rhythm days, but a staggering total of 97.85 percent occurred on critical days—with 46.5 percent on double critical days.

Police records also suggest some interesting connections between biorhythms and crime. With regard to the intellectual cycle, a crime is 4.7 percent more likely to be committed during the negative phase than the positive one. Violent crimes, such as murder or armed robbery, appear to be more frequent when a criminal has been in a positive physical phase, with large amounts of physical energy, and negative emotional and intellectual phases. Several American airline companies have also taken biorhythms seriously and have used them as a tool in planning the work schedules of their pilots and air-traffic controllers.

See also: **Biological clocks; Birth; Menstruation**

Birth

Questions and Answers

Is it true that if I eat well during my pregnancy I will have an easier labor?

It is true that women who are fit and healthy when labor begins are likely to cope better. If a pregnant mother eats a balanced diet, in addition to avoiding cigarettes and alcohol, it is known to result in healthier babies. If a mother and baby are both healthy, there is less chance of complications.

What can I do to make my labor as easy as possible? I feel nervous, and I want to take every precaution.

It helps to understand what is happening to your body in labor. Women who are afraid of birth are usually very tense. In extreme cases, tension can slow labor, and a prolonged labor is more tiring. Most hospitals provide some prenatal education. Even if it is only parenting skills, it will include information about what will happen in labor. Other hospitals or health clinics offer classes in relaxation (psychoprophylaxis) that are invaluable in making labor as easy as possible.

My sister gave birth naturally in a clinic. Can I have a natural birth in the hospital? Will the hospital let my husband stay with me?

Most women will have a natural birth wherever they are. In the hospital, however, fetal monitoring is used to detect fetal distress. While a lot of hospital procedures are unnecessary in normal labor, they are carried out to prevent complications. It is the duty of the hospital to prevent anything from going wrong, but there is no harm in asking why each procedure is to be done and if it is necessary. No procedure in normal labor will warrant your husband's absence, but discuss this with your doctor beforehand.

Childbirth is one of the most rewarding experiences in a woman's life, and if she is well informed about what is involved, she will be more confident and relaxed during labor and the birth itself.

Having a baby usually means going into the hospital, particularly if it is a first baby. Although giving birth almost never has anything to do with ill health, going into the hospital, where specialists are always available, is advised in most cases, just in case a problem or unexpected complication should arise during labor or immediately after birth.

Of course most women are unlikely to have any problems and will enjoy a healthy pregnancy and normal labor. Because a woman does have the right to choose where her baby will be born, if she wishes to have a baby at home, it can usually be arranged.

However, women giving birth in the hospital, especially for the first time, will find it helpful and reassuring to know exactly what happens and what to expect—and so will their partners.

A special experience

Having a baby is a perfectly natural process. A baby is born practically every second somewhere in the world. At the same time, it is an amazing experience for those involved. This is because every human being is unique, and every birth is a very special experience. No one can predict what will happen during labor, but obstetricians—doctors who specialize in the delivery of babies—nurses, and nurse-midwives are able to recognize when things may go wrong and can

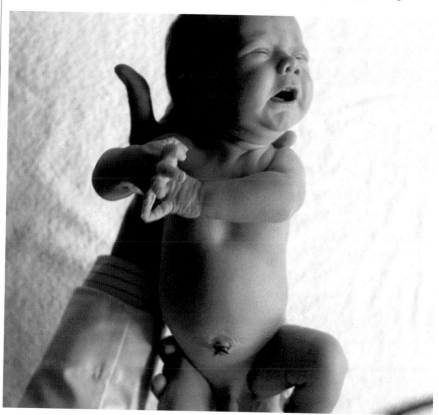

▲ *Even immediately after birth, the baby's flexing and grasping movements are already strongly developed.*

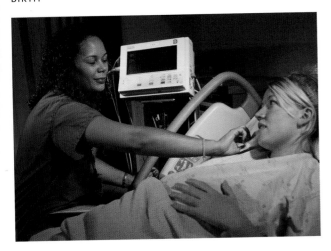

▲ *Reassurance from caring hospital staff is an important feature of a positive experience of labor and childbirth.*

deal with the unexpected with the maximum degree of safety for the mother and her baby. Deaths are now rare in childbirth. Old wives' tales, superstitions, and ancient folklore are best ignored.

The start of labor

No one really knows what causes labor to start, but there are a number of theories. The most recent theory is that the hormone levels in the baby change, triggering labor. Labor can begin in one of three ways. In the last months of pregnancy, the uterus (womb) prepares for birth with a tightening and relaxing of its muscles, known as Braxton Hicks contractions. These are entirely painless. When the tightenings become regular and strong, and last for more than just a few seconds, true labor has begun.

Contractions occur every 10 to 15 minutes, and they remain constant; if they become irregular, it may be a false alarm—simply more Braxton Hicks contractions. This is why it is important not to rush to the hospital at the first hint of contractions.

Signs to watch for

When the baby is due to be born, a woman may notice a small discharge of blood and mucus from her vagina when she uses the toilet, or she may simply find a slight discharge of mucus coming from the vagina. This can occur by itself, followed a little later by contractions. It is usually a sign that labor has begun, for the mucus acts as a plug over the cervix (neck of the uterus). As the uterus opens up, this plug is released from the cervix.

It is common for contractions to begin very soon after the discharge. If the contractions start first, the discharge often follows. Some women do not notice the discharge at all, especially if it is not bloodstained. (Bleeding with contractions is not normal, and in cases of slight bleeding the doctor should be called and asked for advice. With severe bleeding, the woman should go to the hospital immediately.) A third way labor may begin is with the breaking of the water—the rupturing of the membranes of the amniotic sac that protects the baby in the uterus. In most cases, contractions will soon follow. Occasionally, however, the water breaks and nothing happens. In this case, the baby is no longer protected from infection and the woman should contact the hospital immediately.

The obstetrician may, for a number of different reasons, decide to speed things up, in which case he or she administers a drug that stimulates the uterus and causes contractions. If this is done after labor has started, it is called an accelerated labor. When it is done to start labor, usually in conjunction with the artificial rupture of the membranes, it is called induced labor.

When to go to the hospital

If there are no complications, home is the best and most comfortable place to be in early labor. It is extremely rare for a baby to be born within the first hour after the onset of labor. For a first baby, labor may last about 12 hours; for a second or subsequent baby, it is possible that labor will be quicker, lasting about seven hours, but these times are just averages and and can vary enormously. If a woman is able to relax at home between contractions, she can use the time to make last-minute preparations. It is best to eat something; but if she cannot face food, a hot, sweet drink will prevent the blood sugar from dropping as labor progresses (a drop in blood sugar would make her feel tired).

The labor ward staff can be kept informed by phone about the progress of the labor, and they will tell a woman when to come to the hospital. If she is anxious, she may prefer to go in just in case, but there is usually no need to hurry.

Pain relief in labor

Relaxation techniques (psychoprophylaxis) are very helpful. Ask your doctor if there are any classes you can attend.

Meperidine hydrochloride (Demerol) is the drug commonly used for relieving the pain of contractions during the first and second stages of labor. This synthetic opiate acts mainly on the central nervous system and is usually administered by intramuscular injection. Since it crosses the placental barrier, it may cause depressed or delayed respiration in the newborn infant; the effects may be less if less than an hour elapses between injection of the drug and delivery. Transcutaneous electrical nerve stimulation (TENS) is an alternative to Demerol and is just as effective.

Other drugs are sometimes given during labor in conjunction with the primary analgesic. These include a tranquilizer, such as promethazine hydrochloride (Phenergan); a barbiturate, such as secobarbital (Seconal) or pentobarbital (Nembutal), which is a controversial treatment during labor; or an amnesiac, such as scopolamine.

An epidural injection involves injecting a local anesthetic into the epidural space (just behind the spinal cord). A fine needle with a plastic tube inside is placed in the middle of the back. Once the tube is inserted, the drug is given. The area below will be numbed so that contractions are not felt. This drug is very effective, and frequent doses can be given without disturbing the mother. It may cause a drop in blood pressure or other minor reactions such as shivering or faintness, but an intravenous drip is always used to counteract these symptoms. Sometimes the woman's ability to push the baby out is reduced because the pushing sensation is lost. A forceps delivery may be necessary. The anesthetic can cause the baby to be drowsy at birth.

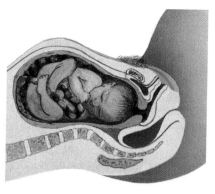
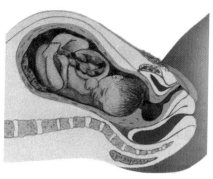
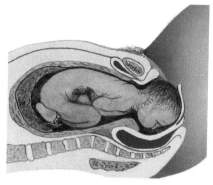

▲ *Before the first stage of labor, the baby's head is resting on the neck of the uterus.*

▲ *Contractions during the first stages of labor are aimed at dilating the neck of the uterus.*

▲ *With its head past the neck of the uterus, the baby twists around to allow the body to emerge.*

Arrival at the hospital

The labor ward in a small hospital is often separate from the admissions ward, and a woman may not see it until she goes into the second stage of labor. In larger, more modern units, the labor wards consist of admission rooms, first-stage rooms, and delivery suites.

On admission, a woman will be asked for details, including a history of her labor up to that point. A vaginal examination (internal) will normally be carried out to confirm the onset of labor by checking for the dilation of the cervix (opening of the neck of the uterus). Someone will check the position of the baby, check the baby's heartbeat, and perform several other routine observations that are designed to prevent any complications at an early stage and also to improve the mother's personal comfort.

Routine tests

Tests will include a check on blood pressure, pulse, temperature, and the length of the contractions. A sample of urine will also be tested to show if the blood-sugar level is satisfactory. Sometimes in labor the blood-sugar level drops, and sugar water is given to prevent exhaustion. Some hospitals do not allow a woman to eat or drink while in labor, not just because some women may vomit but also because if there are problems that require a general anesthetic, the stomach must be empty.

It used to be common practice in hospitals for a woman to be given a warm enema (to clear the bowels of feces); this is sometimes still done if necessary. Occasionally the pubic hair is shaved to reduce the chance of infection.

Monitoring the birth

It is also usual in large maternity units for women in labor to be monitored. This means wearing a rubber belt around the abdomen that has a large knob on the top attached by wires to a machine. The machine records when the contractions begin, when they peak, and when they end. (The doctor or nurse will still feel for the contractions at certain intervals.)

Monitoring does not mean there is something wrong. It is simply a more efficient and safe way of providing a constant check on what is happening to mother and baby.

The progress of labor

To check the progress of labor, a woman will usually be given an internal examination every three or four hours, depending on her individual progress. On one of these occasions, a tiny clip may be attached to the baby's scalp and then to the monitoring machine so that the heart rate can be recorded.

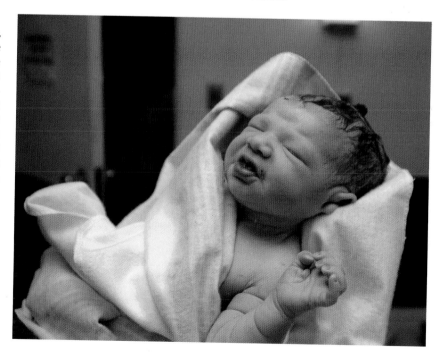

▲ *Even immediately after birth, the baby's flexing and grasping movements are already strongly developed.*

If labor progresses normally and there are no complications, routine observations and monitor recordings will continue with no other interference, except perhaps for some intravenous glucose to keep the blood sugar level up and prevent exhaustion. If the woman decides she would like an epidural or a local anesthetic to prevent pain, the obstetrician or anesthetist may give her one.

If labor is progressing well—most hospitals have a graph that maps out normal progress—and the water is still intact, it will not be broken artificially, since it will eventually break automatically. Occasionally labor slows down and the water will be broken artificially to speed labor up again. This is a fairly painless procedure.

End of the first stage

Toward the end of the first stage of labor, the contractions begin to get much closer together, and they can be very strong. Labor is now fully established, and the baby is almost ready to be born.

This transition period is recognized as a sign that the cervix is more or less fully dilated (usually around 4 inches, or 9–10 cm). A positive sign that the cervix is ready is the gaping of the anus as the head of the baby arrives at the perineum—the area between the anus and the vagina. This area thins and bulges out as the head descends. The woman will feel a very strong urge to push the baby out but may be told to wait a bit if the cervix is not fully dilated.

The second stage of labor

This is when the cervix is fully dilated and the pushing can start. It is now that husbands and partners can be most useful in encouraging the mother-to-be. It is the end of labor, she is tired, and some really hard work is yet to come. Coping with very strong contractions in the late stages of labor may seem almost unbearable, but when pushing begins, it is a relief—all the woman's energy goes into the pushing, the labor feels as if it is progressing, and the contractions do not seem too strong. The urge to push is normally a natural reaction, but if an epidural is being used, the woman may not get this sensation and will need extra help. As the baby's head descends, the perineum will stretch. The doctor or nurse-midwife controls the head as it emerges to prevent the perineum from tearing. If the perineum is too thin and in danger of severe tearing, or if the baby is distressed and needs to be delivered quickly, then the area will be cut with special scissors. This is called an episiotomy and is less painful than one might expect. But most women will stretch adequately without any tearing.

As the baby's shoulders emerge, the birth process is almost at an end. In an ideal situation, the baby should then be lifted onto the mother's stomach while the umbilical cord is cut and a piece of thread tied around the cut ends to stop the bleeding. If the room is warm and all is well, there is no reason for the baby to be taken away for measurements and tests immediately. Instead, time can be

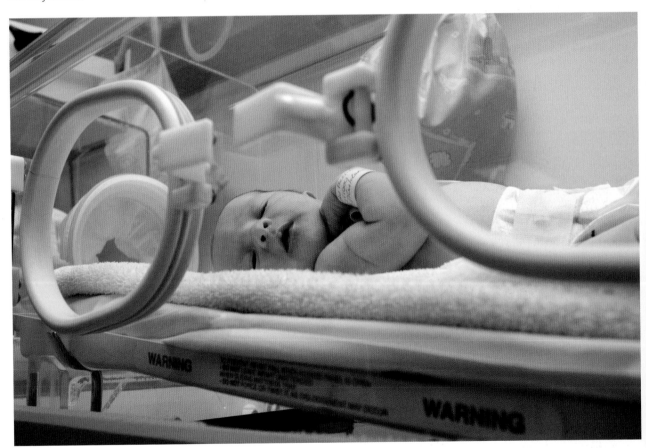

▲ *Babies that are born prematurely may need to be transferred to a specialist neonatal intensive care unit for regular monitoring and possible treatment during the first few days of life.*

Questions and Answers

What will happen if I go into labor early?

If the baby is very tiny, all efforts will be made to delay the progress of labor, and if possible stop it altogether. There are special drugs that can be given to relax the uterus. If this method does not succeed, labor will progress and all precautions will be taken for the birth of a premature baby.

A pediatrician (a doctor who specializes in problems of babies and children) will attend the labor and will probably put the baby in an incubator in the intensive care nursery for observation and possible treatment. Abnormally small babies are vulnerable, and it is much better if a pregnancy goes to its full term.

What will happen if I go past my delivery date?

It is rare for a baby to die because it is overdue. If labor does not start of its own accord, it can be induced with a drug. The water will also be broken artificially. It is difficult to be totally accurate about dates, and for this reason some women will go beyond the time predicted for the delivery by as long as three weeks with no danger to themselves or the baby.

Will the birth of my second baby be easier than the birth of the first?

In a second labor, the muscles are already stretched and the labor is often quicker, especially the second stage. This does not necessarily mean it will be easier.

I am expecting twins. Will the labor be more difficult than with a single baby?

Not necessarily, but it may be a little longer than average. A possible problem is that the babies may be born early because of the extra weight on the neck of the uterus. If there is a risk of their being premature, they may need special attention at the hospital, both before and after birth.

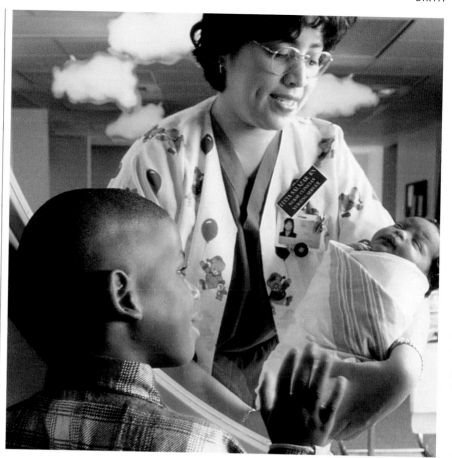

▲ *The arrival of a new baby is an exciting event in the life of the whole family, including older brothers and sisters, who will be eager to meet the new arrival.*

allowed for the bonding between mother and child to begin. Holding the baby as soon as possible is very important. Research shows that mothers who cuddle and fondle their babies from birth are more likely to form early, close relationships with them. The process is called bonding. This, of course, does not mean that if a baby is taken from its mother at birth their relationship is permanently harmed—bonding can take place later. Some women do not possess an inborn maternal instinct, and they may find it more difficult to attach themselves to their babies once they have given birth. But most gradually grow to love their babies over several months.

The third stage of labor

During the third stage of labor, the afterbirth is delivered. Immediately after the birth of the baby, the muscles of the uterus relax. Within minutes, contractions resume to separate the placenta from the uterine wall and expel it. A woman may be unaware of this because the birth area is numb for a while after the delivery of the baby. The delivery of the placenta is painless and over quickly. The doctor or nurse-midwife will gently pull on the cord to ease the placenta out. This process takes from about three to 10 minutes. When it is left to be expelled without help, the process is usually a little longer, lasting from about 20 minutes to an hour. If necessary, a drug may be given to help this process. If the mother puts the baby to the breast immediately after it is born, a hormone is released that causes the uterus to contract and expel the placenta more rapidly.

Cesarean section

The normal process of labor may be bypassed, if necessary, by a cesarean section. This is an abdominal operation in which the baby is born through an incision, either in an emergency, when the life of the baby, the mother, or both are at risk, or as a planned procedure for those

Coping with an emergency delivery

Sometimes a baby is born very quickly. This may happen before medical help can be called. It is vital that any helper should know what to do in this situation.

It should be remembered that a fast labor is generally a normal labor, and the mother should be reminded of this to reassure her.

Cleanliness is essential. The helper must scrub his or her hands thoroughly.

The first stage of labor usually lasts several hours, but it can also be surprisingly brief. The mother should try to relax without worrying between contractions. She should use the bathroom and not be alarmed if there is a small discharge of blood. This is quite normal.

The second stage of labor lasts until the birth. The mother should lie on her back with knees bent up and feet apart. She should push with each contraction, relaxing in between.

At the beginning of each contraction, a deep breath should be taken and released, and then another breath taken, which should be held while the contraction lasts. This helps ease the contractions and provides the baby with oxygen via the mother's bloodstream.

During the first or second stage of labor, there will be a sudden discharge of watery fluid from the vagina—this is from the bag of water that protects the baby in the womb and is completely normal.

The baby is generally born headfirst, and as it emerges the mother should pant. This helps stop violent contractions that would push the baby out too fast.

The helper should now support the baby's head using the palms of the hands, and when the shoulders emerge, lift the baby out by holding it under the armpits. The baby should be put on the mother's stomach. The umbilical cord may be cut with sterile scissors and a knot tied in it to prevent bleeding—but the cord can be left uncut until help arrives.

If the baby is breast-fed immediately following birth, the afterbirth (the placenta and fetal membranes) will be expelled naturally, because breast-feeding releases a hormone that stimulates the process.

▲ *A doctor listens to a premature infant's heartbeat. Regular monitoring takes place in the neonatal intensive care unit.*

however. Most problems that occur can be handled easily by a doctor or a nurse-midwife in a hospital setting. There are also a few nonemergency situations that call for a cesarean section. Some women, for example, have a small pelvic canal, or the afterbirth may have become embedded and more likely to bleed.

Breech babies (this term is used when a part of the baby's body other than the head is born first) are sometimes delivered by cesarean section. The baby can be breast-fed normally after a cesarean section birth. A period of convalescence, however, is required for the mother, to allow rest and healing, as with any other surgical procedure.

Having a baby today is considered a completely safe experience. When everything goes well, there is no need for medical interference, and many hospitals will respect a couple's wish for as natural a delivery as possible. However, sometimes things go wrong, and what may have seemed a normal labor turns out to be not so simple. A nurse-midwife is trained to recognize signs that a labor is becoming complicated and, when she detects them, to call for a doctor. In a hospital, this help is immediately available. All obstetric units have obstetricians on call who frequently check up on normal deliveries and attend abnormal ones.

A home birth can be arranged if there is no likelihood of risk to mother or baby and as long as the mother has no history of difficult pregnancies. The prenatal examinations may be carried out by the nurse-midwife and the doctor, and it is a good idea for the mother-to-be to seek their support for her decision.

Contact with the local hospital's maternity department should be kept up during pregnancy so that it will be familiar if hospitalization is needed. The nurse-midwife will advise on getting the home ready for a delivery. Good lighting and a temperature of around 68°F (20°C) are essential, since newborn babies lose heat rapidly.

The nurse-midwife stays with the mother during labor and birth, and when the placenta is expelled. Most procedures will be the same as in the hospital, although there are not as many painkilling drugs available.

The advantage of a home birth is that a woman is more relaxed and will then find labor easier.

See also: Fetus; Placenta; **Pregnancy**

women who are known to have complications that would prevent them from having a normal birth. A cesarean section may be necessary, for example, if the baby is in distress. Distress can be detected by the doctor or nurse-midwife from the baby's heartbeat pattern, from hypoxia (a lack of oxygen in the baby's bloodstream), or from the release of meconium (a green-black substance) from its bowels. The umbilical cord may be wrapped around the baby's neck, or its head may be tipped so that it cannot fit through the birth canal. Births requiring an emergency cesarean section are very rare,

Birthmarks

Many old wives' tales claim to explain why birthmarks appear, but in fact they have no known cause. All but a few are harmless, and most can be treated or removed.

Questions and Answers

I have heard that birthmarks may sometimes be caused by the mother's receiving a shock during her pregnancy. Is this true?

No—this is an old wives' tale. There is no evidence at all that birthmarks are caused by any external influence, be it emotional or physical.

Can a strawberry birthmark lead to side effects or complications?

The strawberry mark is one of the most common birthmarks. It may become sore and ulcerated, but this can be prevented by carefully washing and powdering it.

If I have a difficult labor or birth, will it give my baby birthmarks?

No—birthmarks are not caused by problems during labor. There may be some bruising or dimpling of the skin if forceps are used for a difficult delivery, but this will soon fade and it should not be confused with a true birthmark.

Can birthmarks disappear by themselves?

The strawberry mark and another common blemish, known as the stork mark, usually fade, but other types tend to be more permanent.

It is true that moles can be malignant?

Almost all moles present at birth are benign, but a few of them may become malignant later in life. Signs that a mole is malignant are bleeding, continuous itching, sudden increase in size, darkening in color, and development of more moles nearby. If you think a mole is malignant, see a doctor immediately. Removal of the mole usually cures the problem, but any delay may be dangerous.

The term "birthmark" is used to describe any noticeable abnormality on a baby's skin, such as a pit, swelling, or mark. It may be present at birth or appear soon after. Such marks can be thought of as defects in the development of part of the skin. They consist of either groups of abnormal blood vessels, collections of cells responsible for the production of pigment (coloring), or groups of cells responsible for the formation of the skin's surface—properly called the epidermis. Most have no known cause, unlike those birth defects that are caused by the mother's catching an infection such as rubella (German measles) during pregnancy, and can thus not be prevented.

The usual word used by doctors to describe one of these marks is "nevus" (plural, "nevi"). Birthmarks composed of abnormal blood vessels are called vascular nevi; those with pigmented cells, melanocytic nevi; and those from the epidermis, epidermal nevi.

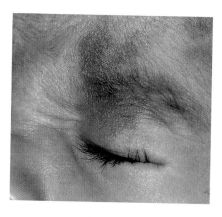

▲ *Port-wine stains are most common on the face and upper body.*

Vascular nevi

There are two types of vascular birthmarks. One, known as nevus flammeus, consists of a defect in the smallest blood vessels in the top layer of the skin. The blood vessels dilate, or widen, causing redness in the overlying skin. The other type of vascular nevus is derived from patches of primitive tissue from which blood vessels are formed. When it is present after birth, it is the blemish known as a strawberry mark or cavernous nevus.

Varieties of the nevus flammeus type include the salmon patch, the stork mark, the port-wine stain, and the capillary nevus. Salmon patches or stork marks are pale pink, with fine blood vessels visible within them. They are often seen on the nape of the neck, the midforehead, and the eyelids and usually fade rapidly, disappearing by the age of one year. They rarely, if ever, need treatment.

Port-wine stains are less common than salmon patches but are usually permanent. They occur on any part of the body, but usually on the face and upper trunk. They range in color from pale pink to deep red or purple and can vary in size from a fraction of an inch to several inches across. Treatment is generally successful, but the outcome depends on size. If the blemish is small enough to be entirely cut out without the need for a skin graft, then a good result is ensured. If a skin graft is necessary, however, scarring usually occurs, which can be unsightly, and the best solution is to hide it with makeup. Occasionally, tattooing can be used to improve the color match with the surrounding normal skin, and freezing the skin and laser treatment can be effective. Similar treatments are used for the more common capillary nevus, a usually strawberry-colored mark that occurs most often on the head and neck but may also appear on the torso and limbs.

The second type of vascular nevi, known as the strawberry mark, can be found on up to 10 percent of babies but is rarely visible at birth. It usually appears on the head and neck as a red blemish during the first month of life, but it may also appear on other parts of the body. Strawberry marks are bright pink, clearly defined, domed swellings that can grow to their maximum size by age six to nine months. They vary in size: most are about 1–2 inches (2.5–5 cm), but in rare cases they can be very large. If located near the mouth, they may cause feeding difficulties, and if they are near the eyes problems can occur with vision. Few cause complications, and they usually disappear in early childhood.

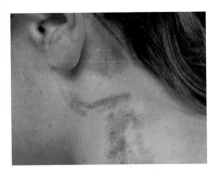

▲ *An epidermal nevus is a light brown or skin-colored permanent stain.*

▲ *A strawberry mark is largest at nine months and then usually fades.*

Pigmented birthmarks

Pigmented birthmarks include café-au-lait patches, moles, and mongolian spots. Café-au-lait patches occur in up to 10 percent of babies. They are not very noticeable, flat, pale brown marks, 2–8 inches (5–20 cm) across, that usually appear on the baby's trunk. More than five patches, especially if large, may indicate the presence of Recklinghausen's disease (neurofibromatosis).

Moles are collections of pigment cells, which give the skin color. They occur more often in childhood and adolescence but, when present at birth, have a small risk of becoming malignant. Moles that develop later on in childhood have little risk of malignancy and can be cut out under local anesthetic.

Mongolian spots are blue-black marks between 0.75–3 inches (2–8 cm) that appear on the buttocks and backs of some African American, Native American, southern European, and Asian babies. They are caused by a local accumulation of the normal skin pigment and usually disappear during early childhood, around the age of four.

Epidermal nevi

Epidermal nevi are linear, raised, sometimes "warty" tan patches that appear on the head, trunk, or limbs at birth or early childhood.

There are several types: most are only a few inches long, but some can cover an entire limb or a side of the trunk. Most are benign, but their presence may point to a more serious syndrome such as epidermal nevus syndrome (accompanied by seizures, mental deficiency, eye problems, and bone malformation) or Proteus syndrome (abnormal growth of the bone, skin, or head)—Joseph Merrick, also known as "The Elephant Man" is a famous case of Proteus syndrome. A biopsy is advisable when epidermal nevi appear.

See also: **Birth; Skin**

Birthmarks

TYPE	TEMPORARY OR PERMANENT	TREATMENT
Salmon patch (small pink mark)	Appears at birth; usually fades quickly, disappearing by the age of one	Rarely needs treatment
Stork mark (red mark on nose)	Appears at birth; usually fades rapidly, disappearing by the age of one	Rarely needs treatment
Port-wine stain (large red patch, often on face)	Remains indefinitely	Cosmetic camouflage; can be cut out, but may require skin graft; laser treatment; occasionally tattooing; freezing
Capillary nevus (red stain under skin, caused by small blood vessels)	Permanent	Cosmetic camouflage; carbon dioxide snow (very cold, semisolid carbon dioxide) may be used to destroy small one by freezing; large one may be cut out; radium therapy
Strawberry mark (red blemish or bright pink domed swelling, usually on head or neck)	Usually temporary	90 percent disappear without treatment. Occasionally steroid drugs are prescribed.
Café-au-lait patch (flat, pale-brown mark, usually on trunk)	Remains indefinitely	Not necessary
Mole (brown or reddish circular mark)	Usually develops in childhood and adolescence; very rarely malignant and unless removed is permanent. Rarely occurs in babies, but if present has small risk of being malignant.	Large one should be removed. A mole that develops later in childhood does not usually require treatment; if it enlarges, it should be removed.
Mongolian spot (blue-black mark, usually on buttock or back)	Usually disappears by the age of four	Not necessary
Epidermal nevus	Present at birth or soon after; permanent	Can be cut out but sometimes returns; freezing can be effective.

Bladder

Questions and Answers

Does urine normally vary in color?

Usually urine is yellow, but it becomes darker when it is more concentrated; this occurs when the body is short of water. The color also varies from person to person and changes when certain substances are eaten or drunk. Beets make urine redder, as do the dyes in some candies and laxatives made from senna. Some medicines also discolor urine.

I always want to urinate more when the weather is cold. Why?

Bladder function is largely governed by the autonomic nervous system. When you are cold this acts to retain body heat by closing the pores. That prevents sweating. The excess liquid must then find another way out of the body, so the system increases the volume of urine produced.

My daughter often holds back from urinating for hours. Is this harmful?

If signals of a full bladder are often ignored, the bladder will eventually stretch. Stretching may weaken the muscles of its walls and prevent complete emptying. A full bladder increases the risk that urine will be forced back up the ureters toward the kidneys, spreading any infection that may exist.

I seem to urinate more often than my friends. Is my bladder weak?

You may have a small bladder, or be more attentive to its signals or less able to override them. People with faster metabolic rates need to urinate more often, as do people who are nervous. In men the most common cause of frequent urination is an enlarged prostate gland. This prevents complete emptying of the bladder, so the bladder fills more quickly and urination becomes necessary.

The bladder acts as a reservoir for urine. Infection, disease, injury, or stress can cause bladder disorders or a loss of control at any stage of life, but most conditions respond well to medical treatment.

The urinary bladder is a hollow, thick-walled, muscular organ that lies in the lower part of the pelvic basin between the pubic bones and the rectum. It is a four-sided, funnel-shaped sac resembling an upside-down pyramid. The base of the pyramid provides a surface on which coils of the small intestine, or in women the uterus, rest. The walls of the bladder consist of a number of muscular layers that stretch while the bladder fills and then contract to empty it. The kidneys pass an almost continuous trickle of urine down the ureters (the tubes between the kidneys and the bladder). Consequently, the pressure increases as the bladder is filled. The muscle fibers allow expansion by adapting to the volume of urine until the bladder is almost full. When the muscle fibers resist, the need to pass urine is felt.

The two ureters enter the bladder near the rear edges at the upper surface. One-way valves in the openings where they join the bladder prevent urine from flowing back toward the kidneys if the bladder becomes too full. Urine is passed out of the body via the urethra, which opens from the lowest point of the bladder. Normally, this opening is kept closed by a sphincter, a circular muscle that contracts to seal the passageway. When someone urinates, this sphincter relaxes at the same time as the muscles of the bladder wall contract to expel the urine.

In women, the urethra is only about 1 inch (2.5 cm) long and does not form a very efficient barrier against the entry of bacteria from outside, especially if the sphincter is weakened by

THE URINARY SYSTEMS

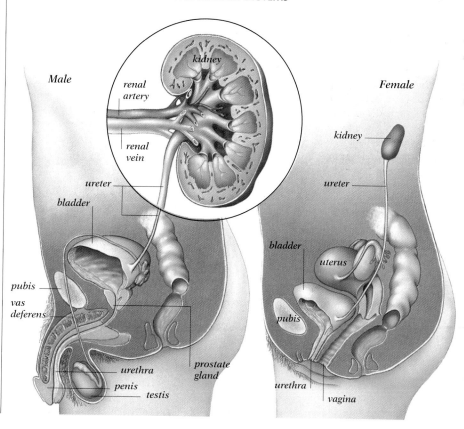

57

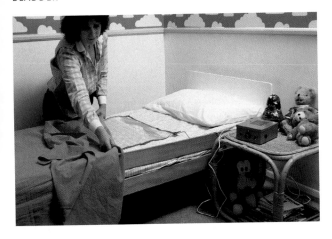

▲ *An alarm unit may help to cure a child's bed-wetting. It is attached to a bed mat that reacts to wetness by setting off the alarm, waking the child and discouraging bed-wetting.*

previous infection, old age, or poor muscle tone. Men are better protected from infection since their urethra has to pass through the penis and the prostate gland to reach the bladder.

Although it is commonly thought that women have bigger bladders than men, since women need to urinate less frequently, this is not the case. If women are able to postpone urination longer, it is because they have gained greater control over the emptying reflex. By suppressing their body signals, some women can increase their bladder capacity until they urinate only once daily. However, it is better to listen to body signals.

Bladder disorders

Cystitis is an inflammation of the bladder. It commonly affects women because their short urethra affords less protection against bacteria. Cystitis may be caused by intestinal bacteria that have passed into the bladder from the anus, from vaginal infections, or from an inflammation originating in an adjacent structure such as the kidney. It can occur because of infections contracted through sexual intercourse or childbirth, when there is damage to the urethra after surgery nearby, or when resistance is low.

Symptoms include a constant need to urinate, with scalding pain in the urethra. The urine is cloudy or may contain pus (pyuria) or blood (hematuria). Pain can also be felt in the pubic area above the bladder, especially before and after urination, and sometimes in the lower abdomen and back.

The condition is not usually serious, but there can be psychological distress in addition to discomfort. Treatment is by antibiotics taken from a few days up to two weeks. The symptoms often disappear within the first few days.

Kidney stones (calculi) that are too large to pass through the urethra can cause an obstruction and prevent people from emptying their bladder freely. The condition brings acute distress with colicky pain and requires urgent medical attention. Smaller stones may pass through, but larger ones may need to be removed by surgery.

Tumors can also be present in the bladder. They are usually benign, though sometimes they may become malignant and will require medical attention. The main symptom is the appearance of blood in the urine; there is no pain.

In middle-aged and elderly men enlargement of the prostate gland can obstruct the flow of urine; straining and a decrease in the strength and force of the urinary stream are common symptoms. Others include pain in the prostate region; burning, painful, and frequent urination; blood or semen in the urine; and difficulties during intercourse and ejaculation. If medical attention is not sought, stones in the bladder or ureter and inflammation of the bladder may develop. A catheter (flexible tube) may have to be passed up the urethra to release the urine that has collected in the bladder.

A fall in the volume of urine passed daily may be due to a narrowing or obstruction of the urethra, an inflammatory condition, drinking less fluid, or spells of hot weather. If obstruction is the cause, the bladder becomes distended and surgery will be needed to relieve the situation. Any difficulties in urinating or any change in the color of urine due to blood or pus are cause for concern. They may be symptomatic of bladder disease or some disorder elsewhere in the body. Seek medical advice as soon as possible.

Bladder control

As the bladder fills, the stretching of the muscle walls passes signals to the spinal cord via nerve endings. Normally the adult bladder will hold up to half a pint (a quarter of a liter) of urine before any discomfort is felt, and emptying (micturition) occurs before a full pint has been stored. In a small child, however, the stretching prompts automatic emptying by reflex action.

With toilet training this reflex is gradually suppressed by control from higher centers in the brain. If signs of fullness occur at an inconvenient time, the brain sends orders to the bladder walls to relax, allowing further filling before the signal is felt again. This process can be repeated several times before the reflex takes over and supersedes voluntary control to prevent damage to the bladder. The muscles then contract until the bladder is empty. Tensing the abdominal muscles aids this process but is normally unnecessary.

Bed-wetting (enuresis) is unavoidable in children until the age of three or four, but older children may begin bed-wetting again after a period of emotional stress, such as the birth of another child. Bed-wetting may also occur in adults as a result of shock or severe emotional problems. It may also be a symptom of physical illness, such as an obstruction or failure of control by the nervous system, although this is much less common. Some children stay dry at night earlier than others, and this difference is probably due to the rate of

BED-WETTING IN CHILDREN

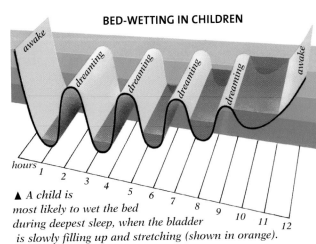

▲ *A child is most likely to wet the bed during deepest sleep, when the bladder is slowly filling up and stretching (shown in orange).*

Bladder problems and their causes

SYMPTOMS	CAUSES	ACTION
Burning pain on urination, cloudy or bloody urine, perhaps pain in lower abdomen and back, no fever.	Cystitis (infection of bladder and urethra)	See a doctor, who may prescribe antibiotics. If recurrent, drink plenty of fluids, urinate before and after intercourse, avoid panty hose and tight trousers, wear cotton briefs, wipe genital area from front to back.
Inability to empty bladder freely; acute, colicky pain.	Stones in urethra	See doctor urgently. Surgical removal of stones may be needed.
In men, straining and decrease in force of flow of urine, blood or semen in urine, problems during intercourse or ejaculation.	Enlarged prostate	See doctor. Treatment by catheter if urine cannot be passed, surgery later. Certain medications may help (shrink the size of the prostate).
Frequent but scanty urination, urgency, possibly blood in urine.	Infection, stones in urethra, tumor	See doctor if condition persists, to establish cause.
Incontinence	Damage to nerves (stroke, multiple sclerosis, spinal cord injury, slipped disk), loss of brain's control over bladder (epilepsy, arteriosclerosis, anxiety), damage to sphincter of bladder (after removal of womb or prostate), damage to pelvic floor muscles in childbirth	See doctor to establish cause.
Need to get up at night to urinate (nocturia)	Warning of prostate, kidney, or heart trouble	Drink less, urinate before going to bed, see a doctor.
Bed-wetting (in children over five)	Small bladder capacity from lack of training in retention, psychological disturbance, heredity, disease	See a doctor to establish cause. If problem is not physical, do not make an issue of it; sensitivity toward child is needed.

bladder growth, family traits, and the home environment. Bed-wetting is also more common in boys, but it has nothing to do with intelligence or lack of development. Over the age of five, confirmed bed-wetting indicates a problem. It may be psychological—resulting from parental conflict, or distress at some aspect of home life—or a call for attention. Alternatively, there may be a physical reason such as chronic urinary infection, injury to the spinal cord, multiple sclerosis, diabetes, or tumors of the central nervous system. Congenital abnormalities of the bladder and urethra account for some cases, but an underdeveloped bladder is no cause for alarm. In some families there is a history of chronic bed-wetting, the reason for which has not been determined.

Coping with bed-wetting

Usually a problem with bed-wetting will resolve itself with sensible care by the parents, but if bed-wetting becomes an emotionally fraught issue it will undermine the child's confidence and result in other problems. Parents should avoid giving the child any fluids at bedtime. They should take him or her to the bathroom before bedtime and again before they themselves go to bed. Various methods of toilet training should also be used to encourage dryness once the child is old enough to learn. An alarm clock, set at two-hour intervals increasing to six hours, can remind the child when he or she should go to the bathroom. A reward system may also be recommended. Otherwise, parents might try a mesh buzzer system—the first few drops of urine set off a buzzer on the bed.

This method may not be successful, however, because bed-wetting occurs during the deepest sleep, and some children do not awaken. In older children, a doctor may prescribe a drug for a short time to help deal with the problem.

Incontinence

Loss of urine without warning, or an inability to prevent urination, often occurs in adults who have suffered a spinal injury or brain damage or who are very old. Injuries received during childbirth or prostate operations may also cause incontinence, and in women it may occur as a result of stress and physical exertion. The inability to control the reflex emptying of the bladder is usually a result of loss of coordination in the nervous system due to injury or disease. The bladder may have to be emptied with a catheter, though sometimes it can be trained to work at planned intervals. Nocturia, which is the need to get up at night to urinate, may simply be a result of having too much to drink in the evening.

See also: **Rectum; Urethra; Urinary tract; Uterus**

Blood

Blood is the vital fluid that circulates around the body and maintains life. It is made up of a complex mix of cells and cell fragments—each of which has a vital function—carried in a colorless, chemical-rich liquid called plasma.

Questions and Answers

My mother thinks I have "tired blood." What is this?

The expression "tired blood" is commonly used to describe anemia, a condition in which the body does not have enough red blood cells. These carry oxygen that provides energy for every part of the body, so you often feel tired and listless when their number is below normal. If this continues, see your doctor. Meanwhile, to help build more blood cells, try eating iron-rich foods, such as liver, and take iron supplements, which you can buy from the drugstore.

My daughter asked me if kings and queens really have blue blood. What is the origin of this saying?

It is just a myth. In diagrams the blood in your veins is often shown as blue because the veins appear blue where they show through the surface of the skin, as on the inside of your wrist. The saying "blue-blooded" came about because, in the past, the aristocracy were not exposed to the sun, so their veins showed up more clearly than those of laborers with weathered skin.

I've been told that I need a blood test. What exactly happens, will it hurt, and how much blood will be taken from me?

The blood is drawn out by a needle and syringe. This feels like a little "pinch" of the skin. The blood is put in a test tube, which holds less than 1 teaspoon of blood, and sent to a laboratory. The number and condition of the cells will be checked, and chemical tests will be performed to check the amounts of certain substances such as sugar. These tests show how the organs are working. Tests will also be run to check for antibodies that form against attacking viruses or bacteria. If an illness is discovered after getting the results, your doctor will be able to treat it properly.

Blood is essential to body functions. It is pumped by the heart around the interior network of arteries and veins from before birth until death, delivering oxygen, food, and other essential substances to the tissues. In return, it extracts carbon dioxide and other waste products that might otherwise poison the system. Blood also helps to destroy disease-producing microorganisms, and through its ability to clot, it acts as an important part of the body's natural defense mechanism.

What the microscope reveals

Blood is not a simple fluid. Its proverbial thickness is due to the presence of millions of cells, the activities of which make it as much a body tissue as bone or muscle. It consists of a colorless liquid called plasma, in which float red cells (also known as erythrocytes), white cells (or leukocytes), and tiny cell fragments called platelets. To get an idea of the size of a single blood cell, first imagine the size of a micron, or micrometer—one-thousandth of a millimeter. Each red blood cell is a flattened, doughnut-shaped disk with a concave center and a diameter of about 7 microns. It is 2 microns thick at its edge and 1 micron thick in the center.

Red blood cells

The red blood cells act as transporters, taking oxygen from the lungs to the tissues. Having done this, they do not return empty but pick up carbon dioxide, a waste product of cell function, and take it back to the lungs, from where it is breathed out. The red cells are able to do this because they contain millions of molecules of a substance called hemoglobin. In the lungs, oxygen combines rapidly with the hemoglobin to give the red cells the bright red color from which their name is derived. Carried in the arteries, this "oxygenated" blood arrives at the tissues. With the help of enzymes in the red cells, carbon dioxide and water (which are other waste products of cell activity) are locked onto the red cells and taken back to the lungs in the veins.

White blood cells

The white cells in the blood are bigger than, and very different from, their red counterparts. Unlike red cells, not all white cells look alike, and the white cells are capable of a creeping motion. Involved in the body's defense against disease, white cells are classified into three main groups, which are known technically as polymorphs, lymphocytes, and monocytes.

The polymorphs, which make up 50 to 75 percent of the white cells, are subdivided into three kinds. Most numerous are those called neutrophils. When disease-causing bacteria invade the body, neutrophils go to work. Attracted by chemicals released by the bacteria, neutrophils "swim" to the site of infection and start to engulf the bacteria. As they do this, the granules inside the neutrophils begin to make chemicals that destroy the trapped bacteria. The familiar pus that collects at the site of an infection is the result of the work of the neutrophils, and is largely made up of dead white cells.

The second kind of polymorphs, eosinophils, are so called because their granules become stained pink when blood

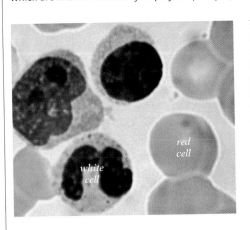

▲ *The blood sample on this slide shows both red (erythrocytes) and white (leukocytes) cells.*

BLOOD CLOTTING

A

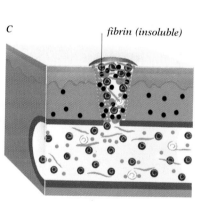

◄ *(A) Injured blood vessels bleed and platelets (small, sticky cell fragments in blood) rush to the site to help seal it.*
(B) Tissue-clotting factors are released and plasma factors enter the area.
(C) The reaction of the platelets, both factors, and other clotting agents convert fibrinogen (a protein) into strands of fibrin that become a jellylike mesh across the break.
(D) Platelets and blood cells are trapped in this mesh, which now recedes, oozing out serum (liquid blood without clotting factors) that helps form a scab. Now bacteria cannot enter the body and cause an infection.

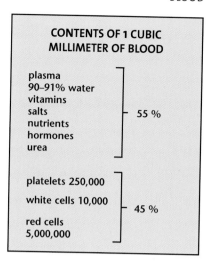

CONTENTS OF 1 CUBIC MILLIMETER OF BLOOD	
plasma 90–91% water, vitamins, salts, nutrients, hormones, urea	55 %
platelets 250,000, white cells 10,000, red cells 5,000,000	45 %

B

C

D

WHERE BLOOD CELLS ARE MADE

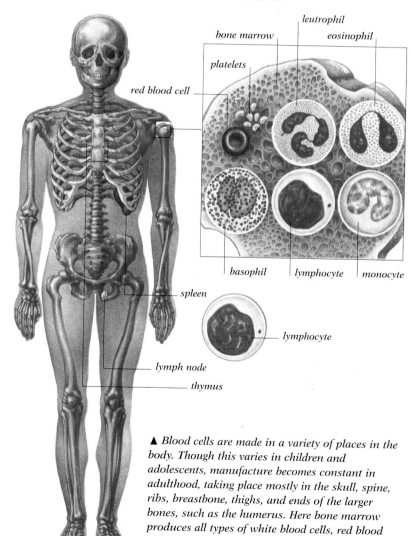

▲ *Blood cells are made in a variety of places in the body. Though this varies in children and adolescents, manufacture becomes constant in adulthood, taking place mostly in the skull, spine, ribs, breastbone, thighs, and ends of the larger bones, such as the humerus. Here bone marrow produces all types of white blood cells, red blood cells, and platelets. Lymphocytes are also produced in the spleen, tonsils, and lymph nodes.*

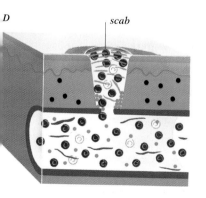

is mixed with the dye eosin. Composing only 1 to 4 percent of the white cells, eosinophils are activated by the antibodies involved in allergic reactions. Eosinophils increase in number during allergic attacks and parasitic infestations, and they release damaging substances that contribute to inflammation.

Basophils are the third type of polymorphs. Although these make up less than 1 percent of all white cells, they are essential to life because their granules make and release heparin, which works to stop the blood from clotting inside the vessels.

Natural immunity

Making up about 25 percent of the blood's white cells are the lymphocytes, which all have dense, spherical nuclei (centers). Lymphocytes give the body its natural immunity to disease in two ways. The B lymphocytes, made in the lymph nodes and spleen, produce antibodies to counteract the damaging effects of bacteria and their toxins (poisons). These antibodies combine with foreign substances to neutralize and render them harmless. The T lymphocytes, made in the thymus, kill foreign tissue elements and infectious organisms and may also protect the body against its own cells when they undergo malignant change. The T lymphocytes also secrete chemicals that help the B cells function. Finally, white cells called monocytes form up to 8 percent of the white cells. The largest monocytes contain large nuclei that engulf bacteria and remove the debris of cell remains after a bacterial attack. Monocytes can develop into macrophages. After the lymphocytes, they are the most important cells of the immune system.

Platelets

The millions of minute platelets in the blood cells have no nuclei. The platelets all have sticky surfaces, and this is a clue to their function. If the minute blood vessels called capillaries are damaged, chemicals are released that make the platelets stick to the broken ends, plugging them to stop the bleeding. The platelets also help trigger blood clotting. The ability of the blood to clot, or coagulate, which prevents fatal bleeding if a blood vessel is severed, comes from the combined action of the platelets and a dozen biochemical substances called clotting factors. Among these is the important substance called prothrombin. Clotting factors are found in the fluid part of the blood—the plasma—and are known as plasma factors. Defects of the clotting process are of two kinds—failure of clots to form, and thrombosis, in which blood clots form in the vessels.

FUNCTIONS OF RED AND WHITE BLOOD CELLS

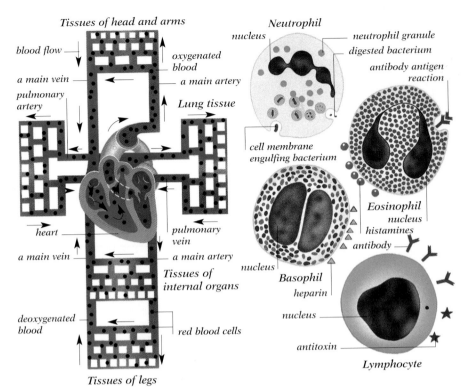

▲ Red cells carry oxygen (oxygenated blood) from lungs to tissues. They then collect carbon dioxide (deoxygenated blood), which is taken to the lungs to be exhaled.

▲ White cells—neutrophils, eosinophils, basophils, and lymphocytes—fight bacteria and prevent allergies and clotting of the circulating blood.

Plasma

Although about 90 percent of it is water, plasma is packed full of vital chemicals that it carries around the body. Among these are vitamins, minerals, sugars, fats, and proteins—in fact, all the constituents needed for cell function and renewal. Just as the blood cells carry away cell waste, so does the plasma. The most concentrated chemical refuse in the plasma is urea, which is ferried to the kidneys for excretion in the urine.

Manufacture of blood cells

The production of red blood cells begins in the first few weeks after conception, and for the first three months manufacture takes place in the liver. Only after six months of fetal development is production transferred to the bone marrow, where it continues for the rest of life. Until adolescence, the marrow in all the bones makes red blood cells, but after the age of 20, red cell production is confined to the spine, ribs, and breastbone. The cells begin life as irregular, roundish cells called hemocytoblasts, with huge nuclei. The cells then go through a rapid series of divisions, during which the nucleus becomes progressively smaller and then is lost altogether. In their travels around the bloodstream, the red cells are subjected to enormous wear and tear and so need constant renewal. Each one has an average life of 120 days; after this, cells made in the bone marrow and spleen attack those that are worn out. Some of the

Causes and treatment of diseases affecting bloodclotting

DISEASE	UNDERLYING CAUSES	TREATMENT
Hepatitis, cirrhosis, and other liver disorders	Deficiency of prothrombin and other clotting factors, made in the liver	Rest, nourishing diet. Avoid alcohol and certain medications, such as acetaminophen.
Gallstones or other interference with bile production or the release of bile	Vitamin K made in the intestine by bacteria is poorly absorbed into the blood in the absence of bile. Lack of vitamin K leads to low prothrombin production.	Vitamin K injections, treatment of underlying cause
Hemophilia	Deficiency of clotting factors. It is an inherited condition that occurs only in males, although women can be carriers.	Bed rest after any injury, blood transfusion, Factor VIII. Preparation containing thrombin applied to wounds.
Purpura—bleeding into the skin or joints that may cause tiny bruises or bleeding spots on the body	Low numbers of platelets. Many causes, including side effects of drugs, viral diseases, and bone marrow disease. Or platelet numbers are normal but function is impaired. Causes include scurvy, prolonged cortisone treatment, or deterioration in old age.	Treatment of underlying cause. Blood transfusions and extra iron may be needed.

White blood cell disorders

DISEASE	TYPES OF DISORDER	CAUSES
Leukocytosis	Overproduction of white cells, particularly neutrophils	Infections, particularly bacterial infections. Aftereffect of heart attacks.
Eosinophilia	Too many eosinophils (type of white blood cell)	Allergies such as asthma and hay fever. Worm infections.
Leukopenia	Too few white cells, particularly all types of polymorphs	Tuberculosis, typhoid, and some viral infections. Side effects of certain drugs including chloramphenicol, sulfonamides. Side effect of radiotherapy.
Leukemia	Too many white cells, and extra cells are abnormal	Cancer of the bone marrow or lymphatic system. Viruses may cause some types.

chemical remains are immediately returned to the plasma for reuse, while others, such as hemoglobin, are sent to the liver for further destruction. The body has a remarkable ability to control the appropriate number of circulating red cells.

If a lot of blood is lost, or parts of the bone marrow are destroyed, or the amount of oxygen reaching the tissues is decreased—owing to heart failure or high altitude, for example—the bone marrow immediately begins to increase its production of red blood cells. Even strenuous daily exercise stimulates extra red cell output, because the body then has a regular need for more oxygen. The bone marrow is also the site of some manufacture of white blood cells. All three types of polymorphs are made here from cells called myelocytes, again by a series of divisions. The average polymorph lives only 12 hours, and only two or three hours when the cells are fighting a bacterial invasion. In such circumstances, the output of all white cells is increased. The lymphocytes, which live an average of 200 days, are made in the spleen, and in areas such as the tonsils and the lymph nodes scattered throughout the body. Both monocytes and platelets are made in the bone marrow. Exactly how long monocytes live is a mystery, for they seem to spend part of their time in the tissues and part in the plasma, but the body replaces all its millions of platelets on average about once every four days. Although bleeding should be taken seriously, the body's survival mechanisms ensure that people can lose a quarter of their blood without long-term ill effects, even if they do not receive a transfusion. Blood is the supply line to and from the tissues, so disorders and diseases cause changes in its makeup. As well as being a reflection of the body's health, the blood can be the site of disorders needing treatment.

See also: Arteries; Cells and chromosomes; Enzymes; Lymphocytes; Spleen; Thymus

Blood-brain barrier

At the interface of the bloodstream and the brain is a protective network that is unique to the central nervous system and is known as the blood-brain barrier. This almost impenetrable, semipermeable barrier is thought to exist to shield delicate neurons in the brain from viruses, toxins, and some poisons, as well as unexpected changes in body chemistry.

Questions and Answers

Which scientist first discovered the blood-brain barrier (BBB)?

The German bacteriologist Paul Ehrlich was the first person to notice that dyes injected into an animal's blood supply stained all the organs except the brain, during experiments in the late 19th century. He thought this was simply because the brain did not absorb the dye.

Not until his student Edwin E. Goldman injected dye into the cerebrospinal fluid (CSF) in 1913, and found that the dye stained the brain but none of the other internal organs, did it became apparent that the central nervous system was separated from the bloodstream by some kind of barrier.

What can cause the blood-brain barrier to be breached?

High blood pressure opens the blood-brain barrier, as well as high concentrations of certain substances in the blood.

Exposure to microwaves, radiation, and infection can also affect the barrier, as can any injury to the brain that is capable of causing inflammation, trauma, pressure, or ischemia (obstruction of the arterial flow of blood).

Why is the BBB of relevance to the HIV virus?

The HIV virus directly infects cells within the brain, but many of the drugs used to combat the virus are excluded by the blood-brain barrier. This means that even if the virus is controlled elsewhere in a person's body, the brain is a source of continuing infection, and dementia may develop.

However, scientists have found ways of overriding the barrier. Either they make drugs imitate substances that can pass through, or they shrink the barrier cells to allow passage of the drug.

The blood-brain barrier (BBB) is present in all vertebrates (animals with backbones). Its function is to protect the brain from harmful substances in the bloodstream such as viruses and toxins, while supplying it with the required nutrients. It also helps to protect the brain from fluctuations in normal blood chemistry, maintaining a constant environment. In humans the BBB forms in the first trimester of fetal life. A BBB that does not form correctly, or subsequently breaks down, is thought to be a major factor in disorders of the central nervous system.

Specific functions of the BBB

Brain cells need a steady supply of glucose to function correctly; even slight variations in concentration can cause cells to malfunction. The BBB transports exactly the right amount from the blood into the fluid of the brain. Another task of the BBB is to maintain the ion balance in the brain, as even minor changes in concentration can affect nerve transmission. The BBB also protects the brain from certain chemicals that could disrupt its function. For example, adrenaline acts as a hormone elsewhere in the body, but in the brain it acts as a neurotransmitter (carrying nerve impulses from cell to cell), altering brain function.

Structure of the BBB

Endothelial cells line all the blood vessels in the body. In most of the organs, these cells are loosely joined, allowing water, ions, molecules, and white blood cells to pass through the gaps. By contrast, single layers of endothelial cells, which constitute the walls of capillaries in the brain, fit tightly together, preventing most substances from passing between them. There are miles of brain capillaries, which all together have a large surface area. To enter the brain a substance must pass through the endothelial cells, each of which contains two membranes, the lumenal and ablumenal membranes, that are a physical and metabolic (enzyme) barrier.

Substances pass through the endothelial cells in several ways. Some molecules diffuse through the membranes when concentrations on the other side are low. This diffusion occurs most readily with small or fat-soluble molecules, such as oxygen, water, carbon dioxide, and barbiturate drugs. Water-soluble molecules cannot usually pass through in this way, but a few, such as glucose, amino acids, and choline, are carried into the brain after they bind to special proteins in the membranes. The transport of certain other substances, such as insulin and some hormones, occurs after an energy-dependent reaction with receptors in special pits.

Molecules that do not normally pass through by any means include proteins, complex carbohydrates, toxins, most antibiotics, many hormones, and substances that act as neurotransmitters (they help to transmit messages between the nerve cells in the brain).

Glial support?

Opinions vary as to whether the glial cells between the brain's capillaries and neurons are a secondary buffer in the BBB. Some researchers believe they do help to obstruct toxins and regulate the flow of nutrients; others believe they do not contribute to the BBB, although they do help to transport ions from the brain to the blood.

The BBB and drugs

The BBB exists to protect the brain, and many drugs are handled like a foreign substance to be excluded so that they will not damage brain function. This can make treating brain disorders difficult, as many of the necessary drugs, such as the chemotherapeutic agents to treat cancer cannot pass across. Chemists have found ways of circumventing this problem, however. One

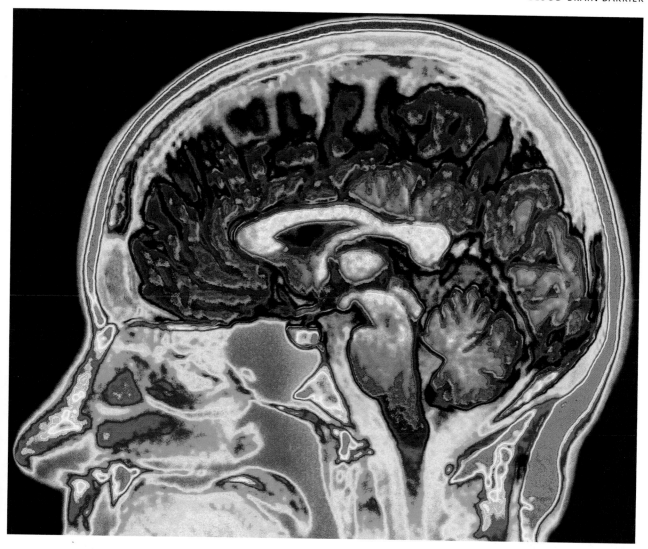

▲ *A magnetic resonance image of a person's head, showing a side view of the brain.*

method is to modify drugs that are usually soluble in water so that they can dissolve in lipids. Changing the solubility of the drug is not without problems, since the activity of the drug can be altered, but researchers have had some success. Another method, called a stealth mechanism, involves creating drugs to mimic compounds that are normally transported into the brain by the endothelial cells. For example, the anti-HIV drug AZT is able to enter the brain because it resembles thymidine, one of the constituents of DNA and RNA. Doctors can also breach the BBB by injecting a strong sugar solution into the arteries that supply the brain. This temporarily shrinks the endothelial cells, allowing drugs to diffuse through the gaps—a method sometimes used at university research centers to treat brain cancer.

Research is also focusing on certain proteins in the BBB. Scientists have identified two proteins, zonulin and zot, that open up the junctions between cells in the intestine. Further research suggests that these proteins may also be involved in the BBB.

Stress and the BBB

Scientists have discovered that stress can affect the blood-brain barrier, causing it to become more permeable. During the Gulf War, Israeli soldiers took advantage of this fact to protect themselves against biological weapons. They took a drug that would not normally breach the BBB, but the stress the soldiers were suffering allowed the chemicals to pass through.

Metals and the BBB

Studies have suggested that certain toxic metals can bypass the BBB by entering the brain via sensory receptors that are exposed to contaminated water. Scientists studying fish and rodents, for example, have found that metals such as mercury are transported along sensory nerves to the brain. From these findings it seems likely that mercury and other toxins could enter the human brain in a similar way.

See also: Autonomic nervous system; Brain; Capillaries; Nervous system

Blood groups

Questions and Answers

If Rh-negative blood is rare, what are the common groups?

You are right in thinking that Rh-negative blood is uncommon—only 15 percent of the population are without the Rh factor.
  A positive and O positive are the most common groups. AB negative—shared by 0.45 percent of the population, or about one in 200—is by far the rarest.

I had a short-lived affair a few months ago but then returned to my husband. I am now pregnant, but my husband does not believe that the child is his. Can a blood test reveal who the real father is?

Blood tests cannot prove that a man is definitely the father, but they can prove that he is not related to the child. It is impossible to predict with absolute certainty what characteristics a child will inherit from the parents, but a child who is group B, for example, will not have two parents whose blood is either group A or O. Proof or disproof of paternity can be established by DNA testing.

I have Rh-negative blood. What would happen if I were given a transfusion of Rh-positive blood?

You could get away with such a transfusion—but only once. Your body would form antibodies to the Rh-positive blood cells, but not fast enough to cause immediate problems. However, those antibodies would react very quickly to a further Rh-positive transfusion, destroying the red cells in the new blood and clogging up your blood vessels and kidneys. The kidneys are the body's principal filtering system, and if prompt action were not taken, such as putting you on a dialysis machine, the situation could be fatal. This is why doctors routinely test patients before giving them any kind of transfusion.

For centuries people assumed that everyone had identical blood. Now, knowledge of the different blood groups enables lives to be saved daily by blood transfusion without the risk of dangerous reactions.

The Viennese doctor Karl Landsteiner first identified the major blood groups in the 1930s, when he experimented by mixing blood from various donors and found that only some people's blood could be mixed satisfactorily. He also discovered that, more often than not, mixing blood from different donors would result in massive damage to the oxygen-carrying red blood cells.

Four groups

Landsteiner identified four blood groups that are now called A, B, AB, and O. What differentiates the groups is the presence or absence of protein coats on the red blood cells, and of antibodies (which are part of the body's defense system) in the plasma, the colorless fluid part of the blood. These proteins act rather like a badge or coat of arms: they enable a cell to "see" or judge whether other cells belong to its own group or whether they are potentially dangerous outsiders. If a cell

HOW BLOOD GROUPS ARE INHERITED

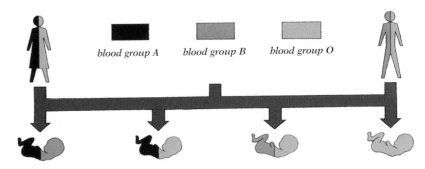

HOW PATERNITY TESTING WORKS

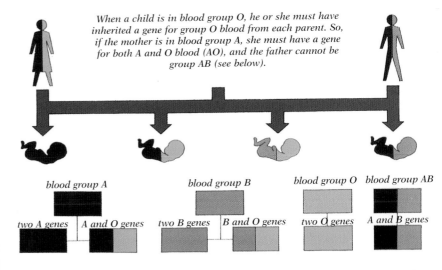

When a child is in blood group O, he or she must have inherited a gene for group O blood from each parent. So, if the mother is in blood group A, she must have a gene for both A and O blood (AO), and the father cannot be group AB (see below).

▲ *Blood tests are often used in paternity cases. But a test will only show that a child is not related to the father; it cannot be used to give positive proof of fatherhood.*

wears a different protein coat from that of the native cells, it will be attacked and neutralized by the antibodies in the plasma. Group A and group B red cells each have their own distinguishing protein on their surface.

Compatible groups

People with group O blood are called universal donors. Their blood contains neither distinguishing proteins nor aggressive antibodies, so everyone will accept it. Group AB people are known as universal recipients, because they have no antibodies to destroy alien red cells. But although it may be relatively safe to mix blood groups under these circumstances, it is not done in practice. It is always safest to transfuse blood of exactly the same group.

If someone is given a transfusion of the wrong kind of blood, the antibodies in his or her own blood will attack the red cells in the transfused blood, and the antibodies in the transfused blood will attack the patient's own red cells.

When an antibody attacks a red cell, they clump together in a sticky mess that clogs up the blood vessels and kidneys. A small transfusion would therefore cause jaundice and fever, and a larger one might block the flow of healthy blood to major organs and eventually lead to the death of the patient. The technical term used for this type of clumping is agglutination.

Heredity

A person's blood is determined by heredity. Broadly speaking, if parents have the same blood group as each other, their child will have the same group, too. If one is A (or B) and the other O, the child will be A (or B). If one is A and the other B, the child will be AB. If it were as simple as this, however, group O would have disappeared

HOW THE RH FACTOR CAN AFFECT A BABY

FIRST PREGNANCY

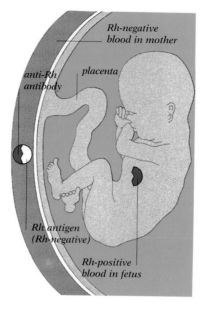

▲ *If blood from an Rh-negative mother and an Rh-positive baby mixes at all during labor, the mother will produce anti-Rh antibodies, but too late to affect a first baby.*

SECOND PREGNANCY

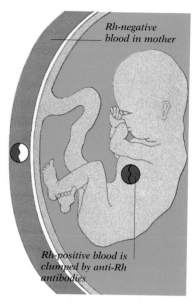

▲ *If the bloods mix during a second labor, the baby's blood will coagulate, because the mother now has anti-Rh antibodies in her blood as a result of the first baby.*

by now, but it is actually the most common group in the world. It survives because anyone with a group O parent retains what is called a recessive gene; if two people with this recessive gene have a child, his or her blood type is as likely to revert to group O as it is to follow the pattern described above.

Certain blood groups are more susceptible than others to particular diseases. For instance, people with group O blood are more likely to develop gastric ulcers, duodenal ulcers, and pernicious anemia. Group A has a higher incidence of stomach cancer, and group A and group AB are more prone to diabetes.

The Rh factor

Another important aspect of blood grouping is the Rh factor, a protein first found in the blood of rhesus monkeys. Eighty-five percent of white people are Rh-positive: that is, they have the Rh factor. It was the discovery of this factor, which can be present in blood of any group, that finally led to an explanation for what was once a common, dangerous, and even fatal kind of anemia in newborn children.

Researchers found that these children invariably had a father who was Rh-positive and a mother who was Rh-negative. The strange thing about the condition was that it affected only the mother's second or subsequent children, never the firstborn. The trouble arises when the Rh-negative mother is

THE CONNECTION BETWEEN BLOOD GROUPS AND DISEASE

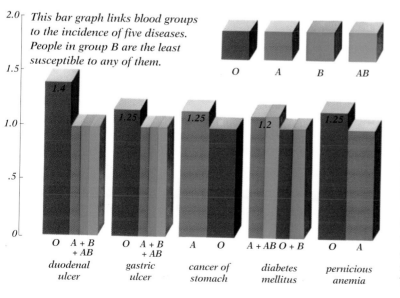

This bar graph links blood groups to the incidence of five diseases. People in group B are the least susceptible to any of them.

How the different blood groups are identified

Whenever red cells from group A blood encounter the anti-A antibody found in group B blood, they will stick together (agglutinate). Similarly, if group B red cells meet the anti-B antibody found in group A blood, they will agglutinate. Group AB blood cells can be agglutinated by exposing them to either of these antibodies. But group O cells will remain unharmed by either antibody. The test is carried out by mixing one drop of blood on a glass slide with a preparation containing anti-A serum and another drop of blood with anti-B serum. The analyst then looks at the samples through the microscope to see if they agglutinate.

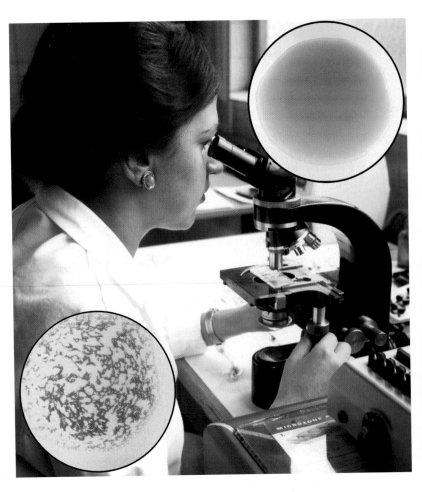

▶ *Group A blood (above right) is identified in a laboratory. It has coagulated (below left) after group B blood, containing anti-A antibodies, has been added.*

carrying an unborn baby with Rh-positive blood. During labor a certain amount of the baby's blood always gets into the mother's bloodstream. The Rh factor is "seen" as a foreign invader by the mother's immune system, which produces antibodies against it. There is some transfer of blood from mother to baby and vice versa through the placenta throughout the pregnancy, but it is mostly in labor that the baby is exposed to the mother's blood. However, there is not enough time for the mother's antibodies to build up and get into the baby's bloodstream during the labor of this first pregnancy.

During subsequent pregnancies, however, the antibodies that have formed do filter through into the baby, where they start destroying its Rh-positive cells. This situation can be compared to the progress of an allergy; no symptoms are produced by a first exposure, but subsequent contacts with the offending substance will produce some kind of reaction. This condition affects one child in 500 and used to lead to stillbirth, severe jaundice, and anemia or serious heart or brain damage. It can be prevented by giving the mother an injection of Rh antibodies shortly after the first delivery. These antibodies mop up any of the baby's blood cells that are floating around in the mother's bloodstream and are themselves consumed before her immune system has had time to produce its own permanent antibodies. So if the woman has another Rh-positive baby, that baby will face no greater risk than the firstborn.

Other ways of grouping blood

In addition to the A, B, AB, and O groups and the Rh factor, there are more than a dozen other systems of grouping blood, all of which rely on identifying some kind of distinguishing protein in the blood cells. Two such blood types, known as M and N, were also identified by Landsteiner. For the practicing doctor worried about blood transfusions or the health of individual babies, these other groups are of no great practical importance. Their principal uses are in the study of heredity, and they have little significance for daily health care. At one time they were also used in paternity disputes, but testing blood groups is no longer considered the most appropriate method for establishing paternity. Although blood groups may be useful in denying paternity in certain cases, the development of DNA fingerprinting is now the definitive method.

Ethnic blood groups

There is a difference in the percentage of blood groups in the various ethnic populations throughout the world. In the United Kingdom, for example, more people have group O blood than any other single group; people with group B blood are a small minority. In India and China, on the other hand, the picture is reversed: in these countries, the largest number of people have group B blood.

In the United States, among the white population, 45 percent are group O, 41 percent are group A, 10 percent are group B, and 4 percent are group AB. Among African Americans, 50 percent are group O, 29 percent are group A, 17 percent are group B, and 4 percent are group AB. Among Asian Americans, 35 percent are group A, 30 percent are group O, 23 percent are group B, and 12 percent are group AB.

See also: **Pregnancy**

Blood pressure

Blood pressure problems affect many people and are a major cause of ill health. But regular checkups can detect warning signs, and treatment does not usually interfere with a person's normal life.

The normal blood pressure is very variable with changes in the level of activity, and during exertion it commonly rises harmlessly above figures quoted as normal. Pressure that is consistently unduly high at rest, however, is called hypertension, and if it is not treated, the chances of disease—or even death—are increased. In fact, the major causes of death in the Western world today are diseases of the heart and blood vessels. Blood pressure is therefore not just a symptom but an urgent warning signal.

Causes

The trouble starts within the arteries (the thick-walled vessels that carry blood from the heart to the tissues of the body). The blood is driven by the main pumping chamber of the heart, the left ventricle, and a great deal of force is required to send the blood out of the heart and into the arteries, through the tissues, and then back into the heart again to be redelivered to the arteries. Therefore, even under ideal conditions, the walls of the arteries are continually under considerable stress. The level of arterial pressure is of great importance. If the constant running pressure within the system is raised for any reason—a condition called hypertension —this stress is increased and paves the way for the development of atherosclerosis, a common arterial disease characterized by stiffening and narrowing from deposits of fatty material. The heart and the arteries can be severely strained and damaged because the blood is pounding through with a very unnatural force.

On the other hand, seriously low blood pressure, called hypotension, is not a common problem and is usually the result of shock from a heart attack, acute infection, or loss of blood following an accident. Occasionally it may occur in people suffering from Addison's disease, a failure of the adrenal glands. This condition is rare and can easily be corrected by drug treatment. Because the maintenance of correct blood pressure is so important, sophisticated

▼ *Blood pressure can be tested quickly and easily at the doctor's office.*

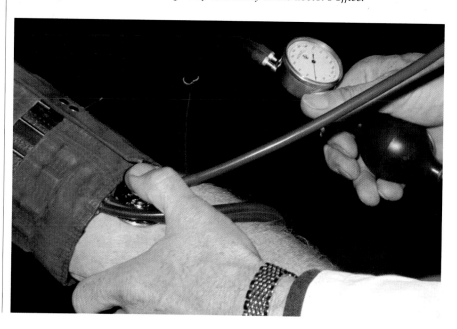

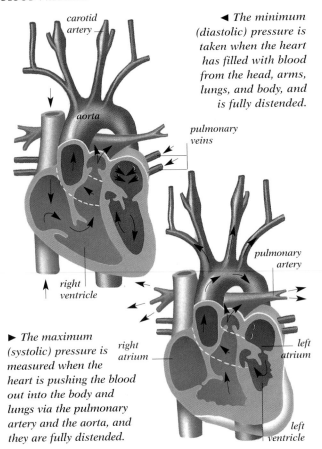

◄ The minimum (diastolic) pressure is taken when the heart has filled with blood from the head, arms, lungs, and body, and is fully distended.

► The maximum (systolic) pressure is measured when the heart is pushing the blood out into the body and lungs via the pulmonary artery and the aorta, and they are fully distended.

which increases blood pressure. Although the average blood pressure in various groups has been found to correspond with their average salt intake, it has been impossible to prove conclusively that higher salt intake necessarily leads to higher blood pressure.

Blood pressure starts to rise when people adopt a more "developed" way of life. Why does blood pressure change in this way? Known factors associated with hypertension include genetic influences, low birth weight, obesity, high alcohol intake, and hardening of the arteries owing to age. It is also believed that both salt intake and an excessively stressful lifestyle are among the factors that can produce essential hypertension.

Control of hypertension

Whatever the causes may be, the tendency to develop essential hypertension is definitely connected with some kind of overactivity of the normal control mechanisms of the body. There is an area in the lower part of the brain called the vasomotor center that controls blood circulation and hence blood pressure. The blood vessels that are responsible for controlling the situation are called arterioles and lie between the small arteries and the capillaries in the blood circuit. The vasomotor center receives information about the level of the blood pressure from pressure-sensitive nerves in the aorta (the main artery of the body) and the carotid arteries (to the head) and sends instructions to the arterioles via the sympathetic nervous system.

Kidney control

In addition to this fast-acting control, there is a slower-acting control operating from the kidneys, which are very sensitive to blood flow. When the pressure falls, they release a hormone called renin, which in turn produces an inactive substance called angiotensin I. This is converted to angiotensin II by an enzyme. That fact was responsible for an important class of drugs, the angiotensin converting enzyme (ACE) inhibitor drugs. The process of converting angiotensin has two effects: first, it constricts the arterioles and raises the blood pressure; second, it causes the adrenal glands to release a hormone called aldosterone, which makes the kidneys retain salt and causes the blood pressure to rise. This interaction of the pressure and salt-control systems does not explain the cause of essential hypertension.

mechanisms have evolved in the body to stabilize it. In the West, however, the level of general stress has led to a situation in which many people develop a level of blood pressure that is much too high for the continuing good health of the arterial system. When this is not the result of disease elsewhere, it is called essential hypertension. The major long-term effect of high blood pressure is on the arteries of the brain, the heart, and the kidneys, with the eventual likelihood of strokes, heart attacks, or kidney failure.

What is normal

The maximum pressure of each heartbeat, or systole, is called the systolic pressure, and the minimum pressure is called the diastolic pressure. These two pressures are measured to determine a person's level of blood pressure. Obviously, some figure has to be adopted as "normal." For young and middle-aged adults, a pressure of 120 (systolic) over 80 (diastolic)—written as 120/80—is considered normal; 140/90 is cause for concern, and 160/95 is definitely high and requires treatment. The difficulty of measuring blood pressure accurately is increased by the fact that in the industrial world pressure rises with age. This is not the case in primitive communities that are untouched by the industrial way of life. In these communities, people enjoy stable blood pressure throughout life; in fact, in some cases pressure even tends to go down with increasing age. There is a significant difference in the type of food that is eaten by developed communities as compared with primitive people. The amount of salt consumed is particularly important, since it has been found that salt tends to increase the volume of blood in circulation,

---

Self-help for high blood pressure

Try to reduce the stress in your life. Much of this is due to worry, frustration, and disappointment. If you have mild hypertension, dealing with the problems causing stress may help lower your blood pressure.

Cut out smoking; the nicotine in cigarettes is rapidly absorbed into the bloodstream and is known to increase blood pressure.

Keep your weight down. If you are overweight, you are making extra work for your circulatory system. Blood pressure can fall within weeks after excess weight is shed.

Take some kind of regular exercise. Jogging and yoga are excellent, although almost any kind of regular exercise will improve the general health and fitness of your body.

HOW THE KIDNEYS CONTROL BLOOD PRESSURE

▼ *Kidneys secrete renin, which produces angiotensin when pressure is low. This constricts arteries and raises blood pressure. Simultaneously, the adrenal glands produce aldosterone, causing salt retention, which also raises pressure and stops renin production.*

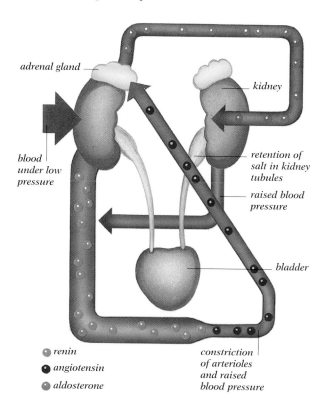

adrenal gland

kidney

blood under low pressure

retention of salt in kidney tubules

raised blood pressure

bladder

🔴 renin
🔴 angiotensin
🔴 aldosterone

constriction of arterioles and raised blood pressure

Diagnosis

Raised blood pressure may be the result of a number of conditions apart from essential hypertension. Many kidney diseases cause high blood pressure. Therefore, when people are suspected of having this problem, their kidneys are usually checked. This is easily done, in most cases, with a single blood and urine test. Only occasionally is it necessary for the person to have more advanced testing such as a kidney ultrasound or other imaging.

The blood test measures the urea in the blood—this is likely to be raised if there are kidney defects. The blood level of various salts (sodium, potassium, and bicarbonate) gives clues to other causes of high blood pressure. The urine is screened for the presence of protein, which also occurs in chronic kidney infection or disease. Many doctors will also perform a cardiogram and chest X ray to see if the raised blood pressure has affected the heart in any way. If the kidneys are found to be working abnormally, it is possible that the raised blood pressure could be the result of renal (kidney) disease. On the other hand, the raised blood pressure can itself cause deterioration in the kidneys. This happens because continuing high blood pressure particularly affects the arterioles, causing their walls to thicken. This obstructs the flow of blood and has an adverse effect on kidney function. Sometimes it may be impossible for the doctor

to distinguish between cause and effect. A vicious circle may be set up that maintains, and even accelerates, raised blood pressure. Kidneys that are short of blood secrete additional renin. This leads to more angiotensin and higher blood pressure.

Essential hypertension accounts for 90 percent or more of people with high blood pressure levels. Most of the remaining sufferers have kidney disease, and a few have abnormalities of hormone secretion such as an overproduction of cortisone or adrenaline. Another condition that may cause high blood pressure is excessive secretion of aldosterone by the adrenal gland. All these forms of hypertension respond to treatment.

Symptoms

Sometimes by the time people feel it is necessary to see their doctor, their blood pressure is already very high indeed. They may have had blackouts, a minor stroke, and swollen ankles or shortness of breath. This situation is likely in malignant hypertension and is fatal if left untreated. The brain disturbances here are due to an increase in the pressures operating inside the skull and pressing on the brain. This can be detected by examining the eye with a strong light; if there is undue pressure, the central nerve at the back of the eye will look swollen. Examination of the eye will also provide the doctor with other useful information. For example, small arteries at the back of the eye, which are the only ones that can be seen without an operation, will also show definite signs of hypertension.

Aside from cases of malignant hypertension, high blood pressure causes no symptoms. People may complain of headaches, but this does not necessarily indicate that they have hypertension. Dizziness and nosebleeds are not symptoms of hypertension. The only real guide to the state of your blood pressure is to have it checked regularly, every year or two.

Treatment

Although hypertension cannot usually be cured, it can be controlled by various treatments. It has been shown that drug treatment improves survival for both young and middle-aged people suffering from hypertension. But since no drug is without side effects, doctors must consider very carefully at what point a drug is justified. This is particularly difficult with elderly patients, who are much more prone to suffer from the side effects of drugs.

Five main types of drugs are used to treat high blood pressure: diuretics, beta-blockers, ACE inhibitors, angiotensin II inhibitors, and calcium channel inhibitors.

Diuretics are widely used in treating high blood pressure. They cause the kidneys to pass more salt and urine, reducing the volume of the blood and thus bringing down its pressure.

Beta blockers lower blood pressure; they get their name from their effect of blocking the action of adrenaline.

Outlook

If a person's high blood pressure is treated, the chance of a stroke is greatly reduced, and the risk of having a heart attack is also lowered considerably. This is why it is so important for people to take the drugs that have been prescribed by the doctor—even when they are feeling perfectly healthy.

See also: **Arteries; Kidneys**

Blushing

Questions and Answers

My face turns red when I eat hot food. Is this the same as blushing?

No. Hot peppers and other spicy foods have this effect on many people, as do alcohol, heat, and sexual arousal. It is called flushing and is one of the body's ways of cooling itself down. A flush can last for up to an hour, while a blush—which stems from emotional causes—is usually over in minutes.

Whenever I have to speak to a large group of people, I can't stop blushing. Would it be better to just avoid these situations?

No. You will blush less frequently if you get accustomed to circumstances that you find awkward or embarrassing. It helps to concentrate on what you want to say rather than on how you look when you say it.

How can I become less self-conscious when I meet strangers?

Try to concentrate on putting the other person at ease. If people seem standoffish, it may be that they are shy, too, and do not wish to show their vulnerability.

How can I increase my confidence?

Rehearse in your mind every day all your achievements, however small, and chalk up to experience the things that went wrong. People often exaggerate the importance of qualities they lack and belittle those they have. Keep things in perspective, and you will gradually become more confident.

Why do I blush more in winter?

The main reason is that a blush shows up more on pale skin than on suntanned skin. Also, in winter we tend to sit next to fires or radiators, which cause flushing.

Blushing has an unfortunate way of occurring when people least want it to, such as when they are embarrassed, anxious, or angry. The reaction is only momentary, however, and many people grow out of it altogether.

Everyone knows what it feels like to blush. A sensation of warmth and of blood rushing to the surface spreads from the neck all over the face. The feeling is often accompanied by a quicker heartbeat, tense muscles, a prickling sensation, sweat on the forehead, and a strong urge to hide the face. This lasts only a minute or two, and for most people it does not happen often enough to be a nuisance. However, a few people blush very easily, and this tendency can make their lives a misery.

Emotional alert

Blushing generally occurs only on exposed areas such as the face and neck. It is usually set off by an unwanted or unexpected feeling or thought that catches a person off guard. The part of the brain that controls emotion reacts by telling the body to pump out adrenaline and prepare for action. This causes some of the blood vessels near the skin's surface to relax. They then dilate as extra blood flows into them. This shows through the skin as a rosy glow. A blush is more obvious in a fair-skinned person, especially a redhead, than in people with more pigment (coloring) in their skin.

▲ *Blushing is especially apparent in people with fair skin, as the dilated blood vessels show readily through the surface.*

Blushing tendencies

Some people are more prone to blushing than others. Women blush more easily than men, and the hot flushes of menopause (when women cease to menstruate) have no counterpart in the middle-aged male. Shy people also blush more than those who are self-confident and extroverted, and some elderly people blush frequently if they are suffering from loss of confidence as a result of the aging process and social pressures put on them. Adolescents are also very susceptible, because they tend to be particularly self-conscious at this stage in their physical and social development. The emotional triggers of blushing can include a feeling that a person may look or sound silly, a fear of disapproval, anxiety, anger, shame, or guilt. It is even possible to blush when experiencing secret thoughts and desires.

Mind over matter

It is impossible to control the physical mechanism that causes blushing, but it is possible to reduce the number of situations that make a person feel like blushing. Practice in handling social situations and exercises to increase confidence are the best ways to do this.

For the few people who blush excessively, medication may prove useful. Most people hardly blush at all by their mid-twenties, but the ability to blush occasionally can indicate a refreshing sensitivity.

See also: Autonomic nervous system

Body odor

Which is better, a deodorant or an antiperspirant?

Deodorants contains pleasant-smelling chemicals that mask the odor of sweat; antiperspirants actually reduce the evaporation of the sweat so that the odor is sealed in. Try a deodorant first, and if this does not control the odor, then try an antiperspirant.

Children do not seem to have body odor. Why is this?

Children's sweat is similar to adults' sweat, but children have smaller bodies that sweat less. Also, they have no underarm hair to trap stale sweat. They still smell when they sweat heavily, however, and the odor is slightly metallic.

Is perspiration just another word for sweat?

Yes, but perspiration generally describes moderate sweating—a steady production of sweat that does not form into heavy droplets.

Why do some people seem to sweat more than others?

The mechanism of sweating is controlled by the nerves and is affected by both a person's excitability and the state of the skin's blood vessels. Everyone is different. For example, heavier people have to sweat more to cool themselves off, and anxious people sweat more because their nerves are more active.

Why is it that women can find men's body odor attractive?

When the smell is mild, it can be sexually attractive. Sweat contains substances that in many animals are used for sexual signals, and humans may react unconsciously in the same way as animals.

Most people worry about their body odor. In fact, every person has an individual smell, but it is what each person does about this odor that determines whether or not the odor becomes offensive.

Body odor is caused by sweat, or perspiration, which helps to cool the body as it evaporates from the skin. The sweat glands—of which humans have about 2 million—can produce up to 6 pints (2.8 l) of perspiration a day. This consists mostly of water and salt, with a small amount of waste products. It has little odor in itself and begins to smell unpleasant only when it becomes old and stale and the skin's bacteria begin to act on it. Sweat from some parts of the body is more likely to cause an odor than that from others. This sweat, called apocrine sweat, is produced by the sweat glands in the armpits and around the genitals and anus, and it is more concentrated than perspiration from the rest of the body.

Individual differences

Sweating varies from person to person and from time to time. Some people sweat so little that their perspiration is almost unnoticeable, while others have to contend with very active sweat glands. Sweating also increases with exercise and with nervous tension. On a cool day a person may hardly sweat at all, but on a hot day the sweat glands work overtime. If the air is very humid, it stops the sweat from evaporating, so people tend to feel sticky and uncomfortable.

Very pungent body odor can occasionally be due to a serious medical condition or skin infection. Certain diseases can cause excessive sweating, but in some medical conditions sweating is reduced to a minimum. A few people suffer from hyperhidrosis—increased sweating—which requires medical treatment.

Body odor can be controlled by regular hygiene. A bath or shower once a day is ideal. Washing more often can disturb healthy skin bacteria and make odor worse. Shaving the armpits does not reduce sweat, but it does reduce trapped sweat that can produce odor. Deodorants and antiperspirants can help by masking or reducing apocrine sweat.

See also: **Glands; Perspiration**

▲ *The best way to prevent odor is regular, thorough bathing or showering.*

How to prevent body odor

Bathe or shower daily, especially after exercise, to prevent a buildup of sweat.

Dry the skin thoroughly, using a clean towel that is not too soft—a slightly rough surface is more effective.

Keep the use of scented talcum powder and perfume to a minimum. These can combine with sweat or bacteria to produce a stale odor.

Apply deodorant or antiperspirant regularly if you sweat heavily.

Wash your clothing regularly.

Wear comfortable, loose clothing when the weather is hot.

Wear cotton underwear; it lets the skin breathe more than nylon.

Body structure

Aside from the obvious sexual differences, do men and women differ in their body structures? And, if so, by how much?

On average, men are 6 in. (15 cm) taller than women and have several internal differences in body structure. Most men have larger hearts and lungs than women, made of 42 percent muscle, compared with 36 percent in women. A woman's body contains about 4 percent less water than a man's, because she has more fat beneath the skin, and fat is a water-free substance.

Is it true that some people have their appendix on the wrong side?

Yes, some people are born with their appendix on the left rather than the right side of the body. Such people have transposition of the whole intestine, a condition known as situs inversus. Their hearts also tend to be reversed so that the apex points to the right rather than to the left. This is called dextrocardia.

My husband and I are both much taller than average. Does this mean that our son will be tall too?

The environment as well as genes has a great effect on height. Your son will probably be taller than average, but his final height will also depend on his diet and the proper functioning of many of his internal organs.

Your son will not grow properly if his diet is deficient in proteins or the mineral calcium, which is needed for building bones. Even with a perfect diet, his growth would be retarded if he had something wrong with his supply of the growth hormone, somatrophin. This is a substance that is necessary if his body is to be able to use the food he eats to enlarge his bones and other internal body structures.

Many people find the human body a fascinating subject, but for a better understanding of the discoveries made about it, a working knowledge of its structure is required.

Structure simply means the way something is put together. In the case of the human body it is possible to talk about an enormously complicated structure by reducing its many parts to a set of simple labels, or medical shorthand. Doctors must do this when they talk to each other. They may say something about a skeletal defect or problems with the digestive system, and unless the patient understands what a doctor means, being treated for an illness can be a mystifying experience. Some of the terms in this basic shorthand are concerned with differences such as race, but most terms make up a picture of what every human body is like under the skin.

For purposes of classification people can be divided into three large groups, or races: the Mongoloids, Negroids, and Caucasians. Each of these groups has certain structural characteristics, and each, in turn, can be divided into many subgroups. People of the Mongoloid group are typified by their yellow skin, straight black hair, and eyes with folded lids that give the eye an almond-shaped appearance. The Mongoloid body has little hair, and the height of the average male ranges from 5 feet 2 inches (1.57 m) to 5 feet 8 inches (1.73 m). The Chinese and the Inuit are typical Mongoloid types.

BODY CAVITIES AND URINARY SYSTEM

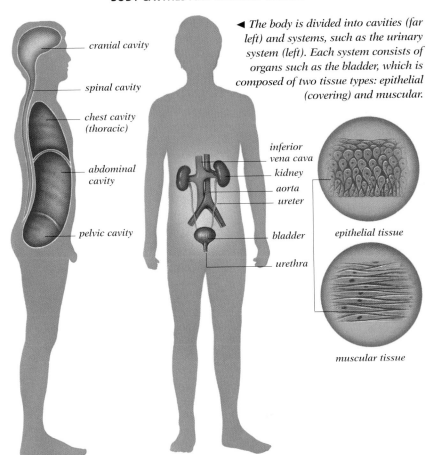

◀ *The body is divided into cavities (far left) and systems, such as the urinary system (left). Each system consists of organs such as the bladder, which is composed of two tissue types: epithelial (covering) and muscular.*

cranial cavity

spinal cavity

chest cavity (thoracic)

abdominal cavity

pelvic cavity

inferior vena cava

kidney

aorta

ureter

bladder

urethra

epithelial tissue

muscular tissue

▲ *This Caucasian female has the pale skin of the Nordic group of the race.*

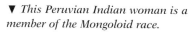
▼ *This Peruvian Indian woman is a member of the Mongoloid race.*

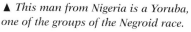

▲ *This man from Nigeria is a Yoruba, one of the groups of the Negroid race.*

▲ *An endomorph tends to have excess body weight and a rounded build.*

▲ *The ectomorph is lean and angular, but not excessively muscular.*

▲ *This athlete is a prime example of the muscular mesomorph.*

Negroids have brown to black skin and woolly hair. Like Mongoloids, they have very little body hair, and they have broad noses, although the Mongoloid nose is narrower at the base. Negroids vary in height—from the Bushmen, averaging 4 feet 8 inches (1.42 m), to the Nilotes, whose adult males average 5 feet 10 inches (1.78 m). Most African Americans are classified as Negroids.

Caucasians, the so-called white races of the world, are actually very mixed, with skin colors ranging from the brown of Indian peoples to the pale coloring of the Nordic (meaning generally found in the north) subdivisions of the race. Their hair may be any shade of brown, black, red, or blond. The body of a Caucasian male is much more hairy than that of members of the other two races, and the face and nose are narrower. One of the least variable Caucasian features is average height, which ranges (for the adult male) from 5 feet 3 inches (1.6 m) in Mediterranean peoples to 5 feet 8 inches (1.73 m) in Nordic types. As well as these three major groupings and their subdivisions, there are many mixed, or composite, races.

Shape

People can also be divided into three groups according to their body shape: endomorph, ectomorph, or mesomorph. A typical endomorph is heavily built and well rounded, with a higher than average proportion of body fat and a tendency toward obesity. The ectomorph is lean and angular with less fat and muscle than average and is capable of a high degree of physical endurance. The mesomorph is muscular and agile—having the typical athlete's body shape—and is the strongest of the three. Not everyone conforms exactly to a standard pattern—most people are a mixture of all three—but each person has an overall leaning toward one of the three body shapes.

Each body type tends be associated with a certain behavior pattern. Extreme endomorphs tend to be very relaxed, with a slow heartbeat and breathing rate. They are tolerant and good company but are inclined to overindulge in food and drink. Extreme ectomorphs are the exact opposite. Hypersensitive and very aware of everything that is happening, they tend to react quickly to new situations. But they are likely to be obsessive and easily thrown off balance by personal setbacks. Between the two are the mesomorphs, who are typically dominating, aggressive, and successful.

Body systems

Whatever a person's race or shape, the body is divided into four sections: the neck and head; the chest or thoracic cavity, which contains the heart and lungs; the abdomen, housing the alimentary canal, kidneys, liver, and various other organs; and the limbs. To understand how these parts link up and work together, the body can be studied as systems—groups of organs that work together. One of these is the digestive system; others include the nervous and reproductive systems. Each organ is made up of tissues formed from cells. These tissues are complex collections of chemicals, including such components as the chromosomes that carry the genes. Many of the body organs have familiar names, such as the adenoids, appendix, or tongue, but the tissues are less well-known. There are four main types, and each organ contains at least one of them. Epithelial tissues cover or line body organs and may secrete substances such as hormones. Connective tissues—which include bones and tendons—link, support, and fill out body structures, including the blood. Muscle tissue enables the body and internal parts to move. Nervous tissue helps the body parts to work in harmony by providing communication and control.

Body fluids

About 60 percent of the body is made up of water, or body fluids, but these are not like tap water. Body water contains a huge variety of dissolved salts that make it more like seawater—where the first living things evolved. A 154-pound (70-kg) man contains an average of 18.5 pints (8.7 l) of body fluids. Of these, 12.5 pints (5.9 l) are inside the cells and are called intracellular fluids. The other 6 pints (2.8 l) are extracellular fluids. Of these, 1.3 pints (0.6 l) are in the blood plasma. The remaining 4.7 pints (2.2 l) are divided between the interstitial fluid, which bathes the cells and body cavities; the lymph fluid; the fluid around the spine and in the joints; and the fluid in places such as the eyes. All this fluid moves constantly in and out of the cells. The body's fluid is even more important to survival than food. It is essential for carrying oxygen, food substances, and hormones around the body and getting rid of wastes. The body maintains a constant check on its fluid level, and if this level becomes too low, endangering body functioning, the water-monitoring center in the brain generates a sensation of thirst, so the person drinks to rebalance body fluids.

SYSTEM	MAJOR ORGANS	MAJOR TISSUES IN TYPICAL ORGAN
Digestive	Mouth, teeth, tongue, salivary glands, esophagus (tube from back of throat to stomach), stomach, small and large intestines, anus, liver, gallbladder, pancreas	Stomach—epithelial, muscular
Urinary	Kidneys, ureters (tubes joining the kidneys and bladder), bladder, urethra	Bladder—epithelial, muscular
Muscular	All body muscles, some under conscious control (skeletal or striped muscles), others working unconsciously (smooth muscles)	Biceps muscle—muscular, connective
Skeletal	All the body's bones and the connecting joints	Connective
Respiratory	Lungs, bronchi (tubes to lungs), trachea (windpipe), mouth, larynx (voice box), nose, diaphragm	Lungs—epithelial, connective
Circulatory	Heart, arteries, veins, capillaries, blood	Heart—muscular
Nervous	Brain, sense organs (eyes, ears, taste buds, smell and touch receptors), nerves, spinal cord	Nervous
Endocrine	Hormone-producing glands: pituitary, thyroid, parathyroid, adrenals, pancreas, thymus, parts of testes and ovaries, and small areas of tissue in the intestine	Pancreas—epithelial, connective
Reproductive (male)	Testes, penis, prostate gland, seminal vesicles, urethra	Penis—muscular, vascular
Reproductive (female)	Ovaries, fallopian tubes, uterus (womb), cervix (neck of womb), vagina	Uterus—muscular, epithelial
Immune	Structures involved in the body's defense against disease, including lymph nodes, lymph vessels, spleen, tonsils, adenoids, thymus (gland in chest)	Spleen—connective, epithelial

A guide to body systems

See also: Blood; Bones; Muscles

WHAT WE ARE MADE OF

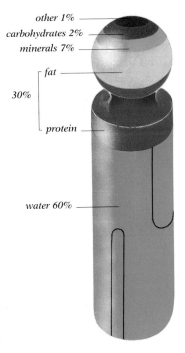

other 1%
carbohydrates 2%
minerals 7%
fat
30%
protein
water 60%

Bones

Bones are light, extremely strong, and joined so that the human body is highly mobile. They are also involved in the production of red blood cells. There are few serious bone diseases, and these are usually treatable.

Most people think of bones as simply a stiffening framework deep inside the body that helps to keep them upright. Although this is basically true, bones are really a reminder of the fight for survival that all animals faced in the earliest stage of life on Earth. Bones protected animals from damage or attack, and almost all successful primitive land creatures carried their bones outside their body—like the bony armor-plating (shell) of a tortoise. Only later did some groups of animals develop so that their shells grew partially, then completely, inside the body. In humans, bones reach a very high form of development, each of the hundreds of different bones being joined to the next to create a fantastically strong, yet agile, framework: the skeleton.

DIFFERENT TYPES OF BONES

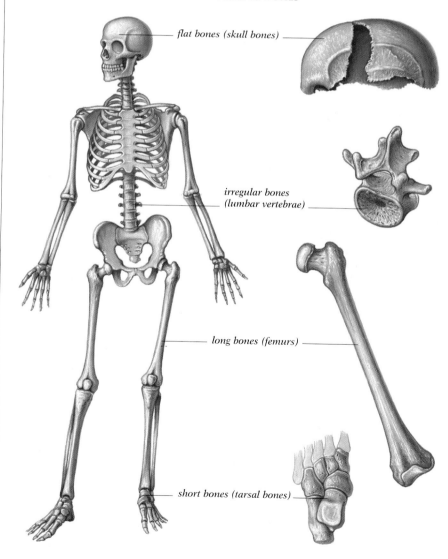

flat bones (skull bones)

irregular bones
(lumbar vertebrae)

long bones (femurs)

short bones (tarsal bones)

Questions and Answers

I have an extra little finger on my left hand. It has never bothered me, but now I'm pregnant. Will my baby have it too?

Yes, probably. This is an inherited characteristic: it can be passed on from parents to children. Your ancestors probably had it too. However, it is a harmless bone defect and can be removed by a simple operation, leaving the hand perfectly normal. You may feel that this is worth doing if the extra finger gets in the way.

My husband has had a bony lump on his skull for years. He's never thought anything of it, but with all the talk of malignant lumps, he now wonders. What should he do?

Don't worry. What you describe is most likely to be an acceptable exostosis—a benign (harmless) bone tumor. Exostoses can occur on bones and elsewhere and are always hard, pain-free lumps that grow very slowly. Your husband's can probably be removed in a minor operation if it is unsightly.

I had a chip fracture in my finger three months ago. My finger was splinted for a few weeks but is still swollen. How long will it be before the swelling goes away?

Any finger injury produces this condition, called spindle finger. It is caused by the body's tendency to produce more new bone than is needed to mend the break. It will slowly return to normal.

My baby has two soft patches on his skull. What are these?

They are known as fontanels and are part of nature's way of ensuring a safe delivery. They allow the skull to be squashed a little as it passes through the pelvis. After birth, rapid bone growth quickly closes the gaps within a few months.

I saw an X ray of my 10-year-old son's arm, and there seemed to be lots of breaks in it. Yet the doctor said that the bone was not broken. Can this be true?

The breaks you saw were gaps between the growing points of the long bones. They join together after puberty, but in a 10-year-old they can give a confusing picture.

The doctor says my brother has osteomyelitis. How did he get it, and is it serious?

Osteomyelitis is a bone infection that seems to occur in areas where there has been a previous injury—perhaps a break. Stray bacteria in the bloodstream from a cut or chest infection multiply in the area where the bone's natural defenses against them have been reduced. Osteomyelitis can usually be cured.

Why do the bones of elderly people break so easily?

As people grow older, their bones have a tendency to become thinner, less solid, and more brittle. The condition is called osteoporosis, and it is defined as decreased density of normal bone. Osteoporosis can be improved by treatment.

Other bone diseases, such as Paget's disease and cancer, are also common in the elderly and predispose them to fractures. Patients on long-term steroid therapy may also develop thinner bones over time.

The skeleton seems to be a remarkable piece of engineering, very light but also strong and maneuverable. So why do humans suffer so much from backache?

When humans began walking on two legs instead of four, they used the same skeleton that had evolved for four-legged animals. Unfortunately, they didn't develop much further. So while we are well adapted horizontally, vertical postures create strain at the center of gravity situated at the bottom of the back, causing pain.

The need for bones

The primitive function of bones as armor-plating is still obvious in certain parts of the human body, such as the skull, which forms a complete protective case around the brain, or the ribs, which do the same thing for the heart and lungs.

Bones also provide the support that keeps the many components of the body together and upright. When the body thinks support is no longer needed—such as in the prolonged weightlessness of space flight, or after a long bed rest—the bones lose their strength and will break easily if put under strain. Another vital use of the bones is as girders to which muscles may attach. Muscles provide the power to move the various limbs and body parts, but they can do so only by moving the bones relative to each other.

The body, with great economy of space, uses the hollow cavities inside bones for the manufacture of blood cells. The bones also store another vital substance for the body—calcium.

The structure of bones

Like everything else in the body, the bones are made up of cells. These are of a type that creates what is technically called a fibrous tissue framework—a relatively soft and pliable material, osteoid. The osteoid becomes a base for the deposition of calcium and phosphorus compounds, which give strength and rigidity to bones.

The growth of bones

When bones begin growing, they are completely solid. Only at a secondary stage do they start to develop a hollow center. Hollowing out a rod of material reduces its strength only slightly but reduces its weight substantially. Nature takes full advantage of this structural law in the bones. The hollow centers are filled with a soft substance, called marrow. The marrow is where the blood cells are manufactured.

Bones start forming in a human baby during the first month of pregnancy, but just like the skeletons of primitive creatures, at this stage they are made of cartilage—a soft material with a rubbery flexibility. As the baby grows, this cartilage frame is replaced by fibrous tissue with little of the hard base that adds strength. Hardening of the bones is a gradual process that takes place throughout childhood and is not completed until the end of puberty.

Keeping in shape

Another important and remarkable feature of bones is their ability to grow into the right shape. This is especially important for the long bones that support the limbs. These are wider at each end than in the middle, so that extra solidity is provided at the joint, where it is most needed. This shaping, which is technically called modeling, is specially engineered during growth and goes on all the time afterward.

Different shapes and sizes

There are several different types of bone, designed to perform in varying ways. Long bones, forming the limbs, are simply cylinders of hard bone with the soft, spongy marrow interior. Short bones, such as those found at the wrist and ankle, have basically the same form as long bones but are more squat to allow a variety of movements without any consequent loss of strength.

Flat bones consist of a sandwich of hard bone with a spongy layer in between. They provide either protection (as in the skull, to protect the vulnerable brain) or a large area

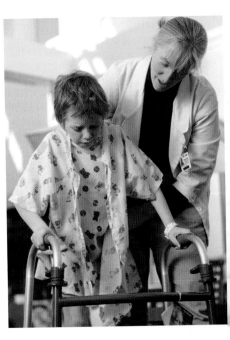

▲ *After a major bone in a leg is broken, the affected limb takes time to heal, and the patient will need patience and perseverance to learn to walk properly again.*

CROSS SECTION OF A LONG BONE

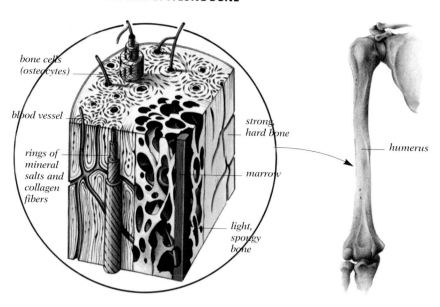

bone cells (osteocytes)

blood vessel

rings of mineral salts and collagen fibers

strong, hard bone

marrow

light, spongy bone

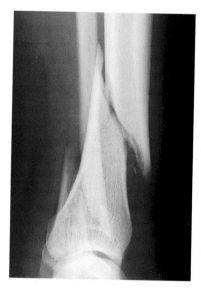

humerus

▲ *This X ray shows clear fractures of both long bones of the lower leg.*

for attaching muscles (as in the shoulder blades). Irregular bones are the final type; they come in several different shapes, designed specifically for the job they do. The bones of the spine, for example, are box-shaped to give strength and plenty of space inside the marrow. The bones that make up the structure of the face are hollowed out into air-filled cavities to create extra lightness.

The joints

Bones have to be joined securely to one another, but some must be able to move extensively in relation to each other. Nature solves this problem with two types of joint—the ball-and-socket type and the hinge type. Within the joints, the ends of the bones are lined with a pad of soft cartilage, so that in movement and weight-bearing they do not damage each other. The joint is also lubricated by specially produced fluids. Tying the whole structure together are tough thongs known as ligaments.

Self-maintenance

Like many other parts of the body, bones have an extraordinary capacity to maintain themselves if infected or damaged. The most obvious example is their ability to repair themselves when broken—even if they are broken completely in two.

People often find it hard to imagine how this can happen. The key to the process in the first place is the fact that when a bone breaks, the blood vessels running through the bone automatically break too.

A large loss of blood results (blood needs to be replaced in many cases), but it is the blood lying around the area of the break that creates the scaffolding for the repair by clotting (hardening) into a solid mass. Next, cells from the broken ends of the bone spread into the clotted area and lay down fibrous tissue. This unites the two broken ends, but before the joint is really complete, the hardening process must take place.

The finished joint is rather large and unwieldy, forming a mass of new bone around the place of the break. But later on the bone's ability to shape itself remodels the area into the original, smooth contours. This process takes place over a period of years after the break is completely mended and the limb is once again in use, so that eventually the place of the break—doctors call it a fracture—is not distinguishable from original, smooth bone, except by X rays.

Congenital diseases

Congenital bone diseases are rare but are severely crippling and in some cases fatal. Some of these conditions are due to a lack of the enzymes needed for formation of the ground substance, with a resultant buildup of the chemicals that should be processed by these enzymes. In other cases the formation of ground substance is defective, for unknown reasons. The result is increased fragility of the skeleton, with a tendency toward numerous fractures and, in some cases, dwarfism and marked skeletal deformities.

Chemical problems

Bone undergoes continual resorption and reformation, a process influenced by numerous hormones and vitamins. Softening of the bone, or osteomalacia, results when these substances are present in abnormal amounts.

The hormone PTH is made by the parathyroid glands and helps maintain normal calcium levels in the blood. Bones are the reservoir for the body's calcium stores, and bone is destroyed when more calcium is needed than is supplied by the diet. This may happen in several ways. Sometimes a tumor forms in one of the glands, which then produces too much hormone, even though the calcium level is normal. More frequently the glands enlarge as a result of a persistently low blood-calcium level due to disease in other parts of the body. Calcium absorption is impaired when a person has chronic kidney disease or severe intestinal malabsorption, and after resection of a large portion of the stomach.

All these factors can cause the parathyroid glands to enlarge and produce excess PTH hormone that will demineralize and weaken the bone. The condition can be treated by removing the excess parathyroid tissue.

REPAIR OF A FRACTURED BONE

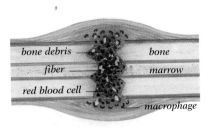

◄ *The site of the break is full of blood (which clots through the action of red cells, platelets, and fiber) and bone debris, which is removed by macrophages (large white blood cells). Surrounding bone produces cells that form the swollen or callus areas, one on either side of the break. These will create new bone.*

bone debris
fiber
red blood cell
bone
marrow
macrophage

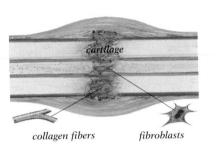

◄ *Fibroblasts (fibrous tissue cells) from intact bone produce collagen fibers, which help to make connective tissue.*

cartilage
collagen fibers
fibroblasts

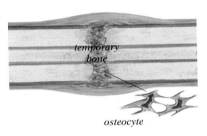

◄ *Cartilage is replaced and made more rigid by osteocytes (bone-producing cells), which create temporary bone at the fracture site.*

temporary bone
osteocyte

◄ *Permanent bone replaces temporary bone, and callus is reabsorbed. This takes a minimum of four to six weeks.*

permanent bone

Deficiency of essential vitamins

Vitamin D deficiency can also weaken bones. The disease that it causes, called rickets, was once widespread but has been virtually eliminated because vitamin D supplements are added to food. Vitamin C deficiency can likewise affect the health of bone by impairing production of ground substance, but this condition—called scurvy—is now practically unknown in the West. Vitamin A deficiency or excess also had adverse effects on bone formation in the past. A diet that is deficient in calcium will produce the same harmful effects as a vitamin D deficiency.

Infections

Infections of bone are called osteomyelitis. They may result from direct invasion of the bone by external trauma or may be spread via the blood from another site of infection in the body. The organisms most frequently involved are *Staphylococcus aureus* and gram-negative bacilli. Tuberculous osteomyelitis was once a common disease, but improved public health efforts to control this disease and the availability of effective antitubercular drugs have largely eliminated it. Osteomyelitis is a serious and debilitating disease, but it can now almost always be cured by treatment involving higher doses of and longer treatment periods with antibiotics.

Bone tumors

The most common primary malignant tumor of bone is osteogenic sarcoma, which predominantly affects people who are under 30 years of age. At one time such tumors were invariably fatal, but now they can often be cured by using combined treatments of X ray, surgery, and chemotherapy.

Bone, because of its rich blood supply, is often invaded by tumors from other body sites, such as the breast, prostate, and lungs. Although they are usually not curable, these tumors can be controlled by hormone therapy, chemotherapy, and X-ray treatment in varying combinations, allowing perhaps a period of reasonable comfort and activity.

Osteoporosis

Osteoporosis is a condition that affects the elderly, especially women, and is characterized by thinning and brittleness of the bones. It has multiple causes, including diminished bone formation, accelerated bone destruction, calcium deficiency, and declining levels of the hormone estrogen in postmenopausal women. It causes almost three-quarters of the fractures in women over 45 years of age, and it has become a greater problem as the elderly population increases.

Hormone replacement therapy (HRT) was widely prescribed after menopause to reduce the risk of osteoporosis but has recently been called into question. Less controversial treatments involve taking calcium supplements and the recently introduced bisphosphonate drugs that prevent bone resorption. The decreased physical activity of old age also contributes to bone loss.

Immobilization osteoporosis is a thinning of bone caused by lack of use. Long familiar as a result of prolonged bed rest, paralysis, or plaster-cast immobilization for fractures, it is now becoming a problem in space exploration, owing to the prolonged period of weightlessness experienced in orbital flights. Its cause is not known, but it may be due to abnormal sensitivity of the immobilized part to thyroid and parathyroid hormones.

Paget's disease

Local acceleration of bone resorption and deposition characterizes Paget's disease, which produces markedly disordered growth. The disease usually involves multiple areas of the skeleton (especially the skull, pelvis, and leg bones) and predisposes the sufferer to develop malignant tumors (osteogenic sarcoma). Most conditions of the bone can be either cured or controlled. Because of modern medical treatments, bone conditions need not be dreaded as they were in the past.

See also: **Cartilage; Hormones; Joints; Skeleton; Tendons**

Brain

Sometimes when I smell a certain smell, old memories come flooding back. How does this happen?

The connections in the brain for the sense of smell are closely linked with the circuits of the limbic system, which deals with emotions. Smells can thus take on emotional significance, whether pleasant or unpleasant. Because events associated with strong emotions are often firmly stored in memory, the smell itself can bring them back.

My friend says he's more intelligent than I am because his head is larger. Is this true?

Probably not. Intelligence has more to do with how you use your brain than how much you have. Some people's brains are probably better at organizing their perceptions, relating them to memories, and forming plans of action on the basis of all this information. The emotional level is also important. Too little emotion deprives the brain of essential psychological energy, while too much causes poor concentration.
  Your personality—especially the way that you react to problems and cope with your emotions—is a big factor in determining how intelligent you become.

My memory is poor. Is there anything I can do to improve it?

It is not your memory that needs improving; it is the way you try to remember things. To make sure that something is firmly stored, it must be presented in a way that arouses the most links with other knowledge already in the brain.
  For example, written directions are more difficult to remember than a map, which will stay fixed in the mind longer. If the map is humorous, the memory will be even stronger, because the emotions are involved.

The human brain is more sophisticated than the largest computer—yet it fits neatly inside the skull. With its many billions of cells, it directs and monitors all of our activities, even when we are asleep or unconscious.

The brain is at the center of the complex network of nerves that run through the body, and together with the spinal cord, it makes up what is known as the central nervous system. The central nervous system controls the whole body by means of messages that are continually passing up and down its nerve pathways. All the information a person receives about his or her surroundings comes from the five senses—sight, hearing, touch, smell, and taste. The nerves carrying this sensory information up to the brain are known as sensory nerves. Once the brain makes a decision, it sends instructions down other nerve cells called motor nerves.

Short circuit

All nerve impulses going to and from the brain have to go up or down the spinal cord. But now and then—for example, if a person touches a flame—such speedy action is needed that there is no time for the message to go all the way to the brain. The message goes only as far as the spinal cord, which processes the message and responds to it on a relatively simple level. The result is known as a reflex. When a doctor taps a patient's knee and makes it jerk, he or she is looking for damage to the spinal cord by testing how fast the reflexes are. Other examples of reflexes are blinking, reactions to pain, and sexual responses.

MAJOR DIVISIONS OF THE BRAIN

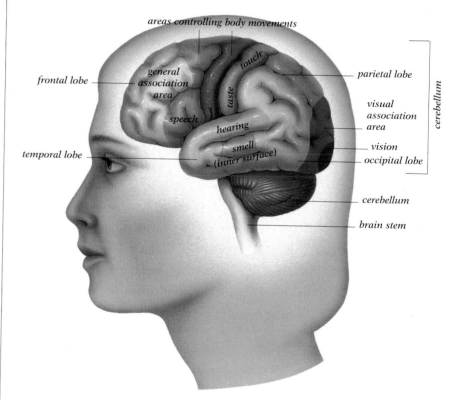

▲ *The brain consists of the brain stem, the cerebellum, and the cerebrum, which has four lobes. Body activities are controlled by specific areas within the brain.*

Inside the skull

All the messages flashing to and from the brain are transmitted by minute electrical impulses. They travel through special nerve tissue cells called neurons. The electrical activity in the brain creates waves that can be picked up on a machine called an electroencephalogram, or EEG, and studied for abnormal or unusual patterns. The brain cells form a mass of soft, jellylike tissue, surrounded by three layers of protective membrane called the meninges, and cerebrospinal fluid. Four arteries in the neck supply the brain with blood, which it needs to survive. The brain can be divided into three regions: hindbrain, midbrain, and forebrain. Each of these is subdivided into separate areas responsible for different functions, all connected to other parts of the brain.

Balance and coordination

The largest structure in the hindbrain is the cerebellum. This is the area that is in charge of balance and coordination, and it works closely with the organs of balance in the inner ear.

Early origins

Also part of the hindbrain is the brain stem, which links the brain with the spinal cord. This is the part of the brain that was first to evolve in primitive humans. It is in

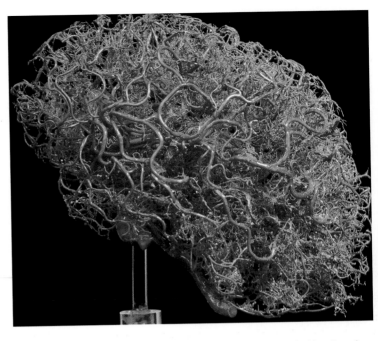

▲ *This resin cast shows the network of vessels that supply blood to the brain. Oxygen is vital to brain function, and brain cells die within a few minutes if they are starved of oxygenated blood.*

INTERNAL STRUCTURES OF THE BRAIN

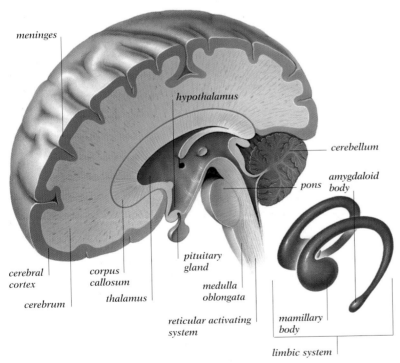

meninges

hypothalamus

cerebellum

pons

amygdaloid body

cerebral cortex

corpus callosum

pituitary gland

cerebrum

thalamus

medulla oblongata

reticular activating system

mamillary body

limbic system

▲ *This cross section highlights the major structures of the brain. The limbic system (inset), located within the thalamus, is chiefly concerned with memory, learning, and emotions.*

the brain stem that all incoming and outgoing messages come together and cross over, since the right-hand side of the brain governs the left side of the body, and the left-hand side of the brain governs the right side of the body. The various structures in the brain stem—including the medulla, pons, and reticular-activating system—control heart rate, blood pressure, swallowing, coughing, hiccuping, vomiting, breathing, and consciousness.

Brain censor and other structures

One of the brain's most crucial functions is controlling the level of consciousness. A mechanism in the brain stem's reticular-activating system sifts through the mass of incoming information and decides what is important. It does this by controlling the amount of electrical activity that each part of the brain receives. In turn the brain's decisions affect the reticular-activating system. It is this interplay that determines consciousness and alertness. Just beyond the brain stem, in the midbrain, is an area devoted to eye movements and pupil size. Beyond this the forebrain begins. Here is the thalamus, which acts as a relay station for incoming information. Just below the thalamus is the hypothalamus. This is involved in such bodily functions as hunger and thirst, body temperature, sex, and sleep, and it works closely with the pituitary gland.

PROPORTIONS OF THE CORTEX DEVOTED TO SENSORY AND MOTOR ACTIVITIES

▼ *The cortex (surface of the cerebrum) has one area that just receives sensory or incoming information (left), and another area that deals with outgoing or motor information (right). The amount of cortex devoted to receiving or giving information depends on the specific part of the body. For instance, a large part of the sensory cortex is given to the lips because they are very sensitive; a large part of the motor cortex is related to sending out information to the hands.*

INCOMING INFORMATION—SENSORY

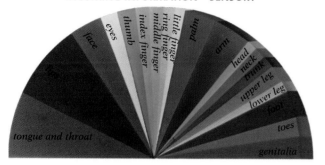

OUTGOING INFORMATION—MOTOR

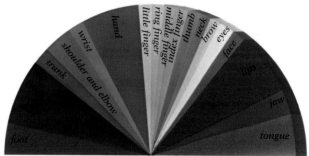

Deep-seated emotions

Nearby is another important system: the limbic system, made up of a number of structures including the hippocampus, amygdala, and septum pelucidum. This is the second most primitive part of the brain and is concerned with deep-seated emotions like rage, excitement, fear, sexual interest, pleasure, and even sociability. The limbic system is closely connected with the smell centers of the brain, and there are also rich connections with areas involved with the other senses, behavior, and the organization of memories.

In addition to information from our sense organs, we also receive messages from our internal organs, and these are relayed to the decision-making part of the brain by the limbic system. This accounts for the fact that these particular sensations are usually tinged with emotion—and that emotions can affect digestion.

Cerebrum

The largest part of the brain is the cerebrum, located in the forebrain. It is divided right down the middle into two halves known as hemispheres, which are joined at the bottom by a thick bundle of nerve fibers called the corpus callosum.

Although the two hemispheres are mirror images of each other, they have completely different functions and work together through the corpus callosum. The cerebrum is more developed in humans than in any other animal and is essential to thought, memory, consciousness, and the higher mental processes. This is where the other parts of the brain send incoming messages for a decision.

Gray matter

The wrinkled layer of gray matter, 0.1 inch (3 mm) thick, folded over the outside of the cerebrum is called the cerebral cortex. This part of the brain has become so highly developed in humans that it has had to fold over and over in order to fit inside the skull. Unfolded, it would cover an area 30 times as large as it covers when folded.

The four lobes

Among all the folds are certain very deep grooves that divide each of the two hemispheres of the cerebrum into four areas called lobes. The temporal lobes are involved with hearing and also with smell,

the parietal lobes with touch, the occipital lobes with vision, and the frontal lobes with movement and complicated thinking. Within each of these lobes, there are specific portions devoted to receiving the sensory messages from one area. The sense of touch, for example, has a tiny area in the parietal lobe devoted to nothing but sensation from the knee, and a large area for the thumb. This is why areas like the thumb are more sensitive than areas like the knee. The same principle applies to the other sensory parts of the cerebrum, and to the motor, or movement, parts as well. Of course none of these areas of the brain can work by itself. Every instruction is, in effect, a joint decision, resulting from the different areas comparing and pooling their information and then coordinating the resulting action. Most human behavior involves movement. In addition, an individual goes over possible behaviors in his or her head and uses language to think as well as act. This gives a sense of an individual's relationship to the outside world, which is felt as consciousness or personality. While the entire brain takes part in this process, it is the frontal lobes that act as the organizing and coordinating areas. They are involved with directing a person's behavior in accordance with the plans the person formulates; they also have the task of ensuring that this behavior remains socially acceptable.

Memory and learning

No particular area of the brain stores all memories; memory is not a bank of information but a process. When an electrical message passes through a brain cell, the cell changes physically. As a person learns something, new electrical pathways are set up that enable it to be remembered for a few minutes. To remember something for a longer period of time, closer attention has to be paid to it, and it has to be gone over many times. As a result, a permanent physical change takes place in the brain cells, which makes the memory a part of the person. Sometimes a special trigger is needed to recall a memory; and some memories seem to be buried so deep that, for all practical purposes, they are lost.

See also: **Autonomic nervous system; Memory; Nervous system**

Breasts

My daughter is 10 and her breasts are already quite well developed. Is this unusual at her age?

No, it is perfectly normal—your daughter has just begun her sexual development a little earlier than the average age. It is important that you don't let her become embarrassed about her body, when it is undergoing so many changes. It may help her self-confidence to let her choose some pretty light-support bras. Girls mature a lot earlier than they did 40 years ago, so your daughter's friends will soon be catching up.

I'm very embarrassed because my breasts are so large. Is there anything I can do about it?

If you are above average weight, losing some weight may help you to lose some of the excess fat in your breasts. In the meantime it might help to wear a well-fitting bra and to choose clothes that flatter your overall figure rather than overemphasizing your breasts. Although doctors are reluctant to recommend surgery, if your breasts are still a problem, particularly if they are causing physical problems, breast tissue can be removed in an operation called a reduction mammoplasty.

My 15-year-old son seems to be developing small breasts. He's thin, so it's certainly not just fat. Is this serious? What can I do about it?

Some boys develop breastlike swellings during puberty, not because they are overweight but because of increased levels of sex hormones in their blood. The swellings may feel very tender and can cause extreme embarrassment, but they usually disappear on their own in 12 to 18 months. However, it might be wise to take him to see your doctor in case it is a symptom of a more serious abnormality or needs surgical treatment.

The breasts play a vital role for women in rearing children, but they are also an important part of a woman's self-image. Regular self-examination of the breasts is essential to ensure that all remains well.

Most people think of the budding of the breasts, which begins before the start of the menstrual periods, as a sign that a girl is moving from childhood to womanhood. Breasts appear in rudimentary form in boys and girls before birth; and for a few days after birth, a baby of either sex may produce from the nipples—as a result of the action of the mother's hormones—a few drops of colostrum, a nutritious fluid that appears before lactation (milk production) begins.

Development

At the start of sexual development, the pituitary gland at the base of the brain stimulates a girl's ovaries and they begin to release large amounts of the hormone estrogen. This hormone travels in the bloodstream to the breast area and triggers the enlargement of the nipples. It also encourages the growth of the lactiferous ducts, the channels through which milk can be released when it is required to feed a baby, and the depositing of fat between and around them.

The completion of breast development, which takes about 18 months from the first appearance of small swellings on the chest, depends on another sex hormone, progesterone, which is produced monthly during the menstrual cycle. Under the influence of progesterone, the ends of the ducts swell out into lobes, each composed of many smaller lobes (called lobules) containing glands that lactate. Meanwhile, as a result of the continued release of estrogen by the ovaries, more fat develops between the lobules, until the breasts have fully developed.

The mature breast is roughly hemispherical in shape, with a tail-like extension toward the armpit. The slightly upward-pointing nipple contains 15 to 20 tiny openings from the ducts. These are too small to be seen with the naked eye and have a ring of darker tissue called the areola. Aside from the tissues directly involved in the production and release of milk, each breast contains nerves and fibrous supporting tissue that gives it firmness and shape. The nipple is particularly well supplied with nerves. These are important in breast-feeding, because it is their

STRUCTURE OF THE BREAST

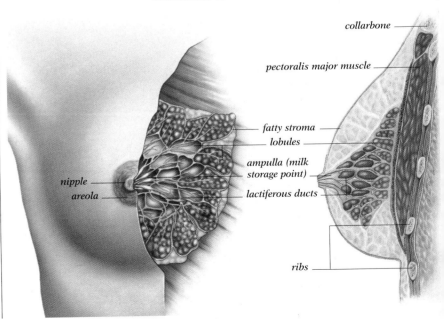

collarbone

pectoralis major muscle

fatty stroma
lobules

ampulla (milk storage point)

nipple
areola

lactiferous ducts

ribs

Breast problems

SYMPTOMS	POSSIBLE CAUSES	ACTION
Fullness and discomfort	Premenstrual changes, effects of the Pill, breast development at puberty.	Try taking extra vitamin B, but if symptoms persist, see your doctor.
Pain in one or both breasts, which may feel lumpy	Abnormal growth of fibrous tissue (fibroadenosis), presence of cyst or tumor.	See your doctor as soon as possible.
Discharge from nipple; may be yellow, greenish, blackish, or bloodstained	As above, or the release of pus from an abscess. A clear discharge is normal in pregnancy.	See your doctor as soon as possible.
One or more lumps in breast	Fibroadenosis, cyst, or tumor.	See your doctor at once.
Inversion or puckering of nipple	May be normal, but it can be a symptom of cancer.	See your doctor at once.
Breasts very small	Probably normal but may be due to failure of hormone activity in puberty, causing underdevelopment.	See your doctor. In cases of hormone abnormality, treatment with hormones is possible for women aged 20 to 30.
One breast bigger than the other	Small variations normal. Larger variations due to abnormal development during puberty.	Hormone treatment not possible. Plastic surgery possible in severe cases. Wear padding in one bra cup.

stimulation that causes the nipple to become erect. Whether the breasts are large or small, they will increase in size when a woman is pregnant and may feel more tender than usual.

What happens during pregnancy

During the course of pregnancy, the placenta surrounding the baby in the womb makes enormous amounts of estrogen, which, together with other hormones from the placenta and secretions from the thyroid and other glands, cause the ducts to grow in size and form more branches. At the same time the hormone progesterone, which is also secreted by the placenta, stimulates the glandular tissue to enlarge.

Sacs, or alveoli, lined with true milk-producing cells are formed, and these produce colostrum, which flows into the ducts and out through the nipples even before birth. Large amounts of fat are also deposited in the breasts during pregnancy, so that the total breast weight increases by about 2 pounds (1 kg). Through the effects of hormones, the areola around the nipple becomes darker.

A hormone called prolactin, responsible for milk production, is produced in increasing quantities throughout pregnancy, but it is prevented from acting because it is antagonized by estrogen. The rapid fall in estrogen level that occurs in the first 48 hours after birth removes this inhibition, and milk production (lactation) begins. For the first two or three days the cells secrete colostrum—a thin, yellowish fluid that contains the protein, minerals, and nutrients necessary for the baby. They then release true milk.

Breast care

Because the breasts contain no muscle, they are naturally likely to sag with age. The fibrous tissue within them becomes less elastic, and the heavier the breasts, the more likely this is to happen. This does not mean that it is essential to wear a bra, particularly if the breasts are naturally small, but many women find it more comfortable to have their breasts supported, if only lightly.

Another much more important part of routine breast care for all women over the age of 20 is to get into the habit of making a thorough examination of each breast. This examination should be done immediately after the menstrual period, when the breasts are smallest or, for women past menopause, at monthly intervals. What must be looked for is any abnormal lump or swelling that may need medical investigation.

It is recommended that all women over the age of 50 have regular breast examinations performed by a doctor, and a mammogram (an X-ray examination of the breasts) on a yearly basis. Whether this should be extended to include women between the ages of 40 and 50 remains a controversial issue.

Changes in the breasts

Just as the breasts enlarge in pregnancy, they become bigger and feel tender just before a period. This is a reaction to the high levels of progesterone present in the bloodstream at this point in the menstrual cycle. It is a healthy sign that the sexual cycle is working normally and nothing to worry about. Because the contraceptive pill contains a synthetic progesterone, it can also cause similar breast enlargement.

Other changes in the breasts can be a source of great anxiety to women, because there is always a possibility that a lump discovered in the breast may turn out to be cancerous. Therefore, it is always wise to report any such changes in the breasts to a doctor as soon as possible after their appearance.

The most usual sorts of lumps, however, are those due to a condition called fibroadenosis, or sclerosing adenosis, but also still sometimes known by its old name, chronic mastitis. Lumps due to this condition can occur in women from puberty onward, but they are most common toward the onset of menopause. What happens

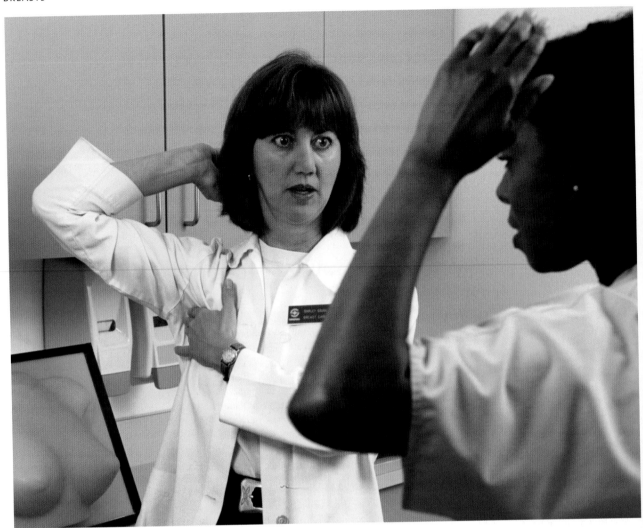

▲ *Breast cancer is one of the leading killers of women. However, a doctor can show a woman how to check her breasts for any lumps that could be malignant and thereby catch the cancer before it develops beyond the stage where it is treatable.*

is that the hormones cause a thickening and growth of the fibrous tissue in the breast. Often the lumps disappear after menstruation, but it is still a good idea to seek medical advice. Although this condition is not serious in itself, it may precede cancer.

In addition to the growth of fibrous tissue, the hormonal changes of the menstrual cycle may result in abnormal collections of fluid in the breasts. Rounded lumps, called cysts, can form. Common after the age of 40, they can cause pain, discomfort, and sometimes a blood-stained discharge from the nipple.

Cysts need medical attention to distinguish them from cancer, and they may need to be removed surgically. Some cysts appear and disappear with menstruation and may completely disappear spontaneously after a few months.

Breast tumors

Tumors of the breasts are of two kinds—benign (those that are not harmful) and malignant (those that are cancerous)—and like other breast problems, they are most likely to occur after the age of 40. All tumors are the result of cells multiplying abnormally, but while benign tumors are confined in a fibrous capsule, malignant ones are able to spread into other parts of the breast.

Typical symptoms of breast cancer include a lump in the breast that is not painful and does not change in size or consistency with the menstrual cycle, a discharge from the nipple, involution (pointing inward) of the nipple, and dimpling of the skin on the rest of the breast so that it resembles an orange peel.

The appearance of any one of these symptoms may have a simple explanation, but because cancer can kill, it is not worth taking even the slightest risk. Therefore, it is advisable for a woman to consult a doctor without delay if she develops any of these symptoms.

See also: **Aging; Birth; Cells and chromosomes; Nervous system; Ovaries; Pituitary gland; Pregnancy; Thyroid**

Breathing

Questions and Answers

My husband smokes 40 cigarettes a day. Does smoking cause cancer, and should I get him to cut down?

Although lung cancer can affect nonsmokers, and not all smokers develop it, there is a strong link. It is the most common form of cancer among smokers, and the more a person smokes, the higher the risk. If your husband is motivated to stop, he may be helped by taking nicotine in the form of chewing gum or a skin patch. This treatment should continue for a few weeks after he quits. Regular smokers are also more liable to chronic bronchitis and emphysema. Every year, some 70,000 people die from these two diseases, and in more than 80 percent of cases the deaths are attributable to smoking. In the United States, smoking causes some 350,000 people a year to die earlier than they otherwise would.

I have a tendency to get bronchitis. Will it help to leave the bedroom windows open at night?

Bronchitis often begins as an infection in the nose and throat, followed by a chest infection. The bronchi are more liable to infection if they are irritated by tobacco smoke, noxious fumes, and cold air. If there are smokers in the house, opening a window may remove some of the smoke. However, cold air and pollution coming in through the window, especially in cities, are just as bad. It is better to leave windows closed at night because a warm atmosphere is best for all bronchial conditions.

My son has been diagnosed as having small lungs. Is this serious?

Each lung has a capacity of about 0.09 cubic ft. (2.5 l), but the amount of air passing in and out is often only a tenth of this, so your son should be all right. Many people manage with one lung.

Breathing is an essential life-supporting process. It is therefore vital to preserve the health of the nose, throat, bronchial tubes, and lungs—and to ensure that any problems in these areas receive early treatment.

Awake or asleep, humans breathe an average of 12 times a minute, and in 24 hours they breathe in and breathe out more than 282 cubic feet (8000 l) of air. During heavy physical exercise a person's breathing rate will increase substantially. The purpose of moving so much air in and out of the body is to enable the lungs to do two things: extract the oxygen needed to sustain life, and rid the body of carbon dioxide, the waste product of internal chemical processes.

Before the air reaches the lungs, it passes through a series of filters that purify it. Even the fresh air of the countryside contains bacteria, fungal spores, and dust; and the town or city dweller has additional pollution to contend with. Although the body is equipped with a series of traps and filters to deal with both situations, it is not a totally fail-safe system; illnesses can and do result, particularly if people smoke or work in industries where they are exposed to certain kinds of dust.

Early development

The respiratory system begins to develop early on in the growth of the fetus, the branching pattern of the airways and arteries being complete 16 weeks after conception. At 28 weeks, cells that secrete surfactant, a fluid, start to develop in the lungs, preventing them from sticking together. The vital gas-exchanging parts, where oxygen is absorbed into the bloodstream and carbon waste dioxide removed, then remain filled with fluid until the baby is born.

This fluid can be a problem in premature babies, who are often not strong enough to do the deep breathing that would inflate the lungs and enable the fluid to disperse. In full-term babies, most parts of the respiratory system are fully developed, but it is not until the age of eight that the gas-exchanging part of the lung in children, whether premature or full-term, is fully formed.

Inhaling

Inhaled and exhaled air enters and exits through the nose and mouth. As air enters the nose, coarse hairs trap dust particles and other foreign bodies. The air continues its passage into the nasal cavity, where the moist membrane that lines the walls warms the air and produces mucus

▲ *The inside of the nasal passages is lined with minute appendages called cilia, which can be seen magnified above, trapping tiny particles of dust.*

HOW OXYGEN IS CARRIED AROUND THE BODY AND CARBON DIOXIDE IS REMOVED

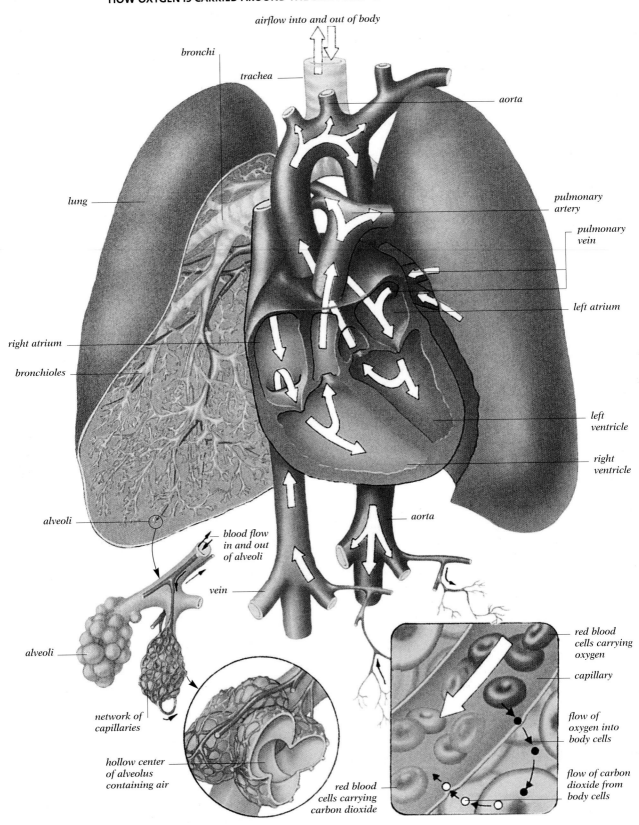

airflow into and out of body

bronchi

trachea

aorta

lung

pulmonary artery

pulmonary vein

left atrium

right atrium

bronchioles

left ventricle

right ventricle

alveoli

blood flow in and out of alveoli

aorta

vein

alveoli

network of capillaries

hollow center of alveolus containing air

red blood cells carrying carbon dioxide

red blood cells carrying oxygen

capillary

flow of oxygen into body cells

flow of carbon dioxide from body cells

◄ *Air inhaled via the trachea, bronchi, and bronchioles reaches the alveoli, where oxygen from the air is transferred to the capillaries surrounding each alveolus. The oxygenated blood is carried to the pulmonary vein and into the left side of the heart and pushed into the aorta. Blood then moves around the body, through the arteries to the capillaries. The oxygen carried by the red blood cells is given to the tissue cells, which transfer their waste product, carbon dioxide, to the red cells. This is carried back through the veins into the right side of the heart, and finally the blood flows out through the pulmonary artery and into the lungs. At the site of the alveoli, the circulating blood gives up its carbon dioxide—which is exhaled—and takes in oxygen again.*

to collect even more particles of dirt. Hairlike projections on the membrane, called cilia, move continually, pushing the film of mucus and its trapped contents back toward the throat to be swallowed.

At the back of the nasal cavity, above the mouth, lie two sets of lymph glands: the tonsils and the adenoids. Their role is to pick up and destroy invading bacteria, so they often become infected and swollen, causing tonsillitis. A sustained buildup of invading bacteria also creates the swelling and irritation of the throat that is a symptom of colds and bouts of flu.

From the throat, the filtered, moistened, and warmed air passes into the windpipe (the trachea), where, as in the nasal cavity, cilia move the mucus layer and its contents toward the throat for disposal by swallowing. Once past the trachea, the inhaled air receives no more screening before it passes into the lungs, where oxygen is absorbed into the bloodstream.

At its lower end, the windpipe divides into two smaller tubes called the bronchi, one leading to each lung. It is here that infections, such as those which cause bronchitis and pneumonia, can build up and cause breathing problems.

The lungs

Within the lungs each bronchus divides into smaller tubes called bronchioles; these in turn branch and form millions of tiny air sacs, the alveoli—each surrounded by a meshwork of fine capillaries—where the exchange of oxygen and carbon dioxide takes place. The branches of the pulmonary artery carry blood rich in carbon dioxide, and this gas is given up in return for the oxygen in the new air that has entered the alveolar sacs. The lungs then exhale the carbon dioxide in the deoxygenated air along with a certain amount of water vapor that comes from the moist membranes of the alveoli. On cold days this water vapor condenses and shows up as steaming breath.

The lungs fill most of the chest cavity. They are inflated and deflated by muscular movements of the chest, and by the rise and fall of a sheet of muscle called the diaphragm, which lies under the lungs and divides the chest cavity from the abdomen.

A double membrane known as the pleura surrounds each lung. The outer pleura is attached to the chest walls and to the diaphragm; the inner pleura is attached to each lung. Between the two membranes of the pleura is a cavity, and as the lungs inflate and deflate, a thin film of fluid on its surface protects the delicate tissues as they move against each other. Inflammation of the pleura, a condition known as pleurisy, causes pain when a person inhales or coughs.

When a person is breathing normally, the diaphragm does most of the work. The muscle contracts, and the volume of the chest cavity increases. The ribs are pulled upward and outward by muscles, called intercostals, that lie between them, and as the pressure within the lungs drops, air is sucked in. Breathing out is a passive process: the diaphragm and the intercostal muscles relax, and the natural elasticity of the lung tissue forces air out.

Breathing rates

Breathing is controlled by the respiratory center of the brain, the medulla oblongata, and is regulated according to the levels of carbon dioxide in the blood rather than the amount of oxygen present. The brain responds to an increased production of carbon dioxide—for example, during a period of physical exercise—and adjusts the breathing rate accordingly. Breathing becomes deeper and faster so that more oxygen is inhaled, stimulating the heartbeat, increasing the blood flow, and burning off carbon dioxide. Once the exercise ceases, the carbon dioxide level falls and breathing returns to normal.

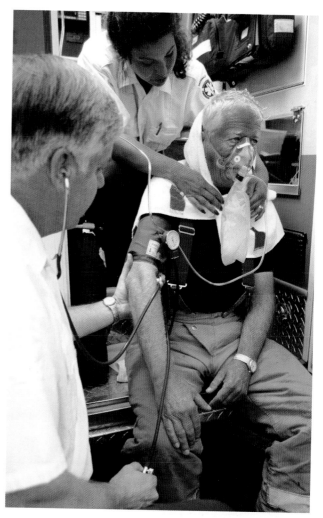

▲ *Oxygen can help relieve the respiratory distress caused by conditions such as bronchitis, emphysema, and asthma.*

Breathing disorders

SYMPTOMS	CONDITION	ACTION
Runny eyes and nose, coughing, sneezing, sore throat, and headache	Common cold	Colds rarely last for more than a week but, if they persist, can lead to complications such as bronchitis. There is no known cure, but aspirin or antihistamine may make the symptoms less uncomfortable. Keep warm and stay indoors. Antibiotics will not help, but drinking fluids may.
Chills, high fever, pain in tonsil area, difficulty swallowing, headache, pain in jaw and neck. Tonsils red, enlarged, spotted.	Bacterial tonsillitis	Antibiotics; cold drinks or food (ice cream). Seek medical attention; if severe or untreated, can lead to other diseases, such as rheumatic fever.
Dry, irritating cough with thick, yellow mucus and light fever. Cough persists for up to two weeks.	Acute bronchitis	Bed rest in a humid room with inhalants, plenty of hot drinks. Stimulant cough medicine by day and a sedative cough medicine at night.
Constant, vigorous cough that is worse in the mornings than in the evenings. Clear sputum that may become yellow if a secondary infection sets in.	Chronic bronchitis	Antibiotics to prevent secondary infection; breathing exercises; clean air; no smoking. Immediate treatment of any other respiratory ailments.
Shortness of breath, difficulty in breathing, tight chest, wheezing, sweating	Asthma	Test for allergies or infection. Treat with appropriate drugs. Breathing exercises and bronchodilators also give relief. Avoid stressful situations and catching colds or other respiratory infections. Prolonged attacks require hospitalization.
Pain caused by coughing, breathing deeply, or moving. The pain may occur in the shoulder if the pleura covering the diaphragm is affected.	Pleurisy	Almost always caused by a virus or bacterial infection; often associated with pneumonia. Appropriate treatment for the infection should be prescribed, along with painkillers.
Sudden fever, chest pains, cough and blood-stained sputum, sweating, shivering, and often vomiting and diarrhea	Pneumonia	Bed rest, antibiotics, and breathing exercises under the doctor's direction
Children: Fever, swollen lymph glands in the chest, weight loss, coughing, and breathlessness Adults: Fever, heavy sweating at night, fatigue, weight loss, coughing up blood-tinged sputum, and possibly pleurisy	Tuberculosis	Effective drugs to treat tuberculosis are available and must be taken for several months or more to cure the patient of TB.
Feeling of constriction of the chest and breathlessness that progresses over the years. Frequent coughing attacks and production of sputum.	Emphysema	Emphysema is a result of cigarette smoking and/or occupational exposures. Treatment consists of clean air, no smoking, and medications (including inhalants). Severe cases may need oxygen before physical activity or during sleep. As the condition worsens, continuous oxygen may be needed.
Spitting blood, coughing, and breathlessness. Possibly asthma.	Pneumoconiosis	Lung disease caused by inhaling certain dust particles, such as silica and asbestos. Early detection is essential.
Coughing with sputum that may be bloodstained. Later pneumonia and partial lung collapse, followed by weakness, weight loss, and lethargy.	Lung cancer	Surgical removal of the tumor is possible, and radiotherapy and drug therapy are the usual treatments.
Sudden onset of high fever, followed by croupy, nonproductive cough, then thick phlegm in the bronchi and trachea. Extreme shortness of breath.	Laryngotracheobronchitis (croup)	If in a child, urgent medical attention advised. Antibiotics given, sometimes hospitalization necessary. Keep room warm, take prescribed drugs. There has been a resurgence of this infection in adults in recent years.

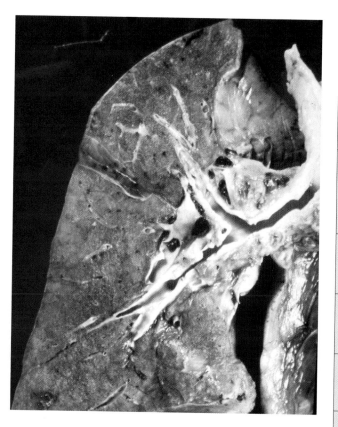

▲ *Smokers are much more likely to suffer from lung problems such as cancer, as shown in this lung.*

Respiratory system health	
DO	**DON'T**
Breathe through the nose, as this gives slightly more protection than breathing through the mouth.	Don't smoke. There is a direct link between smoking and lung cancer and other respiratory infections.
Exercise: walking, running, and cycling increase the blood flow, maintaining the capacity of the lungs and improving the supply of oxygen to the heart.	Don't overfeed babies: overweight babies have an increased risk of developing respiratory disorders.
Wear an approved facial mask if there is asbestos or silica dust in the work environment.	Don't work or live in a dust-laden environment.
Get medical advice for any cold or minor ailment that does not start to improve within a week.	Don't sleep with the windows open in a cold or polluted atmosphere if you have a respiratory infection.
Consult a doctor if a respiratory complaint develops.	Don't spend too much time in smoky atmospheres, especially if you are suffering from cough.
Practice breathing exercises such as those in yoga. Some patterns of deep breathing can help to remove carbon dioxide waste from the bloodstream.	Don't try to exercise strenuously at high altitudes without acclimatizing.

Voluntary alterations in people's breathing rates occur during activities such as talking, singing, and eating. Yawning, sighing, coughing, and hiccuping involve still other kinds of respiration. Laughing and crying, both of which involve long breaths followed by short bursts of exhalation, are respiratory changes caused by emotional stimuli. Holding the breath, either deliberately (when swimming under water) or unwittingly (as a result of an attack of nerves), also alters the breathing pattern. The carbon dioxide level falls after the first few deep breaths, which are then held, and the brain ceases to be stimulated. This can lead to a blackout, and when a person is swimming under water, death by drowning if he or she cannot return to the surface.

Blackouts may also occur when an unacclimatized person climbs to high altitudes and develops altitude sickness. The body tissues become unoxygenated, and the person feels giddy and nauseated. As he or she goes higher, there will be a feeling of sublime indifference called euphoria, loss of muscular coordination, and eventually—if the person does not either receive oxygen or make a rapid descent at the onset of these symptoms—unconsciousness.

Mechanisms of injury

Although subject to many different forms of injury, the lungs have only a few basic ways of reacting. Assaults as varied as inhaled toxic gases, drugs, viruses, and lack of oxygen due to shock, trauma, or mountain sickness all produce a profound outpouring of fluid into the alveoli (pulmonary edema), which may effectively drown the patient if untreated. Bacterial infections cause the alveoli to fill with pus; chronic bronchitis causes progressive destruction of the bronchial walls. All these conditions, if not stopped, progress in a similar fashion, destroying functioning lung tissue and replacing it with fibrous scars. This severely impairs the individual's ability to transfer oxygen to the blood.

Prevention of disorders is therefore important. An individual should be aware of toxic materials in the workplace or living environment and take suitable precautions. A cough that brings up mucus and does not clear up in a week or two should be treated by a doctor. Above all, it is best not to take up smoking, and people who already smoke should make every effort to stop, since cigarette smoke alone (or acting with any of the above agents) is by far the single most destructive factor causing chronic lung disease.

Many therapies and treatments are available that can help curb a smoking habit. Chewing gum (with added nicotine) and skin patches (through which the nicotine is absorbed into the skin) can help wean the body off cigarettes. Acupuncture, hypnosis, and meditation can also be effective to help a person stop smoking.

See also: Coughing; Crying; Diaphragm; Lungs; Throat

Burping

Burping, or bringing up gas, is a normal reflex action in babies after a meal. Adults, too, may suffer from excessive gas in the stomach, but they can learn to control the response—if they so choose.

Questions and Answers

My husband and his family seem to have a burping problem. Can this run in families?

Yes. If it is acceptable behavior in his family to burp, then family members will be unlikely to suppress the habit. However, it should be possible to persuade your husband to control his burping if you want him to conform to the social norm. It may be simply a matter of reviewing his eating and drinking.

My brother is always burping and claims he cannot help it. Why is he doing this?

The most likely cause is that he is swallowing air, although your brother may be unaware of this. If he has no other gastric symptoms, it is most unlikely that this is a medical problem.

My baby cries a lot after he has been fed. I spend ages trying to burp him, but nothing happens. Am I doing anything wrong?

No. He's probably not bringing up gas because he has not swallowed much air. It does not sound as if gas is the problem here.

Can a good, loud burp show appreciation of a good meal?

In American society, a burp is considered antisocial. In some countries where eating is considered a pleasure of life, and the stomach distension of a good meal can be relieved by burping, it can be seen as a compliment.

Do men burp more than women?

No—both sexes have an equal tendency to burp. However, men tend to overindulge more in food and beer, both of which cause an excess of gas.

◄ *Helping a baby to bring up gas is best done by placing the child on a shoulder and firmly patting his or her back.*

Burping (belching) is the involuntary reflex (backward flow) of gas from the stomach and out of the mouth. With every mouthful of food that a person swallows, some air is also taken down into their stomach. Babies swallow a lot of air as they suck their milk—the actual amount varies with how well and how hard the baby sucks and how much milk is available from the breast or bottle.

Adults, too, often swallow excessive amounts of air, and this results in uncontrollable burps. Carbonated (fizzy) drinks contain dissolved gas, which is quickly released in the stomach. This can also cause burping, as can air swallowed to cool the taste of very hot food or to hide the taste of unpleasant food. People who eat too quickly and swallow a lot of air are often prone to burping and sharp stomach pain.

How it happens

Once in the stomach, air can escape in two directions. First, it can pass on with the food into the small intestine. This passage is closed immediately after a meal to ensure that food is adequately digested in the stomach before being allowed to progress down to the gut. Alternatively, the air can return back up the esophagus (the tube that extends from the throat to the stomach) to the mouth. Then, any excessive buildup of gas will put pressure on the valve at the stomach entrance, which is also closed to prevent food from being regurgitated (vomited).

As the stomach churns away, digesting the food, the pressure of the air may get to be too much for the stomach valve. A burst of gas is released up the esophagus without warning.

Prevention

Burping is a natural phenomenon. Children are usually taught to control their burps, since the habit is considered antisocial in most cultures. Eating more slowly and not swallowing too much air, along with avoiding hot or spicy food, will prevent excessive burping.

In babies up to six months old, burping may cause concern. After being fed, babies should not be put directly to bed, because the air in their stomach could lead to pain and discomfort. Cuddling will be appreciated, during which babies may burp a little naturally. Parents will often use traditional methods of bringing up gas if a baby appears uncomfortable, such as placing the baby over a shoulder and patting his or her back gently but firmly. Preparations designed to help eliminate gas, often containing an herb such as fennel, may help in some cases.

> *See also:* Breathing; Digestive system; Esophagus; Reflexes; Stomach

Capillaries

The capillaries are the smallest blood vessels and form a complex network throughout the body. In addition to carrying oxygen and other vital substances to and from cells, they also help regulate body temperature.

The spontaneous appearance of purple patches on the skin always calls for medical attention. The most likely cause is a condition known as purpura, where the capillaries leak blood into the skin. In older people this may be due to loss of collagen support for the fine vessels. However, the condition may also be caused by a drop in the number of blood platelets, which assist in the clot formation necessary to plug small openings in blood vessels. Purpura may be a minor problem but can indicate a serious bleeding disorder.

I have always bruised easily. Do I have weak capillaries?

Yes. The walls of your capillaries could be more fragile than usual, but this is nothing to worry about. You should see a doctor only if you get a crop of tiny bruises without any injury, as these can be a sign that something is wrong with your blood.

Why does drinking alcohol make my face turn pink?

Alcohol dilates the capillaries of the skin. Rapid temperature changes can have the same effect. If this happens often, the capillaries can become permanently stretched, resulting in a pinkness of the skin that does not die down. The best way to prevent this is to avoid excessive alcohol and temperature changes.

Do capillaries occur in all parts of the body?

Yes. Capillaries are essential as the interface between the blood and the tissues and occur everywhere. No part of the body, even the bones, is free from capillaries.

The capillaries form an extensive network of vessels between the arterial system, which takes the blood from the heart, and the venous system, which returns the blood to the heart. Each capillary measures only about 0.0003 inch (0.008 mm), or just wider than one single blood cell. The capillaries' job—one of the most essential in the human body—is to deliver oxygen and other vital substances to the cells and to collect the cells' waste products, which they do through their thin walls.

Structure

Capillaries are simple structures and their walls consist of little more than a single layer of very thin, flattened cells called endothelial cells, which are connected together edge to edge. Each capillary consists of a thin layer of tissue rolled up into a tube and surrounded by an equally thin membrane. All the capillary walls are thin enough to allow certain substances to pass in and out of the blood.

Electron microscopy has shown that in different locations throughout the body, capillary structure varies widely. Those in the kidneys, the lining of the intestines, the endocrine glands, and the pancreas, for instance, are perforated by tiny pores of widely differing size.

Those in the brain differ considerably from those in the rest of the body, especially in thickness. Their thicker walls provide what is known as the blood-brain barrier. This is an effective obstruction to the passage of certain drugs and other substances from the blood to the brain

▶ ▼ *A capillary (right). Transfer of substances from blood to surrounding tissues happens as shown below: water, food molecules, and hormones go through pores; oxygen and carbon dioxide are exchanged via walls; and protein molecules are engulfed by capillary walls, then released outside.*

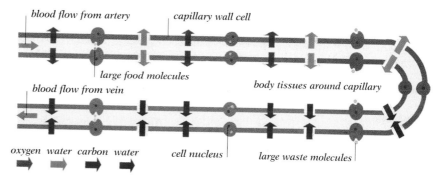

blood flow from artery   capillary wall cell

large food molecules   body tissues around capillary

blood flow from vein

oxygen   water   carbon   water   cell nucleus   large waste molecules

cells and the cerebrospinal fluid (the fluid that surrounds the brain). This barrier offers protection for the brain against many potentially damaging substances commonly found in the blood, but it does have the disadvantage that it can interfere with treatment from antibiotics and other drugs. Doses of antibiotics for the brain may have to be increased to many times those that are needed for infections elsewhere in the body.

Connecting link

As the heart pumps blood through the body, the blood goes first through the arteries. The arteries divide into branches called arterioles. The branches become smaller and smaller, and the smallest are the capillaries. In the capillaries the blood cells jostle along in single file, giving up oxygen, nutrients, and other substances, and taking in carbon dioxide and other waste products from the cells. When this process is finished, the blood needs to return to the heart. As this return journey starts, the capillaries join up to form small veins that gradually grow larger as many branches join them.

When the body is at rest, blood flows through preferential, or preferred, channels. These are capillaries that have become larger than average. However, when extra oxygen is needed by a particular part of the body—for example, by the muscles or the heart during exercise—blood flows through nearly all the capillaries in that area.

Capillary gaps

All capillaries have small gaps between the edges of the cells that form their walls. These gaps are very important. The pressure of the blood in the capillaries is low, but it is highest at the arterial end of any capillary (the part through which blood enters the capillary) and lowest at the venous end (the part through which blood leaves the capillary). As a result, some of the watery part of the blood, but not the red cells or large protein molecules, passes out through the gaps and pores in the capillary walls at the arterial end. Most of this fluid passes back in through the walls at the venous end of the capillary.

Outside the capillaries, this fluid is called tissue fluid. It bathes the cells of the body, allowing diffusion of important dissolved substances from the blood to the cells, and from the cells to the blood. This is how oxygen, nutrients, vitamins, minerals, hormones, and so on are able to get to the cells from the blood, and how waste products of cell metabolism are carried away from the cells to be disposed of via the bloodstream.

There is another important reason for the gaps in the capillary walls. Although they are too small to allow red blood cells to pass, cells capable of changing their shape can do so. Such cells are said to be ameboid, and they pass through the capillary wall gaps in a remarkable way. First the cell pushes a tiny finger through a gap. The substance of the cell then flows along this finger and expands outside the capillary. This process continues until the whole of the cell is outside. Called phagocytes, cells of this kind belong to the immune system. Their function is to combat infection, and millions of them pass through capillary walls at the site of any inflammation.

Sinusoids

There is one particular class of capillaries that differs from all other capillaries. These are known as the sinusoids. The sinusoids are of wider caliber and are more irregular than other capillaries, but the most striking difference is the number and size of the openings in

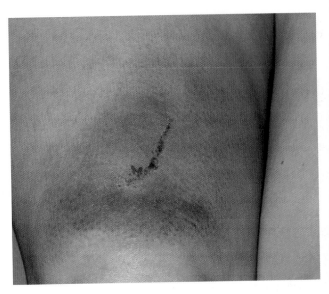

▲ *If the skin receives a heavy blow—in the case above a kick from a horse—the capillaries in the skin's surface break and release their blood. This released blood causes the skin to discolor and form a bruise.*

their walls. The sinusoids in the liver have numerous relatively large openings grouped so as to form sieve plates. Those in the spleen have long, slitlike openings in their walls that, unlike capillaries anywhere else in the body, allow whole blood to pass through into the surrounding space.

Regulating body temperature

In addition to the exchange of substances, the capillaries located in the skin play a special role—they help to regulate body temperature. When the body is hot, the capillaries in the skin get wider, making it possible for a larger volume of blood to reach the skin, where it can be cooled by the outside air. Capillaries widen and narrow passively as a result of changes in the pressure of the blood within them. This, in turn, is determined by the flow rate in the tiny arterioles supplying the capillaries. Arterioles have muscle fibers in their walls, which can tighten to narrow the vessels or relax to widen them. The muscles in the walls of the arterioles are under the control of the autonomic nervous system and the endocrine system.

Damage to capillaries

Since they are thin-walled, capillaries are easily damaged. Those most at risk are the capillaries in the skin. If the skin is scratched, cut, or injured, or if it receives a blow, the capillaries release blood. A bruise results as blood collects in the skin. Capillaries can also be damaged or destroyed by burning, but they have some ability to renew themselves. As a person grows older, or as the result of drinking excess alcohol over a long period of time, the capillaries collapse, leaving visible purple patches or a network of reddish lines on the skin of the face.

> **See also:** Blood; Blood-brain barrier; Cells and chromosomes; Circulatory system; Endocrine system; Immune system; Nervous system; Veins

Cartilage

Questions and Answers

My toddler is always falling over but has never broken any bones. Is she just lucky?

A child's skeleton contains a great deal of cartilage. This is much tougher and more flexible than true bone and has more shock-absorbing power; that is why your daughter can take countless falls without breaking any bones. Falling is also less damaging to the young because they weigh less than adults. Also, the younger they are, the less fear they have, and therefore they do not get tense as they feel themselves falling. This is why you may occasionally hear of babies falling several stories out of buildings and literally bouncing.

My brother had the cartilage taken out of his knee. How can he manage without it?

The cartilage of the knee is part of the movement system of the joint. If the cartilage is removed, the joint is less efficient for a while, but then the muscles learn to compensate for the missing cartilage.

I am 63, and I recently discovered that I am 2 in. (5 cm) shorter now than when I was a girl. Why?

As the body ages, not only do the bones in the vertebral column get smaller, but the disks of cartilage between them get thinner and harder. This makes the disks shrink in size; with bone shrinkage, this makes you shorter.

My joints click loudly whenever I bend down. What is wrong?

Your bones click because the movement releases a vacuum in the joint. A high-pitched clicking sound is nothing to worry about, but if the sound gets deeper and there is pain, see your doctor.

The different types of cartilage form a vital part of the body's framework. Among its many functions, it surrounds the bronchial tubes, supports the nose and ears, and lines bones and joints.

Cartilage, or gristle, is a smooth, tough, but flexible part of the body's skeletal system. In adults it is mainly found in joints and covering the ends of bones, but it forms the complete skeleton of a developing fetus. True bone forms in it. Cartilage that is subject to a great deal of wear and tear may cause problems, particularly in the spine and knees.

Nearly all the bones in the body begin as rods of cartilage, which gradually become hardened by deposits of calcium and other minerals. This process, which is called ossification, begins before birth, in the third or fourth month of fetal development.

Ossification is not fully complete until about the age of 21, since the cartilage not only forms the foundation for bone formation but also allows the bones to grow. In adults, however, cartilage remains at strategic points in the skeleton, where its toughness, smoothness, and flexibility are most needed.

THE DIFFERENT TYPES OF CARTILAGE

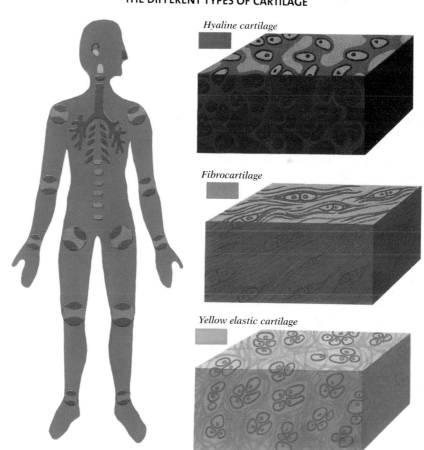

Hyaline cartilage

Fibrocartilage

Yellow elastic cartilage

▲ *Hyaline cartilage lines bones in joints and forms the respiratory tract. Fibrocartilage acts as a shock absorber in the backbone and other joints. Elastic cartilage is extremely flexible and forms the outer ear and the epiglottis in the throat.*

Structure

The structure of cartilage is not the same throughout the skeleton. Its makeup varies according to the specific job it has to do. All cartilage is composed of a groundwork, or matrix, in which there are embedded cells plus fibers made up of substances called collagen and elastin. The consistency of the fibers varies according to the type of cartilage, but all cartilage is alike in that it contains no blood vessels. Instead, it is nourished by nutrients that diffuse through the covering (perichondrium) of the cartilage, and lubricated by synovial fluid, which is made by membranes lining the joints. According to its different properties, cartilage is known as hyaline cartilage, fibrocartilage, and elastic cartilage.

Types of cartilage

Hyaline cartilage is a bluish white, translucent tissue and, of the three types, has the fewest cells and fibers. What fibers there are all consist of collagen. This cartilage forms the fetal skeleton and is capable of the immense amount of growth that allows a baby to grow from about 18 inches (45 cm) into a person up to 6 feet (1.8 m) tall or more. After growth is complete, hyaline cartilage remains in a very thin layer, only 0.039 inch (1 mm) across, on the surface of the ends of the bones in the joints.

Hyaline cartilage is also found in the respiratory tract, where it forms the end of the nose and the stiff but flexible rings surrounding the windpipe and the larger tubes (bronchi) leading to the lungs. At the end of the ribs, bars of hyaline cartilage form the connections between the ribs and the breastbone and play their part in enabling the chest to expand and contract during breathing.

In the larynx, or voice box, hyaline cartilage not only helps to support the structure but is also involved in the production of the voice. As it moves, it controls the amount of air passing through the larynx, and therefore the pitch of the note that is emitted.

The second type, fibrocartilage, is composed of many bundles of the tough substance collagen, which makes it both resilient and able to withstand compression. Both these qualities are essential at the site in which fibrocartilage is most plentiful: between the bones of the vertebral column. In the spine each bone or vertebra is separated from its neighbor by a disk of fibrocartilage. The disks cushion the spine against jarring and help the human frame to remain upright. Each disk is made up of an outer coating of fibrocartilage that surrounds a thick, syrupy fluid. The cartilaginous part of the disk, which has a lubricated surface, prevents the bones from being worn away during movement, and the fluid acts as a sort of natural shock absorber. Fibrocartilage also forms a tough connection between bones and ligaments; in the hip girdle it joins the two parts of the hips together at the symphysis pubis joint. In women this cartilage is particularly important because it is softened by the hormones of pregnancy to let the baby's head pass through.

The third type of cartilage is known as elastic cartilage. It is made up of fibers of elastin, with collagen. The elastin fibers give elastic cartilage a distinctive yellow color. Strong but supple, elastic cartilage forms the flap of tissue called the epiglottis, which snaps down over the entrance to the airway as food is swallowed.

Elastic cartilage also makes up the springy part of the outer ear and supports the walls of the canal leading to the middle ear and the eustachian tubes that link each ear with the back of the throat. Along with hyaline cartilage, elastic cartilage forms part of the supporting and voice-producing areas of the larynx.

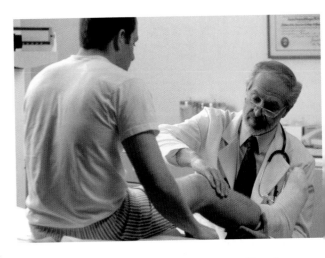

▲ *Torn cartilage in the knee is a common problem for professional soccer players.*

Problems and treatment

The cartilage that has to withstand constant pressure in the knees and vertebral column is the most vulnerable. The knee joint contains a pair of cartilage structures shaped like half-moons, and they are most likely to be damaged by strenuous sports, particularly sports in which the knee joint is frequently twisted. If one or both half-moons become torn, and their frayed ends get caught between the surface of the joint, the knee will lock or give way. In such injuries, which are 20 times more common in men and are a notorious problem for soccer players, the knee should be firmly bandaged. In severe cases a splint may be needed behind the knee. The doctor may try to manipulate the joint to move the torn ends of the cartilage away from the joint surface, and to remove some of the excess fluid that accumulates around the site of the injury. However, the only way to cure the complaint completely is by surgical removal of the offending cartilage. The surgeon makes an incision from the bony lump at the side of the knee to the base of the knee joint and takes out the cartilage. After surgery it is important to get the knee moving as soon as possible, and physiotherapy should be given.

A slipped disk is another common cartilage problem. Strain or sudden twisting movements can tear the disks of cartilage between the vertebrae of the spine, making the pulpy center protrude and pushing the whole disk out of shape. The protrusion can press on nerves, causing pain in the back and possibly in the buttocks, thigh, and one leg. Rest in bed used to be recommended, but now doctors may recommend anti-inflammatory medications along with gentle exercise, manipulation, or the wearing of a corsetlike belt. If these approaches are not successful, surgery may be required.

In old age the joints become stiffer as the cartilage covering the ends of the bones loses its smoothness. This is quite normal, but in a condition called osteoarthritis, the cartilage and part of the underlying bone degenerate. Osteoarthritis is not only a disease of old age; it can affect athletes and people who put excess strain on their joints by being overweight. Analgesics should relieve the pain, and physiotherapy and heat treatment will ease the joints.

> **See also:** Bones; Joints; Ligaments; Skeleton

Cells and chromosomes

Do the number of cells in a child's body increase as he or she grows?

Yes. A person has millions more cells as an adult than at birth, although the increase is not evenly distributed throughout the body. The number of cells in the brain, for example, increases much less as the body develops than the number in the bones, skin, and muscles.

I have heard that certain creams can revitalize my cells. Will they really stop the aging process?

No. There are no preparations that can stop the gradual deterioration and replacement of cells or reverse the changes that take place in cells as they age.

I have read that nuclear radiation can cause birth abnormalities. Why?

Nuclear radiation can kill cells, and it can produce permanent changes called mutations in the chromosomes—the parts of the cell that carry hereditary instructions. If these changes affect the chromosomes of the eggs or sperm, which are brought together at the moment of fertilization to produce a new individual, then an abnormal baby can result. If the baby survives into adulthood, he or she could well pass on the abnormality to future generations.

My twin girls are identical. Are their cells the same, and if so, why?

Identical twins are born when an egg from the mother splits in two after it has been fertilized by a sperm from the father; thus the twins are always the same sex and always carry exactly the same genetic instructions. Scientists find identical twins fascinating because all the differences that exist between them must be due to environmental factors.

Every part of the body is composed of millions of microscopic cells. In the nucleus of each of these are strands of DNA, which contain the genetic instructions that determine the characteristics each person inherits.

The cells are the basic units of life, the microscopic building blocks from which the body is constructed. Within the cells are the lengths of DNA that contain the vast amount of information essential to the creation and maintenance of human life and an individual's personal characteristics. Every adult body contains more than 100 million cells, microscopic structures averaging only 0.00039 inch (0.01 mm) in diameter. No one cell can survive on its own outside the body unless it is cultured (artificially bred) in special conditions, but when grouped together into tissues, organs, and systems of the body, the cells work together in harmony to sustain life.

Types and structures

The body cells vary greatly in shape, size, and detailed structure according to the jobs they do. Muscle cells, for example, are long and thin and contain fibers that can contract and relax, allowing the body to move. Many nerve cells are also long and thin, but they transmit electrical impulses that compose the nerve messages. The hexagonal cells of the liver are equipped to carry out a multitude of chemical processes, doughnut-shaped red blood cells transport oxygen and carbon dioxide around the body, and spherical cells in the pancreas make and replace the hormone insulin.

Despite these variations, all body cells are constructed according to the same basic pattern. Around the outside of every cell is a boundary wall or cell membrane enclosing a jellylike substance, the cytoplasm. Embedded in the cytoplasm is the nucleus that houses the genetic instructions in the DNA. The cytoplasm, although between 70 and 85 percent water, is far from inactive. Many chemical reactions take place between substances dissolved in this water, and the cytoplasm also contains tiny structures called organelles, each with an important and specific task. The cell membrane also has a definite structure: it is porous, and it is rather like a sandwich of protein and fat, with the fat as the filling. As substances pass into or out of the cell, they are either dissolved in the fat or passed through the porous, semipermeable membrane.

Some cell membranes have hairlike projections called cilia. In the nose, for example, these trap dust particles. The hairs can also move in unison to waft substances along in a certain direction.

PARTS OF A CELL

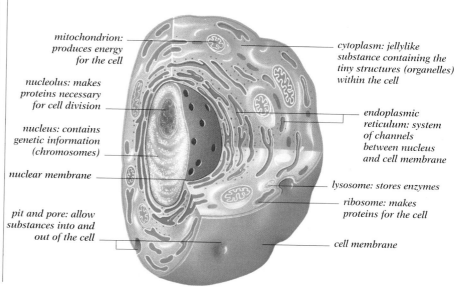

mitochondrion: produces energy for the cell

nucleolus: makes proteins necessary for cell division

nucleus: contains genetic information (chromosomes)

nuclear membrane

pit and pore: allow substances into and out of the cell

cytoplasm: jellylike substance containing the tiny structures (organelles) within the cell

endoplasmic reticulum: system of channels between nucleus and cell membrane

lysosome: stores enzymes

ribosome: makes proteins for the cell

cell membrane

DIFFERENT TYPES OF CELLS IN THE BODY

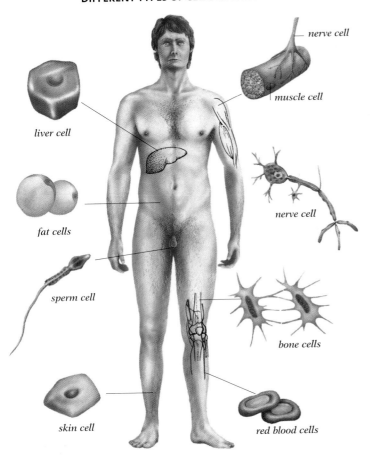

liver cell

nerve cell

muscle cell

fat cells

nerve cell

sperm cell

bone cells

skin cell

red blood cells

tightly to form chromosomes that are visible with an ordinary microscope. Every chromosome contains thousands of genes, each with enough information to produce one protein. This protein may have a small effect within the cell and on the appearance of the body, but equally it may make the difference between, for example, brown and blue eyes, straight and curly hair, normal and albino skin. The genes are responsible for every physical characteristic.

Apart from mature red blood cells, which lose their chromosomes in the final stages of their formation, and the eggs and sperm (the sex cells), which contain half the usual number of chromosomes, every body cell contains 46 chromosomes, arranged in 23 pairs. One of each pair comes from the mother and one from the father. The eggs and sperm have only half that number, so that when a sperm fertilizes an egg, the new individual has the correct number of chromosomes.

At the moment of fertilization, the genes start issuing instructions for the molding of a new human. The father's chromosomes are responsible for determining sex. These chromosomes are called X or Y, depending on their shape. In women both the chromosomes in the pair are X, but in men there is one X and one Y chromosome. If a sperm containing X fertilizes an egg, the baby will be a girl; but if a Y sperm fertilizes the egg, the baby will be a boy.

How a cell divides

In addition to being packed with information, the DNA of the chromosomes has the ability to reproduce itself; without this the cells could not duplicate themselves, nor could they pass on information from one generation to another. The process of cell division in which the cell duplicates itself is called mitosis; this is the type of division that occurs when a fertilized egg grows first into a baby and then into an adult, and when worn-out cells are replaced.

When the cell is not dividing, the chromosomes are not visible in the nucleus; but when the cell is about to divide, they become shorter and thicker and can be seen to split in half along their length. These double chromosomes then pull apart and move to opposite ends of the cell. Finally, the cytoplasm is halved and new walls form around the two new cells, each of which has the normal number of 46 chromosomes.

A huge number of cells die and are replaced by mitosis every day, but some cells are more efficient at this than others. Once formed, the cells of the brain and nerves are unable to replace themselves, but liver, skin, and blood cells are replaced several times a year.

Making cells with half the usual chromosome number, in order to determine inherited characteristics, involves a different type of cell division called meiosis. First the chromosomes become shorter and thicker (as in mitosis) and divide in two; at the moment of fertilization the chromosomes pair up so that the one from the mother and the other from the father lie side by side. Next the chromosomes become tightly intertwined so that when they eventually pull apart, each new chromosome contains some of the mother's genes and some of the father's. After this the two new cells divide again so that each egg or sperm contains the 23 chromosomes it needs. The interchange of

The cytoplasm of all cells contains microscopic, sausage-shaped organs called mitochondria, which convert oxygen and nutrients into the energy needed for all other cell actions. These powerhouses work through the action of enzymes—complex proteins that speed up chemical reactions. They are most numerous in the muscle cells, which need a vast amount of energy to carry out their work. Lysosomes—another type of microscopic organ in the cytoplasm—are sacs filled with enzymes that make it possible for the cell to use the nutrients with which it is supplied. The liver cells contain the greatest number of these. Further minute organs called the Golgi apparatus are necessary to package and store substances made by the cell, such as hormones, that are needed in other parts of the body.

Many cells also possess a network of tiny tubes that are thought to act as a kind of internal cell skeleton, and all of them contain a system of channels called the endoplasmic reticulum. Dotted along the reticulum are tiny spherical structures called ribosomes, which are responsible for controlling the construction of essential proteins. All cells need proteins for structural repairs and, in the form of enzymes, for cell chemistry and the manufacture of complex molecular structures such as hormones.

What a chromosome is

Each nucleus is packed with information, coded in the form of a chemical called deoxyribonucleic acid (DNA), and organized into groups called genes. When cells are about to divide, the DNA coils up

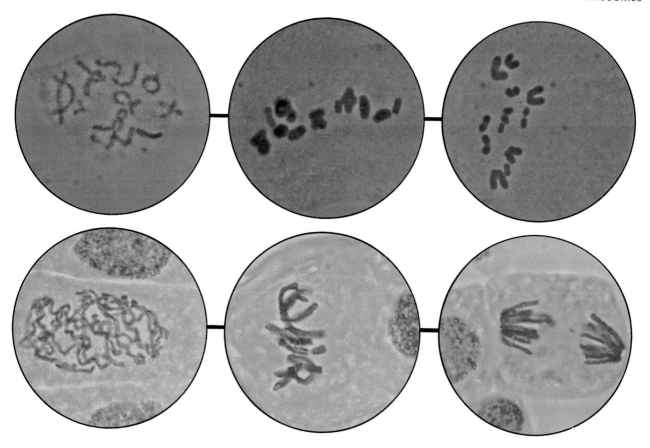

▲ ► *The main difference between the two methods of cell division are shown above, greatly enlarged. In meiosis (top sequence above), the chromosomes are duplicated and then pair up and intertwine, before pulling apart and dividing to produce sex cells containing half the genetic information needed to produce a human (the remaining half is supplied during fertilization). In mitosis (bottom sequence above), pairs of chromosomes separate, and each half divides into two identical parts. These arrange themselves so that when the respective parts move to opposite ends of the cell and the cell divides into two, each new cell will contain the genetic information necessary to replace or duplicate existing body cells. The illustration at right shows the structure of a chromosome in detail.*

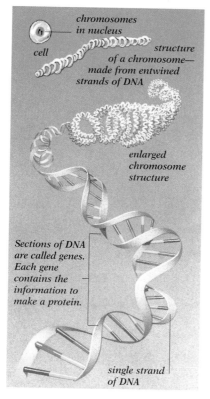

chromosomes in nucleus

cell

structure of a chromosome— made from entwined strands of DNA

enlarged chromosome structure

Sections of DNA are called genes. Each gene contains the information to make a protein.

single strand of DNA

genetic material during the process of meiosis explains why children do not look exactly like their parents and why each person, with the exception of identical twins, has a completely unique genetic makeup.

Problems

Considering the number of cells and the complexities of their structure and chemistry, it is surprising how little goes wrong during the average life, and how few babies are born with deformities. Apart from accidental damage and disease, things go wrong only when there is some abnormality of the chromosomes or of the genes they contain—so that faulty information is sent out to the cell—or if a cell is unable to respond to the messages it receives, although these are correct.

Sometimes entire chromosomes can be responsible for abnormalities. For example, children with Down syndrome are born with one extra chromosome, while extra sex chromosomes can cause abnormalities in sexual development. Many abnormalities at birth are caused by faulty genes. Other problems, such as muscular dystrophy, arise some time after birth, although these, too, are caused by faulty genes. Many aspects of cell life and action have yet to be clarified by science.

See also: Enzymes; Hormones

Cervix

The cervix is the neck of the uterus, which remains closed until a woman gives birth. Cervical smears detect the presence of abnormal cells that can lead to cancer—if cancer is discovered early enough, treatment can provide a cure.

Questions and Answers

Are cervical smear tests really necessary? I don't like the idea of them at all.

The incidence of cervical cancer in American women has fallen by 70 percent as a result of widespread use of the Pap smear test. Many thousands of lives have been saved by the test. In spite of this, over 12,000 American women get cervical cancer annually, and over 4,000 of them die from it each year. These women are mainly those who did not have Pap smear tests, perhaps simply because they didn't like the idea.

Isn't a Pap smear painful and embarrassing?

The doctor or the nurse is not embarrassed—it is just routine—so there is no reason for you to be. You will not be able to see what is happening, so just relax and avoid tensing up, and the procedure should be painless.

What if my Pap smear test result indicates problems?

If you get what is called an abnormal result, don't panic. It may simply mean that the specimen was questionable and that the pathologist couldn't safely grade it as normal. It may have been taken too near the time of your period, it may have been badly preserved, or there may just be inflammatory changes, not cancer. Just make sure that you have a repeat test.

And if the result is truly abnormal?

Mildly abnormal cells call for a repeat test every three months or so. If there is a severe abnormality, you will have to have a biopsy of the cervix, a minor procedure in which a cone of cervix is removed for full microscopic examination. This procedure has saved many women from grave problems.

The cervix is the narrowed lower part, or neck, of the uterus. Although the uterus enlarges greatly during pregnancy, the cervix remains closed until the baby's head descends during childbirth and forces it open. The cervix is the site of various disorders, the most important of which is cancer.

Like the rest of the uterus, the cervix is largely muscular and is lined with a mucous membrane. It is almost cylindrical, about 1 inch (2.5 cm) long, and loosely connected to the bladder in front. The lower part of the cervix, which is somewhat conical and rounded, projects into the vagina so that there is a shallow cul-de-sac (or fornix) all around. This part of the cervix is covered with the same mucous membrane that lines the vagina and is readily accessible for examination. In the center of the vaginal part of the cervix is the tiny, circular external os (mouth), the mouth of the cervical canal that runs down the cervix from the cavity of the uterus.

Infections of the cervix

Cervical infections are fairly common and can cause inflammation (cervicitis). They are often sexually acquired and may be caused by herpes, chlamydia, gonorrhea, or syphilis. All but the herpes virus respond well to treatment with antibiotics. Persistent (chronic) cervical infections can cause pelvic pain and backache, and there may be pain on intercourse. Infection with the human papillomavirus is believed to be an important causal factor in cervical cancer. Because sexually promiscuous females have a higher chance of acquiring this virus, they are more likely to develop cervical cancer. Infection with the herpes virus can make the infected person more susceptible to infection with HIV.

POSITION OF THE CERVIX

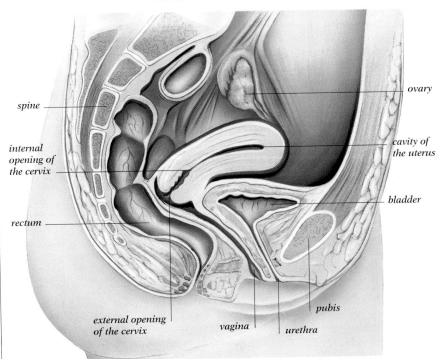

spine

internal opening of the cervix

rectum

external opening of the cervix

vagina

urethra

pubis

bladder

cavity of the uterus

ovary

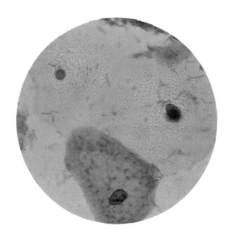

► *For a cervical smear a few cells are scraped off the lining of the cervix near the os. They are then placed on a glass slide, stained, and examined under a microscope to reveal any abnormalities. The cells to the right are normal.*

Cervical erosion

The term "cervical erosion" is a remnant from an earlier misinterpretation of the appearance commonly seen on direct visual examination of the cervix. It is inaccurate, because this appearance is not erosion or any other form of ulceration, nor is it an inflammation or the result of infection. Earlier gynecologists were convinced that cervical erosion was abnormal, and all kinds of symptoms were attributed to it. Many women underwent unnecessary treatment, especially cauterization with a hot probe.

The term refers to a conspicuous dark, raw-looking appearance of the outer part of the cervix, caused by an extension of the columnar epithelium (inner lining) of the canal of the cervix out onto the usually smooth and lighter-colored covering membrane of the vaginal part. The extension of this velvety red area onto the cervix is especially common during pregnancy, when the high levels of estrogen present cause it to expand and extend. Some contraceptive pills also produced well-marked "erosions."

Cervical erosion is seldom seen after a woman goes through menopause. This is a time when estrogen levels are lower, and any vaginal bleeding at this point must be taken more seriously.

Bleeding and discharge

Occasionally this extension of columnar epithelium, with its more profuse blood supply, leads to a slight, intermittent, bloodstained mucous discharge. It is a rule in gynecology that unexplained vaginal bleeding must never be ignored, so even bleeding from this cause should be investigated. If the cervical smear test gives a normal result in a pregnant woman, the condition can safely be ignored.

Cervical incompetence

In some women the upper part of the narrow cervical canal is, for various reasons, abnormally open to just under 0.5 inch (1 cm) at the point at which it joins the cavity of the uterus. During pregnancy the internal pressure from the increasing volume of fluid surrounding the fetus (the amniotic fluid) tends to force this opening wider, so that the outlet of the uterus progressively expands. As a result, such women may repeatedly suffer the misfortune of painless, spontaneous miscarriages, usually around the fourth or fifth month of pregnancy. Miscarriage may also occur because the abnormal widening promotes premature rupture of the membranes.

This condition is known as cervical incompetence, and it is usually, although not always, a result of earlier damage to the cervix during delivery or previous surgery, such as a cone biopsy for suspected cancer, repeated dilatations and curettage, or an amputation of the cervix. When recognized, the condition is easily treated. At some time between the 12th week of pregnancy (when most spontaneous abortions have already occurred) and the 16th week, the cervix is reinforced with a single, strong, purse-string stitch of nonabsorbable material, such as nylon, sewn around it in an in-and-out manner. This procedure, known as the Shirodkar operation, keeps the cervix firmly closed until the baby can be safely delivered. The stitch is then cut and pulled out. This is a simple matter that takes only a few minutes.

Cervical smear

Cancer of the cervix is common. The cancer remains in the surface layer for years before invading the muscle. The smear test can detect it at this harmless stage. For this reason the cervical smear test should, ideally, be done on all females. The test can detect 90 percent of cell abnormalities, and treatment given during this preinvasive stage is simple, safe, and nearly always completely curative.

The cervical smear test, or Papanicolaou (Pap) smear, was instituted not by a gynecologist but by an American physiologist and microscopist, George Nicholas Papanicolaou (1883–1962). While investigating the reproductive cycle using vaginal smears, Papanicolaou recognized cancer cells and realized that this was an important way of diagnosing cervical cancer early. He then promoted the test among the medical profession and the general population.

The procedure

The Pap smear test is a simple, virtually painless, and highly reliable method of detecting cervical cancer in its earliest stages. The vagina is held open by a device called a speculum, and a shaped spatula is used in a rotary manner to gently scrape the area around the os for 360 degrees. The most important cells are those at the junction between the lining of the canal and the lining of the vaginal part of the cervix, known as the transformation zone, where cancer is most likely to start. The almost imperceptible smear of cells is put on a glass slide, fixed, stained, and then examined microscopically by an expert pathologist who has been trained to detect abnormalities in single cells. The whole technique is known as exfoliative cytology.

Suggested frequency

The test should be done regularly on all females, initially at age 21 or within 3 years of first intercourse, whichever is first, and then yearly. If by age 30 all tests are normal, longer screening intervals may be recommended by the doctor. Those with abnormal Pap smears require more frequent follow-up and, in some cases, additional testing.

Abnormal Pap smear tests should ideally be accompanied by microscopic examination of the cervix (colposcopy). This allows accurate localization of the abnormal surface tissue. The instrument used is a long-focus microscope with coaxial illumination that provides an enlarged view of the vaginal part of the cervix.

Cancer of the cervix

Worldwide, cervical cancer is the most common female malignancy. Because of the Pap smear test, however, in the United States it now ranks only number eight in causes of death among women. This is a huge reduction and highlights the importance of the test.

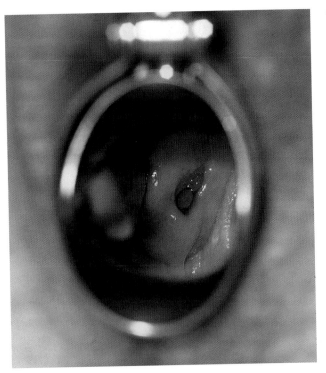

▲ *View of a cervical polyp, as seen through a cervical speculum. Polyps are benign growths that may occur on any mucous membrane; cervical polyps are removed if it is suspected that they may become malignant.*

Even so, the American Cancer Society estimated in 2010 that some 12,200 new cases of invasive cervical cancer would be diagnosed annually in the United States, and 4,210 women would die from it each year. These deaths are particularly tragic when one considers how accessible the site of the cancer is and how easily it can be detected in the early stages if it is looked for. Laboratory tests to detect the types of human papillomaviruses (HPVs) that cause cancer are under development.

The Hybrid Capture HPV Test is useful in determining which women with abnormal Pap results should have a colposcopy. Vaccines against HPV have been developed.

High-risk factors
Risk factors for cancer of the cervix include starting to have sexual intercourse at an early age, having many sexual partners, sexually transmitted diseases (such as genital warts), repeated pregnancies, and cigarette smoking.

As suggested, the sexual factors probably relate to the increased probability of infection with the human papillomavirus. It is probably because of this that sexually promiscuous females are more likely to develop cervical cancer.

Research has shown that the DNA of human papillomavirus types 16 and 18 is found in 62 percent of women with cervical cancer, but in only 32 percent of women without cancer.

Other studies have shown that cervical cancer is more prevalent in those females who use drugs intravenously and in females who have had a positive HIV test.

Cervical intraepithelial neoplasia
Cancer of the cervix is preceded, for a number of years, by a recognizable and easily diagnosable preinvasive condition known as cervical intraepithelial neoplasia (CIN), or carcinoma-in-situ. The epithelium is the mucous membrane lining of the cervix, and neoplasia means a cancerous change in the cells. Intraepithelial means that these changes are still confined to the cells within the lining and thus have not invaded any other tissue. Cancer that remains at this stage is harmless, although potentially devastating. Half of all cancers of the female reproductive system are in the cervix, and so early detection is essential for a course of treatment, which in many cases will provide a complete cure.

Symptoms
Cancer of the cervix often causes no symptoms until it has spread and may cause no symptoms at all before reaching an incurable stage. Sometimes there is bleeding between periods or following sexual intercourse, but there are no dramatic early signs.

Pain and general upset are rare until a late stage is reached and the cancer has spread to other sites. Such pain, which may be felt in the pelvis, buttocks, or lower back, often indicates that the disease is far advanced. It may imply that the cancer has spread widely into the pelvic or abdominal regions.

Involvement of the bladder and the rectum may cause blood in the urine or bleeding from the rectum. The moral is clear. Cancer of the cervix has to be looked for, and the best way to do this is by the Pap smear test.

Complications
Once the cancer has passed from the epithelium into the underlying cervical muscle, the treatment becomes more difficult. If the cancer is confined to the cervix, the choice rests between removing a cone of the cervical muscle and removing the whole uterus (simple hysterectomy). The former has a recurrence rate of about 5 percent; the latter, a zero recurrence rate.

More extensive cervical cancer is difficult to treat successfully, and the choice rests between extensive surgery and radiotherapy. There is no universal agreement on which of these forms of treatment is best. Radiotherapy is widely used—this treatment is usually administered by means of a sealed container of radioactive material placed in the vault of the vagina and in the cavity of the uterus.

Outlook
The cure rate for cervical cancer depends on the extent of its spread at the time of diagnosis. If it has spread to the vagina and surrounding tissues, the cure rate drops sharply to about 50 percent.

Extensive spreading to the organs of the pelvis and remote spreading to other parts of the body has a very poor outlook. In only about 10 percent of such cases is the patient still alive five years later. When the condition is detected in its early stages by means of the Pap smear test, there is an excellent chance of cure and recovery for many women who suffer from this type of cancer.

See also: **Cells and chromosomes; Estrogen; Genitals; Hormones; Pelvis; Pregnancy; Uterus; Vagina**

Chest

Questions and Answers

I am an avid gardener, but I get pains in my chest after I have been digging. Could this mean that I have strained my heart?

If these pains occur only with movement, and your chest wall feels tender or sore, the problem is most likely muscle soreness. However, physical activity such as this could, in some cases, overtax your heart. Without knowing your age, medical history, and so on, it is difficult to say which problem is occurring, so you should see your doctor just to be safe.

I broke my ribs playing basketball, but I was not bandaged or given any treatment. Why was this?

Although broken or cracked ribs can be uncomfortable or painful, the main danger is that the chest movement will be reduced, producing less airflow into and out of the underlying lung. This can cause pneumonia, so it is unusual to bandage broken ribs.

My doctor says I am pigeon-chested. What is this? And am I more likely to get chest infections?

Minor deformities of the chest wall are often referred to as a pigeon chest. The most common form is a hollowing of the center of the chest at the front, but this does not mean that you are more liable to chest infection.

Can people still die of pneumonia as they did in the old days?

Unfortunately, yes. Pneumonia used to be a common cause of death not so many years ago, even in healthy young people, but generally this is no longer the case. However, in people who are seriously ill for some other reason, or are elderly, pneumonia is often the final illness that kills them.

The bony, muscular structure of the chest forms a protective framework around two of the body's most important organs, the lungs and the heart. It is essential to know when a cough or chest pain needs medical attention.

The chest is a bony cage that contains two of the most important organs in the body: the lungs and the heart. The basic function of these organs is to transfer oxygen from the air to the tissues, where it is essential for the continuation of life.

The bell-shaped rib cage is located just under the skin of the chest. It encloses the lungs and heart on all but their lowest surface. It is attached to the spine at the back, and its base is sealed off by the diaphragm, the thick muscular sheet separating the chest from the abdomen. In between the ribs are further muscular sheets called the intercostal ("between the ribs") muscles. The chest wall thus consists of a bell-shaped muscular bag with the ribs as struts. By expanding and contracting it sucks air in and out through the windpipe, or trachea, which emerges into the neck.

A membrane called the pleura lines the whole of the inside of the chest, and similar membranes cover the lungs and the heart. When the pleura becomes inflamed, this leads to pleurisy.

The two lungs fill the bulk of the chest and are connected by their tubes, the main bronchi, to the trachea. Smaller tubes, or bronchioles, split off from each main bronchus like branches of a tree, carrying air to the air sacs (alveoli). Here oxygen is extracted from the air and passed into the blood, while carbon dioxide—the body's waste product—moves in the opposite direction.

The heart lies at the front of the chest between the two lungs, inside its own membranous bag. It receives blood from the body through the pumping chambers on its right side (the right atrium

ORGANS OF THE CHEST

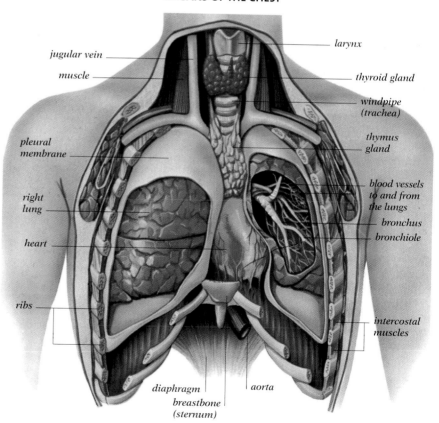

jugular vein

muscle

pleural membrane

right lung

heart

ribs

larynx

thyroid gland

windpipe (trachea)

thymus gland

blood vessels to and from the lungs

bronchus

bronchiole

intercostal muscles

diaphragm

aorta

breastbone (sternum)

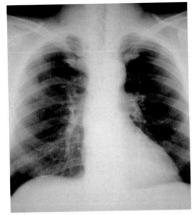

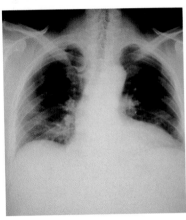

▲ *These X rays show how the muscular walls of the chest expand when air is inhaled (top) and contract when it is exhaled (above).*

Pain in the chest: When to see your doctor

TYPE OF PAIN	OTHER SIGNS	CAUSES
Central pain, pressing and dull in character.	Breathlessness, nausea, or sweating, lasting more than 20 minutes.	Angina (heart disease). Heart attack. Pericarditis (inflammation of the membrane lining the heart). Indigestion.
Central, gripping pain spreading to the neck, shoulders, or arms.	Brought on by exercise or emotional excitement.	Angina. Pericarditis.
Anywhere, worse on inspiration (breathing in) or on coughing.	May be associated with a cough or an attack of bronchitis.	Pleurisy. Pericarditis. Pulmonary embolus.
Central, burning. Worse after food or on bending forward; may be worse at night.	Foods may bring it on, and it may be relieved by milk or indigestion tablets.	Esophagitis (inflammation of the gullet, a form of indigestion).

When a cough needs medical attention

TYPE OF COUGH	CAUSE
Green or yellow sputum coughed up.	Bronchitis (inflammation of the lining of the bronchial tubes in the lungs) or pneumonia (inflammation of the lung).
Cough or wheezing.	This may be true asthma or wheezy bronchitis.
Coughing up blood or bloody streaks in sputum.	There are many causes of this, but the most serious are tuberculosis or lung cancer.

and ventricle) and pumps the blood into the lungs. Blood full of oxygen returns from the lungs to the left atrium and ventricle, from where it is pumped out into the main artery of the body—the aorta.

The chest also contains the gullet, or esophagus, which carries food from the mouth to the stomach. At the top of the chest in front of the windpipe is the thymus gland. It is important in the maturation of immune system cells but virtually disappears before adulthood.

Chest problems

Chest complaints have three main symptoms: pain, coughing, and breathlessness. Pain in the chest may arise from the chest wall, as a result of muscle strain and other conditions, or it may arise from the heart. The esophagus may be the source of pain if the acid contents of the stomach wash back up and cause inflammation. Pleurisy may also be the underlying cause of chest discomfort. Pleurisy occurs when the two layers of pleura lining the inside of the chest and outside of the lungs become inflamed and rub together. The pain of pleurisy is worse on deep breathing or coughing.

Since the lungs do not give direct pain signals, coughing is an important symptom of damage to the lung. Doctors call a cough "productive" when it produces phlegm or sputum. This may indicate infection, particularly if the sputum is green or yellow rather than white. Most coughs, however, do not produce sputum and are simply a result of inflammation of the upper airways rather than a sign of lung disease. Such coughs usually follow a cold.

Breathlessness may result from either lung disease or heart disease. Asthma is a common cause of breathlessness, particularly in younger people. Heart problems lead to breathlessness because the reduced blood circulation, which occurs if the heart is weakened, cannot carry the required amount of oxygen; the lungs move stiffly because they are distended with blood. This situation is referred to as heart failure and is common in the elderly. In its most serious form, known as pulmonary edema, it can be life-threatening.

Urgent medical advice is needed if a person experiences a new, severe pain in the chest, shoulders, or arms, especially if there is breathlessness, nausea, and sweating, or if breathing or coughing makes the pain worse. Prompt advice is also needed if a cough produces phlegm or is accompanied by wheezing, if blood is coughed up, if there are bloody streaks in the sputum, or if there is extreme breathlessness with a bubbling cough.

See also: **Coughing; Esophagus; Heart; Lungs**

Circulatory system

Questions and Answers

My mother sometimes complains of pins and needles. Is something wrong with her circulation?

The pricking sensation of pins and needles is actually the irritation of a nerve, caused when blood supply is restricted for some reason. This often comes about through lying in an awkward position, but it can also be a sign of circulatory disease that has damaged the blood vessels. If the problem persists, your mother should see a doctor.

I am heavy. My doctor has warned me that my weight is harmful to my circulation. Why is this?

When you are overweight, you carry too much fat. The surplus fat must have a blood supply, so additional and unnecessary blood vessels open up. Obesity is also associated with the serious arterial disease atherosclerosis, which can narrow arteries. This causes further problems: to keep the blood flowing around the system, the heart has to do extra work, which may strain it. A heavy body also requires more effort to move around, and this, too, could strain the heart. To avoid these risks, shed the extra pounds.

My teenage daughter keeps on fainting, but she looks completely healthy. Do you think she might have problems with her circulation?

The most usual cause of fainting is a temporary fall in the volume of blood reaching the brain. This is a common problem in adolescents, particularly girls. It is often caused by emotional disturbance, which can make the arteries widen, lowering the blood pressure and preventing blood from being pumped up to the brain. Girls usually grow out of this sort of fainting, but if the fainting spells increase or your daughter is worried that she may be ill, she should see her doctor promptly.

Blood pumped by the heart is continuously circulating around the body. Its journey has several essential purposes—among them, to supply the body's cells with food and oxygen and to clear them of waste products.

The circulatory system is a closed network of blood vessels—tubes that carry blood around the body. At its center is the heart, a muscular pump that keeps the blood constantly moving on its journey.

Arteries and arterioles

Blood begins its journey around the circulatory system by leaving the left side of the heart through the large artery known as the aorta. At this stage blood is rich in oxygen, in food broken down into the microscopically small components called molecules, and in vital substances such as hormones.

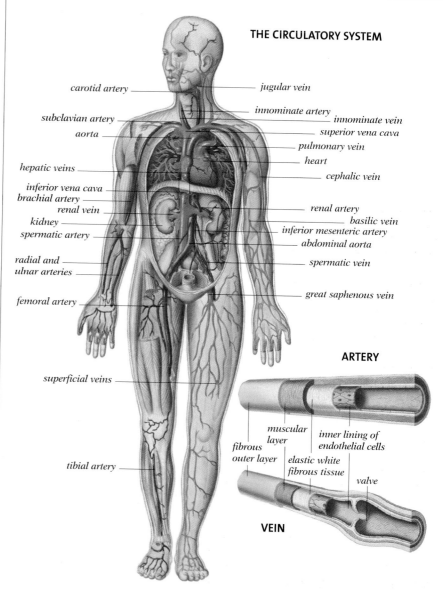

THE CIRCULATORY SYSTEM

carotid artery
jugular vein
subclavian artery
innominate artery
innominate vein
aorta
superior vena cava
pulmonary vein
heart
hepatic veins
cephalic vein
inferior vena cava
brachial artery
renal vein
kidney
spermatic artery
renal artery
basilic vein
inferior mesenteric artery
abdominal aorta
radial and ulnar arteries
spermatic vein
femoral artery
great saphenous vein

superficial veins

ARTERY

muscular layer
inner lining of endothelial cells
fibrous outer layer
elastic white fibrous tissue
tibial artery
valve

VEIN

DISTRIBUTION OF BLOOD IN THE BODY

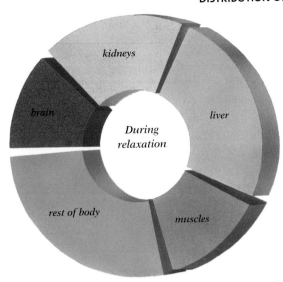

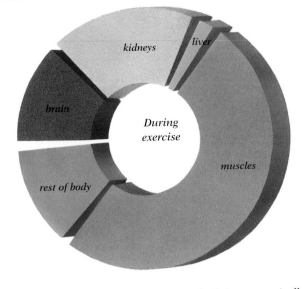

In the early part of its journey, blood flows through fairly large tubes called arteries. It then passes into smaller ones known as arterioles, which lead to every organ and tissue in the body, including the heart. From here, it enters a vast network of tiny vessels called capillaries, where oxygen and vital molecules are exchanged for waste products.

The veins

Blood then leaves the capillaries and flows into the small veins, or venules, where it starts the journey back to the heart. All the veins from the various parts of the body eventually merge into two large blood vessels: the superior vena cava and the inferior vena cava. The first collects blood from the head, arms, and neck, and the second receives blood from the lower part of the body. Both veins deliver blood to the right side of the heart. From here it is pumped into the pulmonary artery (the only artery to carry blood with no oxygen), which takes the blood to the lungs, where oxygen from air breathed in is absorbed into it, and the waste product, carbon dioxide, is released and breathed out.

The journey's final stage is for the now oxygen-rich blood to flow through the pulmonary vein (the only vein to carry oxygenated blood) into the left side of the heart, where it starts its circuit once again.

Distribution

Blood is not spread evenly throughout the system. At any given time there is about 11 percent in the arteries, 12 percent in the lungs, 61 percent in veins and venules, 7 percent in arterioles and capillaries, and 9 percent in the heart. Blood also flows at different rates. It spurts from the heart and through the aorta at 13 inches (33 cm) per second, slows down to 0.1 inches (0.25 cm) per second when reaching the capillaries, then flows back through the veins to gradually reach 8 inches (20 cm) per second when it gets to the heart.

Pulse and blood pressure

One of the main guides to the condition of a patient's circulation is the pulse, for it is produced by each heartbeat. Arteries have elastic walls that expand every time the heart pumps a wave of blood through them. It is possible to feel this happening if an artery near the body's surface is found, such as at the wrist, and pressed against a bone.

▲ *Blood circulation throughout the body is automatically regulated according to certain activities, so that it is always in the places where it is most needed.*

The normal pulse rate is between 60 and 80 beats a minute but varies widely. Thus a doctor taking a pulse is not only counting the beats but feeling for changes in their strength and regularity.

Blood pressure is different from pulse rate—it is a measurement of the force with which the heart pumps blood into the arteries. A blood pressure measurement is often taken on the upper arm because here a suitable artery is in a situation in which it can be readily compressed. The instrument for measuring blood pressure consequently works on the principle that if the flow of blood through the artery is temporarily closed (by inflating a special encircling bag around the arm), the sound made by the initial movement of blood through the compressed artery, as the pressure is gradually released, indicates the point at which the pressure in the artery and in the compression bag are the same. A gauge attached to the bag indicates the pressure.

If the blood pressure is higher than normal, the heart is probably having to work harder to push the blood through the system, perhaps indicating a disease in the circulatory system, such as narrowing of the arteries. The smaller the opening in a blood vessel, the harder the heart has to work to pump the blood through it. If the heart works too hard for a long period of time, its life may be shortened, and this is why doctors constantly look out for high blood pressure.

Further controls

The width of blood vessels also controls circulation. Changes in width are caused by two means: hormones and nerves. If blood pressure drops, the kidneys emit a hormone called angiotensin, which causes the arteries to narrow and the blood pressure to return to normal rate.

See also: Arteries; Blood; Blood pressure; Body structure; Brain; Capillaries; Coronary arteries; Heart; Hormones; Kidneys; Lungs; Pulse; Veins

Colon

I have chronic colitis. Is there a chance my children will catch it?

No. Although the cause of chronic colitis is uncertain—it may be a bacterial infection or a result of psychological disturbance—it is not at all contagious.

If I have chronic colitis, am I likely to get bowel cancer in later life?

It is highly unlikely. Modern drug treatment with salazopyrin is very effective and will usually control the symptoms. If the disease is severe, the colon, or a portion of it, can be removed. Only 5 to 10 percent of those patients who have chronic colitis for 10 years or more develop colonic cancer.

I sometimes have severe diarrhea after I eat certain foods. How can I tell if this is colitis or not?

If you find that specific foods trigger diarrhea, you may be having an allergic reaction and be suffering from acute colitis. See your doctor. If the diarrhea is constant, possibly containing mucus, pus, and blood, and you are anemic and dehydrated, you should be examined by your doctor for chronic colitis. The lining of the colon is examined by a colonoscope: the appearance will resemble red velvet and bleed readily on contact if you have chronic colitis.

My father gets attacks of acute colitis. What should he do to treat the condition?

He should go to bed, drink plenty of fluids (including electrolyte-balanced liquids), and take kaolin to help control diarrhea. If the condition still persists, he should see a doctor. If his colitis is the result of an allergic reaction to certain foods, he should try to avoid them.

Colitis is an inflammation of the colon's mucous membrane, and it is unpleasant and debilitating. However, although chronic colitis can become serious, and a long-term problem, acute colitis is usually easy to treat.

The function of the colon is to move solid material from the small intestine to the anus and to absorb salt and water delivered to it from the small intestine. Colitis is an inflammation of the colon's mucous membrane. There are two kinds of colitis. Acute colitis is often a result of an infection—usually viral but may be bacterial (e.g., salmonella, shigella, E. coli) or parasitic—and lasts only a short while. Food allergy or lactose intolerance can also cause this problem. Chronic colitis, such as ulcerative colitis or Crohn's disease, is much more serious and requires prolonged treatment. The cause of chronic colitis is unknown but is thought to be due to a malfunction of the immune system in which the basic controls go awry and the system attacks the body (autoimmune). Stress may play a role as well.

In both acute and chronic colitis, the symptoms are abdominal pain, followed by an explosion of watery diarrhea. In chronic colitis, there may be as many as 15 to 20 bowel movements each day. Large quantities of mucus, pus, and sometimes blood are passed with the movements. On occasion there is rectal tenesmus (an ineffectual urge to defecate). In more severe cases dehydration, anemia, loss of appetite and weight, vomiting, and high fever may be present.

For acute colitis, bed rest is advisable. Kaolin may slow diarrhea. The mainstay of treatment for chronic colitis is salazopyrin (a combination of antibiotic and aspirinlike drugs), taken three times a day. A liquid preparation of hydrocortisone can be given as a suppository; this has a marked soothing effect. Diet should consist of bland, high-protein food, with only a small amount of fruit and roughage.

With acute colitis, the outlook is excellent as long as the cause of the illness is treated or removed. Various treatments are available for chronic colitis disorders, with the goal being to control the symptoms as well as possible.

See also: Anus; Digestive system; Excretory systems; Membranes; Mucus

HOW COLITIS AFFECTS THE BODY

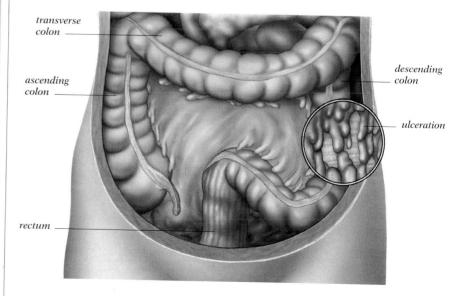

transverse colon

ascending colon

descending colon

ulceration

rectum

▲ *Here, chronic colitis has caused ulcers (inset) in the mucous membrane of the colon.*

Conception

Knowing what happens during conception can help couples understand this complicated process and may help them overcome any problems they experience when trying to plan a family.

No. The sperm determines whether the child is female or male, and there is nothing a couple can do to make sure that either a sperm carrying a male chromosome or one carrying a female chromosome gets to the egg first. It has been suggested that the acid-alkali balance of a woman's vagina and cervical canal has some influence on sperm, and that an acid or alkali gel inserted into the vagina before intercourse may make it possible for the couple to choose a boy or a girl. There is no reason to suppose this method can work. Its reputation arises from the common fallacy of accepting evidence that supports the desired outcome and ignoring contrary evidence.

How long is the period during which conception is possible?

The egg lives for less than 24 hours after ovulation, which takes place in the middle of a woman's menstrual cycle; it is during this time that conception is possible. Sperm can live in a woman's body for 24–48 hours, or very rarely up to five days, so the egg may be fertilized by sperm as long as five days after ovulation.

My husband has a low sperm count. What can be done to help men with such a problem?

The sperm count is the number of sperm in 0.061 cu. in. (1 ml) of seminal fluid. Sperm cannot survive in high temperatures, so hot baths, tight underwear, or being overweight can cause a low sperm count. Too much alcohol or cigarette smoking are also likely to have a significant adverse effect on male fertility, as may the use of anabolic "muscle-building" steroids. Referral to a fertility specialist may be needed.

Conception is a great deal more complicated than the simple joining of a sperm and an egg. It is a complex process, and various conditions have to be right to ensure that it is successful.

Every time a man ejaculates, he produces sperm, but a woman is physically able to conceive only once during each menstrual cycle. Approximately 14 days before her period, she produces a single egg or occasionally two eggs from one of her two ovaries. The egg is then drawn into the fallopian tube. It lives for less than 24 hours, and if it is not fertilized, it dies and is absorbed into the cells lining the tube. The menstrual period follows 13 days later, and the cycle then begins again. The average cycle lasts 28 days, although some women find that their cycles are longer or shorter. Most women release approximately 12 eggs per year, if they have a regular 28-day cycle.

The sperm and ejaculation

If intercourse takes place around the time of ovulation, conception is likely. A man produces about 400 million sperm in each ejaculation. These are surrounded by seminal fluid, which protects the sperm from the acidity of the vagina. Once deposited, the sperm immediately start their journey up

HOW AN EGG IS FERTILIZED

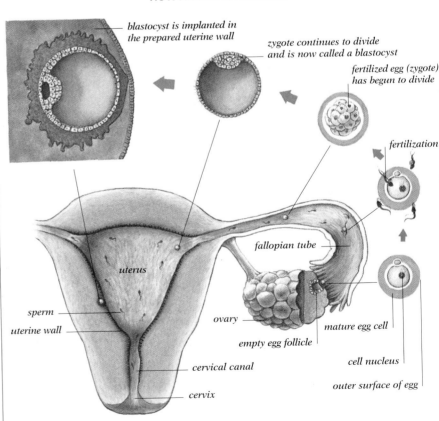

- blastocyst is implanted in the prepared uterine wall
- zygote continues to divide and is now called a blastocyst
- fertilized egg (zygote) has begun to divide
- fertilization
- fallopian tube
- uterus
- sperm
- uterine wall
- ovary
- empty egg follicle
- mature egg cell
- cell nucleus
- outer surface of egg
- cervical canal
- cervix

▲ An egg is released from one of the two ovaries, is fertilized by a sperm as it travels down the fallopian tube, and then implants itself in the wall of the uterus.

the vagina, through the cervix (neck of the womb), and into the uterus. They move by vigorously lashing their tails. Many of the sperm do not make this journey successfully and wither and die in the acidic conditions. This is nature's way of ensuring that damaged or unhealthy sperm do not fertilize the egg.

Fertilization

The millions of sperm that reach the uterus are nourished by the alkali mucus of the cervical canal. They then travel up into the fallopian tubes, a journey of about 8 inches (20 cm) that takes approximately 45 minutes. Only about 2,000 sperm may actually survive, but those that reach the fallopian tubes may stay alive within them for up to five days, ready to fuse with an egg if ovulation takes place. If an egg is already present within the tube, fertilization takes place immediately.

Fertilization occurs when a sperm penetrates the surface of the egg. Each sperm carries an enzyme (a substance responsible for activating biological chemical processes) that helps liquefy the outer surface of the egg to make penetration by a single sperm easier. Once the egg is fertilized, a membrane forms around the egg that prevents other sperm from entering. The egg and sperm (which has now discarded its tail) then fuse together to form a single nucleus (center), which then begins to divide into two cells. Within 72 hours, the cells divide five times to produce a 64-celled egg.

The fertilized egg travels down to the uterus in approximately seven days (day 21 of a 28-day cycle). During this time it grows tiny projections that help it burrow into the lining of the uterus, where it can be nurtured and a pregnancy can start. This process is called implantation; once it has occurred, conception is complete.

The egg can now be nourished by the rich blood supply present in the uterine lining. From the moment of fertilization, the egg produces a hormone called human chorionic gonadotropin (HCG), which informs the ovary that fertilization has taken place and maintains the blood flow to the lining of the uterus so that the egg can continue its development. (Pregnancy tests work by detecting this hormone in blood or urine.) The body therefore knows that the menstrual cycle must not continue, since a period would remove the fertilized egg.

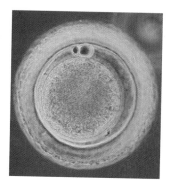

► *This photograph of a living egg, a few hours after fertilization, has been magnified 200 times.*

Not every conception occurs in this way. If the fertilized egg begins to divide and the two cells separate, they form two embryos, which become identical twins because they came from the same egg and sperm. Nonidentical twins occur when two separate eggs are released at ovulation and are fertilized by two separate sperm. Multiple births (three or more babies) occur for the same reasons, but although the use of fertility drugs has increased their number, they are still rare.

Difficulties in conceiving

Although most couples conceive within six months, it can take up to two years for a woman to become pregnant. Even after two years some still find difficulty in conceiving. The cause of the problem may lie with either the man or the woman, or with both. The ease with which a woman becomes pregnant depends on her age and the age of her partner and the state of their health. Women are most fertile up until the age of 25; from 35 onward their chances of becoming pregnant decline rapidly until menopause. A man's decline in fertility is more gradual; at 60 he is still fertile, though to a lesser degree.

In spite of sex education, sexual intercourse for some couples may be unsatisfactory, although sex manuals and clinics may be helpful. Other couples have difficulties because they are unsure of the best time in the menstrual cycle to conceive; and when this uncertainty is combined with infrequent intercourse, the chances of conceiving are low. Emotional factors such as stress, anxiety, tiredness, or overwork can also play a large part in conception—intercourse may be less frequent, or the man may be incapable of maintaining an erection or of ejaculating during intercourse.

Women with infrequent periods ovulate less often, but they are still fertile, although their chances of conceiving may be reduced to only three or four times a year. A failure to ovulate, caused by thyroid problems, drastic weight loss, or premature menopause, is another common cause of infertility. There may also be a delay in the return of a woman's fertility after she comes off the Pill. Damage to the fallopian tubes, arising from an infection in the tubes themselves (salpingitis) or other causes, will prevent them from functioning normally. Adhesions may form, which prevent the egg from passing down to the open end of the tube.

In men difficulties may arise when the testicles are not working properly, or there is a low sperm count. Diseases such as mumps may cause an inflammation of the testes, or there may be a testicle failure or a blockage in the vas deferens (down which the sperm travel) or epididymis (where sperm mature). There are many techniques to discover the cause of infertility and to help circumvent it.

▲ *Identical twins are born when one egg is fertilized and splits into two as it begins to develop in the womb.*

See also: Menstruation; Ovaries; Pregnancy; Semen; Sperm; Testes; Uterus; Vagina

Coordination

Why do some people seem to be natural athletes, dancers, or gymnasts, while others seem to be less gifted? The answer lies in the complicated processes of coordination that begin in the brain.

The supple movements of a champion gymnast reveal, in their flowing patterns, how delicately the human brain can control the hundreds of muscles in the torso and limbs. To achieve such intricate sequences of action, the human brain has evolved a complex system of control and guidance that makes even the most sophisticated computers look primitive.

Babies are born with many reflexes (muscular responses that occur without conscious thought). To visualize these reflex actions in an adult, imagine how quickly a person would withdraw his or her hand from a hot saucepan. The movements that are directed by the brain (voluntary movements) are superimposed onto these simple reflex actions. For every action that is performed, some muscles contract, others relax, and still more maintain their contraction to stabilize the rest of the body. The process by which all the individual muscle contractions are carefully synchronized to produce a smooth order of activity is called coordination.

How coordination works

To understand this process, consider an everyday action such as leaning over a table to pick up a cup of coffee. How does the brain direct this apparently simple task? Before someone can pick up a cup of coffee, a series of events must happen.

First the person must know where the cup and his or her hand are and the relationship between them. This means that the brain must be able to generate a "map" of the space for necessary movement to be planned. This is called spatial perception. The brain must then interpret this internal "map" of the outside world so that the problem of getting the coffee cup from the table to the hand can be solved. It must then generate a plan of action that can be translated into a detailed set of instructions to the muscles so that they will contract in the right order.

During the movement, started by the planning parts of the brain (the premotor area), continuous streams of information pour in from all the sensors (nerves) in the muscles and joints. This information, which has to be organized and relayed back to the brain, describes the positions of the muscles and joints as well as their states of contraction.

In order to move the hand to pick up the cup of coffee, the person also needs to lean slightly toward it, and this alters the center of gravity of the body. All the reflex balance mechanisms must be controlled to ensure that the correct changes in muscle tone are made, allowing the movement across the table that the brain has ordered. This means that the background tone of many other muscles has to be monitored and coordinated.

First stages of coordination

All intentional movements need to be practiced before they become coordinated. Even such ordinary actions as walking are

▲ *A game such as golf requires a great deal of coordination between the hands and eyes.*

HOW THE BRAIN ENABLES US TO PICK UP A CUP OF COFFEE

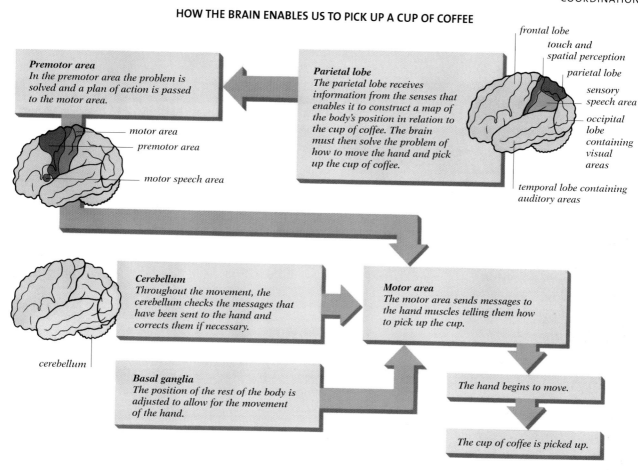

Premotor area
In the premotor area the problem is solved and a plan of action is passed to the motor area.

Parietal lobe
The parietal lobe receives information from the senses that enables it to construct a map of the body's position in relation to the cup of coffee. The brain must then solve the problem of how to move the hand and pick up the cup of coffee.

frontal lobe
touch and spatial perception
parietal lobe
sensory
speech area
occipital lobe containing visual areas
temporal lobe containing auditory areas

motor area
premotor area
motor speech area

Cerebellum
Throughout the movement, the cerebellum checks the messages that have been sent to the hand and corrects them if necessary.

Motor area
The motor area sends messages to the hand muscles telling them how to pick up the cup.

cerebellum

Basal ganglia
The position of the rest of the body is adjusted to allow for the movement of the hand.

The hand begins to move.

The cup of coffee is picked up.

---

major problems for every developing child. As a baby's brain matures and the interconnections increase, the primitive reflexes with which he or she was born (such as the startle reaction, which causes the hands to be outstretched when a baby feels he or she is falling) are overlaid with progressively more complicated ways of moving.

A toy might attract the baby's eye, because its bright color causes a strong signal in the visual centers, but the baby finds that reaching out is not enough to touch this object, so he or she is impelled to move toward it. The first attempts to move are not coordinated: the limbs just thrash around. But these initial attempts enable the necessary brain connections to develop for the set of actions that make up a coordinated crawl. Once crawling has been achieved, the messages sent from the brain to the muscles can be improved on until nothing at ground level is safe from the child's grasp.

When the baby discovers that he or she can get into an upright position, the cerebellum (the part of the brain responsible for coordinating voluntary muscular activity) has to analyze new instructions coming from the balance centers in the brain stem. Walking is another new skill to learn, requiring many attempts during which the cerebellum cooperates with the motor area to develop efficient messages to send to the muscles.

The separate parts of each action learned in this way are reprogrammed into the spinal cord, but they must form a coherent pattern to produce a coordinated movement, in the same way that an orchestra must have a conductor before it can produce a tuneful sound from all its instruments. Once these "simple" skills have been perfected, the brain has been programmed so that no concentration is necessary—the premotor area says "walk," and the right set of instructions go into action to produce the complicated mechanical actions that are involved. The cerebellum monitors the progress of the action, but this is less and less a conscious event. If a "problem" is introduced into the system, such as the change in the foot's posture caused by wearing high heels, some reprogramming is necessary, and concentration is needed while the motor cortex learns this new "tune."

Advanced coordination

In complex movements, the movement of the eyes is coordinated with the visual receptive centers of the brain and then with the movement of the rest of the body. This coordination is the last to mature. It forms the basis for learning the type of complex movements that are needed in sports or in skills such as playing a musical instrument.

Some brains seem better equipped from birth to develop in particular ways. However, to a large extent, abilities in complex types of coordination depend on how much the individual can concentrate to build up these specialized "programs" for complex movements.

> *See also:* Balance; Brain; Movement; Muscles; Reflexes

Cornea

The cornea—the eye's main lens—is the shining, transparent bulge at the front of the eye. If it is scarred or damaged, sight is impaired, but a graft operation from a donor eye can sometimes be performed to improve vision.

Questions and Answers

Is having an eye graft painful?

The procedure, lasting about half an hour, is performed under a local or general anesthetic. The eye may feel a little bruised, but there is seldom more than mild discomfort, and usually the patient is too happy with the result to notice any soreness. If sutures (stitches) have to be removed, a short stay in hospital may be required, but a buried continuous suture is often left.

Will my eye look different in shape or color if I have an eye graft?

If the eye needing the graft is scarred, a white area will show up on the cornea. After a successful graft, your eye will look much more normal, since the operation will bring back the color of the iris. The scar from the operation is visible only at very close range.

Can I become a cornea donor?

Anyone can become a donor, as long as the eyes are free from disease or injury. Contact your nearest eye hospital, and tell your doctor and your nearest relatives.

How can I take care of my eyes to keep them functioning well?

Let nature do the work. Tears contain a natural antiseptic, which eye baths can wash away. Do not use drops or put anything in your eye unless a doctor prescribes it. If you get grit in your eye, wash it out with a mild, "balanced" salt and water solution.

Is it true that a disease of the cornea can cause blindness?

A chlamydial infection, trachoma, is still common in underdeveloped countries, and it can cause blindness even in children.

The cornea, together with the fluid behind it, forms the powerful, fixed-focus lens of the eye. It measures just 0.02 inch (0.5 mm) thick at the center and 0.04 inch (1 mm) thick where it joins the white of the eye, the sclera. The optical power of the cornea accounts for about two-thirds of the total eye power.

The cornea consists of five layers. On the outside is a five-cell layer called the epithelium, which corresponds to the surface skin. Underneath this is an elastic, fiberlike layer called Bowman's layer. Next comes the tough stromal layer, made up of a protein called collagen. The stromal layer is the thickest part and contains various infection-fighting antigens that help keep the cornea free from infection. It is also thought to help control inflammation in the cornea.

After the stroma comes a layer called the endothelium, which is only one cell thick. This layer keeps the cornea transparent and maintains a balance of water flow from the eye to the cornea. Once formed, the cells of this layer cannot regenerate, so injury or disease to the endothelium can cause permanent damage to sight. The final layer, called Descemet's membrane, is an elastic one.

A tear film covers the epithelium. Without tears the cornea would have no protection against bacterial microorganisms, pollution, or dust. The tear film is also essential for vision, since without tears the epithelium would lose its transparency and become opaque.

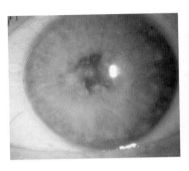

▲ *This diseased cornea has lost its transparent quality, affecting the focus of the eye.*

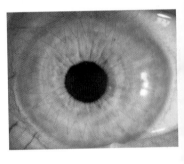

▲ *The same eye five weeks later. Note the fine zigzag line holding the clear grafted cornea in place.*

Corneal grafting

This operation is done on diseased or injured corneas when the central portion is scarred or the curvature is deformed. Where the cornea is deformed, the irregular surface causes gross distortion of the image on the retina, and very poor vision.

A disk is cut out of the cornea, removing the diseased area or scar tissue. Most grafts done to restore sight are around 0.27 inch (7.5 mm) in diameter. There are two procedures: one is to cut the full thickness of the cornea, the other is to cut only part of the thickness. The latter operation, called a lamellar graft, is usually done to replace diseased surface tissues, when the deeper tissues are still in good shape. The donor eye has a similar-size disk cut from it, and this is placed in position over the living eye and sewn into place using an operation microscope and fine nylon or collagen thread mounted on a curved needle 0.16 inch (4 mm) long.

In most instances the patient is allowed to leave the hospital within a few days, and many notice an improvement in their sight even in this short time. If the graft starts to be rejected, anti-inflammatory drugs are used, and in many cases the graft survives. Stitches may be removed several months later, and then, provided that the graft is clear, the sight is corrected by glasses or contact lenses. At this point, if the operation has gone well, the patient should notice a great improvement in his or her sight. The graft will leave only a faint scar.

See also: Eyes; Optic nerve; Retina

Coronary arteries

Questions and Answers

Could I have had a minor heart attack without knowing it?

Yes. It is fairly common to find clear evidence of a previous heart attack on the electrocardiogram of a patient who has never had any symptoms. These "silent" heart attacks are most common in the elderly and in diabetics.

How old do I have to be before I am at risk of a heart attack?

Heart attacks occasionally occur in the teens and twenties, but most of those who suffer heart attacks are middle-aged or older.

Does jogging lessen the risk of coronary disease?

Exercise requiring stamina gives some protection, and jogging, cycling, swimming, or walking are good for you. Muscle- or body-building exercises are unlikely to give protection unless they are part of a controlled gymnastic program. If you have been sedentary all your life, do not suddenly start exercising; build up gradually.

Why is my angina much worse in cold weather?

If you exercise in the cold, the circulation in the skin shuts down and increases resistance to the blood flow and heart action. Exercise only when warmed up.

Should I take beta-blockers for my angina if I am asthmatic?

If your asthma is under control, then treatment of angina with "selective" beta-blockers is acceptable. Only the "nonselective" beta-blockers should be avoided in patients with asthma.

What happens when people experience a coronary thrombosis, or a heart attack? And what are their chances of leading an active life again? Advances in medicine can do much to help, but prevention is better than treatment.

▲ *A heart attack causes severe, crushing pain in the chest.*

The coronary arteries are the vessels that supply blood to the heart itself. They are particularly prone to partial or total obstruction by atheroma—a buildup of fat that is caused by many factors, but principally by excessive stress, sedentary living, smoking, and an unhealthy diet. Obstructed coronary arteries are the cause of heart attacks; disease of the coronary arteries is the most common cause of death in the majority of Western countries.

The three arteries

The heart is a muscular bag that pumps blood around the body. Like any muscle, it has to be supplied with oxygen and food to continue working. This supply is carried in the right and left coronary arteries, which are the first vessels to leave the aorta (the body's main artery) as it emerges from the heart. Almost as soon as it branches off the aorta, the left coronary artery splits into two big branches. So there are, in effect, three coronary arteries: the right plus the two branches of the left. They go on to completely surround and penetrate the heart, supplying blood to every part of it. The coronary arteries are particularly affected by any obstruction because, like the heart itself, they are always in motion and the resulting strain on their walls hastens the buildup of atheroma. Atheroma is the principal feature of the arterial disease atherosclerosis. Except in a tiny proportion of cases, the disease process is always the same. Fatty deposits build up on the wall of the artery, narrowing the whole artery and creating the risk of a total blockage.

Heart attack

If a coronary artery becomes completely blocked, the blood supply to an area of heart muscle is shut off. There is an intense, heavy pain, often lasting for hours or even days, and described by the patient as resembling a viselike grip. The patient also experiences shortness of breath, cold sweats, and heart palpitations, and he or she looks very pale. Eighty-five percent of those who have a heart attack recover, but in some patients there is another, sometimes disastrous, attack in the first hour. After the attack, the area of heart that was affected eventually heals into a scar. That particular part of the heart muscle will never work again. But with careful treatment the patient will, in most cases, be able to lead a healthy, active life once more.

The blockage itself usually comes about as a result of what is known as a thrombosis (blood clot). The artery, narrowed by atheroma, restricts the flow of blood to such a slow pace that its natural tendency to clot or thicken begins to operate. This clot makes the final obstruction. A heart attack is often referred to as a coronary thrombosis, or simply a coronary.

The other problem caused by coronary artery atherosclerosis is angina. In this case there is a partial block that allows the heart to function normally during rest but does not allow the extra blood flow necessary for exercise. In some patients the pain of angina is caused by intense

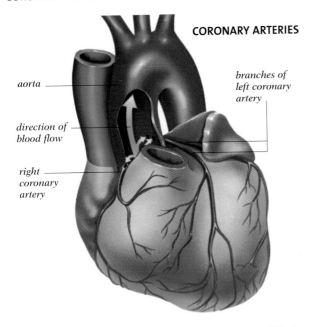

CORONARY ARTERIES

aorta

direction of blood flow

right coronary artery

branches of left coronary artery

◄ *Three main coronary arteries supply the heart muscle with oxygen and nutrients: the two branches of the left coronary artery, and the right coronary artery.*

Treatment

Angina can be a crippling disease, even when the patient has not suffered a heart attack. At his or her worst, the patient may not be able to move more than a few yards without pain. Fortunately modern drug treatments are now available that temporarily widen the coronary arteries.

Patients can carry these drugs with them. When slipped under the tongue or taken by inhaler, they quickly stop attacks; but they are not very good at preventing an attack, because their effect lasts only for a few minutes.

Beta-blockers, however, have been a more lasting treatment for angina since their development in the mid-1960s, a great medical advance. These drugs block some of the effects (the so-called beta effects) of adrenaline. In doing so they also reduce the amount of work the heart has to do and, therefore, its need for oxygen. Taken regularly, not just when there is pain, they reduce the number of angina attacks and help prevent heart attacks.

For people who cannot take beta-blockers, there is an alternative treatment with calcium channel blockers, which interfere with the movement of calcium through cells to help reduce the workload on the heart.

Surgery

There have recently been considerable advances in surgery for the treatment of coronary artery disease. In bypass grafting, the surgeon removes a length of vein from the leg and uses it to connect the diseased blood vessel directly to the aorta so that blood bypasses the

spasm of the coronary arteries rather than actual physical blockage. This type of pain manifests itself at rest rather than during periods of exercise.

This relative lack of blood flow produces pain. The typical angina chest pain spreads to the arms, shoulders, or neck. It is usually brought on by exertion or excitement and lasts only a few minutes. Patients who have angina may develop a full-blown heart attack, and, conversely, patients who have had a heart attack may get attacks of angina.

Symptoms of heart attack and angina		
	HEART ATTACK	**ANGINA**
Type of pain	Dull, crushing, or heavy pain, "a tight band around the chest." Patient often describes the pain by clenching a fist.	May be heavy or dull pain, or may be sharp.
How long the pain lasts	More than half an hour, often much longer.	Minutes only.
What brings it on	May come on at rest or during sleep, but may be precipitated by exertion, excitement, or a heavy meal.	Almost always brought on by exertion or excitement, but also by sudden exposure to extreme cold.
What stops it	Usually nothing. If a heart attack is suspected, and if there are no contraindications, the affected person should immediately chew an aspirin tablet.	Stopping the exertion; nitroglycerin spray or tablet placed under the tongue.
Sweating	Usual.	Rare.
Nausea or vomiting	Usual.	Rare.
Breathlessness	Common.	Uncommon.
Patient's appearance	Often very sick-looking with grayish skin.	Most patients know they have angina and may not give any signs of pain or distress.

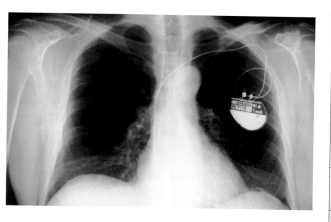

▲ *If the heart's timing system is severely impaired, a pacemaker can be implanted in the chest that sends out an electrical impulse to make the heart beat at the correct pace.*

obstruction. Successful bypass grafting depends on sophisticated surgical techniques. The joins must be able to withstand high pressures, and the blood vessels are only a few hundredths of an inch (a few millimeters) wide. There is no doubt that such surgery can be effective at relieving angina pain. It is also possible to attach an artery from the chest wall to the blocked artery.

Another important procedure to relieve angina is angioplasty. This is done using a special tube called a balloon catheter. The catheter is passed along the coronary artery until the tip, bearing the balloon segment, lies in the narrowed section. The balloon is then inflated with fluid under pressure so that the atheromatous plaque is compressed into the wall and the vessel widened. The results of balloon angioplasty are usually excellent.

Treatment problems

The reason why blood flow in coronary arteries becomes obstructed is still being extensively investigated. But one fact seems clear: blockage nearly always occurs when there is atheroma. Death from heart attacks occurs for two basic reasons. First, the death of an

---

Coronary disease—Who is at risk?

Smoking 10 cigarettes a day doubles the chances of a coronary, since nicotine in the bloodstream causes the arteries to go into spasm, narrowing them and making thrombosis more likely.

People who are more than 20 or 30 percent above the usual weight for their age, height, and sex are two to three times more prone to heart disease than people of average weight.

Anyone who has been under work or family pressures for a long time is at definite risk of a heart attack.

People in desk jobs who do not exercise are certainly more at risk than active people.

People who eat large amounts of cholesterol in dairy foods or animal fat are at risk and should adjust their diet.

People from families in which there is a history of heart disease should be sure to exercise and eat a healthy diet.

---

area of heart muscle, caused by the blocked artery, causes a major disturbance of heart rhythm, which reduces the efficiency of the heart so severely that it may stop working. Second, if too much heart muscle is destroyed, the heart is simply not powerful enough to pump an adequate amount of blood around the body. In contrast, relatively minor disturbances of heart rhythm—known as arrhythmias—can usually be treated with drugs or by giving electrical shocks. If the timing sequence becomes totally interrupted, however, and the heart slows or even stops—a condition known as a heart block—it may be necessary to use a pacing system.

Pacemaker

A wire is passed into a vein and threaded in the direction of the blood flow, until it becomes lodged against the wall of the heart. The other end of the wire is connected to a pacemaker implanted in the chest. This sends out a regular electric impulse that drives the heart at the correct speed.

Recovery

After one or two days in a coronary care unit, heart attack patients usually spend about 10 days to two weeks in the hospital. During this time they gradually regain their strength and resume normal activities as much as possible.

After leaving the hospital and returning home for a period of recuperation, most people are well enough to go back to work within two or three months of a heart attack. In general, patients are encouraged to resume an active and normal life. There is no need for anyone to overprotect the heart or to consider the patient a permanent invalid. In fact, a lack of exercise or activity may probably have been a major cause of the heart attack in the first place.

Preventing coronary disease

Exercise regularly: swimming, walking, and jogging are ideal, although people who have been sedentary for a long time should avoid suddenly starting vigorous exercise. Build up gradually, and if in doubt, ask the doctor's advice. When exercising, think about the exercise, not personal or work problems. If work involves sitting at a desk all day, walk as much as possible instead of driving; take the stairs, not the elevator; and seek medical advice about an exercise program that can be done while sitting.

Eat a sensible diet: cut down on potentially harmful substances such as animal fats. Replace butter with certain types of margarine and use sunflower or canola oil for cooking. Cut down on sugar and starch, and avoid large, heavy meals.

Reduce mental stress: stress is part of living, but the body is not designed to put up with it constantly—so slow down!

See also: Arteries; Circulatory system; Heart; Muscles

Coughing

Coughing is not dangerous, or a disease, in itself. However, it may be an indication of something more serious. A respiratory infection will cause an acute, short-lived cough, whereas smoking will cause chronic coughing.

Questions and Answers

Do cough medicines really work?

Cough medicines do work in that they can suppress a cough, but there is much debate about whether or not they are necessary. The best cough suppressant is one that contains codeine. Specially formulated expectorants will help to dilute and loosen the mucus that has pooled in the respiratory tract, making coughing easier. A hot drink with honey can be similarly soothing.

I coughed up yellow phlegm when I had a bad cold recently. Why?

Phlegm is normally clear mucus and indicates that the secretions of the mucous membrane are normal. If the color changes to either yellow or green, it implies that there is an infection present. Since this cleared up by itself in your case, it could not have been serious. For more serious infections, a visit to the doctor is necessary—he or she may prescribe a course of antibiotics.

Does coughing really spread infection?

Yes. Although coughing and sneezing are reflex responses to outside stimuli, such as dust or gas, they can also transmit germs if a person has a respiratory infection. This is why it is so important to cover your nose or mouth and to avoid coughing or sneezing directly onto anyone.

Why is it that I sometimes can't control my coughing?

Coughing is the body's way of dealing with a foreign body in the upper airways or inflammation in the trachea. It is a reflex action— the messages to and from the brain are extremely rapid and not under voluntary control.

A cough is an explosive current of air driven forcibly from the chest. It forms part of a protective reflex to clear the air passages of any obstruction. Irritation of the upper airways by noxious gases and inflammation by infections cause coughing by a similar mechanism, but in this case the coughing is persistent. Coughing is an essential protective mechanism designed to get rid of potentially harmful substances in the lungs and air passages. Using medicines to suppress a cough may sometimes do more harm than good.

Symptoms

The most important symptom is not the cough itself, but rather the material that is coughed up, the frequency of the coughing, and whether there is any pain. A persistent cough caused by smoking, especially when accompanied by sputum production, indicates chronic bronchitis. The person is in danger of developing COPD, lung cancer, and other disorders if he or she does not quit smoking.

Coughs are not dangerous. However, exhausting coughs—accompanied by hoarseness, chest pains, breathlessness, fever, fatigue, and weight loss—should always be treated by a doctor.

In adults, a dry, persistent cough without any phlegm may be a symptom of pneumonia or heart disease, though an inflammation of the trachea (windpipe) or bronchi (air passages in the lungs) is more likely. If the cough produces phlegm and the phlegm changes from white to yellow or green, this is a sign of infection. In asthma without infection, the phlegm is white and frothy. Bloodstained phlegm may indicate lung cancer, pneumonia, tuberculosis, or even severe bronchitis. Coughing that becomes painful can be a symptom of the development of pleurisy. In children, croup might be the cause of a cough with noisy, labored breathing that is dry and then produces mucus. Coughs that sound like crowing, with heavy phlegm, might be whooping cough.

Treatment and outlook

Diagnosis is based on phlegm color and patient symptoms. Antibiotic drugs may be given to treat certain infections. Bronchodilators are usually used to relieve asthma, and surgery may be necessary in cases of cancer. Stethoscope examination of the chest, possibly followed by an X ray or certain laboratory tests, enables the doctor to determine the cause.

Minor coughs can be treated at home. The best cough suppressant is one that contains codeine. Most other cough medicines are ineffective. Rather than buying expensive expectorants to loosen the cough, it is probably wiser to drink plenty of fluids, such lozenges, and stay in a humid atmosphere.

Simple cough remedies	
SYMPTOMS	REMEDY
Postnasal drip (mucus dropping down from back of nose).	Ephedrine or similar drops, 3 or 4 times a day. See doctor after 5 days.
Inflammation of the back of the throat or larynx.	Inhalations of menthol or eucalyptus vapor, several times a day.
Irritating, throaty cough.	Cough lozenges.
Dry cough or cough interfering with sleep.	Cough syrup or suppressant. Take 2 teaspoonfuls, up to 3 times a day and at night.
Thick, sticky phlegm that will not come up easily.	Expectorant cough syrup, as directed on the bottle.

See also: Chest; Lungs; Mucus; Reflexes

Crying

My relatives keep telling me that I will spoil my son, now two months old, if I pick him up every time he cries. Should I just let him cry? Or is it okay for me to continue picking him up?

At that age, crying is the only way a baby can get love and attention. To ignore it would be distressing for both of you. Small babies need lots of physical comfort and contact, and it is impossible to spoil them during their first few months. As your son grows older, you may decide to let him cry for a while sometimes. Many children develop a "testing cry" at bedtime. But at no age should a child be left to cry for more than five or ten minutes.

My friend and I have babies of similar ages, but mine seems to cry much more easily than hers. Is there any particular reason for this? Or is it natural for some babies to cry more than others do?

Babies cry for various reasons, and one of those reasons may be colic. Although the term and concept may be controversial, the assumption is that the baby is suffering from abdominal pain. The problem seems to wear off by the age of three or more months. On the other hand, some babies just tend to fuss more than other babies for no particular reason, and in this case, there is no need to worry, because the fussing will pass.

Men don't seem to cry as easily as women. Why would there be any difference?

From a physical point of view, crying operates equally in both sexes. Boy babies cry just as much as girls do. But rightly or wrongly, boys are more often encouraged not to cry; thus girls and women appear to cry more easily.

Crying is a spontaneous and necessary expression of human emotion. It is perhaps easier to respond to tears in children, but people of all ages are usually asking for love, understanding, and reassurance when they cry.

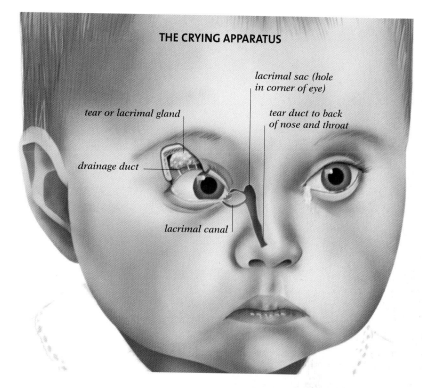

THE CRYING APPARATUS

lacrimal sac (hole in corner of eye)

tear or lacrimal gland

tear duct to back of nose and throat

drainage duct

lacrimal canal

All humans cry—in fact, crying is often the first sound a newborn baby makes. Usually it is a way of expressing grief or pain when words cannot be used; at other times it may be a natural and often involuntary reaction, caused by an emotional state or as a response to pain. In this case the purpose is to release tension.

How crying occurs

Tears, in the form of a watery, salty fluid, are produced from the tear (lacrimal) glands, situated above the outer corner of the eye. This flow, medically known as lacrimation, serves to keep the eye clean and germ-free and to

▲ *For crying, tears from the tear glands, which normally lubricate the eye, drain into the tear ducts. The excess flow of tears pours down the cheeks.*
▶ *Before they are able to show their dissatisfaction with words, children cry to show distress.*

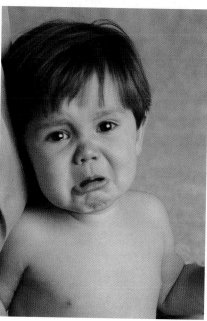

◄▲ Although babies and children are more apt to cry in order to express even slight discomfort or unhappiness, people of all ages and of both genders are likely to cry during times of great pain, stress, grief, or depression.

When a person cries loudly, the face becomes flushed, the forehead wrinkles, the corners of the mouth may turn down, and there is a marked change in breathing. The rate of respiration gets much faster—a deep initial breath can be followed by a series of sobbing or wailing sounds.

The length of time that anyone cries varies. Older children and adults can get their tears under control fairly quickly, but a baby may continue to sob long after he or she has been soothed.

Why babies cry

Infants spend a great deal of time crying, and during the first weeks of life crying is the only way in which they can communicate. Some of the most common causes are hunger, gas, general discomfort and pain, boredom, pain from teething, loneliness, and insecurity.

A sudden change or unexpected noise can also set off crying, as can tiredness or overstimulation. Strangely, a wet or dirty diaper is unlikely to produce much distress, but careless handling or being dressed or undressed can cause tears very easily. Other possible reasons may be the onset of illness such as a cold or an earache.

lubricate the movement of the eyelid over the eyeball. Every time the eyelid blinks, the fluid drains away into small holes in the inner corner of the eye, down the tear duct, and into the back of the nose and throat. But when a person laughs or cries, or if the eye is irritated by a foreign body, such as dust or grit, lacrimation increases, and there is an overflow of tears.

The reasons for this are not fully understood, but the lacrimal glands are controlled by the parasympathetic nerve fibers of the central nervous system. These act automatically, without a conscious decision on a person's part. Under certain emotional influences, these nerves, which in turn are controlled by the brain, stimulate tear production.

Soothing techniques can be used. Some babies like being swaddled tightly in a blanket or being rocked. Others relax to soft music or singing or enjoy being carried around a room. Pacifiers, too, can provide comfort, but they should never replace affection or physical reassurance.

See also: **Eyes; Glands; Nervous system; Teeth**

Diaphragm

The diaphragm is a sheet of muscle and fibrous tissue that forms a complete wall between the chest and the abdomen. Although it is only a thin layer, it is an important part of the anatomy, helping a person to breathe.

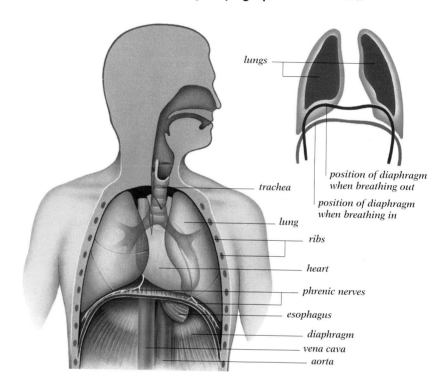

lungs

position of diaphragm when breathing out

position of diaphragm when breathing in

trachea

lung

ribs

heart

phrenic nerves

esophagus

diaphragm

vena cava

aorta

▲ *The diaphragm controls the volume of the lungs as air is inhaled and exhaled.*

The diaphragm forms the bottom of the cage that encloses the heart and lungs, while the ribs form the upper part. Seen from above, the diaphragm would appear as a large central fibrous portion, connected by muscle fibers to the inside of the lower six ribs. This resembles a sun, with rays spreading out toward the rib cage to anchor it. From the front, the diaphragm looks like a dome, attached by muscular strings to the inside of the ribs.

When a breath is taken in, the diaphragm does most of the work. The muscle fibers contract, flattening the diaphragm by drawing the highest central part down into the abdomen, and this flattening increases the volume of the lungs and draws air into the chest through the trachea (windpipe). Simply relaxing the muscles causes air to be breathed out, in much the same way as releasing the contents of a balloon into the air.

Like any other muscle, the diaphragm receives its instructions to contract or relax from the nervous system. The nerves that control it are the left and right phrenic nerves, and these arise from high in the spinal cord and make their way from the neck down to the bottom of the chest. The phrenic nerves can be damaged by injury or disease, but although the diaphragm appears to be an essential part of the anatomy, most people manage to breathe quite adequately if it stops working. People cannot deliberately turn the activity of their diaphragm on and off, and most are unaware of diaphragmatic movements, but it is possible to learn to breathe either mainly from the abdomen or mainly from the chest. Breathing can then be consciously controlled during such activities as singing, playing a musical instrument, or swimming.

See also: Breathing; Muscles; Nervous system; Ribs

Digestive system

Is it true that babies cannot digest cow's milk?

Compared with human breast milk, cow's milk contains large amounts of a protein called casein, which babies do not have the equipment to digest properly. The curds that can be seen in the stools of a baby fed on cow's milk are the result of the incomplete digestion of casein. The many powdered formula milks for babies are made from cow's milk and thus contain more casein than breast milk, but they are specially treated to make the casein more digestible. However, breast-feeding is better for a baby's digestive system, and better in other ways, too.

Why do some foods, such as baked beans, give me gas?

Gases in the alimentary canal, known medically as flatus, form both as a result of swallowing air during eating and by the action of intestinal bacteria on the undigested remains of food.
    The lining of the large intestine absorbs many of these gases, but some foods are an ideal medium for the growth of bacteria. They tend to irritate the intestinal lining, causing food to be moved along quickly, so that there is less time for the gases to be absorbed. If gas is painful or embarrassing, it is best to avoid foods that cause it.

My three-year-old daughter loves spicy foods. Won't they upset her stomach and her digestion?

There is no reason why they should. By the age of three, a child's digestive system can cope with a very mixed diet, and the degree of tolerance to spicy foods will vary, just as it does in adults. As long as these foods do not give her diarrhea or make her vomit, it is much better to give her foods that she likes, rather than having constant battles at mealtimes.

Digestion is the process that breaks down food into substances that the body can absorb and use for energy, growth, and repair. By understanding this process, a person can learn to recognize the signs of any potential problems.

The digestive system depends on the action of substances, called enzymes, on the foods eaten. These enzymes are produced by the organs attached to the alimentary canal, and they are responsible for many of the chemical reactions involved in digestion.

The changes begin in the mouth. When the teeth chew food, the salivary glands beneath the tongue increase their various secretions, one of which, the enzyme ptyalin, starts breaking down some of the carbohydrates into smaller molecules, known as maltose and glucose. The food then travels down the esophagus and into the stomach, where a mixture of chemicals—including mucus, hydrochloric acid, and the enzyme pepsin—pours onto it. Ptyalin stops working, but a new series of chemical reactions begins, triggered by a set of nerve impulses.

Nerve impulses, the presence of food, and the secretion of hormones govern the quantity of juices in the stomach and intestine. The hormone gastrin stimulates the stomach cells to release hydrochloric acid and pepsin once food is in the stomach, so that it can be broken down into substances called peptones. Mucus secretion prevents the acid from damaging the stomach lining. When the acidity reaches a certain point, gastrin production ceases.

The food leaving the stomach—a thickish, acidic liquid called chyme—enters the duodenum, the first part of the small intestine. The duodenum makes and releases large quantities of mucus, which protects it from damage by the acid in the chyme and by other enzymes. The duodenum

HOW A CHEESE SANDWICH IS DIGESTED

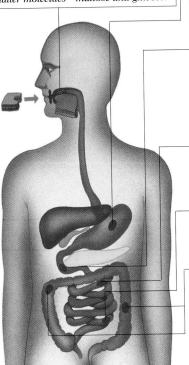

Saliva contains an enzyme called ptyalin, which breaks some carbohydrates into smaller molecules—maltose and glucose.

In the stomach an enzyme called pepsin begins to break protein into smaller molecules—peptones.

The gallbladder releases bile into the duodenum. This breaks fat into small droplets so that an enzyme called lipase can break fat into smaller molecules—glycerol and fatty acids. Lipase is made in the pancreas, as are two other enzymes—trypsin and amylase. Trypsin breaks peptones into smaller molecules—peptides—and amylase breaks carbohydrates into maltose.

In the jejunum and ileum, fat, carbohydrate, and protein are broken into the smallest molecules; peptidases break peptides into amino acids, lipases reduce the remaining fats to glycerol and fatty acids, and other enzymes break down the remaining carbohydrate.

Now the molecules can begin to pass into the capillaries in the villi (small protrusions from the wall of the ileum).

The residual waste matter continues through the colon, where water is taken from it into the bloodstream. This makes the feces semisolid when they are finally expelled from the body through the anus.

◄ *A cheese sandwich contains fat, protein, and carbohydrate, which must be broken down into very small molecules to be absorbed into the bloodstream and used by the body.*

Digestive problems

SYMPTOMS	CAUSES	ACTION
Dry mouth, swelling and pain below ears and in neck particularly after eating (which stimulates salivation).	Lack of saliva due to infection of salivary glands or blockage of the duct carrying saliva to the mouth.	Take plenty of fluids. See a doctor, as drug treatment, and possibly surgery to remove blockage, may be needed.
Pain in back or upper abdomen or chest (heartburn) up to two hours after a meal. Possibly vomiting and loss of appetite.	Excessive secretion of acid, overactivity of stomach, destruction of mucus-secreting cells in stomach; due to stress, anti-inflammatory medicines, autoimmune disorders, smoking, and drinking liquor.	Take antacids for immediate relief. Stop smoking, drinking alcohol, and using anti-inflammatory medicines. See a doctor if the condition persists.
Burning or gnawing pain high in abdomen two to three hours after meal. Is relieved by eating. Vomiting and lack of appetite not unusual.	Most peptic ulcers are caused by H. pylori infection.	See a doctor.
Jaundice, pale feces, possibly pain high in abdomen on the right and in the right shoulder. Symptoms made worse by fatty foods. Possibly fever and vomiting.	Lack of bile secretion due to blockage or inflammation of bile duct or gallbladder. Gallstones or liver disease may also be present.	Take analgesics for immediate pain relief. See a doctor as soon as possible, as drug treatment and possibly surgery may be needed.
Diarrhea, rapid pulse, possibly fever, crampy abdominal pain.	Overproduction of mucus and water by large intestine due to infection or disease; or lack of water absorption due to food passing too quickly down intestine. Commonly due to gastroenteritis.	Take only fluids for 24 hours. If symptoms do not improve in 48 hours, see a doctor in case there is something seriously wrong.
Persistent indigestion, vomiting of blood or dark "coffee grounds" material, weight loss, abdominal pain.	May indicate cancer of the stomach or ulcers.	See a doctor, who will order further tests.

also receives digestive juices from the pancreas and large amounts of bile, which is made in the liver, stored in the gallbladder, and released into the duodenum to break down fat globules.

Two hormones trigger the release of pancreatic juices. The first, secretin, stimulates the production of large quantities of alkaline juices, which neutralize the acidic, partially digested chyme. A second hormone, pancreozymin, causes pancreatic enzymes to be produced, which help the digestion of carbohydrates and proteins, in addition to fats. These enzymes include trypsin, which breaks the peptones into smaller units known as peptides; lipase, which breaks fat down into smaller molecules of glycerol and fatty acids; and amylase, which breaks down carbohydrates into maltose.

The digested food then enters the jejunum and ileum, farther down in the small intestine. Cells in small indentations in the walls of the jejunum and ileum, called the crypts of Lieberkuhn, release more enzymes, causing the final stages of chemical change to take place.

Most food absorption occurs in the ileum, which contains millions of tiny projections called villi on its inner wall. Each villus contains a small blood vessel (capillary) and a tiny, blind-end branch of the lymphatic system called a lacteal. When digested food comes into contact with the villi, the glycerol, fatty acids, and dissolved vitamins enter the lacteals, are carried into the lymphatic system, and then pour into the bloodstream. Amino acids from protein digestion and the sugars from carbohydrates (plus vitamins and important minerals, such as iron, calcium, and iodine) are absorbed directly into the capillaries and transported into the hepatic portal vein, which takes food directly to the liver. The liver filters out substances for its own use and storage, and the remainder pass into the general circulation.

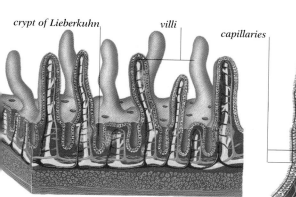

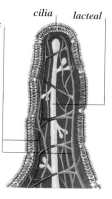

crypt of Lieberkuhn   *villi*   *capillaries*   *cilia*   *lacteal*

Small projections called villi cover the surface of the ileum. These increase surface area, enabling food to be absorbed quickly into the capillaries.

See also: Bile; Duodenum; Enzymes; Hormones; Lymphatic system; Stomach

Duodenum

Questions and Answers

I have heard duodenal ulcers described as an executive disease. Are all executives prone to them?

While it was once thought that stress and diet caused duodenal ulcers, we now know this is not the case. A bacterium called *Helicobacter pylori (H. pylori)* is the major cause of this problem. Emotional stress perhaps plays a role in the process, but certainly not a major one. On the other hand, bodily stress (such as an ICU hospitalization) does increase the risk for the development of duodenal ulcers. Patients in such situations are treated prophylactically with anti-ulcer medications.

Do children get duodenal ulcers?

It is very rare for children to get duodenal ulcers, and when this does happen, there may be predisposing factors involved. Examples would include stress resulting from hospitalization due to severe trauma or perhaps treatment with medications known to interfere with the protective coating of the stomach and duodenum, namely corticosteroids, aspirin, or other anti-inflammatory medicines. The major risk factor by far, though, is infection with *H. pylori*.

My husband has a duodenal ulcer and is frightened that it will lead to cancer eventually. Is there any possibility of this?

Duodenal ulcers do not increase one's risk of developing duodenal cancer. However, if the cause of the ulcer is *H. pylori* infection, there is an increased risk of gastric cancer. Evidence shows that this bacterium is associated with gastritis, gastric ulcers, duodenal ulcers, and gastric cancer. It is unknown how a person contracts this organism, but theories include person-to-person spread or perhaps spread via food and water.

The duodenum, a small section of the alimentary canal that is joined to the lower part of the stomach and curled around the head of the pancreas, plays a vital part in digestion by neutralizing the acid in partially digested food.

The duodenum is a horseshoe-shaped tube about 10 inches (25 cm) long, located in the first part of the small intestine. It is vital in the efficient digestion of food. Two layers of muscle in the wall of the duodenum alternately contract and relax and so help move food along the tube during digestion. Between the muscle layers is the submucosa, containing many glands, called Brunner's glands, which secrete protective mucus. This helps to prevent the duodenum from digesting itself and from being eaten away by the acid mixture arriving from the stomach.

In the duodenum's innermost layer, the mucosa, lie glands that secrete an alkaline juice containing some of the enzymes needed for digestion; this juice also neutralizes stomach acid.

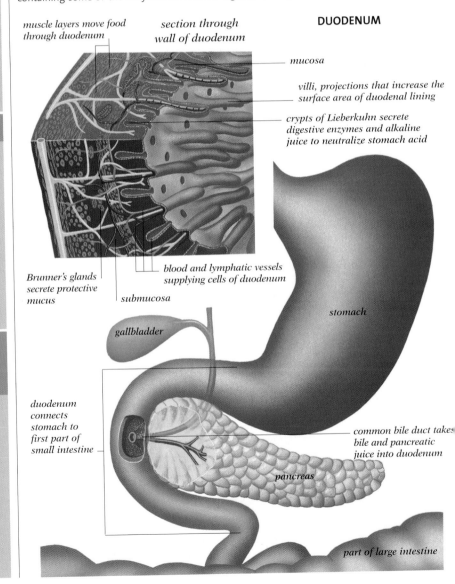

muscle layers move food through duodenum

section through wall of duodenum

DUODENUM

mucosa

villi, projections that increase the surface area of duodenal lining

crypts of Lieberkuhn secrete digestive enzymes and alkaline juice to neutralize stomach acid

Brunner's glands secrete protective mucus

blood and lymphatic vessels supplying cells of duodenum

submucosa

stomach

gallbladder

duodenum connects stomach to first part of small intestine

common bile duct takes bile and pancreatic juice into duodenum

pancreas

part of large intestine

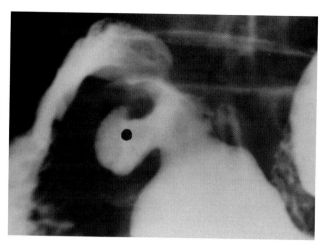

▲ *A light coating of barium makes the pouch caused by this duodenal ulcer (the red dot) clearly visible on an X ray.*

The cells of the mucosa constantly need replacing and multiply faster than any other cells in the body; of every 100 cells, one is replaced every hour throughout life.

Digestion

The partially digested, liquefied food reaching the duodenum contains hydrochloric acid from the stomach. The duodenum neutralizes this acidity by its own secretions, and by the actions of bile and the pancreatic juices that pour into the duodenum from the gallbladder and pancreas, through the common bile duct. These juices continue the process of digestion.

Duodenal ulcers

The duodenum's most common problem is the formation of ulcers—an eating away of small patches of the mucosa. Duodenal ulcers are often called peptic ulcers because they involve the combined corrosive action of hydrochloric acid and the enzyme pepsin. Pepsin breaks down proteins and, with ulcers, it works on the mucosa instead of on food.

Peptic ulcers in the duodenum are usually caused by the overproduction of hydrochloric acid and pepsin in the stomach. They may also be due to lowered mucus production, which can be caused by, among other things, reduced blood supply to the duodenum, or drugs such as aspirin, which are thought to alter the mucus chemically. This will, of course, reduce the degree of protection the mucus gives.

Duodenal ulcers are up to 12 times more common in men than women and typically occur between the ages of 30 and 40. Women tend to develop ulcers later in life, often between the ages of 45 and 60.

It is now known that 95 percent of people with duodenal ulcers are infected with the spiral-shaped bacterium *Helicobacter pylori*. It has also been established that if this infection is eradicated duodenal ulcers are very unlikely to recur. Surprisingly, about 60 percent of all adults worldwide carry this germ, but only 15 percent develop duodenal ulcers. One explanation is that the bacteria are known to vary widely in their virulence.

There is evidence, although it is not clear-cut, that duodenal ulcers run in families, and strong statistical proof that people whose blood type is O are more likely to be affected than those with A, B, or AB blood. Stress may play a role in the development of duodenal ulcers, albeit not the major role once thought. (A bacterium called *Helicobacter pylori* is now the major cause.) Both anxiety and bad eating habits tend to increase the amount of acid and pepsin produced in the stomach at times when there is no food present for these substances to work on. Cigarette smoking is another important causal factor of duodenal ulcer.

Heavy drinking is a more likely cause of duodenal ulcers than hot, spicy food. Alcohol, especially in large amounts, is known to impair the digestive process; conversely, people with ulcer-prone personalities tend to drink to relax and to reduce the effects of stress.

Symptoms and treatment

The symptoms of a duodenal ulcer include a pinching pain in the abdomen, or sometimes in the back—its precise position varies with the position of the ulcer. The pain, accompanied by a feeling of fullness and discomfort in the abdomen, occurs between meals, usually between two and three hours after eating. It can be eased by eating and by alkaline indigestion medicines. There is rarely a loss of appetite. Sometimes the first sign of an ulcer is the sudden vomiting of blood or the appearance of black feces, black indicating the presence of blood.

Most duodenal ulcers are now diagnosed by an endoscopy—the insertion of a flexible, lighted tube into the duodenum so that the ulcer can actually be seen. Endoscopy of this sort involves an injection to make the patient sleepy, and so a short stay in the hospital is necessary.

It is currently recommended that all patients with duodenal ulcers should have *H. pylori* infections eradicated. Treatment for eradication is 90-percent effective and reinfection occurs in only about 1 percent of cases. The *Helicobacter* is best eradicated with a one-week triple therapy using two antibiotics such as clarithromycin and metronidazole in combination with one of the recent proton pump inhibitor antacid drugs such as omeprazole. The results of such a regimen are excellent. Bismuth chelate is also widely used. The earlier antacid drugs have now all been superceded by proton pump inhibitors or H2-receptor antagonists such as ranitidine.

As far as diet is concerned, the patient should try to eat small, regular meals frequently. The only foods that should be avoided are highly spiced foods, large amounts of alcohol, and anything known to have given the patient indigestion in the past. The patient is also advised to stop smoking.

Other problems

Other problems of the duodenum occur if the pancreas or gallbladder, whose secretions are essential to digestion, are not working properly. Problems can also occur if the duodenal lining is irritated by the toxins produced by food-poisoning bacteria; this usually produces vomiting and abdominal pains. When the duodenum is irritated in this way, its muscles contract while those of the stomach relax. As a result, partially digested food that has already passed through the stomach is pushed back again, and this may trigger vomiting. As for all intestinal infections, the best treatment is to take nothing but fluids for 24 hours. If the condition has not improved within a couple of days, the patient should see a doctor.

See also: Abdomen; Bile; Blood groups; Cells and chromosomes; Digestive system; Enzymes; Mucus; Pancreas; Stomach

Earwax

Is it harmful to use cotton swabs to clean the inside of my ears?

Cleaning the inside of the ear is unnecessary because it has its own self-cleaning mechanism. What's more, cotton swabs can cause minute skin abrasions, exposing the deeper layers to bacteria. You also interfere with the natural cleaning process by pushing wax deeper into the ear canal, resulting in blockage and deafness. Should a doctor need to examine your ears, this impacted wax will block his or her view of the eardrum and delay diagnosis.

Can earwax cause deafness?

Yes, earwax can cause temporary deafness. A small amount does no harm, but if the canal is blocked completely over a period of time, or if the wax swells suddenly because water enters the ear canal, moderate hearing loss, discomfort, and pain may result.

Do some people's ears have more wax than others?

Yes. Some people are more likely to have a buildup of excess earwax, especially those who work in a dusty environment or who have eczema of the ear skin, profuse hair growth in the ears, or infections of the ear canal skin.

Does ear syringing hurt?

Ear syringing carried out by skilled medical personnel should be pain-free, but no one would say that it was an enjoyable experience.

Is it safe to syringe your own ears?

No. If you find that your ears are blocked, consult your doctor. Hearing is precious, and any attempt to remove wax by yourself could be dangerous.

Earwax has a useful and important function in the human body: to provide a barrier against infection. However, wax sometimes overaccumulates in the ear canal, and medical help is then required to remove it.

Earwax, or cerumen, consists of a mixture of oily secretions from the modified sweat glands situated in the outer third of the ear canal, scales from the skin, and dust particles. It is sticky and water-resistant and forms a natural barrier against infection.

In normal circumstances wax does not accumulate in the ears, because it is continually being pushed outward by the movement of the jawbones and the natural shedding of the skin. However, some people produce too much wax. This may be because of an inflammation of the skin or scalp, excess hair in the ears, or employment in occupations in which they encounter a lot of dust.

Accumulated wax may cause a variety of symptoms. The wax may become impacted at the narrowest part of the ear canal by unskilled attempts to remove it with matches, hairpins, cotton swabs, or other implements. This accumulation may cause irritation and noises in the ear, but it is rarely painful. Although harmless, however, earwax may occasionally accumulate to a point where temporary deafness occurs. Medical treatment should be sought when this happens.

A person may experience a sudden deafness, together with a feeling of pressure, after he or she takes a swim or shower; this is because water has entered the ear and caused the wax to swell. Any attempts to clear the ears will only tend to push the wax deeper into the ear canal, causing pain, noises in the ear, or, more rarely, dizziness.

SYRINGING THE EAR

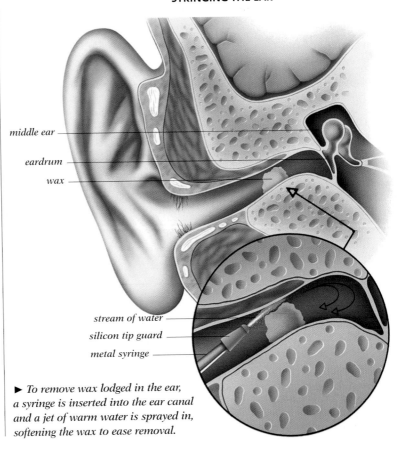

middle ear

eardrum

wax

stream of water

silicon tip guard

metal syringe

▶ *To remove wax lodged in the ear, a syringe is inserted into the ear canal and a jet of warm water is sprayed in, softening the wax to ease removal.*

General points

Never insert hard objects like safety pins, matches, or paper clips into the ears to clean them.

If ears are inflamed or the skin is sensitive, do not allow water to enter the ears. When showering, swimming, or washing the hair, protect the ears by placing a plug of cotton covered with petroleum jelly in the opening of the ear canal.

If dry skin is causing itching in the ears, a few drops of warm olive oil once a week helps to keep the skin soft and moist and will ease the itching.

Always seek medical attention for an ear complaint.

Treatment

Impacted earwax is one of the most common conditions seen in a doctor's office. A doctor can remove the wax safely by picking it out with very fine forceps or by a blunt hook using an otoscope (a viewing instrument that illuminates and magnifies the inner ear).

Alternatively, the wax can be removed by a syringe. This is a safe and painless procedure. Before syringing, the doctor may advise the patient to put drops, such as bicarbonate of soda in solution, warm olive oil, or a wax solvent, in the ears to soften the wax and make it easier to syringe. In some cases such preparations, available in drugstores, may themselves be enough to treat the problem.

Where syringing is still necessary, a jet of warm water is forced into the ear canal, without touching the skin with the metal tip of

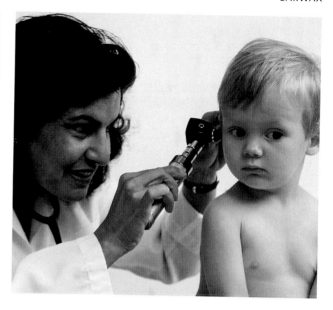

▲ *A pediatrician examines a child's ear. Wax, a foreign body, or debris may either be removed with a fine forceps or aspirated by a suction tube attached to a vacuum pump. The experience should be free from pain and discomfort.*

the syringe. The ear is then carefully and thoroughly cleansed and dried. Relief from deafness in the patient is immediate and dramatic. Special care and gentleness are required in syringing a child's ear. If the child is reluctant or uncooperative, it may be better for him or her to have the wax removed under anesthesia.

If the patient has a previous history of ear trouble or has had surgery on his or her ears, there may be perforation of the eardrum. The normal eardrum is not easily ruptured, but where there has been a perforation that has healed, the scar tissue is vulnerable to injury and great care must be taken during syringing.

Where wax has accumulated in large amounts within a person's ear and become solid, it should be removed by an otolaryngologist. The procedure is carried out by a suction machine, together with an operating microscope that provides high magnification and powerful illumination. The patient lies on his or her back on a couch, with the head turned to one side. The otolaryngologist, using a metal or plastic speculum, examines the ear with a microscope. Wax, any foreign body, or debris can then be removed with fine forceps. If there is a discharge of soft wax, then this can be aspirated (drawn out) using a suction tube attached to a vacuum pump. Suction clearance is very safe in experienced hands and is a most satisfactory way of clearing accumulated wax from the ears.

That people should never attempt to clear the wax out of their own ears cannot be overstated. Far too much damage is done by people poking about in their own ears with matchsticks, hairpins, and cotton swabs. In particular, these objects should never be used on babies or children. If people cannot remove what they see at the opening of the ear and wipe it away easily, they should leave it alone.

See also: Glands; Hearing

Care of the ears

DAILY CARE

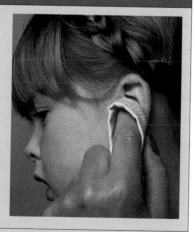

Wash the outer part of the ears with soap and water, and dry them gently with a soft towel. Do not rub the skin.

Should water enter the ears, hold the earlobe and shake it gently after turning the head sideways. Usually the water will pour out.

If water is trapped in an ear and the wax expands and causes discomfort, do not try to remove it. Consult a doctor.

If wax removal is necessary, the doctor is the expert. One specialist warned: "Never put anything smaller than your elbow in your ears."

Elbow

Why do you get such an odd sensation in your elbow when you bang your funny bone?

One of the nerves of the arm, the ulna nerve, lies in a groove of bone near the tip of the elbow. When the elbow is hit, the nerve becomes pinched in the groove, generating the tingling, painful sensation that gives the funny bone its name.

I injured my shoulder recently and had difficulty moving my elbow. Why?

Many of the elbow's movements depend on the actions of muscles that are attached at one end to the shoulder and at the other end to the bones of the forearm. Because these muscles are involved in elbow movement, any injury to the shoulder can affect the elbow, too.

My elbow joints seem rather weak. Are there any exercises I can do to help strengthen them?

As long as your joints are not painful—if they are painful, you should see your doctor—a few simple push-ups each day will help strengthen your elbows by giving added power to the muscles that move the joints in the elbows.

What is the best kind of first aid to give for elbow injuries? Should you use some type of sling?

If someone injures an elbow in an accident, first try bending his or her elbow very gently. If the person feels any pain when this is done, the arm should not be put in a sling, since bending the elbow in this position might do further damage to the joint. To keep it immobile, straighten the arm and tie it to the body with bandages or strips of material, placed above and below the elbow. Then take the patient to the hospital emergency room as quickly as you can.

The elbow plays a vital, complex part in a whole range of intricate movements of the arm and hand, yet its structure is based on two simple joints: the hinge and the ball-and-socket. Any injury to the elbow can be painful and debilitating.

The elbow is the meeting place for the ends of the three main bones in the arm: the humerus (the single bone in the upper arm), the radius, and the ulna (the two bones of the forearm). These three bones are connected in two joints, each making a different range of limb movements possible.

Between the ulna, which is the longer of the two forearm bones, and the humerus bone of the upper arm, there is a hinge joint that allows the elbow to be bent up and down. In this joint, a rollerlike extension called the trochlea, at the lower end of the humerus, moves within a hollow near the head of the ulna. Beyond this hollow is a projection of bone, the olecranon, which forms the point of the elbow and prevents the elbow from being overstraightened.

STRUCTURE OF THE ELBOW

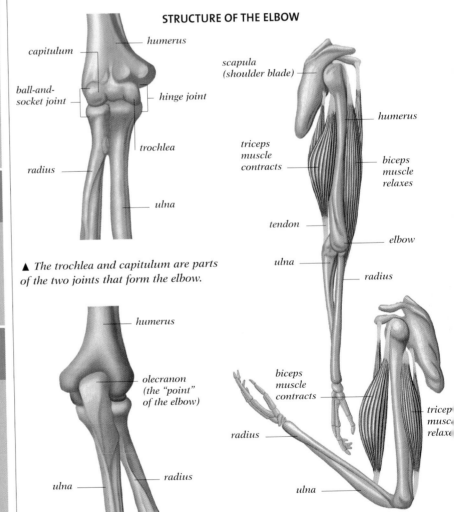

▲ The trochlea and capitulum are parts of the two joints that form the elbow.

▲ The olecranon, the bony projection beyond the hollow at the head of the ulna, forms the point of the elbow.

▲ The biceps muscle runs from the shoulder to the top of the radius, the triceps from the shoulder to the humerus and ulna.

▶ *Bursitis of the elbow results from excessive leaning, a direct blow, or a fall on the tip of the elbow that causes inflammation of the bursas protecting the tendons that join muscle to bone.*

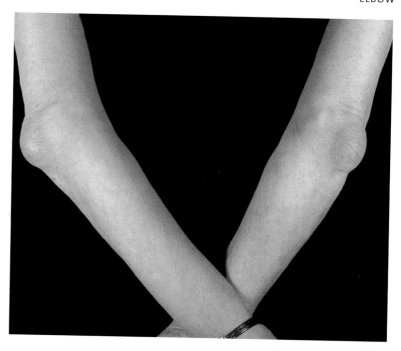

Between the radius and the humerus is the ball-and-socket joint, the socket lying on the radius, the ball formed by a spherical projection of bone on the outer end of the humerus, the capitulum. This second joint is also used in bending movements, but its main function is to make rotation of the elbow possible so that the palm of the hand faces outward, although the extent of the movement is rather limited.

Movement

During movements of the elbow, the radius and the ulna always move together, because they are joined just below the elbow itself and at the wrist. When the palm is facing outward, the two bones lie parallel to one another, but when the back of the hand is uppermost—its usual position—the bones are crossed over one another.

If the arm is held straight down and the elbow is then turned so that the palm of the hand faces outward, the part of the arm below the elbow automatically moves outward at an angle from the side of the body. This angle, formed by the positioning of the bones in the elbow, is known as the carrying angle and is vital to prevent the arms from banging into the sides of the body when it is in motion. The angle also helps to increase the precision with which a tool can be held in the hand while the arm is being straightened out. The carrying angle is larger in women than in men; women have wider hips than men.

To allow smooth movement, the ends of all bones that meet in the elbow are covered with translucent hyaline cartilage (connective tissue) that provides a low-friction surface. The synovial membrane, the lining of the elbow, secretes a clear liquid called synovial fluid that lubricates the whole joint. Covering the membrane is a capsule of fibrous tissue; surrounding the elbow, and between the tissue and the membrane, are several small pads of fat that help to cushion the joint against injury.

If it were not well supported, the elbow would dislocate every time the arm moved. Strong ligaments joining the humerus with the radius and the ulna keep it in place and bind the latter two bones together.

Movement of the elbow is made possible by various muscles, joined to the bones by tough, sinewy tendons. These are protected by fluid-filled sacs or bursas, which lie between tendon and bone and prevent friction. The bursas are similar in structure to the capsule joint of the whole elbow, and fluid also flows between them.

Injuries

The most common wear-and-tear injury is called tennis elbow. This is caused by tiny tears in part of the tendon and in muscle coverings. Like other joints, the elbow may be affected by osteoarthritis (degeneration of the cartilage lining the joint, and of the bones themselves); it is most common in people who have put great strain on their elbows in the course of a lifetime. For example, men who have regularly used pneumatic drills are particularly prone to osteoarthritis of the elbow.

Damage to the elbow is likely to result from falls or other accidents. It may involve tearing or rupturing the tendons, ligaments, or muscles;

dislocating the bones, so that they are moved out of their correct positions; or fracturing the ends of the bones that form the joint.

Since the injured elbow may have to be manipulated to get the bones back into their proper positions, all such injuries need prompt medical attention. If the ends of the bones have been broken, surgery may be required to pin the bones back into place. Whatever the injury, it is usual for the elbow to be immobilized in a plaster of paris cast until it has healed. This is followed by physiotherapy to help get the joint mobile again.

Even with expert medical attention, elbow injuries often result in permanent stiffness of the joint. One cause of this is that, for some unknown reason, injury to the internal tissues of the elbow may be followed by an abnormal growth of bony tissue within the injured muscle, at the site of the injury. This growth, called myositis ossificans, is difficult to treat; if it is removed surgically, the surgery may stimulate the abnormal bone growth again, and it may reform.

In children, a fracture of the elbow is serious. The break, which occurs near the top of the humerus, is usually the result of a fall onto outstretched hands. The danger of the injury is that it may cut off the blood supply to the forearm and also damage the nerves of the arm.

Treatment

If there is severe swelling and the injured elbow cannot be moved, the patient should be taken immediately to the hospital emergency room for treatment. Even if the elbow injury does not seem serious, the pulse in the wrist should be checked, and if it is weak or absent the patient should be taken to the hospital at once. If there is any chance that the circulation in the arm has been impeded, it will be necessary for the patient to stay in the hospital until the surgeon is absolutely sure that all is well.

> *See also:* **Bones; Cartilage; Hand; Hip; Joints; Ligaments; Muscles; Shoulder; Tendons; Wrist**

Endocrine system

Questions and Answers

My teenage sister is very fat. Are her glands faulty?

Probably not. It could be that she is simply eating too much or is not active enough. Several hormone conditions can cause obesity, including overactive adrenals or an underactive thyroid. Weight gain is not a common feature in the few young people who get this thyroid disease. An overactive adrenal gland is a rare condition and is always accompanied by other signs.

My friend's doctor advised her that it may be a good idea for her to begin hormone replacement therapy. What do you think?

Current research indicates that hormone replacement therapy (HRT), usually given to women after menopause, is not as effective as had previously been thought for treating postmenopausal symptoms. Although HRT has been found to reduce hot flashes in women in the first few years after menopause and may reduce the risk of developing osteoporosis, it is now thought to increase the risk of cardiovascular disease and stroke. Women considering HRT should consult with their doctor, who should treat each case individually.

I am an 18-year-old female and hair has started growing on my face and around my nipples. Can anything help?

The growth of hair on the face and body is a response to male hormones. In most cases there is no definite abnormality in the hormone balance, since it is normal and desirable for girls to have small quantities of male hormones circulating in addition to female ones. If hairiness is excessive, however, there may be a definite hormone imbalance that can be treated medically.

The body is like a finely tuned musical instrument, with the glands of the endocrine system keeping all the parts working in harmony with each other. The pituitary gland plays a particularly important role.

Many of the vital functions of the body are controlled by the endocrine system, which consists of glands that secrete hormones (chemical messengers) into the blood. Hormones are concerned with metabolism (the many processes that keep the body functioning) and tend to interact. They are are responsible for balancing the levels of basic chemicals such as salt, calcium, and sugar in the blood.

MAJOR HORMONES SECRETED BY THE ENDOCRINE SYSTEM

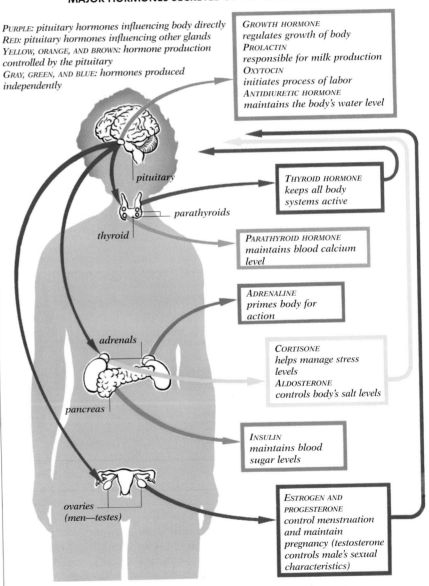

PURPLE: *pituitary hormones influencing body directly*
RED: *pituitary hormones influencing other glands*
YELLOW, ORANGE, AND BROWN: *hormone production controlled by the pituitary*
GRAY, GREEN, AND BLUE: *hormones produced independently*

GROWTH HORMONE
regulates growth of body
PROLACTIN
responsible for milk production
OXYTOCIN
initiates process of labor
ANTIDIURETIC HORMONE
maintains the body's water level

THYROID HORMONE
keeps all body systems active

PARATHYROID HORMONE
maintains blood calcium level

ADRENALINE
primes body for action

CORTISONE
helps manage stress levels
ALDOSTERONE
controls body's salt levels

INSULIN
maintains blood sugar levels

ESTROGEN AND PROGESTERONE
control menstruation and maintain pregnancy (testosterone controls male's sexual characteristics)

pituitary

parathyroids

thyroid

adrenals

pancreas

ovaries
(men—testes)

▲ *The pituitary gland controls its own hormone level, in addition to those of other glands*

FEEDBACK MECHANISM

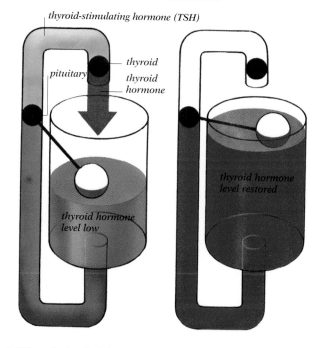

thyroid-stimulating hormone (TSH)

thyroid

pituitary

thyroid hormone

thyroid hormone level restored

thyroid hormone level low

▲ *When the level of thyroid hormone is low (left), the pituitary gland secretes TSH (thyroid-stimulating hormone), which sets off its production. When there is enough thyroid hormone (right), the pituitary stops producing TSH.*

They are also essential to the reproductive system, controlling sperm production in men, and ovulation, menstruation, pregnancy, and milk production in women. They are also critical to the growth of children and to development during puberty.

Other glands, such as the sweat glands in the skin and the salivary glands in the mouth, open onto an organ's surface and are known as exocrine glands. The pancreas has both endocrine and exocrine activity, since it secretes hormones into the blood but also produces alkali and digestive substances that are secreted directly into the intestine.

Overall control

A small gland in the base of the skull, called the pituitary, controls much of the hormone system. It acts as a conductor, secreting hormones that turn the other glands on and off. The pituitary interacts with three important groups of glands. The first, the thyroid, is situated in the neck just below the voice box. The pituitary also directs the activities of the two adrenal glands—which are found in the abdomen lying just on top of the kidneys—and affects the sex glands, the two ovaries in a woman and the two testes in a man.

The pituitary gland exerts its control in a simple process called feedback. An example of this occurs in the thyroid gland. When the gland is stimulated by TSH (thyroid-stimulating hormone) made by the pituitary, it produces the thyroid hormone. When this hormone's level in the blood rises, the pituitary stops secreting TSH. But when the level starts to fall again, TSH is produced once more. In this way, a relatively constant level of thyroid hormone is maintained. It may be raised or lowered by the pituitary, which is itself controlled by the part of the brain immediately above it, the hypothalamus.

The pituitary also controls the adrenal cortex (part of the adrenal gland), where it stimulates cortisone production with ACTH (adrenocorticotropic hormone). It controls the action of the ovaries with FSH (follicle-stimulating hormone) and LH (luteinizing hormone) and that of the testes with ICSH (interstitial cell-stimulating hormone). The balance of the ovarian hormones, estrogen and progesterone, affects a woman's menstrual periods.

As well as controlling other endocrine glands, the pituitary secretes several other important hormones, including the growth hormone, which is essential for normal development in children, and prolactin, which plays a role in the production of breast milk.

All these hormones come from the front part of the gland, or anterior pituitary. The back, or posterior pituitary, secretes only two hormones: oxytocin, which causes the womb to contract in labor, and ADH (antidiuretic hormone), which is concerned with maintaining the correct amount of water in the body.

Acting independently

Various other glands and their hormones act more or less independently of the pituitary. Perhaps the most important of these is the pancreas, which secretes insulin. Insulin controls the sugar level in the blood; a lack of insulin causes diabetes.

Also important are the four tiny parathyroid glands, each about the size of a pea, that lie behind the thyroid gland. These control the level of calcium in the blood, which is essential to good health.

Finally there are the hormones secreted by the walls of the gut and by the pancreas, which direct the processes of digestion. Many of these gut hormones have been discovered only recently, and a complicated system of hormone-controlled mechanisms affect such things as acid production by the stomach, alkali secretion into the intestine by the pancreas, and bile excretion by the gallbladder.

Scientists are still investigating the precise way in which the various gut hormones work. Perhaps one of the best-known and best-understood is the hormone gastrin, which is secreted by the pylorus and causes acid to be produced in the stomach.

Problems and their treatment

The endocrine system can go wrong in many ways, but luckily most endocrine diseases can be cured or at least alleviated. The two main problems are over- and undersecretion of hormones.

Excess hormone production is probably commonest in the thyroid gland, but the reasons for this are still being researched. In other glands, excess production of hormones is usually the result of a tumor. Since these tumors are usually benign (not cancerous), surgery is often successful in curing these endocrine conditions.

Underactivity can be dealt with by giving replacement doses of the relevant hormone by mouth. This is particularly successful in failure of the adrenal glands (Addison's disease).

When the pituitary gland causes trouble, the situation can become complicated, since the failure of hormone secretion leads to a failure of the thyroid and adrenal glands. Nowadays, however, medical investigation is more straightforward than it was in the past, although many blood tests may be required to measure the levels of the various hormones in the body.

See also: Adrenal glands; Bile; Birth; Gallbladder; Glands; Hormones; Insulin; Metabolism; Pancreas; Pituitary gland; Pregnancy; Testes; Thyroid

Enzymes

Enzymes are proteins that speed up reactions in the body. They are known as catalysts, and they make possible almost all of the body's vital processes, including digestion, growth, and reproduction.

Questions and Answers

I have stomach ulcers. Are they affected by enzymes?

Yes. When food enters the stomach, the stomach wall secretes two substances—pepsinogen and hydrochloric acid. Together they form an enzyme called pepsin that helps digest food. Ulcers are caused when pepsin begins to digest the stomach wall after excessive acid secretions have destroyed the protective lining of mucus. Stress and smoking are thought to be largely responsible for increasing acid secretion. Stomach ulcers can usually be treated by taking medicines, by not smoking, and by avoiding alcohol consumption. Aggravating medicines should be stopped. In some cases surgery may be necessary.

My daughter has an enzyme deficiency called phenylketonuria and follows a special diet. Will all of our children be affected?

Any further children you and your husband have will have a one-in-four chance of being affected. Hospitals screen for the deficiency at birth, and if the test is positive, the child needs to be put on a diet that cuts out proteins containing the substance phenylalanine. However, consult your doctor before you decide to have another child or if you become pregnant again.

I use biological detergents for washing clothes. Can the enzymes they contain damage my hands?

Some detergents contain protease enzymes, which help dissolve protein stains, such as blood or egg. If you hand-wash clothes, the detergent removes oil from the skin. The enzymes can attack skin proteins and cause irritation or damage. Always wear rubber gloves and rinse clothes and hands after using such detergents.

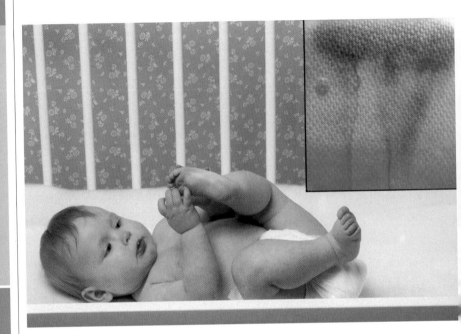

▲ *Enzymes activate processes that create and sustain life. Here (inset), it is possible to see their action in a biological detergent, as the detergent digests a bloodstain.*

Enzymes accelerate chemical reactions that allow any living organisms, including humans, to function normally. Enzymes speed up some chemical reactions several thousandfold. Thousands of varieties of enzymes are found throughout the body; each promotes a different chemical reaction. The majority of genes in our DNA genome function to provide the genetic code for the synthesis of enzymes. That process demonstrates the importance of enzymes in the functioning of the human body. Hereditary diseases are the result of gene mutations that cause defective enzymes. Each cell in the body makes the enzymes it needs to function. Some enzymes are so important that they are produced by every cell; they are usually used within the cell itself. Others are produced only by specialized cells; they may be secreted and used outside the cell. For example, enzymes involved in the digestive process act outside cells, in the gut itself. Some substances can deactivate enzymes by occupying their working areas. Cyanide has this effect on an enzyme called cytochrome oxidase, which is involved in the production of energy, and small quantities of cyanide can quickly cause death. Many disorders are caused by enzyme deficiencies. Albinos lack an enzyme called tyrosinase, which helps make the skin pigment melanin. Other enzymes convert the amino acid tyrosine to thyroid hormone: if one of the enzymes is missing at birth, the child will be mentally retarded. Enzyme deficiency disorders can be traced back to defects in genes. Problems can be alleviated if the missing enzyme is replaced.

Many prescribed drugs work by altering enzyme activity. Aspirin inhibits the activity of cyclooxygenase, an enzyme involved in producing prostaglandins, which are thought to cause painful inflammation of the joints, so aspirin is a useful treatment for disorders such as rheumatoid arthritis.

See also: **Melanin; Thyroid**

Esophagus

Questions and Answers

If I drink something that is too hot, can I burn my esophagus?

In theory, yes, you can; in practice it rarely happens. The nerve endings in your lips, your tongue, and the lining your mouth would have warned you that the liquid was painfully hot and you would have stopped drinking and spat out any of the drink already in your mouth before it was swallowed.

If my son accidentally swallows something sharp, such as a piece of broken glass, what damage could he do to his esophagus?

Usually, amazingly little damage is done. Sometimes the glass scratches the lining of the esophagus and causes a little bleeding, but seldom more than that. If something like this does happen, however, it would be wise to let your doctor know. It may be necessary to follow the foreign body's progress through the gut, possibly by taking X rays, until it is excreted.

Can anything be done to treat cancer of the esophagus?

Yes, a great deal. Many people now survive for years after being successfully treated for cancer of the esophagus. It is important that the diagnosis is made at an early stage when a curative operation is usually possible.
   If the cancer is in the lower part of the esophagus, an operation will be carried out to remove the cancerous section, bringing up the stomach to be joined with the cut end. Growths in the upper part of the esophagus used to be much more difficult to deal with, because there is not much room to maneuver. It is sometimes possible, however, to transplant part of the colon to act as a replacement for the diseased section of the esophagus.

Most food is swallowed so easily that we are unaware of the esophagus. It is only when we eat something too large, too hot, or too cold that we notice this vital link between mouth and stomach.

The esophagus or gullet is the tube that connects the back of the mouth to the stomach. Its only function is to carry food from the mouth to the stomach, where the food is broken down into chyme by various digestive processes and passed into the duodenum.

Structure

The esophagus is an elastic tube about 10 inches (25 cm) in length and about 1 inch (2.5 cm) in diameter. It extends from the back of the throat (the pharynx)—where it lies immediately behind the windpipe (trachea)—down through the diaphragm and into the upper part of the stomach. The top and bottom ends of the esophagus are each controlled by a strong muscular ring, called a sphincter, that can open and close to allow the passage of food.

▲ *This X ray shows how The Amazing Stromboli swallowed a sword. He kept his throat open to allow the blade into his esophagus without swallowing.*

HIATUS HERNIA AND THE ESOPHAGUS

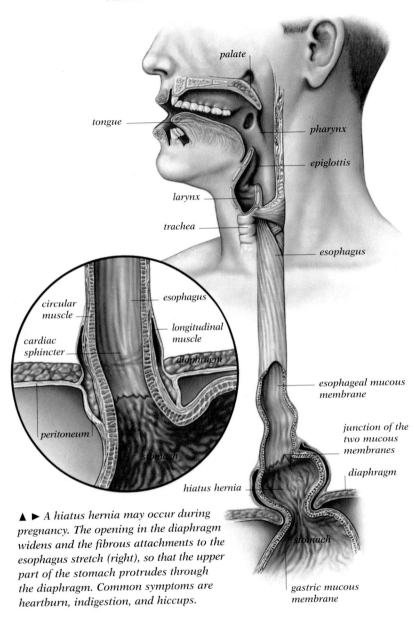

palate

tongue

pharynx

epiglottis

larynx

trachea

esophagus

circular muscle

esophagus

longitudinal muscle

cardiac sphincter

diaphragm

esophageal mucous membrane

peritoneum

junction of the two mucous membranes

stomach

diaphragm

hiatus hernia

stomach

gastric mucous membrane

▲ ▶ *A hiatus hernia may occur during pregnancy. The opening in the diaphragm widens and the fibrous attachments to the esophagus stretch (right), so that the upper part of the stomach protrudes through the diaphragm. Common symptoms are heartburn, indigestion, and hiccups.*

same time, other throat and face muscles raise the tongue up against the roof of the mouth so that food does not pass back into the mouth; these muscles also move the palate upward to prevent food from entering the space at the back of the nose, and close the epiglottis (a flap of cartilage) over the raised larynx so that food cannot enter the windpipe (trachea) and lungs.

Occasionally the epiglottis does not close in time and food or liquid does get into the larynx—a sensation that people often refer to as something "going down the wrong way." When this occurs the swallowed substance is immediately expelled by forceful coughing—a reflex action that is not under the person's conscious control.

The first part of swallowing is a voluntary act that is consciously controlled. Once the food has passed the back of the tongue, however, the continuation of the act of swallowing is an involuntary, automatic act that cannot be stopped consciously.

The bolus of food does not just slide down the esophagus into the stomach. It is actively pushed down by a series of wavelike muscular contractions of the esophagus wall, a process known as peristalsis. The passage of food is, therefore, an active process and not just a passive mechanism that depends on gravity. People could eat and drink just as well standing on their heads as sitting down.

Common problems

A surprising variety of things can pass down the esophagus without damaging it, and people have even been known to swallow broken glass without being harmed. However, if a large object is swallowed the esophagus may become blocked, for example when a child swallows a small toy. If this happens any swallowed food is brought back up rapidly, and the person will retch persistently and feel pain behind the breastbone. Surgery may be required to remove the blockage. Similarly, if a person swallows a corrosive substance, such as a strong acid or a strong alkali such as bleach, the esophagus can be seriously damaged and surgery may be required to repair it.

Inflammation of the esophagus (esophagitis) is one of the most common problems in this area. It is usually caused by acid passing up into the esophagus from the stomach, resulting in a burning sensation behind the breastbone after eating. The condition, which is often called heartburn, occurs frequently during pregnancy, and is a common symptom of a hiatus hernia (see diagram). The treatment for esophagitis normally depends on antacids and drugs to reduce acid production in the stomach.

Like the rest of the alimentary canal, the wall of the esophagus is made up of four layers: a lining of mucous membrane to enable food to pass down easily, a submucous layer to hold the mucous layer in place, a relatively thick layer of muscle consisting of both circular and longitudinal fibers, and an outer protective covering.

Swallowing

When food enters the mouth, it is chewed by the teeth and lubricated by saliva until it forms a smooth, slippery mass called a bolus. The act of swallowing then enables the bolus to enter the esophagus. Swallowing is a complex activity involving several groups of muscles. The muscles of the throat, or pharynx, contract, forcing the food toward the upper end of the esophagus. At the

See also: **Alimentary canal; Coughing; Digestive system; Muscles; Pharynx; Stomach**

Estrogen

Estrogen is a general term for a group of female sex hormones. The hormones in this group have slightly different names (such as estrone or estrol), and they perform essential functions in a woman's reproductive life.

Estrogens first achieve prominence in a woman's life at puberty. About four years before a girl has her first period the hypothalamus in the brain begins to secrete substances called releasing factors. These act on the pituitary gland and stimulate it to produce the hormones responsible for a girl's sexual development. One is a growth hormone; two others—follicle-stimulating hormone (FSH) and luteinizing hormone (LH)—are responsible for controlling the various changes in the monthly menstrual cycle.

Follicle-stimulating hormone

FSH has the effect of stimulating the growth of egg follicles (Graafian follicles) to maturity in the ovary. Millions of follicles are present at birth. Only a few follicles grow at first, but as they do, the layers of cells surrounding them begin to secrete estrogen. The follicles produce estrogen for about a month and then fade. Each month, however, more and more follicles are stimulated by FSH until eventually between 12 and 20 become active at a time. Thus there is a gradual increase during puberty in the amount of estrogen that stimulates follicles to mature, causing the release of an egg or eggs every month (ovulation).

I have been taking the Pill for years. Will it upset my hormones?

In some women the Pill does upset normal estrogen and progesterone levels, and it is probably responsible for the side effects, such as cramps, mood changes, headaches, and fluid retention, which some women experience. These side effects usually disappear when the Pill is discontinued. In a few women the natural cycle is disturbed so much that periods cease and it is difficult to restart them. Fertility, although altered by the Pill, returns when it is stopped.

I am just starting menopause. Will the lack of estrogen in my body cause any physical differences?

The most obvious difference is the cessation of periods. Also, the breasts tend to become smaller and less firm. In some women, the vagina may become less moist and supple. This can encourage urinary tract infections. Finally, loss of estrogen may lead to progressive bone weakness (osteoporosis). These effects can be substantially improved by taking hormone replacement medication.

Does estrogen have any effect on pregnancy?

After stimulating the uterus to prepare for a possible pregnancy, estrogen and progesterone continue to play an important part. Estrogen levels increase through pregnancy, and reach a peak just before birth, then rapidly drop to normal levels after it. Much of the estrogen is made in the placenta rather than the ovaries. Estrogen does not seem to play a part in the onset of labor, but high estrogen levels during pregnancy keep the enlarging breasts inactive until milk production is needed.

ESTROGEN PRODUCTION AND THE DEVELOPING EGG

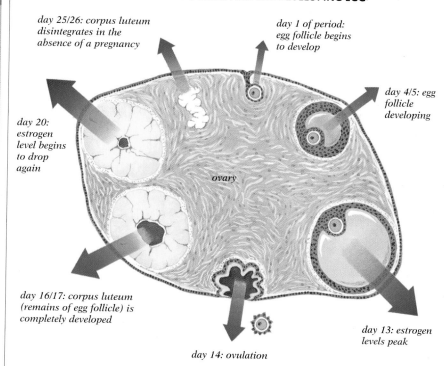

day 25/26: corpus luteum disintegrates in the absence of a pregnancy

day 1 of period: egg follicle begins to develop

day 4/5: egg follicle developing

day 20: estrogen level begins to drop again

ovary

day 16/17: corpus luteum (remains of egg follicle) is completely developed

day 14: ovulation

day 13: estrogen levels peak

▲ *The quantity of estrogen produced by the egg follicles varies during the menstrual cycle. Initially the follicles produce very little estrogen, but the level gradually builds up as the follicles develop and reaches a peak on day 13. At ovulation the estrogen level drops dramatically. It rises once more as the corpus luteum develops, and drops after day 20, unless the egg has been fertilized.*

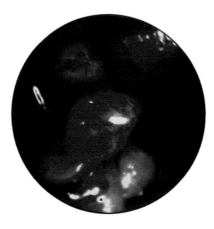

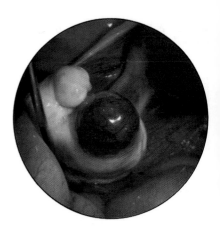

▲ *The egg is just beginning to leave the Graafian follicle, where it has been maturing for at least 10–14 days.*

▲ *The egg is expelled from the ovary toward the fallopian tube, surrounded by a jellylike fluid from the follicle.*

▲ *Here, artificial hormones have been used to stimulate ovulation. The egg can be seen lying on a small follicular cyst.*

Effects of estrogen

As the months pass, the amount of estrogen circulating in the body increases more rapidly. This has the effect of stimulating growth in the lining of the uterus, the endometrium, in preparation for its role of accommodating a fetus. Feedback to the pituitary causes it to secrete less FSH, with the effect that the ovaries in turn produce less estrogen. Estrogen support for the thickening of the lining of the uterus is withdrawn, and the lining begins to break down and flow out through the vagina as a mixture of blood and debris, the first period or menstruation. During a period, the hypothalamus stimulates the pituitary to produce more FSH, which triggers the ovary to mature another batch of egg follicles and manufacture more estrogen; and so the cycle is repeated.

The amount of estrogen rises steadily all through the first (or proliferative) phase of the menstrual cycle, during which the lining of the uterus is growing thicker, until by the 13th day after the onset of the last period the lining is six times its original thickness.

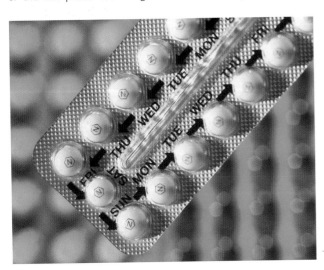

▲ *The combined contraceptive pill containing estrogen and progestin is the most commonly used and is very reliable.*

Feedback to the hypothalamus and pituitary causes the production of FSH to slow down once the estrogen peak has been reached, but it also stimulates the manufacture of the other major pituitary sex hormone, luteinizing hormone. This also acts upon the ovary, stimulating one of the egg follicles to break open and release the egg, which then makes its way into the fallopian tube, where it can be fertilized. Once this monthly cycle is established, it continues until a woman reaches menopause.

Estrogens are necessary to initiate puberty and the succession of menstrual cycles, and to maintain the woman's other sexual characteristics. If, for any reason, the supply of estrogen fails, the menstrual periods cease, the fertility rate falls, and the woman may begin to take on a more masculine appearance.

The activity of the hypothalamus gland, which is the initiator of the chain of events that results in estrogen production, can be influenced by emotional factors. It is because of these influences that anxiety and depression can affect estrogen secretion as well as fertility and the cycle of menstruation.

Uses of estrogen

A small daily dose of synthetic estrogen, called hormone replacement therapy (HRT), can be taken as a substitute for the ovaries' naturally produced hormones. HRT may slightly alleviate some of the effects of menopause, such as hot flashes and the risk of developing osteoporosis, but it may also contribute to the risk of stroke. The discontinuance of HRT brings about an artificial menstruation, called withdrawal bleeding. This technique is sometimes used to provoke regular periods in patients when periods are erratic or absent.

One of the changes that may accompany menopause is a drying up of vaginal secretions; this symptom can often be improved by using estrogen preparations prescribed by a doctor. Most estrogen preparations are chemically (synthetically) manufactured rather than prepared from natural sources. They are most widely taken as a constituent of the contraceptive pill.

See also: Hormones; Hypothalamus; Menopause; Menstruation; Ovaries; Uterus

Excretory systems

The human body constantly produces a variety of waste products that have to be gotten rid of if it is to remain healthy. A number of organs throughout the body are responsible for this vital process.

Excretion is the process by which the body gets rid of waste products. The different constituents of the body continuously produce their own by-products. These must be eliminated or the body will effectively poison itself. It is the job of various organs—including the lungs, the kidneys, the liver, and the intestines—to ensure that this does not happen.

The function of the lungs

Waste products are produced by the process in which body cells burn up fuel with oxygen to produce energy for the body's functions. When glucose, the most common body fuel, is burned, the waste product is the gas carbon dioxide. This dissolves in the bloodstream and is carried to the lungs and exhaled.

It may seem odd to think of the lungs as organs of excretion, but carbon dioxide is the most important waste product excreted by the body. If carbon dioxide starts dissolving in the blood in greater quantities than normal, the blood becomes very acidic. This in turn prevents many essential chemical activities in the body and death can result. This is known as respiratory failure and may occur as the final stage in severe emphysema.

The kidneys

Most cells in the body use some form of protein in their chemical activities. The breakdown of protein results in waste products that contain nitrogen. The kidneys are responsible for filtering this nitrogen-containing waste, the most common compound of which is urea, out of the bloodstream. The kidneys also regulate the amount of water passed out of the body and keep the correct balance of salt in the body.

▲ *As this child breathes out to make bubbles, he gets rid of carbon dioxide, the main waste product excreted by the body.*

The kidneys work in a complicated way. They receive about 2 pints (1 l) of blood every minute. The blood passes through filters at the end of each of the two million tubules in the kidneys. The filters separate out the watery part of blood (the plasma, but no proteins), which passes into the tubules while the rest stays in the bloodstream. As the filtered fluid moves down the long kidney tubules, the majority of the water, salt, and other valuable substances is absorbed back into the bloodstream. Water, urea, and other waste substances make up urine, which passes down two tubes into the bladder.

The kidneys produce urine continually day and night. About 4 pints (2 l) of urine are passed in 24 hours, but this can vary a lot. The kidney tubules control the body's delicate water balance by the amount of filtered fluid they absorb. If the body is becoming dehydrated, the pituitary gland in the brain secretes antidiuretic hormone (ADH) to instruct the kidneys to absorb more water. The total amount of urea passed remains around the same, but it is dissolved in a greater or lesser proportion of water and so leads to weaker or stronger urine.

A similar system exists to manage the balance of salt in the body. A hormone called aldosterone, secreted by the adrenal glands just above the kidneys, sends a message to the kidney tubules to reabsorb more or less salt according to the body's needs.

The liver

The liver is like a high-powered chemical factory and storehouse combined. Its cells are grouped in clusters around veins, into which they pass their waste products. The waste products are excreted in the form of bile into nearby bile ducts. The bile ducts come together into a large duct that drains into the small intestine.

The gallbladder

Bile is stored and concentrated in a sac called the gallbladder, which squeezes it out into the duodenum, the first part of the intestine. Bile is involved in the process of digestion; bile breaks down large droplets of fat into smaller droplets by a process called emulsification, which makes fat easier to absorb. The biliary system therefore not only provides a useful way of eliminating waste products from the liver, but also plays an important role in the digestion of food. Bile is a bright green color when it is secreted, but it changes color after bacteria in the bowel break it down into compounds that are yellow to dark brown. It is the bile that gives the feces their usual brownish color.

The bowels

When food enters the stomach, it is churned up and broken down by acid until it is liquid. It then enters the small intestine, where the real process of digestion takes place and all the desirable nutrients in the food are absorbed. Finally, what is left of the digested food enters the colon or intestine. This is a long, wide tube that starts in the lower right-hand corner of the abdomen, then works up and around in a horseshoe shape before coming to an end at the anus. During this passage through the large intestine, the remaining waste matter gradually solidifies as water from it is absorbed into the bloodstream through the intestinal wall. The final consistency of the food waste, or feces, depends upon how much water is absorbed. Most of the substance of the feces consists of bacteria and food residue after the nutrients have been removed. It is debatable

How the body clears itself of waste

The body has several systems for ridding itself of waste products, which are mainly the result of digestion and other chemical processes necessary to maintain life.

The skin excretes water and salt (derived from food) through the pores via the sweat glands.

The lungs excrete carbon dioxide (from the burning of glucose as fuel) and some water via the windpipe (trachea) and mouth.

The liver and gallbladder excrete bilirubin (from the breakdown of hemoglobin from red blood cells in the liver) via bile passed out with the feces.

The kidneys excrete urea (from the use of proteins by the cells), water, and mineral salts via the bladder and urethra.

The intestines excrete feces (the remains of food after the nutrients have been removed) via the anus.

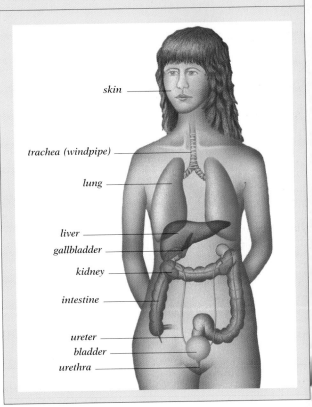

skin

trachea (windpipe)

lung

liver

gallbladder

kidney

intestine

ureter

bladder

urethra

whether this should be called excretion, but the intestine certainly does contain some true excretions, because it contains the waste products of cell chemistry in the form of bile.

The skin

On a hot day the body loses a large amount of salt and water in sweat. Sweat is the product of sweat glands in the skin, and its sole

Questions and Answers

What's the difference between excretion and secretion?

Excretion is the process in which waste products are lost. Secretion, on the other hand, is the process in which glands produce substances for a particular purpose. The tear glands, for example, secrete tears that lubricate the eyes and to some extent protect them from infection.

My grandfather gets up once or twice a night to urinate. Does this mean he has a weak bladder?

For many people it is normal to urinate once or twice during the night. Urinating at night is unlikely to be due to a disease, unless your grandfather has noticed any change in the pattern of urination. If there is a change or he is having pain or difficulty urinating, he should see his doctor.

Is it true that senna pods are a good remedy for constipation?

No. Senna and similar laxatives move the feces by irritating the intestine. It is better to take the natural approach and give the intestine more work to do by eating plenty of unrefined food like whole-grain breads and bran and increasing your intake of fresh fruit.

I find that whenever I eat beets, my urine turns red. I thought this was just the color in the beets, but my friend tells me that I may be bleeding from the kidneys. Should I see my doctor?

No, you are right. It is common for food coloring to come out in the urine. This happens with beets and with asparagus in some people. Some drugs may also color the urine.

Is my baby's diaper rash due to his urine's being strong?

Urine can become ammoniacal by bacterial action. The best thing to do is to change his diapers frequently and use a barrier cream.

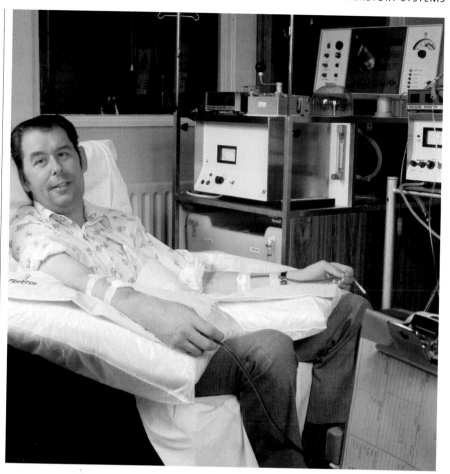

▲ *This patient is undergoing dialysis: a machine is taking over the excretory function of the kidneys and removing waste products from his body.*

purpose is to regulate body heat, because heat is lost as the sweat evaporates from the skin. However, if a person did not sweat for a day, then any excess salt or water could easily be excreted by the kidneys. So sweat does not fulfill any essential function in the clearing of waste products.

Tests

Many of the widely used blood tests are designed to find out whether the main parts of the excretory process—the lungs, kidneys, and bile system—are working properly.

Many patients who are treated in the hospital will have preliminary tests to measure the levels of urea in the blood. The level of bilirubin, one of the main constituents of bile, is also commonly measured in the hospital. If either of these substances is present in excess, it is an indication that either the kidneys or the bile system is not working fully. This type of malfunction may not produce any symptoms in the patient, but it is essential for doctors to know whether these organs are functioning because all the drugs used in treatment have to be eliminated through the kidneys or the liver, and hence through the bile. If either system is not working properly, drug levels may build up in the system and so smaller doses must be given to the patient.

It is slightly more difficult to measure carbon dioxide levels because blood has to be taken from an artery, usually at the wrist, rather than from a vein. However, the level of carbon dioxide can and must be measured when an overall assessment of respiratory function is required in medical tests.

See also: Bile; Digestive system; Gallbladder; Kidneys; Liver; Lungs; Perspiration; Pituitary gland; Skin; Urinary tract

Eyes

Often compared to a versatile and accurate movie camera, the eye should provide a person with a lifetime's service. It needs the minimum of attention, is self-repairing, and is prone to relatively few diseases.

I am very nearsighted. Will the problem get worse until I go blind?

Neither nearsightedness nor farsightedness leads to blindness. Nearsightedness usually progresses during body growth and then stabilizes in adulthood. Progress may continue in severe cases, increasing the risk of retinal damage or detachment, but effective treatment is available.

In young people the internal lens focuses automatically to correct farsightedness (accommodation), but the power to do so decreases steadily with age.

My son has measles. I heard it is dangerous for him to read now, and for a while after. Is this true?

No. Although measles causes inflammation of the eyes and photophobia (discomfort in strong light), there is no danger to the sight. It is simply an old wives' tale.

My eyesight became worse after I started wearing glasses. Why?

This is a common experience, especially for farsighted people who have to use reading glasses. Before the glasses are worn, the focusing muscles in the eyes have to work hard and are kept strong by exercise. With glasses they can relax and may weaken slightly. The muscles will still weaken eventually, however, so there is little point in not wearing glasses.

What causes a black eye?

The skin of the eyelids and around the eyes is usually the thinnest in the body. Beneath it is an elaborate complex of thin-walled veins that tend to bleed after a blow to the eye region. The blood then spreads under the skin and darkens because of loss of oxygenation, causing a black eye. The blood will be absorbed in about two weeks.

When people want to explain in simple terms how the eye operates, they usually compare it to a brilliantly designed camera. To understand fully how the outside world can be viewed inside the tiny chamber of the eye, however, it is necessary to go back to basics.

The role of light

Light is essential to people's ability to see. Whether it comes from the sun or from an artificial source, it can be thought of as a transmitting medium that bounces off objects in countless directions, making it possible for the human eye to see them.

Light usually travels in straight lines, but it bends if it passes through certain substances such as the specially shaped glass of a camera lens or the lenses in a human eye. The degree of bending can be precisely controlled by the shape of the lens. If the light is bent inward, or concentrated, it forms tiny but perfect images of much larger objects. The light-sensitive area at the back of the human eye, the retina, detects these images.

The cornea and the anterior chamber

When a ray of light strikes the eye, the first thing it encounters is a round, transparent window called the cornea, which is the first of the eye's two lenses. The cornea is fixed in position, is always the same shape, and does a major part of the light bending. It is surrounded by the white of the eye, a tough substance called the sclera that excludes light. A protective membrane called

▲ *A doctor examines the internal structures of the eye by looking through the lens of an ophthalmoscope. Eyedrops are used to dilate the pupils.*

STRUCTURE OF THE EYE

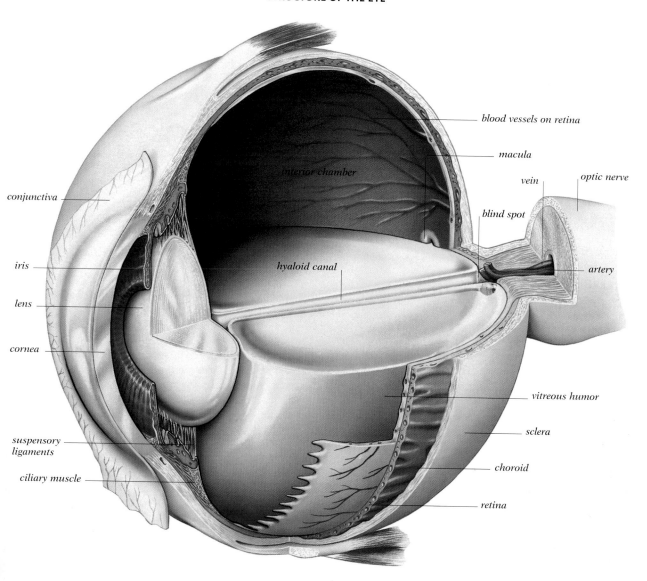

conjunctiva

iris

lens

cornea

suspensory
ligaments

ciliary muscle

interior chamber

hyaloid canal

blood vessels on retina

macula

vein

optic nerve

blind spot

artery

vitreous humor

sclera

choroid

retina

The conjunctiva covers the white, and this membrane, together with the lacrimal glands behind the upper eyelids, produces a constant film of tears essential to the health of the eyes.

After a ray of light has passed through the cornea, it enters the outer of two chambers within the eye, properly called the anterior chamber. This chamber is filled with a watery fluid called the aqueous humor, which is constantly drained away and replaced. Together, the cornea and the aqueous humor form a powerful lens.

The iris

The iris forms the back of the anterior chamber, with the second lens just behind it. It is a round, muscular diaphragm—in other words, a disk with an adjustable hole in the center. In the eye this hole is called the pupil, and its size is altered by two sets of muscles.

The iris is pigmented; this pigmentation gives the eye its color when seen from the outside (the word "iris" is derived from the Greek word meaning "rainbow"). Different eye colors are due not to different pigments but to varying amounts of a pigment called melanin. Dark eyes have a lot of pigment all the way through the iris. Lighter eyes have less melanin, concentrated toward the back of the iris, so the color is seen through a layer of aqueous humor and cornea and tends to look blue or green.

The iris controls the amount of light entering the eye, rather like the aperture of a camera. If a strong light falls on it, the pupil automatically gets smaller. In dim light, the pupil grows larger. Emotions such as excitement and fear and the use of certain drugs also make the pupil widen or contract.

The lens

Just behind the iris is a soft, elastic, transparent lens. It provides about a third of the eye's focusing power and is responsible for the fine focusing of light rays. For this reason, it is adjustable. The lens is

Questions and Answers

My brother is color-blind but says he doesn't see in black and white. What is color blindness?

There are three types of light-sensitive cells in the retina, each type specializing in seeing either red, green, or blue light. Color blindness results from having too few of one of these types of cell. A color-blind person does not see in black and white but is unable to distinguish between certain colors, the most common ones being red and green.

Will taking extra vitamins improve my eyesight?

Only severe vitamin deficiency affects the eyesight, and this happens only when people are on the verge of starvation. In the days when sailors got scurvy (caused by lack of vitamin C) from being without fresh vegetables for long periods, their eyesight suffered, too, resulting in night blindness. The condition improved when they returned to a balanced diet. We now know that severe vitamin A deficiency can cause night blindness. Carrots are a rich source of this vitamin, but people with a normal diet cannot, as some believe, improve their night vision by eating large numbers of them.

Severe vitamin A deficiency can also cause disease of the cornea, the transparent outer window of the eye, ultimately resulting in total blindness. Lack of vitamin B may cause blindness by interfering with the normal nutrition of the optic nerve. Eyesight is restored if vitamin B is put back in the diet.

What should I do if I get grit in my eye?

A foreign body in the eye is always worrying because it may scratch and damage the cornea. A simple remedy is to rinse the eye with a warm solution of salt and water, or eye drops, or to blink repeatedly while rolling the eyes. If this fails, go to the doctor. A penetrating eye injury, such as occurs when people fail to wear protective glasses at work, is more serious and requires hospital attention.

▶ *Light rays from a near object diverge (top), so the surface of the lens becomes more curved to focus them. Light rays from a distant object are almost parallel (below), so the lens has less focusing to do.*

held around its outer edges by the ciliary muscle, which can change the shape of the lens to alter the angle at which the light rays passing through it are bent.

The vitreous humor

Behind the lens is the main interior chamber of the eye. This is filled with a substance called the vitreous humor, which has a jellylike texture. Running through its center is the hyaloid canal, the remains of a channel that carried an artery during the eye's development in the fetus.

FOCUSING

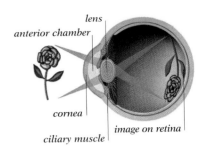

lens
anterior chamber
cornea
ciliary muscle
image on retina

The retina

A light-sensitive layer called the retina lines the curved interior of the back chamber of the eye. It is made up of two different types of light-sensitive cells, which are called rods and cones because of their shapes. Rods are sensitive to light of low intensity and do not interpret color, which is picked up by the cones. The cones are also responsible for clarity, and they are most plentiful at the back of the eye in an area called the macula. That area is where the lens focuses its sharpest image, and that is therefore the place where vision is at its best.

In the area around the macula the retina still registers images with clarity, but out toward its edge, where images are less clear, is the area called peripheral vision. Together, central and peripheral vision make up a complete view of the outside world.

How objects are viewed

The eyes work by rapid scanning, each eyeball moving around by means of a specially arranged set of muscles attached to it (see diagram, page 141). When people look at someone's face, for example, their eyes tend to concentrate on that person's eyes, darting glances from one eye to the other and occasionally moving down to look at the mouth.

From optic nerve to visual cortex

After the information that comes through the eyes has been adjusted to a manageable level of intensity by the iris, focused by the two lenses, and recorded by the light-sensitive surface of the retina, it is sent to an area of the brain called the visual cortex to be processed and understood. The information from each eye passes along a cable called the optic nerve, which runs from the eye to the brain. At the optic chiasma, which is situated at the base of the brain, the nerve fibers from the inner half of the retina cross over to the optic nerve from the other eye, while those from the outer half continue directly into the brain.

The visual cortex has two halves, one in each hemisphere, or side, of the brain, and each with its own separate blood supply. Consequently, it is rare for a stroke—which almost always involves a single artery—to cause complete blindness, although severe damage to one half of the visual cortex always causes the loss of part of the field of peripheral vision. In most cases stroke

▲ *To discover your blind spot, close your right eye and look at the cross (above). Continue looking at the cross, and slowly move the page toward your face. There is a point at which the dot will vanish—this is your blind spot.*

HOW THE EYEBALL MOVES

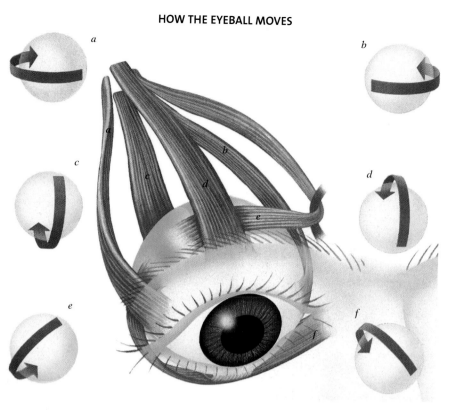

a

b

c

d

e

f

◄ *Six main muscles move the eyeball. Muscle (a) swivels it away from the nose; (b) toward the nose; (c) rotates it downward; (d) upward; (e) moves it down and outward; and (f) moves it upward and outward.*

blank space or a dark patch; they do not see anything at all. The blind spots are also concealed because the area of visual field covered by the optic disk of each eye is only small, and the eyes move constantly, even when a person is concentrating on a single object.

The arteries of the eye

A large artery leads into the eye beside the optic nerve, and many smaller blood vessels branch out from this main artery and spread over the retina like the tributaries of a river. Their vital job is to supply the nerve cells of the retina with nutrients and drain away waste products.

The presence of this network enables a doctor using an ophthalmoscope to examine some of the body's blood vessels without any skin in the way.

damage causes the complete loss of the corresponding halves of the visual fields of the two eyes—in other words, the outer half of the field of vision of one eye and the inner half of the field of vision of the other eye. This is called homonymous hemianopia.

Although the loss of visual field is considerable, stroke sufferers do not experience a sense of blackness. They are often unaware that they have any visual loss at all until they discover that they are unable to read properly or find themselves having accidents with objects in the blind part of the field.

Damage to the optic nerve at the point where the nerve fibers cross causes a different kind of visual loss. In this case, because the crossing fibers come from the inner halves of the two retinas, both outer halves of the fields of vision are lost. This is called bitemporal hemianopia. The condition is usually caused by upward pressure from a tumor of the pituitary gland, which lies immediately under the point where the optic nerve fibers cross.

The blind spot

At the point where the million or so nerve fibers from the retina emerge from the eyeball through the sclera to form the optic nerve is a small, insensitive area called the optic disk. There are no rods or cones on the optic disk, so any part of an image that falls on it is simply not perceived.

People are not usually aware of the blind spot in each eye because the fields of vision of the two eyes overlap, each one compensating for the other's blind spot. However, it is possible to demonstrate the existence of the blind spot by trying the simple experiment at the bottom of the opposite page.

Another reason that people are unaware of their blind spot is that visual field loss is an absence of perception; people do not see a

Common problems

Few eye disorders end in blindness. One of the most common minor problems at the front of the eye is conjunctivitis: an inflammation of the conjunctiva, the thin membrane covering the whites of the eyes. It is especially common among babies and young children and is caused by bacteria, viruses, irritation, or an allergy such as hay fever. Treatment with antibacterial drops or creams usually clears up the condition if it is due to bacterial infection.

The cornea is subject to various types of inflammation. Pain in the eyeball, redness, blurred vision, and clouding of the cornea are all typical symptoms. These may be caused by minor conditions but may also indicate more serious problems, so anyone experiencing such symptoms should get immediate medical attention. The cornea contains no blood vessels, so if grafting a new one in its place is necessary, the graft is usually successful.

Glaucoma

To work properly the eyeball must stay the same shape. Its normal shape, and the relationship of the cornea to the retina, is maintained by the constant secretion of water (aqueous humor) within the eye. The water fills the interior, including the vitreous gel, and passes out by a restraining meshwork near the root of the iris. The balance between the secretion and outflow of fluid normally maintains the pressure in the eye within narrow limits. If the pressure falls, the eyeball indents and vision is lost. If the pressure rises too high, the small vessels that supply blood to the retina and optic nerve fibers are constricted and closed off, causing serious damage to the nerve fibers. The condition in which pressure rises is glaucoma and is due to interference with the free outflow of fluid from the globe of the eye. This can occur in several ways.

NEARSIGHTEDNESS

FARSIGHTEDNESS

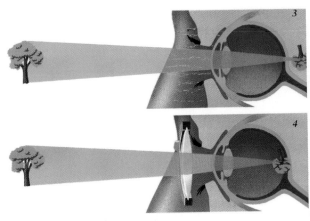

▲ *One cause of nearsightedness (1) is an unusually long eyeball, causing light rays from a distant object to form an image in front of the retina. The defect is corrected (2) by a concave lens. If the eyeball is too short for the strength of the corneal lens (3), the image cannot be formed within the eye, causing farsightedness. A convex lens (4) focuses the image on the retina.*

The most common kind of glaucoma, chronic glaucoma, is entirely painless and insidious and is one of the major causes of blindness in the West. Affected people are often unaware that anything is wrong until a late and irreversible stage of the disease. This is because the optic nerve fibers damaged initially are almost always those concerned with peripheral vision. Serious loss in the peripheral visual field can occur before any harm is suspected.

For these reasons, people should have their eye pressures tested routinely at about the age of 40, and afterward at intervals of a few years. People with a family history of glaucoma should be tested at an earlier age and more often. Raised pressures can easily be brought down, usually by the use of appropriate eye drops. In some cases surgery may be needed to correct the condition.

Acute glaucoma, which causes pain, misting of vision, and the perception of colored haloes around lights, is much less common and is readily detected because of the symptoms. It also requires urgent treatment by an ophthalmologist.

Defects of the lens

The human eye commonly suffers optical defects that cause unclear vision or a sense of strain during prolonged visual work. These defects result from two principal causes: a lack of correspondence between the curvature of the cornea and the length of the eyeball, and a failure of the internal focusing lens to adjust adequately to the optical requirements for clear vision of near objects, such as print. The latter defect is almost always age-related.

Refraction, or the bending of light rays, occurs when the rays pass from one medium to another—such as from air into glass. Contrary to popular belief, the main refracting lens of the eye is not the internal crystalline lens, because it is bathed in fluid and has a refractive index (the degree to which light rays are bent) similar to that of the fluid. The main refracting lens is the outer lens, the cornea, and most of the bending of light rays occurs at the interface between the air and the corneal surface. The power of this lens depends on its curvature, the radius of curvature ranging from about ¼ inch (6.5 mm) to about ⅓ inch (9 mm).

If the corneal lens is too steeply curved relative to the inner length of the eye, either because the lens is particularly curved or because the eyeball is particularly long, it will be too strong, causing myopia, or nearsightedness. Rays from close objects diverge and can focus sharply on the retina, but the parallel light rays from distant objects form an image plane that lies in front of it. Concave (minus) eyeglass lenses are needed to correct this defect.

If the cornea is too flat relative to the inner length of the eye, either because the lens is particularly flat or because the eyeball is particularly short, it will be too weak, causing hyperopia or farsightedness. In this case the image plane falls behind the retina, so clear vision at any distance is possible only by strengthening the power of the internal lens system with convex eyeglass lenses.

Some people have an imperfectly curved cornea that throws a distorted image on the retina, causing their vision to be partly blurred. This is called astigmatism, and glasses or certain types of contact lenses can correct it. With this condition, the cornea has a maximum and a minimum curvature at right angles to each other—the surface is toric, like the surface of a football. Eyeglass lenses of corresponding opposite power are required and must be set accurately in the correct orientation.

Aging

Many eye problems are caused by getting older. The focusing ability of the internal lenses of the eye lessens with age, owing to loss of elasticity. Weakness of fine focusing (presbyopia) is universal and progressive and is most commonly apparent from about age 45. It is caused by deterioration of the internal lens. Convex lenses are required to bring near objects into focus for reading and other close work. Other problems may occur if the area near the optic nerve gets an insufficient blood supply; sight in the center may be lost.

Retinal detachment

Sometimes the retina can lift from the blood vessels and supporting tissues underneath. This condition is known as a detached retina and usually begins with a small tear in the retina. It can happen as a

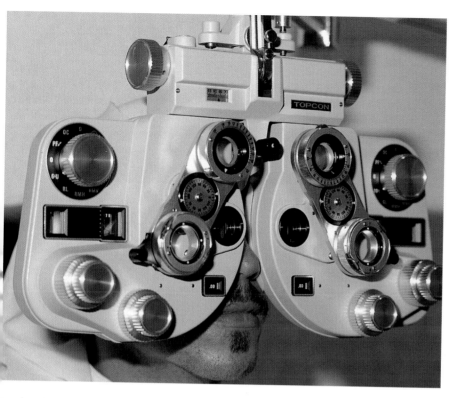

◄ This instrument, called a refracting head, is used to perform eye tests. The patient looks through various lenses at an eye chart on the wall opposite. Different lenses are selected until the correct prescription for the patient is found. This system has replaced the older method of testing with single lenses in trial frames.

apparent until the damage is severe, although there may be earlier blurriness. Later, rupture of one of the new blood vessels or separation of the retina from the underlying tissues may cause sudden blindness in an eye. The risk is minimized by tight diabetic control, and laser surgery can be used to destroy abnormal areas of the retinal periphery, preventing further loss of sight and reducing the risk of blindness from this cause.

Squinting

Squinting is an inability to focus both eyes on the same spot at the same time. It is caused by a defective balance of opposing eye-moving muscles, and it can be corrected by a course of expert ophthalmic and orthoptic treatment. Surgery is often necessary.

Other visual disorders

Hereditary degeneration of the retina, macular degeneration, which usually affects old people, and accidental injury to the eye can also damage vision. In addition, certain diseases can lead to serious, permanent vision disorders. For example, multiple sclerosis may inflame the optic nerve, leading to rapid reduction of central vision. If the herpes simplex virus gets into the eye it may cause a dendritic ulcer, which will cause gradual loss of vision and may permanently damage the cornea if left untreated.

Infections

Eye infections can develop if the cornea is scratched, and sometimes an ulcer may occur. In such cases the eye becomes red and painful and vision is blurred. Medical treatment is vital because scarring may form and permanently affect vision. In cases of severe damage a corneal transplant may be possible. Other eye infections include trachoma and stys (hordeolum, an infection of the eyelid).

Outlook

Near- and farsightedness, although common, can be easily corrected. Blindness is a much more terrible affliction, but it is relatively rare and becoming rarer with advances in medicine.

result of an eye injury or an eye disorder such as severe nearsightedness. Although retinal detachment is painless, certain visual symptoms may be experienced. The person may see dark spots or "floaters" in the field of vision, flashing lights in the corner of eye, and possibly a black area across the field of vision. If any of these symptoms are noticed the person should immediately go to a doctor or the emergency room of a hospital.

Retinal detachment can be diagnosed by ophthalmoscopy. If only a small area of the retina has detached, the tear can usually be repaired by laser treatment. A larger area will need surgery under general anesthetic. Immediate treatment is vital to restore normal vision, since any delay may make the treatment less effective.

Cataracts

Cataracts are a clouding of the internal focusing lens. For the person affected this is like looking through a window that is slowly frosting up. The condition may also alter color vision, making objects appear more red or yellow. Cataracts are painless, develop slowly, and are commonly caused by advancing age or diabetes—an abnormally high level of blood sugar. They may also develop because of an eye injury or prolonged exposure to sunlight. They are treated by removing the opaque contents of the lens capsule and inserting a tiny, optically perfect plastic lens. The power of the plastic lens is calculated from ultrasound measurements of eye length and optical measurements of corneal curvature.

Diabetic retinopathy

Long-term diabetes is also likely to cause damaging changes in the retinal blood vessels, leading to a condition called diabetic retinopathy. Initially the vessels leak; later, delicate new blood vessels grow into the vitreous humor. Symptoms may not be

> *See also:* **Aging; Arteries; Cornea; Optic nerve; Retina**

Feet

Questions and Answers

My son's feet turn inward. Does this mean he is pigeon-toed?

Yes, it may. Pigeon toes are not usually a big problem, and the condition often cures itself. It can result from either deformities of bones in the foot arch or from weakness in the leg muscles attached to the foot. Take him to the doctor if you are worried.

The verrucae on my feet continue to disappear and return again after a time. Why can I do keep them from returning?

Verrucae are a type of wart on the sole of the foot and are caused by a viral infection. All warts are prone to reappearing, making it hard to assess whether treatment has been effective. It is best to have verrucae removed by a doctor.

My husband has been told that he will not be able to play soccer for a while because of damage to his Achilles tendon. What is this?

The Achilles tendon is attached to the heel and is the link between the foot and the powerful calf muscles at the back of the leg. These muscles provide the power for walking and running; when they contract, they pull on the tendon and the heel is raised. This tendon is extremely strong, but occasionally it gets torn. Surgery is the only means of repair, but a fairly long convalescence is needed before the tendon can again take the great stress placed on it when it is in action.

My friend's feet smell a lot. What can he do about it?

He can wear moisture-wicking or wool socks and leather shoes, which allow some ventilation and minimize sweating. Regular washing, foot deodorants, and clean socks daily also help.

Babies tend to be born with smooth, beautiful feet, but those feet will have to endure a lifetime of wear and tear, bearing the brunt of the body's weight and its contact with the outside world. Commonsense care can help keep feet healthy and in good working order to meet the demands put upon them.

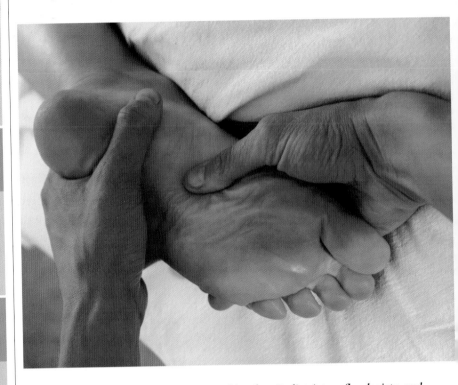

▲ *A foot massage can bring relief from aching feet. Podiatrists, reflexologists, and general massage practitioners may provide foot massage as part of their therapy regime.*

Each human foot is a mechanical masterpiece that consists of 26 bones, 35 joints, and more than 100 ligaments. In many ways, the formation of the bones, blood vessels, and nerves of the foot resembles the structure of the hand, although the proportions are different; for example, the phalanges, the bones making up the toes, correspond to the phalanges in the fingers of the hand.

It is not surprising that the hand and the foot have many similarities in structure, because they both developed over centuries of evolution from limbs that were originally designed for climbing and grasping. The human ability to make controlled movements of the hand was gradually refined, while the foot's function changed as people began to walk on two legs instead of four.

The foot has two important tasks: to support the weight of the body, and to act as a lever to move the body forward in walking or running. However, people who have lost the use of their hands through an accident or disease are able to achieve a surprising degree of dexterity with their feet. Some can even use them for writing or painting.

Walking

The foot is made up of many small parts and joints, so it is flexible and adapted to walking on uneven surfaces. Most of its power is derived from the strong muscles in the leg, with a series of smaller muscles in the foot adding support.

The weight of the body is supported on the largest bone in the foot, the calcaneum, which forms the heel, and by the heads of the metatarsals, which are the longest bones of the foot

The other bones are raised from the ground in the form of an arch, because this is the only way that a segmented structure (like the arch of a bridge) can hold up any weight.

The foot has three arches: the medial longitudinal arch, which runs along the inside of the foot; the lateral longitudinal arch, which runs along the outside of the foot; and the transverse arch, which crosses the foot. The arches can be seen by looking at a wet footprint (of someone who is not flat-footed) on the bathroom floor. The foot will appear narrower in the middle, because the inner arch is higher than the arch on the outside of the foot. All three arches are supported by an elaborate system constructed of ligaments and muscles.

When a person stands still, his or her body weight is supported on the heel and the metatarsals, but when he or she begins to walk, the load is borne first by the lateral margin (outside edge) of the foot, and then by the heads of the metatarsals (toes). The toes are extended as the heel rises and the contracting muscles shorten the longitudinal arch of the foot. The body itself is thrown forward by the action of the powerful muscles in the calf (the gastrocnemius and soleus), which pull on the ankle joint using the ankle as a lever, while the flexor muscles of the foot flex the toes for the final thrust forward. The toes are kept extended to prevent them from folding underneath the foot upon the next step. The big toe is the most important of all, but loss of the other toes does not affect a person's ability to walk.

Why feet ache

Feet may ache after a long walk; after a long time spent standing; or if a person is overweight, ill, or employed in an occupation, such as waiting on tables, that requires being on the feet all day. Aching feet result when the supporting muscles tire and the ligaments become stretched from supporting the load on the arches. Flat feet result when the ligaments become permanently stretched.

Children's feet

Most babies are born with what appear to be flat feet, and the arch develops as the tendons, muscles, and ligaments of the feet strengthen with use. For this reason it is essential that young children should always wear well-fitting shoes, so that their feet can develop correctly.

It is not a good idea to give children shoes at too early an age. Let them run around barefoot as long as there is no danger from sharp objects. Once a child has begun to wear shoes, parents should make sure that his or her feet are measured for length and width about every two months. Most reputable shoe stores provide this service and will offer advice.

As soon as shoes become too small, they should be discarded; and the reason that parents should keep an eye on this themselves is that young children are unlikely to complain, even if their shoes are too tight. Cramped shoes are the cause of many deformities, such as bunions, ingrown toenails, and splayed toes, which will be difficult to correct later in adult life.

Preventing foot problems

Adults' shoes should be chosen with the same care as children's regardless of the dictates of fashion. Most foot deformities are acquired as the result of wearing badly fitting shoes; problems can be accentuated by platform soles or stiletto heels, because they tilt the body at an unnatural angle.

THE MECHANICS OF WALKING

▼ ► *The flexibility of the human foot is due to its intricate anatomy. Shown here are the many bones, ligaments, and powerful muscles of the foot and leg that go into action with every step that a person takes.*

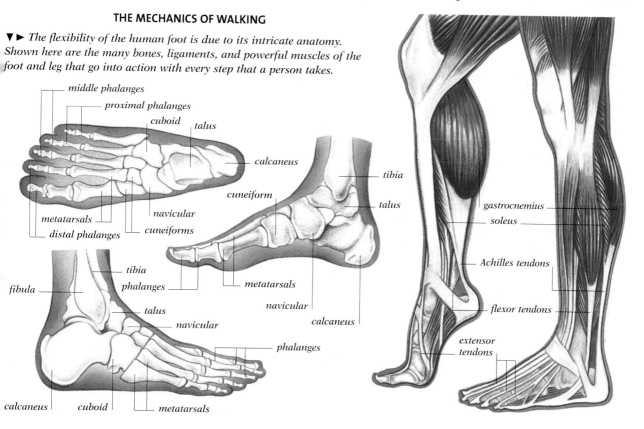

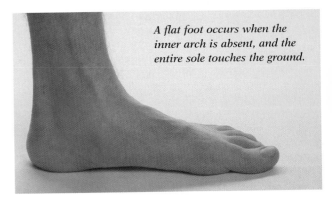

A flat foot occurs when the inner arch is absent, and the entire sole touches the ground.

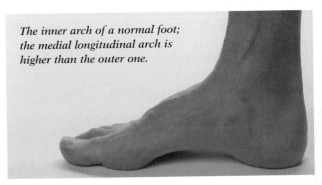

The inner arch of a normal foot; the medial longitudinal arch is higher than the outer one.

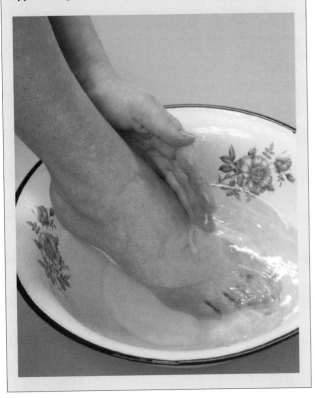

The same pressures that cause feet to ache can also produce calluses and corns: hard areas of skin on areas of the feet that can irritate the underlying nerves. Although it is tempting for people to try to treat corns themselves, it is always advisable to consult a podiatrist.

Good foot hygiene is important in preventing infection. Because feet are mostly enclosed in socks and shoes all day, they tend to sweat. Stale sweat not only smells very unpleasant but also creates a suitable environment in which infections can easily grow.

The chances of contracting skin infections such as athlete's foot can be reduced by a careful choice of footwear. Socks made of moisture-wicking materials or wool absorb sweat better than human-made fibers such as nylon. Leather shoes allow the foot to have some ventilation, whereas plastic ones trap sweat and bacteria.

Foot injuries

Breaking a bone in the foot is a fairly common accident, usually caused by dropping a heavy object on the foot. Fractures of the toes (excepting the big toe) are generally not a problem; the fractured toe can be splinted to the sound toe alongside it and be left to heal like this. However, fractures of the tarsal bones are serious, because it is difficult to align the broken bones. Damage to the joints sometimes occurs after a bone has been broken. Often the victim has to learn to live with a painful foot; however, surgical operations can be undertaken to seal the joint by locking the bones together, providing a joint that is less mobile but painless.

Flat feet

The condition known as flat foot, or *pes planus*, can happen at any age. As the name suggests, the foot loses the arch on the inside, with the result that the entire sole is in contact with the floor. The bathroom test will show a person if his or her feet are flat; wet footprints made by normal feet are definitely narrower in the middle than those made by flat ones.

Feet have to serve their owner for many thousands of miles during a lifetime; it is no wonder that the mechanics go wrong occasionally. The inside arch of the foot acts as a spring that helps to distribute a person's weight evenly through the heel and toes. If the feet are flat, this bracing is not necessarily absent; it is merely not shown by the presence of an arch. There are many athletes who have flat feet and perform perfectly well. The only trouble they may suffer from is uneven wear of their shoes.

Perhaps the most common cause of flat feet in adults is laxity of the ligaments holding the forefoot bones together. The forefoot splays and the arch drops, and consequently there is often a generalized aching of the foot.

Questions and Answers

Can flat feet be found in several members of the same family?

Yes, they can often be traced back through the family tree. In some ethnic groups up to 35 percent of the population has flat feet.

My son has knock-knees and flat feet. Are the two often associated?

Yes, they are. What happens is that the knock-knees force the feet to tilt inward, so that the feet's arches come into contact with the ground. However, this condition often clears up by itself. If your son is unhappy or being teased about his condition, take him to the doctor for advice.

I have recently been sick and have spent a lot of time in bed. Now that I'm up again, I notice that I have flat feet, which are very painful to walk on. Is the condition permanent?

This is a temporary condition. The muscles in the feet and lower leg have become flabby and weak with disuse. It will take some time before they regain their ability to support the arch. You should do some exercises to regain the strength in the arch, and the rest of the foot too; but do this gradually, not all at once.

My son is badly pigeon-toed. Can this condition cause flat feet?

No, pigeon toes do not cause flat feet. However, if your son stands the opposite way, with his feet turned outward, Charlie Chaplin-style, that may cause flat feet. Ask your doctor to advise you if you need reassurance.

My teenage daughter is becoming flat-footed. Will her career as a dancer be affected?

It depends upon exactly what is causing the problem. You should take her to a doctor as soon as possible to see if any remedial measures can be taken.

Foot exercises

The time when exercise can help your feet most is after a period of illness, when your muscles are weak and slack. Even before you are up again, you can practice these exercises regularly in your bedroom. Do them on a hard floor, with bare feet. They will help to retone and strengthen your muscles.

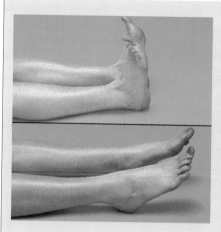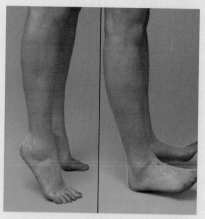

1. Sit on the floor, leaning back very slightly, with your hands supporting your body. Stretch your legs straight out in front of you, keeping them slightly apart. Starting with your feet and toes pointing upward, stretch your feet down toward the floor as far as you can, feeling the stretch in your instep. Relax, then repeat the exercise 10 times.

2. Stand upright with your feet slightly apart. First, stand on tiptoe, stretching upward as high as you can. Hold for a count of five. Now lower your feet to the ground and then stand on your heels and raise your toes. Hold this position (if you can) for a count of three. Lower your toes to the ground. Now repeat the whole exercise slowly five times.

Flat feet in children

Many parents worry about their children's feet. Babies appear to have flat feet, but the feet contain a large quantity of fatty tissue in the soles, as well as a prominent pad of fat in the front of the heel that fills the arch. By the age of three, when children are walking properly, their feet will no longer have such a flat appearance. Parents who are concerned about their children's feet should look at the soles of their shoes. If the sole is markedly worn on the inside, then their feet are probably flat. If parents think that this is affecting their child's walking, it would be worth consulting the doctor.

Treatment

Pain in the feet can happen for various reasons, such as infection of the soft tissues caused by an abscess or boil, abnormality of the bones of the foot, or an acquired deformity such as a bunion, in which the anatomical changes can also lead to flat feet. Surgery cannot correct flat feet, but discomfort can be relieved by wearing shoes or sandals with instep inserts for extra support. These can be bought from drugstores and some department stores.

Opinion is divided as to whether remedial exercises can help flat feet. Exercises are definitely beneficial when the condition is due to a prolonged period of rest in bed.

Choice of footwear

No shoes are really good for the feet, but the best are low-heeled or flat shoes with plenty of room for the toes and good ankle support. A light, flexible sole is also important for easy walking. In suitable climates and inside the house, go barefoot whenever possible.

See also: Bones; Joints; Ligaments; Muscles; Skin; Tendons

Fetus

Questions and Answers

I am pregnant and have a vaginal yeast infection. Will this damage my fetus?

No, it will not affect the fetus. In fact, yeast infections are common during pregnancy. Since yeast is harmless, and inserting pessaries or creams into the vagina might start a miscarriage, many doctors prefer to leave such infections untreated during pregnancy. Although it is uncomfortable, it is nothing to worry about.

Can sexual intercourse during pregnancy damage the fetus?

No. You can continue to have sexual intercourse for as long as you wish and as long as you are comfortable. You may need to find new sexual positions, because any weight on the woman's abdomen can be uncomfortable. However, a woman who has miscarried during a previous pregnancy should not have sexual intercourse during the first trimester. Intercourse should not take place at all if there is a threat of a miscarriage. Consult your doctor if you are anxious.

I saw a road accident when I was nine weeks pregnant, and it really upset me. Will the experience mark my baby in any way?

No. It is a myth that a disturbing or frightening experience could result in a baby with an abnormality or disfigurement.

I love swimming, but a friend told me that I should stop exercising while I'm pregnant. Is this true?

No. You should stay as fit as you can throughout your pregnancy. As long as you don't overdo it, exercise such as swimming, tennis, and walking is perfectly safe. Doctors advise pregnant women to avoid only horseback riding and diving.

A pregnant woman is often eager to understand the stage-by-stage development of the fetus within her during the nine months of her pregnancy, as well as the changes that her own body will go through.

A doctor will date the start of a woman's pregnancy from the first day of her last menstrual period, adding nine months plus seven days to give the estimated delivery date.

Pregnancy is divided into trimesters (periods of three months in the life of the fetus). In fact, conception will probably take place between the 10th and 14th day of the menstrual cycle, when a woman is most likely to be ovulating and at her most fertile. Therefore, the pregnancy is likely to begin during the second week of the first trimester. At this early stage, just after conception, the pregnancy consists of a single fertilized cell, or egg. For three days after fertilization, the cell moves along the fallopian tube toward the uterus (womb), dividing and redividing to form a small group of cells called the morula.

The fetus in the first trimester: Weeks 1–14

Week 1: For three more days the morula floats in the uterus. It divides and redivides to form a hollow clump of cells called the blastocyst, which is just visible to the naked eye.

Week 2: The blastocyst embeds itself in the endometrium, the lining of the uterus. This stage is called implantation. Chorionic villi—projections from the blastocyst's covering—burrow into the endometrium to secure nourishment. The outer lining of the blastocyst—called the trophoblast—begins to develop into the placenta, which provides the vital link between the fetus and the mother. Blood cells start to form and the first heart cells are laid down.

Week 3: Hormonal changes in the mother's body cause the endometrium to thicken; the blood from it nourishes the blastocyst.

Week 4: The amniotic sac is by now well developed. The fetus will stay in the sac throughout the pregnancy, suspended in the amniotic fluid. Here it is kept at a constant temperature and is well buffered against any shocks. The fetus's heart is already beating, irregularly at first, but soon steadily and faster than the mother's.

The spine and the beginning of the nervous system are starting to form in the fetus, which is now about ⅕ inch (5 mm) in length.

Week 5: The first organs form. The head is growing, enclosing the developing brain, which is linked to a rudimentary spinal cord.

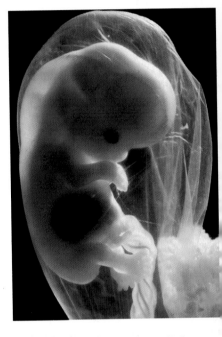

▲ *In this nine-week old fetus all the parts of the body are present, even though they are not all fully formed.*

▶ *Between the third and the seventh week of pregnancy, the fetus increases in length from about ¹⁄₁₀ inch (2.5 mm) to ¾ inch (20 mm).*

3 weeks, ¹⁄₁₀ inch (2.5 mm)

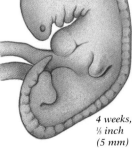

4 weeks, ⅕ inch (5 mm)

The fetus's arms and legs begin to show as little buds, and the heart and blood circulatory systems are well established.

Blood vessels from the fetus join with others in the developing placenta to form the umbilical cord. The chorionic villi continue to increase in number and to branch, attaching the fetus firmly to the wall of the uterus.

In the fetus, now about ⅖ inch (10 mm) long, the digestive system has begun to form, starting with the stomach and parts of the intestines. Although there is as yet no recognizable face, there are small depressions where the eyes and ears will be. The mouth and jaws are also starting to form, and the brain and spine continue to develop.

Week 6: The head continues to develop. The internal parts of the ears and eyes continue to form (the latter covered with the skin that will become the eyelids). The holes that later become the nostrils start to develop. The brain and the spinal cord are nearly formed. The development of the digestive and urinary systems continues, although the liver and kidneys are not yet able to function. The arm and leg buds have grown and it is now just possible to see the rudiments of hands and feet. By the end of week 6 the fetus is about ½ inch (13 mm) long.

Week 7: The placenta—through which the fetus takes nourishment from its mother's circulation and passes back waste products to be excreted—is now well developed. This is an important time for the growth of the eyes and parts of the inner ear, and the heart beats more powerfully. The digestive system continues to form, and many of the internal organs, although still in a very simple state, now exist. The lungs are growing but are solid at this point. There are small spinal movements, and the face continues to form, to the point where it is possible to see the beginnings of the mouth. The arms and legs are growing and have developed hip, knee, shoulder, and elbow joints.

Week 8: The eyes are almost fully developed but are still covered with half-formed eyelid skin. The face continues to form and now has the beginnings of a nose. It is possible to see

FORMATION OF THE BLASTOCYST

fallopian tube

fertilized egg

fertilized egg dividing

blastocyst embedded in uterine lining

morula (solid ball of cells)

sperm fertilizing egg

empty egg follicle

ovary

mature egg follicle

stages in maturation of egg

uterus

chorionic villi (projections from blastocyst lining)

trophoblast (blastocyst lining)

fluid-filled ball of cells

uterine lining

◄▲ *The fertilized egg divides into the morula. This then divides to form the blastocyst, which embeds itself in the lining of the uterus, the endometrium.*

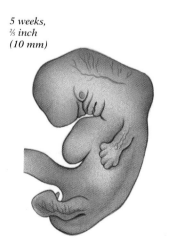

5 weeks,
⅖ inch
(10 mm)

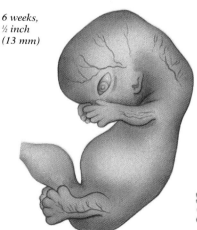

6 weeks,
½ inch
(13 mm)

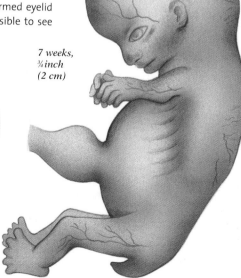

7 weeks,
¾ inch
(2 cm)

separate toes and fingers, and the limbs are able to move a little. The head, large in comparison with the rest of the body, leans downward over the chest. The fetus is now about 1½ inches (3.8 cm) long.

Week 9: The umbilical cord is formed completely and nourishes the fetus's circulatory system with blood. The inner part of the ear is complete, while the outer part is starting to form. All the major inner organs of the body continue to develop, and the mother's uterus has increased in size. The fetus is now about 1⅞ inches (4.8 cm) long.

Week 10: The circulatory system is now pumping blood around the body of the fetus. The reproductive system has begun to form, but only inside the body; the external genitals are not yet visible. The face continues to develop, and the arms and legs are now clearly formed, with tiny webbed fingers and toes. Movements of hands and feet are more vigorous but still cannot be felt by the mother. By the end of week 10, the fetus measures 2 inches (5 cm).

Week 11: The face is almost formed, and the eyelids have developed from the skin covering the eyes. Muscles are starting to form, and the development of the external sexual organs has begun. The placenta by now is a separate organ, a soft pad of tissue. The volume of fluid in the amniotic sac increases continually between the 11th and 40th weeks of the pregnancy.

Weeks 12–14: Nearly all the internal organs are now formed, but they cannot yet function independently of the mother. The obstetrician will now be able to just feel the uterus rising above the mother's pelvic bones, but she does not yet show her pregnancy.

The mother in the first trimester
Since this is the three-month period in which the basic formation of the fetus takes place, it is important for the mother to avoid anything that could cause fetal malformation. The doctor should be consulted before any drugs are taken, and all women are advised to give up smoking as soon as they become pregnant.

▶ *Between the eighth and 40th weeks of pregnancy, the developing fetus increases in length from about 1½ inches (3.8 cm) to 20 inches (51 cm).*

Before starting a pregnancy, a woman should make sure that she is immune to rubella (German measles). If she is not, the doctor will vaccinate her against it. Contracting rubella during pregnancy might cause the baby to be born with a number of grave abnormalities.

It is important for the mother to see a doctor at the beginning of the pregnancy for a thorough physical checkup and to arrange for prenatal care. Checking on the progress of the fetus is an important part of this and can range from simply measuring the size of the mother's abdomen to the use of ultrasound.

About a week before the normal menstrual period would start, there may be a little bleeding as new blood vessels are forming to nourish the growing embryo. The doctor should be told of this and of any other symptoms. He or she will advise on diet and the extra vitamins and iron that may be needed throughout the pregnancy. The doctor will also carry out regular blood pressure and urine tests.

Morning sickness (nausea and vomiting) is common in the first trimester. It is usually of no concern, although very severe morning sickness can lead to dehydration.

The fetus in the second trimester: Weeks 14–28
Weeks 14–16: The limbs continue to form; the joints are able to move; fingernails and toenails develop; and a soft, fine hair, called lanugo, covers the whole fetus. After week 14 the placenta is fully formed. Growth begins to be rapid—the fetus now weighs about 4⅞ ounces (138 g) and is approximately 5 inches (12.7 cm) long. After week 16 or thereabouts, the kidneys begin to produce dilute urine.

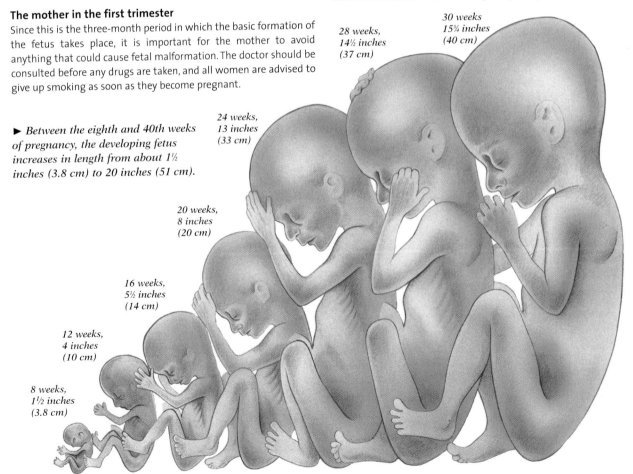

30 weeks
15¾ inches
(40 cm)

28 weeks,
14½ inches
(37 cm)

24 weeks,
13 inches
(33 cm)

20 weeks,
8 inches
(20 cm)

16 weeks,
5½ inches
(14 cm)

12 weeks,
4 inches
(10 cm)

8 weeks,
1½ inches
(3.8 cm)

Week 20: By now the fetus is able to make vigorous kicking movements, which the mother will be able to feel. The muscles are developing fast, and hair has begun to grow on the head. The fetus will now be about 8 inches (20 cm) long.

Week 24: The muscles are almost completely formed. The placenta is growing continually—all the necessary nutrients, including oxygen, pass through it from the mother to the fetus, and waste products go back through it into the mother's circulation and she excretes them. The fetus is still not able to exist independently of the mother, although in rare instances babies born prematurely at this point and given expert care have survived. The fetus weighs about 20 ounces (567 g) and is about 13 inches (33 cm) long.

Week 28: This is the point at which the fetus is said to be viable, because it would have a 5 percent chance of survival if it were to be born prematurely. It is now 14½ inches (37 cm) long and is covered with a grease (vernix) to protect it from the fluid in the amniotic sac.

The mother in the second trimester

The mother will feel the fetus moving inside her, particularly just before she falls asleep. Her own blood circulatory system has changed, with a continual increase in the production of blood cells.

Many women find that they drink larger quantities of liquid than usual during this period, and some may need iron supplements to help in the increased production of blood.

By week 20, the breasts are ready for breast-feeding. Some women find that their nipples produce a yellow fluid, called colostrum, but not all pregnant women experience this. Those who do not should not worry that their ability to breast-feed will be affected.

At this stage of the pregnancy, some women have indigestion, heartburn, and constipation, and they need to take these things into account when they are planning their diet. As the pregnancy advances, the increase of weight and pressure on the woman's internal organs can cause hemorrhoids in the rectum and varicose veins in the legs. The hemorrhoids can be prevented partly by avoiding constipation, and the irritation they cause can be relieved by ointment or suppositories, obtained from the doctor. Elastic support stockings or panty hose, put on before getting out of bed in the morning, help to prevent varicose veins.

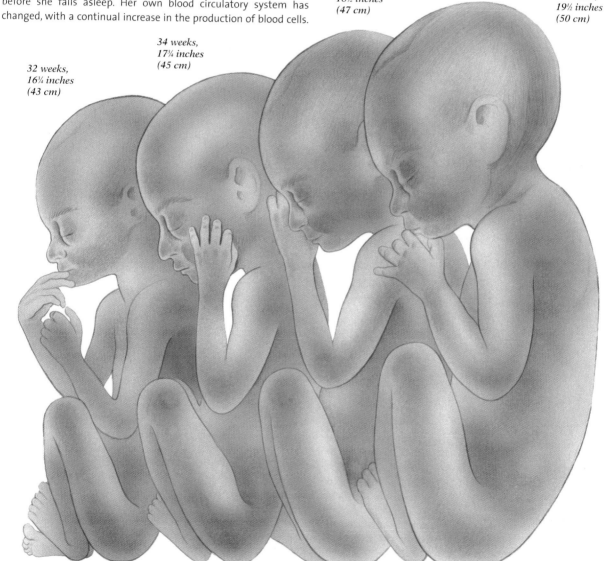

32 weeks, 16¾ inches (43 cm)

34 weeks, 17¾ inches (45 cm)

36 weeks, 18½ inches (47 cm)

38 weeks, 19½ inches (50 cm)

Questions and Answers

My sister had a baby with spina bifida. Will I have one too?

There is some evidence that spina bifida (a spinal abnormality) occurs more in some families than others. Supplementary folic acid, taken from the start of the pregnancy (or preferably before), can significantly reduce the risk of spina bifida. In the United States, breakfast cereals are enriched with folic acid for this reason.

Do I need to take vitamin pills during pregnancy?

If you are following a good diet, you may not need to, but many women find that because of the demands of pregnancy, they do need extra vitamins. Most doctors recommend vitamin supplements.

Is it dangerous to drive a car during pregnancy?

No, as long as you are still able to concentrate (some women find concentration difficult in the later months). By the seventh or eighth month, wearing a seat belt may be uncomfortable and you should ask someone else to drive.

Can I find out the sex of my baby through amniocentesis?

In amniocentesis, a fine needle is passed through the woman's abdomen to draw off a sample of amniotic fluid. A chorionic villus biopsy samples cells surrounding the embryo. Both techniques are used to test for abnormalities in the fetus and can also detect its sex, although amniocentesis carries a slight risk of miscarriage. Later in pregnancy, an ultrasound scan can reveal the sex, depending on the baby's position.

Should I stop smoking while I'm pregnant?

Yes, definitely. If you smoke, your baby may have a low birth weight and be more vulnerable to illness.

The development of the hands

▲ *By the sixth week, the arm buds are growing and the fetus has rudimentary hands.*

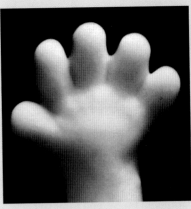

▲ *In the seventh week, the structure of the hand begins to form: the finger ridges are visible.*

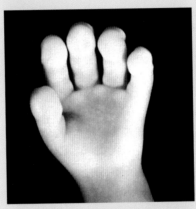

▲ *In the eighth week, the fingers and thumb, with their broad pads, are separate from one another.*

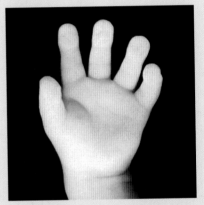

▲ *By the 13th week, the pads are smaller, the nail beds have begun to develop, and the hands curl.*

The fetus in the third trimester: Weeks 29–40

The growth of the fetal body has now caught up with that of the head, and the fetus has the physical proportions of a baby. He or she is much thinner, however, because the subcutaneous fat (fat under the skin) has not yet developed. The amount of vernix has increased. The length of the fetus at 34 weeks is about 17¾ inches (45 cm). A baby that is born prematurely at this stage has a 15 percent chance of survival.

Week 36: By this point, the chance of survival is increased to 90 percent, because the lungs are fully formed. In many cases, the baby has turned to rest head downward in the uterus, but in women who have already had a child, this may not happen until later. The testicles of the male baby have come down into the scrotal sac, and the vernix has increased. The baby's weight is increasing by about 1 ounce (28 g) per day at this stage.

Some babies are born with the fine lanugo hair still on their arms, legs, or shoulders, but it usually disappears in the final weeks of pregnancy.

Birth will take place at about the 40th week, although some women go into labor later or earlier than this. When the baby is born, there will still be patches of vernix on the body but not on the eyes and mouth. At birth the child will be about 20 inches (51 cm) long and will, on average, weigh about 7¾ pounds (3.5 kg).

The development of the feet

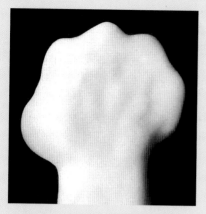

▲ *By the seventh week, clefts have formed at the ends of the leg buds, which will separate into toes.*

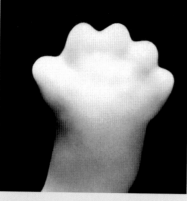

▲ *Early in the eighth week of pregnancy, toe ridges develop. The rudimentary toes are webbed.*

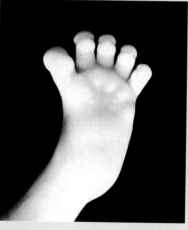

▲ *By the ninth week, the pads of the toes are visible, and so is the heel. The legs are lengthening.*

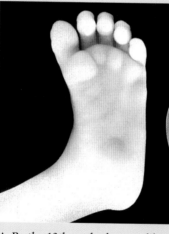

▲ *By the 13th week, the toes, like the fingers, are separate from one another and their pads are smaller.*

the mucus plug in the cervix—the neck of the uterus—are among the first signs that the baby is about to be born. The woman's cervix then starts to dilate, and the baby begins his or her journey to the outside world.

Although at birth the average baby weighs approximately 7¾ pounds (3.5 kg), the size of a newborn baby is determined by many factors. These include the length of the pregnancy, the size of the child's parents, and how well the fetus was nourished during the pregnancy.

The hair on the newborn baby's head varies from down that is hardly visible to thick hair that is about 1½ inches (3.8 cm) in length. The nails reach to the ends of the fingers and toes, or even a little beyond them, and the baby's eyes are almost always dark blue in color, or have a bluish tinge, because the eye coloring is not yet fully formed.

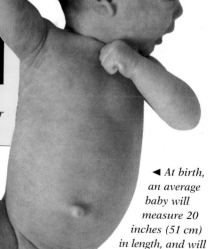

◄ *At birth, an average baby will measure 20 inches (51 cm) in length, and will weigh about 7¾ pounds (3.5 kg).*

The mother in the third trimester

By the third trimester, the uterus has expanded a great deal, and many women find that it is difficult to walk without leaning back a little, which can cause backache. It is likely that the mother will experience occasional, painless contractions of the uterus, which are normal and help with circulation through the placenta.

The mother will find that lying on the stomach has become uncomfortable. However, once the baby's head has engaged—descended into the pelvis—many women feel a great deal more comfortable, because the pressure on the stomach and diaphragm will now be reduced substantially.

Sometime around the 40th week, labor will begin. The mother's pelvic bones have already become more separated in readiness for the delivery of the baby. Powerful contractions, rupture of the amniotic sac (and the sudden loss of amniotic fluid), and the loss of

See also: **Birth; Ovaries; Placenta; Pregnancy; Uterus**

Gallbladder

Gallstones are made from bile pigment and cholesterol. They sometimes form in the gallbladder, where bile, produced by the liver, is stored. They may not cause trouble, but if they do, treatment is straightforward and effective.

Bile is made in the liver and drains into a channel called the common bile duct. During digestion bile is secreted into the intestine via an opening in the side of the duodenum, which is the first part of the small intestine. Bile plays an important part in the digestion of fatty foods, then passes on to join the feces (solid waste matter).

Spouting from the side of the bile duct is a channel leading to a bag called the gallbladder, where 1/4 pint (0.12 l) of bile can be stored. Water in the bile is absorbed gradually through the walls of the gallbladder so that eventually the bile becomes up to as much as 10 times more concentrated than it was in its original form. Exactly why our bodies store bile remains something of a mystery. If a person's gallbladder is removed by surgery, he or she can do very well without it. The gallbladder empties its bile into the intestine whenever fatty food arrives from the stomach. Bile has the ability to break down or emulsify fat, much like detergent.

Gallstones

There is a tendency for stones to form from the concentrated bile in the gallbladder. These are exactly what they sound like: hard, stony objects that can vary in size from that of a pigeon's egg to a tiny bead. About one in 10 people over the age of 50 probably has a stone or stones in the gallbladder, and these may cause trouble at some time.

There are three kinds of gallstones. The most common variety are called mixed stones because they contain a mixture of the green pigment in bile and cholesterol, one of the chemicals synthesized in the liver from available materials. They develop in multiples of up to 12

POSITION OF THE GALLBLADDER

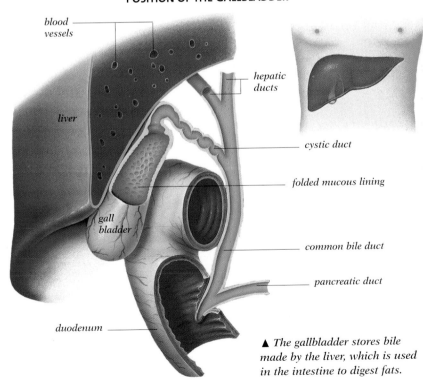

blood vessels

liver

hepatic ducts

cystic duct

folded mucous lining

gall bladder

common bile duct

pancreatic duct

duodenum

▲ *The gallbladder stores bile made by the liver, which is used in the intestine to digest fats.*

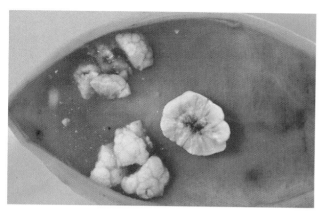

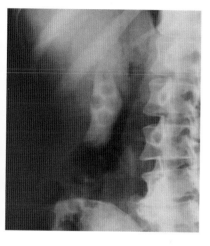

▲ *This opened gallbladder contains cholesterol stones.*

▲ *These stones are composed mainly of green bile pigment.*

◄ *A dye, either injected or swallowed, makes gallstones visible on an X ray.*

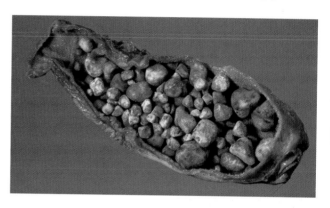

► *Mixed stones containing bile and cholesterol can occur in large numbers.*

at a time and are faceted so that they fit together in the gallbladder. Cholesterol stones, as their name implies, are formed largely from cholesterol. They occur singly or in pairs and can grow up to ¹/₂ inch (1.27 cm) in diameter, becoming large enough to block the common bile duct. Pigment stones are largely made of green bile pigment. They are small and occur in large numbers. They tend to form owing to illnesses affecting the composition of the blood.

Inflammation of the gallbladder

The gallbladder itself can become inflamed. This inflammation, called cholecystitis, is caused by a bacterium. It can develop as a complication of stones in the gallbladder. Symptoms vary, but there is usually pain and vomiting. Acute cholecystitis is sudden, with agonizing pain in the upper right side of the abdomen accompanied by fever and vomiting. Chronic cholecystitis is a long-standing inflammation causing an ache, nausea, and flatulence.

Biliary colic

Occasionally a large stone will find its way out of the gallbladder and become stuck in the common bile duct. The resulting complaint, which is called biliary colic, causes severe pain in the abdomen, high fever, and sweating. The stone prevents bile from reaching the intestine, so that the feces turn putty-colored. This damming-up of bile also causes the pigment normally excreted in bile to enter the bloodstream, and the patient turns a yellowish color and becomes jaundiced. At the same time, the body tries to compensate by

getting rid of the excess pigment in the urine, which turns dark brown or orange. It is vital that this condition is relieved promptly, otherwise the obstruction can cause severe damage to the liver, eventually leading to cirrhosis.

Diagnosing gallstones

Anyone suffering from cholecystitis, biliary colic, or jaundice is examined for gallstones to find the cause of the problem. Various techniques are used. Not all gallstones show up on X-ray pictures; however, certain dyes may be introduced into the gallbladder or its adjoining areas to give a clearer indication. The fiber-optic tube, which transmits an image down its length, and a technique similar to echo-sounding, using high-frequency sound waves, are both used.

Treating gallstones

Some doctors think it is better to leave gallstones alone if they are not causing problems; others prefer to remove them.

A single attack of cholecystitis can be settled by antibiotics, and it may be possible to dissolve stones over a period of months by means of drugs. Obstruction of the common bile duct, however, requires a surgical operation to find and remove the stone then relieve the blockage. The gallbladder is now frequently removed through a laparoscopy, using a tube inserted through a small incision.

See also: **Bile; Digestive system; Liver**

Genitals

Questions and Answers

My husband and I have been attending a fertility clinic, and he has been told that he has a low sperm count. Could this be caused by wearing nylon underwear?

Unlike any other body organ, the testes, which produce sperm, cannot work properly at body temperature; that is why they are located outside the body. Nylon does not absorb perspiration and may therefore raise the temperature of the testes. Nylon underwear alone should not cause infertility, since it is not worn 24 hours a day (for instance, it is probably not worn in bed). However, if your husband's sperm count is low naturally, nylon underwear may affect his fertility.

My mother uses a douche for personal cleanliness. Is this necessary to keep the vagina clean?

No. The vagina is well lubricated with its own secretions from the uterus, and this natural flushing of the vaginal canal keeps it clean and free from germs. Removing these secretions by douching can make it more vulnerable to infection.

My dark-haired daughter's pubic hair is red. Is this normal?

Hair color is a result of pigment in the hair cells. It is normal for people to have pubic hair a different color from the hair on their heads. Men with dark hair may also have red beards.

Why did the school doctor check to see if my testicles had dropped?

During fetal development the testes lie in the abdomen, descending into the scrotal sacs before birth. Sometimes they don't descend, however, and if this condition is not corrected surgically before puberty, sterility may be the result.

Owing to the social embarrassment people often feel about discussing the male and female genitals, many people still do not fully understand the structure and working of their own reproductive organs or those of the opposite sex.

The genital organs in men and women are those designed for the purpose of sex and reproduction. An understanding of how the male and female genitals function can help people who are engaging in these activities become more aware of their natural physical processes.

The male genitals

In men the genitals consist of the penis and the testes (which are situated outside the body) and the prostate gland, the seminal vesicles, and the two vasa deferentia (each one is a vas deferens) through which the sperm pass up from the testes to the ejaculatory area.

The testes are enclosed in a pouch of loose skin called the scrotum. This consists of two compartments called scrotal sacs, one for each testis. The testes have two functions: to produce sperm (the male reproductive cell) and to produce the male hormone testosterone. The hormone is responsible for male genital development and has a vital role in the secondary sexual changes of puberty in boys, such as the growth of body hair and the changing and deepening of the voice.

The male genital system is designed to produce millions of sperm in a continual process and deposit them in the female. They are first formed in a mass of small tubes, the seminiferous tubules. As new sperm cells are produced, they push the older ones along to an area where the tubules join together into a coiled tube called the epididymis; there is one of these in each testicle. Here the sperm mature and gain the ability to move. The sperm are stored in the epididymis and vas deferens until they are ejaculated. If the sperm remain in the epididymis and are not emitted for several weeks, however, they break down, turn into liquid, and are then reabsorbed by the system.

MALE GENITAL ORGANS

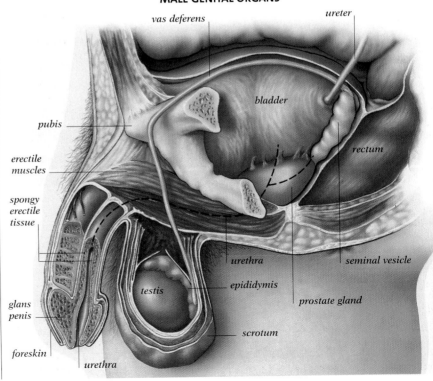

FEMALE GENITAL ORGANS

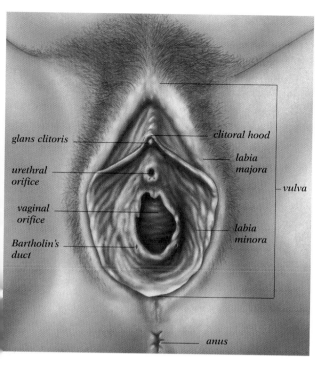

glans clitoris

clitoral hood

urethral orifice

labia majora

vaginal orifice

vulva

Bartholin's duct

labia minora

anus

When a man is aroused sexually and ready to ejaculate, the sperm travel up from the epididymis and along the vas deferens on each side until they arrive at the prostate gland. Here the sperm become mixed with seminal fluid, which is secreted by the seminal vesicles and the prostate itself, to form a sticky, gel-like fluid called semen. This passes into the penis via the urethra, a tube running down through the middle of the penis that also carries urine from the bladder.

During sexual excitement the urethra is shut off from the bladder so that the two functions are kept separate. When the man has an orgasm, the semen spurts out of the tip of the urethra in an ejaculation that contains more than 300 million sperm. Besides providing an outlet for urine, the main function of the man's penis is to deposit semen in the woman's vagina. The penis consists mainly of spongy erectile tissue and is normally soft. When a man is aroused sexually, however, a large amount of blood flows into his penis under pressure, and muscles at the base of the penis tighten, trapping the blood. This makes the penis longer, wider, stiffer, and erect so that it can be inserted into the vagina. The glans penis, a bulge at the end of the penis, is sensitive to stimulation. This area is covered by a collar of skin called the foreskin, which is often removed for cultural or medical reasons in a minor operation called circumcision.

Male problems

Because the urethra serves the dual purpose of carrying semen and urine, pain in the penis can be caused by problems with the kidneys and bladder, such as kidney stones that lodge in the penis.

Infection of the testes can be due to viruses or bacteria. Painful inflammation of the testes is a known complication of infection with the mumps virus. Bacteria that cause sexually transmitted diseases can also cause testicular inflammation. In the latter case,

symptoms include burning with urination, urethral discharge, and possibly rash. Prompt medical evaluation is essential. The prostate gland usually increases in size with age. This can cause difficulty in urinating, with failure to empty the bladder in turn causing increased frequency of urination. Once an enlarged prostate is diagnosed, treatment with medicines to shrink the size of the prostate may be of benefit. Some men will require surgery. Prostate cancer is a common cancer in men. Inflammation of the prostate, usually from infection, is called prostatitis.

The female genitals

The woman's reproductive system produces ova, or eggs, to be fertilized by the sperm. It is also equipped to nurture the fertilized egg so that the egg can develop into a fetus. Strictly, the egg-producing ovaries, the fallopian tubes leading out of them, and the uterus, where the fetus grows, are all part of the female genitals. However, the term is used most often for the parts of the reproductive system below the cervix, or neck of the uterus—the external genitals plus the vagina.

Below the cervix is the vagina, a muscular canal 4–5 inches (10–13 cm) long. Normally the walls of the vagina lie flat against each other but open to accommodate the penis during sexual intercourse (and enlarge wide enough for the baby to pass down during birth). The vagina is lubricated by secretions through its walls and from Bartholin's glands, located on the wall near the opening of the vagina. These secretions increase with sexual excitement and at different times during the menstrual cycle. The continual secretion passing down the vaginal canal keeps the vagina free from infection.

The vagina's entrance is covered and protected by the external genitals, known as the vulva. The vulva consists of two folds of flesh, the outer and inner labia. The outer labia are padded folds of muscle and tissue that become covered in pubic hair at the onset of puberty. Between them lie the smaller inner labia, which are sensitive to touch and swell up during sexual excitement, when they change from pale pink to a darker pink.

The inner labia join together in front of the clitoris, a small fleshy organ covered by a flap of skin called the clitoral hood. The clitoris corresponds to the male penis and is made of erectile tissue that swells up when the woman is excited sexually. It is extremely sensitive to touch and for most women is the center of orgasm. Unlike the penis, however, the clitoris plays no other part in reproduction and does not carry urine; the urethra opens directly into the vulva.

In virgins the entrance to the vagina is partly closed by a thin membrane called the hymen. Rupture of the hymen may result from trauma. However, it is often stretched and torn by the first experience of sexual intercourse, resulting in temporary pain and bleeding. Rarely, the membrane is so thick that penetration is impossible and the hymen has to be removed surgically.

Female problems

There are many minor ailments of the female genitals that can cause discomfort or pain. They include bacterial, fungal, and viral infections, most of which can be treated with medicines. If left untreated, sexually transmitted diseases can have serious consequences. Regular Pap smear tests can detect the precancerous stage of cervical cancer, and the outlook is good if treatment is begun early.

See also: **Birth; Conception; Penis; Prostate gland; Sperm; Uterus; Vagina**

Glands

Glands secrete substances that perform a wide variety of special functions, such as producing sweat, regulating the metabolism, aiding the digestive processes, and controlling the reproductive system.

Sweat does not smell offensive until it begins to decompose. The solution is frequent washing and using a deodorant. Also, consult your doctor—he or she can recommend other treatments.

If one of the glands stops working, can another be transplanted?

No, but other treatment can be given. If the endocrine glands (which form the hormone system) stop working, a hormone substitute can be given by mouth. If the pancreas stops producing insulin, diabetes results. This is treated with insulin by injection.

My aunt is tired all the time. Could there be something wrong with her glands?

An underactive thyroid gland can cause tiredness, but it would also cause other symptoms. Her doctor will be able to tell her if her thyroid gland is working properly.

I had infectious mononucleosis, and my friend said it was a glandular disease. Was she right?

A common feature of infectious mononucleosis is a swelling of the lymph nodes, and people often refer to this as a swelling of the "glands." However, infectious mononucleosis is a disease that affects the whole body, rather than the lymph nodes alone.

Why, when my son gets tonsillitis, do the glands in his neck swell?

The tonsils are part of the lymph defense system. The tonsils communicate with the lymph nodes (which are not actually glands) in the neck, which swell up in response to the infection.

The body contains many types of glands—organs that manufacture and secrete substances with various functions.

Glands fall into three groups based on the way that they secrete substances. Endocrine glands are located in sites throughout the body. They pass secretions into the bloodstream, which then carries them to the specific body organs that they stimulate. Endocrine glands collectively form the hormone system. They include the pituitary gland and the thyroid.

Exocrine glands include those glands that release their secretions to the surface of the body—such as the sweat glands—or through the large ducts: for example, the pancreas, which secretes juices into the small intestine.

Finally, there are the so-called glands of the lymphatic system. In fact, lymph nodes are not glands but were so named when their function was unknown. They produce the special type of blood cell called lymphocytes, and antibodies against infections. The thymus gland at the base of the neck prepares the lymph cells to deal with infection.

▲ *The swollen "glands" that an infection may produce are, in fact, lymph nodes.*

Functions

The endocrine glands form a system that controls the way the body handles its basic chemical processes. The hormones produced by these glands are responsible for controlling such things as the levels of steroids, sugar, and calcium in the blood. Hormones also influence nearly every aspect of life, including growth, reproduction, sexuality, metabolism, and mental function.

The pituitary gland, part of the endocrine system, is the body's master gland. It produces its own hormones but also influences the hormone production of other glands. These include the thyroid gland, which stimulates metabolism and regulates the heart rate; the parathyroid glands, which control the amount of calcium and phosphorus in the blood; the adrenal glands, which produce steroid hormones for metabolism and adrenaline for energy; and the sex glands, which control the production of testosterone in men and estrogen and progesterone in women.

The pancreas has both exocrine and endocrine functions. It secretes enzymes for digestion into the small intestine, and as an endocrine gland, the pancreas produces, monitors levels of, and secretes two hormones—insulin and glucagon—from the islets of Langerhans, which are even smaller glands found throughout the pancreas. Insulin aids the body in utilizing the sugar in the blood for energy, and glucagon promotes the release of sugar from the tissues.

Exocrine glands do not form a coordinated system; each type usually works independently. One group of exocrine glands is responsible for breaking down food into basic substances that can be absorbed into the bloodstream via the intestines. This group includes the salivary glands and the glands of the stomach.

The largest exocrine gland is the liver. Among its other functions, the liver secretes bile, which is carried to the small intestine, where it breaks up fat.

Another group of exocrine glands is the collection of cells that produce substances to protect the body surfaces. The inside of the intestines, the airways in the lungs, and the nasal passages are all lined with mucus-secreting glands that form a protective layer. The skin is protected by its specialized surface, but it too has glands opening onto it. These include the sweat glands, which help control the body's temperature, and the sebaceous glands, which produce an oily or waxy secretion. Wax in the ear is from modified sweat glands.

Glands and their functions

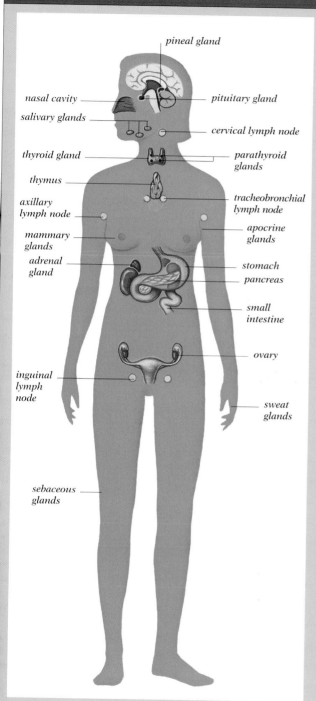

GLAND	FUNCTION
ENDOCRINE Pituitary Adrenal Thyroid	Combine to control the life-supporting processes of the body. The pituitary is the master gland with which the other glands interact.
Parathyroid	Controls calcium and phosphorus levels in the blood.
Pancreas (islets of Langerhans)	Produces insulin and glucagon to aid the body in using sugar and promoting the release of sugar from the tissues.
Ovaries/Testes	Interact with the pituitary gland to control reproduction and sexual characteristics.
EXOCRINE Salivary	Produce saliva that lubricates food to be swallowed.
Pancreas	Produces alkali and digestive enzymes that drain into the small intestine.
Glands in the lining of the stomach and the intestines	In the stomach the glands produce acid as well as mucus to aid digestion. In the small intestine they produce alkali and digestive enzymes. Throughout the intestines they produce protective mucus.
Mammary glands	Produce breast milk.
OTHER Sweat glands	Produce sweat that helps the body to lose heat by evaporation.
Apocrine sweat glands	Function uncertain, but they produce sweat that is odorous as it breaks down.
Sebaceous glands	Produce sebum that is waxy to help protect the skin.
Lacrimal (tear) glands	Produce tears that lubricate the eyes and wash away any particles of dust.
Mucous glands	Produce mucus to help protect the air passages from infection.
Pineal gland	Produces melatonin to control body clock.
LYMPHATIC Thymus	Concerned with programming lymph cells to fight infection.
Lymph nodes	Produce large amounts of lymph cells and antibodies to fight infection.

The mammary glands, which supply breast milk, are another example of exocrine glands. Finally, there is a structure deep in the brain known as the pineal gland. It produces a secretion called melatonin, which synchronizes the body's diurnal ("24-hour") clock.

See also: Adrenal glands; Endocrine system; Hormones; Liver; Parathyroid glands; Pituitary gland; Sebaceous glands; Thyroid

Growing pains

Questions and Answers

My son is tall for his age. Is he more likely to suffer from growing pains than boys who are shorter?

No. As far as we know, growing pains have nothing to do with growth, and they are no more common in children who are tall or going through a spurt in growth. There is no need for you to worry about this.

Are growing pains similar to the muscular aches you get after strenuous exercise?

Yes. Growing pains usually occur in the calf and thigh muscles and do not affect the joints. They often happen when a person is fatigued after strenuous exertion.

Do growing pains mean that my child is likely to have rheumatoid arthritis when she grows up?

Children with bad growing pains often come from families who suffer minor, vague rheumatic pains. However, there is no connection between growing pains and severe rheumatoid arthritis. Your child is no more likely than any other to have it in adult life.

My son has just entered junior high school and often complains of growing pains. He says they're so bad that he doesn't want to attend classes. What should I do?

Growing pains are not caused by illness; your son may be using them as an excuse not to attend classes. Many children find it difficult to go from the juvenile atmosphere of grade school to the more adult atmosphere of junior high school, and they should be treated with understanding. Try to get him to discuss what is really bothering him, and help him to deal with it. You may need to talk to his teacher if there seems to be a real problem at school.

In their preadolescent and early adolescent years, children often complain of mysterious pains in their limbs. These are known as growing pains. Their cause is unknown and may be psychological; loving reassurance is the cure.

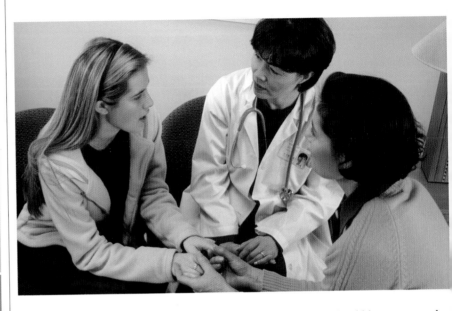

▲ *Anxiety may cause growing pains, so a child who has them should be encouraged to discuss his or her problems.*

"Growing pains" is the name given to some of the more indefinite pains of childhood, which are not related to any kind of physical disease. They are most common between ages eight and 12, and almost all children have them at some time or other.

Causes
Growing pains are not connected with physical growth, nor are they caused by allergies, infections or any other disease. They may be an expression of emotional growth, which can be painful for children who have difficulties adjusting to new phases of life. High-strung, nervous children complain more often about growing pains, perhaps because they pay more attention to their bodies than children who are busy playing. Children with growing pains are more likely to belong to families where relatives suffer from rheumatic aches and pains, which may be hereditary.

Symptoms and dangers
Growing pains occur mainly in the long bones of the arms and legs, and in the calf muscles. They occur more often when the child is tired or fatigued after a prolonged bout of exercise. Severe pains in the arms or legs should always be taken seriously. Growing pains are not likely to be confused with any serious illness, but if they are extremely painful, the child should see a doctor.

Treatment and outlook
Growing pains should not be dismissed just because they are not threatening to health. A little extra mothering or fathering is soothing at a time when there may be stress in a child's life, such as starting a new grade at school or changing a circle of friends. The best remedy is a hug, patience, and encouragement. All children with growing pains will grow out of them.

See also: Bones; Growth; Muscles

Growth

Is there any food I can give my sons to make them grow tall?

No. While it is certainly true that a lack of food will stunt growth, any child who eats a balanced diet will achieve his or her full height. However, children whose parents are both tall are likely to grow into tall adults.

Do growing pains exist?

It is not usual to suffer any pains from normal growth, but growth can cause great psychological problems for a late developer. A teenager may find it distressing to see everyone else grow up while he or she remains small and childlike. Provided that there is no abnormality causing the child's slow growth, there is no cause for concern; any growing pains are psychological in origin and are likely to be cured by love and reassurance from the parents.

My brother is very tall for his age. Is he likely to become a bully?

No, being tall is not going to make him into a bully. If he has that sort of personality it will happen anyway, no matter how tall or short he is. However, tall children, particularly boys, seem to be better at games and sports. This makes them feel more secure and they are less likely to bully others. There is also a tendency toward greater height in people who are successful in business.

Do all children have a growth spurt?

Yes. The rate of growth tends to be fairly steady until the adolescent growth spurt, which occurs at about the age of 11 in girls and later in boys, although the timing and extent are variable. The average 13-year-old girl tends to be slightly taller than a boy of the same age; he will catch up later.

Growth is an amazing process, from the development of a baby to the sudden spurt toward maturity that occurs in adolescence. What is normal growth—and what happens inside a child when he or she is growing?

▲ *The exact timing of the adolescent growth spurt varies a great deal among any group of teenagers, with the girls usually starting before the boys.*

It is striking to notice how much people vary in height and shape, and these variations make it difficult to provide measurements that could indicate normal growth. However, to monitor healthy growth in children, a range of normal heights has to be established for different age groups. The way doctors express the range of heights is to talk about percentile values.

Percentile values for heights are established by grouping children of the same age and sex (but of different heights) and arranging them in height order. The 50th percentile refers to the height of the child in the middle; thus half of the children will be taller and half shorter than this measurement. With the 75th percentile, three-quarters of the children will be shorter and a quarter taller than this measurement. Similarly, a quarter will be shorter and three-quarters taller than the 25th percentile. The same principle applies in working out percentile values for weights.

Doctors do not line up groups of children every time they establish percentiles; they use standard height charts, based on surveys of a wide range of children. In general they regard anyone who is above the 3rd percentile and below the 97th percentile as normal. Thus 94 percent of the population is viewed as normal.

Provided that a child's height falls between these two benchmarks for his or her age group, growth is considered satisfactory. If a child is outside these limits the doctor will look for some explanation, but this does not mean that there is a problem necessarily. The reason for the child's height could simply be that he or she has very tall or very short parents.

Growth spurts

The rate at which children grow depends very much on their age. They grow quickly during the first months of life and even by their first birthday they are still growing at the rate of 6 inches (15.2 cm) per year. After about the age of three, growth settles down to a steady rate of between 2 inches (5 cm) and 3 inches (7.6 cm) per year, until the adolescent growth spurt toward maturity.

Throughout childhood the average girl is about ½ inch (1.3 cm) shorter than the average boy. However, at about the age of 11 the girl will overtake the boy, because the adolescent growth spurt comes earlier in girls. This means that at the age of 13 the average girl is about 1 inch (2.5 cm) taller than the average boy. By the time they are both 14, the boy's growth spurt will be taking place while the girl's has almost finished. Over the next two or three years, the boy will continue to grow, and he will end up approximately 5 inches (12.7 cm) taller than the girl.

In reality there is a great variation in the ages at which the growth spurt starts, and this can cause considerable heartache in late developers. The spurt may start at any time between the ages of 9½ and 14½ years in girls, and between 10½ and 16 in boys. Some boys may not have even started their growth spurt by the end of their sophomore year in high school, and it is easy to imagine how they must feel if all their friends are fully developed by this stage.

The growth spurt begins before the other changes associated with puberty in both sexes. Sometimes this is very obvious, particularly in girls, who may grow to nearly their full height without much change in the shape of their breasts or hips. Normally girls start their periods, which mark the beginning of puberty, after the growth spurt has reached a peak. Although a girl's development in height precedes that of a boy by about two years, the changes in sexual characteristics happen only about six months earlier in a girl than in a boy.

The rate at which these changes occur and the degree of sexual development vary. Children whose sexual development lags behind feel left out, and their parents may worry about whether development is occurring normally. The factor that seems to have the most influence is the age at which puberty occurred in the parents; if both parents had a late puberty, then the children can also be expected to experience a late puberty.

Until the growth spurt, children tend to follow the lines on a percentile chart closely. If a three-year-old is of exactly average height on the 50th percentile, he or she would be expected to be within 1½ inches (3.8 cm) of that average height by the age of 10.

Changes in shape

As children grow, their proportions change because various parts of the body grow faster than others. The most striking example is the head; at birth it is already about three-quarters of its final size, and by the first birthday it has grown to almost full size. By adulthood the head is a much smaller proportion of the total body size.

The order in which parts of the body grow during adolescence also varies. As a general rule leg length peaks first, followed about

Growth and facial changes

Compared with an adult's, a child's head is big in relation to his or her body. At the age of 1½, this boy's head is already almost full-size.

By the age of four, the boy's growth rate has slowed from the rapid development that takes place in infancy. He will grow gradually and steadily until adolescence.

At 10 the boy is on the verge of the adolescent growth spurt that starts shortly before puberty. His face still has the soft, rounded look of childhood.

At 13 the boy's height is shooting up and his face has changed considerably. The changes are most noticeable around the brows and chin.

nine months later by the length of the torso. Chest and shoulder width, particularly in boys, is the last dimension to reach a maximum size. An interesting difference between the two sexes is in the development of the face during puberty. In girls there may be little change, while in boys there is forward growth of the chin, which becomes more pointed, and a forward growth of the eyebrow ridges, which make the face look more rugged.

How growth takes place

All the bones enlarge as children grow, but the most dramatic growth takes place at each end of the long bones. These bones end in growth plates made of cartilage, and for this reason the plates do not show up on an X ray. These plates continue to grow until the end

My 15-year-old cousin seems to be growing and growing. Can you tell me when he is going to stop?

The average age for upward growth to stop is about 17 for boys and 15 for girls, but there is often a variation of at least two years on either side. In your cousin's case it all depends on the time when his growth spurt started. On average the rate of growth will fall to about 1 in. (2.5 cm) a year within three years after the growth spurt starts.

My 14-year-old daughter complains that she is badly proportioned and ungainly. Do different parts of the body grow at different speeds?

Yes. The brain and head have practically reached their full size by the time a child is about eight. Hands and feet can appear large in a young teenager because they reach their full size early in the adolescent growth spurt; they will appear in proportion when the rest of the body catches up. The reproductive organs, however, grow very slowly before puberty and then take a few years to reach their full size.

My son is 13, and I am very worried because he doesn't seem to be growing at all. What should I do to encourage his growth, and should I take him to the doctor?

It is unlikely that anything is wrong, particularly if he has been at a reasonable height during childhood. You have probably become worried because all his friends are getting taller.
Start measuring his height carefully every three months. Within the next year you will almost certainly see that his rate of growth has increased enormously. If this doesn't happen within a year or so, it is worth seeing your doctor.
Keep all his measurements, along with any from his younger days, and take them with you to the doctor.
If your son has always been very small, you could see your doctor now to make sure that nothing is wrong.

HOW LONG BONES LENGTHEN

Cartilage cells in the growth plate multiply, move down the bone, and produce a calcified matrix. The cells die, leaving spaces. Osteoblast cells produce bone to fill the spaces and replace the matrix.

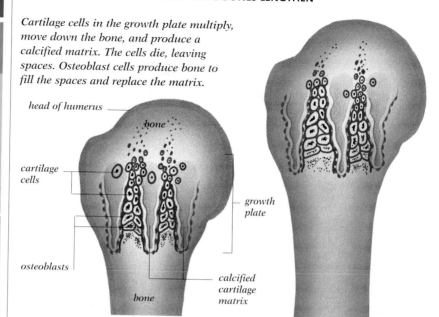

head of humerus — bone — cartilage cells — growth plate — osteoblasts — bone — calcified cartilage matrix

of adolescence, by which time all the cartilage has turned, or fused, into solid bone. When this has happened, usually by the early twenties, the bones are no longer able to increase in length.

The growth plates in the various bones turn into solid bone in a particular order. By X-raying the bones and looking for fused growth plates, it is possible to deduce the bone age. This may differ by several years from the real age and is a better guide to a child's state of growth. Growth is regarded as delayed if the bone age lags more than two years behind the real age.

How growth is controlled

An interaction of various hormones controls growth. The most important is growth hormone, which is secreted by the pituitary gland in the base of the skull. Children who lack the hormone will not reach their full height unless the hormone is replaced early on in life. Overproduction of the hormone, often caused by a pituitary tumor, results in gigantism, a condition that causes a child to grow to well above average height. Other hormones are essential for normal growth. Lack of thyroid hormone also stunts growth, but its replacement will rectify matters.

The sex hormones are essential to changes in shape at puberty. The wider hips of girls, and their breast development, depend on estrogen, and the increase in muscle bulk and strength in boys depends on testosterone.

Growth problems

Delayed growth may be present in some of the small number of children who are born with genetic defects. One example of this is Turner's syndrome, which occurs in girls who have only one female chromosome instead of two. These children remain short and require sex hormone replacement in pill form to acquire normal female characteristics.

Occasionally poor development starts in the uterus. Disease, starvation, or smoking by the mother may cause this, but it can occur for unknown reasons. Such children are born small and remain shorter than would be expected, although otherwise they grow up healthy.

Disease of all sorts may stunt growth, and recording growth can be a useful way of monitoring the general health of a child with a disease such as diabetes. Rare cases of bone disorders may also produce a short stature; for example, achondroplasia stops the growth of the long bones of the arms and legs.

A child's growth may also be slowed by nutritional deficiencies or by severe psychological deprivation—for example, a lack of love and attention.

> See also: **Bones; Growing pains; Hormones; Puberty**

Gums

My gums often bleed when I brush my teeth. Does this matter?

Healthy gums do not bleed easily, so something is probably wrong. Your gums are likely to be inflamed because of a buildup of plaque. This is a sticky substance made of bacteria that live in the mouth, food debris, and dried saliva. Plaque can be removed by brushing in the correct way—a circular, up-and-down rather than a side-to-side motion of the brush. You must keep this up for several minutes each day, even if it makes the gums bleed more at first. If your gums are still giving you trouble after a week or so, visit your dentist.

My husband eats in the restaurant at work, where I suspect that the dishes are not washed very well. Is it possible to catch gum disease from another person's cutlery?

With the exception of one rare condition, gum disease is not infectious. The bacteria that usually cause gum disease are present in small numbers even in a healthy mouth. If the bacteria are not removed by brushing, their numbers increase and cause serious damage to the gums.

I have irregular teeth. Does this mean I will get gum disease?

No, but it may make gum disease more likely, since your teeth will be difficult to clean where they overlap. Your dentist can advise you on cleaning.

Do we have to lose our teeth as we grow old?

No. There are a few diseases that tend to make the teeth fall out, but they can be treated. For almost all of us, the simple way to keep our teeth for life is by proper brushing and flossing.

More than half the population over the age of 35 in Western countries has gum disease, a condition that can be easily prevented in most cases by correct brushing and oral hygiene.

It is impossible to have good teeth without healthy gums. Disease of the gums leads to disease of the whole supporting structure of the teeth, and this is in turn the most common cause of tooth loss in people over the age of 35.

The supporting structure

Teeth must withstand the strain of biting and chewing and the knocks of daily life. They are held in the jaws by a complex mechanism that keeps them firm and resistant to stress. This mechanism, called the periodontium, has four essential components: the gingiva (gums), the cementum (between the enamel and the gums), the peridontal ligament, and the alveolar and the supporting bone.

The first component is what people see from the outside—the gums. They are actually a sleeve that forms a tight collar around the neck of each tooth and covers the bony bed in which the roots are fixed. When it is healthy the outer layer of gum is pink, flat, firm, and curved to the shape of the teeth. It fits tightly onto the enamel (the outer surface) of each tooth, providing an excellent seal.

The teeth are anchored in place by fibers called the periodontal ligaments. These are tough but they have a certain amount of elasticity, which works like a shock absorber when the teeth come under pressure as they bite and chew.

The ligaments are attached to the teeth by a hard substance called the cementum, which is out of sight beneath the gums. The other ends of the periodontal ligaments are attached to the alveolar bone, which is part of the jawbone. The roots of the teeth are embedded in this.

STRUCTURE OF THE GUMS

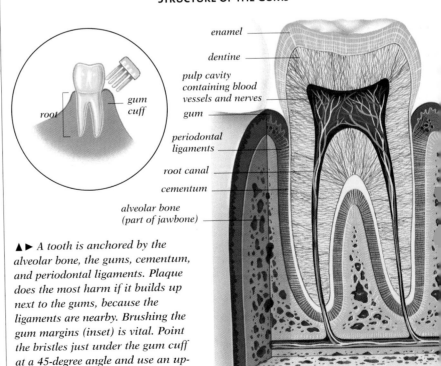

enamel

dentine

pulp cavity containing blood vessels and nerves

gum

periodontal ligaments

root canal

cementum

alveolar bone (part of jawbone)

root

gum cuff

▲▶ *A tooth is anchored by the alveolar bone, the gums, cementum, and periodontal ligaments. Plaque does the most harm if it builds up next to the gums, because the ligaments are nearby. Brushing the gum margins (inset) is vital. Point the bristles just under the gum cuff at a 45-degree angle and use an up-and-down, circular motion.*

▲ *Gum disease results from the accumulation of plaque. People should use dental floss or dental tape to clean areas that are difficult to reach with a toothbrush.*

Plaque

No matter how healthy a person's mouth is or what he or she eats, a mass of bacteria and their secretions continually accumulate on the teeth, along with particles of food. This accumulation, called plaque, damages and eventually destroys any tissue with which it comes into contact. Plaque is invisible in small amounts; this invisibility makes it all the more harmful.

Plaque is the main cause of gum disease. If it is not cleaned off regularly by proper brushing, a damaging layer of plaque can build up within days, causing a condition called gingivitis.

Gingivitis and periodontal disease

Gingivitis causes red swollen gums that tend to bleed easily. It can be prevented or cured by cleaning the teeth properly. However, if it is allowed to continue, gingivitis will result in serious damage.

As the gum continues to swell, it starts to form a pocket or crevice at the neck of the tooth. The bacteria that form plaque can multiply easily in this crevice because they are out of reach of the toothbrush. It is only a matter of time before the periodontal ligaments are damaged. This is followed by a loss of the alveolar bone itself.

Over the years this deterioration process causes the teeth to loosen until they either fall out or have to be removed because they are loose. In addition, the gums often recede and abscesses may form in the diseased gum tissue. Gingivitis is the first stage in this disease, when only the gums are affected (*gingiva* is Latin for "gums"). However, if the deterioration progresses to affect the whole supporting structure, it develops into chronic periodontal disease.

Diabetes, leukemia, and the changes that take place during pregnancy can also make people susceptible to gingivitis, as can certain vitamin deficiencies. Plaque, however, is easily the most important source of gum problems.

Other problems

Compared with gingivitis and chronic periodontal disease, other problems of the supporting structure of the teeth are comparatively rare. One exception is a condition called acute ulcerative gingivitis, which affects about five people in 100 and tends to occur when a person's general health is poor. This disease damages the parts of the gums between the teeth. It is painful and progresses rapidly. Acute ulcerative gingivitis can be treated with a mouthwash and antibiotic to counter the action of the bacteria that cause it.

Careless brushing can also tear the gums, leaving them open to infection, followed by general damage to the periodontium.

Prevention

Proper oral hygiene, the technical term for cleaning the teeth, is the simple way to prevent gingivitis. This involves using a medium-hard brush, circling repeatedly up and down from the gums to the teeth and back to the gums. This brushing needs to be done repeatedly because plaque is sticky. It takes at least three minutes of brushing correctly to remove plaque properly.

Places that cannot be reached with a toothbrush, the sides of the teeth that butt each other, need to be cleaned with tools such as dental floss or dental tape.

Most people are able to brush their teeth well enough with a plain brush, but an electric one that brushes automatically with the correct action can be more thorough. Electric toothbrushes are particularly suitable for people with weak wrists.

To ensure that all of the plaque has been removed, a person can rinse his or her mouth out with a harmless solution that stains any remaining patches of plaque red. Dentists and parents often find this useful when trying to prove to children that care of their teeth is important, or when they want to check the progress of treatment.

More serious cases

If gingivitis has developed into chronic periodontal disease, other treatments may have to be used. Some of these are uncomfortable or painful. For example, the dentist may have to cut away gum to eliminate the pockets that hold plaque. This prevents further damage to the ligaments by making adjacent tooth surfaces accessible to cleaning. The gum contours can also be reshaped to make cleaning easier, and gum tissue may be repositioned to replace gum that has been destroyed. The dentist will also urge better oral hygiene as an essential element for future health.

Outlook

In the past people used to expect to lose their teeth in old age. However, provided that people brush their teeth properly and go to the dentist every six months for checkups, there is a good chance that they will enjoy healthy gums and teeth for life. Good oral hygiene is a small price to pay to avoid needing dentures.

See also: **Bones; Ligaments; Mouth; Teeth; Tongue**

Hair

I had to have all the hair shaved off my arms for an operation. Will it grow back much longer?

No. This is one of those myths, like the myth that if you tweeze out a hair, two will grow in its place. Cutting or shaving off hair has no influence on the rate or length of growth. Hair tissue is dead, and the nerves in the follicle do not know when hair is cut.

My mother has recently had a major depression and lost all her hair. Will it ever grow back?

Probably. The reason for hair loss following a bout of depression or a sudden psychological upset is not known, but it is thought that all the hairs go into a resting phase, stop growing, and may then fall out during washing or brushing. Normally only about 10 percent of hairs rest at any one time, and so when they fall out the loss is unnoticeable. The hair follicles remain normal, and in time all the hair grows back.

Why do albinos have white hair?

Albinos lack the pigment melanin that is normally present in the hair, skin, and eyes and provides their coloring. Without any pigment albinos inevitably have white hair as well as very fair complexions and pink or red eyes.

When I was pregnant my hair became really thick, but after my daughter was born it began to fall out. Why was this?

During pregnancy, all of a woman's hair grows at once; there is no resting or shedding phase, so your hair became thicker. However, once your baby was born, the normal cycle of growth and rest resumed, and the hairs that should have fallen out did so. This is normal and there is no need to worry.

Hair has no clear survival purpose for people, but it is of great psychological importance to them, whether they worry about having too much or too little. Excessive growth of hair or hair loss can also be indicators of ill health.

Hair of some sort covers most of the human body, except the palms of the hands and the soles of the feet. It is most noticeable on the scalp, armpits, pubic area, arms, legs, eyebrows, and eyelid margins, and in men on the lower part of the face and on the chest.

The visible part of a hair is called the shaft. This is formed from a protein called keratin and is composed of dead tissue. The shaft is rooted in a tubelike depression in the skin called the follicle. The hair develops from a root, the dermal papilla, which is at the bottom of the follicle and is nourished by the bloodstream. If the root is damaged, hair stops growing and may never regrow.

The follicle also contains a gland, the sebaceous gland, which secretes a greasy substance called sebum. This lubricates the hair shaft and surrounding skin and can give hair a greasy appearance. Finally the follicle contains tiny muscles called erector pili. When a person is cold, afraid, or alarmed, the muscles contract, the hair stands on end, and the skin around the shaft bunches to form what are known as goose pimples.

Types of hair

Adults have about 120,000 hairs on the head, but this number varies with a person's hair type; redheads have fewer, blonds more. Hairs in different parts of the body have different structures: there are fine, soft baby hairs that grow on some areas of the body; long hairs that grow on the scalp; and short, stiff hairs that compose the eyebrows. Blond hair is the finest; black is the coarsest.

The type of hair shaft determines whether hair is straight or curly. Straight hair has a cylindrical shaft; curly or wavy hair has an oval shaft. The hair that is characteristic of African Americans has a flattened or kidney-shaped shaft.

Growth and color

The cells that make keratin for hair are among the most rapidly dividing cells in the body. Scalp hair grows by an average of $\frac{1}{2}$ inch (1.27 cm) a month. Hair growth is not continuous, and every five or six months the hair goes into a resting phase, during which no growth takes place. The

▼ *Race and genetic inheritance determine both the type of hair and its color.*

STRUCTURE OF HAIR

► *Each hair consists of a shaft (visible hair) and a follicle containing a root (dermal papilla) that is nourished by venous blood. A sebaceous gland lubricates the follicle. Erector pili muscles contract in response to cold or fear.*

▼ *A cross section of scalp, with hairs, follicles, and roots.*

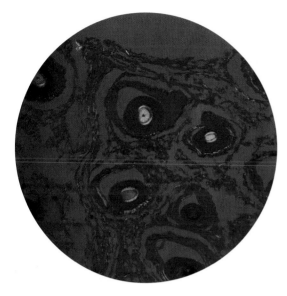

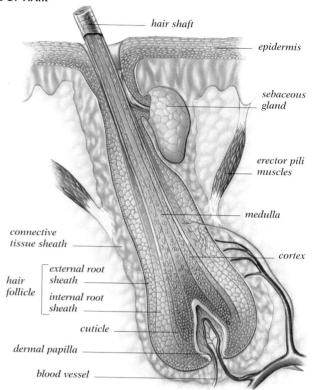

hair shaft

epidermis

sebaceous gland

erector pili muscles

medulla

cortex

connective tissue sheath

hair follicle
- external root sheath
- internal root sheath

cuticle

dermal papilla

blood vessel

roots of resting hair become club-shaped—thus the name "club hairs"—and lose their normal pigmentation. Up to 10 percent of human scalp hairs are in the resting phase at any one time.

These club hairs are the ones that seem to come out in handfuls when people wash their hair. No damage is done to the follicles, and normal hair growth begins again when the root has finished its rest.

Lining the follicles and mixed in with the cells making keratin are other cells, containing a pigment called melanin. Melanin stains the keratin and gives hair its color; red hair has an extra pigment. As people get older the melanin-producing cells in each follicle gradually stop working, turning the hair gray, and finally white. The timing of this process as well as hair color are due to heredity.

Excessive hair

Hirsutism, or excessive hairiness, is caused not by an increase in the number of hairs, but instead by a change in the character of the hair.

Our bodies are covered with soft, downy hairs called vellus hairs. In certain parts of the body, notably the scalp, the armpits, and the groin, these hairs become thicker and are called terminal hairs. At puberty, especially in males, terminal hairs become more abundant on the face, in the armpits, in the groin and pubic area, and on the legs. This change is brought about by hormones called androgens.

Hairiness is inherited, and certain types of people are naturally hairier than others. People of Latin descent, for example, tend to be much more hairy than people from Scandinavia.

Excessive hairiness is only rarely a sign of disease. It can result from excessive production of androgen hormones, either from the adrenal glands or from the ovaries. A benign condition in which multiple cysts appear on the ovaries is a common cause of hirsutism. It is treated with medicines. Development of hairiness in children or excessive hairiness in women, particularly if associated with male characteristics such as a deepened voice, requires investigation.

Hair loss

Alopecia (baldness) can be a major worry for men and, occasionally, for women. It is often due to excessive sensitivity of the follicle to androgen hormones, and is usually hereditary and irreversible. Baldness is characterized by a steady recession of the hairline from the front of the head, progressing back to the crown. It can start in the late teens in men or at any time thereafter, or in women in their forties or fifties.

Occasionally the timing of the resting phase of the hair happens all at once, instead of taking place haphazardly at different parts of the head. This occurs frequently in babies or in mothers who have just delivered, and their hair usually grows back normally. Hair may be lost in patches, which can be extensive. The cause of this pattern of hair loss, known as alopecia areata, is unknown. In adults the hair often grows back, but in children the loss is usually permanent.

Glandular diseases occasionally cause hair loss, owing to a lack of hormones such as the thyroid and pituitary hormones. Severe illness or high fever can also cause hair loss, as can radiation and some cancer drugs.

A type of alopecia that develops along the front of the scalp can result from constant pulling of the hair, and it is usually seen in girls or women who do this as a nervous habit, or who wear very tight braids or ponytails. Hair growth usually returns once this is corrected.

See also: **Baldness**

Hand

When I move my ring finger, the fingers on either side move a little also, but I can move my index finger on its own. Why is this?

The three fingers farthest from the thumb all share the same muscles to flex them into a fist and straighten them out again, so they cannot move totally independently. The index finger has some independent muscles, so it is able to move on its own.

Is it true that no two people ever have the same fingerprints?

Yes. The tips of the fingers have a precise pattern of lines, in the center of which is a loop, a whorl, or an arch. The pattern may be the same on all the fingers; it may be different on all of them; or two or three fingers may be the same.
　The pattern regrows exactly the same even after an injury, so your fingerprints never change. It is not known how the pattern evolves, but there are so many combinations that no two people have ever been found to have the same set of fingerprints.

I am a little nervous, and I find that I get breathless sometimes. When this happens my fingers start to tingle, and once my hand cramped. What causes this?

You have a complaint that is common among nervous people. It is called hyperventilation tetany and is harmless. Hyperventilation (overbreathing) takes place and alters the balance of acid and alkali in your blood. This makes the fingers tingle and your hand go into tetany—the fingers and thumb get stuck in a position with all of them straight but bunched up together. Treatment is simple: breathe in and out of a paper bag for a few minutes. This will bring the carbon dioxide in your blood back to its correct level and your hand will return to normal.

Human hands are versatile and very powerful, especially when combined with tools. However, they are also vulnerable to injury and infection, and people should take particular care when using them.

STRUCTURE OF THE HAND AND LOWER ARM

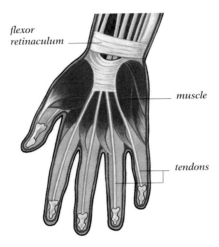

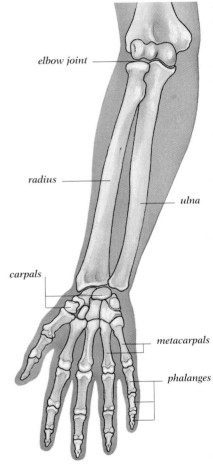

The human body has evolved over the ages. From their tree-dwelling ancestors, humans have inherited arms and hands that were developed for climbing and holding onto branches, yet were also versatile enough to pick fruit and berries. Thus the thumbs developed so that the four fingers and thumb could encircle a branch and make delicate movements. Fingernails developed to provide a hard backing for the sensitive surfaces of the fingertips.
　The human hand, like that of the monkey, is highly versatile. However, humankind has outstripped all the animals by inventing tools that the hand is ideally equipped to use.

How the hand works

To see how the hand works, a person must look at the structure of the arm from the elbow downward. In the forearm (elbow to wrist) there are two bones, the radius and the ulna. The ulna is slightly longer. When a person holds his or her hands palms upward, the two bones lie side by side in the forearm, the radius on the thumb side. If he or she then turns the hands over, palms down, the head of the radius rotates in a cuff at the elbow, crossing over the ulna.
　The wrist is a hinge between the forearm and the hand. It consists of eight cubelike bones, or carpals, joined by fibrous ligaments passing from bone to bone. The carpals are joined to the metacarpals, five bones radiating out from the wrist. They are buried in layers of flesh in the palm of the hand. Each metacarpal has a finger (or in one case, a thumb) at the end of it. The fingers consist of three bones, or phalanges, with a joint between each two. The thumb has only two phalanges in it, and so only one joint.
　The muscles that move the wrist and the fingers are mainly in the forearm. The part of the forearm that leads to the palm of the hand contains the muscles that bend (flex) the wrist and

fingers forward. The muscles in the part of the forearm leading to the back of the hand straighten (extend) the fingers and wrist.

Each muscle has a tendon (a long fibrous cord) that stretches from the body of the muscle to one of the bones. The tendon that bends a finger, for example, passes from the muscle in the underside of the forearm over the wrist joint and palm, and is attached to one of the phalanges. A strap of fibrous tissue, the flexor retinaculum, holds down the tendons and prevents them from springing away from the wrist when it is bending.

There are a number of small muscles in the hand. Some are in between the bones, but others form the fleshy bumps at the base of the palm just below the wrist. The bigger bump that lies under the thumb consists of the muscles that move the thumb across to make a pinching movement (opposition) with the fingers.

The lines that can be seen on the palm of the hand (which, some people believe, can be used by palm readers to read the future) simply bind the skin down to a flat sheet of fibrous tissue. This tissue is bonded to the bone at the base of each finger. It enables the fingers to flex and grip without a large wad of skin bunching up in the middle of the hand.

Possible problems

The hand has a large number of joints and may be particularly prone to arthritis (inflammation of the joints) as a result. Rheumatoid arthritis, which is quite common, causes inflammation in the membrane that lines the joint cavities. Arthritis may affect many joints in the body, but it seems to affect particularly the joints at the base of the fingers, between the metacarpals and the phalanges, and also the joint between the radius and the carpal bones at the

▲ *Because the thumb can move in the opposite direction from the fingers, humans are able to carry out delicate and skillful tasks with their hands.*

wrist. In a hand that is affected severely, the joints at the base of the fingers become very swollen and the fingers turn away from the thumb.

Osteoarthritis, which results from wear and tear on the joints, is not as serious as rheumatoid arthritis. It often affects large joints such as the hips and knees, but it may also affect the joints between the phalanges in the fingers, particularly the joint of the fingertip. When this happens, the fingers become slightly crooked. A small swelling (Heberden's node) often develops, one on either side of the outer joint.

A variety of treatments, including painkillers, anti-inflammatory drugs, and physical therapy, can help relieve these conditions.

Accidents

People doing jobs involving physical labor usually use their hands; therefore, the hands can be prone to accidents. Fingers may be lacerated, crushed, or even cut off.

If a person suffers anything more serious than a minor cut or bruise to the hand, it is important for him or her to see a doctor or to go to the local emergency department. Such action is necessary because deep lacerations can sever nerves and tendons in the hand. Permanent loss of function may result if the hand is not treated quickly. When the tendon to a finger is severed, for example, it is gradually pulled back toward the wrist. Surgery can repair the damage easily and effectively if it is carried out immediately, but may be impossible within a few hours.

Repair work

In recent years there has been a tremendous advance in the surgical techniques required for delicate repair work, and nowadays severed arms and fingers can sometimes be sewn back on.

Such an operation requires careful surgery using a microscope, because the stitches must be tiny so that all the blood vessels, nerves, and tendons are joined properly.

Fractures in the bones of the hand are generally less serious than bad cuts. Broken fingers are fairly common among athletes and are often treated simply by strapping the broken finger to the one next to it until it has healed.

Caring for your hands

Wear rubber gloves when you wash dishes or clothes. However, because wearing rubber gloves in hot water causes sweating and can cause irritation, it is better to use a long-handled brush when the water is hot.

Use hand cream or lotion daily to help replace lost moisture.

Generally, stains on the hands can be removed by rubbing them with a slice of raw potato or lemon. For really persistent stains, leave lemon juice on for five minutes, then rinse it off and apply some hand cream or lotion.

For cracks on the skin, sprinkle 1 teaspoon (5 ml) of sugar in 1 tablespoon (15 ml) of olive oil; mix together and rub into the skin for five minutes before rinsing off. Dry the hands and apply hand cream.

Nervous tension can cause sweaty, clammy hands. Wash the hands frequently in cool water, and keep some packaged hand-wipes or tissues nearby for use during the day.

To prevent chilblains (redness and swelling of the skin due to cold), wear warm gloves in cold weather. Do not warm cold hands close to a fire; rub them to improve the circulation instead.

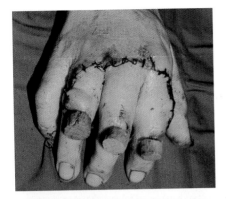

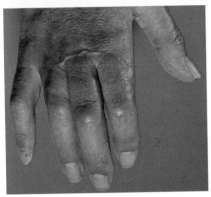

▲ *Three fingers of this woman's hand, severed in a factory accident, were successfully sewn back on within hours. The blood vessels and nerves were rejoined and steel pins secured the bones.*

Health warnings in the hand

SYMPTOM	SIGNIFICANCE
Clubbing (swelling) of the fingers, causing the nails to curve over the ends of the fingers	May be an indication of chronic infection, cancer, or chronic insufficiency of oxygen in the blood
Pale nails and pale creases in the palm	Anemia (lack of red blood corpuscles)
Koilonychia (spoon-shaped nails with prominent ridges)	Deficiency of iron
Pigmentation (coloring) of the skin creases	Addison's disease—failure of the adrenal gland to produce enough cortisone. This is extremely rare.
Liver palms (red areas on the palm, particularly at the base of the thumb and along the edge of the palm below the little finger)	Fairly common in healthy people, but may be associated with liver disease
Onycholysis (when the nail comes away from the finger)	Overactive thyroid gland, other medical conditions
Tremor of the outstretched hand	May indicate an overactive thyroid but may also be seen in alcohol withdrawal
Leukomychia (white nails with enlargement of the crescent at the bottom of the nail)	Liver disease

These symptoms may be an indication of trouble elsewhere in the body, so you should check with your doctor if you have any of them.

Dermatitis and infections

The hand is often affected by contact dermatitis (inflammation of the skin) as a result of allergy to, for example, detergents or rubber gloves. Treatment involves topical corticosteroid preparations and soothing or drying lotions.

Because the skin of the hand is exposed, it is also prone to infections. Warts, which result from a virus infection, are a common problem. Whitlows (paronychia) are infections beside the fingernails, resulting when bacteria get into the nail folds. If an abscess forms, it may need to be drained.

Dupuytren's contracture

A common condition of the hand is Dupuytren's contracture, which affects the skin of the palm and the tendons that run underneath it. The tendons, skin, and fibrous tissue of the palm cramp into a thickened mass, and the fingers become rigid, as though they were gripping something. The condition usually affects the ring and middle fingers and is especially common in older men. Treatment is not usually necessary unless the condition is very bad. In such cases some of the tissue may have to be removed.

Raynaud's phenomenon

A condition of the hand that affects younger women is called Raynaud's phenomenon. The condition involves the circulation of the fingers and makes them turn white, or even blue, in response to cold temperatures. Even minor degrees of cold may produce this reaction, and it is painful. Keeping the hands warm is often sufficient to prevent it, but sometimes drugs or surgery may be needed.

Raynaud's phenomenon occurs on its own and is not usually serious. Occasionally, however, it may be a sign of a disease that affects other parts of the body, such as scleroderma (a thickening and contraction of the skin of the hands and face).

Carpal tunnel syndrome

Carpal tunnel syndrome is caused by undue tightness under the ligamentous band called the flexor retinaculum, with compression under it of the median nerve to the hand. Treatment involves a non steroidal drug such as ibuprofen, a wrist splint, and physical therapy, or an injection of steroids into the carpal tunnel itself. It may be necessary to cut the flexor retinaculum surgically. The affected hand may need to be rested for months to recover.

Signs of disease

Because signs in the hand can tell your doctor that there is trouble elsewhere in the body, a physical examination may start with the hands.

See also: Bones; Joints; Ligaments; Muscles; Tendons; Wrist

Head

Questions and Answers

My son is always fighting. If he was knocked on the head, which part of the skull would be particularly vulnerable?

The temples, because the temporal artery runs inside the skull just beneath that area. A fracture of the temporal bone could tear the artery and result in bleeding, and cause pressure on the brain. Instead of worrying, why not talk to your son's teacher, and discuss why he gets into fights and what can be done about it.

My boyfriend is an amateur boxer. Is he running the risk of permanent brain damage?

Repeated, heavy blows to the head over a prolonged period of time will cause small hemorrhages throughout the brain, resulting in diminished mental ability, slurring of speech, and general slowness.
    Many doctors say boxing is an unacceptably dangerous sport and would like it banned. Medical advisers for the sport, however, claim that if proper precautions are taken, there is not much risk.

One of my children's friends fell out of a tree and fractured his skull. His doctor said he didn't need any treatment. Is this right?

Yes, the doctor's conclusion that additional treatment is not needed is correct—assuming that the imaging study (CT scan) showed only a fracture and no bleeding and that the fractured bone did not compress the brain.

Is phrenology a true science?

It is the so-called study of the mind and character revealed by the shape of people's heads, and it is not a science at all.

Because the brain controls everything the body does, it is the most important organ, so it is vital to protect the head from injury and to seek medical help should an accident involve the head.

The head comprises the skull, a rigid, bony structure that houses the brain and the organs for hearing, balance, sight, smell, and taste. The skull consists of two parts: the cranium, which encloses the brain; and the face, which provides a bony framework for the eyes, nose, ears, and mouth. The eight bones of the skull are joined together by cartilage tissue before a baby is born. This permits the bones to move over one another, if necessary to allow the head to pass down the mother's birth canal. The cartilage is gradually replaced by bone during the first 18–24 months of life; after that the skull becomes rigid.

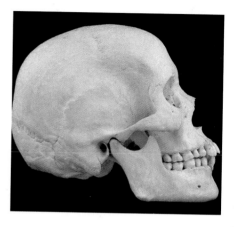

HOW THE SOFT TISSUES OF THE BRAIN ARE INJURED

▶ *Although the jaw is well protected by skin, bone, membranes, and fluid, if it receives a blow, the blow can shake the brain and compress the part of the brain farthest from external contact.*

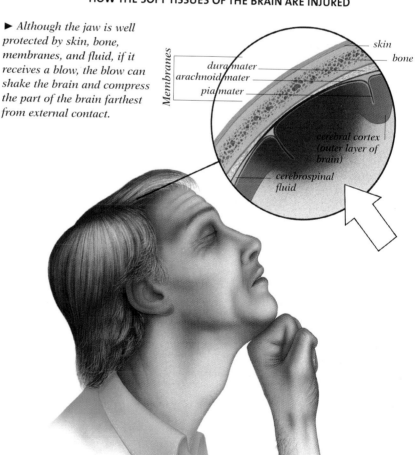

Membranes

skin
bone
dura mater
arachnoid mater
pia mater
cerebral cortex (outer layer of brain)
cerebrospinal fluid

▲ *BMX riders are vulnerable to accidents; hence they are required to wear crash helmets for protection.*

The brain and spinal cord

The brain fills the cranial cavity and consists of a soft jellylike substance that is easily compressed, torn, or crushed. It is wrapped in an extremely tough layer of tissue called the dura mater. This tissue and the bony part of the skull protect the brain.

If the brain is damaged and swells, however, its tissue can be further damaged by being crushed against the bony outer layer.

At the base of the skull are several openings, enabling arteries, nerves, and veins to pass through. The largest of these spaces, the foramen magnum, is the outlet for the spinal cord. This passes down through the vertebral (spinal) bones in a channel called the spinal canal, which is widest at the top end and narrowest at the bottom.

The skull rests on the first cervical vertebra, called the atlas, which articulates with the second cervical vertebra, the axis. The two vertebrae allow the head to move backward and forward and to rotate. If the spinal cord is cut across or damaged at this level, death will result almost instantaneously.

Injuries

Almost two-thirds of the people who die under the age of 35 do so because of head injuries. These are not necessarily from skull fractures but also result from serious damage to the soft tissues of the brain caused by a violent blow on the head.

Injuries result from two causes: deceleration, when a head that is moving is brought suddenly to rest (for example, by striking the windshield of a car or hitting the ground in a fall); and acceleration, when a head that is stationary is violently struck (as from a blow with a baseball bat). The result of both these shocks is that the soft tissue of the brain is shaken.

Two typical injuries result. One is immediately beneath the point where the head has come into contact with something solid and is called the coup injury. The other is inside the part of the head farthest away from the blow, where the brain is either pushed against the skull or pulled away from it by the force of the blow; this is called the contrecoup injury.

The most critical factor in every case is whether damage has been done to the blood vessels. If bleeding begins, blood spills into the brain substance, spreading through it with relative ease and occupying a great deal of space. Because the brain is contained within the bony skull, the pressure rises rapidly and some of the brain substance is crushed. Much of this damage can be prevented if the bleeding can be stopped or the rise in pressure relieved.

RELATIVE POSITIONS OF THE SKULL, BRAIN, AND SPINAL CORD

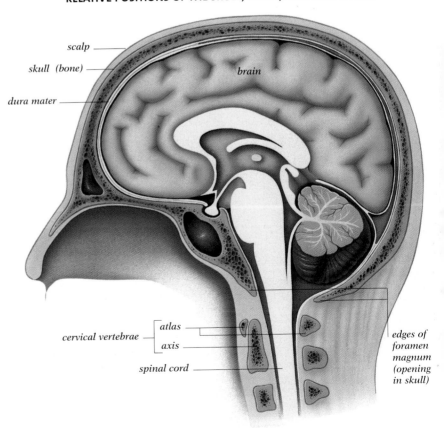

Questions and Answers

Why do injuries to the head bleed so much? I was shocked by this when I witnessed a road accident.

There are two reasons why any injury to the skull or face will bleed profusely. First, this part of the body has a very rich blood supply; second, skin over the face and scalp differs from that in every other part of the body (except the scrotum in men), in that the skin is attached to muscle. This is the reason why cuts on the face or scalp tend to gape open, and why the only way such wounds will heal properly is if the edges are stitched together.

My husband tried to photograph our baby in her crib. To get a good view, he lowered the side and the baby fell out and landed on her head. Will she be brain damaged?

Babies are amazingly resilient, and their heads are fairly elastic. It is most unusual for any brain damage to occur. If, however, a fall is followed by unconsciousness or a convulsion, seek immediate medical evaluation.

My father received a head injury three months ago and has not been himself since. What are his chances of recovering fully?

Very good. It is important to remember that confusion, dizziness, headaches, and depression often follow head injuries. They almost always get better in time; sometimes it takes up to a year after the injury.

Why do Inuit wear fur hats? Is it because it's harmful to let the head get cold?

If the body is very cold, heat is conserved when the blood vessels in the skin constrict. This happens over the entire surface of the body, but much less on the head. For this reason up to 90 percent of body heat can be lost via the head in very cold weather. The Inuit realized this long ago and started wearing fur hats.

▲ *Small children are agile and have no sense of fear. Well-planned play areas have sand laid down to make accidental landings less hazardous.*

Diagnosis of head injuries

Doctors divide people with head injuries into three groups: those who are fully conscious at the time they are seen; those who have been unconscious since the time of the accident; and those who have been conscious for a period of time following the accident but are slipping into unconsciousness again. Doctors have to distinguish between concussion (from which the patient recovers, usually without brain damage), and bleeding inside the skull, which may cause further damage and eventual death. Once brain tissue is destroyed, it never regenerates.

Minor injuries

Minor head injuries are most often the result of an accident in the playground or on the sports field, or of a fall at home in which the head is knocked against a solid object. The usual symptoms are those of concussion: unconsciousness seldom lasts for more than a few seconds; the victim often sees stars and feels dizzy; sometimes he or she has a headache that is made worse by exertion, stooping, or emotional excitement; the events that happened in the minutes immediately following the injury may be forgotten (a state known as post-traumatic amnesia), but it is most unusual for the events leading up to it to be forgotten (retrograde amnesia). These symptoms seldom last for more than a few hours. You should see your doctor after such an accident, especially if the period of unconsciousness has lasted much more than five seconds.

Serious injuries

A more serious state of affairs exists when the victim has been unconscious since the time of the accident. As with any unconscious patient, the first priority is to ensure that he or she is able to breathe. The person must be placed on one side to lessen the risk of inhaling any vomit and therefore suffocating. A head injury can be complicated by bleeding inside the brain, and there are certain warning signs that the doctor looks for. One of these is an increase in size in one or both pupils of the eyes, caused by stretching and paralysis of the nerve that supplies the muscles of the iris. At the same time, when a bright light is shone into the eye, the pupil does not constrict. These symptoms occur because the brain swells from the increase in pressure caused by internal bleeding. Another well-known warning sign is the direction in which the big toe points when the sole of the foot is stroked. In a healthy person the big toe usually points downward when a blunt object is rubbed along the sole of the foot. In the case of a serious brain

▲ *Bicycling is fun for children, but falls can result in serious injuries. This child is well protected by a safety helmet; elbow and knee pads would also be a sensible precaution.*

▲ *Old people may not be very steady on their feet; a helping hand when they are getting out of a vehicle will be appreciated and could prevent a bad fall.*

injury, however, the direction is reversed and the toe may point upward. This is known as Babinski's sign, for the man who first described it. Patients with this kind of head injury need to be taken to the hospital immediately.

Sometimes after a brief period of consciousness, perhaps less than an hour, the patient lapses into unconsciousness again. Typically this is the result of a blow in the region of the temples in the skull, beneath which there is a vulnerable artery called the temporal artery. This condition is usually caused by bleeding between the tough covering of the brain and the skull, which is called an epidural hematoma. The trouble begins when the bleeding occupies enough space to compress the brain underneath it. This pressure can be relieved almost at once if a surgeon drills a hole in the skull over the site of the hematoma.

Problems

Doctors encounter a more difficult problem when bleeding occurs between the tough covering of the brain and the brain tissue itself; this is known as a subdural hematoma. In this case the skull injury causing the trouble may be so trivial that the patient may not remember it. What can complicate matters further is that the period of unconsciousness may occur several weeks after the original injury. Both epidural and subdural hematomas may occur at the nearest and farthest point of the skull from the blow, just like coup and contrecoup injuries. Again, drilling a hole in the skull to remove the blood clot will rapidly remove the pressure.

CT and MRI scanning have greatly eased the problems of accurate diagnosis of the extent of head injuries and have saved many lives. Most patients with a severe head injury will have a brain scan.

Treatment

In many cases of minor head injuries, only first-aid measures will be needed, but in more serious cases the patient will need hospital care.

As described, subdural and epidural hematomas may be removed by a relatively simple operation, but there is not much that can be done for bleeding deep within the brain, although everything possible will be tried. To ensure normal breathing, a tube may be placed in the patient's windpipe (trachea) and a mechanical ventilator used. There then follows a period of waiting to see whether the brain recovers from the injury. This may last for months, but full recovery has occurred after very long periods of unconsciousness, during which time the patient is on a life support machine to help him or her breathe, and the electrical activity of the brain is monitored.

Outlook

An aftereffect that occurs in a small proportion of cases following head injuries is epilepsy, which is usually the result of damage to the brain tissue. The interval between the head injury and the first attack is usually several months, but it may be much longer, even years, particularly in children. Concussion itself seldom leads to epileptic seizures.

Patients may also suffer from headaches, moodiness, depression or excessive nervousness, which may also have a psychological basis; however, complete recovery is still possible in many cases.

See also: Brain; Mind

Hearing

Hearing is one of the most complex senses in the human body. Nowadays it can be tested by sophisticated machines so that hearing problems can be accurately diagnosed and treated.

My son was recently blinded in an accident. Will his hearing improve to compensate for the loss?

No, but blind people train themselves to be more discerning in picking out sounds, because they must rely more on their hearing.

Can a person with hearing in only one ear tell which direction a sound is coming from?

To some degree, yes. Sounds made behind the head have a slightly different quality or timbre from sounds made in front of the head. However, such a person will be much less accurate in pinpointing the direction of the sound than someone with hearing in both ears.

My brother has measles. Is it true that measles can cause deafness?

Yes, but rarely. Measles can be complicated by an inflammation of the middle part of the ear, which contains bones that transmit sound from the eardrum to the inner ear. If the infection is not properly treated, this part of the conducting system may become damaged.

Why can I pick out one sound from many others when I concentrate?

This is the result of selective interpretation by the brain. By making a special effort, your brain can screen out all the unwanted sounds and tune itself to notice one particular frequency.

My ears ring after a rock concert. Could the noise damage my ears?

Prolonged exposure to very loud noises can cause irreversible ear damage. It starts with deafness in the highest frequencies and moves on to the lower ones, but avoiding prolonged exposure to loud noises can stop its progress.

The human ear hears sound waves that are produced by pressure vibrations in the air. The size and energy of these waves determine loudness, which is measured in decibels (dB). The number of vibrations or cycles per second make up the sound's frequency: the more vibrations, the higher the pitch of the sound. Sound frequency is expressed in cycles per second, or hertz (Hz).

In young people the range of audible frequencies is about 20–20,000 Hz per second, though the ear is most sensitive to sounds in the middle range of 500–4,000 Hz. As people get older, or if they are exposed to extremely loud noise over a period of time, hearing becomes less acute in the higher frequencies. To measure the extent of hearing loss, normal hearing levels are defined by an agreed-on international standard. The patient's level of hearing is the difference in decibels between the faintest pure note perceived and the standard note generated by a diagnostic machine.

How people hear

Sound waves transmitted through the air are collected by the outer ear (pinna), and are funneled through the ear canal. When they reach the eardrum (tympanic membrane), they cause it to vibrate. This vibration is passed on to the three bones of the air-filled middle ear—the malleus, incus, and stapes—which are arranged in a chain and transmit the vibration to the inner ear.

Diseases that affect the middle ear and result in the nonfunctioning of this bony chain will cause conductive hearing loss. If the normal pathway to hearing is obstructed, transmission of sounds across the bone of the skull is also possible, but there is nevertheless a considerable hearing loss of about 25–30 dB. This type of hearing loss occurs in otosclerosis, where the stapes becomes rigidly attached to the inner wall of the middle ear and is unable to vibrate.

Sound vibrations are transmitted through the oval window to the inner ear, which is filled with a fluid called endolymph. The inner ear contains two separate organs: the vestibular apparatus (including the three semicircular canals), which provides the sense of balance; and the organ of Corti, located in the cochlea, which is shaped like a snail. The fluid transmits sound waves to the cochlea. The organ of Corti cells, distributed along the length of the cochlea, bristle with specialized hairs that are sensitive to movement. As the waves pass, they are sensed as pressure and transmitted along the cochlear nerve to the auditory area of the brain, where they are interpreted. High-frequency tones are sensed only by the specialized hair cells in the first part of the cochlea, whereas medium and low tones are sensed farther along the chain.

The cochlear hair cells are fixed to a membrane, the basilar membrane, that is vibrated by the sound waves. As the hair cells move, their hairs wipe against another membrane. This causes the cells to generate nerve impulses that are then transmitted to the brain. Many diseases of the inner ear result in a disordered interpretation of sound.

Hearing tests

The simplest hearing test involves whispering standard phrases at varying distances from the patient's ear and measuring the distance at

▲ *A doctor examines a patient using an instrument called an otoscope, which illuminates and magnifies inside the ear.*

THE STRUCTURE OF THE EAR AND HOW SOUND IS HEARD

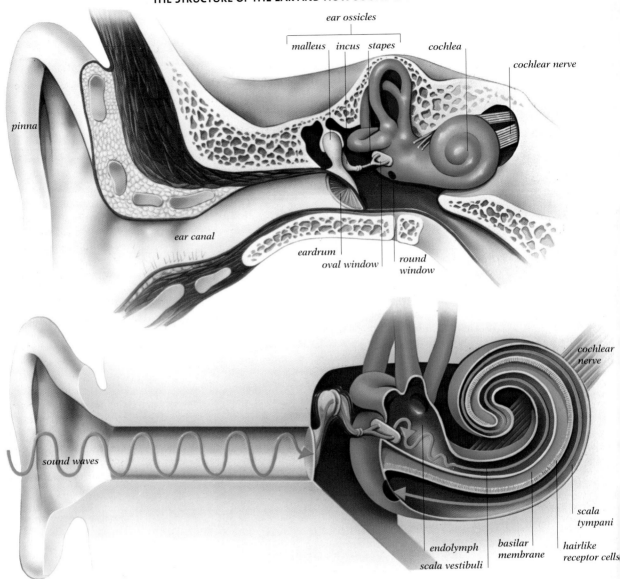

▲ Sound entering the ear canal causes the eardrum (tympanic membrane) to vibrate. The vibrations are transmitted through the ossicles (small bones). These intensify the pressure of the sound waves and transmit vibrations to the oval window, a membrane over the entrance to the cochlea. The simultaneous pulsating movements of the round window stabilize pressure within the inner ear. The fluid (endolymph) that fills the cochlea transmits the waves along the scala vestibuli and around into the scala tympani, making the basilar membrane separating them vibrate. This membrane contains hairlike receptor cells (organs of Corti) that produce nervous impulses that are sent along the cochlear nerve to the brain, which interprets them.

which the patient stops hearing the whispers. To conduct a test with a tuning fork (an instrument pitched at a certain frequency), the doctor taps the fork and holds it against the patient's ear, then moves the fork away until the patient can no longer hear any sound.

More sophisticated tests are conducted with an audiometer machine in a soundproof room with the patient wearing earphones. The instrument emits notes at different volumes, beginning with sounds beyond the hearing range and gradually lowering the frequency. When the patient can hear the sounds, he or she presses

a button. Not only does the audiometer measure air conduction from the sound emitted and heard in the earphone; it also tests bone conduction, when the sound is played into the bone of the skull bypassing the external and middle ear. The results are presented graphically on a scale measuring frequency and hearing loss and show the number of decibels to be generated above those considered just audible for the patient to hear the tone.

See also: Brain; Earwax

Heart

My husband's heart beats at a much slower rate than mine. Could his heart actually stop completely?

Usually not. Lower heart rates, around 50 to 60 beats per minute, are generally found in athletes. However, those lower heart rates do not always mean that someone is physically fit.

Can a healthy person strain the heart through exercise in the same way that you can pull a muscle?

No. In a healthy person, normal exercise would not result in the heart being strained in the same way as a muscle being pulled. Coronary artery disease is increasingly common as we age and may limit the work the heart can do. People in their forties and fifties are likely to have some degree of artery disease. Everyone should build up their endurance gently to avoid putting the heart under undue stress.

I am nervous, and when I get a shock my heart feels as if it will jump out of my chest. What causes this feeling? Is it serious?

Don't worry, your heart is quite normal. The shock makes your adrenal glands pump out adrenaline, which suddenly drives your heart to beat very fast and forcefully, causing the symptoms.

I am worried about my son, who is 14. I felt his pulse and it was very irregular. Is this normal?

The pulse rate can vary in response to breathing, since the heart speeds up on breathing in and slows down on breathing out. This irregularity, called sinus arrhythmia, is normal and is probably what you found in your son. It diminishes with age.

The heart is one of the most important organs for the maintenance of life. Situated in the middle of the chest, it is a powerful pump that drives blood around the body, carrying vital food and oxygen to the tissues.

The heart is a large muscular organ in the middle of the chest. Although it is often thought of as being on the left-hand side of the body, it actually straddles the middle with more of it on the left side than on the right. It weighs about 12 ounces (340 g) in men and a little less in women.

The right-hand edge of the heart lies behind the right-hand edge of the breastbone. On the left side of the breastbone, the heart projects out as a kind of rounded triangle with its point lying just below the left nipple in men. At this point, the apex beat can be felt pulsating with each heartbeat.

The job of the heart is to pump blood through two separate circulations. First it pumps blood out into the arteries via the aorta, the central artery of the body. This blood circulates through the organs and tissues, delivering food and oxygen, and then returns to the heart in the veins, after all the oxygen has been absorbed from it. The heart then pumps the blood on its second circuit—this time to the lungs to replace the oxygen. Blood is then returned to the heart with its oxygen renewed. The circulation to the lungs is called the pulmonary circulation and the one to the rest of the body is called the systemic circulation. Pulmonary and systemic arteries carry the blood outward from the heart, and pulmonary and systemic veins return it.

Structure of the heart

The heart is almost all muscle and is divided into four chambers: two atria and two ventricles. These chambers are arranged in pairs, with the two atria situated behind and above the two ventricles. Both the atria and both the ventricles lie side by side, and the portions of their walls that separate them are called the interatrial septum and the interventricular septum, respectively.

When the muscle tightens, blood is squeezed onward in a direction determined by the valves. The thickness of the muscular wall depends on the amount of work the chamber has to do, the left ventricle having the thickest walls because it does the most work. Each thin-walled atrium receives blood from the veins, then pumps the blood through a valve into the thicker-walled ventricle below, which then pumps the blood into a main artery.

POSITION OF THE HEART

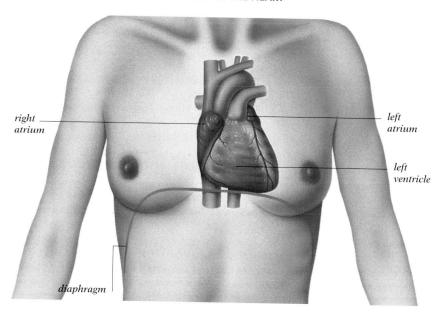

right atrium

left atrium

left ventricle

diaphragm

STRUCTURE OF THE HEART

▼ *Arrows show the direction of blood flow through the four chambers and main blood vessels of the heart.*

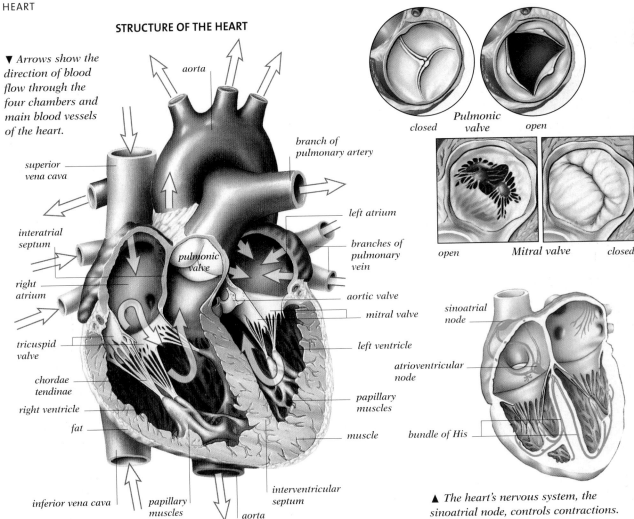

Pulmonic valve

closed   open

Mitral valve

open   closed

- aorta
- branch of pulmonary artery
- superior vena cava
- interatrial septum
- left atrium
- branches of pulmonary vein
- pulmonic valve
- right atrium
- aortic valve
- mitral valve
- tricuspid valve
- left ventricle
- chordae tendinae
- right ventricle
- papillary muscles
- fat
- muscle
- sinoatrial node
- atrioventricular node
- bundle of His
- inferior vena cava
- papillary muscles
- interventricular septum
- aorta

▲ *The heart's nervous system, the sinoatrial node, controls contractions.*

Blood returns to the heart from the lungs in the pulmonary veins with its oxygen store renewed. It enters the left atrium, which contracts and pushes the blood through a valve called the mitral valve into the left ventricle. As the left ventricle contracts, the mitral valve shuts so that the blood can go out only through the open aortic valve into the aorta. The blood then goes on into the tissues.

Blood returns to the heart from the body in a large vein called the inferior vena cava, and from the head in the superior vena cava. Blood enters the right atrium, which contracts, squeezing the blood through the tricuspid valve into the right ventricle. A right ventricular contraction then sends the blood through the pulmonic valve into the pulmonary artery, and then through the lungs, where it has its oxygen renewed. The blood then returns to the heart in the pulmonary veins, ready to start all over again.

This process is repeated 60 to 80 times every minute at rest. During exertion the heart rate will be much higher.

The valves

The pulmonic and aortic valves are very similar, each having three leaflets of yellowish membrane which open upward to allow blood to flow forward, but which fall back into place to stop any backward flow. The mitral and tricuspid valves are more complicated, although they are similar in structure. They also have leaflets: two for the mitral and three for the tricuspid valves. The bases of these leaflets form a ring-shaped border between the atrium and the ventricle. Cords called chordae tendinae extend from the leaflets to part of the ventricle wall known as the papillary muscles. By pulling down on the chordae tendinae during each ventricular contraction, the papillary muscles make sure that no blood flows backward into either atrium.

Timing system

With each heartbeat the two atria contract together and distend the ventricles with blood. Then the ventricles contract. This ordered series of contractions depends on a sophisticated electrical timing system. The basic control comes from the sinoatrial node in the right atrium, which sends impulses through both atria to make them contract.

Another node, the atrioventricular node, lies at the junction of the atria and ventricles. This delays the impulse to contract and passes it down through a bundle of conducting fibers in the intraventricular septum called the bundle of His. After passing through the bundle, the impulse spreads out into the ventricles, causing them to contract.

See also: Arteries; Blood; Circulatory system; Muscles; Veins

Hip

Questions and Answers

Why is it so easy for young children to do the splits, but almost impossible for most adults unless they have trained as dancers?

The ligaments and muscles around the hips are much more elastic in a child than in an adult. However, as any ballet dancer will testify, if the ligaments are kept supple by constant exercise, then there is no reason why adults should not retain the mobility of childhood.

My six-year-old son complains that his hips ache. Are these aches just growing pains?

Probably not. If pain is present in a particular part of the body, there may be an underlying cause, even if that underlying cause is just a mild muscle strain. You should take your son to see his doctor as soon as possible, just in case something is wrong.

Why is it that elderly women seem to break their hipbones so often?

In old age, all the bones in the body become more brittle. Bone weakness (osteoporosis) in older people affects women more so than men due to estrogen deficiency after menopause. The hipbone is likely to be broken in a fall because it is often the part that hits the ground first and so bears the whole weight of the body. Broken hipbones in the elderly involve a prolonged period of immobility, which can diminish an older person's zest for living, so it is important for older people to be extra cautious.

My mother is very overweight. Am I right to worry that this is putting too much strain on her hips?

Yes. Being overweight affects many internal organs, but the hips are particularly vulnerable, since they bear most of the body's weight.

The hips are designed to bear the weight of the body and enable a person to stand, walk, or run. Although they are very strong, problems still arise, especially in the very young and the elderly.

The hip joint is the largest in the body. It has two main functions: to carry the considerable weight of the trunk and, with the help of the spine, to enable a person to walk upright. The description "ball and socket" that is given to the hip joint is particularly appropriate because a large hemispherical projection at the top of the thighbone (the femur) fits into a cup-shaped socket (the acetabulum) in the iliac bone of the pelvic girdle.

Smooth movement is ensured by the shiny cartilage that covers the head of the femur and lines the acetabulum. Lubrication is provided by the synovial fluid secreted by the synovial membranes that surround the joints. These membranes form the inner layer of the tough fibrous capsule that makes a hermetic seal around the joint. This seal is very strong and airtight. The capsule is thin at the back of the joint but very thick at the sides and the front of the body, where the greatest pressure is exerted on the hip joint. Several accumulations of fatty tissue, which lie inside the tight seal provided by the capsule, pad the inside of the joint and act as shock absorbers.

Added stability is provided by a set of ligaments that bind the head (top) of the femur to the pelvic girdle, crisscrossing over one another to maximize their strength and to prevent the ball from slipping out of its socket. One of these, the iliofemoral ligament, toward the outside of the joint, is thought to be the strongest in the body. For both walking and standing, it is essential that the thigh does not move farther back than an imaginary straight line drawn vertically down the side of the trunk. This constraint is provided by the ligaments. When the leg seems to be pulled back beyond this line—for example, before kicking a football—it is the whole pelvic girdle that moves, not the hip joint.

If a person stands and moves the legs into different positions, it can be seen how many actions the hip joints allow. These movements are made possible by the muscles that lie over the ligaments. The action of pulling the knee up toward the head involves bending, or flexion, of the hip joint. Such a movement is

MUSCLES THAT MOVE THE HIPS AND LEGS

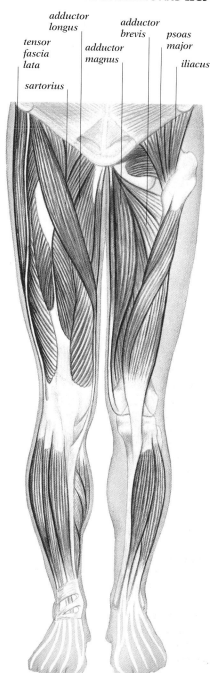

adductor longus

tensor fascia lata

sartorius

adductor brevis

adductor magnus

psoas major

iliacus

179

HOW LIGAMENTS STABILIZE THE HIP JOINT

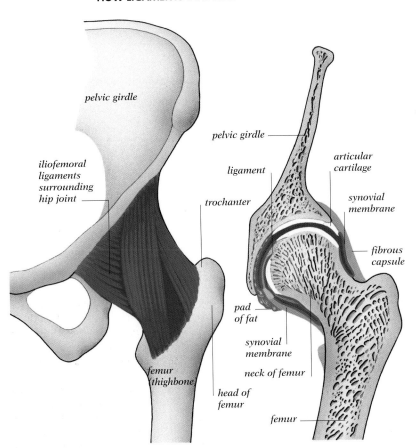

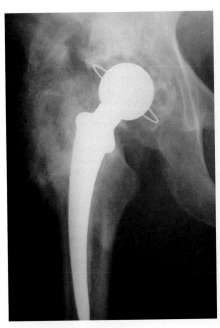

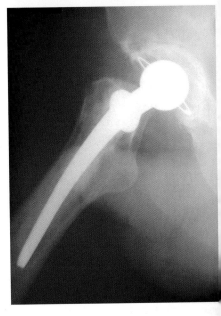

brought about by two main muscles: the psoas major, which runs from the base of the spine across the front of the hip to the femur; and the iliacus, a flat, triangular muscle that is attached to the pelvis at one end and to the femur at the other. These two muscles are assisted by others in the thigh.

Straightening a bent leg is called extension. To perform this action the outermost and biggest buttock muscle, the gluteus maximus, is used, plus a group of muscles at the back of the thigh called the hamstrings. It is these muscles that are actually at work when a person is standing still.

Moving the leg out sideways is called abduction, and bringing it back again is known as adduction. The first of these movements involves two of the muscles that lie beneath the gluteus maximus: the tensor fascia lata, which joins the pelvis with the femur; and the sartorius, the longest muscle in the body, which runs from the pelvic girdle to the knee. A group of muscles including the adductor longus, adductor brevis, and adductor magnus provide the pulling power between the pelvis and the femur needed for adduction. Many of these muscles are also used to rotate the leg, although this movement is limited by the binding of the ligaments.

Congenital dislocation

Because of its complex structure, the hip joint is extremely stable compared with its counterpart, the shoulder. This stability means that in adult life, dislocation of the hip joint is a rare injury. Paradoxically,

▶ *A hip replacement (right and top right) allows for full movement of the patient's leg. The new joint is made of stainless steel and titanium.*

it is not only the most common hip problem of childhood but also the most common site of joint dislocation at birth.

Examination of the hips is carried out as a routine precaution immediately after the birth of every child. The baby is placed on his or her back, with the legs wide apart, and held firmly by the feet. If the hip joints are dislocated, the doctor will hear a deep clicking sound as he or she bends the child's knees and hips. The sound comes from the head of the femur as it goes into its socket. When the baby's legs are extended, another click will be heard as the femur disengages. The doctor will also place his or her hand so that the thumb is on the baby's groin and the middle finger is on the projection of the thighbone. If the head of the femur is easily moved out of its socket, the doctor will suspect that the hip joint is potentially unstable, or

► *The ability to do splits so easily is lost unless muscles are trained for it.*

irritable. If this is the case, these tests are followed by X rays to check the diagnosis.

Nowadays the treatment for a congenital dislocation of the hip joint is extremely effective. Usually the infant is put into a plaster splint that keeps the legs wide apart in a froglike position. After a few months the joint is X-rayed again to check progress, and in most cases the splint is then removed. Whether or not the baby has to stay in the hospital during this period depends on the domestic circumstances of the family. The child can be taken care of at home if this is possible.

If the infant's hip joints are not dislocated but simply irritable, the doctor may advise the mother to put a very thick cloth diaper on the baby to keep the legs wide apart until the ligaments and muscles around the joint have developed enough strength to overcome the problem naturally.

Diseases and injuries

Problems may arise in the hip joint later on in childhood, owing to diseases that cause inflammation of the synovial membranes or, rarely these days, as a result of tuberculosis. An unusual but serious hip condition is Perthes' disease, which is much more common in boys than girls. In this condition, for some unknown reason, the bony tissue in the head of the femur begins to disintegrate. The child will complain of aching or of sharp pains in the hip—and possibly in the knee on the same side, too—and he or she may have difficulty walking. With early X-ray diagnosis

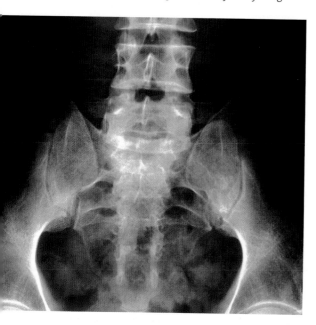

This X ray shows the hip joints that attach the pelvis and thighbone (femur) to the spinal cord.

and rest in bed, often for several years, the condition corrects itself and the bone grows back normally. However, if it is not treated in the early stage, it can lead to a permanent deformity of the joint. For this reason alone it is most important for any child who complains of hip pains to be seen by a doctor, who will arrange for X rays.

Accidental injury to the joint is most common in old age and often follows heavy falls. The most likely place for the bone to break is across the neck of the femur. Symptoms of hip fracture include unbearable pain on trying to walk, bruising, swelling, tenderness, or deformity of the hip. In cases of suspected hip fracture, call an ambulance first, then check for signs of shock or internal bleeding. You should not attempt to move the patient or splint the bone unless it is absolutely essential—for example, if the patient is in harm's way, such as in the middle of the street.

Because of the great weight the hip joint has to bear and the many muscles that surround it, the most usual method of treatment is orthopedic surgery in which metal nails or screws are used to keep the bones in place.

After surgery, although the patient is allowed out of bed a few days after the operation, he or she will not be permitted to put any weight on the joint until the surgeon is satisfied that the fracture has begun to heal properly. One of the problems of hip surgery is that the blood vessels may be damaged so that there is an insufficient supply to the fracture for mending to take place. The worse the break, and the older the patient, the more likely it is that such problems will arise.

A fracture of the neck of the femur is commonly associated with loss of the blood supply to the head of the femur. This leads to death of bone known as avascular necrosis. In this case, the treatment is replacement of the whole joint with a metal prosthesis.

Hip joint replacement is one of the most common major operations on elderly people. Restoration of hip function is almost immediate and the stay in the hospital is short.

See also: **Bones; Cartilage; Joints; Ligaments; Membranes; Muscles; Pelvis; Skeleton**

Homeostasis

Questions and Answers

Does homeostasis have anything to do with homeopathy?

No. The term comes from two Greek words: *homeo,* meaning "the same," and *stasis,* meaning "standing still." Homeostasis is the maintenance of a stable state in a dynamic system that can readily deviate from normal.

What is negative feedback? Has it anything to do with homeostasis?

Yes. The action of a homeostatic system is to provide a response to a deviation from the normal that is proportional to it but that acts in the opposite direction. This is called negative feedback and is highly stabilizing. Almost all homeostatic systems display negative feedback.

Is hunger part of a homeostatic mechanism?

Yes. Hunger is induced when brain cells detect low levels of blood sugar (glucose) and emptiness in the stomach. The sensation of hunger prompts the individual to seek and eat food. If this is successful, the blood sugar rises and the stretch receptors in the stomach signal that no more food is currently required.

Can homeostasis break down?

Yes. This is common and is one of the causes of human disease and disorder. Examples of the breakdown of a homeostatic mechanism are obesity and anorexia nervosa. Both are results of the failure of a homeostatic mechanism that normally keeps body weight within reasonably normal limits. In both cases the negative feedback loop is broken. Obese people fail to respond normally to satiety, and anorexics have an abnormal perception of their own body shape.

The human body is a remarkably complex machine requiring many built-in control systems to maintain its stability and constancy. Underlying all these systems is the principle of homeostasis by negative feedback.

It is essential for the healthy functioning of the body that the composition and state of the internal environment should be maintained within narrow limits. This applies, at a higher level, to all the systems of the body, but it also applies at the level of individual cells. Cells are highly sensitive to changes in the composition of the fluid in which they are bathed, and any large deviation from normal will adversely affect cell function, often fatally. The constancy of composition of the internal

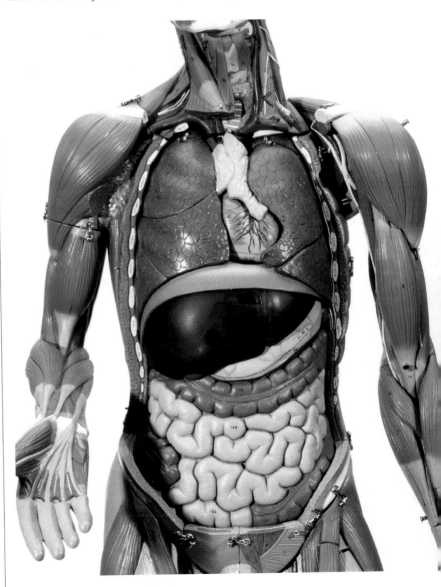

▲ *This model of a male torso shows all the major body organs held in homeostasis— the process by which the body keeps all the different metabolic processes in balance.*

► *People who live in high altitudes, such as the Quechua Indians of Peru, have a homeostatic response when their bodies adjust to the lower oxygen levels in the atmosphere.*

environment is essentially a chemical matter, and the maintenance of this vital constancy is called homeostasis.

How does homeostasis work?

In broad general terms, whenever a significant change occurs anywhere in the body, a mechanism comes into action to correct that change and restore normality. Such a mechanism requires the presence of a receptor that can detect the change and that can, directly or indirectly, bring about an effect to counter it. Such a correction system is called a homeostatic control system, and body function (physiology) incorporates many such systems.

To give a simple example: if a person exercises strenuously, muscle contractions produce heat and the person's temperature begins to rise. This rise is detected by cells in the brain, internal organs, and skin that are sensitive to heat. The output from these cells relaxes the skin blood vessels and stimulates the sweat glands. The skin flushes so that more blood flows near the skin surface and is cooled. Evaporation of the sweat causes cooling.

The same mechanism can work in the opposite direction. A person starts to get too cold, and the heat-sensitive cells detect the internal drop in temperature. This immediately causes muscles to start contracting. If the drop in temperature is not quickly corrected, however, the muscles go into a mode of rapid contraction and relaxation, the process familiar to us as shivering. This is a highly efficient heat-generating mechanism, and all the heat is produced internally. If the cause of the cooling is removed, the body temperature will quickly rise to normal.

Chemical homeostasis

The body's endocrine system is a striking example of homeostasis. The endocrine glands are under the direct control of the pituitary gland, located at the base of the brain and attached to the hypothalamus with nerve tracts.

The pituitary releases into the blood a range of chemical substances called hormones; their function is to stimulate the various endocrine glands into action. The thyroid-stimulating hormone, for instance, causes the thyroid gland to secrete the hormone thyroxine, which boosts metabolism and growth.

At the same time cells in the hypothalamus are monitoring the levels of thyroxine in the blood. If the level of this hormone rises too high, this is detected by these cells and a signal is sent to the pituitary to prompt it to secrete less thyroid-stimulating hormone. If the level of thyroxine in the blood drops too far, the pituitary is directed to produce more stimulating hormone. Thyroid-stimulating hormone is only one of several pituitary hormones that take part in a homeostatic feedback loop. A similar process controls the levels of cortisol from the adrenal glands, estrogens and testosterone from the sex glands, prolactin to control milk production from the breasts after pregnancy, and growth hormone to control protein synthesis and metabolism.

Other examples of homeostasis

Blood pressure is controlled by a homeostatic mechanism. In this case the receptors are sensitive to pressure and are cell groups called baroreceptors, situated in the large arteries near the heart. From these, nerve fibers run to a heart and artery control center in the brain stem, where the information from the baroreceptors is analyzed. If the baroreceptors report a rise in blood pressure, the brain stem center acts via the autonomic nervous system to reduce the heart output and relax the arteries. In this way the blood pressure is reduced.

The principal fuel of the body is glucose, and a minimal level of this fuel must always be present in the blood. If the level of glucose falls too far, this is sensed by cells in the liver where a form of concentrated glucose called glycogen is stored.

When the need arises, the liver splits glucose molecules off the glucogen store and releases them into the bloodstream. When there is plenty of glucose in the bloodstream, glycogen is resynthesized.

Some homeostatic mechanisms work more slowly. The process of acclimatization (the body's response to changed climatic conditions) and the process of adaptation to high altitudes and lower atmospheric oxygen levels are also examples of homeostasis. In these cases the responses are, respectively, earlier onset of, and more profuse, sweating, and an increase in the number of oxygen-carrying red cells in the blood.

See also: **Autonomic nervous system; Blood pressure; Endocrine system; Glands; Hormones; Hypothalamus; Metabolism; Perspiration; Pituitary gland; Shivering; Thyroid**

Hormones

I've been told that estrogen cream can cure vaginal dryness and shrinkage and that HRT by mouth is not necessary. Is this true?

It's true that estrogen vaginal cream can treat postmenopausal vaginal changes effectively, and that if these changes are the main problem, such creams may be preferable to oral HRT. Treating vaginal atrophy with local estrogen preparations also has the advantage that it can cure the persistent cystitis that often results from local infection encouraged by the atrophy.

Is it true that hormones cause morning sickness in pregnancy?

Morning sickness is one of the early symptoms of pregnancy. It coincides with a big upsurge in the ovaries' production of the hormone progesterone, which helps bring about physical changes in the uterus lining to prepare it for nourishing the fetus. The effects of progesterone may produce feelings of nausea and tenderness of the breasts. These discomforts pass about 14 weeks after the beginning of the pregnancy, when the placenta takes over much of the hormone production.

Since having treatment for problems with her thyroid gland, my wife is much calmer and more even-tempered. Why is this?

Your wife was probably producing an excess of thyroid hormones, which have many physical and emotional effects. Many people who produce an excess become jumpy or overanxious, and this situation is not helped by the fact that they know they are not reacting normally. The physical symptoms are relieved when the thyroid problem has been dealt with, and also the emotional problems are often brought back into balance.

Hormones are chemical messengers. They are made and released in one part of the body before being circulated in the blood to target cells, where their effects are brought about. The correct balance of hormones is essential to good health.

Compared with nerve impulses, which carry information around the body in the form of electrical charges, hormones tend to act slowly and spread their activity over a much longer period. Not all of them act so slowly, but many of those that do influence fundamental activities such as growth and reproduction. They tend to work by controlling or influencing the chemistry of target cells: for example, by determining the rate at which the cells use up food substances and release energy; or whether or not they produce milk, hair, or some other product of the body's metabolic processes.

The organs responsible for making most of the hormones are the ductless, or endocrine, glands, so called because they discharge their products directly into the blood and not elsewhere in the body via a tube or duct. Hormones made by the major endocrine glands have the most widespread effects. They include insulin and the sex hormones. The body also makes local hormones, or transmitters, which act much nearer to their point of production. One example is secretin, which is made in the duodenum in response to the presence of food. Secretin travels a short distance in the blood to the pancreas, stimulating it to release a flood of watery juice that

THE BODY'S CHEMICAL MESSENGERS

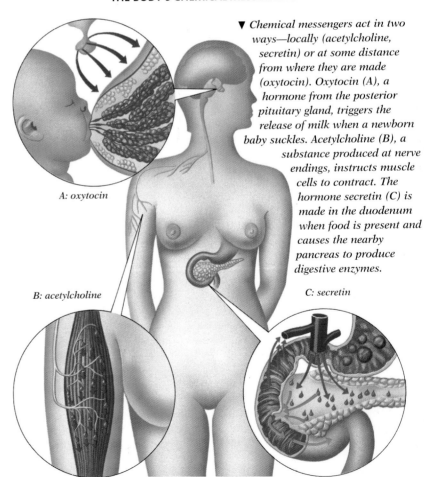

A: oxytocin

B: acetylcholine

C: secretin

▼ *Chemical messengers act in two ways—locally (acetylcholine, secretin) or at some distance from where they are made (oxytocin). Oxytocin (A), a hormone from the posterior pituitary gland, triggers the release of milk when a newborn baby suckles. Acetylcholine (B), a substance produced at nerve endings, instructs muscle cells to contract. The hormone secretin (C) is made in the duodenum when food is present and causes the nearby pancreas to produce digestive enzymes.*

Hormone-related disorders and their treatment

HORMONE	DISORDER	SYMPTOMS	TREATMENT
Growth hormone	Too little	Failure of growth, often linked with failure of sexual maturity	Administration of growth hormone
Growth hormone	Too much	Excessive growth in childhood leading to very long limbs (gigantism). In adults causes acromegaly—excess growth in skull, feet, and hands; enlarged larynx; deepening of voice; thickened skin.	Treatment of pituitary gland by radiotherapy, by removal of part of the gland (the other pituitary hormones may then need to be replaced), or with hormone antagonist drugs
Prolactin	Too much	Periods stop; breasts may produce milk and become tender; infertility	Drug treatment to reduce production
Antidiuretic hormone	Too little (or kidneys fail to respond to hormone produced)	Production of large quantities of very dilute urine (diabetes insipidus)	Synthetically produced hormone usually given as a nasal spray. Hormone then absorbed into blood.
Thyroxine	Too much	Weight loss; large appetite; excess body heat; periods may stop in women. One form (Graves' disease) also causes popping eyes.	Antithyroid drugs; radioactive iodine by mouth to destroy overproducing cells; surgery to remove part of thyroid gland
Thyroxine	Too little	Weight gain; loss of appetite; general body swelling. Lassitude; constipation. In infants produces condition called cretinism, associated with failure of physical and mental development.	Replacement of missing hormones at carefully controlled doses needed for life. It is now standard practice in the United States to screen newborn infants for cretinism.
Parathormone	Too much (usually owing to tumor)	Passing a great deal of urine; kidney stones; indigestion; feeling of malaise	Removal of parathyroid tumor
Parathormone	Too little	Muscular spasms; convulsions; lassitude; mental disturbance	Administration of vitamin D pills that mimic the action of the missing hormone
Hormones of adrenal cortex (e.g., cortisol, aldosterone)	Too much	Muscle wasting and weakness, leading to thin limbs but obese trunk. Fragile bones and blood vessels; purple stretch marks on skin. Diabetes; high blood pressure (Cushing's syndrome).	Drug treatment to block cortisone production. Where only one adrenal gland is involved, it is removed. Usually both are involved as a result of a tumor in the pituitary or elsewhere.
Hormones of adrenal cortex	Too little	Faintness; nausea; vomiting; loss of weight; low blood sugar; increased pigmentation on skin (Addison's disease); vague abdominal pain	Cortisone pills taken for life at carefully controlled dosage
Adrenaline	Too much	Episodes of palpitations; fright; raised blood pressure; fast pulse, leading to permanently raised blood pressure; pale face or occasional flushing	Removal of adrenaline-secreting tumor (usually found in the adrenal medulla)
Insulin	Too little	High blood sugar that may lead to loss of weight, thirst, and passing large quantities of urine (diabetes mellitus)	Diet is the cornerstone of treatment, with a reduction in the amount of sugar. This may be supplemented by antidiabetic tablets or insulin injections.
Male sex hormones	Too little	Failure of growth and sexual development or, in adulthood, impotence, infertility, mood changes, and sleep disturbances	Replacement of missing hormones by monthly injections, topical gels, or transdermal (skin) patches
Female sex hormones	Too little	Failure of growth and sexual development; menstrual periods do not start. Later in life, menopause (a normal event) due to reduced hormone levels.	Replacement of hormones by pills, transdermal (skin) patches, lotions, or gels

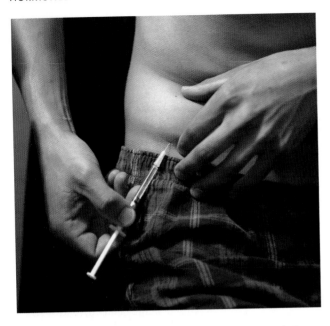

▲ *Injections of the hormone insulin can control type I diabetes.*

contains enzymes essential to the digestive processes. Another example is acetylcholine, which is made every time a nerve passes a message to a muscle cell, to tell it to contract.

All the hormones are active in tiny amounts. In some cases less than a millionth of a gram is enough for a task to be carried out.

Proteins and steroids

Chemically, hormones fall into two basic categories: those that are proteins or protein derivatives, and those that have a ring, or steroid, structure. Insulin is a protein, and the thyroid hormones are manufactured from a protein base and are protein derivatives. The sex hormones and the hormones made by the outer part, or cortex, of the adrenal glands are all steroid hormones.

When each hormone reaches its target, it will work only if it finds itself in a correctly shaped site on the target cell membrane. Once it has become locked into this receptor site, the hormone stimulates the formation of a substance called cyclic adenosine monophosphate (AMP). Cyclic AMP is thought to work by activating a series of enzyme systems within the cell so that particular reactions are stimulated and the required products are made.

The reaction of each target cell depends on its chemistry. Thus the hormone insulin—made by the pancreas—triggers cells to take up and use glucose. The hormone glucagon causes glucose to be released by cells, and to build up in the blood, to be burned off as energy-giving fuel for physical activity.

After they have done their work, the hormones are either rendered inactive by the target cells themselves, or are carried to the liver for deactivation. Once they are broken down, they are either excreted or used to make new hormone molecules.

Controlling the system

The complex system of hormone production and use in the body is governed by the pituitary gland. This lies at the base of the brain, connected to it by a short stalk, and has two parts. The front portion (or anterior pituitary) secretes many hormones that stimulate other endocrine glands to release their products. These trophic, or stimulating, hormones include the adrenocorticotrophic hormone that stimulates the adrenal cortex to make cortisol (assessment of cortisol levels in the blood helps measure the function of the pituitary and adrenal glands), the hormone that triggers the thyroid gland to release thyroid hormone, and the sex hormones (follicle-stimulating hormone that controls the release of hormones by the ovaries and testes and luteinizing hormone that triggers the output of testosterone in males and progesterone in females).

Other hormones made by the anterior pituitary exert their influence directly. These include growth hormone, which acts on cells throughout the body to promote normal growth and cell replacement, and prolactin, which stimulates milk production and inhibits menstruation while a woman is breast-feeding.

The posterior pituitary makes two important hormones: antidiuretic hormone, which travels to the kidneys and helps to maintain a correct fluid balance within the body; and oxytocin, which triggers the flow of breast milk when a baby starts to suckle. The release of these posterior pituitary hormones is directly controlled by nerve impulses generated in the hypothalamus (the part of the brain to which the pituitary gland is attached).

The only other hormones directly controlled by nerve impulses are adrenaline and noradrenaline, which are made and released by the inner medulla of the adrenal glands. There is a less direct link between the nerve cells of the hypothalamus and the anterior pituitary. Special nerve cells in the hypothalamus make releasing factors, which must act on the cells of the anterior pituitary before they can send out their hormones.

Effects on emotions

The strong link between the brain and the pituitary helps explain the connection between hormones and the emotions. For example, many women find that the timing of their periods alters if they are anxious or upset. The levels of estrogen and progesterone, which control the periods, can also affect a woman's mood. The sudden fall in their levels before menstruation is thought to play a vital part in creating the symptoms of premenstrual tension, whereas high levels at midcycle give many women a sense of well-being. This may be why women are most fertile and most responsive sexually at this time. Emotional factors can also alter the levels of these hormones. During sexual foreplay, for example, it is thought that estrogen and progesterone levels rise as a result of pleasurable impulses to the brain. Conversely, the thought of having sexual intercourse with someone who is repulsive is a turnoff because it inhibits hormone production.

At menopause a woman may experience great emotional ups and downs when the ovaries stop making estrogen and progesterone. Similarly, the sudden withdrawal of these hormones from the system after a woman has given birth may have emotional effects.

Hormonelike substances are found in many preparations, from the contraceptive Pill to the ointments used to treat eczema, but if a doctor recommends these types of drugs, it is important for him or her to point out the potential dangers. The drugs must be used with care and exactly as prescribed.

See also: **Endocrine system; Enzymes; Growth; Insulin; Menopause; Pituitary gland; Thyroid**

Hymen

Questions and Answers

Will my daughter be able to use a tampon if her hymen is not broken?

Hymens vary in the extent to which they cover the vaginal entrance, and in their elasticity, so it is often possible to use a tampon without damaging the hymen. It is incorrect to say that only women with sexual experience are able to use tampons successfully.

Can a girl be born without a hymen? I am sure I never had one.

Although most women do have a hymen, some are born without one. The hymen has no biological function, so its absence does not create any physical problems.

Could a woman reach old age without breaking her hymen?

A woman rarely reaches old age with her hymen intact, since most women will have had sexual intercourse. Also, the hymen may tear as a result of physical exercise or even for no apparent reason.

If my hymen were to break naturally, would this mean I was no longer a virgin?

No. Unless you have had sexual intercourse, you are still a virgin. The unbroken hymen as proof of virginity is a myth that has caused a lot of anxiety for many women.

Is it possible to have an artificial hymen made?

It is possible for a doctor to stretch a piece of tissue across the vagina, but this is an unusual practice. It may be done in places where for cultural reasons it is essential for blood to be present after the first intercourse. This is an unfortunate consequence of the myth that links virginity with the presence of an intact hymen.

The hymen is a thin fold of mucous membrane lying just within the vaginal orifice. It rarely covers the entrance of the vagina completely, and the belief that an unbroken hymen is proof of a woman's virginity is a myth.

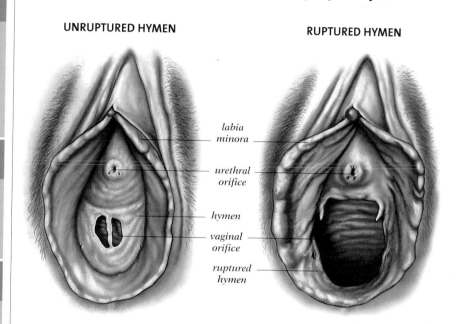

UNRUPTURED HYMEN — RUPTURED HYMEN

labia minora · urethral orifice · hymen · vaginal orifice · ruptured hymen

The hymen, also known as the maidenhead, is named after the Greek god of marriage, Hymen. It has no known physiological function but has achieved great importance in most cultures as a sign of virginity. However, hymens come in all shapes and sizes, and although they are often broken during the first experience of sexual intercourse, there is no way in which one can be a reliable indication of virginity. The membrane is usually thin and punctured by holes; rarely it is complete (imperforate). It can easily be broken by strenuous physical exercise such as running and horseback riding, by heavy petting, by masturbation, or by the insertion of tampons.

When the hymen is first torn, there may be some pain and slight bleeding. Both of these last for only a short time and rarely cause much discomfort. If a woman has a thick hymen and is worried about pain during her first experience of sexual intercourse, her gynecologist can show her how to stretch her hymen with her fingers so that it will be less of a potential problem.

Problems

On rare occasions the hymen forms a complete barrier across the vagina, and this can cause problems because it may restrict the menstrual flow. When this happens the hymen can be removed by a simple operation called a hymenectomy, which causes very little pain.

Sometimes this operation is necessary if the normally thin and flexible hymen is thick and fibrous and resistant to penetration of the penis during sexual intercourse. However, if the hymen is so elastic that it stretches rather than tears during sexual intercourse, it may need to be removed from a pregnant woman before her child is born.

Pregnancy

Contrary to popular belief, an intact hymen does not prevent pregnancy. Sperm that come into contact with the genital area, perhaps as a result of heavy petting, may still travel through a gap in the hymen and up into the vaginal canal.

See also: Genitals; Menstruation; Vagina

Hypothalamus

Many life-maintaining processes, including eating, sleeping, and drinking, are controlled by one small area of the brain—the hypothalamus—which plays a vital role in both the nervous system and the endocrine system.

The hypothalamus lies at the base of the brain, under the two cerebral hemispheres and immediately below the thalamus, which acts as a link between the spinal cord and the cerebral hemispheres. Like the thalamus, it is an important structure in the nervous system, consisting of a collection of specialized nerve centers that connect with other important areas of the brain and with the pituitary gland. The hypothalamus plays a vital role in controlling functions such as eating and sleeping and is closely linked with the endocrine system.

Exerting control

Some of the nervous pathways of the hypothalamus connect with the limbic system, which influences smell and sight. This portion of the brain also deals with the formation of memories, so abnormalities here may cause memory loss.

A vital function of the hypothalamus is temperature control. Substances that act directly on the hypothalamus can cause high temperatures, and when serious head injuries affect the hypothalamus, temperatures of over 105°F (40.5°C) may prove fatal.

The hypothalamus also connects with the reticular activating pathways, which run up and down the central core of the brain and spinal cord. The reticular activating formation is concerned with sleeping and waking. The condition called narcolepsy, in which a person falls asleep suddenly at any time and cannot be roused, is usually due to disease of the hypothalamus.

As if this were not enough, the hypothalamus is also the controlling center for both appetite and thirst. Although problems in this area are rare, one of the more common syndromes is gross

▼ *It is the hypothalamus that controls a person's appetite and thirst.*

obesity due to massive overeating, often with excessive sleeping. This is called the Pickwickian syndrome after a fictional character in Charles Dickens's *Pickwick Papers*.

The control of thirst and water balance is handled separately. The hormone that influences water loss from the kidneys is called antidiuretic hormone, or ADH. It is produced by the hypothalamus but secreted from the posterior part of the pituitary gland, which is closely connected to the hypothalamus by the pituitary stalk. ADH passes from the hypothalamus to the posterior pituitary via the specialized nerve cells of this stalk. However, the receptors that measure the concentration of the blood (and thus help the hypothalamus to regulate the body's water balance) are found in the hypothalamus itself. Excessive ADH activity causes the kidneys to retain

HOW THE HYPOTHALAMUS CONTROLS BODY TEMPERATURE

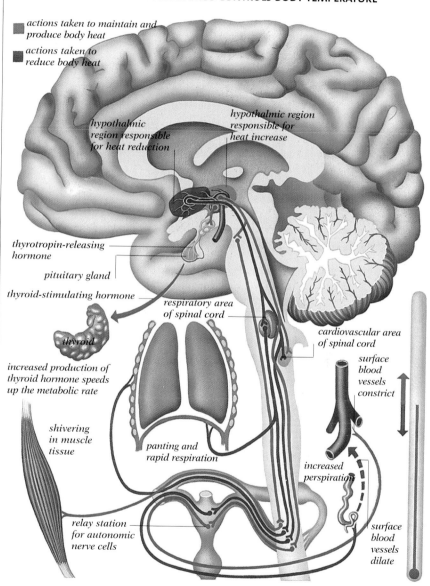

◼ actions taken to maintain and produce body heat

◼ actions taken to reduce body heat

hypothalmic region responsible for heat reduction

hypothalmic region responsible for heat increase

thyrotropin-releasing hormone

pituitary gland

thyroid-stimulating hormone

respiratory area of spinal cord

cardiovascular area of spinal cord

surface blood vessels constrict

thyroid

increased production of thyroid hormone speeds up the metabolic rate

shivering in muscle tissue

panting and rapid respiration

increased perspiration

relay station for autonomic nerve cells

surface blood vessels dilate

▲ *The hypothalamus receives information about the body's temperature both from the heat of the blood passing through it and from the messages sent by the temperature-sensitive nerve endings in the skin surface. One region is sensitive to an increase in body heat and another reacts to a decrease. The nerve pathways from these regions pass through the spinal cord via the autonomic nervous system and control a number of bodily activities, such as sweating, which increase or decrease the temperature of the body.*

▲ *A defective hypothalamus may cause narcolepsy; the person affected keeps suddenly falling asleep at any time.*

too much water. A loss of ADH secretion causes diabetes insipidus, in which the kidneys produce large quantities of dilute urine. The latter condition may also result from head injuries that damage the pituitary stalk.

There is no doubt that the hypothalamus is involved with the control of behavior and emotion. Withdrawn and confused behavior with some aggressive outbursts may happen in cases of disease. Sexual overactivity may also occur.

Links with hormones

Apart from its role in the nervous system, the hypothalamus is an integral part of the endocrine system. Although the posterior part of the pituitary gland is actually part of the hypothalamus, the anterior part of the gland is anatomically separate, and most of the important interactions occur between the hypothalamus and the anterior pituitary. It is really a little endocrine system of its own. Hormones—either inhibitory or releasing factors—are produced in the hypothalamus and are carried to the pituitary in blood vessels that course through the pituitary stalk.

The hormone TSH (thyroid-stimulating hormone) or thyrotrophin, produced by the anterior part of the pituitary gland, stimulates the thyroid to release thyroid hormone. In turn the production of TSH is stimulated by thyrotrophin-releasing hormone (TRH) from the hypothalamus.

Similar systems exist to stimulate the release of corticotrophin, which controls adrenal activity, and FSH (follicle-stimulating hormone) and LH (luteinizing hormone), which stimulate the ovaries and the testes.

The anterior pituitary also produces two hormones that act directly upon the tissues. Growth hormone is one of these. It is essential for normal growth and is also involved in the control of blood sugar. Growth hormone seems to be controlled by two factors, one of which stimulates release while the other inhibits it.

In contrast, the hormone prolactin—which stimulates lactation (milk production) and inhibits menstruation—seems to be largely controlled by a substance called prolactin-inhibiting factor, which restrains rather than stimulates secretion.

All these releasing and inhibiting factors are comparatively simple chemicals, and the formulas for many of them have been figured out. Nowadays they can be made and used not only to test pituitary function but occasionally for treating patients.

One of the most fascinating developments in this field has been the realization that some of these compounds, particularly somatostatin (which affects growth hormone release) and TRH, are actually widespread in the nervous system and act as transmitters.

Problems

Diseases of the hypothalamus are rare. Tumors may occur in this area, and infection of the nervous system in the form of encephalitis or meningitis may also affect the function of the hypothalamus. In a similar way, head injuries or surgery may cause problems that are difficult to treat.

It is possible that some of the small benign hormone-producing tumors that occur in the pituitary actually result from overstimulation by the hypothalamus.

See also: **Appetite; Brain; Endocrine system; Estrogen; Excretory systems; Growth; Hormones; Kidneys; Memory; Nervous system; Pituitary gland; Spinal cord; Temperature; Thyroid**

Immune system

Questions and Answers

Can people's state of mind affect their immune system by making them prone to illness?

Yes. Current medical thinking holds that the interaction of the mind and all the systems of the body is so intimate and interdependent that hardly anything can happen in one area without affecting another. This certainly holds true for the emotions and the immune system. There is a discipline called psychoneuroimmunology that studies this relationship.

Can a child inherit immunity to any disease from its mother?

Yes. There are some globulins that cross the placenta and so give the baby some protection for the first six weeks to three months of life while his or her immune system is still immature. An example of this is protection of the baby from infection by the measles virus. However, if the mother hasn't suffered from measles, she won't pass on the antibody created to fight the infection.

Another important source of immunity is from globulins in the mother's breast milk. Even if a mother breast-feeds her baby for just the first few weeks, this will still provide some extra protection from disease to her child.

After having an illness once, can I get it again?

Yes, it is possible for you to have some illnesses again once you have already had them. This is especially true with the common cold, simply because there are so many different viruses that cause this illness that a person could never be immune to all of them. Another issue is that the level of a specific antibody to a disease can wane with time, leaving the person vulnerable to that disease in the future.

The body has its own internal defense system; the immune system, which is designed to protect it from illness. How does the immune system work, and why does it sometimes fail?

The immune system protects people from infection and invasion by all sorts of bacteria, viruses, and other microbes. It has two major weapons to accomplish this. The agents of the humoral (antibody) system are plasma cells, special kinds of white blood cells called B lymphocytes that secrete antibodies (immunoglobulins). The cellular immune system works through a second kind of lymphocyte, the T cell, which in early childhood migrates from the bone marrow to the thymus to mature, then settles in the lymph nodes and other lymphatic tissues or circulates in the lymph and blood. The T cells either kill invaders directly or act as helper cells.

▼ *People who are physically active are generally less vulnerable to infection.*

B LYMPHOCYTES

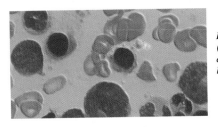

▲ B lymphocytes, or plasma cells, produce immunoglobulins that enter the bloodstream to ward off attack by bacteria, microbes, and viruses.

T LYMPHOCYTES

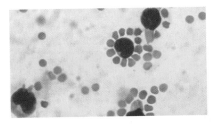

▲ T lymphocytes, stored in the lymph nodes, are alerted to attack foreign tissue and viruses by helper cells, which carry messages about any invasion.

HOW IMMUNITY-PRODUCING CELLS FUNCTION

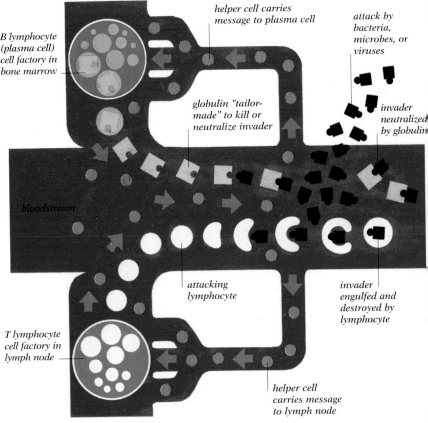

helper cell carries message to plasma cell

attack by bacteria, microbes, or viruses

B lymphocyte (plasma cell) cell factory in bone marrow

globulin "tailor-made" to kill or neutralize invader

invader neutralized by globulin

bloodstream

attacking lymphocyte

invader engulfed and destroyed by lymphocyte

T lymphocyte cell factory in lymph node

helper cell carries message to lymph node

How immunity works

Immunity by B cells: The B cell makes substances called antibodies or immunoglobulins (Ig). The globulin molecules consist of two long chains of amino acids, called heavy chains because of their large size, flanked by two shorter strands of amino acids, the light chains.

There are five classes of immunoglobulins: IgM, IgG, IgA, IgD, and IgE. The most primitive is IgM, which occurs in lower vertebrates. This is the globulin made by all B cells at first, some of them changing later to make one of the other classes. IgM is the largest globulin molecule and makes up 10 percent of the total plasma globulin concentration. IgM is the first antibody to appear after an infection starts; it may assist B cells to remember invaders they have met before.

IgG is the most important globulin, making up 80 percent of the total concentration. It plays the major role in neutralizing bacterial toxins and in coating the bacteria so that they can be engulfed by other blood cells. Being a small molecule, it is able to cross the placenta and transfer a mother's immunity to disease to her baby.

IgA is the principal antibody in mucous secretions and is found in saliva, in tears, and along the respiratory and intestinal tracts. It plays a major role in defending the body against bacteria. The function of IgD is unknown; it may help B cells to grow and develop. IgE is responsible for hypersensitivity in hay fever and hereditary allergic disease.

Immunity by T cells: The precursors of T cells migrate from the bone marrow to the thymus, where they are programmed for the number of antigens they will recognize. Leaving the thymus, they settle in the lymphatic tissues or circulate in the lymph and blood, where they constitute 70 percent of all lymphocytes. Some T cells are killer cells that attack bacteria and viruses directly. They also detect and destroy foreign tissue and so are responsible for rejecting transplanted organs. The same quality may protect people against cancer: T cells recognize developing malignant cells and destroy them before they spread. Other T cells are thought to regulate B cell antibody production, either by increasing it (helper cells) or turning it off (suppressor cells).

What can go wrong

Immunity in the system depends on the ability of B and T cells to recognize an invading bacterium or virus. Once the B and T cell remember an invader, they will attack and repel it, sometimes having been boosted in number by a vaccine.

There are some rare inherited diseases that result in a deficiency in production of globulin antibodies, of T lymphocytes, or of both. In certain cases one particular class of globulin is missing; for example, the IgG antibody is absent in a disease called agammaglobulinemia.

It is possible to survive a severe deficiency of immunoglobulin because globulin from other people can be given by a monthly injection. A complete deficiency of white cells is more difficult to treat and sufferers usually succumb to an infection early in life.

The AIDS virus, HIV, acts on a type of T lymphocyte (T4), causing it to die and so crippling the immune system.

See also: **Lymphatic system; Marrow; Thymus**

Insulin

Insulin is a hormone produced in the pancreas to keep levels of sugar in the blood within normal levels. Insulin was one of the major discoveries of the early part of the 20th century and is now the main treatment for diabetes.

Questions and Answers

I have been taking insulin for two years since I was diagnosed as a diabetic. Will I have to take it for the rest of my life?

Probably. However, insulin is sometimes needed only to tide diabetics over a difficult period, after which they may manage with just a special diet. Some diabetic women need insulin only while they are pregnant. In some cases pregnancy actually brings out a tendency to diabetes, and the woman is given insulin until after the birth, when she can return to normal and no longer needs it.

Can you have too much insulin in your body as well as too little?

Yes. In a rare condition called insulinoma, too much insulin is secreted by an insulin-producing tumor of the pancreas. Symptoms may include loss of consciousness and seizures. The most common cause of excess insulin, however, occurs in diabetics who eat less, or exercise more, than is appropriate for the insulin dosage taken. The result is a dangerous drop in blood sugar known as hypoglycemia.

My son is diabetic; should I encourage him to give himself his own insulin injections or should I continue to do them for him?

This depends on his age. Obviously, very young children need to have the injections given to them by an adult, but many specialists believe that children do better if they can take control of the condition and give their own injections right from the start or as soon as possible. Even children as young as seven or eight may manage their own injections with parental help and encouragement. Learning to do it at an early age will also help your son to be independent and can give him some pride in the achievement of managing to inject his insulin himself.

HOW INSULIN IS MADE BY THE BODY

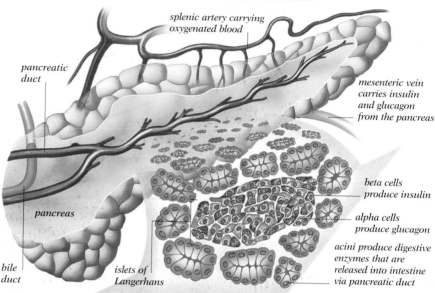

splenic artery carrying oxygenated blood

pancreatic duct

mesenteric vein carries insulin and glucagon from the pancreas

pancreas

beta cells produce insulin

alpha cells produce glucagon

acini produce digestive enzymes that are released into intestine via pancreatic duct

bile duct

islets of Langerhans

▲ *The hormones insulin and glucagon are produced in the islets of Langerhans. They enter the bloodstream via the mesenteric vein and balance the body's sugar level.*

Insulin is made by the islets of Langerhans, a cluster of hormone-producing cells in the pancreas. If the level of sugar in the blood begins to rise above a certain level, the islets of Langerhans release insulin into the bloodstream. The insulin acts to oppose the effects of hormones such as cortisone and adrenaline, which raise the level of sugar in the blood. Insulin allows sugar to pass from the bloodstream into the body's cells, where it can be used as fuel. If the body produces only a very small amount of insulin or no insulin at all, there is no mechanism for controlling the blood sugar level, because the sugar in the blood cannot be converted into fuel for the cells. The lack of insulin causes a condition known as diabetes mellitus. The pancreas also produces another hormone called glucagon, which helps to break down glycogen that is stored in the liver and muscles. This has the effect of raising blood glucose levels.

Treatment for diabetes

The treatment for diabetes is to replace the insufficiently produced or absent hormone with either insulin from animals or genetically engineered human insulin. Because insulin is destroyed by gastric juices if taken by mouth, it is given by self-injection. In general, diabetics have two injections of insulin a day, one before breakfast and one before their evening meal.

Diabetics who accidentally take too much insulin may develop hypoglycemia, in which the blood sugar is drastically reduced. The symptoms include agitation, trembling, sweating, speech difficulty, pallor, and unconsciousness. Prompt action is necessary to prevent possible death. People with early symptoms are given sugar in the form of candy or fruit; unconscious sufferers are given an injection of glucagon into a muscle or a glucose solution into a vein.

> **See also:** Hormones; Pancreas

Joints

Joints, which connect the bones that make up the human skeleton, give the body a full range of movements. However, the individual joints vary in the amount of flexibility they allow, and in their vulnerability to injury.

There are two main types of joints—synovial joints and fibrous joints. Synovial joints are designed to allow a large range of movements, and they are lined with a slippery coating called synovium. Fibrous joints have no synovium; the bones are joined by tough fibrous tissue, permitting very little movement.

Synovial joints

The synovial joints can be subdivided according to the range of movement of which they are capable. Hinge joints, such as the ones at the elbow and knee, allow bending and straightening movements. Gliding joints allow sliding movements in all directions because the opposing bone surfaces are flattened or slightly curved. Examples of these joints are found in the spinal bones, the wrist, and the tarsal bones of the feet. Pivot joints in the neck at the base of the skull and at the elbow are special types of hinge joints that rotate around the pivot. The pivot joint in the neck

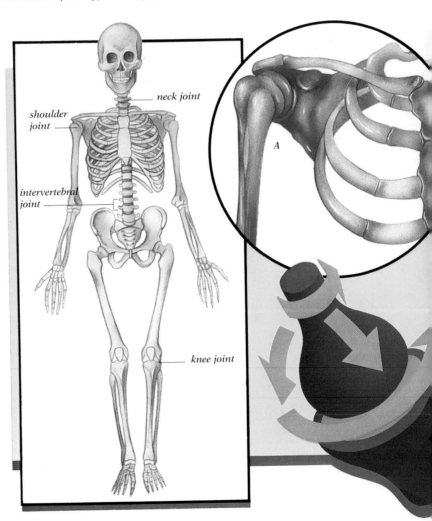

neck joint

shoulder joint

intervertebral joint

knee joint

A

allows the head to turn, and the joint in the elbow allows twisting of the lower arm to enable movements such as turning a doorknob. Joints that can be moved in any direction, such as the hip and shoulder, are called ball-and-socket joints. The joints in the fingers are typical examples of hinged synovial joints. The bone ends are covered in a tough elastic material called articulating cartilage. The entire joint is enclosed in a very strong coating of strong fibrous tissue called the joint capsule. This holds the joint in place and thereby prevents any abnormal movement. Lining the inside of the joint, but not running over the articulating cartilage, is the synovium. This layer of tissue is often only one cell thick and provides fluid that lubricates the joint and prevents it from drying up. It is not absolutely essential for the normal functioning of the joint, and in certain conditions in which the synovium becomes diseased, such as rheumatoid arthritis, some of it may be removed without damaging the joint in the short term. A healthy synovial membrane is thought to be essential to help prevent wear and tear of the joint.

Knee joint

The knee joint is a much more complicated structure than it would first appear, and it is commonly injured by athletes. When a person extends his or her knee beyond the straight position, it locks. This locking is brought about by a very slight but important twisting movement of the large bone in the lower part of the leg (the tibia). There are two pieces of gristle that pad between the cartilage linings of the thighbone (the femur), the largest bone of the body. One piece of cartilage wedges the outside of the joint and the other

▼ *People can bend and move many of the 206 bones of the human skeleton because the bones have joints at each end. The synovial joints allow the greatest range of movements, and there are several types. A: Ball-and-socket joints, like those in the shoulder, allow movement in any direction. B: Hinge joints, like those in the knee and elbow, allow only bending and straightening movements. C: Pivot joints are hinge joints, as in the neck, that rotate around a central pivot. These types of joints allow movements such as turning the head from side to side. D: Gliding joints, which are found in the spine, wrists, and feet, allow a sliding movement, which happens in twisting the body from side to side.*

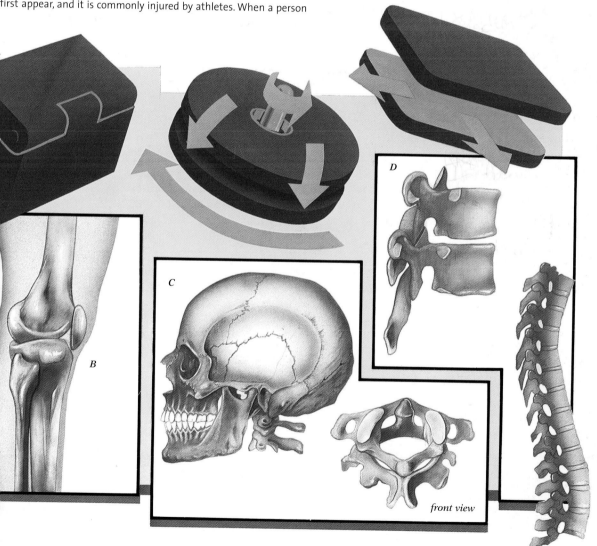

front view

Is it possible to get osteoarthritis of the jaw?

Yes. It occurs more often in people who have lost their teeth and in those who clench or grind their teeth while sleeping. It may also cause pain elsewhere, for example a headache. The most effective treatment is with an anti-inflammatory painkiller, such as aspirin or ibuprofen.

My uncle has suffered from back pain for years. The doctors now tell him he needs an operation to set his back solid. What does this mean and how will it help relieve my uncle's pain?

Pain in an arthritic back is caused by joint movement. If the joints can be prevented from moving, then the pain is relieved. An operation called a spinal fusion can be performed to fuse the bones together, usually just at the lower part of the back.

My aunt has arthritis. She has read that aspirin is bad for her stomach. Is there a painkiller she can take that doesn't damage the stomach?

All anti-inflammatory drugs cause a certain degree of irritation of the stomach. Aspirin is particularly bad in this respect, but it is also one of the most effective drugs. It is always better to use a soluble type of aspirin than one that doesn't dissolve in water and better still to have food in the stomach at the time the aspirin is taken.

My sister can make amazing movements with her thumb. Could she be double-jointed or is there a problem with her joints?

There are no truly double-jointed people, although some people can move all their joints through a very large range of movements; this is usually because they have trained themselves to do so. However, an ordinary person could simply have greater laxity of a joint or two and thus have greater mobility.

the inside of the joint. These are called semilunar cartilages, since they are shaped like a crescent moon, with the points toward the interior of the joints. The outer edge of each semilunar cartilage is attached to two very strong ligaments (elastic tissue that connects muscles), one running down the outside of the leg and one down the inside. These form the lateral and medial collateral ligaments of the knee, with the cartilage forming a sandwich between them. Since the cartilages are attached at their points and their edges, a twisting movement of the knee that pulls the ligaments can tear the cartilages. A torn cartilage—common among football players—causes severe pain and sometimes an inability to bend the joint. If the condition persists, the only treatment is to remove it. The cartilage may grow back, but if not, the muscles strengthen to compensate for the missing cartilage.

Both the knee and the hip joint have ligaments in the center of the joint for added stability. In the case of the knee, there are two of them that cross, one from front to back and the other back to front (for this reason they are called anterior and posterior cruciate ligaments). These too can be torn, particularly when the knee is dislocated.

The synovial membrane lining the inside of the knee is very extensive and runs under the kneecap (patella) for about 1 inch (2.5 cm) down the front of the shin. There are several bursas (small enclosed cavities) at the knee that are lined with synovial membrane and filled with synovial fluid. Constant kneeling or a bang on the knee can result in inflammation of the synovium, causing painful swelling due to increased production of fluid by the synovium. This condition is called bursitis.

One of the most common diseases of the joints is rheumatoid arthritis. This is an autoimmune disease and is characterized by an inflammation of the synovium, and, as a result, accumulation of fluid in the joints. This condition is quite distinct from wear-and-tear arthritis.

Fibrous joints

The fibrous joints are those of the back and the sacrum (the triangular bone between the hipbones), and some of the joints in the ankle and the pelvis. The joints of the spine are

SYNOVIAL JOINTS

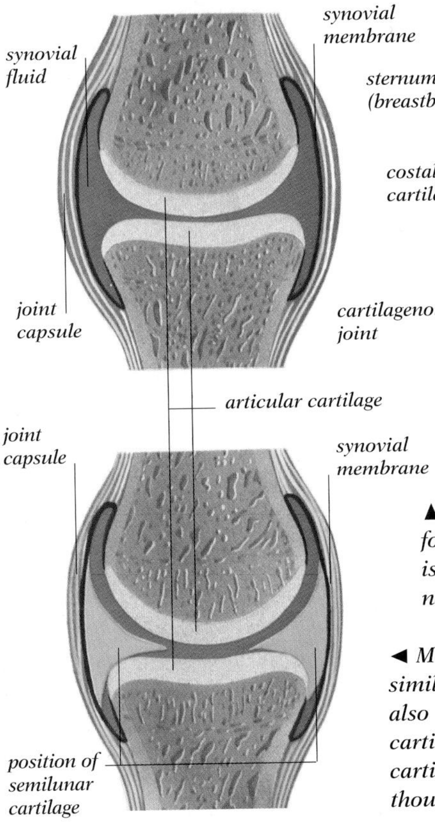

synovial fluid

synovial membrane

joint capsule

joint capsule

articular cartilage

synovial membrane

position of semilunar cartilage

CARTILAGINOUS JOINTS

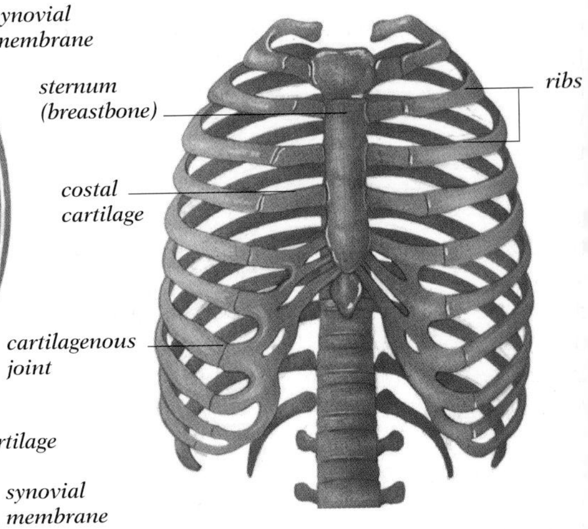

sternum (breastbone)

ribs

costal cartilage

cartilagenous joint

▲ Certain joints on each of the ribs are formed from bone and cartilage. The cartilage is flexible, allowing movement without the need for fluid-filled synovial membranes.

◀ Most synovial joints, such as the elbow, are similar to the one above left. The knee joint is also a synovial joint, but it has semilunar cartilage between the two layers of articular cartilage, as seen on the left. This cartilage is thought to act as a shock absorber.

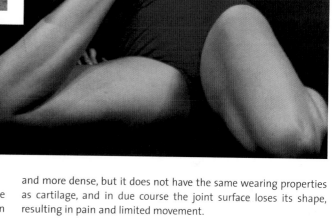

▲ ▶ *Games like those above would be impossible without the joints that allow the limbs to bend. Some people show a greater range of flexibility than others, as in the case of the woman on the right, who is practicing yoga.*

different from the other fibrous joints. Besides being joined by fibrous tissue, they are cushioned by a disk of pulpy material surrounded by a strong ring of fibrous tissue. This acts as a shock absorber, and there is one between each of the spinal bones (the vertebrae).

If the vertebrae are crushed together, as during heavy lifting, the pulpy material can be forced out through the fibrous tissue. When this happens the material may push backward and press against the spinal cord or nerve root, causing severe pain. This is called a slipped disk. Often the pulpy tissue pops back by itself after plenty of rest, or the body simply breaks the tissue down over time (resorption). Occasionally, surgery is needed, and the damaged disk has to be removed from between the vertebrae.

Cartilaginous joints

Some joints in the body are formed between bone and cartilage. Cartilage is very flexible, and this flexibility allows a good deal of movement with no need for fluid-filled synovium. The joints between the ribs and breastbone at the front of the chest are cartilaginous joints. The joints attaching the ribs to the back are synovial joints. These allow the ribs to move up and down freely when a person breathes.

Damage

Damage to the joints may give rise to problems many years later. People who injure their joints regularly are more prone to osteoarthritis. The cartilage is worn away from the end of the bone so that bone rubs against bone. The exposed bone becomes stronger

and more dense, but it does not have the same wearing properties as cartilage, and in due course the joint surface loses its shape, resulting in pain and limited movement.

Treatment

The development of spare-part surgery has transformed the treatment of severe osteoarthritis. At one time the only treatment that offered relief from constant pain was aspirin. Now complete replacement joints are fashioned out of plastic or metal, and they will last for up to 20 years or more.

Pain relief is dramatic following joint replacement, and with regular physical therapy a full range of movements is likely to be restored. Many joints can be replaced, but the hip is the most common joint to be treated using a prosthesis.

Rheumatoid arthritis features inflammation of the synovium, which itself begins to eat away the cartilage at the end of the bone, leading to an increased chance of osteoarthritis. In severe cases of rheumatoid arthritis, when inflammation cannot be controlled by anti-inflammatory drugs, the synovium can be surgically removed to provide some relief and to slow down the rate at which osteoarthritis develops.

See also: Back; Cartilage; Elbow; Hand; Hip; Knee; Movement; Neck; Shoulder; Skeleton

Ketones

Questions and Answers

When I had a stomach upset, my doctor asked for a urine sample so he could test for ketones. Why?

Any illness that causes loss of appetite or vomiting is likely to result in increased levels of ketones in the blood and urine because the body is breaking down fat to produce energy. A positive test for ketones in the urine just confirms that.

Whenever I am overweight I go on a crash diet. Will I start to produce ketones?

Mostly you will lose water. If you do not exercise as well as diet, your body will metabolize more muscle than fat. A crash diet will slow your overall metabolism; although you eat less, your body will use proportionately less energy and store more fat. If you reduce your glucose intake, your body will metabolize muscles and fat and produce ketones.

My husband has diabetes, and sometimes his breath smells bad. What causes this?

Ketones give the breath a characteristic sickly odor. In diabetics this is a warning sign of ketosis (the presence of excessive ketones in the body), which may lead to a diabetic coma. In this condition, there is insufficient insulin to enable the glucose in the blood to be used as fuel. As a result, an excess of ketones is released to give energy.

I've heard that children are more likely than adults to suffer from an excess of ketones. Is this true?

Yes. Children often develop ketosis when they are sick, especially if they have been vomiting. It is usually mild and is the body's initial attempt to provide energy.

An overproduction of ketones is called ketosis. It may occur in diabetes when the body's tissues cannot take up glucose because of an inadequate production of insulin. The condition is a medical emergency.

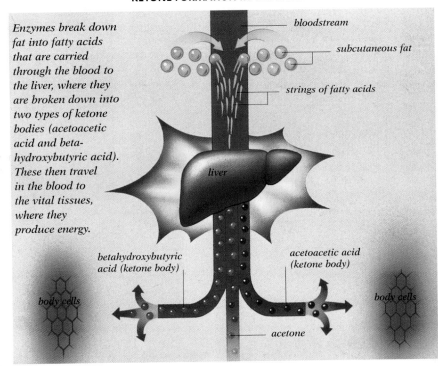

KETONE FORMATION IN THE LIVER

Enzymes break down fat into fatty acids that are carried through the blood to the liver, where they are broken down into two types of ketone bodies (acetoacetic acid and beta-hydroxybutyric acid). These then travel in the blood to the vital tissues, where they produce energy.

bloodstream

subcutaneous fat

strings of fatty acids

liver

betahydroxybutyric acid (ketone body)

acetoacetic acid (ketone body)

body cells

body cells

acetone

Nearly all a person's energy reserves are stored in fatty tissue created and maintained by glucose and carbohydrates. A constant supply of glucose in the bloodstream is necessary to carry out the body's functions and to supply tissues with energy, so when glucose intake is low, proteins and carbohydrates are broken down to make it. Metabolism occurs in every cell. Ketones are produced at a stage of fat metabolism, and insulin metabolizes them into glucose.

There are three types of ketones: acetoacetic acid, betahydroxybutyric acid, and acetone. Acetone, a waste product of fat metabolism, is produced at the same time as the ketone bodies but has no useful function. Ketone bodies, on the other hand, are readily utilized as a source of energy. When glucose is scarce, fatty tissue is broken down into fatty acids and carried in the bloodstream to the liver, where ketone bodies are formed. The ketones are released into the circulation and used for energy by the muscles, the heart, the brain, and by many other tissues and organs.

Ketosis is a potentially serious condition in which there is an abnormal accumulation of ketones in the body due to a carbohydrate deficiency or inadequate carbohydrate metabolism. As a result, fatty acids are metabolized, and the end products, ketones, accumulate. The underlying causes of ketosis include starvation, alcohol consumption, inadequate intake of protein and carbohydrates and, most commonly, diabetes mellitus. Women in labor can also become ketotic, and high levels of ketones in the blood may delay labor by interfering with the uterus's ability to contract. Symptoms are sweet, sickly-smelling breath, loss of appetite, and vomiting. Untreated ketosis may lead to unconsciousness and death. A glucose IV will stop the formation of ketones.

See also: **Insulin; Metabolism**

Kidneys

I know that alcohol causes damage to the liver, but does it affect the kidneys? And how much fluid should I drink each day?

Alcohol tends to have no detrimental effect on the kidneys. The amount of fluid you need to drink each day depends on how much you lose and the amount of physical work you do, since this will increase the volume lost in sweat. Even office workers should drink at least 2.12 pt. (1 l) of water a day. People with urinary infections should drink large amounts of water to flush out the infection.

My uncle has high blood pressure. Will this harm his kidneys?

High blood pressure causes the small blood vessels to thicken. This causes damage to the kidneys' filtering units (nephrons), and as these are lost, so is the kidneys' ability to remove waste products. This may result in kidney failure unless the condition is treated.

If one kidney fails, can the other one cope on its own?

Yes. In fact, we have so many nephrons in each kidney that we can easily do without not just one, but almost half of the other kidney as well. For this reason it is perfectly reasonable to remove a kidney from a healthy person and donate it to someone else. The donor can live with one kidney for the rest of his or her life, provided that it remains healthy.

I have diabetes mellitus. Is it likely to damage my kidneys?

Four out of 10 people who have had diabetes mellitus for more than 15 years develop diabetic kidney disease. However, most people with diabetes mellitus are monitored to detect kidney damage at an early stage.

The main purpose of the kidneys is to extract impurities from the urinary system. The kidneys are amazingly efficient, and because there are two it is possible to live normally even if one fails.

The kidneys contain a complicated system of filters and tubes. Apart from their main function of filtering impurities from the blood, the kidneys also absorb many essential nutrients back into the bloodstream from the tubes. Another important function performed in the tubes is balancing the amount of salt and water that is retained.

The kidneys lie on the back wall of the abdomen. A tube called the ureter runs from the inner side of each of the two kidneys down the back of the abdominal cavity and into the bladder. The tube leading from the bladder is called the urethra. In women its opening is in front of the vagina, and in men it is at the tip of the penis. The urethra is much shorter in women than men; therefore women are more prone to bladder infections than men are.

Functions

The kidneys contain thousands of tiny filtering units or nephrons. Each nephron can be divided into two important parts; the filtering part (glomerulus) and the tubule, where water and essential nutrients are extracted from the blood.

▲ *Drinking plenty of water helps keep the kidneys healthy. People with diabetes, dehydration, or urinary infections need to drink even more to replenish the lost water.*

199

Questions and Answers

Do some forms of sexually transmitted diseases (STDs) lead to kidney trouble?

STDs are a rare cause of kidney disease, although occasionally gonorrhea can lead to kidney infection by spreading up the ureter from the bladder. A much more common cause of kidney trouble is bacterial, fungal, or viral infection, or infection of the urinary tract.

The doctor has told me that I have bacteria in my urine and have to take antibiotics. Why does this happen, and is it easy to cure?

Urine is normally sterile. However, some people shed bacteria in their urine, and this is a sign of infection somewhere in the urinary tract. As long as the bacteria persist, you have a chance of developing acute inflammation of the kidney. Doctors may switch from one antibiotic to another over several weeks until the most effective is found and the infection is completely cleared.

I have read that drinking cranberry juice helps the kidneys to function better. Is there any truth in this?

Yes, a number of studies have concluded that cranberries are beneficial to the kidneys and the urinary tract. One study reported that out of 60 patients with urinary tract infections who were given 2 cups (473 ml) of cranberry juice per day for 3 weeks, 70 percent showed moderate to excellent improvement. The *Journal of Urology* has stated that cranberry juice is a potent inhibitor of bacterial adherence in the urinary tract. Cranberry juice also helps dissolve kidney stones, according to an article in the *New England Journal of Medicine* that stated, "An 8-fl. oz. (240-ml) glass of cranberry juice four times daily for several days, followed by a glassful twice daily, is valuable therapy for stone-forming patients." Cranberry taken in tablet form is also thought to be beneficial to the kidneys.

▶ *Magnified view of a kidney, showing its components and how they work. The renal artery carries blood to the kidney (bottom left), splitting into arcuate arteries and finally into afferent arterioles (top left). Each of these ends in a glomerulus (inset). As blood passes through the glomerulus, it is filtered through the glomerular wall and enters the renal tubule. The basic components of blood (such as plasma, protein, and red and white corpuscles) are too large to cross the glomerulus's semipermeable membrane, but most of the other material that the blood carries around the body (such as water, salts, and hormones) can pass across it. The next stage, selective reabsorption, is seen in the detailed enlargement (bottom, far right). Materials essential to the body, including almost all of the filtered water, are reabsorbed into the efferent arterioles, across the tubule wall. Once the blood has been thoroughly filtered and all the required materials have been reabsorbed, the blood leaves the kidneys in the renal vein while waste products are excreted in the urine.*

The glomerulus consists of a knot of tiny blood capillaries that have very thin walls. Water and the waste dissolved within it can pass freely across these walls into the collecting system of tubules on the other side. This network filters about 4.4 fluid ounces (130 ml) from the blood each minute. Most of this fluid is reabsorbed so that urine production is about 0.3 fluid ounce (1 ml) per minute.

The holes in the capillary wall form a biological sieve and are so small that molecules beyond a certain size cannot pass through. If the kidneys become infected, the glomeruli inflame and the sieve fails to be so selective, allowing larger molecules, such as proteins, to escape into the urine. One of the smallest protein molecules to find its way into the urine is albumin (the main protein in the blood). When a doctor tests a person's urine for protein, he or she is checking to see if the kidneys (and liver) are working properly.

The tubules run between the glomeruli to a collecting system that ultimately drains into the bladder. Each glomerulus is surrounded by a structure called Bowman's capsule, which is the beginning of its tubule. It is in the tubules that almost all the filtered water and salt is reabsorbed so that the urine is concentrated. To reabsorb all this water, the body has a highly sophisticated system in which a hormone secreted into the blood from the pituitary gland in the brain changes the permeability (ability to reabsorb water) of the tubule.

While the hormone is in the blood, the tubule allows a great deal of water to be reabsorbed. When the hormone is turned off, however, the tubule becomes less permeable to water and more is lost in the urine; this situation is called diuresis, and the hormone concerned is antidiuretic hormone (ADH). In certain conditions, such as diabetes insipidus (not to be confused with

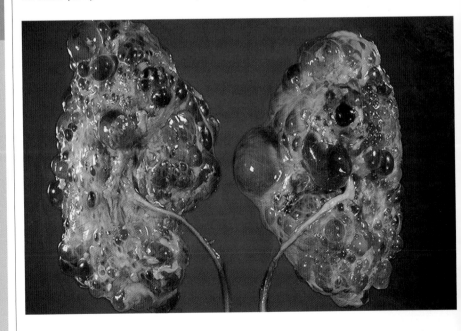

▲ *In polycystic kidney disease, which is caused by an inherited abnormal gene, the normal tissue of the kidneys is replaced by many cysts that look like fluid-filled bubbles.*

THE KIDNEY'S FILTERING SYSTEM

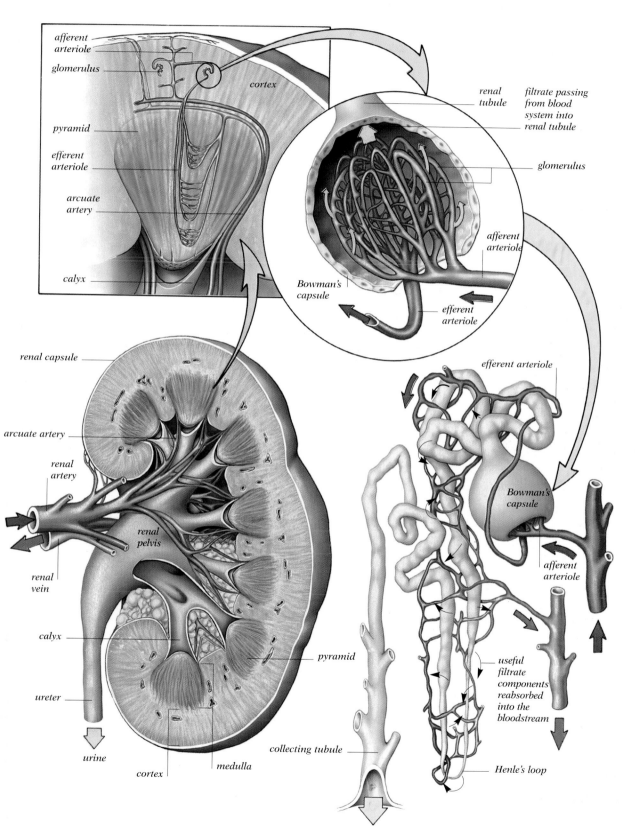

afferent arteriole

glomerulus

cortex

pyramid

efferent arteriole

arcuate artery

calyx

renal tubule

filtrate passing from blood system into renal tubule

glomerulus

afferent arteriole

Bowman's capsule

efferent arteriole

renal capsule

arcuate artery

renal artery

renal pelvis

renal vein

calyx

ureter

urine

cortex

medulla

pyramid

efferent arteriole

Bowman's capsule

afferent arteriole

useful filtrate components reabsorbed into the bloodstream

collecting tubule

Henle's loop

Questions and Answers

Are there any drugs that are likely to cause damage to the kidneys?

Yes. There is a dangerous form of kidney disease caused by certain drugs that results in damage to the duct system. The drug Phenacetin has been linked with this kind of kidney disease, but other anti-inflammatory drugs, like aspirin and ibuprofen, may also cause trouble if large amounts are taken regularly. Thus the people most at risk are those who take tablets of this type habitually, especially if they live in hot climates where the amount of the drug concentrated in the urine is likely to be high, making kidney problems more probable.

My father's doctor has told him he needs an IVP. What is this, and how is it carried out?

IVP, or intravenous pyelogram, is a test done to check the function of the kidneys and to identify any obstruction to urine flow (such as occurs with prostate enlargement). A dye that concentrates in the kidneys is injected into the bloodstream, and an X ray is taken soon after the injection. The kidneys, ureters, and bladder can be seen outlined by the dye. Men with an enlarged prostate tend to be unable to empty their bladder completely, so any residual dye left in the bladder will show on the X ray and confirm the diagnosis. This test is painless.

My uncle has kidney stones and his doctor has suggested that he be treated by lithotripsy. What is this?

Lithotripsy is a technique usually carried out in a hospital in which a machine called a lithotripter sends high-energy ultrasonic shock waves onto the stones to pulverize them. Once the stones have disintegrated, they will pass out through the urine. Before receiving lithotripsy your uncle will be given an analgesic to ease any discomfort caused by the procedure, and the affected area may feel bruised and tender for a few days afterward.

▲ *Various types of kidney stones, including the staghorn formation.*

diabetes mellitus), this hormone may be totally lacking. When this happens the patient cannot conserve water and loses large quantities in the urine, which have to be replaced by drinking.

Another hormone, aldosterone, secreted by the adrenal glands just above the kidneys, is responsible for exchanging sodium salt for potassium salt—thereby helping to control blood pressure and the balance of salt in the body. Parathormone, a hormone made by four small glands buried behind the thyroid gland, regulates the reabsorption of the essential mineral calcium, from which our bones and teeth are made.

Kidney disorders

Once a nephron is destroyed, it seldom regrows. From birth onward the body slowly loses nephrons, but it has so many that this loss seldom becomes a problem. In fact, about one-third of all nephrons do very little. These redundant nephrons are gradually brought into use as other nephrons are lost through wear and tear.

Doctors can get a good idea of the number of nephrons working by measuring the level at which certain waste substances are present in the blood and removed in the urine. One such waste product is called creatinine. Its level is kept very low in the blood but rises when the blood becomes concentrated, as when a person is thirsty. When doctors see a rise in the level of creatinine in the blood, it is a warning sign that the kidneys are failing. The rate at which creatinine can be removed from the blood is a better indication of renal (kidney) function. If a patient's kidneys fail to clear more than 0.34 fluid ounce (10 ml) of blood of creatinine each minute, the assistance of a dialysis machine, or artificial kidney, will be needed soon.

Inflammation of the glomeruli (glomerulonephritis) leads to sudden kidney failure. The early signs are the appearance of the protein albumin in the urine and sometimes particles of red blood cells. If a person's urine becomes pink or rust in color, he or she should see a doctor as soon as possible. In kidney failure excess water and salts cannot be excreted and so collect in parts of the body, either at the base of the spine in patients who are lying in bed, or around the ankles in patients who are up and about, causing swelling. In most cases acute kidney inflammation clears up completely and full recovery is normal. Rarely, however, damage to the glomeruli is progressive, and the patient gradually develops renal failure. This is called chronic glomerulonephritis. Some streptococci carry markers (antigens) very similar to kidney tissue. So a streptococcal infection may cause the immune system to produce antibodies that end up attacking the kidneys, causing glomerulonephritis. In the days before penicillin, streptococci caused epidemics of sore throats followed by acute glomerulonephritis, and a high proportion of the victims developed chronic renal failure. However, this type of disease is seldom seen today.

Serious injury of the tubules resulting in kidney failure can be caused by severe hypotension as well as certain drugs (aminoglycoside antibiotics, contrast dye used for studies). Inflammation of

glomerulonephritis and may cause the urine to look cloudy. There are a number of other causes of cloudy urine. The first amount urinated in the morning may be cloudy because it contains a deposit of undissolved salts that have collected overnight; however, a person who frequently has cloudy urine should see his or her doctor.

Kidney stones, called calculi, are composed of the salts of calcium and phosphorus. Major causes of kidney stones include infection and an excessive excretion of calcium into the urine. Stones associated with infection are more common in women and can reach very large sizes, so that they cannot pass down the ureter. They lie inside the kidney, and as they grow they take on the shape of the collecting duct system in which they lie. When an X ray is taken, they show up as a branched lump of chalk, looking similar to the antlers of a stag (see photograph, page 202). Smaller stones may be passed down the ureter, causing severe pain.

Hypertension and its effect on the kidneys

High blood pressure, left untreated, can damage the kidneys. The risk increases depending on how long the condition has been present. If the hypertension is severe and of long duration, it can damage the arteries in the kidneys, eventually leading to chronic kidney failure.

> See also: Adrenal glands; Hormones; Urinary tract

See your doctor if:
Your urine is discolored, particularly if it is red.
You have pain on urinating, lasting more than two days.
You urinate frequently for more than three or four days.
You have accompanying pain in the abdomen.
Your ankles are swollen.

the kidneys resulting from infection (pyelonephritis) usually responds well to treatment with antibiotics. Patients who have certain deformities of the kidney (and are thus at risk of repeated infections) can develop chronic pyelonephritis and, ultimately, kidney failure. Pus is present in the urine of patients with

▼ *A patient receives lithotripsy to treat kidney stones. An X-ray beam locates the stones, and then shock waves generated by the lithotripter machine are sent to pulverize the stones through a water- or gel-filled cushion placed under the patient's back.*

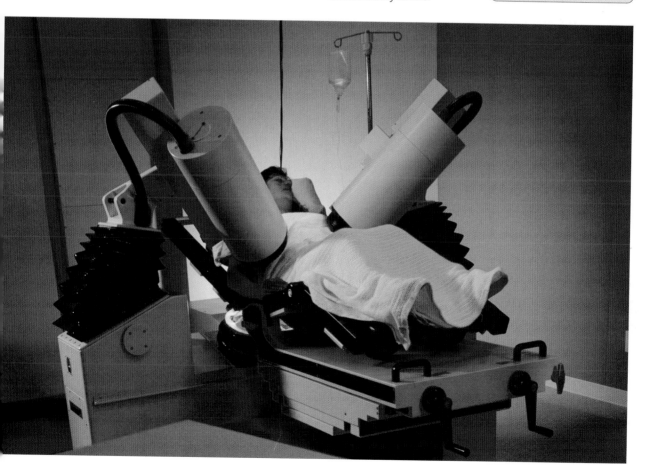

Knee

Why do I get a tender swelling in front of my knee after kneeling to wash the kitchen floor?

This is a form of bursitis, an inflammatory swelling of the bursas at the front of the knee. You have probably noticed that the inflammation dies down with rest but flares up again whenever you kneel for any length of time. When you do any work that involves kneeling, use rubber kneeling pads or towels to ease the pressure on your knees.

My son injured his knee while playing football. It swelled and became very stiff. Why was this?

The inflammatory response to injury involves a large outpouring of fluid into the tissues, resulting in a swollen joint and stiffness. A doctor may decide to drain some of the fluid to relieve the condition.

Sometimes my knee won't straighten. I'm told I have a "loose body" in the joint. What is this?

Any solid material in a joint can limit its range of movement. In the case of the knee, the loose body can be a bit of cartilage, or a chip of bone that may have broken free. Diagnosis is made by X-ray examination and arthroscopy. If necessary, loose bodies can usually be removed during arthroscopy.

When I had a checkup, the doctor tested my knee reflexes. How do they work?

The tendon over the kneecap has tension-sensing nerves in it. These run up to the spinal cord, then directly down to stimulate the muscles in the thigh. This feedback ensures that the nerves work properly. If the tension is increased by tapping the tendon below the kneecap, the muscles will contract, causing the lower leg to jerk.

Because the knees are so important for movement, any injury or disorder can render a person immobile. However, with modern surgical and medical treatment, permanent disability is rare.

The knee is the joint situated between the thighbone (the femur) and the shinbone (the tibia). It is designed to allow great flexibility of movement, yet still to be strong and stable enough to hold the body upright.

Structure and function

To allow easy movement, the knee joint is constructed something like a hinge. The end of the femur is smoothly rounded off and rests comfortably in the saucer-shaped top of the tibia. The surfaces of the bones are covered with a shiny, white, hard substance called cartilage.

To stabilize the joint further, yet still allow flexibility of movement, two leaves of cartilage lie in the joint space on either side of the knee. These are the bits of cartilage that sometimes get torn in sports injuries and may be removed in an operation. Without them the knee can still function, but wear and tear can increase so that a person may develop arthritis later on in life.

To lubricate the joint the surfaces are bathed in a slippery liquid called synovial fluid. This is made, and held, in the joint by a capsule that surrounds the whole knee area. There are also bags of fluid called bursas that lie in the joint and act as cushions against severe stresses.

Strength and stability are provided by fibrous bands called ligaments. Without hindering the hinge movement of the knee, these ligaments lie on both sides and in the middle of the joint and hold it firmly in place.

The movements of the knee joint are governed by muscles within the thigh. Those at the front pull the knee straight and those at the back hinge it backward. At the top these muscles are attached to the hip and the top of the thighbone (the femur). Farther down the leg they

STRUCTURE OF THE KNEE

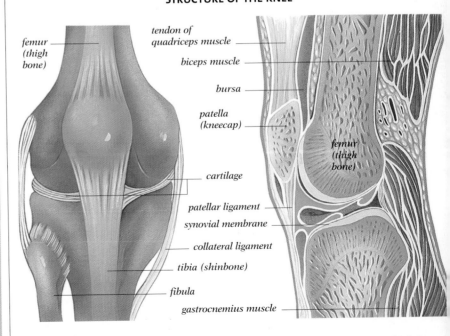

▲ *The front view (left) shows the bones of the knee joint. The cross section (right) shows the structure in detail, including the bursas.*

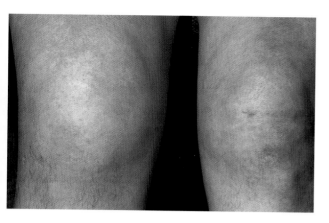

◄ Increased fluid in the knee, or knee effusion, often affects those who play contact sports, such as football players. There may be swelling of the joint and loss of bone tissue around the kneecap.

The knee joint is traversed by blood vessels that nourish the ligaments, cartilages, capsule, and bones. As with all important structures, nerves run to the knee so that the brain knows its exact position to coordinate balance and movement.

The knee is important principally for locomotion. At every step it bends to allow the leg to move forward without hitting the ground; otherwise, the leg would have to be swung outward by tilting the pelvis, as in the typical stiff-leg walk. To complete the step, the knee is straightened and the foot brought back to the ground by movement at the hip.

condense into fibrous tendons that cross over the knee and are then attached to the shinbone (the tibia).

To prevent the tendon at the front from rubbing the joint as it moves, a bone has been built into the tendon. This bone is called the kneecap, or patella, and lies within the tendon itself; that is, it is unattached to the rest of the knee. The kneecap runs up and down the bottom of the femur in a cartilage-lined groove and is lubricated by synovial fluid. There are also two additional bursas that act as the shock absorbers for the kneecap.

Treatment of injuries

Twists and sprains are common knee injuries caused by excessive forces stretching the knee joint ligaments. The knee swells and becomes stiff, and the injured ligament feels tender. On the whole, sprains get better by themselves with rest and gentle remobilization. An elasticized bandage, applied in a figure eight, will give good support to the knee initially, and a walking stick can help take the weight off the leg. In a severe ligament injury, the knee may need to be put in a plaster cast.

Either of the two leaves of cartilage in the joint space may get torn by a sudden twist of the knee; this tearing is a common sports injury. The knee locks, will not straighten, and also tends to give way.

▼ The knee can be prone to injury during sports, particularly those such as running, which put a lot of strain on the joint.

◄ *The fine motor skills in ballet depend on the smooth movement of all the joints, especially the knees. Injuries must receive prompt treatment, followed by physical therapy.*

Torn cartilage will often heal by itself, but if the symptoms persist or recur, the doctor will refer the patient to an orthopedic specialist. In the hospital the diagnosis is usually confirmed by arthroscopy. An arthroscope is a fine telescope that is fed through a small cut into the joint. The doctor can see the torn cartilage through the instrument, and he or she can perform minor repairs by remote control. For more serious injuries it may be necessary to remove the cartilage completely. In such cases the muscles will gradually strengthen to compensate for this loss.

Dislocation of the kneecap may result from a sideways knee injury. The kneecap is pulled out of its groove, and the patient is unable to straighten the knee. Treatment consists of pushing the patella back into place, followed by a period of rest. However, the dislocation may recur with only minor stresses on the joint, in which case surgery may be necessary.

Severe injuries can break the bones at the knee, with the result that the knee cannot bear weight and becomes very swollen and painful. A definitive diagnosis is made by X-ray examination. Surgical correction may be necessary, followed by a prolonged period in a cast.

Prevention of injuries

Knee injuries are very disabling, so it makes sense for people to take precautions to avoid them. Before athletes undertake any sporting activity, it is essential that they warm up by stretching the major muscle groups before starting each exercise program. Jogging and bicycling—short distances at first—are good for building up the muscles in the knee. Although jogging can put considerable stress on knee joints, the risk of injury can be minimized by a very gentle buildup to the exercise.

Other problems

Rather than suffering a sudden loss of function following injury, many people find their joints getting stiff and painful with progressing age. In general, stiffness is a sign that the muscles need to be stretched. Morning stiffness that improves as the day progresses is typical of rheumatoid arthritis. It may also be caused by a previous injury.

Anti-inflammatory drugs should relieve pain, but if the arthritis becomes crippling, a knee replacement operation may be necessary.

After persistent, unexpected activity the bursas may get inflamed, a condition called bursitis. The knee swells in the area of the affected bursas. This condition usually settles with rest, but it does tend to recur from time to time.

Injury or surgery will put the knee out of use for a time, with the risk of muscle wasting. As a result, physical therapy forms an important part of aftercare in such cases. It is this aftercare, together with modern advances in orthopedic surgery, treatment, and diagnostic instruments, that makes it unnecessary for anyone to be permanently disabled by a stiff knee.

See also: **Balance; Cartilage; Joints; Ligaments; Movement; Muscles; Tendons**

Larynx

Questions and Answers

My husband keeps coughing and clearing his throat. He says it is a smoker's cough, but he is also hoarse. What should he do?

It sounds as if your husband has chronic (long-term) laryngitis from smoking. He should cut down on smoking, and preferably quit altogether. Also, because he has had the condition for some time, it is essential that he see a doctor. It is possible he is developing cancer of the larynx; to postpone treatment is to risk the spread of a disease that is curable in its early stages.

I've just heard a person who had his larynx removed talking on the radio. How can he speak?

He is using esophageal speech. The esophagus is the tube that takes food down to the stomach. What this person does is swallow air, so that it passes down to the stomach; he then belches it up, controlling the amount released, and the sounds produced, by means of the esophagus. With practice people can become skillful at this type of speech, although they do tend to have a rather unusual voice.

Can my larynx get tired? My voice seems to crack after a lot of conversation.

The larynx does get tired, like any other organ, especially if it is overused by shouting, screaming, or singing for too long, or talking for hours, particularly in a dry or smoky atmosphere.

In fact, what you describe is a mild laryngitis or irritation of the vocal cords within the larynx. The cure is to stop talking completely for a few hours. You could also try inhaling water vapor, with a soothing preparation added. Boil water, then pour the hot water into a bowl. Cover your head with a towel and inhale the vapor.

The larynx contains the vocal cords, so any disorders of this organ will affect speech. However, most conditions of the larynx can be completely cured if they are diagnosed and treated early enough.

▲ *The larynx cannot be seen easily because of its position in the throat. Doctors use a special technique in order to view it.*

The larynx, or voice box, contains the vocal cords that vibrate to produce speech. Thus it is an extremely delicate instrument. Together with the epiglottis it also guards the entrance to the lungs. When a person eats or drinks, the epiglottis closes tightly over the opening to the larynx so that food or liquids slide over it down into the esophagus, which leads into the stomach. When a person breathes, the larynx is open.

Position and structure

The larynx is placed at about the center of the neck, at the top of the windpipe (trachea), out of sight around the corner of the back of the throat.

The larynx is essentially a specialized section of the windpipe with an external sheath of cartilage. Positioned over it is the epiglottis, the flap-valve that comes down from the back of the throat to cover the opening into the larynx, called the glottis. The action of the epiglottis is controlled automatically by the brain, but it sometimes fails, and then liquids or food particles go down the wrong way. Unless a lump of food is so large that it sticks in one of the passages below the larynx, it will be coughed up.

The voice

The vocal cords are mounted on specially shaped pieces of cartilage within the larynx. Air breathed out over them makes them vibrate, producing sound. The cartilages move in order to tighten or relax the cords, producing high- or low-pitched sound.

However, the kind of sound produced by the vocal cords is also strongly influenced by the nature and action of the tongue, lips, and jaw, because the sound waves resonate (bounce) around the adjacent passages.

Disorders of the larynx and their treatment

DISORDER AND CAUSE	SYMPTOMS	TREATMENT
Injury by burns from scalding liquids, inhalation of hot gases, or acute allergies causing swelling of the lining.	Sudden pain; difficulty in breathing; harsh, hoarse cough or croup.	Urgent medical attention. Tracheostomy (an opening made in the windpipe) may be needed to save life in severe cases.
Acute laryngitis (short-term hoarseness and soreness) caused by a bacterial or viral infection.	Sore larynx; difficulty in speaking or loss of voice.	Strict voice rest; water vapor inhalations; antibiotics may be needed if a bacterial infection is suspected.
Chronic laryngitis (any hoarseness that lasts more than 10 days), usually caused by too much smoking, by dust, or by overuse of the voice.	Hoarseness; constant "frog in the throat"; occasionally tenderness in the larynx or adjacent areas.	Specialist examination to exclude the possibility of cancer; strict voice rest; removal of the irritant.
Diphtheria (a severe, mainly childhood disease that has now been virtually eliminated in the West by mass vaccination).	Sore throat, fever, coughing; the telltale sign is a web over the larynx caused by white blood cells and other substances produced in response to inflammation.	If severe, a tracheostomy is required, and antibiotics as well as antitoxin are given. Immunization would have prevented the disease.
Laryngotracheobronchitis. Caused by a virus or bacteria, or occasionally by an allergy.	The larynx suddenly becomes swollen; severe cough or croup; difficulty in breathing.	Urgent hospital treatment; possibly antibiotics; steamy inhalations and moist atmosphere; tracheostomy in some cases.
Benign tumor (a lump or growth that will not spread). Such tumors are called polyps or nodes, and their cause is abuse of the voice: too much singing, shouting, or even loud talking, as, for example, in teaching.	Hoarseness that develops gradually.	Removal of polyp or node. This provides a complete cure, but speech training may be required to prevent recurrence.
Cancer of the larynx. The majority of cases are males aged between 50 and 69; a high proportion of sufferers are heavy smokers.	Hoarseness becoming progressively worse over several weeks. Difficulty in breathing or swallowing. Coughing blood and often a sticky mucus that necessitates constant clearing of the throat. Pain in the throat and ears and enlarged glands in the vicinity of the larynx occur when the condition is established.	Radiotherapy, or surgery if the case requires it. Recovery excellent; speech therapy, if the larynx is removed, will be needed to teach the patient to speak with the esophagus (the tube leading down to the stomach) and to breathe through a tracheostomy.

▲ *People who use their voices a great deal, such as singers or teachers, are likely to get laryngitis or, in some cases, to develop polyps or nodes (tissue lumps) on their vocal cords.*

Injury to the larynx

If the larynx is injured by a blow or a knife wound, for example, may become blocked. Similarly, scalding liquids or poisons can bur and irritate its lining, causing swelling and blockage. The difficulty breathing and the pain of such injuries usually severely disable th patient. In such cases it is vital to restore the airflow to the lung and this is usually done in a hospital. It may be possible for a doct to pass a tube down the patient's windpipe (trachea), creating a ne artificial passage; or it may be necessary for him or her to make hole, known as a tracheostomy, in the trachea below the larynx that air can pass directly to the patient's lungs. The actual procedu is known as a tracheotomy.

Acute laryngitis

Acute laryngitis means inflammation of the larynx caused by infection, such as a cold or influenza; by overuse of the voice, such from shouting or singing; or by irritation, often through smoking.

Either the sufferer has difficulty speaking and has a hoarse a throaty voice, or else his or her voice disappears completely. There pain in the larynx, and tenderness is often felt in the region.

Questions and Answers

Is it true that cancer of the larynx can now be completely cured?

Yes. If the cancer is detected early enough it can be cured with radiotherapy—a controlled dose of radioactive rays delivered to the area of the cancerous tumor. This procedure kills the malignant cells, and, provided the cancer has not spread too far, the treatment can effect a complete cure. In such cases there is no need to remove the larynx. However, if the condition is comparatively advanced, removal of the whole larynx may be necessary.

My neighbor's three-year-old contracted croup the other day and was rushed to the hospital, because it was feared he would be unable to breathe. Why was this?

Your neighbor's child was almost certainly rushed to the hospital because he was so young. In children under the age of five, croup can be dangerous because the larynx is so small. If it becomes inflamed, as happens in croup, the airway narrows, and there is a danger of suffocation. This is why emergency treatment is needed.

My husband has been diagnosed as suffering from laryngitis. However, his doctor has referred him to the hospital for a specialized examination of the larynx. Why is this necessary?

Your husband's doctor is acting properly. Laryngitis is not a serious condition and usually clears up of its own accord if the voice is rested. However, its initial symptom, hoarseness, is also a symptom of cancer of the larynx.

Although a serious disease, this cancer can be cured if diagnosed early. This involves examining the larynx closely. Since the organ is situated out of sight around the corner of the throat, a special technique and skill are required to view it. Sometimes, too, a person cannot tolerate anything near the back of the throat, in which case an anesthetic may be required if the larynx is to be seen at all.

POSITION AND STRUCTURE OF THE LARYNX

▶▼ *Front and side views of the larynx*

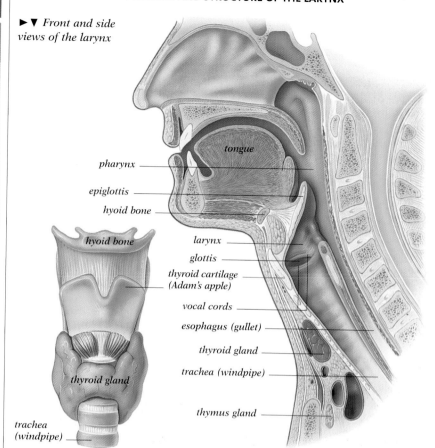

TRACHEOSTOMY

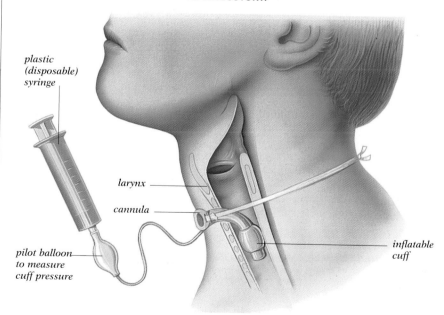

▲ *The tracheostomy is an emergency operation that restores breathing if the larynx becomes blocked or is injured in some way.*

Acute laryngitis is an irritating but not a dangerous condition, and it passes if the voice is given a complete rest; this involves not speaking for as long as it takes for the condition to correct itself—normally a day or two. Inhaling hot vaporous air with a soothing additive can help; antibiotics help only rarely.

Chronic laryngitis

Chronic laryngitis occurs over a long period of time. Its cause is irritation, mostly of the sort caused by violent shouting. Tobacco smoke, dust, and overuse of the voice are also common causes. However, continued hoarseness may be a symptom of cancer of the larynx; chronic laryngitis should always be reported to a doctor.

Cancer of the larynx

This type of cancer is 10 times more common in men than in women, and is most common in heavy smokers. It arises in the lining cells of the larynx, and its initial symptom is hoarseness lasting several weeks. A dry cough follows, and occasionally coughing up flecks of blood, followed some weeks later by swallowing and breathing difficulties.

Untreated, cancer of the larynx can end in death. If detected soon enough, it can be cured by radiotherapy. More severe cases require removal of the larynx. The patient breathes with the help of a permanent tracheostomy and learns to speak again using his or her esophagus. Alternatively, an electronic artificial larynx may be used.

Polyps and nodes

Polyps are tiny lumps attached to tissue by a stalk. If they develop on the vocal cords as a result of long-term abuse of the vocal cords, they cause hoarseness and a breathy quality of the voice. Vocal cord nodes (sometimes called singer's or teacher's nodes) are also lumps, and develop from long-term abuse. Both need to be removed surgically.

Laryngeal diphtheria

Diphtheria used to be greatly feared, but today it has been almost eradicated by vaccination. However, it can still occur, particularly among young children, and is extremely contagious. If contracted,

▲ *Mild laryngitis can be relieved by inhalation of water vapor from a soothing preparation in boiling water.*

diphtheria tends to spread via the throat; this, with the larynx, becomes sore and inflamed. In response to the inflammation, a sticky, grayish, false membrane or skin forms across the larynx and obstructs the airway.

Without an emergency tracheostomy and appropriate antibiotics and antitoxin given urgently, death can occur from suffocation.

Laryngotracheobronchitis

Laryngotracheobronchitis is the technical name for croup, the dry, distinctive-sounding cough caused when inflammation narrows the larynx. Such inflammation is frequently caused by bacteria, but it may also be caused by a virus. The larynx, trachea, and bronchi (the airways leading to the lungs) become inflamed and swollen—and a sticky, heavy mucus is formed—so that the air supply is in considerable danger of being cut off.

If the case is severe, a tracheostomy is necessary. Milder cases respond to oxygen, antibiotics, and ensuring that the atmosphere around the patient is warm and humid to soothe the inflamed areas.

Hoarseness

The nerve serving the vocal cords travels down into the chest before ascending to the larynx; therefore, injury or disease affecting the chest can paralyze the vocal cords or produce hoarseness. For this reason hoarseness is a symptom that may be difficult to interpret; a person should not be unduly alarmed if a doctor refers him or her to a hospital for simple tests.

The number of disorders involving the larynx is relatively high but they can all be cured if the main symptom, hoarseness, is recognized early enough. For this reason any hoarseness lasting more than 10 days should be reported to a doctor without delay.

See also: Vocal cords

Home treatment of laryngitis
If you are suffering from a short-term hoarseness, loss of voice, or pain felt deep in the throat:
Don't speak any more than is absolutely necessary. Rest your voice; carry a notepad and pencil with you, and write messages to people, rather than talk to them.
Avoid dusty or smoky atmospheres. If you smoke, quit, if possible permanently.
Try inhaling vapor from hot water in a jug or bowl (don't inhale from a boiling kettle, since this could burn). An additive, such as lemon, may make this more pleasant. A pharmacist will advise on other preparations available.
If you have been hoarse for more than 10 days, report this to your doctor without delay.

Ligaments

Questions and Answers

My wife keeps dislocating her shoulder while playing squash. Can ligaments be strengthened?

Ligaments are not elastic, so it is a question of their being shortened, not strengthened. Recurrent dislocation can also be due to muscular deficiencies or problems in the joint, so your wife should see her doctor for a checkup.

Stretched ligaments can shorten, but they need rest; playing squash may be aggravating the problem. If necessary, some ligaments can be shortened surgically.

Does exercise strengthen the ligaments?

No, but exercise does strengthen the muscles; this takes the strain off the ligaments, so it amounts to the same thing. Exercises that demand a full range of movement from the joints particularly benefit ligaments.

My son has terrible posture. Could his ligaments be loose?

Many ligaments become stretched as a result of long-term poor posture, but this is an effect, not a cause. Unless there is a structural abnormality, poor posture is usually due to poor habits. As people allow their postural muscles to weaken, it puts all the weight on the ligaments and gradually stretches them. Postural reeducation and exercise will help.

Is it true that movement can be brought to a paralyzed joint by transferring a ligament from another part of the body?

No. Ligaments are passive structures that can only prevent movement at the joints; they cannot initiate it. In paralysis the muscles do not work, but the cause lies in the nervous system, not in the muscles or ligaments.

Like car doors, which have checkstraps to prevent them from being opened too far and torn off their hinges, joints in the human body have ligaments to restrict movement and prevent dislocation.

The bones at a joint are moved by muscles. These are joined to the bones by tendons, which cannot stretch. Ligaments, which can stretch very slightly, join the two bones that form the joint and keep them in place by restricting the amount of movement they can make.

People are often unclear about the difference between ligaments and tendons. In a ligament, bundles of collagen form sheets of tough fibrous connective tissue. These sheets prevent dislocation by holding bones together at the joints to restrict any inappropriate movements. In tendons similar bundles are twisted together to form a cordlike structure of white inelastic collagenous tissue that runs between the ends of a muscle and the bones to which the muscle is attached. Tendons transfer the pull of a muscle to a bone. Many ligaments become very closely associated with tendons when tendons are joined to bones near joints; sometimes tendons reinforce or even replace the ligaments that form the capsule (covering) of a joint.

Other locations

Ligaments are also found in the abdomen, where they hold in place organs such as the liver and uterus. They allow the degree of movement necessary for adopting different postures and for the changes that accompany eating, digestion, and pregnancy. There are also ligaments made up of very fine fiber strands in the breasts; these provide support and prevent the breasts from sagging.

People are not usually aware of the existence of ligaments until they injure one. A strained or torn ligament can be just as painful as a broken bone.

Tensile forces

The pulling (tensile) forces applied to many ligaments are considerable, and their slight elasticity would not, in itself, provide sufficient protection against tearing. However, an additional protective mechanism comes into play.

ANATOMY OF LIGAMENTS

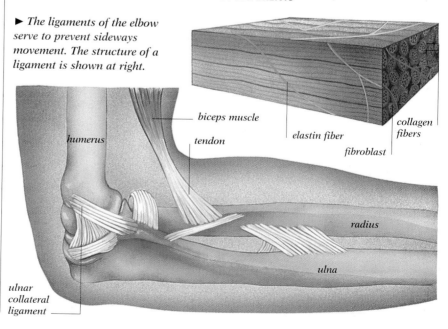

▶ *The ligaments of the elbow serve to prevent sideways movement. The structure of a ligament is shown at right.*

biceps muscle

humerus

tendon

elastin fiber

fibroblast

collagen fibers

radius

ulna

ulnar collateral ligament

When a ligament is suddenly and excessively stretched, nerve endings in it are stimulated strongly. As a result, nerve messages are sent to the appropriate muscles, which, on contraction, move the part (or even the whole body) in such a way as to relieve the pull on the ligament. Even this reflex mechanism is not always capable of preventing ligament damage. Some ligaments, such as those on the inner and outer side of the ankle that help to bind the foot onto the lower leg, are extremely susceptible to sudden stretching and tearing.

Structure

The body's bones and internal organs are kept in place by connective tissue. This consists of cells and protein fibers gelled together; without it the whole framework would sag and collapse. The connective tissue in ligaments is mainly made up of a tough white protein called collagen, with some yellowish and more elastic protein called elastin. In most ligaments this tissue is arranged in sheets of fibers.

These fibers run in definite directions, depending on the type of movement they resist. In ligaments that are arranged as a long cord, the fibers run longitudinally down the length of the cord and resist stretching along the length. Others, which help prevent joints from moving sideways, are arranged as a flat band of crisscross fibers that prevents movement through the band.

Between the fibers there are specialized cells, called fibroblasts, that are responsible for the creation of new collagen fibers and the repair of damaged ones. Between the fiber bundles there is a spongy tissue that carries blood and lymph vessels and provides space for the nerves to pass through.

Ligaments are attached to the bones they unite by fibers that penetrate the outer covering of the bone (the periosteum). The periosteum is supplied with nerves and blood vessels so that it can nourish the bone as well as provide an attachment for the ligaments and muscles. The ligament and periosteum grow together so perfectly that the periosteum is often affected if a ligament is injured.

Specialized ligaments

There are various types of joints in the body, and specialized ligaments exist for each. Major joints such as the knees, hips, elbows, and spinal joints are completely enclosed by a fibrous joint capsule containing synovial fluid for lubrication. Parts of this capsule are thickened for strength and are known as intrinsic (capsular) ligaments. There are also ligaments inside and outside the joint capsule that play individual roles in restricting movement. These are known as extrinsic (accessory) ligaments.

B

humerus

annular ligament

ulnar collateral ligament

coracoacromial ligament

A

muscle

clavicle

humerus

coracohumeral ligament

coracoclavicular ligament

scapula

Purpose

The variety of movements that the body can make depends on two factors: the shape and design of the bone surfaces at the joint (the articulating bone surfaces), and the ligaments.

In some joints the bones are the most important factor. In the elbow joint the ulna (in the forearm) forms the lower half of the joint and is a hooklike shape that allows only backward and forward movement, similar to a hinge. Here the ligaments serve to prevent side-to-side rocking, and a specialized ligament (the annular ligament) fits like a collar around the head of the radius (the outer bone in the forearm) to attach it to the ulna; it still allows for rotation.

In the knee joint the shape of the bones offers no resistance to movement. So, although the knee is also a hinge joint, it is controlled by specialized (cruciate) ligaments that prevent it from bending backward and help lock the joint while a person is standing still.

The muscles at the joints work in groups, some contracting while others relax, to enable the bones to move. The ligaments work in concert with the muscles, preventing them from making excessive movement. Ligaments have no ability to contract. However, they can be stretched slightly by movement in the joint; as this happens they gradually tighten until no more movement is possible.

There are also ligaments that pass between two points on a bone and remain unaffected by any movement. Their function is to protect or hold in position structures such as blood vessels and the nerves.

Movement

Every joint relies on some form of control by ligaments for its stability. This is especially important where the bones themselves cannot restrict movement. The mobility of joints such as shoulders, hips, and knees relies on the fact that ligaments are progressively tightened at the extremes of any movement.

▼ *Each movement that a trained athlete makes depends on the interplay of joints, muscles, tendons, and ligaments. Shown below are the most important ligaments needed for actions involving (A) the shoulder, (B) the elbow, (C) the hip, and (D) the knee when Daley Thompson, an Olympic medalist, goes into action.*

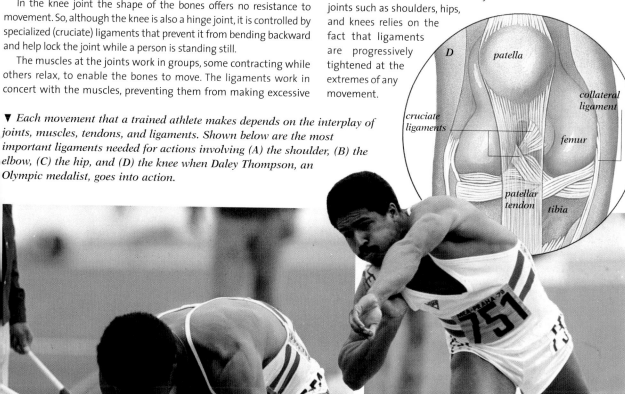

◄ Habitual slumping in a chair when watching television can eventually cause the back muscles to become flabby and prone to ligament strain.

Strains

The most common cause of pain from ligaments is either a strain or a sprain. A strain is caused by prolonged or violent stretching of the ligament. Many cases of pain in the lower back and hips are due to ligament strain that arises if the muscles are allowed to lose their tone.

If faulty posture develops through habitual slumping in chairs or lack of exercise, the normal curves of the spine are increased, because the muscles are not taking the load sufficiently. Thus the ligaments are having to continually restrict excess movement; eventually they stretch, leaving the joints more prone to injury. Pain may start in the ligaments, the surrounding structures, or both.

A strain may be brought on by unaccustomed vigorous activity, such as lifting heavy boxes or skiing. The demands of the particular form of exercise often exceed the fitness of the muscles and ligaments; they protest with pain when they are stretched.

Sprains

Sprains often occur in joints such as the ankle, shoulder, knee, and wrist following a fall or sudden blow. The joint is twisted or forced so quickly that the muscles do not have time to react and guard the joint against impact.

If it is a severe sprain, some of the ligament's fibers may be torn and the periosteum may be lifted from the bone underneath. This type of injury is extremely painful and is accompanied by inflammation and swelling. The most serious form of injury occurs if the ligament is completely separated from the bone or if it takes part of the bone with it. This is called an avulsion fracture.

Ruptures and tears

A ruptured ligament is one that has broken completely; a torn ligament results when some of the fibers twist and tear. Complete rupture of a ligament will produce an unstable joint. An avulsion fracture, where the ligament remains intact but the whole of its attachment to the bone has been pulled off, is easier to deal with than a complete rupture of the ligament. Such injuries commonly occur at the ankle, knee, and finger joints. There is severe pain and often bleeding under the skin; this can cause swelling in the joint.

Treatment

In an avulsion fracture an X ray will show a detached flake of bone near the point of insertion of the ligament. This bony flake makes the reattachment of the ligament to the bone easier and makes for better healing than in the case of a torn ligament. Flakes can be put into place using stainless-steel wire or some other strong suture.

Torn ligaments heal well in four to six weeks if they are held without tension. In young people with severe joint instability, such injuries are best treated by surgery. This must not be delayed, owing to the tendency for torn ligaments to shorten fairly quickly. Once this has happened it may be impossible to bring the edges together.

After an operation the joint is immobilized, often in a cast, to ensure that there is no pull on the ligament. The cast will usually be removed after three or four weeks to allow exercising of the muscles, but wrapping the joint in an Ace bandage (a bandage of woven elastic material) will be continued for several more weeks.

Most strains and sprains heal by themselves, although an injured ligament needs to be protected from further strain by rest in bed or a supporting bandage. Ace bandages are preferable because they give support and warmth without restricting circulation. If the sprain involves a limb, further support from a sling or crutch may be needed.

After the injury some of the pain is caused by the pressure of the swelling; cold compresses and massage will help to relieve this.

Ligaments have a poor blood supply, are slow to heal, and can take several months to mend completely. Although gentle use promotes healing, it is important not to use the joint too vigorously too soon.

If a ligament is badly torn or ruptured from the bone, surgery will be necessary to reconnect it before it can heal and its function can be restored. A new technique involves replacing a length of tendon with a specially-designed and extremely strong carbon-fiber substitute.

Dangers and outlook

Permanent instability at a joint may result if a severe ligament injury is not treated properly. When a ligament is torn and the free edges are brought firmly together by stitching, the scar tissue that forms between the edges is minimal; the tensile strength of the ligament will be slightly, if at all, impaired. If, however, the ligament is allowed to heal slowly with a gap between the torn ends, this gap will fill with fibrous tissue that will stretch under tension. The result will be an incompetent ligament and a joint that is more prone to injury.

Another important point is that since ligaments require the cooperation of muscles to work correctly, the treatment of a severe ligamentous injury must not be allowed to interfere with the health of the relevant muscles. Therefore, treatment must ensure that the torn ends of the ligament are brought together securely so that exercises can be undertaken to strengthen the muscles. However, such exercising must not be started too soon and should be deferred until all acute pain has subsided.

See also: **Bones; Elbow; Joints; Knee; Movement; Muscles; Reflexes; Tendons**

Liver

Questions and Answers

Is it only heavy consumption of alcohol that damages the liver, or can any amount cause damage?

A regular intake of 1½ fl. oz. (40 ml) of pure alcohol per day over the course of a decade is a serious health hazard. This is roughly the equivalent of five 16-fl.-oz. (2.36-l) cans of beer or 10 single shots of liquor per day. However, women's livers contain about half the amount of the enzyme ethanol dehydrogenase that breaks down alcohol than men's livers, so women suffer liver damage at only half the level of consumption.

For years I avoided aspirin for fear that it would damage my stomach. I now hear that acetaminophen, which I have been taking instead, can harm the liver. Is this true?

Yes. Acetaminophen taken in large quantities can cause severe and sometimes irreversible damage to the liver, and even death. A safe dose is considered to be less than two extra-strength tablets four times per day, or 4 g. For alcoholics, even smaller doses can be toxic. It is very dangerous to take an overdose of acetaminophen; when this occurs, the patient must have emergent evaluation and treatment with the antidote for this poisoning. Tragically, by the time the overdose damage is evident, it is usually too late to help the patient.

Are liver transplant operations performed often?

Yes. They were first performed at least 20 years ago and are now common. Yet while 90 percent of kidney transplant patients survive their first year, only 80 percent of liver transplant patients, who are usually much sicker, do. However, improvements in the drugs used to control the body's tendency to reject the new organ may increase the operation's success in future.

The liver has such amazing powers of regeneration that only the most persistent abuse with alcohol or drugs can really damage it. Otherwise, there is very little that can go wrong with this vital organ of the human body.

The liver has two vital roles: making (or processing) new chemicals, such as proteins, enzymes, and hormones, and neutralizing poisons (toxins) and wastes. The liver filters blood flowing away from the intestines; this blood carries all the nutrients absorbed from food. Blood can return to the heart and lungs from the stomach only by first passing through a system of veins into the liver, called the portal system.

Location and appearance

The liver is the largest organ in the body, weighing between 3 and 4 pounds (1.36–1.80 kg). Tucked underneath the diaphragm, it is protected from damage by the lower ribs. There are two projecting parts, or lobes, called the left and right lobes. The right is the larger and occupies the whole of the abdomen's upper right side. The left is smaller and reaches the midpoint of the left side. It is not usually possible to feel the liver, but when it enlarges, as a result of disease, it protrudes from behind the rib cage and can then be felt if the abdomen is pushed in.

Functions

As in any other part of the body, the liver cells do the real work of maintaining life's processes at a microscopic level. Hepatocytes are the creative cells of the liver. They are specialized to handle the basic substances the body runs on—proteins, carbohydrates, and fats.

Protein processing: Proteins are essential for the renewal and creation of cells all over the body; for making enzymes (substances that are secreted by cells to bring about chemical changes); and for the formation of hormones (the body's chemical messengers). Protein is consumed by the body in various forms, both vegetable and animal in origin. The liver breaks down and rebuilds raw proteins into a form the body can use.

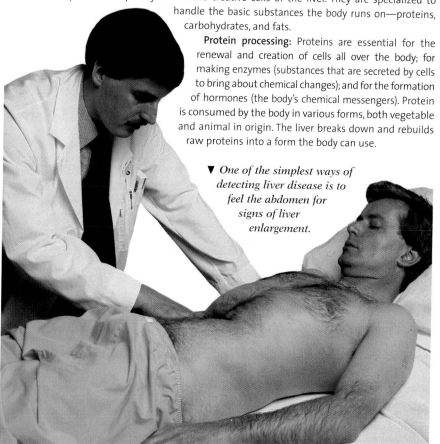

▼ *One of the simplest ways of detecting liver disease is to feel the abdomen for signs of liver enlargement.*

HOW THE LIVER WORKS

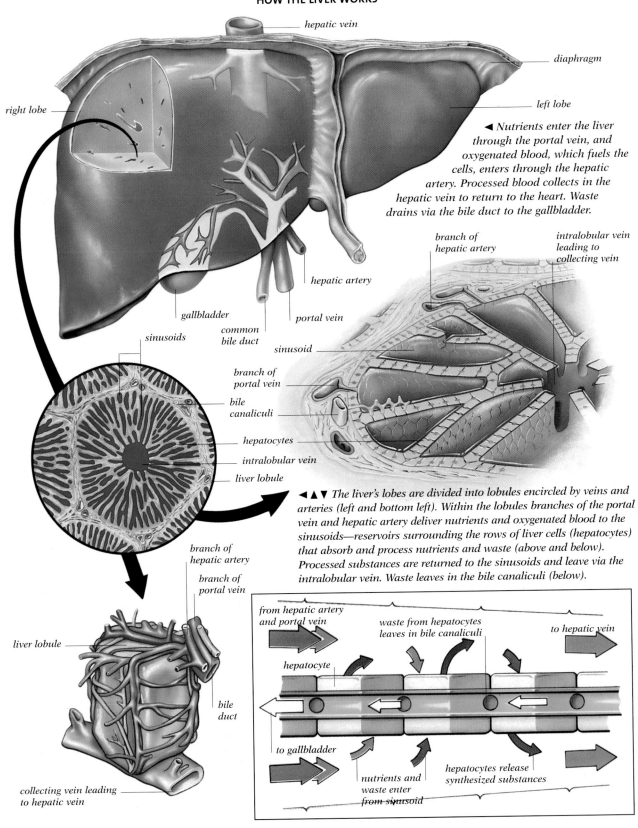

hepatic vein

diaphragm

right lobe

left lobe

◄ *Nutrients enter the liver through the portal vein, and oxygenated blood, which fuels the cells, enters through the hepatic artery. Processed blood collects in the hepatic vein to return to the heart. Waste drains via the bile duct to the gallbladder.*

branch of hepatic artery

intralobular vein leading to collecting vein

gallbladder

common bile duct

hepatic artery

portal vein

sinusoids

sinusoid

branch of portal vein

bile canaliculi

hepatocytes

intralobular vein

liver lobule

◄▲▼ *The liver's lobes are divided into lobules encircled by veins and arteries (left and bottom left). Within the lobules branches of the portal vein and hepatic artery deliver nutrients and oxygenated blood to the sinusoids—reservoirs surrounding the rows of liver cells (hepatocytes) that absorb and process nutrients and waste (above and below). Processed substances are returned to the sinusoids and leave via the intralobular vein. Waste leaves in the bile canaliculi (below).*

branch of hepatic artery

branch of portal vein

liver lobule

bile duct

collecting vein leading to hepatic vein

from hepatic artery and portal vein

waste from hepatocytes leaves in bile canaliculi

to hepatic vein

hepatocyte

to gallbladder

nutrients and waste enter from sinusoid

hepatocytes release synthesized substances

Questions and Answers

I was told that the liver has great powers of self-healing. Is this true?

Yes. Liver cells are among the most rapidly dividing in the body, and it is well known that if a portion of the liver is removed during an operation, that piece is completely regenerated within a few weeks. It is not known why certain cells in the body are capable of this rapid replacement while others, such as those that make up nerve and muscle tissue, are not.

I have heard that eating shellfish may cause hepatitis. Why is this?

Shellfish, such as oysters and mussels, commonly grow near sewage outlets because of the rich supply of food they provide. Sewage usually contains a virus that causes hepatitis; if this is absorbed by the shellfish, it may infect humans, especially if the shellfish are eaten raw. If you go to collect mussels, check whether it is the right season. Leave them in fresh water for 12 hours before cooking. Mussels constantly pass fluid through their bodies, and the fresh water will help clean them.

I have a busy social life that involves a great deal of entertaining. A few weeks ago I had hepatitis with slight jaundice. How long must I refrain from drinking alcohol?

To be absolutely safe it is wise to avoid any kind of alcohol for six months after the jaundice disappears. Some doctors recommend abstaining for six months after the liver tests have returned to normal. It is important to do this because acute liver failure may occur if alcohol is taken when the liver is inflamed.

Can a person's body survive without a liver?

No. If you had no liver, your body would have no usable fuel and would be unable to break down poisons and waste. Diseases such as viral hepatitis can lead to liver failure, which is fatal.

▲ *The greater the alcoholic intake, the greater the risk of liver damage. Experts agree that a daily intake of 5 pt. (2.4 l) of beer—whether it be of ale (left) or of lager (right)— or the alcoholic equivalent, is highly dangerous.*

In this process, which is sometimes called synthesis, the raw proteins are taken or absorbed from the blood flowing through the portal veins into the surrounding hepatocytes. In the hepatocytes the proteins are rebuilt by the liver's enzymes and then finally handed back to the blood in their new form. Waste material, however, does not return to the bloodstream.

Carbohydrate processing: Carbohydrates are a large class of chemical substances in the body made up of three atoms (basic building blocks of all physical matter)—carbon, hydrogen, and oxygen. They occur typically in sugary or starchy foods, and the body needs carbohydrates for energy. The muscles "burn" sugar or sugarlike substances whenever they work, a process for which oxygen is essential. The liver plays a vital role in organizing this fuel into usable forms. It does this by turning carbohydrates into two forms closely akin to pure sugar. One form produces instant energy: glucose. The other is storable energy, a substance similar to glucose called glycogen. A shortage of sugar that is severe and sustained rapidly causes brain damage, and so the level of sugar in the blood must be precisely maintained.

It is therefore necessary to store sugar for times of need, such as sudden exertion or starvation. Equally, if too much sugar is present in the blood, a hormone made by the liver can store the excess as glycogen.

Conversion of fats: Fats are essential to the body, too. The liver turns fats into new forms that can be built into or renew existing fatty tissue, typically the subcutaneous layer beneath the skin. Fat insulates the body, absorbs shocks, and stores energy.

Waste disposal: Lining the veins of the liver are highly specialized cells—called Kupffer's cells for the man who identified them—that clean the blood of impurities such as bacteria. These cells also deal with the debris from red blood cells that have reached the end of their 120-day life and have broken up.

By-products are produced from all the substances mentioned—blood itself, proteins, fats, and to a much lesser extent carbohydrates—during the rebuilding that goes on in the hepatocytes. Some chemicals, such as ammonia (produced during the breakdown of protein), are toxic.

Liver diseases and their treatment

DISEASE AND CAUSE	SIGNS AND SYMPTOMS	TREATMENT
Acute liver failure, brought on by hepatitis virus; alcohol; acetaminophen poisoning	Jaundice and coma	Low-protein diet; vitamin K; antibiotics; blood transfusion; intravenous glucose
Hepatitis A, B, or C; infectious mononucleosis; other viruses	Abdominal pain; nausea; distaste for cigarettes; loss of appetite; jaundice	Rest in bed and avoid alcohol for at least six months after the jaundice has gone
Cirrhosis—alcohol and hepatitis B and C are the best-known culprits, but there are some unknown causes	May be none until quite late in the disease when signs associated with hepatitis and finally liver failure develop	Stop drinking alcoholic beverages and avoid acetaminophen
Kernicterus, severe brain damage caused by jaundice in newborn babies	Severe neurological symptoms, lethargy, and sometimes death	Jaundice is common in babies, but if it is exceptionally severe it is treated by complete replacement of the blood, called exchange transfusion

The liver cells neutralize toxins, sending the harmless waste product, urea, back into the main circulation. Fat and blood waste products pass out as bile.

The same process applies to actual poisons that are consumed, like alcohol, and also to medicines. If a drug is to be long-lasting in its effects, it needs either to be resistant to the liver's enzymes or to bypass the liver completely.

What can go wrong

The liver has an extraordinary capacity to renew itself; a whole lobe cut away in an operation can be replaced in a few weeks. However, on rare occasions, the destruction of liver cells outstrips the rate of replacement, and this leads to acute (immediate) liver failure.

The most common cause of this situation is viral hepatitis. Acetaminophen poisoning, often caused by a deliberate overdose, is also a common cause.

The results of liver failure are easy to imagine by considering the various tasks that the liver performs. Blood sugar falls, and without

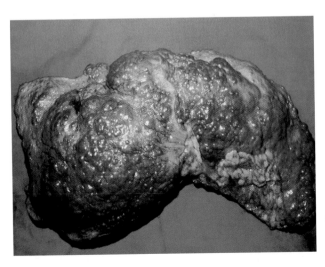

▲ The deadly effect of cirrhosis of the liver. The main cause of cirrhosis is a high level of frequent alcohol consumption.

a proper level, brain damage can result. The failure of protein production, including the manufacture of those proteins that cause clotting in the blood, can make the patient bleed easily; it can also lead, for various technical reasons, to complications such as the accumulation of fluid in the abdomen, called ascites.

The failure to eliminate wastes causes jaundice—the yellowish tinge of the skin, mucous membranes, and eyes that is caused by too much bile pigment in the blood—and also coma.

Other conditions

Certain inborn defects can cause long-term liver problems, most notably the failure of the liver enzymes to remove excess bile. This condition is called kernicterus and occurs almost exclusively in premature babies.

The enzymes that make and store glucose may also be deficient, causing a rapid fall in the blood sugar level.

The liver is a common site for cancer to develop after spreading from the primary site in the body through the bloodstream. However, primary cancerous liver tumors are relatively rare in developed countries.

Poisons such as acetaminophen and alcohol, or infections such as hepatitis, can cause such rapid destruction of liver cells that repair is not possible. This then causes scarring, and if severe, the hardening and shrinking of the liver and a failure to properly regenerate. Eventually this condition, known as cirrhosis of the liver, becomes irreversible; this is the most common cause of liver failure and coma in the West.

Treatment

Liver tumors can be treated by either surgery or radiotherapy. Hepatitis usually clears up after a few weeks. However, if a liver disease has progressed to the stage where coma is likely, emergency action is required. Protein intake must be restricted and essential substances should be given intravenously. In severe cases, a liver transplant may be the patient's only hope of recovery.

See also: Abdomen; Bile; Digestive system; Enzymes; Excretory systems; Hormones; Transplants

Lungs

Questions and Answers

I get short of breath when I run. Is there something wrong with my lungs?

Perhaps there is something wrong with your lungs, but maybe you are just not particularly fit. If the problem continues once your endurance has improved, see a doctor. Some forms of asthma can act like this (exercise-induced asthma).

Can nerves cause asthma attacks?

Anxiety can trigger an asthma attack in some people, but asthma is a medical condition, not proof of someone's mental state.

My friend told me that measles can also cause serious lung disease. Is she right?

Yes, but it is extremely rare. The measles virus itself may infect the lungs, but it may also reduce the patient's resistance to other infections, and it is usually these that cause the trouble. If a child develops a persistent cough after a measles infection, take him or her to a doctor for a checkup.

I've heard that just one cigarette can damage the delicate tissues of the lungs. Is this true?

It is doubtful if one cigarette would cause any damage, but smoking is now recognized to be the single most dangerous avoidable health hazard. It is especially likely to damage the lungs.

Does it hurt to have the inside of your lungs examined?

No. The instrument used is a bronchoscope: a flexible tube containing a bundle of glass fibers that can be passed down the windpipe, enabling the doctor to see the deeper parts of the lung.

The lungs are essential to life. They allow life-maintaining oxygen from the air that people breathe to be absorbed into the bloodstream, and waste carbon dioxide to be removed from the body.

Structurally, the two lungs are little more than a dense latticework of capillaries and tiny tubes containing air. The whole structure is suspended on a framework of elastic strands and fibers. The right lung is slightly larger than the left one because the heart takes up more room in the left side of the chest cavity.

Each lung is divided into lobes, which are supplied with air by divisions of the bronchus that lead down from the trachea (windpipe). The right lung has three lobes: upper, middle, and lower. The left has only two: upper and lower. These lobes are separate from one another and are marked by grooves on the lung surface called fissures. The fissures can give important information to doctors, since they can be seen on a chest X ray. By looking carefully at their position and observing, for instance, whether they have moved up or down, doctors can tell whether someone has suffered a collapse of part of the lung.

The entrance to the trachea is guarded by a flap valve called the epiglottis. When people swallow, this shuts, preventing food from entering the lungs. Should this mechanism fail, allowing food to get into the trachea, violent coughing results.

How the lungs work

If the lungs were removed from the chest they would shrink like deflated balloons. They are kept inflated by negative pressure in the chest between two membranes called the pleura; the

▼ *It is possible to screen for lung cancer by taking an X ray of the lungs.*

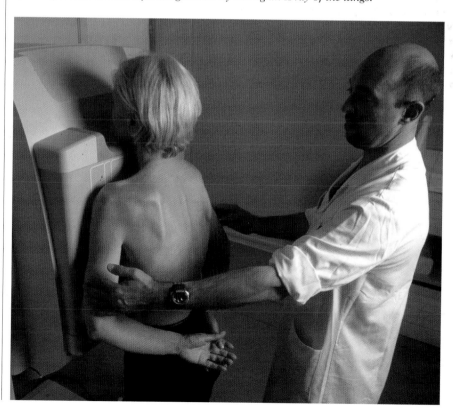

219

LUNG—STRUCTURE AND DISORDERS

▼ ▶ *In a normal lung, oxygen from the air is transferred to the capillaries that surround each alveolus (main picture). Lung disorders (insets) include bronchitis, in which the bronchus fills with mucus; a cancerous tumor of a bronchiole; asthma, in which the muscular walls of the bronchioles are narrowed; emphysema, in which the walls of the air sacs break down; and pneumonia, in which the air sacs fill with fluid.*

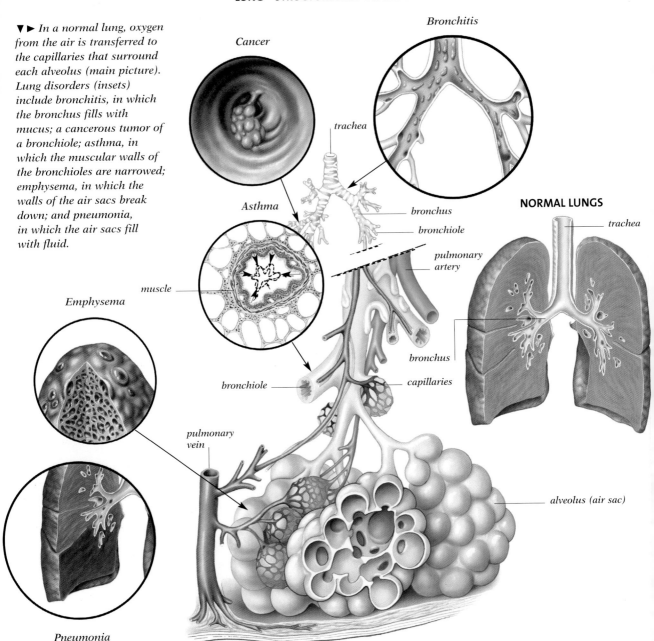

Cancer

Bronchitis

Asthma

muscle

Emphysema

bronchiole

pulmonary vein

Pneumonia

trachea

bronchus

bronchiole

pulmonary artery

bronchus

capillaries

alveolus (air sac)

NORMAL LUNGS

trachea

---

visceral pleura coats the outside of the lungs, and the parietal pleura lines the inside of the chest. Surface tension usually keeps the two membranes clinging together, and as long as they remain like this, the lungs stay inflated.

If air or liquid gets into the space between the two membranes (the pleural cavity), it breaks the surface tension between them and causes the lungs to collapse. Such a collapse may occur if the membranes become inflamed or irritated, producing an excess of fluid that accumulates in the pleural cavity. This is commonly called fluid on the lungs, although the medical term for the condition is a "pleural effusion."

Breathing and gas exchange

When the chest expands, the lungs are pulled out and air is taken into the alveoli—tiny air sacs in the lungs, each surrounded by fine capillary blood vessels where the exchange of oxygen and carbon dioxide takes place.

Chest expansion occurs by virtue of the shape of the ribs and the way they are attached and move. When the muscles between the ribs contract, the whole rib cage is pulled upward, but since the ribs are secured so that their front and rear ends cannot rise, this movement causes the curved ribs to swing outward, thereby increasing the volume of the chest. At the same time, the

How the lungs start working

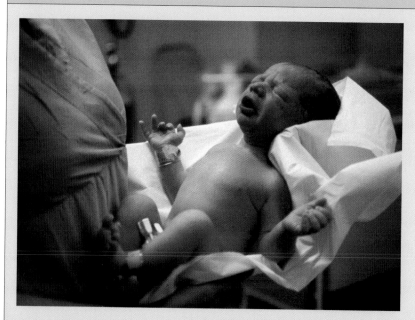

Breathing begins after a baby is born. While the fetus is still in the uterus, it derives oxygen from the mother's blood via the placenta (below left). Although blood passing to the liver or the inferior vena cava via the ductus venosus is fully oxygenated, the blood in the rest of the body is neither oxygenated nor deoxygenated. Much of the blood reaching the right atrium of the heart is shunted to the left atrium through an opening that closes after birth; most of the blood reaching the right ventricle leaves through the ductus arteriosus. Very little blood passes through the lungs, which contain no air and have yet to function.

After the baby is born, the oxygen supply from the placenta is cut off, but the baby's first cry expands the lungs so that they start to work (below right). The ductus arteriosus becomes redundant and closes naturally after a few days.

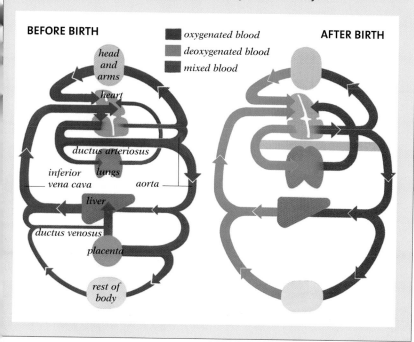

diaphragm—the upwardly domed sheet of muscle that forms the floor of the chest cavity—flattens, increasing the volume of the chest even further.

In addition to the surface tension effect, the expansion of the lungs is brought about in another manner. As the total volume of the chest increases, the pressure of the air within the lungs is reduced below atmospheric pressure. Since the interior of the lungs is directly connected with the outside atmosphere via the bronchial tubes, trachea, and nose or mouth, the greater pressure outside causes air to be forced into the lungs.

The exchange of oxygen and carbon dioxide takes place in the alveoli in less than one-tenth of a second. Hemoglobin in the red blood cells combines with fresh oxygen and discharges its load of carbon dioxide for the lungs to exhale.

Carbon dioxide is much more soluble in water than is oxygen, and most of it is dissolved in the plasma, where it is converted into bicarbonate. Some of the carbon dioxide is transported in the form of compounds with protein constituents.

When a person exhales, the rib muscles relax gradually. Because the lung walls are extremely elastic, like a rubber band, this relaxation allows the lungs to decrease in volume and return to a resting state.

How babies begin breathing

A fetus spends the whole of its existence before birth completely immersed in amniotic fluid; oxygen is supplied by the mother through the placenta. Because the fetus is not breathing air, the alveoli never fully collapse but contain small amounts of fluid, which are absorbed into the rich capillary network with the first breath after birth. Very little blood is pumped around the lungs, and this puts the main pump to the lungs—the right ventricle of the heart—under a great deal of pressure. To relieve this pressure, a large artery called the ductus arteriosus connects the pulmonary artery directly to the aorta, bypassing the lungs.

At birth a startling transformation occurs as the placenta separates from the uterus and cuts off the previous oxygen supply. This prompts the baby to take his or her first breath—first with little gasps, followed by a prolonged cry. Both the rapid breathing and the long cry expand the alveoli and the lungs. At first only a few alveoli expand, then more, until after a few hours both lungs are

Lung disorders

CONDITION	SIGNS AND SYMPTOMS	CAUSES	TREATMENT
Lung cancer	Weight loss, coughing up blood, persistent cough, hoarseness of the voice.	Smoking, exposure to asbestos dust, exposure to silica dusts (as in mining), industrial pollution.	Surgical removal or radiotherapy and anticancer drugs. Painkillers. Oxygen.
Chronic bronchitis	Persistent cough with phlegm for more than three months of the year for two years in succession. Phlegm changes from white to yellow or green when infected.	Smoking, industrial pollution, inheritance (the English have 20 times more chronic bronchitis than Scandinavians).	Stop smoking. Wear an approved face mask for dust protection or protection from chemicals, whichever the case may be.
Emphysema	Breathlessness. Enlargement of the chest.	Smoking. Rarely, long-standing chronic bronchitis may be inherited.	Stop smoking, breathing exercises, use bronchodilators.
Pneumonia	Cough with green or yellow phlegm, occasionally streaked with blood. Fever, sweating, loss of percussive resonance (sounds solid when tapped).	Bacteria (pneumococcus or hemophilus influenza), viruses (whooping cough or measles). Smoking makes adults more susceptible.	Antibiotics (ampicillin or tetracycline). Stop smoking.
Asthma	Breathlessness, cough, and wheezing.	Both genetics and environment. Allergies, often to house mite or pollens, household pets, irritants.	Avoidance of environmental triggers. Prescription medications.

fully expanded. As the lungs expand, the pressure drops, the resistance to blood flow in the pulmonary artery falls sharply, and instead of flowing through the ductus arteriosus into the aorta, the circulation changes so that blood flows through the baby's lungs to be oxygenated.

For the first few days the ductus arteriosus remains open, and blood from the aorta, where the pressure is high, comes back the other way into the pulmonary artery, where the pressure is low. Usually, the vessel closes off naturally in the first week or so of the baby's life, but sometimes the ductus remains open. When this happens, an abnormal sound called a murmur is produced, which can be heard with a stethoscope when the child is examined by a doctor. This is one of several congenital heart conditions and is comparatively easy to correct—more recently using laparoscopic (keyhole) surgery.

Bronchitis and emphysema

In Western countries, the most common chest conditions affecting adults are chronic bronchitis and emphysema. People with chronic obstructive pulmonary disease usually have both.

In bronchitis, the bronchial tubes become chronically inflamed and produce an excess of mucus, resulting in a cough that produces phlegm. Smoking is the most important causal factor, especially in the persistent form known as chronic bronchitis. Acute bronchitis is usually caused by a viral infection, but chronic bronchitis is not primarily an infective condition, and between 80 and 90 percent of cases are caused by smoking, usually of at least 10 years' duration. Environmental pollution is another contributory factor, and the condition is invariably worse in winter.

People with emphysema have lost the elasticity of their lung tissues so that the alveoli are permanently distended; this distension causes breathlessness. Coughing is less common and rarely brings up phlegm. As with bronchitis, smoking is the major causative factor. Chronic bronchitis and emphysema are among the leading causes of disability and death in the United States; they are the fifth largest cause of death and are second only to coronary artery disease as a cause of disability. In 2009, the National Center for Health Statistics estimated that 9.9 million Americans suffered from chronic bronchitis and 4.9 million suffered from emphysema. In 2007, it was reported that 667 died from bronchitis (chronic and unspecified) and 12,790 died of emphysema.

Asthma

Impairment to breathing caused by obstruction in the bronchial tubes may be persistent and difficult to treat, but when the obstruction is easily reversed by medication or rapidly gets better by

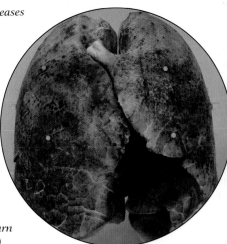

► *Smoking increases the chance of developing disabling and potentially fatal lung disorders, including cancer, bronchitis, and emphysema. Typically, after years of smoking, pink healthy lungs turn black (see right).*

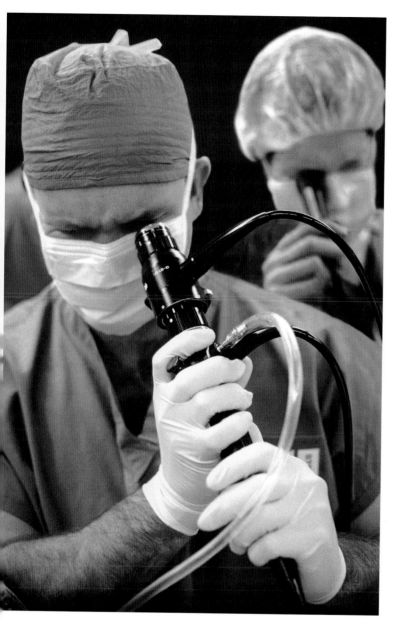

◄ A doctor uses a bronchoscope to examine a patient's windpipe (trachea) and bronchi. The instrument allows the doctor to check for tumors and to take tissue samples for biopsy.

lung, leading to bronchopneumonia. Most cases of bronchopneumonia are caused not by bacteria, however, but by viral infections. Although these types of pneumonia do not respond to antibiotic treatment, antibiotics are commonly given because secondary bacterial infection is often present.

Viruses weaken patients and make them more susceptible to bacterial infections; consequently, serious bacterial pneumonia can complicate the viral diseases of childhood, such as measles, whooping cough, chicken pox, and influenza.

Lung cancer

There has been an enormous increase in cases of lung cancer in the last 50 years and this, too, is entirely due to cigarette smoking. In the United States lung cancer is the second most common form of cancer in men, and causes more deaths among men than any other form of cancer.

After breast cancer, lung cancer is the most common form of cancer among women, and the incidence in women continues to rise. According to statistics, lung cancer is the leading cause of cancer death in both sexes.

Of all patients with lung cancer, 87 percent are, or were, cigarette smokers. Moreover, the enormously increased risk among cigarette smokers is not reduced to the same level as that of nonsmokers until 10 to 15 years after quitting.

There are many different types of lung cancer, and they each present differing chances of survival. In general, however, the outlook depends less on the type of cancer than on the stage to which it has progressed at the time of diagnosis; the outlook in most cases of established lung cancer is very poor. The early stages of the disease are often insidious, and the person concerned may simply consider the symptoms to indicate no more than a slight worsening of smoker's cough.

It has now been proved without doubt that smoking cigarettes has dire consequences. It is tragic that so many young people should be induced to smoke by peer pressure and the advertising of the tobacco industry.

Swell of opinion

Public opinion is now being mobilized against smoking, and governments are, at last, taking heed of the medical advice that they have for so long been ignoring.

tself, the condition is called asthma. Asthma attacks are intermittent and may be triggered by exposure to an allergen. Symptoms include hortness of breath, wheezing, and an excess of mucus.

Pneumonia

Pneumonia is an infection of the deepest parts of the lungs, the alveoli, and interferes with oxygen exchange. It typically affects the weak, the sick, the old, or the very young. Adolescent or middle-aged people who develop pneumonia are often smokers. The condition may be caused by a bacterial infection, and before the discovery of antibiotics, it caused many deaths.

If the infection is confined to one lobe of the lung, it is called lobar pneumonia, and it is most likely to be caused by the pneumococcus bacterium. Other bacteria such as staphylococcus, streptococcus, klebsiella, and hemophilus may invade the whole

See also: **Blood; Breathing; Chest; Coughing; Diaphragm**

Lymphatic system

The lymphatic system—comprising a network of fine lymphatic vessels, clusters of lymph nodes, and other areas of lymphatic tissue—collects surplus fluid from the tissues and blood vessels, filters it, and returns it to the blood. It also makes lymphocytes, which provide the body with immunity to disease.

Questions and Answers

My son has repeated attacks of tonsillitis, which are always accompanied by swellings in his neck. Why does this happen?

The tonsils and adenoids are small patches of lymphoid tissue in the upper part of the throat. Both can become infected by the bacteria and viruses that we inhale, causing inflammation. The infection then spreads along the lymphatic channel to other lymph nodes lower in the neck, so that the neck appears to swell. If the situation recurs and there are no other alternatives, your son may need to have his tonsils removed.

I recently had my right breast removed because I had cancer. Now, my right arm tends to swell. Is there anything I can do?

It is common for an arm to swell after such an operation, especially if the surgery involved both the breast and the lymph nodes in the armpit. The swelling should gradually settle down. However, if a large number of lymph nodes have been removed as a precautionary measure, it is unlikely that the swelling will get much better. In this case a surgical sleeve may help.

My daughter had influenza, and the doctor told me that her spleen was enlarged. Is this serious?

No. Enlarged lymph nodes and an enlarged spleen commonly occur when children are either suffering from or convalescing after a viral infection. The enlargement may become very marked in certain diseases, especially infectious mononucleosis, an infection of all the lymphatic tissues. It may be wise to stop your daughter from participating in contact sports for a while; otherwise, her spleen, which is not completely protected by her ribs, might be struck and then rupture.

The smallest vessels of the lymphatic system are each about the diameter of a needle. They crisscross the blood vessels and body tissues collecting surplus fluid, called lymph, which contains protein, fats, salts, white blood cells called lymphocytes, and other substances. Lymph is returned to the bloodstream via a myriad of larger vessels called lymphatics. Some of these contain an involuntary muscle that contracts rhythmically and drives the lymph forward, with valves set at intervals of 1.5 inches (3.8 cm) to prevent the lymph from flowing backward.

Lymphatics are found in all parts of the body except for the central nervous system, and the constituents of the lymph they contain depend on their location. The vessels draining the limbs contain fluid in surplus to body needs, which is leaked from cells or blood vessels; the lymph is therefore rich in protein, especially albumin (the main protein in the blood). The lymph in the intestines is full of fat, which it has absorbed from the intestines during digestion, so it has a distinctive milky appearance.

The lymph vessels join together to form two large ducts: the right lymphatic duct, which collects lymph from the right arm and the upper right side of the body; and

▲ *Swollen nodes in the neck can occur with tonsillitis or infectious mononucleosis.*

▼ *Nodes in the armpits may also swell.*

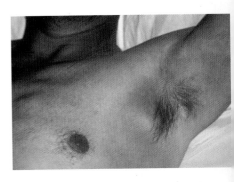

the thoracic duct, which collects lymph from the rest of the body. The lymphatic ducts empty into large veins on the right and left side of the neck respectively, returning the lymph to the blood.

Lymph nodes

At various points around the body, the lymphatics join together to form knots of tissue called lymph nodes, which are most numerous around the major arteries. It is sometimes possible to feel these nodes at points where the arteries run close to the skin, for example in the groin, in the armpits, and in the neck. Although the lymph nodes are sometimes referred to as lymph glands, the term is medically incorrect. They have no glandular function but are part of the immune system, being packed with immune cells called lymphocytes. As the lymph passes through the nodes, these cells trap and destroy infectious organisms and cancerous cells, filtering the lymph before it is returned to the bloodstream.

Lymphocytes

The bone marrow makes the precursor cells of all the immune system cells, including the lymphocytes. There are two main types of these: B lymphocytes, which mature in the bone marrow; and T lymphocytes, which mature in the thymus. Although they look identical under a microscope, the two types are controlled by different mechanisms and serve different functions in developing immunity. B cells are activated by the foreign proteins of disease organisms that invade the body. These stimulate the B cells to multiply and to make antibodies, which are proteins that inactivate the disease organisms, either directly or by marking them for attack by T cells. There are three types of T cells: some help to activate the B cells and other T cells; some attack disease organisms and infected body cells directly by releasing toxic proteins; and some suppress the immune response of other cells to the invading organisms.

Since T cells recognize any abnormal cells that enter the body, they may cause grafted tissue to be rejected. They are also thought to identify and eliminate abnormal cells produced by the body itself—for example, cancer cells. Many doctors now believe that the T cells are responsible for preventing people from developing cancer, since the incidence of cancers rises enormously if the T cells are diseased or destroyed by radiation or drugs.

Questions and Answers

My father has cancer of the lymph nodes, but my doctor says this is now curable. Is she correct?

Nowadays, using a combination of modern medicines, it is often possible to cure lymphomas. Even in advanced cases, life can be prolonged greatly by combining drug therapy with radiation treatment.

When my daughter had mumps, her glands became swollen. Was this something to do with her immune system?

It is unlikely. Mumps is a viral infection that spreads mainly to the salivary glands beneath the tongue and to the parotid gland, which also makes saliva and lies between the ear and the angle of the jaw. The swelling and pain that your daughter experienced were probably confined to those glands, since the lymph nodes in the surrounding area are seldom involved during mumps.

When my brother became ill at college, the doctor told him that he had infectious mononucleosis. What is this?

Infectious mononucleosis is a disease caused by the Epstein-Barr virus. The infection invades all the lymphatic tissues and may cause lymph node enlargement in the groin, the armpits, and the neck, and enlargement of the spleen and liver. The mainstay of treatment is general rest.

I've been told that swellings in the neck and armpits caused by infections are swollen lymph glands? Is this true?

Although you may hear some people talk about lymph glands, the term is medically incorrect. True glands secrete substances for use by the body: for example, saliva, hormones, digestive enzymes, or sebum. Lymph nodes have no glandular function; they are part of the immune system, and this is why they may become enlarged during an infection.

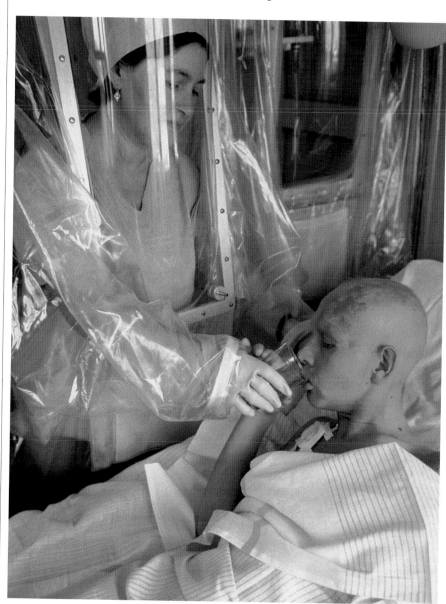

▲ *This 27-year-old man is a victim of the disaster at the nuclear plant in Chernobyl. He is receiving chemotherapy for cancer, which was caused by nuclear radiation. Affected lymph nodes often need to be removed to keep cancer from spreading further.*

LYMPHATIC SYSTEM

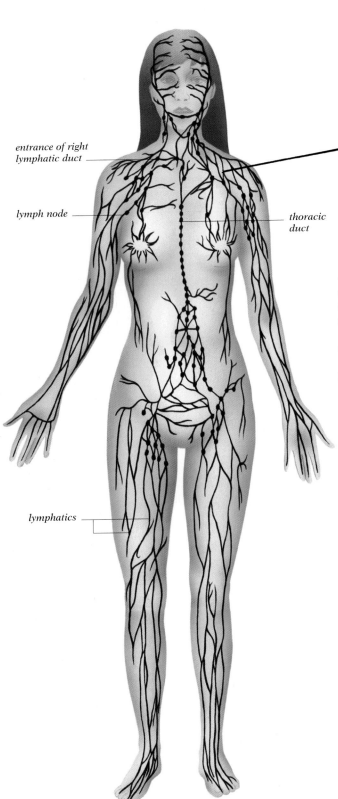

entrance of right
lymphatic duct

lymph node

thoracic
duct

lymphatics

◄▲ The lymphatic vessels are a network of fine tubes that collect surplus fluid (lymph) from the blood and tissues and return it to the bloodstream. Ultimately, the lymph drains into either the right lymphatic duct or the thoracic duct (left). The view above shows a section of a lymph node, where white blood cells called lymphocytes proliferate.

Infection and disease

Unless an infection is contained at the point of entry, invading bacteria or viruses gather in the lymph and pass to the nearest lymph node. Sometimes, a bacterial infection of this type can cause a condition called lymphangitis. The connecting lymph vessel becomes inflamed and can be seen through the skin as a red line that is tender to the touch. Once the infection reaches the lymph node, it may swell up; thus an infected wound in the finger will produce a swollen lymph node in the armpit if it is not treated promptly.

Besides acting as a filter for bacteria and viruses that invade the body, the lymph nodes trap cancer cells that are carried away from a tumor in the lymph. Once in the lymph nodes, the cells may lodge and form a secondary deposit. Cancers of the breast, for example, may drain toward the lymph nodes of the armpit, and cancers of the lung may spread to the nodes in the neck. When the lymph node is invaded, it enlarges and hardens. Thus the neighboring lymph nodes often have to be removed during surgery for cancer.

Not only do the lymph vessels and nodes form a means by which cancer is spread, but the nodes may be the site of primary cancers called lymphomas or lymphosarcomas, which consist of abnormal lymphocytes. The latter are more serious. Lymphomas can usually be cured if they are caught early enough.

Lymphatic obstruction

Sometimes lymphatic vessels can become blocked. This blockage seldom causes a problem in the chest or abdomen, because the lymph will usually drain through an alternative pathway in the lymphatic network. If an obstruction arises in the limbs, however, this choice of routes is not available and the lymph dams up. In this case the lymphatic blockage causes fluid to collect outside the cells, usually resulting in a type of swelling called lymphedema.

Diseases affecting the lymphatic system

DISEASE	CAUSE	SYMPTOMS	TREATMENT
Tonsillitis	Bacterial or viral infection of the lymphoid tissue of the tonsils.	Chills; high fever; headache; sore chest. Swallowing difficult. Sometimes painful swollen nodes under the jaw.	Antibiotics for bacterial infection. If severe and recurrent, removal of tonsils may be necessary.
Infectious mononucleosis	Viral infection spread by close contact.	Listlessness, fatigue, headache, chills, high fever, and sore throat. Swollen lymph nodes in the neck, armpits, and elsewhere.	Treat each symptom separately: mouthwashes; painkillers; bed rest; hot water bottle to relieve swollen glands. Convalescent period of six to eight weeks.
Lymphangitis	Bacterial infection from a cut or abscess that has spread to the nearest lymph nodes.	Chills; fever; swelling of the lymph nodes. Red lines spread up the arm or leg as the channel becomes inflamed.	Antibiotics. Elevate affected part and apply hot wet or dry compress.
Lymphedema	Blockage of the lymph vessels; can be congenital or follow childbirth or surgery for breast cancer.	Swelling of affected area.	Usually disappears in time or with treatment for the underlying cause. Elastic sleeve for limb if swelling does not resolve.
Lymphomas, lymphosarcomas	Tumors of the lymphatics; no known cause in most cases.	Tiredness; loss of energy, appetite, and weight. Lymph nodes in neck swell and feel rubbery. Unusual fever pattern. Itchiness of skin.	Radiotherapy in early stages; radiotherapy and anticancer drugs in later stages.
Elephantiasis	Filarial (worm) infestation, leading to blockage of lymph vessels.	Swelling of affected area, sometimes difficulty passing urine, skin thickens and becomes discolored.	Combination of two anti-filarial drugs. Elevate leg and wear compression stocking.

Lymphatic obstruction can have various causes. Occasionally people are born with deficient lymphatic vessels—a deficiency that may affect just one limb or part of a limb or may be generalized. This deficiency seldom causes problems in the major organs, however, because the network of vessels is so extensive that the organs usually continue to receive lymph.

Lymphedema commonly results and may be persistent when lymph vessels have been damaged by surgery. In women, this may be seen in one arm after a mastectomy operation for breast cancer in which the lymph nodes have been removed from the armpit on that side. This is a precautionary measure, in case the lymph nodes contain cancer cells that may spread.

In women, obstruction of the lymphatic vessels in the legs may also follow normal childbirth. Although this obstruction is usually temporary, swelling of the legs may persist in some cases.

Tumors may sometimes obstruct the lymphatic vessels and cause lymphedema, and since many tumors are invasive, they may also result in the lymph vessels becoming perforated, leading to a leakage of lymph. Where there are large numbers of lymphatic channels, as in the abdomen and chest, the leakage may be considerable and contain a high proportion of fat. Lymph may then accumulate, either between the lungs and the chest wall to form an effusion, or in the abdomen to form ascites (edema within the membrane of the abdominal cavity). If a creamy white fluid is obtained when these accumulations are drained, the patient is likely to have a malignant disease.

Certain diseases are also the result of lymphatic obstructions. Perhaps the most well-known is the condition called elephantiasis, even though it never occurs in the Western world. It is the result of an infection caused by tiny parasitic worms called filariae that gain entry to the lymphatic system through the feet. The ensuing blockage of the lymph nodes results in permanent damage and unsightly swelling. The condition may also occur in the arms.

Diagnosis

For less serious illnesses the doctor can feel the armpits or groin or around the neck to see if the lymph nodes are enlarged, but in other instances special techniques may be required.

If the disease involves the lymph nodes near the surface, a local anesthetic is given, a small incision is made through the skin, and the node is removed for study under a microscope: this procedure is called an excisional biopsy.

To check for abnormalities in the deeper nodes around the aorta, either a biopsy is done by endoscopic surgery under local anesthesia, or a lymphangiogram is carried out. This involves injecting a dye directly into the lymph vessels (usually in the foot); the dye is then carried to the lymph nodes in the groin and in the abdomen. Any abnormality will then show up on an X ray.

See also: Blood; Immune system; Lymphocytes; Marrow; Spleen; Thymus

Lymphocytes

Human immunity to disease depends on lymphocytes—white blood cells that protect people from attack by bacteria, viruses, and other microbes. There are two main types, B cells and T cells, which each have specialized functions.

Questions and Answers

My son has to have his tonsils removed. Could the number of white blood cells in his body be affected? I'm told that the tonsils are made of lymph node tissue, and that this processes these cells.

Removing the tonsils does not affect the number of white blood cells, nor does it interfere with the normal immunity to diseases that these support. Both types of lymphocytes (the white blood cells that provide immunity) are produced in the bone marrow, and T lymphocytes mature in a chest structure called the thymus. So you need not worry.

Can a viral infection lead to a rise in the number of lymphocytes in the blood?

Yes. Since lymphocytes are specially adapted for dealing with viruses, changes in their number generally occur after a viral infection. One example is infectious mononucleosis: not only does the number of lymphocytes increase, but their appearance under the microscope changes. Rest is the treatment.

I often hear of certain types of diseases described as autoimmune. What does this mean?

Autoimmune diseases develop when white cell antibodies attack tissues instead of fighting outside infection. The lymphocytes make such antibodies in error in diseases such as rheumatoid arthritis. Most treatment involves relieving the symptoms. In some cases, depending upon the underlying autoimmune disorder, immunosuppressive medicines are used to control the disease. These drugs include the corticosteroid prednisone. Such treatment becomes a balancing act of suppressing the disorder without compromising the immune system.

TYPES OF LYMPHOCYTES

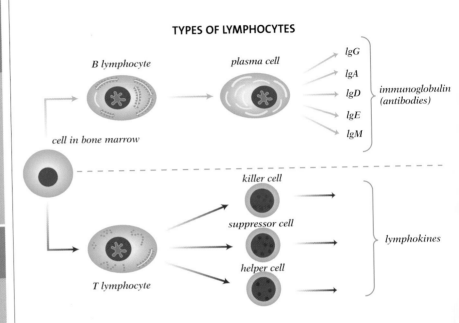

▲ The two types of lymphocyte react in different ways when exposed to disease organisms.

Lymphocytes are specific types of white blood cell devoted entirely to immunity. They circulate in the blood and lymph, and accumulate in the lymph nodes and other areas of lymphatic tissue. Unlike other elements of the immune system, which react indiscriminately when they encounter a foreign substance, lymphocytes form a coordinated attack against specific disease organisms (antigens), giving the body acquired immunity.

B lymphocytes

B lymphocytes (B cells) make antibodies: protein molecules called immunoglobulins that tag, destroy, or neutralize bacteria, viruses and other toxins. Most antibodies are made by a mature offshoot of B cells called plasma cells, which develop when the B cells are alerted to the presence of an antigen. B cells can be identified using a fluorescent stain that shows up the antibodies.

T lymphocytes

T lymphocytes (T cells) are customized by the thymus gland in the chest, which shrinks after puberty. It is not known exactly how the customizing works, but researchers do know that T cells make chemicals called cytokines, which are released into the circulation.

T cells are capable of recognizing foreign organisms and tissues; they attack foreign tissue and help to fight infectious microbes by releasing substances called lymphokines. There are three main types of T cell: helper cells assist the B cells and other T cells by stimulating them to fight specific infections; suppressor cells either slow or help the immune reaction; and killer cells produce toxins that destroy specific organisms or body cells that they identify as foreign.

T cells can be identified by their capacity to attract the red blood cells of sheep; when mixed with lymph, the red blood cells gather around the T cells in rose-shaped clusters called rosettes.

See also: Blood; Immune system; Lymphatic system; Thymus

Marrow

Our young daughter needs a bone marrow transplant. Can doctors be absolutely sure that the marrow donor isn't suffering from a disease that will be transmitted to her?

Yes. Potential donors are carefully questioned about previous illnesses, such as AIDS and jaundice, and have a series of tests on their blood before being accepted by a donor bank.

Can an unborn baby's bone marrow be diseased, and if so would it show up in prenatal tests?

Usually not. Standard prenatal testing would not be able to identify this problem. In many cases, genetic testing is needed. A possible sign of such problems in the unborn might be failure to grow at the expected rate.

After his marrow transplant, will immunosuppressive drugs be given to our son for the rest of his life?

Yes. Although these drugs do have side effects, they are minimal compared with what would happen without the transplant.

Is it painful for a donor to have bone marrow taken?

About 17.6 fl. oz. (500 ml) of marrow is taken from the donor under a general anesthetic, using a suitable needle, from several sites on the pelvis and the breastbone. There is very little pain afterward.

What is the outlook for someone who has had a marrow transplant?

This depends on the original disease and the match between the patient's tissue type and the donor's. Bone marrow transplants have not been performed long enough to say for certain what the long-term results are.

The soft marrow in the center of the bones is where the blood and the blood cells are manufactured. This marrow is essential to the body's health, and when problems develop a bone marrow transplant may be the only solution.

At birth all the bones in the body contain red bone marrow. However, at the age of about six or seven the marrow in the outer areas of the body stops producing blood cells and becomes yellow. This process continues until, at the age of 20 or so, the red marrow is found only in the skull, ribs, vertebrae, breastbone (sternum), and pelvis, together with small islands in the head of the long bone in the upper arm (humerus) and the head of the thighbone (femur). At the age of 70 the area of red marrow is about half that of a young adult.

If demand for the manufacture of red blood cells increases (as it increases in a person with continuous blood loss, for example), the yellow marrow has the ability to turn back into red marrow, acting as a reserve site for the synthesis (composition) of red blood cells.

Three types of cells are made in the bone marrow: the red blood cells responsible for carrying oxygen; the various types of white blood cells, whose main function is concerned with immunity (or defense of the body against invaders); and the platelets, the tiny fragments of cells that are a vital part of a blood clot.

Bone marrow diseases

Bone marrow may be defective at birth or may become so later on as a result of acquired disease.

Congenital defects: A child may be born with the inability to make one or more types of blood cell. He or she may be anemic because of an inability to make red blood cells, succumb easily to infections because of a lack of white cells, or bleed easily because of a lack of platelets.

Acquired diseases: Later in life the bone marrow can be affected by various diseases. There

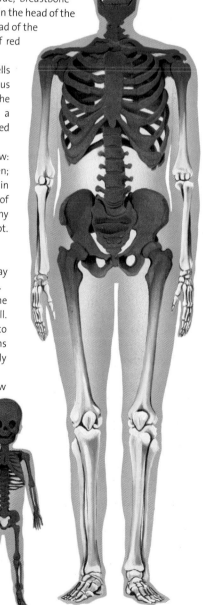

BONE MARROW IN A CHILD AND YOUNG ADULT

▶ *At birth all the bones in the body contain red marrow. At the age of 20 or so, the red marrow is mainly in the skull, ribs, vertebrae, breastbone, and pelvis. At 70 the area of marrow is about half that of a young adult.*

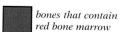

bones that contain red bone marrow

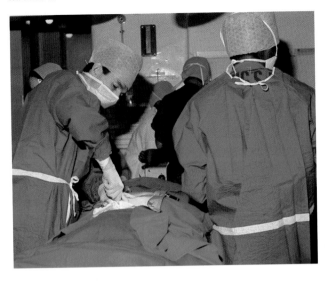

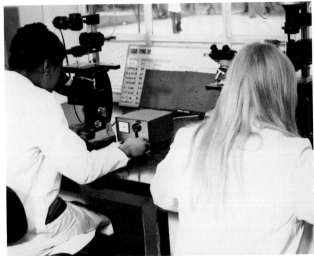

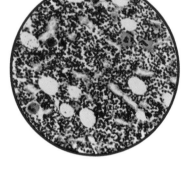

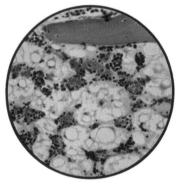

▲ *A successful marrow transplant (left) requires a close match between the tissue types of the donor and the recipient. Computerized registers (above right) help to achieve this by collecting a database of volunteer donors and their tissue types.*

▶ *As a person ages, red bone marrow (near right) is gradually replaced in the shafts of the long bones in the limbs by yellow marrow (far right). If necessary, however, yellow marrow can turn back into red marrow in the liver and spleen as well as in the bones.*

can be a sudden failure of the bone marrow to produce the different types of cells. This can happen for no obvious reason or as a result of allergy to certain substances, notably various drugs (such as some antibiotics and chemotherapeutic agents); the marrow usually recovers when the drug is withdrawn. Large doses of X rays can have the same effect, and cancer cells can infiltrate the bone marrow from a tumor elsewhere in the body. Cancer cells multiply in the marrow and prevent it from functioning.

The cells in the marrow can themselves become cancerous, leading to the production of vast amounts of unwanted cells. A form of leukemia can occur if the white blood cells are affected in this way.

Bone marrow transplants

Transplants are done for two main groups: patients born with defects in their own marrow, and patients with leukemia. In the latter the leukemic cells are destroyed by X rays and drugs, and the transplanted marrow takes over.

There are two types of marrow transplants. In allogenic transplant a matching donor is used. In autologous transplant marrow is removed from the patient, and both the patient and the marrow are treated to kill cancer cells. The marrow is then reinfused into the patient. The success of an allogenic transplant depends on various factors, the most important being the tissue types of the donor and the recipient. If the tissue types match closely, the recipient's body does not try to reject the donor's marrow. Because marrow contains cells concerned with immunity,

the reverse can happen: the donor marrow reacts against the recipient, in what is called the graft versus host reaction. There are about 100 different tissue types, and usually about 10 of these in any one individual. There is about a one-in-four chance that two members of a family will have the same tissue type; identical twins will have exactly the same type. The chance that two unrelated people will have the same type is remote, and so centers where bone marrow transplants are performed try to get as many donors as possible to build up a tissue bank, one that is cataloged via a computerized register of donors.

Carrying out a transplant

The donor has a blood test to determine his or her tissue type, and various other tests to exclude the presence of certain diseases that would make him or her an unsuitable donor. The marrow is then removed from the bones of the donor's pelvis and from the breastbone under general anesthetic, using a large needle. The recipient is treated with large doses of X rays to the bone marrow, and with drugs to remove any remaining cells, so that the transplanted marrow will not be rejected by the recipient's own marrow. The transplant is injected into the recipient and enters the bones via the bloodstream. Transplant drugs must be taken to prevent the graft versus host reaction.

See also: **Bones; Cells and chromosomes; Immune system; Transplants**

Mastication

Mastication—or chewing—is an everyday action we take for granted, but it is a complicated process in which the facial muscles work with the teeth and salivary glands. Biting and chewing are the first stage in the digestive process.

Questions and Answers

As a child, I was told to chew my food 32 times. Is this necessary?

Not really. The number 32 corresponds to the number of teeth in the full adult set. No doubt the idea was that the food should be chewed once for each tooth. It is important to chew your food well, since this makes swallowing easier and helps the digestive enzymes work more efficiently.

Do babies have to learn to chew?

Babies are born with a sucking reflex for suckling at their mother's breast. Teeth develop later, and when babies get to solid food they have no difficulty in chewing it.

I have problems chewing my food because my mouth is very dry. What could cause this?

Older people may suffer from this because the salivary glands become less efficient with age. Other causes may be lack of the B vitamins, blockage in one of the salivary glands, or Sjogren's syndrome, or the effect of drugs and smoking.

I have to have dentures. Will I be able to chew as well as with my natural teeth?

Today's dentures fit so well that you should be able to eat most things, except perhaps very sticky foods, like toffee. If your dentures do slip, buy a dental fixative from your local drugstore.

My young son gobbles down his food. Should I try to make him eat more slowly?

Some children do bolt their food. In adults this would probably cause indigestion, but children seem to get away with it. However, it might be wise to try to slow him down, if only to get him to savor his food.

THE CHEWING PROCESS

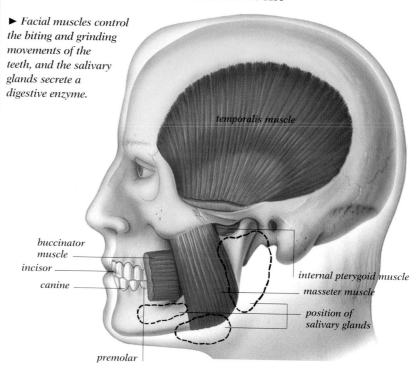

▶ *Facial muscles control the biting and grinding movements of the teeth, and the salivary glands secrete a digestive enzyme.*

temporalis muscle

buccinator muscle
incisor
canine
internal pterygoid muscle
masseter muscle
position of salivary glands
premolar

"Mastication" is the medical word for chewing. The purpose is to reduce food to pieces that can be swallowed easily; it also creates large surface areas on which enzymes can begin to work, breaking down the substances in the food that can be used by the body for its maintenance.

An adult has 32 teeth. The eight incisors are shaped like chisels to cut food into small pieces; the four canines are longer and are used for tearing; the eight premolars slice food by shearing the top teeth against those in the lower jaw; the 12 molars have flattened surfaces that pulp food.

Biting is controlled by two muscles, the masseter and the temporalis, working together. The masseter stretches from the cheekbones to the base of the lower jaw and raises the lower jaw, while the temporalis on the side of the head acts to clench the teeth.

Chewing consists of three actions: positioning the food in the mouth (assisted by the tongue), side-to-side grinding movements, and back-and-forth grinding movements. Contractions of muscles called the buccinators compress the cheeks and position the food in the mouth. The internal pterygoid muscles move the jaw sideways, while the external pterygoids move it back and forth.

As a person chews, three pairs of salivary glands secrete saliva. This lubricates food to make it easier to swallow, and enables it to be tasted, since taste buds will not work on dry substances. Saliva also contains the enzyme ptyalin, which begins the digestion of starches in food by converting them into simple sugars. Once food has been sufficiently chewed and lubricated, it is pushed to the back of the mouth by the tongue, where it enters the esophagus to begin its journey through the gastrointestinal tract.

See also: **Digestive system; Enzymes; Esophagus; Glands; Mouth; Saliva; Teeth**

Melanin

Melanin, a dark-colored pigment, protects a person's skin from the potentially harmful ultraviolet rays of the sun. The amount of melanin produced varies greatly between a dark- and fair-skinned person.

Questions and Answers

I am fair-skinned, and however hard I try, I can never seem to get a good tan. Why is this?

Fair-skinned people tan more slowly than darker-skinned people. The reason is that although all people have the same number of melanocytes, the cells in fair people respond more slowly and take longer to form melanin, the pigment that darkens the skin.

I have heard that women tan more quickly if they are on the Pill, although I have never noticed it myself. Is this true?

The contraceptive pill does change the level of all the major hormones in the body, including the melanocyte-stimulating hormone. Although this increased tanning effect is not noticeable in most women, it is said to exist. It is thought to be harmless.

I am convinced that my friend's teenage daughter is getting darker every month. She says that she has not been sunbathing, so what could be causing this change?

Because of her age this darkening of the skin could be caused by puberty. However, it could be a sign of illness, so a doctor should be consulted. Melanin disorders are quite uncommon, are rarely life-threatening, and can usually be treated.

If melanin protects us from the harmful rays of the sun, why is sunbathing so dangerous?

The sun ages the skin; it also reduces its elastic properties and increases the risk of skin cancer. Although sunlight also makes vitamin D in the skin, which is essential for healthy bones, it is best to limit the time you spend in strong sunlight, particularly if you are fair-skinned.

▲ *Exposure to sunlight stimulates the formation of melanin, the dark-colored pigment that protects the skin from harmful ultraviolet rays; this causes the skin to darken and helps prevent it from burning.*

Melanin is a dark-colored pigment found in the skin, the hair, and the iris of the eye. It is formed in melanin-making cells, called melanocytes; these are situated in the basal, or lower epidermal, layer of the skin.

The same number of melanocytes can be found in the skin of every human being, no matter what his or her racial type. However, the amount of melanin produced by these cells varies greatly. In dark-skinned people the melanocytes are larger and produce more pigment.

Function and formation

The function of melanin is to protect the skin from the harmful rays of the sun; the darker a person's skin, the less likely it is to suffer from sunburn. The complex chemical process of the body that creates melanin takes place on the outer part of each melanocyte. Once the dark pigment is formed, it moves to the center of the cell to cloud over, and thereby protect, the highly sensitive nucleus.

Exposure to ultraviolet light, from either sunlight or artificial sources such as sunbeds or sunlamps, stimulates melanin production by the normal process of tanning. Melanin is formed, the cells expand, and the skin darkens in color. The response varies from individual to individual, but all people except albinos can eventually become pigmented when they are exposed to enough sunlight. In addition, sunlight can stimulate pigmented skin overgrowth.

Melanin and maturity

In both men and women some areas of the body become pigmented naturally as a result of sexual development. In men the scrotum and penis become slightly pigmented. In women, parts of the genitals and also the nipples and areolae tend to darken; this is particularly true during pregnancy and after the birth of children.

Skin pigmentation can also change as a result of age; the elderly are more likely to develop liver spots than younger people.

Melanin disorders

Disorders of melanin are rare and mostly concerned with absent pigmentation or increased pigmentation.

Skin conditions such as eczema and psoriasis, both of which cause severe itching and lead to scratching and skin abrasions, can cause increased melanin production. Even friction from tight clothing can stimulate melanin formation.

A rare medical condition called Addison's disease occurs when the adrenal glands do not function properly. This causes the oversecretion of melanocyte-stimulating hormone (MSH), which is produced in the pituitary gland. The patient gradually becomes darker-skinned as melanin forms in response to the hormone; the pigmentation of the skin can also be patchy.

In an inherited condition called albinism, the enzyme that creates melanin is absent. Although the correct number of melanocyte cells are present in the skin, no melanin can be produced. Albinos (people who suffer from albinism) have snow-white or pale yellow hair and pink or light blue eyes. In addition, because the protective function of melanin is lost, albinos are at a much greater risk from sun-induced skin cancers in hot, sunny climates.

In the condition called vitiligo, the cause of which is unknown, there is a spontaneous and patchy loss of skin pigment. Areas of a

▲ *People of all racial types have the same amount of melanin-making cells. In darker-skinned people, however, the melanocytes are larger and produce more melanin*

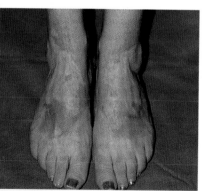

◄ *Vitiligo causes a patchy loss of skin pigmentation. The condition may occur anywhere on the body and can appear and disappear spontaneously.*

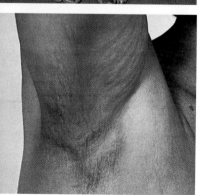

◄ *The increased pigmentation found in Addison's disease is a symptom of a complex condition that results in an overproduction of melanocyte stimulating hormone.*

sufferer's skin remain unpigmented for months or even years in some cases, and then may suddenly return to normal. Vitiligo can be particularly noticeable in dark-skinned people.

Treatment and outlook

Most melanin disorders are not life-threatening. The treatment usually depends on identifying the cause of the increased or decreased pigmentation and then giving the correct treatment.

There is no treatment for patients with albinism, but steroid creams can be applied to a person's skin to help in cases of vitiligo. When irritant skin conditions such as eczema and psoriasis are treated, skin pigmentation will gradually return to normal over a period of a few months. In Addison's disease, however, when pigmentation is caused by the abnormal activity of the pituitary gland, the patient is treated with drugs.

See also: Adrenal glands; Skin

Melatonin

Melatonin is a hormone produced by the pineal gland—a tiny structure about the size of a pea situated deep in the brain. Melatonin controls daily rhythms of sleeping and waking and may hold the key to the aging process.

Questions and Answers

I've seen both natural and synthetic melatonin on sale in health food stores. What's the difference between the two?

Supplements that contain "natural" melatonin use melatonin extracted from animal pineal glands, whereas synthetic melatonin is manufactured using the same chemical constituents as the melatonin that occurs naturally in our bodies. The synthetic variety is generally recommended because the supplier can ensure consistent quality and purity.

I've been taking melatonin to relieve my insomnia. As a result I've been sleeping really well, but I wake up in the morning feeling drowsy and sluggish. Is this an inevitable side effect of melatonin?

You are probably taking a higher dose than you need to. The tablets and capsules on sale typically come in 2-mg, 2.5-mg, and 3-mg doses. Break up the tablets or divide the contents of the capsules so that you can begin with a 1-mg dose. This may be enough to help you get a good night's sleep—if not, gradually increase the dose by 1 mg each day to no more than 3 mg in total. See your doctor.

My 43-year-old mother has started taking melatonin because she believes it will stop the effects of aging. If this is true, then isn't it best to start as young as possible? However, she says that, at 18, I'm too young to benefit. Is she right?

Yes. Even the most ardent champions of melatonin as an antiaging treatment (a claim that is still controversial) do not recommend it for young people, whose bodies produce high levels of it naturally. They suggest that those who hope to slow the aging process should not normally start taking melatonin before the age of about 40.

Melatonin is a hormone secreted by the pineal gland. This gland used to be called the third eye and was believed by some to be the seat of the soul. Until recently, however, scientists considered it of little importance and its function was not understood. Now it is believed to be closely linked with other endocrine (hormone-secreting) glands, such as the pituitary, adrenal, and thyroid glands.

The role of the pineal gland is thought to involve regulating, or fine-tuning, the other glands' production of hormones. In addition, through the production of melatonin, the pineal gland controls the circadian rhythm—the internal clock that tells the body when to sleep and wake.

Melatonin's function in the body

Researchers in the 1950s and 1960s who discovered melatonin and explored its properties found that its production followed a daily cycle, with levels of melatonin in the blood 10 times higher at night than in the day. People who were given extra melatonin started feeling drowsy, and it was found that the production of melatonin was the body's way of regulating sleep. The pineal gland reacts to light by producing more or less melatonin, cuing us to sleep during the night and to be wakeful and alert during the daytime. The light detected by the pineal gland varies in amount and intensity as the seasons change, and this affects the seasonal production of melatonin in many animals, timing their breeding seasons, migrations, hibernation, and fur growth. Such seasonal fluctuations are not so obvious in humans, although one study has shown that women tend to have lower melatonin levels in the summer. Melatonin imbalance may also be a factor in seasonal affective disorder (SAD), a condition that involves depression, carbohydrate cravings, and excessive sleep in the winter.

Newborn babies do not start producing melatonin until they are two or three days old, and their melatonin cycle does not settle down to a regular daily rhythm until they are about a year old. Melatonin is present in nursing mothers' breast milk, and this could be a reason why bottle-fed babies are sometimes troublesome sleepers. Young children have high levels of melatonin relative to adults and also sleep far more, with regular naps in the daytime. Melatonin production at this time may be associated with the pituitary gland, which is responsible for stimulating growth. During sleep the pituitary gland secretes growth hormone, and the period of highest growth rate coincides with

▲ *Air travelers often experience jet lag when flying across different time zones, and it may take several days for them to resume their normal sleep-wake patterns.*

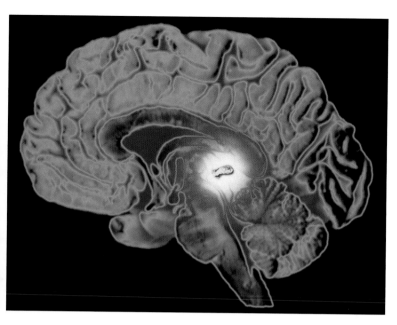

◄ *The pineal gland (highlighted) deep in the brain secretes melatonin, the hormone that sets the body's biological clock and controls sleep–wake patterns.*

proposed that the pineal gland is not just a circadian clock but also an aging clock, timing the body's natural life span by tailing off melatonin production. One way in which this may work is that the melatonin shortage that comes with old age has a critical effect on the thymus gland, which controls the functioning of the body's defenses, the immune system. With age the immune system works less efficiently, laying the body open to viruses and diseases such as cancer and heart disease, the incidence of which increases as the body grows older. Some researchers claim that this decline in immune functioning can be reversed by boosting melatonin levels.

Experiments done to test this hypothesis have involved transplanting the pineal glands of old mice into young ones and vice versa. The results were dramatic—the young mice with older pineal glands (and hence the lower melatonin levels) aged quickly, whereas the older mice with high melatonin production seemed rejuvenated and lived, on average, 30 percent longer than normal.

Those who champion melatonin as an antiaging treatment for humans suggest that taking regular supplements may help people live to over 100 without significant deterioration in mental or physical capacities. However, to date there is insufficient evidence to confirm whether this theory is applicable to humans.

If melatonin does indeed turn out to protect against aging, it may do so because of its action as an antioxidant. The production of energy in our cells gives rise to a by-product called free radicals, unstable atoms that can damage the cells' ability to repair themselves. Melatonin is one of the many natural substances called antioxidants—others include vitamin E and vitamin C—that prevent this damage by neutralizing the action of free radicals.

The effect of melatonin supplements on the treatment of various cancers is also under investigation. Abnormal levels of melatonin have been found in men suffering from prostate cancer, for example, and links have been indicated between breast cancer and disruptions in melatonin production. Melatonin may have a positive influence on the progress of cancer by limiting the production of sex hormones such as estrogen, high levels of which are thought to increase the risk of breast cancer. Melatonin supplements have been given to cancer patients, together with lower than normal doses of conventional chemotherapy, with some success.

Melatonin may also restrict cancer growth because of its properties as an antioxidant. Both this and its influence on the balance of the endocrine system have led to the idea that melatonin supplements may have a role in treating or preventing a range of diseases, including high blood pressure, Parkinson's disease, diabetes, and Alzheimer's disease.

Only time will tell whether melatonin is a genuine medical breakthrough.

the time of most sleep, during the first three years of life. The rise in melatonin levels slows at about age seven and starts its decline at adolescence, which seems to cue the production of the sex hormones in both boys and girls, who start their sexual development as a result. Melatonin levels are lower in adults and begin to decline steeply around the age of 45.

Melatonin supplements

Because of its role in regulating the body's circadian clock, some people use melatonin to help treat sleep disorders, which include insomnia and jet lag. Melatonin supplements are available in pharmacies and some health food stores. However, the long-term effects of melatonin are unknown, and because it is marketed as a dietary supplement and not a drug, it is not subject to FDA approval.

Melatonin in tablet or capsule form taken at night in very low doses (1 mg to 5 mg) is enough to bring the blood melatonin level up to the normal nighttime levels, leading to drowsiness and sleep that follows the body's natural rhythms. Synthetic melatonin does not seem to have any of the side effects associated with drugs such as sedatives or tranquilizers, which can lead to a hung-over feeling the following morning and may be addictive.

People with insomnia may find it useful to take other measures to help them sleep besides melatonin supplements, such as sticking to a fixed sleeping routine, drinking warm milk, and avoiding stimulants such as smoking and exercising just before bedtime.

Jet lag happens when people cross time zones too fast, for example on a transatlantic flight. The body's time clock still believes it is in Los Angeles when the actual time is that of London, eight hours ahead, and it can take several days to adjust. The result of this bodily confusion is a feeling of exhaustion, irritability, and lack of concentration. Taking a small dose of melatonin before bedtime for a few days in the new time zone can help reset the body's internal cues.

A miracle cure for aging?

The fact that melatonin production falls sharply as people get older has led to research into its associations with aging. It has been

See also: Aging; Biological clocks; Endocrine system; Glands; Hormones; Pituitary gland; Thymus

Membranes

Questions and Answers

Is it true that babies are sometimes born with a membrane covering the head?

Yes. A baby develops within the uterus encased in a membrane or sac that contains the amniotic fluid, or waters. This usually ruptures in early labor; if it does not, the baby is delivered with the unruptured membranes covering his or her head and face. This unusual occurrence, called being born with a caul, used to be thought lucky, a sign that the baby would never die by drowning.

Why is the term "membrane" used in connection with so many medical problems?

Membranes cover or line many parts of the body, and they occur throughout the human anatomy. They are not especially susceptible to disease, but they tend to be in the front line, in cases of injury and attack by germs. This is true of the membranes of the throat, which are exposed to germs in the air, and of the membranes that surround the lungs and brain.

Is meningitis a disease of the brain or its membranes?

It is a disease of the membranes that surround the brain. Bacterial cases can usually be cured with drugs if caught early enough.

Is the snuff my grandfather takes likely to cause him any damage?

Yes. Snuff, which is powdered tobacco, contains nicotine and other substances that have numerous adverse health effects. The most serious consequences are an increased risk of cancers of the mouth and esophagus, kidneys, and pancreas. Use of smokeless tobacco increases risks of heart disease as well.

Doctors use the term "membranes" in connection with tissues throughout the human body, from the joints to the nostrils, the stomach, or the brain. They have the basic functions of covering, lining, or dividing.

Membranes are simply layers of body tissue that cover, line, or divide. There are five main types. Mucous membranes are principally found as the lining of tubes, as in the alimentary canal (the system that functions in the absorption and digestion of food). Synovial membranes cover joint surfaces and tendons. Serous membranes surround organs in the chest and the stomach. The meninges cover the brain and spinal cord. Finally, at a microscopic level, all the many millions of cells that make up the body, and the tiny compartments (organelles) within those cells, are enclosed and divided by membranes.

The mucous membranes

As the name suggests, mucous membranes contain specialized cells that secrete the slimy fluid called mucus. Among the many functions of mucus are fighting infection (it contains antibodies) and keeping the throat, the nostrils, and indeed the whole gastrointestinal tract moist and pliable. Some of the mucous membranes, notably those in the respiratory tract (the passages that lead to the lungs), contain cells with additional functions. Sprouting from them are hairlike projections called cilia that move in concerted waves to push harmful foreign bodies, such as dust, back upward to the throat to be coughed out of the body.

The membranes lining the intestines are folded into fingerlike projections called villi to increase the surface area available for digestion of food. There are also mucous membranes in the reproductive systems, notably the lining of the uterus, which is shed each month during menstruation.

Synovial membranes

These occur at moving joints and take the form of bags or sacs containing the lubricating fluid called synovial fluid. Tendons, the tough bands of tissue that connect muscle to bone, are also surrounded with a sheath of protective synovial membrane.

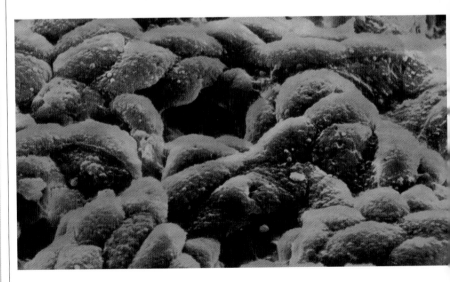

▲ *This colorized electron micrograph scan shows the folds of the stomach mucous membrane magnified 360 times. The folds increase the surface area; these, along with the digestive juices secreted by the membrane, aid food digestion.*

TYPES OF MEMBRANE

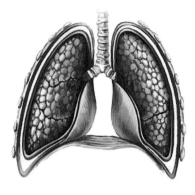

▲ *Serous membranes called the pleural membranes (shown here in purple) prevent friction between the lungs and the rib cage.*

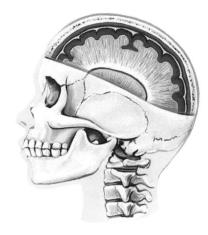

▲ *The meninges (shown in orange) surround the delicate brain tissue and cushion it from the hard bony skull.*

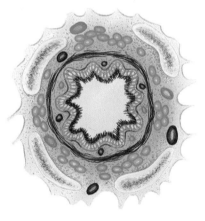

▲ *The mucous membrane lining the respiratory tract supports cilia: tiny hairs to waft foreign bodies away from the lungs.*

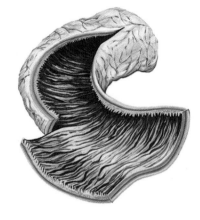

▲ *Villi of the mucous membrane (pink) lining the small intestine increase the surface area available for the digestion and absorption of food.*

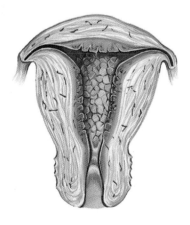

▲ *The mucous membrane (red) lining the uterus provides nutritive secretions and prevents friction.*

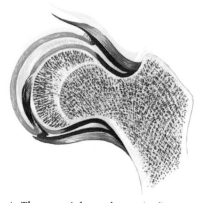

▲ *The synovial membrane (red) lining a joint capsule secretes fluid to lubricate all movements of the joint.*

Serous membranes

The coverings for organs in the chest and stomach are called serous membranes. They provide protection against disease and reduce friction with neighboring parts. In the chest cavity there are two membranes called the pleura. In the abdomen all organs are covered by the membrane called the peritoneum.

Problems with membranes

In the digestive system the mucous membranes have the important function of secreting digestive juices, which contain acid. Excess acid is produced during indigestion. Acid indigestion can usually be relieved with a counterbalancing, or antacid, medicine.

Because the mucous membranes in the mouth and throat are in such close and constant contact with the air, and because so many germs are airborne, these membranes are frequent victims of infections. Familiar infections that start in the membranes of the throat and nose include the common cold, influenza, and measles.

Inflammation of the synovial membranes, because of infection or wear and tear, causes a condition called synovitis. Unaccustomed pressure caused by too much kneeling, or even awkward movements, such as the incorrect use of a screwdriver, can cause irritation of the whole synovial sac, or bursa, resulting in bursitis.

The meninges are vulnerable to meningitis. Bacterial meningitis can in most cases be cured with modern antibiotics.

The problems of the serous membranes are mainly pleurisy (inflammation of the pleura) and peritonitis (inflammation of the peritoneum). In the latter case, oftentimes the patient requires abdominal surgery to repair the underlying problem.

Membranes in pregnancy

Membranes have a special temporary function at the beginning of every new life. The developing baby, or fetus, is surrounded in the uterus by a special membranous bag called the amniotic sac. This contains the fluid in which the fetus floats, creating a protective shock-absorbing system. After the baby's birth, the amniotic sac and placenta are automatically expelled; together they make up the afterbirth.

See also: Alimentary canal; Cells and chromosomes; Digestive system; Joints; Mucus; Peritoneum; Placenta

Memory

Memory is an essential part of day-to-day life. Driving a car, remembering a telephone number, putting a name to a familiar face—all would be impossible without it. How does this intricate mental process work, and is there any way that the memory can be improved?

Everyone's memory is like a personal filing system that contains many pieces of information about his or her life. Memory has many facets, from remembering a telephone number long enough to dial it to recalling an event from childhood accurately.

Scientific research has found that there seem to be at least two types of memory storage. The first contains a small number of very exact memories that fade away very quickly; the second contains a large number of general memories that can go back many years.

Human beings are also said to possess other forms of memory; for example, the so-called photographic memory (eidetic memory). A very small proportion of young children can, for

▲ *The memory functions as an individual's personal filing system. Some types of memory are filed alphabetically; others may be filed aurally or visually.*

Questions and Answers

My elderly grandmother has vivid memories of her childhood but tends to become confused in day-to-day matters. Why?

Her brain is less efficient at creating new memories because she is elderly. However, her older memory paths are probably still working almost as efficiently as when she was young.

Why do people forget what happened to them just before being knocked unconscious?

There are two reasons. The first is that they often want to forget what has happened. The moments before a knockout are usually confused (making memory difficult) and unpleasant (making forgetfulness a relief). Second, memories are not made instantly; they take a few seconds to be organized within a person's brain. If that person has been knocked unconscious by then, the memories just before the knockout are not recorded properly.

Why do we find it so hard to remember our dreams?

The answer is not clear. The most recent theory holds that when we sleep, we watch our dreams go by and make very little attempt to process them into memories. Only when we wake immediately after a dream does normal memory recall take place.

Why do I sometimes find it hard to remember things after I have been drinking a lot of alcohol?

This is mainly because alcohol is a depressant drug, even though it may temporarily make you feel happy. In this depressed state, your attention will wander easily and you will become distracted; memories will tend to get filed less often as a result.

HOW MEMORIES ARE STORED

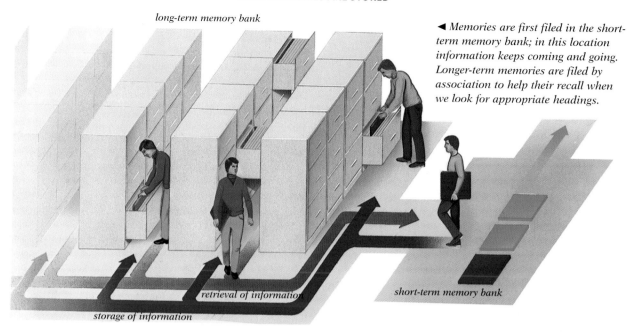

long-term memory bank

◄ *Memories are first filed in the short-term memory bank; in this location information keeps coming and going. Longer-term memories are filed by association to help their recall when we look for appropriate headings.*

retrieval of information

short-term memory bank

storage of information

Most people can recall telephone numbers up to seven digits; a longer list of numbers is more difficult to retain. Numbers that are used frequently are easier to remember.

example, describe in minute detail a picture that they have looked at for only a few seconds; or they may be able to remember exactly a long list of words or figures.

The ability to remember something depends on the efficiency of the mental filing system that services each individual's memory.

With both short- and long-term memory, the filing system performs three functions: the information is first converted into the kind of code that the memory accepts. Next, the coded fact is held in the appropriate memory storage—just as an accounting balance sheet would be placed in an appropriate filing cabinet. Finally, the memory will be retrieved, the way an entry on the balance sheet would be, either by using general association or by pulling out a bundle of possible documents and searching for the right one.

The two main memory storages

Any new piece of information to which a person pays attention goes first into his or her short-term memory storage. The size of the short-term memory is very small; an average person can immediately recall only seven unrelated items. For example, most people can remember a seven-digit telephone number for long enough to dial it but tend to forget one or two digits if the number is any longer.

To preserve a new memory, people often repeat the information to themselves a few times. At the same time they search for ways to remember it, generally by associating the new idea with material already in the memory.

Recalling a memory

The short-term memory is not very well organized, and the last piece of information is usually the easiest to recall.

In the long-term memory, however, a faster recall system is needed to extract a memory from the thousands each person has in storage. Research has shown that if an individual's search mechanism is working well, it races through a series of file headings, giving access to subheadings and further subheadings until the memory required is found. Sometimes, this process is not needed because a person's very vivid memories or pieces of information that he or she uses constantly are available immediately.

Faults in memory

There are five types of memory faults from physical causes. Such faults can occur because of brain damage caused by concussion or skull fracture, brain tumors or hemorrhages, vitamin and other chemical deficiencies, chemical poisoning, and bacterial infection. Memory impairment from these causes is sometimes permanent because in the brain (unlike in other parts of the body) new cells are not created when the old cells die.

Old age often produces failure of memory, because the brain has become less efficient at creating new memory pathways. That is why old people can often recall childhood events yet may be very vague about recent experiences. Such common forgetfulness is different from that caused by Alzheimer's disease, which is not a normal aging process.

Psychological causes

Other memory faults stem from psychological causes. For example, a memory associated with a very unpleasant emotion may be seemingly obliterated completely from the memory.

▼ *Elderly people find it more difficult to form new memories; for example, remembering to take medication each day.*

This condition is called repression and, according to one theory, it occurs because the mind sometimes censors life's most difficult experiences. It prevents the traumatized individual from recalling events that are too emotionally painful for him or her to bear.

By contrast, in many cases of amnesia it appears to be unnecessary for a counselor or therapist to try to deliberately unearth the lost memories. Removal of the stress that produced the amnesia will often enable the memory to return by itself.

Everyday memory failings

The last category of possible memory faults is experienced by everyone, young and old.

When people are in a hurry or distracted, the information they receive is not stored properly in the memory. Similarly, in those individuals who are habitually stressed or in a rush, the memory storage system has never been properly organized in the first place.

Recalling a memory is probably the stage at which most failings occur; it has been said that people have far more in their heads than they can ever get out and use again. There is a retrieval difficulty in the "tip of the tongue" phenomenon, when a word or concept is almost recalled but lies just out of reach. Although a person often knows the concept or word's first letter, its length, its rhythm, or its last syllable, he or she simply cannot remember it.

At other times people may suddenly forget what they were going to say or do next; they cannot remember because agitation interferes with the retrieval process. Only when they switch off and turn their attention to something else does the desired memory return.

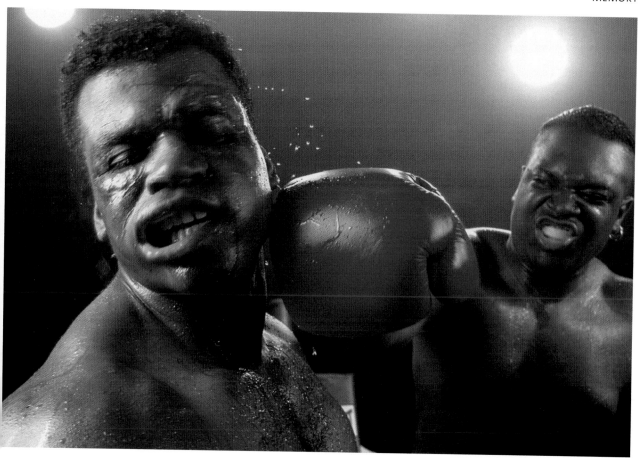

▲ *When a boxer receives a blow from his opponent, events prior to the knockout are often confused. Because the mind has had insufficient time to process them, such events are not stored properly in the memory.*

Variation in memory efficiency

Many individuals who say that they have a poor memory make the mistake of thinking that memory is equally efficient in all fields of knowledge or experience.

Such people may complain that they can never remember dates, but they may remember musical themes with great ease. Other people may get annoyed with themselves because they can easily remember a face but can never attach the correct name to it. This means merely that in some instances they have more file index headings for one field of memory than for another.

Memory and intelligence

The development of intelligence is heavily dependent on memory because much of it is concerned with learned facts and with the knowledge of meanings, logic, and reasoning. However, some facets of intelligence involve common sense more than anything else.

Improving memory

Not much can be done to improve short-term memory. Even with practice it seems impossible to increase the number of separate items the human brain can hold at any one time in the memory.

Improving long-term memory, on the other hand, is both possible and useful. Material can be put more effectively into memory by memory-aiding systems called mnemonics.

Perhaps the oldest known system is called the method of places. With this system a list of objects can be memorized by putting each object (in the imagination) in a different room of a house. The list can then be recalled by walking (in the imagination) through each of the rooms and seeing the object required as it appears.

Another system involves the color-coding of different sections of material—for example, with a colored highlighter. Because the learned information is linked to a color, the mind sorts and remembers it more easily. Finally, information can be remembered by association (linking the word *blanco*, which is Spanish for "white," with a blank sheet of paper) or by anagram (Roy G. Biv for the seven colors of the rainbow—red, orange, yellow, green, blue, indigo, and violet).

General memory tasks

More general memory tasks—for example, remembering the plot of a play or a newspaper story—can be done by organizing the material under small, easily remembered headings and associating the details with similar events. Even though only about one-tenth of human memory capacity is ever used, it is worth putting in the time and effort that is required to make better use of people's memories.

See also: Aging; Brain; Mind

Menopause

The end of a woman's menstrual periods—menopause—is a natural event, not an illness. It can cause problems, but it can also mark the beginning of a life that is free of many physical and emotional pressures.

Why do some women go through menopause without any trouble?

Some women have the good fortune to have a happy and calm personality, or perhaps life has been very kind to them. The level of hormones circulating in the blood does play a large part in health. One of the problems in deciding on treatment for a woman whose menopause is causing her problems is that the level of hormones in the blood is not always consistent with the troublesomeness of the symptoms. Women with low blood levels of hormones may have no symptoms.

Also, sometimes women who have no discomfort at menopause still resent it, perhaps forgetting that it is a change, not a decline. Menopause can mark the end of many pressures and the start of greater vitality and freedom.

I think I have started menopause, and I constantly have to urinate during the day and the night. Should I ask my doctor for hormone pills?

If you have stopped having periods, you have passed your menopause, but you will probably have to wait for a year before you can be sure that this is so. Menopause causes thinning and dryness of both the vaginal tissues and those lining the urethra, leading to an increased risk of urinary tract infections in the latter case. Seek medical attention if you are suffering with increased frequency of urination. Also, if you have had babies, you may have a prolapse (dropping) of your uterus, so that there is weakness of the muscles of your bladder and vagina. This can sometimes be corrected by a set of exercises, but if it does not respond, you may have to have a small operation to correct it. Be aware that the value and safety of hormone pills are now considered doubtful.

Menopause, or "change of life," is the time in a woman's life when her menstrual periods stop. It generally comes between the ages of 45 and 55, but it may happen a good deal earlier or may be delayed until she is in her late fifties. The physical changes that it brings about are a process, not a sudden event, and the time during which they happen—a matter of a number of years—is also called the climacteric.

End of childbearing

The main effect of menopause is that a woman can no longer become pregnant, because it marks the end of her fertility. In the past, for women who had spent much of their lives pregnant, menopause must have been welcome. Today it can still mark a new beginning. The problems it may cause are real but so are the positive aspects.

What is menopause?

At birth, a girl baby's ovaries contain thousands of immature egg cells. At puberty (the age of 12 or thereabouts), the first ovum ripens and moves from the ovary through one of the fallopian tubes into the uterus. At the same time the ovaries are manufacturing two hormones: estrogen, which stimulates the lining of the

▲ *Eating a balanced diet and following a program of regular exercise will benefit a woman before, during, and after menopause.*

uterus to thicken; and progesterone, which encourages the body to prepare for pregnancy. The secretion of both estrogen and progesterone is controlled by the pituitary gland, situated in the base of the brain.

When a pregnancy does not occur, the thickened tissue lining the uterus is shed. This passes out of the body, along with the unfertilized egg, in the menstrual blood. This process is repeated month after month, except during pregnancy—or when it is interrupted by tension, extreme weight loss, prolonged strenuous exercise, or severe illness—until menopause, when it stops.

Signs of menopause

In many women menopause happens suddenly. Others may find their periods becoming irregular, scanty, or brief—or a combination of all three. If a woman finds her periods becoming more frequent or heavier during the middle years, it is important for her to get medical advice.

Hot flashes and vaginal atrophy

The most common physical problems associated with menopause are hot flashes and a decrease in the moisture and elasticity of the vagina (called vaginal atrophy). A hot flash is a sudden feeling of feverish heat in the body, which is sometimes accompanied by patches of redness on the skin. When the hot flash is over, the woman often feels chilled. About 80 percent of women experience hot flashes within three months of menopause. Among women who have hot flashes, 85 percent will experience them for more than a year, and 25 to 50 percent will have them for up to five years.

Flashes may happen infrequently or many times in the course of 24 hours—often they happen at night, when they may disturb a woman's sleep. This sleep deprivation can lead to chronic fatigue, depression, and poor concentration.

Dryness and inelasticity of the vagina, which generally occur in the latter part of the climacteric, are caused by a reduction in the secretions of the vaginal walls. They can produce irritation and increase the likelihood of vaginal infection, and can also at times make sexual intercourse slightly uncomfortable.

Other problems

Medical research has shown that only the end of monthly periods, and subsequent hot flashes and vaginal dryness, are symptoms that are definitely caused by menopausal changes in a woman's body. Other troublesome physical problems—including brittle bones (osteoporosis), swollen ankles, headaches, palpitations, and dizziness—may be connected to the body's adjustment to a change in the level of estrogen, but so far, the link is not clear.

Hormone levels

At the menopause estrogen levels drop virtually to zero, leading to hot flashes, vaginal dryness, and a greater risk of developing osteoporosis. While it is true that both the hot flashes and the vaginal dryness can be corrected by prescribed hormones, women are advised to discuss the risks and benefits with their doctor. Recent studies suggest that taking hormone replacement therapy (HRT)

▼ *Despite many unpleasant physical changes, menopause can be a time when a woman pursues her own interests; the possibilities are endless and she may discover hidden talents.*

can increase the risk of heart attacks, strokes, and breast cancer. The effect of menopause on women is complex, and the psychological elements may be just as important in causing symptoms as the hormonal ones.

Dealing with difficulties

The help that is appropriate for menopausal problems depends on the nature of the difficulties (if any) a woman experiences. HRT does relieve vaginal atrophy, but this symptom can also be relieved by creams containing estrogen. Once it ceases to be a problem, discomfort during sexual intercourse is removed, too.

Like vaginal atrophy, osteoporosis is a condition that may appear later in the climacteric in older women. The reduction in estrogen levels can cause the thinning of the body's bone structure; and this, in turn, can lead to stiff joints, backaches, and an increase in the risk of broken bones. Again, hormone replacement therapy can help, but should be considered only if there are other risk factors for osteoporosis, and other treatment cannot be used.

There are more than 250,000 hip fractures in the United States each year in women over age 65, many of which are debilitating, and about 40 percent of women have osteoporosis of the vertebrae (often leading to vertebral fractures) by age 80. Hormone replacement therapy may prevent osteoporosis and related fractures but the risks involved must be considered in relation to any benefit gained.

It would be a mistake to think of hormone replacement therapy as the one solution to the whole range of menopausal difficulties, since the connection of such problems with estrogen levels is not yet clear. It is encouraging for women to be able to feel that there are a number of solutions to the problems and that some of these are in their own hands.

Questions and Answers

My periods began when I was 10 years old. Does this mean that my menopause will come early?

No, there is no rule about this. It makes no difference whether a woman's periods start early or late: the time of menopause is not related to this.

I am 52 years old and still have my periods. Is there something wrong?

Not at all. Menopause tends to occur sometime between the ages of 45 and 55 years, so you are well within the normal range. Some women's periods continue into their late 50s.

I went through the menopause at 48 without any difficulty, but now I seem to have lost all interest in sex. Will this come back in time?

Have you lost interest in anything else, apart from sex? It may be that you are depressed, and this could result in your feeling a lack of pleasure in sex—and perhaps life in general. Perhaps you no longer find your husband or partner sexually attractive, and you may need to think seriously about your relationship to see whether anything is wrong. Another possible explanation for your lack of pleasure in sex is that intercourse has become difficult or uncomfortable. If this is so, talk to your doctor, who may be able to prescribe an estrogen cream that would help.

My wife is normally an optimistic, calm person; however, since she has started going through menopause, she has become an irritable worrier. This is having a serious effect on our relationship. Will she ever be herself again?

The change you have noticed occurs at menopause but need not be caused by it. If your wife is normally a calm person, it is unlikely that a change in hormones would alter her personality. Talk to her about the change you see in her; perhaps that will help.

CHANGING EFFECTS OF HORMONES

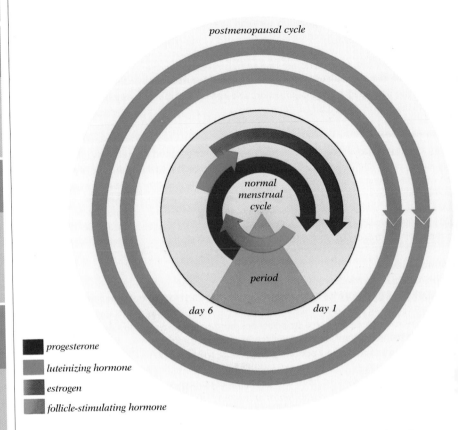

postmenopausal cycle

normal menstrual cycle

period

day 6   day 1

■ progesterone
■ luteinizing hormone
■ estrogen
■ follicle-stimulating hormone

▲ *During the normal menstrual cycle, follicle-stimulating hormone (FSH) and luteinizing hormone (LH) activate the ovary's production of progesterone and estrogen. During menopause FSH and LH are still made, but the ovaries do not respond.*

Heart disease

Heart disease is the leading cause of death in women over the age of 60. Menstruating women have low rates of heart disease, but its incidence increases rapidly after menopause. This suggests a protective role for estrogen. The ability of HRT to protect against postmenopausal heart disease has been the subject of several large studies. The largest study involved 48,000 women, and found that those who took estrogen were half as likely to have major heart disease.

However, recent studies have thrown doubt on the results of the research. In 2002, the U.S. Women's Health Initiative found that the risks were greater for women taking HRT, whereas early research reported that HRT protected women against heart disease and thrombosis. This has reversed accepted thinking and risks and benefits must be weighed with caution.

A full, active life

Menopause is a natural stage in life, not an illness. A tranquil disposition is an ally in almost all situations in life, especially in those having to do with health. However, for those women who tend to be more anxious, a substitute is activity: a woman whose life is busy and satisfying will not only gain confidence but will have little time to dwell on the changes her body is going through.

Far from narrowing the scope of a woman's life, menopause can put new pleasures and achievements within her reach. She no longer has to take into account the inconvenience, and often the pain, of monthly periods. She also need have no more worries about contraception and childbirth. She has come to a turning point in her life and has acquired a new freedom to choose what she will do with the future.

See also: Estrogen; Menstruation; Ovaries; Pituitary gland

Menstruation

Questions and Answers

I am 13, and I worry about getting my first period. What should I do?

Menstruation is a completely natural part of being a woman. The more you understand about your period, the more comfortable you will feel about menstruating.

I become constipated just before my period. Why does this happen?

High levels of progesterone in the blood tend to make the intestine less mobile at this time, causing constipation. Make sure you eat fresh fruits and vegetables and drink plenty of fluids; if this fails, talk to your doctor.

Can I become pregnant during my period?

This is the least fertile part of the cycle, but pregnancy can occur.

After giving birth, can I become pregnant before I have a period?

Yes; 78 percent of women release an egg from an ovary before their first period after giving birth.

Since I started taking the Pill, I have suffered from premenstrual tension. Is it possible to cure this without stopping the Pill?

The symptoms may be relieved by taking vitamin B6 (pyridoxine). Your doctor will advise you further.

I stopped taking the Pill so that I could start a family; but now I get painful periods. Can this be cured without taking the Pill again?

Painful periods can be caused by a substance produced by the lining of the uterus. Medication that blocks its production may cure or help the pain, so consult your doctor about it.

Menstruation is the natural loss of blood and cellular debris from the uterus. It usually occurs at monthly intervals throughout a woman's reproductive life, and is commonly referred to as a period.

The time from the first day of one period to the first day of the next is called the menstrual cycle. During this cycle a woman's reproductive organs undergo a series of changes, during which an egg is released from one of the ovaries and travels to the uterus. If the egg is fertilized by a sperm, it is nourished by secretions from the cells lining the uterus until it burrows its way into the lining, where it is nourished from the blood supply. If the egg remains unfertilized, the lining of the uterus is shed in the menstrual flow, allowing a new lining to grow ready to nourish the next egg.

The menstrual cycle

This intricate cycle of activity is controlled by a center in the brain called the hypothalamus, which acts as a menstrual clock. The clock operates through a small gland called the anterior pituitary gland, situated at the base of the brain. This gland releases several hormones, two of which are particularly important for reproduction: follicle-stimulating hormone (FSH) stimulates the growth and maturation of eggs in the ovary, and luteinizing hormone (LH) stimulates the release of these ripened eggs. One egg matures fully during each menstrual cycle; this egg is surrounded by hormone-producing cells, and together these are called a Graafian follicle. The main hormone produced by the Graafian follicle is estrogen, and during the cycle a surge in estrogen production stimulates the growth and formation of glands in the uterus lining. This surge also changes the secretions at the neck of the uterus, making it easier for sperm to swim into the uterus and meet the egg.

Approximately 15 days before the period is due, the pituitary gland releases a large amount of luteinizing hormone, which stimulates the release of the egg from the follicle about 36 hours

▲ *There is no basis for myths about not bathing or not washing hair during menstruation. Even swimming while using a tampon is perfectly all right.*

If I forget to take the Pill, I have a period the next day. Have I caused some damage?

No; the bleeding is similar to the false period you get when you stop taking the Pill for seven days between sets (a true period occurs by natural cycle without chemical interference). If you forget to take your pill for longer than the safe time recommended by the manufacturer, you must take other contraceptive precautions until you have finished the packet.

I bleed from the vagina after intercourse. I have been told that I have an erosion on the neck of my uterus. Is this causing the bleeding?

An erosion is a harmless extension of the uterus lining onto the vaginal surface, but it can often cause bleeding after intercourse. However, there are other possibilities, such as a polyp or growth at the neck of the uterus.

My grandmother tells me that she has started having periods again. Is this possible?

This bleeding cannot be a true period. It may be caused by a growth in the uterus, which could be cancerous, so she should see a doctor as soon as possible. However, the bleeding is more likely to be the result of a condition called atrophic vaginitis, in which the vaginal walls have become thinned, dry, and sore. This is easily treated by applying an estrogen cream or a slightly acidic gel.

I had a D & C because my periods are heavy, but it does not seem to have cured the problem. Can anything else be done for me?

A D & C (dilatation and curettage) is not usually performed to cure a menstrual problem. The main reason for a D & C is to exclude the possibility of a serious disease in the uterus before treatment. Since you are still suffering from heavy bleeding, you should consult a gynecologist.

later. The egg then travels down a fallopian tube into the uterus. Fertilization usually takes place in the fallopian tube.

The cells in the ovary that had formed the Graafian follicle now undergo changes, which include taking up fat. They are now referred to as the corpus luteum. They still produce estrogen but now also produce a hormone called progesterone. Progesterone has two main functions in the menstrual cycle: the first is to alter the mucus at the neck of the uterus, making it too thick for sperm to swim into the uterus; the second is to make the glands lining the uterus secrete a fluid that will nourish the newly fertilized egg.

If the egg is not fertilized, the corpus luteum degenerates. Small blood vessels in the area go into spasm so that cells lining the uterus no longer receive oxygen and die. They are then shed, together with some blood, as menstruation, and the cycle is complete.

Duration of the menstrual cycle

The normal menstrual cycle can vary from 21 to 90 days, although emotional problems such as stress may affect the menstrual clock, and stringent dieting may stop periods entirely. Women who take the Pill—although they do not undergo the same cyclical changes as described above—have a cycle of 28 days, the lining of the uterus being shed in the same way as it would be at menstruation when they stop taking the pills for a seven-day break each month.

Menstrual flow

Average blood loss during menstruation is between 0.7 and 2.7 fluid ounces (20–80 ml) over a few days. The amount can vary, and it is not really understood why periods are heavier at certain times than at others. Women who take the Pill tend to lose less blood. Very heavy periods (menorrhagia) can be caused by inflammation of the uterus; the use of intrauterine contraceptive devices; certain endocrine diseases, such as an underactive thyroid gland; growths called fibroids, which increase the surface area of the uterus lining and make it irregular; and menopause.

Menstrual flow sometimes has an unpleasant odor, largely as a result of the action of bacteria on the blood and cells that have been expelled from the uterus, so it is important to pay particular care to personal hygiene. If a tampon is left in the vagina for too long or forgotten completely, it can lead to a foul-smelling vaginal discharge.

Blood loss from the vagina is not always due to menstruation. Bleeding may also occur at the neck of the uterus if the tissue is very weak, from a polyp in the uterus, or from a cancer. A woman should consult her doctor if she experiences bleeding after intercourse or between periods.

Premenstrual syndrome

Women are subject to huge hormonal changes during each menstrual cycle, and some experience premenstrual syndrome (PMS), which may occur up to 14 days before the period. The symptoms vary but may include unpleasant mood changes, bloating, tenderness of the breasts, muscle stiffness, and sleep disruption—all of which disappear shortly after menstruation finishes. PMS usually coincides with the time when progesterone production is at its maximum, and there is some evidence to suggest that PMS is due to insufficient progesterone production. Various self-help measures can be tried, such as avoiding stress, eating little but often, and avoiding excessive amounts of sugar and caffeine. Diuretics and painkillers may also help, and in extreme cases a doctor may prescribe antidepressants or the Pill.

Period pains and heavy bleeding

There are two types of menstrual pain (dysmenorrhea). Primary dysmenorrhea usually begins in the early teens and has no obvious cause. It may be caused by the muscle of the uterus going into spasm, or by substances called prostaglandins, which are released from the uterus lining after ovulation and which make the uterus contract. Painkillers such as NSAIDs, or a hot-water bottle may help. Particularly painful periods can sometimes be cured by taking the Pill or drugs that block the production of prostaglandins. In the past, many women had a dilatation and curettage

▶ *The chart shows the changes that take place on each day of the menstrual cycle. Triggered by follicle-stimulating hormone (FSH) and luteinizing hormone (LH) from the pituitary gland, an egg matures and is released from one of the ovaries. Hormone levels then alter, secretions in the neck of the uterus change, and body temperature rises. If the egg is not fertilized, the lining of the uterus is shed around the 28th day.*

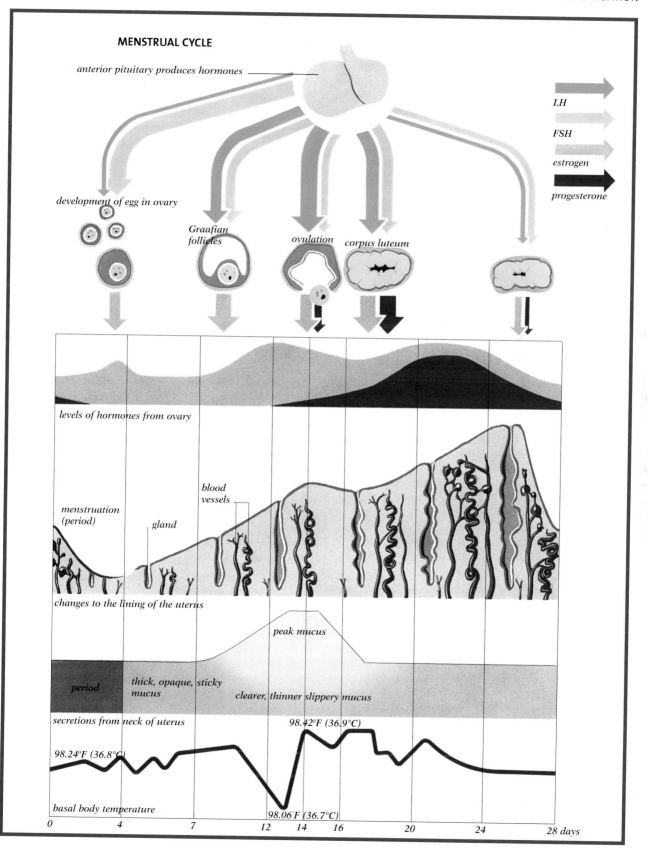

MENSTRUAL CYCLE

anterior pituitary produces hormones

LH

FSH

estrogen

progesterone

development of egg in ovary

Graafian follicles

ovulation

corpus luteum

levels of hormones from ovary

menstruation (period)

blood vessels

gland

changes to the lining of the uterus

peak mucus

period

thick, opaque, sticky mucus

clearer, thinner slippery mucus

secretions from neck of uterus

98.42⁰F (36.9°C)

98.24⁰F (36.8°C)

basal body temperature

98.06°F (36.7°C)

0   4   7   12   14   16   20   24   28 days

(D & C) to cure the problem, with the idea that the stretching of the neck of the uterus made expulsion of the menstrual blood easier. However, there is no evidence to support this theory, and painful periods often return a few months later.

Secondary dysmenorrhea usually affects women aged 20 to 40 and may be caused by conditions such as fibroids, chronic infection of the reproductive organs, or the diseases called endometriosis and adenomyosis—in which pieces of the uterine lining become attached to other organs or to the middle muscular layer of the uterus. Surgery may be necessary in the last two cases, but this is rare.

An unusually heavy or painful period could also mean that the woman is having a miscarriage, especially if the period is a few days late. She should seek her doctor's advice if this happens, because a D & C may be necessary to clean out the uterus.

Menstrual disorders

PROBLEM	CAUSE	TREATMENT
Menstruation does not start at puberty	Genetic abnormalities; hormonal problems	Special investigation
Menstruation stops	Pregnancy; emotional problems; excessive dieting; heavy exercise	Usually returns spontaneously, but may need treatment. If periods are infrequent for more than six months, get medical help.
Periods absent or irregular, in irregular amounts	Premature menopause	None, or hormone replacement medicine
	Disease in genital tract or hormone system	May need evaluation and treatment by a specialist
	Liver disease (liver is unable to break down estrogen)	Difficult to treat
Infrequent light periods	Overactive thyroid	Treat thyroid disease
Irregular periods	Failure of ovulation; most common during puberty	No treatment strictly necessary, but hormonal treatment may be given
Irregular periods, often heavy	At menopause	Dilatation and curettage to exclude serious disease; birth control pills
Irregular periods; possibly mild obesity, hirsuitism, acne	Unclear	Birth control pills to regulate periods; hormonal therapy
	Multiple cysts in ovaries	Removal of diseased ovary
Infrequent periods; secretion of milk from breasts; infertility; headaches; visual problems	Tumor of pituitary gland	Bromocriptine tablets or removal of tumor
Bleeding between periods	Taking the Pill incorrectly	Taking the Pill regularly
Vaginal bleeding between periods	Uterine or vaginal polyp	Removal of polyp
Occasional heavy periods	Fibroids	If bleeding is severe, removal of fibroids (myomectomy)
Heavier periods and a little bleeding between periods	Insertion of intrauterine contraceptive	Usually settles within one to two months; consult your doctor if you continue to bleed between periods or get pain in abdomen
Bleeding after sexual intercourse; heavy prolonged periods; bleeding between periods	Could indicate cancer of uterus	Radiotherapy or surgical removal of uterus; or combination of both treatments if cancerous
Heavier periods and lower abdominal pain	Pelvic infection	Antibiotics
Heavy regular periods	Blood abnormalities stopping clotting	Treatment of blood disease
	Underactive thyroid	Treat thyroid disease

MENSTRUAL DISORDERS

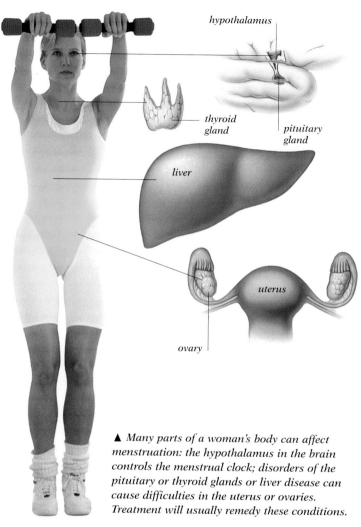

▲ *Many parts of a woman's body can affect menstruation: the hypothalamus in the brain controls the menstrual clock; disorders of the pituitary or thyroid glands or liver disease can cause difficulties in the uterus or ovaries. Treatment will usually remedy these conditions.*

Menstrual problems

Many menstrual problems resolve themselves, but if they persist a woman should consult her doctor. For example, a woman who has heavy periods may become anemic; anemia is diagnosed with a simple blood test and is usually cured by taking iron tablets.

When periods do not occur and pregnancy is ruled out; the condition is called amenorrhea. This takes two forms: primary amenorrhea, when menstruation fails to start at puberty; and secondary amenorrhea, when the periods stop. Both types may be the result of hormonal imbalances caused by factors such as stress or depression, or by extreme weight loss and excessive exercise—as is common in women runners. Primary amenorrhea may also be caused by a chromosomal abnormality, a hymen that is completely intact, or the absence of a uterus. Secondary amenorrhea may be caused by polycystic ovaries or a pituitary tumor.

Infrequent periods are generally of no concern. In an established menstrual cycle, periods may be uneven in quantity or prolonged, or may occur at differing intervals—a pattern of menstruation described as metrorrhagia. However, a diagnostic D & C may be necessary if the periods suddenly change, becoming either very heavy or very light.

Effects of contraceptives

During the reproductive years some menstrual problems may be due to a woman's choice of contraceptive. The Pill may cause bleeding between periods, but this usually settles down within a month or two—if not, a different Pill containing a higher dose of the class of hormones called progestogens will often cure this problem.

Women who use an intrauterine device (IUD) may find their periods become a little heavier than usual, although many are prepared to tolerate this for the sake of the convenience of this form of contraception. However, if the IUD causes severe discomfort it may have to be removed and a different contraceptive tried.

When to see a doctor

Four problems are of particular importance and justify an early visit to the doctor. These are heavy, prolonged periods that occur at the usual time (menorrhagia); bleeding after intercourse; vaginal bleeding between periods; and vaginal bleeding after menopause. Although all of these symptoms are usually caused by minor problems, they may be early signs of cancer of the uterus, which can be cured if it is treated early.

In younger women serious diseases of the reproductive system are uncommon. After examination (which may include a vaginal check) has conclusively established the cause, most problems, such as heavy or painful periods, can be treated with drugs. However, as women become older, menstrual disorders are more often associated with diseases of the reproductive organs. Diagnostic tests may be performed to exclude this possibility; otherwise the condition may be treated with synthetic sex hormones.

During menopause periods become infrequent before they finally stop; this is normal and needs no further treatment. Heavy periods are common at this time, and doctors may prescribe hormone pills or carry out a D & C to control their severity. This treatment does not always work, however, and a hysterectomy (removal of the uterus) may be suggested as a final alternative. A woman then has to decide if her symptoms are severe enough to warrant such a major operation or if she is prepared to tolerate her heavy periods until the onset of menopause.

The intricacies of the menstrual cycle are still not entirely understood by medical researchers and doctors, but a doctor will usually be able to diagnose a problem by performing a pelvic examination and doing a Pap test. Many menstrual conditions can be successfully treated with synthetic hormones or other drugs, and some obvious conditions can be dealt with by surgery.

See also: **Cells and chromosomes; Estrogen; Hormones; Hymen; Hypothalamus; Liver; Menopause; Ovaries; Pituitary gland; Prostaglandins; Sperm; Thyroid; Uterus; Vagina**

Metabolism

"Metabolism" is the collective term for thousands of chemical processes that occur continuously within the body to allow an individual to grow, to survive healthily, and to reproduce.

Questions and Answers

Is it true that if you have a high metabolic rate, you won't gain weight; but if you have a low metabolic rate, you will?

Generally this is true, but weight gain or loss depends on energy input or food intake as well as on energy expenditure or metabolic rate. An excess of input over expenditure will result in weight gain, and more expenditure over input will result in weight loss. If you have a low resting metabolic rate, the food you eat is burned up slowly, so you may have to eat less or exercise a lot to avoid weight gain.

Can metabolic disturbances cause mental illness?

Metabolism is the buildup and breakdown of the body's tissues and the biochemical reactions that these changes involve. In this sense it is not related to mental illness. However, a number of metabolic diseases are linked to mental disturbances. An underactive thyroid gland will reduce metabolism and can cause retardation of the mental processes. Serious metabolic disturbances from severe vitamin B deficiencies can lead to brain damage with mental disturbances.

If you have a fever, does your metabolism increase?

Increases in temperature stimulate chemical reactions in the body, which can result in a rise in the basal metabolic rate (rate of heat produced by the body's cells). A fever can be dangerous because increased heat production by the body's cells may cause a further rise in temperature and heat production. The increased temperatures stimulate sweating, so that the temperature usually stabilizes just above normal until the cause of the fever subsides.

The complex processes that help keep the body functioning normally are efficiently controlled by body chemicals called enzymes and hormones. Enzymes influence chemical conversions so that necessary substances are made available to body cells, while hormones control activities such as growth and the utilization of energy reserves. Efficient metabolism, which is essential to normal body functioning, requires a suitable supply of raw materials; thus a bad diet, overeating or undereating, the heavy consumption of alcohol, and even certain drugs can cause troublesome or potentially dangerous metabolic disturbances.

Metabolism is the product of two processes called catabolism and anabolism. Catabolism is the breakdown of carbohydrates, fats, and proteins to glucose and amino acids to provide energy to the cells for the purpose of renewing cell structures. The energy released is converted into useful work through muscle activity, and some is lost as heat. Anabolism involves the constructive processes by which food materials are adapted to be stored as energy, used to repair or make new cells, or used for reproduction and for defense against infections.

Body inputs and outputs

In a growing child or adolescent, the energy input derived from the breakdown of the food consumed outweighs the energy output to provide for the energy requirements of growth. In adults any excess of energy intake is converted into fat; conversely, excess energy expenditure results in weight loss. All energy content and output is measured in kilojoules (1 kilojoule is equivalent to approximately 4.2 kilocalories).

Metabolic rate

The metabolic rate, the amount of energy a person burns up, has two components. The energy a person expends when he or she is completely at rest is called the basal (resting) metabolic rate, and consists entirely of energy needed to keep the body at a basic level of functioning. An average-size man has a resting metabolic rate of about 7,110 kJ./day (1,700 kcal/day) and an average-sized woman has a rate of about 5,850 kJ./day (1,400 kcal/day), but the rate can vary by up to 1,460 kJ./day (350 kcal/day) either way between different adults. The second component of metabolic rate consists of the amount of energy expended in muscular activity and varies from close to zero for someone perfectly motionless to 2,920 kJ./hour (700 kcal/hour) for someone

▶ *The body burns up energy in kilojoules (1 kJ. = 4.184 kcal), and more energy is expended with more exercise. The rate at which energy is expended is called the metabolic rate; it varies between men and women and between individuals. An average woman burns 210–315 kJ. (50–75 kcal) an hour relaxing; 1,040–1,250 kJ. (250–300 kcal) an hour doing strenuous housework; 1,250–1,670 kJ. (300–400 kcal) an hour playing tennis; and 3,340–4,180 kJ. (800–1,000 kcal) an hour walking up stairs.*

HOW FOOD IS USED IN THE BODY

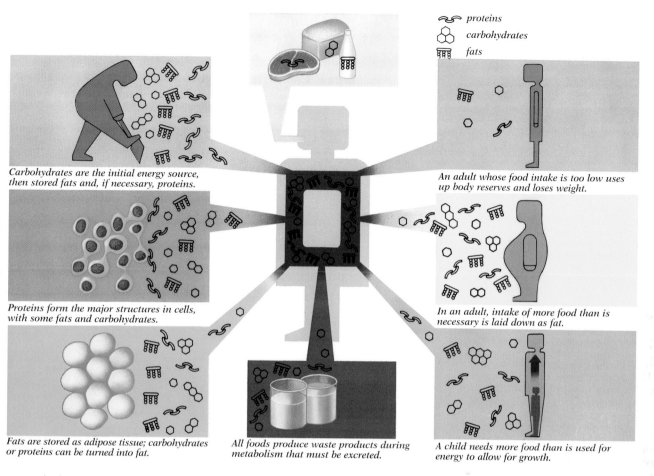

proteins
carbohydrates
fats

Carbohydrates are the initial energy source, then stored fats and, if necessary, proteins.

An adult whose food intake is too low uses up body reserves and loses weight.

Proteins form the major structures in cells, with some fats and carbohydrates.

In an adult, intake of more food than is necessary is laid down as fat.

Fats are stored as adipose tissue; carbohydrates or proteins can be turned into fat.

All foods produce waste products during metabolism that must be excreted.

A child needs more food than is used for energy to allow for growth.

engaged in heavy work such as laying bricks, or a strenuous sport such as racket ball, football, fast swimming, or jogging. Because of these different components of metabolic rate, people vary widely in how much energy they burn up in a day. A woman engaged in light secretarial work may expend about 8,360–10,460 kJ./day (2,000–2,500 kcal/ day), while a man engaged in heavy manual labor may expend 16,730–20,900 kJ./day (4,000–5,000 kcal/day). This is why some people must eat much more than others to provide needed energy.

Much of the body's energy requirement is provided by the breakdown of carbohydrates such as bread and potatoes into sugars. The most common sugars obtained from food are glucose, fructose, and galactose. These are first transported to the liver, where fructose and galactose are converted into glucose. Some of the glucose in the liver is used for energy, and some is stored as glycogen. Once the glycogen storage areas are filled up, the glucose is converted into fat. Stored glycogen or fat can be converted back to glucose to meet extra energy requirements. Cells obtain energy from glucose by breaking it down into a substance called pyruvic acid. The energy released by this process is temporarily stored as a high-energy compound called ATP.

Fat and protein breakdown

Fats and proteins are an important part of the food a person eats, and if carbohydrate intake is too low for energy requirements, fats (and perhaps proteins) may be used as an energy source.

When carbohydrate sources of energy run out, fat molecules are split again into glycerol and fatty acids, which are catabolized separately. Glycerol is converted in the liver into glucose and metabolized as described above. Proteins contained in the diet are broken down into amino acids, which are required for growth, and the enzymes needed to accelerate each cell's metabolic processes.

Many hormones regulate the metabolism of food; they act to maintain a steady state of important chemicals, increasing the production of a chemical if its level falls too low but directing the chemicals into a catabolic or anabolic pathway if its level rises too high.

Metabolic disorders

Many metabolic disorders are caused by an enzyme deficiency at birth, which can cause toxins to accumulate. Disturbances in hormone production also cause metabolic disorders. Diabetes, for example, is caused by decreased production of the hormone insulin by the pancreas. Without insulin, body cells cannot absorb and break down glucose, which then accumulates in the blood. Instead, fat stores are broken down to provide energy, but these cannot be fully catabolized either, and toxic by-products accumulate and poison the body. The disease is treated with insulin.

See also: **Enzymes; Hormones; Insulin; Liver; Pancreas**

Mind

Questions and Answers

Why is the phrase "he's out of his mind" used?

The activities of a person's mind produce patterns of behavior that are usually consistent enough for others to recognize them as typical of that individual. When he or she acts contrary to these patterns, it is not unreasonable to think that the person must have broken away from the mind's control; thus the individual is out of his or her mind.

Can someone really control my mind?

There are examples in which people seem to have lost their free will and are at the command of others—religious cults, for example. However, that involves the use of certain psychologic techniques over a period of time.

Does the mind stop working when we go to sleep?

No. Even if we regard the brain (rather than the mind) as being responsible for all the automatic processes—such as breathing, digestion, and so on—while we are asleep, the fact that we dream shows that the mind is very active. Some research has shown that in addition to anything else the mind may do during dreams, it also uses this time to rearrange and consolidate recent memories.

Do animals have minds?

Many species do. Any activity that is not just an automatic response to an outside event shows the workings of a mind. Cats, dogs, and monkeys are all able to solve problems, and dolphins can even teach each other tricks. In some cases they appear to be able to do so under conditions in which demonstration and copying are impossible. This implies that dolphins have a real language.

Each person has a unique and invisible possession—his or her mind. What it is, where it is, and how it works have intrigued human beings for thousands of years; there is still an immense amount to be discovered about the mind.

▲ *There are many examples of apparent "mind over matter"; however, claims that some people are able to bend metal by the power of the mind are rejected by most scientists.*

The mind is the brain in action. A brain can exist without a mind, but as far as is known, a mind cannot exist without the brain in which it has its being. The brain may supply all the information that enables us to perceive the surrounding world—processing and storing data, and supplying concepts and ideas. How each individual selects that information, reacts to it, feels about it, and thinks about it, however, involves the operation of the mind.

The activities of the mind

In many ways one person's mind is very much like another's. Virtually every mind learns quickly at an early age to recognize objects in the external world and—more important—to make sense of those objects. The mind learns that an object is still there even when it is not in view, and that it will (in most cases) have the same properties when it is seen again as when it was last in view. A young child's mind learns that things actually stay the same size when the child gets closer to them, in spite of what the eyes say. Inhabiting a body that is at first unwieldy and helpless, the mind soon learns to control it and to be independent of it. It discovers how to organize itself and how to think about itself, and in doing so it develops memory, reason, and logic.

▶ *Optical illusion is one of the deceptions practiced by the mind. It can give us evidence of a world in which the laws of gravity and perspective do not apply.*

The unknowable mind

Because of the complexity of the nerve-web that makes up a brain and the brain's capacity to develop in unique ways, it seems unlikely that humans will ever know exactly how the mind inhabiting that brain works. Scientists cannot see a thought, cannot trace its electrical passage through the brain, and cannot deduce why an increase in the concentration of a chemical substance in one part of the brain can produce, say, a feeling of anger within a person.

Despite this, the distinction between the abilities of the mind and those of computers is narrowing. It is no longer possible to claim that all aspects of the power and function of the mind exceed those of computers. Any personal computer (PC) can produce new information that has not been previously put into it by processing input information. This is exactly what the human brain-mind does. A human brain that has not been supplied with information via the eyes, ears, nose, mouth, and skin can produce nothing new. Well-designed and well-programmed computers can read handwritten text better than many humans. Small computers can store far more information than any human and can retrieve it far more accurately. A desktop machine can store the entire text of a library of 100,000 books and can find any word in any of them in seconds. Programmed with artificial intelligence software, a PC can perform a wide range of highly sophisticated "mental" tasks better than any human can.

However, in spite of all this, many experts still believe that there is a qualitative difference between the human mind and a computer. The complexity and scope of brain functions make mind more prone to error and to mental disorder. Some people believe that as computers grow in complexity and application and begin to acquire information comprehensively without human intervention, they too will begin to develop errors, neuroses, and even psychoses.

The deceptions of mind

People generally believe that what their mind tells them is correct, because it is the only means by which they can know anything. In fact, there is evidence to suggest that, in certain situations, the mind does deceive. It deceives when people look at optical illusion drawings; it deceives when people press their fingers over closed eyes and see constantly changing patterns of light and darkness that they know do not exist. It deceives when people dream and believe the dream to be real enough so that they are terrified. It deceives when people are hypnotized and then do not feel any pain when stuck with a pin.

Mind and mental disorder

It is through the mind that humans see the world, and so its powers of deception are of special importance in the treatment of mental disorders of the type not caused by actual brain malfunction.

The information the mind gives to a neurotic person is perfectly straightforward. However, he or she feels the anxieties that arise from that information too intensely, remembers them too easily, and reacts to them too sensitively.

▼ *Young children seem to have an inborn capacity for trusting in what their minds tell them—believing, for example, that when they cover their eyes and cannot see anything, they cannot be seen themselves.*

Questions and Answers

Do scientists locate the mind in the brain?

Yes, insofar as anything nonphysical is located anywhere. Destruction by disease or injury of parts of the brain results in the loss of various functions or features of the mind. Many brain and mind functions such as vision, hearing, taste, smell, and word comprehension are localized in the brain, and their sites are known. Memory is a function of most of these sites.

Is walking on hot coals achieved by mind over matter, or is it a trick?

Neither. A combination of factors goes into the ability to walk across hot coals without being injured. These include the protection of harder skin on the soles of the feet, the slow rate of heat transfer through that skin, the vaporization of perspiration from the foot sole, the brief contact of the foot with the coals at each step, and the confidence not to panic and stay still for even an instant.

How powerful is the influence of the subconscious mind?

Modern psychologists are wary of the old idea of the unconscious, arguing that if a person is not conscious of something, it can't be measured or even proved to exist. However, the power of repressed desires can be just as strong as any desire we freely admit—if not stronger.

My wife and I often think of the same thing without having spoken a word. Does this mean that our minds are in communication?

People who live together develop similar habits of thought. If you both independently pick up a stimulus, reminding you both of the same thing, you will both then speak about it. You may also communicate an idea by quite ordinary means: by humming a tune or making a gesture, which acts as a cue without either of you realizing you have picked it up.

The hysterical person's mind sees the world as paying too little attention to that person, and as a result the mind clamors for attention. Anxious depressives, on the other hand, believe—wrongly—that the world has paid them lots of attention and has judged them unworthy.

What is psychotherapy?

Psychotherapy is intended to correct the wrong interpretations produced by the mind. It may correct them by supplying fresh evidence and asking the mind to notice the good results of acting upon that evidence—this is the technique called behavioral therapy. Another method—psychoanalysis—is used to find out how the mind came to make the wrong constructions and associations. There is no suggestion that the brain's physical activity and information supply are at fault; only the interpretation of that information is in question.

Unusual states of mind

Using electronic equipment that is able to measure the patterns of electrical activity in a person's brain, scientists have become aware of the many different types of mind activity that can go on in the brain. When people go to sleep, the pattern of brain waves becomes different from that recorded while they are awake. When they begin to dream and the mind becomes very active, the pattern changes again. During dreaming the pattern of brain waves is almost the same as that shown during wakefulness, in spite of the fact that people are at that point most difficult to rouse from sleep. Mind and brain at this stage are virtually awake, but the body stays asleep.

▲ *During rituals people can tolerate physical pain if that is what is required. For the duration of the ritual the extent to which the participants feel it is limited by their minds.*

When someone is hypnotized the brain-wave pattern is that of a person who is awake. The body, too, is often awake, especially when it is commanded by the hypnotist to perform an active task. Hypnosis is a state of intense concentration of the mind on the information provided by the hypnotist, so that this is regarded as exclusive of almost all other sensory input. It is for this reason that such information is so influential.

Mind control

Hypnotism is an area in which the mind of one person seems to be controlling the mind of another. It is also possible for a person to achieve an unusual control over his or her own mind so that it can perform feats of mental skill. Some people develop a specialized ability to memorize vast quantities of information. For example, actors who work in a repertory company where the plays change once a week memorize their lines for a whole play and repeat them each night while learning next week's material during the day.

Another mental skill is the ability to do two things at once. Many people believe that they can do this, especially when one of the tasks is automatic—such as driving a car while holding a conversation. However, careful testing nearly always reveals that the person is actually switching attention very quickly from one task to the other and back again, leaving the unattended task to freewheel until attention is returned to it a few seconds later. Only occasionally is it possible to perform two tasks at once. For example, a test of an experienced musician found that she could read music by sight and play it on the piano while repeating a passage of prose being relayed to her through headphones.

Mind over matter

Instances of mind over matter fall into two groups. The first contains examples of ways in which the mind can influence what is sensed or done by its owner. In the second group are instances of the mind's apparently influencing objects or people at a distance.

Many examples exist of the way in which the mind can control bodily activity and sensation. It exerts a great deal of control over the extent to which pain is felt, for example, often blotting pain out completely. Boxers seldom feel the pain of their injuries until after the fight is over; someone injured in a car or train crash may be too shocked to feel any pain, however horrific the injuries.

In spite of the many reported instances of telepathy, clairvoyance, and telekinesis (the ability to move objects by willpower), science has found no examples of these phenomena that can be produced under experimental conditions. In addition, the events themselves cannot be reproduced to order—sometimes they happen, sometimes they do not—and this makes the reality of such happenings suspect. There have been many reports of out-of-body experiences from people who have been near death and who have afterward remembered looking down on themselves from above. However, there is no way of checking the truth of these reports: they are entirely dependent on the honesty and observational powers, under very unusual circumstances, of the people concerned and therefore the phenomena is not able to be verified scientifically, nor can it be considered as evidence.

Dolphins not only possess a mind but are thought to be able to communicate with one another using language.

See also: Brain; Head; Memory; Nervous system

Mouth

The mouth is extremely versatile: it plays a key role in speech, the taking in and digestion of food, and breathing. Despite a great deal of wear and tear over the years, it is resilient, too, and recovers quickly from most common ailments.

The human mouth is a cavern containing the tongue and the teeth. It is the point of entry to the body for food. Its opening is surrounded by the lips, and at its exit it links with passages leading to the digestive tract and lungs. Because of this connection with two vital body systems, the mouth is involved in both digestion and breathing. It is also concerned with speech.

The lips and the lining

The lips give the mouth its expression. They are made up of muscle fibers interspersed with elastic tissues and copiously supplied with nerves; these give the lips their extreme sensitivity.

The lips are covered in a modified form of skin; this is structurally halfway between the true skin that covers the face, and the membrane that lines the inside of the mouth. Unlike true skin the skin of the lips has no hairs, sweat glands, or oil-secreting glands.

The mouth is lined with a tissue called mucous membrane; this contains glands that produce the slightly sticky clear fluid known as mucus. The continuous secretion of these glands keeps the inside of the mouth permanently moist, and this is helped by the action of the salivary glands. The membrane lining of the cheeks receives an enormous amount of wear and tear, and the cells have a great capacity for regeneration.

At the top of the mouth, toward the front, is the hard palate; the soft palate is toward the back. The hard palate—formed by the bottom of the upper jawbone (or maxilla)—allows the tongue to press against a firm surface so that food can be mixed and softened.

The softness of the soft palate is an essential feature, since this enables it to move upward as food is swallowed, and so prevents the food from being forced up into the nose, the passages of which enter the back of the mouth.

Hanging down from the center of the soft palate is a piece of tissue called the uvula. Its true function is a mystery, but some anatomists think that it helps to form an effective seal at the top of the air passages when food is swallowed, so preventing choking.

▲ *The lips are the entry to the mouth, and are extremely sensitive to both physical sensations and differences in temperature.*

STRUCTURE OF THE MOUTH

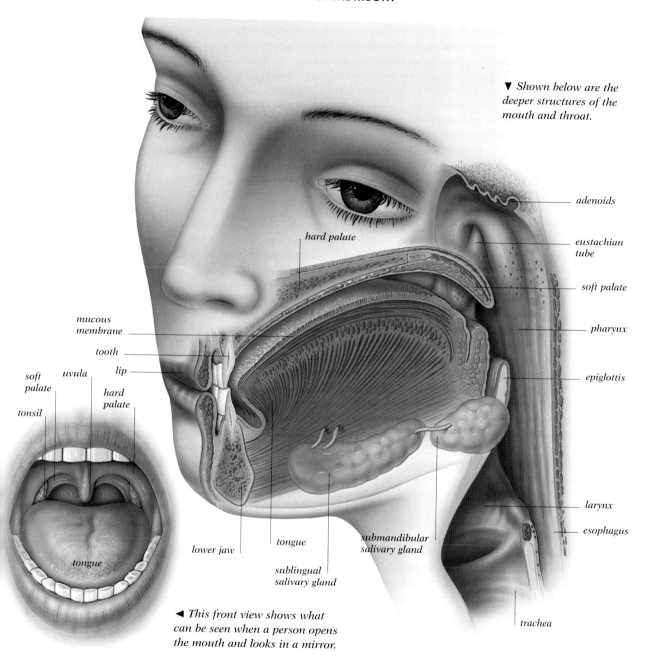

▼ *Shown below are the deeper structures of the mouth and throat.*

adenoids

hard palate

eustachian tube

soft palate

mucous membrane

pharynx

tooth

soft palate

uvula

lip

hard palate

epiglottis

tonsil

lower jaw

tongue

submandibular salivary gland

larynx

esophagus

tongue

sublingual salivary gland

trachea

◄ *This front view shows what can be seen when a person opens the mouth and looks in a mirror.*

The mouth in action

When a mouthful of food passes between the lips and into the mouth, it is chewed in the mouth. Chewing involves the teeth and the muscular tongue, which moves food around the mouth.

In the mouth, food is moistened by both mucus and saliva, a watery fluid that pours into the mouth from three pairs of salivary glands. The largest of these are the parotid glands, which lie in front of and below the ears. The sublingual glands lie below the tongue, and the submandibular glands under the lower jaw.

Saliva is produced when food is put into the mouth, and also in response to the sight, the smell, or even just the thought of food.

Saliva is slightly alkaline and contains the biological catalyst (enzyme) amylase. This enzyme begins the breakdown of starchy foods that is the first step in chemical digestion.

When food has been chewed and softened, it is then swallowed. During the swallowing process food is pushed by the tongue to the back of the mouth into a cavity called the pharynx. The tongue pushes against the hard palate, and the soft palate dilates to let food pass. At the same time the epiglottis, a flap of tissue at the entrance to the airway, is pushed across to seal off the passage to the lungs and to prevent choking. Contractions of the esophagus finally push food into the stomach to continue digestion.

Questions and Answers

My toddler daughter holds each mouthful of food in her mouth for ages and then often spits it out rather than swallowing it. What should I do to try and stop her from wasting her food?

Toddlers often enjoy testing their parents, but the more irritated you get, the more likely she is to continue doing it. Try to make sure that your daughter is really hungry at mealtimes and has not filled up on snacks. Also, do not give her more food than she is likely to want; she can always have a second helping. Above all, do not lose your temper, since this will make matters worse.

Why is it that my mouth always feels so dry when I'm nervous, tense, or frightened?

When you are anxious or afraid, your adrenal glands start to pump out the powerful substance adrenaline into your bloodstream to prepare your body for a fight or flight response. One of the many actions of adrenaline is to divert blood away from areas such as the mouth to the muscles. This lack of blood, which contains a large proportion of water, is the main reason why a dry mouth and feelings of nervousness go together. You can solve the problem by taking sips of cold water when you are experiencing these feelings or emotions.

Why does my baby drool so much? Is he just teething or is there something else wrong?

The salivary glands that lubricate the mouth become stimulated to overproduce when a baby is teething, and so this probably does explain the drooling. Also, infections of the respiratory passages increase the amount of mucus produced by the glands in the mouth, and this may add to the drooling problem.

To protect your baby's clothes, put a simple terry cloth bib on him during the day, not just at mealtimes. If you suspect that he has some kind of infection, take him to a doctor.

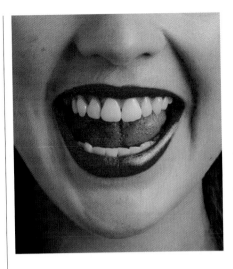

▲ *The letter A is made with the mouth open and the tongue on the bottom teeth.*

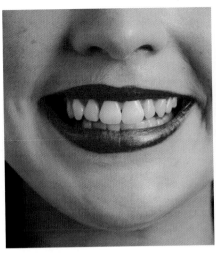

▲ *An E is made with the tongue near the roof of the mouth while the lips smile.*

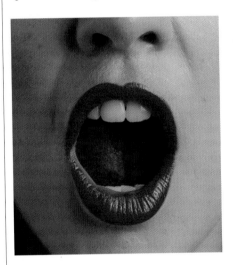

▲ *The I sound is made with the lips wide open and drawn back from the teeth.*

▲ *With the U sound the lips are pursed together. Tongue position is not vital.*

As a person swallows, particularly if he or she is going up a steep hill, or ascending or descending in an aircraft, the ears are often felt to pop. The reason for this is that the eustachian tubes of the ears lead into the back of the mouth. The purpose of these tubes is to help equalize the pressure between the middle ear and the atmosphere, to prevent the eardrum from bursting; swallowing does just this.

Breathing and speech

In breathing, the mouth is involved as an airway. However, breathing through the mouth is not as safe as breathing through the nose, because the air is not as thoroughly moistened or temperature-corrected. Also the mouth, unlike the nose, does not contain any hairs (cilia) to trap dust and other potentially irritating particles.

The mouth is intimately involved in speech because it helps to shape sounds emanating from the voice box, or larynx. Making the sounds of consonants such as K or T, for example, demand that the air coming from the larynx is cut off sharply by the tongue and palate. Vowel sounds such as A and E, on the other hand, need no truncation but require certain positions of the tongue and teeth. Each sound in any language is determined by a slightly different movement

Problems of the mouth

CONDITION	CAUSES AND SYMPTOMS	TREATMENT
Aphthous (viral) stomatitis	Small but extremely painful canker sores (mouth ulcers) in the mouth.	See your doctor. He or she may prescribe lozenges to relieve the pain.
Cheilosis	Lips sore, swollen, and inflamed. Corners of mouth red, fissured from licking lips, drooling.	Soothe pain with analgesics. See your doctor; the problem is usually cleared up with antibiotics and/or topical antifungals.
Herpes virus infection	Cold sores, inflamed blisters on mouth and lips. May be associated with common colds, hot and cold weather, and mouth ulcers.	Aciclovir, a prescription medication sold as Zovirax in tablets or in cream form, can stop an attack of cold sores if used early enough.
Drooling (overwatering of mouth)	Causes range from teething in babies to pain from mouth ulcers. Neurologic disorders can cause this also.	Must be treated according to underlying cause. If persistent, or accompanied by severe pain, seek medical help.
Gingivitis	Bleeding and soreness of the gums, especially when teeth are brushed. Bad breath, excess watering of the mouth.	See your dentist as soon as possible.
Halitosis (foul-smelling breath)	Causes include cigarette smoking, indigestion, tooth decay, and poor dental hygiene.	Stop smoking. Try antacids. Practice good dental hygiene. See your doctor.
Acidic diet	White or yellowish patches in mouth, surrounded by red, sore area. Breath may be bad.	Avoid acidic foods such as oranges, tomatoes, and chocolate. Apply ointment containing choline salicylate or something similar, or suck lozenges to numb the pain. If problem persists, see your doctor or dentist.
Stomatitis	Inflammation, swelling, and redness of the lining of the mouth. Breath may smell foul.	If the mouth is very dry, chew gum or suck fruit candy to stimulate flow of saliva. Wash out the mouth two or three times a day with a medicated mouthwash.
Stone (calculus) in salivary duct	Painful swelling in mouth when food is smelled or put into mouth.	See your doctor. Minor surgery may be needed to remove the blockage.
Yeast infection (*Candida albicans*)	Ulcers covered with white spots in mouth. Can be intensely painful. Patients taking prednisone and those who are immunocompromised are at risk.	See your doctor, who may recommend nystatin lozenges or liquid medicine or systemic (oral, e.g. fluconazole) antifungal medications to help kill the fungus causing the problem.

of the lips, tongue, and teeth. The ability of deaf people to lip-read is proof of the mouth's crucial role in speech.

Disorders of the mouth

The mouth is subject to many disorders, but most are not serious and are quick-healing.

Because the mouth has such a rich blood supply, any injury—no matter how slight—tends to bleed copiously. Similarly, the mouth is so richly supplied with nerve endings that any slight swelling inside the mouth feels enormous. This is also because when a person tries to assess the size of a thing by putting it in the mouth, the sensory data from the tightly packed nerve endings is not corrected by visual cues; he or she assumes the object is much larger than it is.

The worst mouth problems are the gross congenital deformities such as cleft lip and cleft palate, in which the two halves of the roof of the mouth are not properly sealed together.

Mouth infections are fairly common. The mouth can be subject to viruses such as those that cause cold sores and ulcers, bacteria that cause tooth decay and gum infections, and fungi that cause the white-spotted sores that are typical of a yeast infection.

The mouth, like much of the digestive system, is full of bacteria. In fact, it is estimated that there are as many as a few hundred different species of bacteria in the human oral cavity.

These bacteria seem to do no harm as long as they stay in the mouth, but if they get into other parts of the body, particularly into a skin wound, they can cause serious infections that take a very long time to heal.

See also: Breathing; Digestive system; Hearing; Mastication; Mucus; Nose; Saliva; Taste; Teeth; Tongue; Tonsils

Movement

Questions and Answers

Why can't I control my shivering when I get cold?

Shivering is a reflex action—a movement of the skeletal muscles that happens automatically. It is caused by the autonomic system, which maintains a constant body temperature. Whenever the heat lost from your body is greater than the heat being produced by body cells, your temperature drops slightly. The fall is signaled to a part of the brain called the hypothalamus; this automatically sends messages to your skeletal muscles, making them shiver. The shivering causes your muscles to increase their heat production and so brings the body temperature back to normal.

Does everyone have the same potential for fine, accurate hand movements, such as those required for handwriting?

In general, yes, although people do vary in the speed with which they pick up such skills. This variation is in part related to different people's ability to form mental associations between the signals being sent from the brain to the hand and sensory signals coming from the eyes.

How do we control the power of our movements?

Each of our muscles is supplied with many motor (movement-producing) nerve fibers; these branch out and come into contact with a number of muscle cells. Each time a motor nerve fiber fires, it causes several muscle cells to contract. However, one nerve firing is not enough to contract the muscle; this requires the firing of several nerves together or in rapid succession. The power with which a muscle contracts is varied by increasing the number of nerves firing or the rate at which they fire. Signals from the brain control this.

Most people take movement for granted and never think of the sophisticated mechanics involved. Even one small step involves a complicated series of activities in the brain, nerves, bones, muscles, joints, and sense organs.

The most familiar movements are those people are able to stop and start at will as a result of a decision and program of action formed within the brain. Such voluntary movements can involve transport, for example, walking; carrying out a piece of work, such as turning a screwdriver; gathering information, such as turning the eyes to look at a page; and a variety of other actions.

Reflexes are less obvious movements that are sometimes responses to an unpleasant stimulus—blinking when a bright light shines in the eyes or rapidly withdrawing the hand from a hot stove, for example. Reflexes help protect the body from damage and occur automatically, bypassing the conscious areas of the nervous system. They involve much simpler nerve circuits than those involved in voluntary movements.

▲ *To perform their complicated movements, gymnasts require fine coordination of numerous muscles, together with great skill, balance, and concentration.*

HOW MOVEMENT IS SIGNALED

▼ *Voluntary movement is initiated by signals sent from the cerebral cortex of the brain; they are sent from the opposite side of the brain to the side of the body where movement will occur. The signals travel down the spinal cord and along motor nerves to the skeletal muscles. Some cause the muscles to contract; others inhibit motor nerves and ensure relaxation of the antagonistic muscles.*

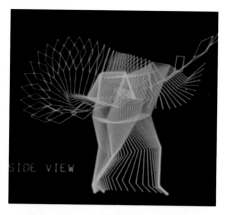

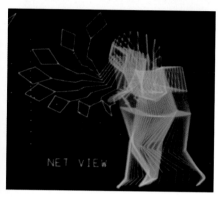

▲ *A computer scan illustrates a tennis player practicing strokes.*

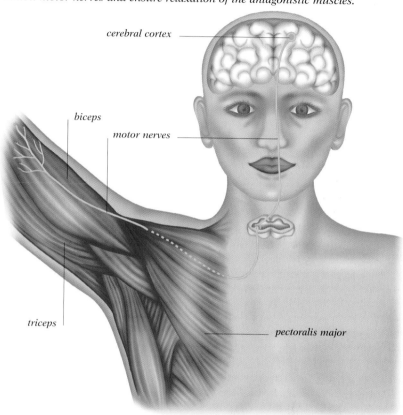

cerebral cortex

biceps

motor nerves

triceps

pectoralis major

Both voluntary movements and reflexes work through the activity of what are called skeletal muscles—muscles that are attached at one or both ends to bones of the skeleton. Skeletal muscles are activated by the nerves of the somatic nervous system. Internal muscles, however, such as the heart muscle and the muscles in the walls of the stomach, are controlled by a separate system called the autonomic nervous system. These muscles can also bring about movement, but their activity is not normally controlled at will; for this reason they are called involuntary muscles.

Skeletal movement

There are three basic moving parts of skeletal movement: bones, muscles, and joints.

Bones: If bones did not exist, humans would be capable only of wobbly, jellylike movements. Two important features of the skeleton are, first, that it provides a framework for the rest of the organs so that the body has a definite shape and structure; and second, that it is not a rigid structure but a jointed one, and therefore allows the parts of the body to be moved into a large number of useful positions. To fulfill these two functions, the components of the skeleton must be rigid. Bones are the hardest and most rigid parts of the human body.

Muscles: Bones are moved in relation to each other by skeletal muscles that are like springs attached at various points to the skeleton. They are made of parallel bundles of cylindrical cells packed into a spindle shape, with a thick central belly and two tapering ends, or tendons, each attached to a bone. Each muscle cell contains many long fibers, called myofibrils. The myofibrils contract when the muscle cell is activated by a nerve. At any given time many muscle cells will be contracting, providing the muscle with a degree of tension, or tone. When enough muscle fibers contract, the whole muscle shortens, reducing the distance between its attachment points so that two or more bones move in relation to others.

Individual muscles can act only to shorten, not to lengthen, the distance between two attachment points; they can pull but not push. For movement in the opposite direction, therefore, another muscle must be activated. For example, the biceps in the upper arm can flex the elbow, but extension is brought about by another muscle, the triceps, on the underside of the upper arm. Muscles such as biceps and triceps are called antagonistic because, in a way, they work against each other.

Joints: Joints are the points at which bones meet and may be the points at which movements of the bones take place. Some joints, such as those between the bones in the skull, permit no movement, whereas others allow a range of movements in various directions.

Questions and Answers

My son keeps shrugging his shoulders for no reason. Why?

This is an example of a tic, which is literally a nervous habit. Your son probably picked up the habit by mimicking an adult who frequently shrugged his or her shoulders to express feelings. However, if such a movement is overused, it can become an unconscious habit. The tic may indicate that your son is a little insecure, but the habit is likely to disappear with time. If it does persist or gets worse, you should consult your doctor, since it may be a symptom of a more serious nervous disorder.

Can we consciously control the movements of internal organs such as our heart or stomach?

The movements of the heart muscle and the smooth muscles in the walls of the stomach are usually thought to be involuntary. However, some experiments suggest that we may be able to learn to control these organs through a process known as biofeedback. For this to work, the person practicing the technique has to receive information about whatever it is he or she is trying to control; the person would have to listen to his or her heart beating if trying to control its rate.

Some people use biofeedback techniques to control stress-related conditions such as high blood pressure and migraines.

My husband suffers from Parkinson's disease. Can you explain why he seems so clumsy?

The movements of the skeletal muscles are controlled by part of the brain, which varies the balance between the stimulating and the inhibiting nerve activity. In people with Parkinson's disease, the part of the brain concerned with controlling this balance is impaired. The unwanted stimulatory messages are no longer canceled by the accompanying relaxing messages, so producing the unintentional, uncoordinated movements that are characteristic of Parkinson's.

▲ *This stroboscopic (time lapse) photograph of a gymnast performing on the beam demonstrates the range of mobility and flexibility the human body is capable of.*

Most movable joints, such as the knee, elbow, and hip, are called synovial joints. They contain a space between the bones, called the synovial cavity, which is filled with a lubricating fluid, and the bone ends are covered with a smooth, tough, layer of gristle called cartilage.

The movements that can take place at a joint depend partly on the shape of the two bone ends and partly on the various muscles, tendons, and ligaments in and around the joint. The most versatile joints are those such as the hip or shoulder joint, where one bone end is ball-shaped and the other forms a socket; this arrangement allows the limb to rotate so that it can move to the front, back, side, down, and up, thus turning inward or outward.

Ligaments are the slightly elastic fibrous bands of tissue that connect bones together at the joints. They cannot contract like muscles, but once they are pulled taut by bone movement, they cannot be stretched further, so they set a limit to the possible movements at a joint. All synovial joints are crossed by muscles or ligaments; both play a part in binding and stabilizing joints.

Nerve control of movement

Skeletal muscles are activated by nerves in the spinal cord—the bundle of nerve fibers running down from the brain through a channel in the vertebral column (spine). These motor nerves split into several strands where they enter a skeletal muscle, the fibers of each strand making contact with a different muscle cell. When a signal passes down the motor nerve, each tiny nerve end releases a chemical called acetylcholine that causes the individual muscle cells to contract.

The most simple reflex movements occur through direct activation of motor nerves by signals arriving at the spinal cord from sensory receptors (the nerves that receive sensations). For example, in the knee-jerk reflex, a tap just below the kneecap is sensed by receptors inside one of the tendons that run across the knee joint. These receptors send signals to the spinal cord; these signals activate motor nerves running from the spinal cord to the thigh muscles. As a result, the thigh muscles rapidly contract, and the lower part of the leg jerks forward.

Voluntary movements

Voluntary skeletal movements, in contrast to involuntary reflex movements, are set off by signals sent down the spinal cord from the brain. Some of these signals act to stimulate

REFLEX MOVEMENT OF THE KNEE

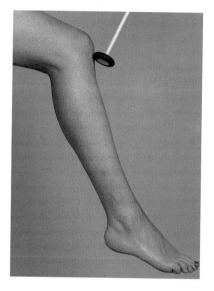

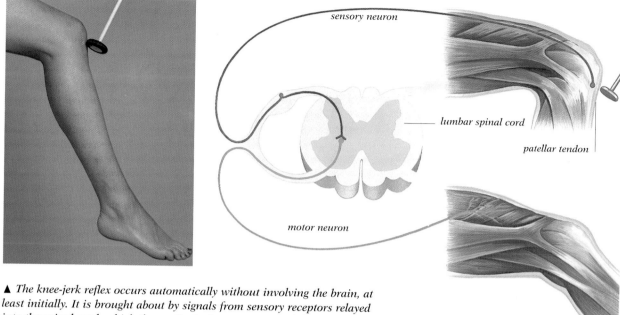

sensory neuron

lumbar spinal cord

patellar tendon

motor neuron

▲ *The knee-jerk reflex occurs automatically without involving the brain, at least initially. It is brought about by signals from sensory receptors relayed into the spinal cord, which then activate the motor nerves (above right).*

▲ *To activate well-rehearsed movements, the brain recalls the program of nerve activity and sends signals down the spinal cord to activate the relevant motor nerves.*

particular motor nerves, and others act to inhibit them; in this way a pattern is worked out that will cause some muscles to contract and other antagonistic muscles to relax.

In the case of a well-practiced movement—for example, picking up a glass and taking a drink—the program of nerve activity that will set off the particular movement is stored in the brain's memory. When a person decides to make that movement, the brain recalls this program from memory and sends the pattern of signals down the spinal cord to activate the motor nerves in the correct order. The movement is monitored by sensors, called proprioceptive receptors, in the muscles and joints. These give information about how much the muscles are contracting and the amount of joint movement. This information is analyzed by the brain, which also keeps track of the body's posture while the movement is taking place and adjusts it accordingly.

Movement of the internal organs

The muscles that move the internal organs are different in structure and method from skeletal muscles.

One of these muscles, the heart muscle, is continuously active. It consists of a matrix of muscle cells that contract in a way similar to the contraction of skeletal muscles. However, contraction is made possible not by motor nerves but by pulses from a special pacemaker tissue within the heart. These pulses pass over the heart about 72 times every minute, causing the heart to contract and to squeeze its blood content into the blood vessels.

The muscles in the walls of organs such as the stomach and intestines are called smooth muscles. They consist of spindle-shaped cells arranged haphazardly and supplied with motor nerves belonging to the autonomic nervous system. Stimulation spreads in a wave over several cells, and this wavelike action helps to move food through the intestines.

CONTROLLING THE HEARTBEAT

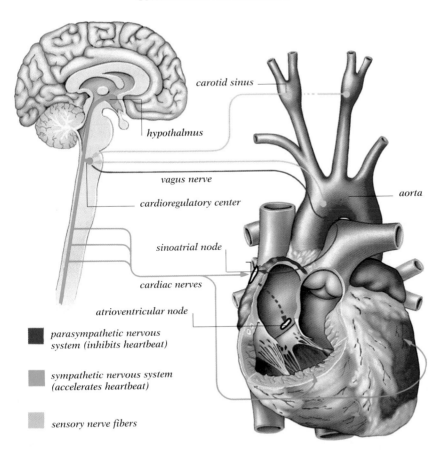

carotid sinus

hypothalmus

vagus nerve

cardioregulatory center

sinoatrial node

cardiac nerves

atrioventricular node

aorta

■ parasympathetic nervous system (inhibits heartbeat)

■ sympathetic nervous system (accelerates heartbeat)

□ sensory nerve fibers

◄ *Sensory nerves send information to the regulatory center in the spinal cord. The heartbeat is then adjusted accordingly by the sympathetic or parasympathetic system.*

The autonomic nervous system, which controls the activity of the heart and smooth muscles, consists entirely of motor nerves arranged in relays from the spinal cord to the various muscles. Some of these nerves bring about muscle contraction; others work to relax the muscle and slow movement.

The role of the brain

The whole autonomic nervous system is controlled by the hypothalamus (an area of the brain). This receives information about any variations—for example, in the body's chemical makeup—and adjusts the autonomic system to bring the body back to normal. If the body's oxygen levels are lowered by exercise, for instance, the hypothalamus instructs the autonomic nervous system to increase the heart rate in order to supply more oxygenated blood.

> *See also:* **Autonomic nervous system; Bones; Brain; Heart; Hypothalamus; Joints; Ligaments; Muscles; Nervous system; Posture; Shivering; Spinal cord**

▼ *The smooth muscles of the intestines move food along with a wavelike action.*

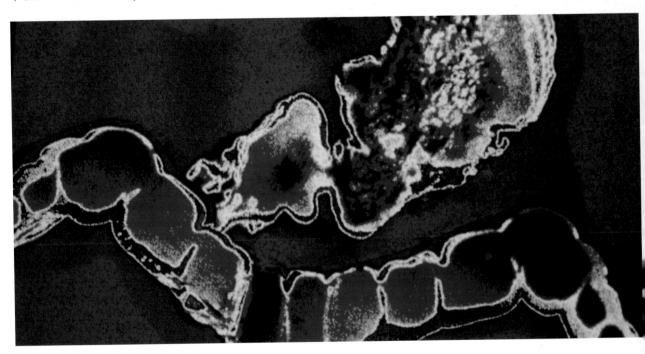

Mucus

Mucus is a clear, sticky lubricant that is secreted by glands in the mucous membranes that line the body cavities. It is normally produced in small amounts, but infections can cause a dramatic increase in its volume.

Questions and Answers

Why do my nose and eyes run when I have attacks of hay fever?

Hay fever is an allergic response to pollen in the atmosphere, and the mucous membranes of the nose and eyelids become irritated and swollen, causing the body to produce more mucus.

I find sexual intercourse painful, even after a lot of foreplay, because my vagina doesn't produce much lubrication. Is this a physical or a psychological problem?

The problem could be due to either of these issues. Perhaps the easiest way to overcome it is to use K-Y jelly for lubrication during intercourse. You may want to see your doctor if lubrication does not help.

Is sinus trouble caused by mucous membranes?

Sinusitis is an inflammation of the mucous membranes of the sinuses in the head. They become swollen, blocking the small passages that lead to the nose. A chronic runny nose may develop. Prolonged attacks of sinusitis may lead to chronic sinusitis.

Is saliva the same thing as mucus?

No. Saliva is a mixture of mucus and fluid containing the digestive enzyme amylase that begins the digestion of starches.

Does indigestion have anything to do with the mucous membranes?

"Indigestion" is a term used to describe almost any stomach complaint. Gastritis, an inflammation of the lining of the stomach, can cause indigestion. This often results from excessive smoking, alcohol consumption, or use of anti-inflammatory drugs.

The mucous membranes are made up of two layers. The outer layer is like skin, and underlying this is a tougher layer that sometimes contains muscular fibers. Mucous membranes line the respiratory tract from the nose to the lungs, the alimentary canal from the mouth to the anus, the reproductive tracts, and the urinary tract. Each membrane is modified to its particular task in the body.

In the respiratory tract, air first comes into contact with mucous membranes in the nose, which are richly supplied with blood vessels to warm and moisten cold air. When an infection takes hold in the body, the mucous membranes themselves suffer and respond by producing more mucus. The mucous membranes of the respiratory tract are involved in colds, bronchitis, hay fever, inflammation, sinusitis, and influenza.

The gastrointestinal tract

In the gastrointestinal tract, mucus is first produced by glands in the membranes lining the mouth. This mucus mixes with saliva and enzymes, and coats the ball of food, making it slippery and easy to swallow. The stomach is lined with a

▲ *When a person has a cold, the swollen, infected nasal membranes produce excessive mucus, which results in a runny nose.*

mucous membrane containing glands that secrete gastric juices and digestive enzymes that also break down food. Problems associated with disorders of the stomach membrane include gastritis and ulcers.

From the stomach food moves down into the small intestine. The membrane lining this tube is covered with fingerlike projections (villi) that absorb nearly 95 percent of the nutrients in the food. The remnants of the food pass into the large intestine, where water is absorbed across the membrane.

The female reproductive tract

The vaginal lining is more like skin than a mucous membrane. It contains no mucus glands but is lubricated by mucus from the cervix and from two external glands called Bartholin's glands. These secretions of mucus increase with sexual excitement and at different times during the menstrual cycle. Mucous membranes are even more closely connected with menstruation, because the endometrium, the lining of the uterus, is also a mucous membrane.

Higher up in the genital tract the fallopian tubes are also lined with mucous membrane. The mucus secreted by this lining is essential for the survival of the fertilized egg following conception.

See also: Alimentary canal; Cervix; Digestive system; Membranes; Menstruation; Nose; Stomach; Urinary tract

Muscles

My son has torn a leg muscle playing football. Will the injury trouble him in the future?

The best way to avoid trouble later on is to build up strength in the injured leg gradually, exercising and training carefully for the first two to three weeks until he can do things without pain in the muscle. The tendency of recurrent injury to the same muscle is much greater if people rush back to being fully active.

What actually happens to a muscle as it is built up and strengthened?

Adults don't increase the number of fibers in a muscle as they train it; the muscle grows bigger and stronger through an increase in the size of each individual fiber. Whether children can make more muscle fibers by exercise or whether the number of fibers is set by genetic makeup is unclear.

Why can't women build up their muscles as much as men can?

Men are more heavily muscled than women because of their genetic makeup. We can all develop muscle by training with weights and other exercises, but most men start off with much more muscle than women. The hormonal makeup of the body also has a lot to do with this: male sex hormones are more anabolic than female sex hormones.

Should my son eat lots of red meat to build up his muscles?

Red meat is the muscle of animals, so it does contain the building blocks for human muscle. However, the protein from which human muscle is made is formed from basic material, and your son's muscles will develop just as well if he consumes protein from fish or vegetable sources.

All the movements of the body, from the twitch of an eyebrow to a high jump, are made possible by muscles: people depend on muscles in everything they do. To stay in shape, muscles need only a reasonable amount of exercise.

▲ *The balance and coordination required of a gymnast demand enormous skill, strength, and almost perfect muscle control.*

There are three different types of muscle in the body. The first, called voluntary muscle, is under the control of the brain. Together with the bones and tendons it is responsible for all forms of movement, from a smile to running up a flight of stairs. The second is smooth muscle—so called because of the way it looks under a microscope; it is concerned with the involuntary movement of internal organs. The third is the cardiac muscle, which makes up the main bulk of the heart.

Types of muscle

Voluntary muscles are also known as striated (striped) muscles because they have a striped appearance under a microscope. They produce their effect by shortening their length, a process called contraction. They have to be able to produce sudden, explosive contractions—of the kind that the muscles of the legs make when a person jumps into the air—and to maintain a constant contraction to keep the body in a particular posture.

Involuntary—or smooth—muscle is made up of long spindly cells instead of fibers and is not under the conscious control of the brain. This type of muscle is responsible for automatic contractions such as those that occur in the process of digestion, when peristaltic waves of smooth muscle contractions move food along the intestine.

Cardiac muscle, found only in the heart, has a structure similar to that of voluntary muscle.

The location of muscles

Voluntary muscle is distributed throughout the human body, making up a very large proportion of its weight—up to 25 percent, even in a newborn baby. This type of muscle controls the movement of the different parts of the skeleton, from the tiny stapedius muscle—which works on the stapes, a minute bone in the inner ear—to the huge gluteus maximus, which forms the bulk of the buttock and controls the hip joint.

The muscles are attached to the skeleton by means of tendons. The end of the tendon nearest to the center of the body is called the origin of the muscle, and it is generally shorter than the tendon of insertion at the other end. It is usual for the muscle origin to be on one side of a joint and the insertion to be on the far side, so that, by shortening, the muscle brings about a movement of the joint.

The structure of muscles

Voluntary striped muscle can be visualized as a series of bundles of fibers gathered up to make a complete unit.

The smallest of these fibers—and the basic working unit of the muscle—are the actin and myosin filaments; these are so tiny that

SOME OF THE BODY'S MAIN VOLUNTARY MUSCLES

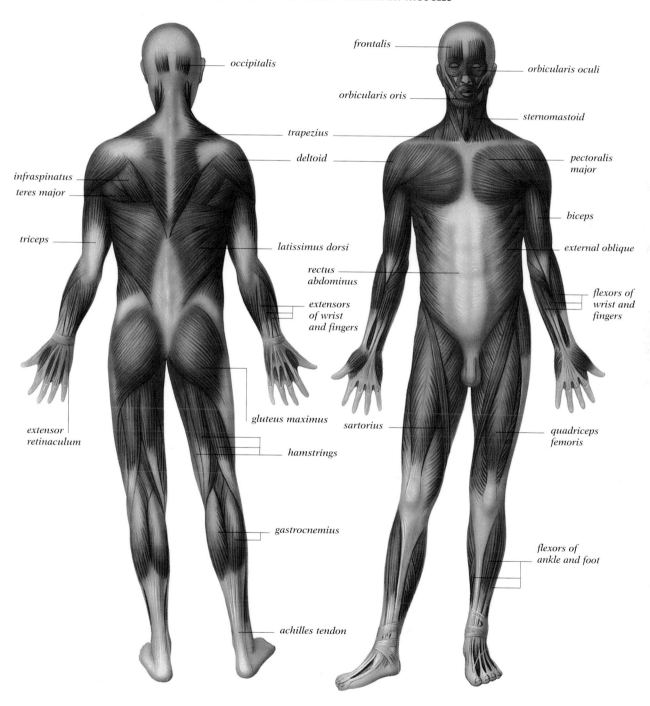

occipitalis

frontalis

orbicularis oculi

orbicularis oris

sternomastoid

trapezius

deltoid

pectoralis major

infraspinatus

teres major

triceps

biceps

latissimus dorsi

external oblique

rectus abdominus

extensors of wrist and fingers

flexors of wrist and fingers

gluteus maximus

sartorius

quadriceps femoris

extensor retinaculum

hamstrings

gastrocnemius

flexors of ankle and foot

achilles tendon

Muscle disorders

DISORDERS	CAUSE	SYMPTOMS	TREATMENT
Pulls, tears, strains, and sprains	All caused by tearing of the muscle fibers, followed by bleeding in and swelling of the affected muscle.	Pain, which may become worse during the day or two after the original injury. Movement is limited.	Cold (ice packs, water) and pressure with bandages help immediately. Later the plan is to get the muscle moving. Heat treatment may help.
PAINFUL MUSCLES WITHOUT INJURY			
Cramps	The muscle goes into spasm. Dehydration, electrolyte abnormalities, certain medications.	Painful tightening of the muscles, often in the calf.	Salt and water replacement are vital when dehydration is the cause of cramps. Muscle relaxants may help.
Polymyalgia rheumatica	Inflammation of the blood vessels supplying blood to the muscles.	Pain and stiffness around the patient's neck, shoulders, and hip regions. Usually affects the elderly.	Low doses of cortisone-type drugs.
Myositis	Painful inflammation of the muscles, many associated conditions. Can be caused by viruses.	Pain and weakness. Blood tests may show that muscle tissue is being broken down.	Treatment depends on the cause; rest alone will improve many types of myositis, but steroid pills may help.
MYOPATHIES—MUSCLE WEAKNESS WITHOUT PAIN			
Endocrine disorders	Endocrine disorders such as underactive pituitary, thyroid, adrenals; also overactive thyroid.	Muscle spasm, stiffness. Weakness.	Treat the underlying hormone problem (over- or underactive).
INHERITED MUSCLE DISEASE			
Muscular dystrophy	Inherited abnormalities in the working of the muscle cell.	Progressive weakness of muscles affecting different areas according to the form of muscular dystrophy.	None, but splints and other aids can be of considerable value.
Hypokalemic paralysis	A low level of potassium in the blood during attacks.	Periodic attacks of paralysis.	Avoiding heavy meals and not exercising before sleep help to prevent attacks. Potassium supplements may also help.
Enzyme defects	Inherited lack of one of the enzymes that control the handling of food and energy by the muscle cell.	Floppiness of the body. In affected babies, delayed walking; a tendency to tire when exercising; cramps.	Avoiding strenuous exercise usually helps.

they can be seen only with the assistance of an electron microscope. The filaments are made of protein and are called the contractile proteins. The muscle shortens in length when all the myosin filaments slide past the actin filaments.

The filaments are gathered into bundles called myofibrils. Between these are the deposits of muscle fuel in the form of glycogen and the energy factories of the cell, the mitochondria, where oxygen and food-fuel are burned to make energy.

The myofibrils are gathered into further bundles called muscle fibers. These are really the muscle cells with cell nuclei along their outside edge, and each one has a nerve fiber coming to it to trigger it into action when necessary.

The muscle fibers themselves are grouped together in bundles in an envelope of connective tissue, a little like the insulation that surrounds the copper strands of an electric cable. A small muscle may consist of only a few such bundles of fibers; by contrast, a huge muscle, such as the gluteus maximus, is made up of hundreds of fiber bundles.

The whole muscle is contained in a fibrous tissue covering, again similar to the insulation around a wire.

STRUCTURE OF VOLUNTARY MUSCLE

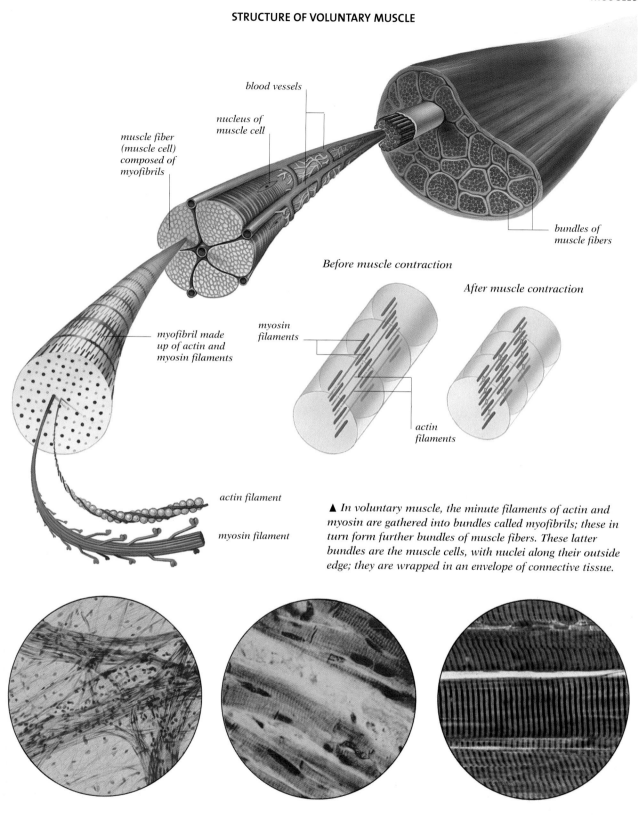

blood vessels

nucleus of
muscle cell

muscle fiber
(muscle cell)
composed of
myofibrils

bundles of
muscle fibers

Before muscle contraction

After muscle contraction

myosin
filaments

myofibril made
up of actin and
myosin filaments

actin
filaments

actin filament

myosin filament

▲ *In voluntary muscle, the minute filaments of actin and myosin are gathered into bundles called myofibrils; these in turn form further bundles of muscle fibers. These latter bundles are the muscle cells, with nuclei along their outside edge; they are wrapped in an envelope of connective tissue.*

▲ *Involuntary—or smooth—muscle is composed of long, spindly cells.*

▲ *Cardiac muscle is made up of fibers in an orderly crisscross pattern.*

▲ *In voluntary muscle, myosin and actin fibers meet like two sets of comb teeth.*

269

The structure of smooth (involuntary) muscle does not show the same orderly arrangement of filaments and fibers as voluntary muscle, which are built into a complicated geometric pattern. However, the contraction of smooth muscle still depends on the myosin filaments sliding past the actin filaments.

The structure of cardiac muscle tissue, when it is viewed under a microscope, is the same as that of voluntary muscle except that the fibers form a crisscross pattern with each other.

How a muscle works

Nerves run down the spinal cord from the motor (movement-controlling) parts of the cerebral cortex in the brain and pass out of the cord into the individual nerves to the muscles. If a muscle is deprived of its nerve supply, it not only loses its ability to contract but also starts to waste away.

There is an area on the surface of the muscle fibers where the nerves are plugged into the muscles. The electrical force of the nerve impulse when it arrives at this junction with a muscle is tiny, whereas the electrical changes that take place in muscles when they contract are, by comparison, quite large—so considerable energy must be supplied.

The transmission of the impulse to contract takes place at a special motor end plate at the point where the nerve fiber meets the muscle fiber. The electrical impulse traveling down the nerve does not stimulate the muscle directly but releases a transmitter chemical called acetylcholine, which causes the muscle to contract.

Filament interaction

The sliding of the myosin filaments over the actin filaments is a complicated process in which a series of chemical bonds between them is continually formed and broken. This requires energy, provided by the burning of oxygen and food-fuel in the mitochondria.

This energy is stored and transferred as a compound called ATP (adenosine triphosphate), which is very rich in high-energy phosphate. The process of muscular contraction is started by a flow of calcium (one of the minerals of the body) in the muscle cells through a series of small tubes—called microtubles—running between the myofibrils.

Contraction

The muscles also contain a set of fibers that registers the force of the contraction. Another set of fibers inside the tendons leading from muscle to bone gauges the stretch. The information is relayed via the nerves back to the brain and is vital in preventing injuries; for example, a cat without such regulating fibers would be able to leap 100 feet (30 m) into the air.

Injuries to muscles

Pulled, strained, and torn muscles are all similar injuries; they all give rise to pain and difficulty in using a muscle. These injuries arise from torn muscle fibers and from bleeding inside the body of the muscle.

▶ *There are about 250 voluntary muscles in the body; many of them must work together to keep a dancer's body poised.*

Muscular weakness

Diseases that lead to muscle problems are rare. Most people who have muscular weakness have a disorder of the nervous system—it is the nervous system that actually controls muscle movement. Muscular weakness may be caused by a stroke, for instance.

Occasionally the muscles themselves are weak; this condition is called myopathy. The most common cause of myopathies is disease of the hormone system, but they can also be brought about by vitamin D deficiency and by some forms of cancer.

The inflammation of the muscles, which is called myositis, can have a viral cause, but more often it seems to result from an abnormal reaction of a person's immune system. Like myopathy, it may be linked with a tumor.

The remainder of the major muscle disorders are inherited. The best-known example of an inherited muscle disorder is a condition called muscular dystrophy.

Muscle health

Good muscle tone is simply the normal firmness of muscles that are used regularly. When they are not in constant use—such as when an individual sustains a serious fracture or is forced to remain in bed for a long period of time—the muscles start to atrophy (waste away). In the process they become weak and flabby; once the person recovers, he or she must build up muscle tone again through physical therapy and consistent exercise.

Regular exercise is necessary for keeping the body in good shape. Running and swimming are particularly good in this respect, since they involve most of the muscles of the body. In addition, most people find it easier to keep up these types of exercise (or another sports activity such as tennis or football), rather than a set exercising regimen.

Weight training

Special weight training that is used for building up extra strength in the muscles is necessary, and probably desirable only for those people who have a particular athletic goal in mind. There is certainly no reason to build up the muscles in this way for the sake of general health; it is much more valuable to increase the efficiency of the heart and lungs by such activities as walking and bicycling.

Early exercise habits

The exercise that children get through ordinary play and activity is usually vigorous enough to give them a good physique. Organized exercise is important, too, as it will encourage them to continue with exercise later in life when health is not so easy to maintain with an adult lifestyle.

See also: **Autonomic nervous system; Digestive system; Heart; Joints; Movement; Nervous system; Posture; Tendons**

Nails

The condition of the nails can reflect someone's general health. Changes in the appearance of the nails can be due to injury, infection, or disease. Good nail hygiene and care are essential to protect the nails from injury.

Nails are made of dead cells that grow from living skin cells in the tissues of fingers and toes. The part of the nail that shows itself is dead and so will not hurt or bleed when it is cut or if it is damaged—but any injury to the living roots can be painful.

Nails can be an indicator of someone's state of health. For example, a change in the appearance of the nails can indicate a deficiency of iron or cysteine.

Infection can occur in the nails, resulting in conditions such as paronychia, a common complaint among people who frequently immerse their hands in water. Such problems can be prevented by regular, thorough care of the hands and feet and by wearing cotton-lined rubber gloves when the hands have to be in water. It is also important to keep nails clean, because dirt can harbor germs that may spread diseases.

How nails are formed

Each nail is made up of a horny, proteinaceous substance called keratin. The visible part of the nail is called its body. The shape of the body can be determined, at least in part, by genetic factors, so nail shapes tend to run in families. The bottom part of the nail, which is implanted in a groove in the skin, is called the root. Overlapping the root are the cuticles (eponychia). These layers of skin cover the white crescent or lunula found toward the base of the nail. The lunula, usually most clearly visible on the thumb, is slightly thicker than the rest of the nail, and so looks white because it hides the blood vessels beneath.

The lowest layer of cells in the skin composing the nail folds is known as the general matrix. The cells of the matrix divide, and the upper ones become thickened and toughened with keratin. When the cells die they become part of the nail itself. If the matrix is seriously damaged, the whole nail will be lost.

Rates of growth

Nails grow an average of about 0.03 inch (0.5 mm) a week, but the actual rate of growth will vary between individual fingers and toes. Studies of nail growth have shown that growth can be slowed down by a hormonal imbalance, by mineral deficiencies, and even by a psychological

ANATOMY OF A NAIL

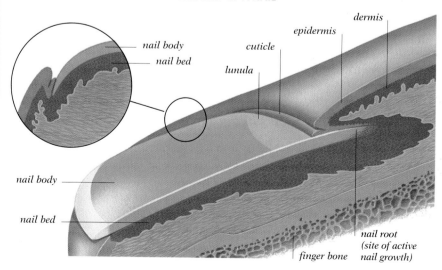

nail body
nail bed
dermis
epidermis
cuticle
lunula
nail body
nail bed
finger bone
nail root (site of active nail growth)

▲ *Dirty, bitten, or ragged nails can become infected. Good hygiene and regular care of the nails will keep them healthy. General care of the hands need take only a few minutes each day and can prevent injury and trauma to the nails.*

disturbance. Poor health is also reflected in the nails, and about 40 different disorders, including anemia, lung disease, heart disease, infection, and hemochromatosis all cause changes in the nails that can help in the diagnosis of disorders.

Functions of the nails

Nails are used mainly to manipulate objects. The living tissue around and beneath them is well supplied with nerve endings, so the nails are also involved in the sense of touch.

Nails also support the tissues of the fingers and toes, particularly when the hands or feet are pressed onto a firm surface when a person is standing and walking.

Nail problems

The most common problem is that the skin around or beneath the nails becomes infected with bacteria. The infection may be caused by incorrect growth of a nail or by a small injury at the side of the fingernail. Acute infections are commonly caused by small pieces of nail, called hangnails, that grow from the side of the nail into the skin, or by bacteria entering skin through minor wounds. This kind of infection, known as paronychia, is easily treated with antibiotics. If pus forms, it may need to be released by

surgical means. Another problem that may occur is a fungal infection of the nail. These can be difficult to treat, requiring many weeks of oral antifungal medicines.

When nails have split horizontally, they can be very difficult to repair. However, a special glue will hold the nail edges in place until the nail grows enough to be filed down. White spots may be due to minor damage from a knock or blow. Ridges in the nails are common in elderly people and can be a sign of rheumatoid arthritis.

Too little iron tends to make the nails concave. This condition, known as koilonychia, is due to softening and thinning of the nail plate. Everyday splitting of the nails does not seem to have any underlying cause. The nails can be treated with nail hardener, but a well-balanced diet and general care should keep them healthy.

The condition known as ingrown toenail does not always involve nail growth into the tissue as is commonly believed. Infected and inflamed tissue along the nail edge can swell over the edge, causing the typical appearance. It is seldom necessary to remove the whole nail. Treatment is directed at the control of the infection in the surrounding tissues. Sometimes, one or both edges of the nail may have to be cut away.

Nail biting

Nail biting can start at the toddler stage and develop into a habit. In older children and adults, the same habit is probably related to anxiety or boredom. Bitten nails look unattractive and can be a source of infection if they are bitten down to the nail bed or beyond

Questions and Answers

One of my fingernails grows very little, if at all, but the others grow normally. Why does this happen?

The cells that produced this nail have been damaged or even destroyed. As a result, the nail is not being continually renewed as it should be. This is nothing to worry about as long as the nail does not fall off altogether, indicating that the matrix from which the nail grows has been destroyed. If this happens, consult your doctor.

Will my son get ingrown toenails if he wears tight shoes?

Shoes that are too tight are a common cause of ingrown toenails, because the soft tissue is pushed onto the edge of the nail and damaged by the shoe. Badly-fitting shoes can also cause a host of other foot problems, so it is not worth taking the risk of letting your son wear shoes that are too small for him.

I am determined to break my nail-biting habit. How can I do this?

Try painting your nails with white iodine, or use a commercial preparation. You could also try treating yourself to expensive nail polish and telling yourself you will use it when your nails grow to a reasonable length. If neither of these measures works, see your doctor about behavior therapy.

My mother is disgusted by the way my children bite their nails and says the only way to stop it is by making them go to bed wearing gloves that are tied to their wrists. Is this a sensible approach?

Most people now agree that it is best not to make an issue of nail-biting. After all, it is not a major failing. Instead, try to persuade your children to overcome the habit by pointing out how nice unbitten nails look and praising them if they succeed in growing their nails. Whatever you do, do not nag them.

Nail biting

If your child is a persistent nail-biter, he or she may be upset about something. Try to find out if there is unhappiness at school or distress caused by some event in family life such as a new house or a new baby. If it just seems to be a habit, try not to nag—this will only make the problem worse. Try to persuade the child to stop nail biting by the promise of a treat or a reward. A daughter may be encouraged to give up nail biting if you give her her own manicure set.

If everything else fails, there are unpleasant-tasting substances, which you can buy from the drugstore to paint on the nails. However, do not worry too much about this minor habit—most children grow out of it.

Some people bite their nails until they bleed and become sore and infected. It is best to avoid making nail biting an issue with children. If the habit is too difficult to break, commercial preparations are available from drugstores. These can be painted on the nails, and they taste so unpleasant that fingers are usually kept well away from the mouth. Adults may try painting their nails twice a day with white iodine, which is also very unpleasant to taste.

If the habit is very ingrained and commercial preparations are not a deterrent, some doctors may recommend behavior therapy. Therapy involves habit-reversal training, which teaches awareness of the habit, how to relax, how to perform a competing response, and that motivation can overcome any habit.

Nail care

Nails should be kept clean so that they do not attract bacteria. Nails should never be filed to sharp points, because this shape weakens them and can cause them to break: a gentle oval is better. Colored nail polish should never be used to cover up dirty nails. Massaging cream or warm olive oil rubbed into the cuticles will aid healthy nail growth, but a diet high in protein and rich in vitamins is the best guarantee of healthy nails. Try to protect the nails by avoiding physical trauma to the fingertips.

See also: **Feet; Hand; Skin**

Neck

Questions and Answers

My son's Adam's apple sticks out a long way at the front of his neck. Is this a symptom of some disease?

Almost certainly not. The Adam's apple is a piece of cartilage that supports the vocal cords. It can be any size, in the same way that ears and noses vary in size and shape.

My wife is having her thyroid gland removed. Why is this being done?

If her thyroid is overactive, the body's metabolic rate may increase, causing heart problems, weight loss, and hyperactivity. Sometimes an overactive thyroid responds to drugs; in other cases surgery is necessary to remove part of it and reduce its action.

My friend broke his neck in a car crash. Will he die?

The extent to which the fracture affects the spinal cord is very important. Severe damage can cause paralysis. If the top part of the cord is torn, death will result. Other fractures of the bones in the neck may be no more serious than any other broken bone.

Can a stiff neck be the sign of a serious disorder?

Perhaps. Consult your doctor if the stiffness persists for longer than a week and does not respond to warmth, massage, and rest. The stiffness could be from arthritis in the neck vertebrae or perhaps from muscle strain in the neck area. Of greater concern is a stiff neck due to meningitis—other symptoms such as severe headache and fever are often present. See emergent medical care in this situation.

The neck not only supports the skull but also contains life-supporting parts of the human body—the vital windpipe, the spinal cord, and important veins and arteries.

The neck connects the chest with the skull. It runs from above the clavicles, or collarbones, upward to the lower jaw at the front and the base of the skull at the back. The neck is built around the seven cervical vertebrae of the spinal cord and has powerful muscles at the back that support and move the head.

At the front of the neck is the cavity that contains the throat, leading downward to the trachea (windpipe) for the intake of air to the lungs. Also in the throat are the esophagus, which takes food down into the stomach; the thyroid gland; four parathyroid glands; the vocal cords; the thyroid cartilage, or Adam's apple; a collection of lymph glands; and the major blood vessels that supply and drain the whole head including the brain—the carotid arteries and the jugular veins. At birth, the neck also contains the thymus, but this gradually shrinks after puberty.

Seven cervical vertebrae are found in all mammals (even giraffes, despite their long necks), and the upper two are the most mobile of those found in the spinal column. The topmost vertebra is called the atlas and joins the skull. It allows the head to nod. The second vertebra is called the axis and allows the head to rotate. The mobility of the head is extremely important.

Humans rely greatly on their hands and feet, coordinated by highly developed binocular vision. It is important, therefore, to be able to move the head without having to make exaggerated movements of the whole body. People who have suffered from a stiff neck know how apparently simple actions such as crossing the road become difficult when they have to change the whole body position quickly.

SKELETON AND MAJOR BLOOD VESSELS OF THE NECK

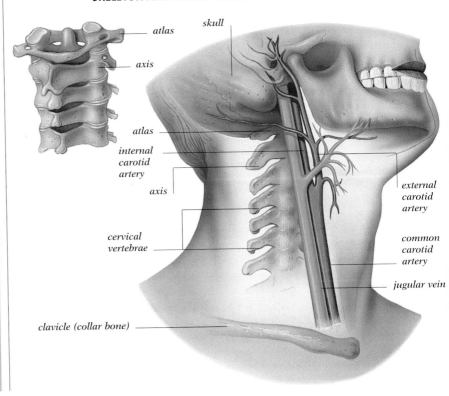

atlas

axis

skull

atlas

internal carotid artery

axis

cervical vertebrae

clavicle (collar bone)

external carotid artery

common carotid artery

jugular vein

IMPORTANT MUSCLES AND NERVE SUPPLY OF THE NECK

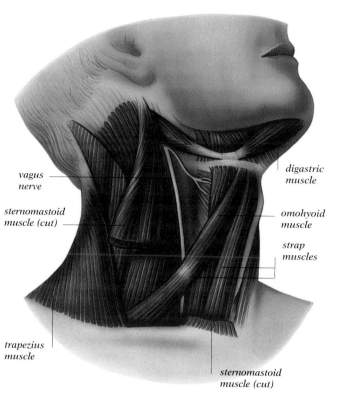

vagus
nerve

sternomastoid
muscle (cut)

trapezius
muscle

digastric
muscle

omohyoid
muscle

strap
muscles

sternomastoid
muscle (cut)

CONTENTS OF THE THROAT

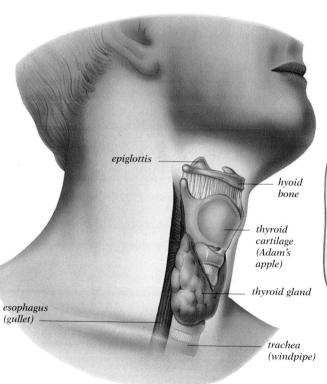

epiglottis

hyoid
bone

thyroid
cartilage
(Adam's
apple)

thyroid gland

esophagus
(gullet)

trachea
(windpipe)

The front of the neck is extremely vulnerable, since it contains so many important organs. The trachea, or windpipe, is at the front of the throat and consists of hoops of cartilage that hold open elastic tissue. It is easy to feel the trachea with the fingers through the skin at the base of the neck. At the upper part of the neck, the windpipe is covered by the thyroid cartilage, or Adam's apple.

Above the trachea lies the larynx, the narrowest part of the windpipe. If the larynx becomes blocked, air cannot get into the lungs, and an operation called a tracheotomy may need to be performed. This procedure involves inserting a pipe into the trachea to let air into the lungs and suck fluid out of the lungs. After the obstruction has been removed, the opening heals.

Not much can go wrong with the trachea, because it is lined with a mucous membrane that contains cells with tiny hairs called cilia, which waft invading germs and dust back up into the throat to be swallowed. If there is an infection of the tracheal membrane (tracheitis), it is usually part of a general infection of the respiratory tract and often accompanies bronchitis.

The vocal cords

At the top of the trachea is the larynx, or voice box, which contains the vocal cords that allow humans to speak. The box comprises a series of cartilages—the thyroid cartilage at the front; a ring of cartilage that forms the top of the trachea at the bottom; and finally, a pair of movable cartilages, the arytenoids.

The vocal cords are stretched across from the thyroid cartilage to the arytenoids, which can move apart to open or close the cords. The vocal cords are generally kept apart, but when a person swallows food they close, and a flap of tissue, the epiglottis, covers the entrance.

CROSS SECTION OF THE THROAT

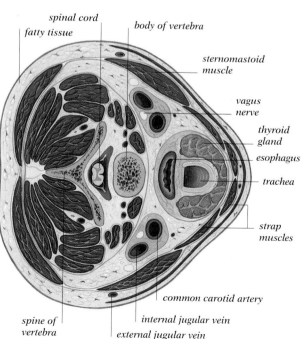

spinal cord
fatty tissue
body of vertebra
sternomastoid
muscle
vagus
nerve
thyroid
gland
esophagus
trachea
strap
muscles
common carotid artery
internal jugular vein
external jugular vein
spine of
vertebra

Questions and Answers

Why do people who live in certain geographical areas seem particularly prone to goiter?

A goiter is a swelling of the thyroid gland. This gland needs a supply of iodine from food or water to produce its hormone, thyroxine. If there is a deficiency of iodine, the thyroid swells up. Certain areas of the world, particularly inland mountain regions, lack iodine in the soil and water, so the inhabitants lack iodine too. In such areas one must use iodized table salt. This type of goiter is called endemic goiter and is five times more common among women than men.

I hurt my neck in an accident and was told that this was a whiplash injury. What is that?

A whiplash injury occurs when the head is suddenly jerked, much as a whip is cracked. This may happen when an automobile suddenly collides with something just in front of it. Even if they are wearing seat belts, passengers' heads are still violently jerked. The injuries sustained may range from two or three days' pain from strained muscles and ligaments to much more serious damage involving the vertebrae and the spinal cord.

Can people lose their voices through shock?

Yes. This is possible in two ways. You may be so taken by surprise by something that you cannot collect your thoughts and thus "lose your voice"; or a longer period of muteness may follow a profound emotional disturbance. This is a psychological problem, like the loss of memory that may occur in similar circumstances.

Why do some people have longer necks than others?

Simply because they have inherited a long neck. We all have the same number of vertebrae in our necks—some are just longer than others.

▶ *A soft cervical collar is typically used on the neck when muscles and ligaments have been strained.*

For speaking the cords are closed but the epiglottis is open and air is passed over them to create speech. Varying the tension in the cords changes the pitch of the voice; the amount of air pushed up from the lungs determines the amount of sound that is made; and intonation, or the sound of the words, is created by the nose and adjusted by the shape of the mouth.

Swallowing

The esophagus lies behind the trachea and is a muscular tube that conveys food from the throat to the stomach. When it is empty it lies flat, but when food passes through it the muscles force the ball of food downward. Food usually passes through in only about five seconds. When vomiting takes place the muscular contractions that force the food down are reversed and food is brought up instead.

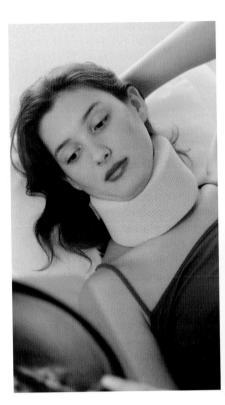

Glands of the neck

People are born with two glands in the neck—the thyroid gland and a set of four parathyroids.

The thyroid is found just below the level of the larynx and can be seen or felt just like the Adam's apple. It has two lobes and is shaped like a butterfly. The two lobes lie just in front and at either side of the windpipe, or trachea, as it passes down the front of the neck. The two lobes are connected by a small bridge of tissue. In an adult the gland will weigh about 0.66 ounce (19 g).

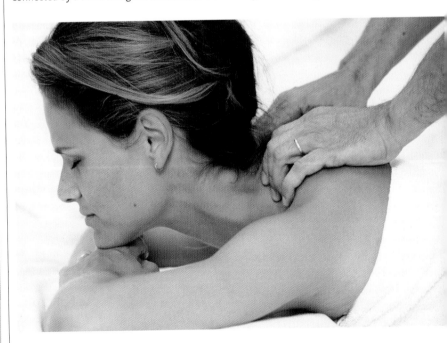

▲ *The muscles in the neck and shoulders can retain a lot of tension and become very stiff and sore. Regular massage can help relax these muscles and release the tension.*

The hormone thyroxine is made and secreted by the thyroid gland. The thyroid requires iodine to function; it is the only organ of the body that requires iodine. Since it is vital for activity, the thyroid is very efficient at trapping all available iodine from the blood. The four parathyroid glands lie behind the thyroid and have one function—the production of a hormone called parathormone, which regulates the concentration of calcium in the blood.

Besides containing particular organs, the neck has important carotid arteries and the jugular veins passing through it that convey blood to and from the brain. The vagus nerve, which is an important part of the parasympathetic nervous system that controls unconscious actions such as breathing, runs close to the internal jugular vein.

Neck problems

Stiffness of the neck can occur after sleeping in a cramped position or in a cold draft, or after some unaccustomed exercise, such as playing tennis or digging in the yard or garden. The pain will often disappear on its own: otherwise two days of rest, massage, keeping the neck warm, and taking anti-inflammatory painkillers such as ibuprofen or aspirin should clear up the trouble.

Other causes of a stiff neck, involving the front and sides mostly, may be the swelling of the lymph nodes that drain the neck. This probably is from an infection of the throat or ears, which should be treated. Pain and difficulty in moving the neck also result from arthritis of the joints between the vertebrae. This effect can occur with rheumatoid arthritis or, more commonly, with osteoarthritis. The arthritic process may cause pressure on the nerves leaving the spinal cord and produce numbness or paralysis in the hands, which are controlled by these nerves.

The voice can be lost by damaging the vocal cords. This can happen after a long bout of shouting—for example, at a football

▲ *Men tend to have larger Adam's apples than women. This photograph shows the characteristic shape—but its size can vary considerably between men.*

game—or after a long singing session. Trauma of this kind can cause a small swelling or polyp. Professional singers sometimes suffer from polyps, but these can usually be easily removed without any permanent damage.

Cancer can affect the larynx and cause hoarseness. Eventually an operation is needed to remove cancerous growths. Sometimes the whole larynx must be removed, but it is possible to learn to speak by using a different method. One way is by swallowing air and releasing it in small bursts to create enough sound to be heard. Almost all cases of cancer of the larynx are caused by smoking.

Laryngitis is an inflammation of the throat that is caused either by a virus or by the inhalation of harmful chemicals. Very rarely, inflammation may be a symptom of tuberculosis.

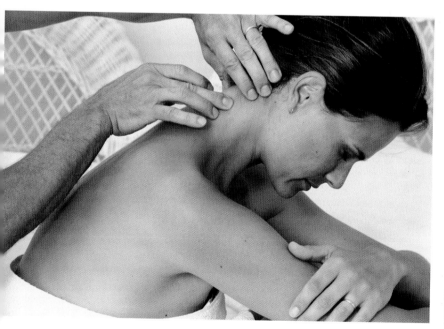

The neck consists of seven cervical vertebrae. Damage to any of these vertebrae, as through a whiplash injury, can cause neck pain that will require treatment.

See also: Arteries; Esophagus; Glands; Larynx; Lymphatic system; Parathyroid glands; Skeleton; Spinal cord; Throat; Thymus; Thyroid; Veins; Vocal cords

Nervous system

My friend says I'm a nervous wreck. Might I have a nervous system disturbance or problem?

This is unlikely—although there are a few physical illnesses that can lead to generalized anxiety.

Your nervousness is probably caused by fear. This may be unconscious—for example, a fear that you are about to lose your job. Being anxious is a normal response to such stresses and need be a cause for concern only if the nervousness is ever-present.

Many forms of anxiety can be relieved by, for example, changing your job, living conditions, or lifestyle, or by seeking medical help or therapy. However, first you may have to identify exactly what it is you are worried about.

Do "pins and needles" have anything to do with nerves?

Yes. If a nerve containing sensory fibers is slightly compressed, its individual fibers may fire off a number of random signals; these signals are perceived by the brain as a tingling sensation.

For example, pressure on the nerves running from your foot up the back of the thigh—perhaps caused by sitting cross-legged—may be felt as the familiar pins and needles in the foot. Continued compression may prevent the nerve fibers from transmitting signals altogether, resulting in a loss of sensation, or numbness.

Can heavy drinking damage nerves?

Yes. Heavy drinking over many years may cause a permanent disturbance in the conduction of signals by nerve cells. It can also speed up the rate at which nerve cells in the brain die, causing mental deterioration. Alcoholics often have vitamin B1 deficiency because of poor diet, and this also disturbs nerve functioning.

Every time a person does anything—literally anything at all—the nervous system is intimately involved at every stage. The nervous system is the body's most complex and important network of control and communication.

▲ *The action of the nervous system allows several activities to be carried out at the same time. This man is using his cell phone and his computer, as well as feeding a baby.*

The nervous system is essential to sight and hearing; the perception of pain and pleasure; the control of movements; the regulation of bodily functions, such as digestion and breathing; and the development of thought, language, memory, and decision making. Basically, the nervous system collects and receives information from the outside world and uses it to frame the body's response.

The working parts of the nervous system include millions of interconnected cells called neurons. Their function is similar to the wires in a complex electrical machine: they pick up signals in one part of the nervous system and carry them to another; here the signals may be relayed to other neurons, or they may initiate some action, such as the contraction of muscle fibers.

Neurons are delicate cells that are damaged easily or may be destroyed by injury, infection, pressure, chemical disturbances, or lack of oxygen. Since neurons cannot be replaced after they have been destroyed, any damage to them tends to have serious consequences.

The nervous system has two interdependent parts. One of them, the central nervous system, consists of the brain and spinal cord. The other, the peripheral nervous system, consists of all the nerve tissue outside the central nervous system. The central and peripheral nervous systems are each further divided into a number of components.

Peripheral nervous system

The peripheral nervous system has two main divisions: an outer system called the somatic nervous system and an inner one—the autonomic nervous system.

The somatic system has a dual role: first, it collects information from the body's sense organs and conveys this information to the central nervous system; and second, it transmits signals from the central nervous system to the skeletal muscles, thus initiating movement.

The autonomic nervous system is concerned with the regulation of the internal organs and glands, such as the heart, stomach, kidneys, and pancreas.

What is it about the "funny bone" that causes such a strange and painful sensation if I knock it?

The sensation you describe has nothing to do with bone but is due to the ulnar nerve. This passes behind the elbow on its way to the forearm and is prone to injury here. A slight knock can cause a volley of signals in the nerve's sensory fibers, and these can be excruciatingly painful.

A friend said the pain I have in my hands and arms could be caused by a pinched nerve. What is this?

At some point along their length, many nerves have to pass through a restricted space—especially near joints. Any displacement or swelling in this space may squeeze or pinch the nerve, causing pain, muscle weakness, numbness, or a tingling sensation.

The median nerve running through the wrist is the nerve most often affected. It may be squeezed between the ligaments and tendons in the wrist and wrist bones, causing numbness and tingling in the index finger, middle finger, and ring finger; pain in the hand and forearm; and weakness in the thumb. The condition, called carpal tunnel syndrome, can affect people who use keyboards frequently. If you have any of these symptoms, you should see your doctor—an operation to free the nerve may be necessary.

Two months ago I had a foot amputated. Why do I still feel that the foot is there and even have pain from the missing toes?

Although your foot has been amputated, the sensory fibers that used to send messages from the foot to the brain are still present in the remaining part of your leg and have their endings in the stump. If these endings are stimulated, the fibers send messages via the spinal cord to the brain, which from past experience interprets the message as having come from the foot. It takes some time for the brain to realize that the foot is not there.

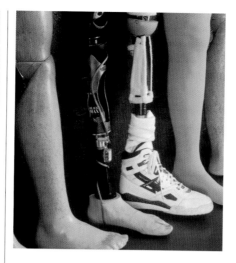

◀ *People with artificial limbs can be trained to use them skillfully. However, they may sometimes experience phantom pain—pain that seems to come from the lost limb.*

The somatic nervous system has two main components: the sensory and motor systems. Information is picked up by sensory organs such as the eyes, which contain receptor cells. There are similar cells for hearing, taste, smell, pain, touch, and skin temperature.

Signals from the receptor cells are carried toward the central nervous system in the sensory nerve fibers. The pattern of signaling in these fibers, which may amount to millions of impulses every second, gives the body and mind essential data about the outside world. Just as the sensory fibers carry information toward the central nervous system, so the motor fibers transmit signals away from the central nervous system toward the skeletal muscles.

Both sensory and motor fibers are themselves part of the sensory and motor neurons. All neurons have a cell body, as well as a number of projecting fibers. The motor and sensory fibers of the peripheral nervous system, called axons, are merely the longest fibers of their respective neurons. The sensory fibers have their cell bodies just outside, and the motor neurons have theirs inside, the brain or spinal cord. An axon may be 10,000 times the length of the nerve cell body.

The motor and sensory fibers that carry messages to and from a particular body organ or area are gathered together in a bundle called a nerve. Different nerves are said to supply a particular area or organ. All together, 43 pairs of nerves emerge from the central nervous system: 12 pairs of cranial nerves from the brain and 31 pairs of spinal nerves from either side of the spinal cord.

The cranial nerves mainly supply sense organs and muscles in the head, although a very important cranial nerve—the vagus—supplies the digestive organs, heart, and air passages in the lungs. Some cranial nerves, such as the optic nerve to the eye, contain only sensory fibers.

The spinal nerves emerge at intervals from the spinal cord and always contain both motor and sensory fibers. They supply all areas of the body below the neck. Each spinal nerve is attached to the spinal cord by means of two roots, one of which carries motor fibers and the other sensory fibers. At a short distance from the spinal cord, each spinal nerve splits into a number of branches.

In effect, the peripheral nervous system acts only to relay sensory and motor messages between the central nervous system and the body's muscles, glands, and sense organs. It plays almost no part in the actual analysis of those sensory signals or the initiation of motor signals. Both of these activities, and much in between, occur in the central nervous system.

The central nervous system
The brain and spinal cord form the nervous system's central processing unit. They receive messages via the sensory fibers from the sense organs and receptors, filter and analyze them, and then send out signals along the motor fibers that produce an appropriate response in the muscles and glands. The analytical, or processing, aspect may be relatively simple for certain functions performed in the spinal cord; however, analysis in the brain is usually a highly complex process that involves thousands of different neurons.

The spinal cord
The spinal cord itself is a roughly cylindrical column of nerve tissues about 16 inches (40 cm) long, which runs inside the backbone from the brain to the lower back. It has two main functions. First, it acts as a two-way conduction system between the brain and the peripheral nervous system. This action is achieved by means of sensory and motor neurons, whose fibers extend in long bundles from parts of the brain. They run varying distances down the spinal cord, and at the ends that are farthest from the brain they come into contact with the fibers or cell bodies of sensory and motor neurons that belong to the peripheral nervous system. Messages can be transmitted across the gaps—called synapses—between the peripheral neurons and the spinal neurons.

LAYOUT OF THE NERVOUS SYSTEM

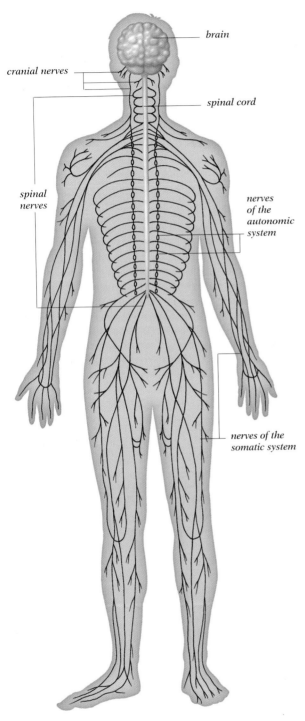

▲ *From the central nervous system (spinal cord and brain), pairs of nerves radiate all over the body to form the peripheral nervous system. This in turn has two main subdivisions: the autonomic system, responsible for unconscious control of functions such as breathing; and the somatic system, responsible for conscious control.*

The second function of the spinal cord is to control simple reflex actions; this control is achieved by neurons whose fibers extend short distances along the spinal cord, and by interneurons, which relay messages directly between the sensory neurons and motor neurons.

If a person accidentally puts his or her hand on a hot stove, pain receptors in the skin send messages along sensory fibers to the spinal cord. Some of these messages are relayed immediately by neurons to motor neurons that control the movements of the arm and hand muscles, and the hand is quickly and automatically withdrawn. At the same time, other messages travel up the spinal cord and are relayed by interneurons to motor neurons that control the neck's movements. In this way, the head is automatically turned toward the source of the pain. Further messages are carried all the way up to the brain and cause the conscious sensation of heat and pain.

The brain

The brain has three main parts. The stalk (brain stem) is a continuation of the spinal cord and supports the brain's large cap—the cerebrum. Below the cerebrum is the cerebellum. Although many sensory neurons terminate, and many motor neurons originate, in the brain, the majority of the brain's neurons are interneurons, whose job it is to filter, analyze, and store information.

One of the brain's most important functions is to memorize information from the sense organs. Later that information may be recalled and used in decision making. For example, when a person touches a hot stove, the pain is memorized; that memory will affect the decision whether or not to touch other hot stoves in the future.

Most of the conscious activities of the brain take place in the upper part of the cerebrum, the cerebral cortex. Some parts of the cortex are involved in the perception of sensations such as hearing, sight, taste, and smell. Others are involved in speech and language, and yet more are the starting point of motor pathways and govern the movements of muscle. Between the motor, sensory, and language areas of the cortex are associated areas that consist of millions of interconnected neurons. These neurons are associated with reasoning, the emotions, and decision making.

The cerebellum is attached to the brain stem just below the cerebrum and is concerned mainly with motor activities. Its job is to send out signals, which produce unconscious movements in muscles so as to maintain posture and balance. It also acts in concert with the motor areas of the cerebrum in order to coordinate body movements.

The brain stem contains different structures with a variety of roles. By far the most important of these structures are the centers that control the lungs, heart, and blood vessels. Functions such as blinking and vomiting are also controlled here. Other centers are concerned with the perception of sensations such as pain, or act as relay stations for messages that arrive from the spinal cord or cranial nerves.

One of the smallest parts of the brain stem, the hypothalamus, controls the body's chemical, hormonal, and temperature balance.

The neurons

Neurons are central to the working of the whole nervous system. However, they are not the only type of cell to be found in the nervous system; another type, called the neuroglias—literally "nerve glue"—are present in large numbers. Their job is to bind, protect, nourish, and provide support for the neurons.

Neurons come in various shapes and sizes; however, they all have the same basic structure. They have a nucleus, or center, which is

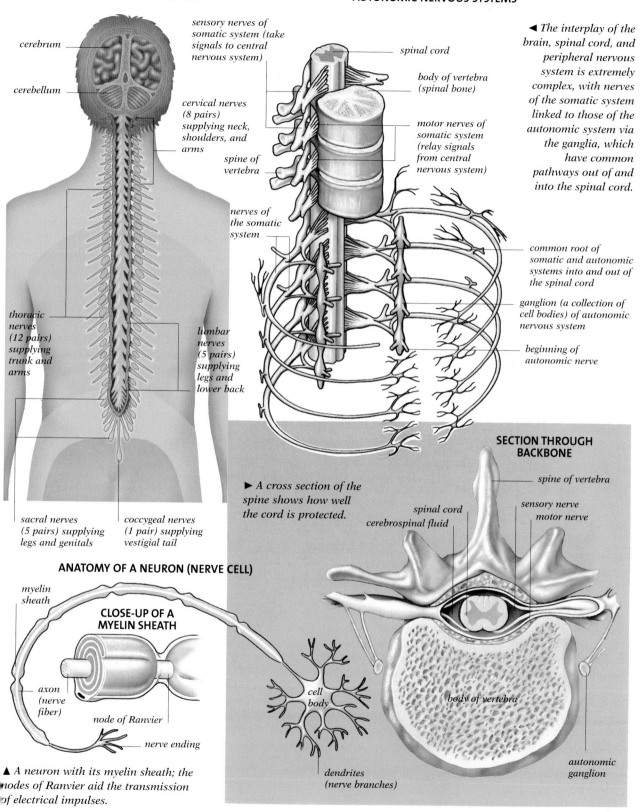

ARRANGEMENT OF THE CENTRAL AND PERIPHERAL NERVOUS SYSTEM

cerebrum

cerebellum

sensory nerves of somatic system (take signals to central nervous system)

cervical nerves (8 pairs) supplying neck, shoulders, and arms

spine of vertebra

nerves of the somatic system

thoracic nerves (12 pairs) supplying trunk and arms

lumbar nerves (5 pairs) supplying legs and lower back

sacral nerves (5 pairs) supplying legs and genitals

coccygeal nerves (1 pair) supplying vestigial tail

INTERPLAY BETWEEN THE CENTRAL, SOMATIC, AND AUTONOMIC NERVOUS SYSTEMS

spinal cord

body of vertebra (spinal bone)

motor nerves of somatic system (relay signals from central nervous system)

◄ *The interplay of the brain, spinal cord, and peripheral nervous system is extremely complex, with nerves of the somatic system linked to those of the autonomic system via the ganglia, which have common pathways out of and into the spinal cord.*

common root of somatic and autonomic systems into and out of the spinal cord

ganglion (a collection of cell bodies) of autonomic nervous system

beginning of autonomic nerve

SECTION THROUGH BACKBONE

spine of vertebra

sensory nerve

motor nerve

spinal cord
cerebrospinal fluid

body of vertebra

autonomic ganglion

► *A cross section of the spine shows how well the cord is protected.*

ANATOMY OF A NEURON (NERVE CELL)

myelin sheath

CLOSE-UP OF A MYELIN SHEATH

axon (nerve fiber)

node of Ranvier

nerve ending

cell body

dendrites (nerve branches)

▲ *A neuron with its myelin sheath; the nodes of Ranvier aid the transmission of electrical impulses.*

281

Diseases of the nervous system

DISEASE	SYMPTOMS	TREATMENT
Brain tumor	Severe headaches; projectile vomiting; neck pain; fits; personality changes; progressive paralysis.	Surgical excision of tumor; radiotherapy.
Dementia	Memory loss; inability to concentrate; confusion; loss of interest; untidiness.	No cure except when a specific cause is known. Vitamin therapy sometimes helps.
Epilepsy	Convulsive fits or temporary loss of consciousness.	Anticonvulsant drugs.
Ménière's disease	Ringing in ear; giddiness; nausea; vomiting.	Antinausea drugs.
Meningitis	Fevers; headaches; neck and back pain; intolerance of bright lights; convulsions; vomiting.	Antibiotic drugs for bacterial meningitis.
Multiple sclerosis	Weakness in one or more limbs; numbness; pins and needles; visual disturbances; walking difficulties. Symptoms vary; may improve for a time, then reappear.	No cure. Various drugs may bring about a temporary recovery.
Neuropathy	Muscle weakness; numbness; pain; pins and needles.	Treat the underlying cause.
Parkinson's disease	Tremors; uncoordinated movements; facial rigidity.	Anti-Parkinsonian drugs.
Sciatica	Back and leg pain along path of sciatic nerve.	Limited bed rest; painkilling, anti-inflammatory drugs.
Shingles	Fevers; pain; skin blistering along the course of affected nerve fibers.	Oral aciclovir taken early; analgesic drugs for pain and fever.
Spastic paralysis (very rare)	Spasms; partial paralysis; lack of coordination; uncontrolled movements.	No cure; drugs to treat specific symptoms; physical and speech therapy.
Stroke	Effects depend on area of brain affected: partial paralysis; speech impairment; severe headaches; visual disturbances; deafness. Sometimes fatal.	Clot-busting drugs within 3 hours (certain situations). Surgery for brain aneurysms.
Trigeminal neuralgia	Severe pain in the side of face lasting for about a minute. Recurs every few hours, days, or weeks.	Injection of alcohol into nerve; drugs; surgery to allow more room for nerve.
Vestibular neuronitis	Vertigo; vomiting; uncontrolled eye movements.	Treatment with drugs.

contained in a roughly spherical part of the neuron called the cell body. In addition, a number of fine, rootlike fibers project from the cell body. These are called dendrites. Also projecting from the cell is a single, long fiber called the axon. At its far end, the axon divides into a number of branches, each of which ends in a number of tiny knobs.

Each knob is in close proximity to, though not actually touching, a dendrite from another neuron. The gap between the two is known as a synapse, and messages are transmitted across it by chemicals called nerve transmitter substances.

Each neuron is bounded by a a thin, semipermeable wall, and this membrane plays an important part in the transmission of signals. The signals are always started by the excitation of one or more of the dendrites and are first carried toward the cell body. They are then transmitted away from the cell body along the membrane of the axon.

When a signal reaches the knobs at the end of the axon, under certain circumstances it may jump across the synapse to the dendrite of an adjacent neuron and so continue its journey.

To speed up the transmission of these signals, there is a covering along many axons—like insulation on electrical wires—called myelin. The areas of the brain and spinal cord that are insulated in this way are called white matter; the remainder is known as gray matter.

The central nervous system is maintained with a plentiful supply of blood, which provides oxygen and nutrients. The system is also protected by two kinds of covering. The first is bone and consists of the skull, which encloses the brain; and the backbone, which encloses the spinal cord. The second covering consists of three membranes of fibrous tissue: the meninges. These cover the whole of the brain and the spinal cord. Fluid circulates through the brain and the spinal cord and acts as a shock absorber. The cerebrospinal fluid contains nutrients and also white blood cells, which fight infection.

See also: Autonomic nervous system; Brain; Hypothalamus; Memory; Mind; Movement; Reflexes; Spinal cord

Nose

There could be many causes, most commonly an allergen such as pollen. Other common irritants include dust, tobacco, cosmetics (especially face powders, talc, and perfume), smoke, and gases. You should consult your doctor, who may refer you to a specialist.

What causes a runny nose when I have a cold?

The inside of the nose is constantly washed with mucus and swept with cilia (the "brush border" on the lining cells). A cold increases watery mucus production in an attempt to get rid of the infection, and a runny nose results.

My son has begun to pick his nose. When should I start being severe with him?

Nose-picking is unhygienic and antisocial, and you should put your foot down now. Train him to carry a hankie or tissue and to always use it when he sneezes or wants to get rid of excess mucus.

My neighbor says that a nosebleed is a sign of pressure on the brain. My son has had several nosebleeds in quick succession. Is this serious?

Nosebleeds are common in children, perhaps because they are so active and thus are likely to have many minor injuries. A frequent cause of nosebleeds is that blood vessels just inside one or both nostrils have burst, after becoming weakened and enlarged through rubbing and picking, or because of inflammation from an upper respiratory infection. Pressure on the brain is not a cause. However, recurrent bleeding can be a symptom of disease, so you should consult your doctor.

In addition to being a distinctive facial feature, the nose is a highly sensitive organ. It detects odor, then sends nerve impulses to the brain. Together with the eyes, ears, and throat, the nose can influence a person's health.

The nose is an important organ that has three main functions. First, it is the natural pathway by which air enters the body through breathing. The air is warmed, moistened, and filtered there before entering the lungs. Second, the nose acts as a protective device—if irritants such as dust enter, they are expelled by sneezing and do not have a chance to pass into the lungs, where they may cause damage. Third, the nose is the organ of smell. The nose also acts as a form of resonator, helping to give each person's voice its individual characteristic tone.

Structure

The external part of the nose consists partly of bone and partly of cartilage. The two nasal bones, one on each side, project downward and also form the bridge between the eyes. Below

SIDE VIEW OF THE NOSE

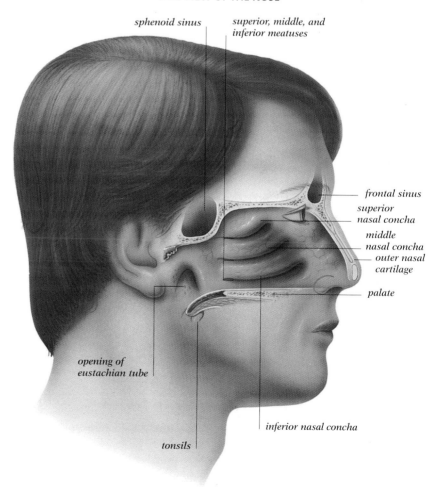

sphenoid sinus

superior, middle, and inferior meatuses

frontal sinus

superior nasal concha

middle nasal concha

outer nasal cartilage

palate

opening of eustachian tube

inferior nasal concha

tonsils

▲ *This cross section of the face shows the important parts of the nose and how it is linked to other areas of the face.*

Questions and Answers

I have had a crooked nose ever since I broke it playing football in school. Could it be the reason that my left nostril is permanently blocked, and can anything be done about it?

Your crooked nose is almost certainly the cause of your blocked nostril. An operation to reset your nose will entail only a day or two in the hospital and should improve your breathing.

I have been advised that my daughter, who is 12 years old, should have an operation for adenoids. A friend of mine says that she will grow out of the condition without needing an operation. What do you think?

Your daughter may grow out of it in time. However, if adenoids are causing her a lot of trouble, an operation may have an immediate and beneficial effect.

Why is it that when my nose is blocked, I get a pain behind my eyes and cannot speak properly?

When your nose is blocked, pressure builds up inside your sinuses, which may also contain some mucus. The pressure and the inflammation cause the pain. The sinuses also affect the way sounds vibrate, because they act as vibrators when we speak.

I keep getting polyps in my nose. Are they cancerous?

Nasal polyps are benign tumors. Treatment is with corticosteroids, such as prednisone. If treatment fails, then surgery is an option.

My grandmother used to put a cloth soaked in witch hazel across my nose when it bled. Is this an effective cure?

No. Although some herbs may have properties that help to stem the flow of blood, it is more likely that the treatment acted as an effective cold compress.

the nasal bones the nasal cartilages and the cartilages of the nostrils give the nose its firmness, shape, and pliability.

Inside the nose are two narrow cavities divided by a partition running from front to back. This partition, called the septum, is made of bone and cartilage. It is covered with a soft, delicate membrane called a mucous membrane, which is continuous with the lining of the nostril. The nostrils themselves are lined with stiff hairs that grow downward and protect the entrance of the nose.

The twin cavities created by the septum are called nasal fossae. They are very narrow, measuring less than ¼ inch (6 mm) in width. At the top of the fossae are thin plates of bone. These contain numerous small receptors from the olfactory nerves, which are responsible for the sense of smell. When a person has a cold, the receptors become covered in thick mucus, which reduces the sense of smell and also the sense of taste.

Warming and moistening the air

The cavity at the back of the nose is divided into sections by three ridges of bone called the nasal conchae. They are long and thin and run lengthwise, sloping downward at the back. The passage between each concha and the next is called a meatus. The conchae are covered

SECTIONS OF THE NOSE

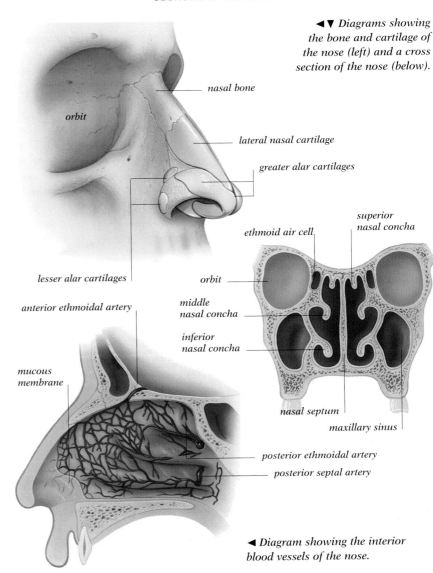

◄▼ *Diagrams showing the bone and cartilage of the nose (left) and a cross section of the nose (below).*

nasal bone

orbit

lateral nasal cartilage

greater alar cartilages

superior nasal concha

ethmoid air cell

lesser alar cartilages

orbit

middle nasal concha

anterior ethmoidal artery

inferior nasal concha

mucous membrane

nasal septum

maxillary sinus

posterior ethmoidal artery

posterior septal artery

◄ *Diagram showing the interior blood vessels of the nose.*

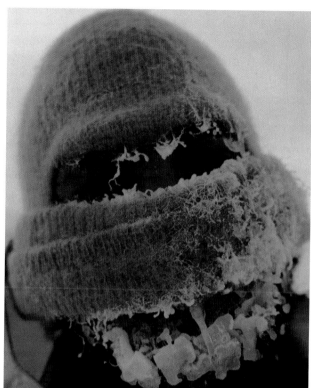

▲ *The nose is a sensitive organ. Here, an explorer in Alaska has his nose and lower part of his face fully protected against freezing temperatures.*

▲ *Sneezing is the natural way of clearing the nose—but the nose should be covered by a tissue or handkerchief. Coughing and sneezing can spread infection, so children must be taught good manners and healthy habits concerning hygiene.*

with a mucous membrane, which helps increase the surface area of the nasal passage. The membrane has a very rich blood supply that moistens and warms the inhaled air.

The mucous membrane inside the nose, which is covered with thousands of tiny hairlike bodies called cilia, secretes just under 1 pint (0.5 l) of mucus every day. The mucus traps dust particles and any harmful microorganisms that enter the nose. The mucus is then carried inward, taking any foreign bodies with it. These harmful bodies are then swallowed and destroyed by the stomach's gastric juices.

Sinuses and tear ducts

The sinuses—spaces in the front of the skull—are connected with the inside of the nose. They are located behind the eyebrows and behind the cheeks, in the triangle between the eyes and the nose. Sinuses help cushion the impact of any blows to the face.

Two other passages lead off the meatuses. Tear ducts carry away tears from the eyes—that is why it is usually necessary for people to blow their noses when crying takes place. The other duct, the auditory tube, is at the back of the nose near the junction with the throat. It leads from the nose to the middle

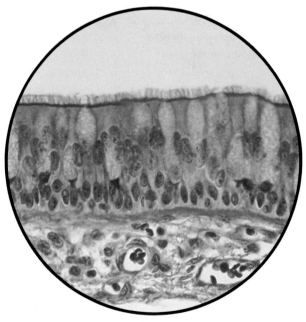

▲ *This image of a nasal membrane shows the microscopic "brush border" of cilia that waft out dust particles and so protect the lungs from pollution.*

Treating nosebleeds

A nosebleed occurs when a small blood vessel inside the nose is ruptured by a blow, by picking the nose, or by a bout of sneezing. It can be a consequence of inflammation from a cold or sinus infection and can also be caused by excessive dryness. People who have hay fever or a nasal infection may get nosebleeds. Although the blood loss looks great, it is not, in fact, copious, and it is rarely very serious. It will generally clear up in five to 15 minutes—in other words, in the time it usually takes blood to clot. Very heavy bleeding may follow damage to an artery at the back of the nose. In this case the nose has to be packed in a special way by a doctor. Nosebleeds that occur within a week of a tonsil or adenoid operation are particularly serious and should be treated immediately, as should nosebleeds that follow a blow to the head, which could indicate a fractured skull. Tell your doctor if the bleeding was caused by a blow. Otherwise:

- Sit the patient down, loosen his or her clothes around the neck, and incline the head slightly forward so that the blood drips into a bowl or any other receptacle that will catch the flow of blood.
- Try to prevent the patient from swallowing too much blood.

- Get the patient to breathe gently through the mouth and lightly pinch the nostrils closed for approximately five minutes.
- Apply ice wrapped in cloth to the nose.

- When bleeding stops, make sure the patient leaves the nose alone.
- If the bleeding persists for 20 minutes, call your doctor or take the patient to the hospital.
- If packing the nasal passage is required, it should be done by a doctor. When this procedure is performed by a physician, antibiotics are usually prescribed because this intervention creates a risk of infection.

ear; that explains why an earache sometimes occurs with a sore throat.

Colds and hay fever
The common cold is caused by a virus. It leads to acute inflammation of the nose, and excessive production of watery mucus, which causes nasal congestion.

Hay fever, or allergic rhinitis, is an allergic form of head cold, which is generally more unpleasant and longer-lasting than a cold caused by a virus. However, the hay fever disappears if the patient is able to pinpoint the cause and then avoid it in the future. Dust, animal dander, and irritating smells are also common triggers of nasal inflammation.

With colds, prevention is often better than treatment. For this reason, it is best to avoid being around people with colds whenever possible. People who have colds themselves should stay at home until they feel better. Acetaminophen and commercial cold medicines are soothing and relieve the discomfort. A nasal spray will relieve the pain, and inhalation of menthol is also beneficial. Douching the nose is particularly helpful if dry crusts have formed inside it.

Repeated colds, or colds that linger, may be caused by a polyp, a deviated septum, or sinus trouble, all of which can be corrected by minor surgery.

Injuries and malformation
A broken nose is one of the most common sports injuries and requires immediate medical attention. Unless a marked deviation in the underlying structure of the nose results from trauma, most surgeons wait until the swelling subsides before performing surgery—if in fact surgery is necessary.

The cavities on either side of the septum are rarely of equal size, because the septum usually leans more to one side. If the septum actually touches the conchae, irritation may arise and perhaps lead to a runny nose. This type of problem can be corrected by an operation.

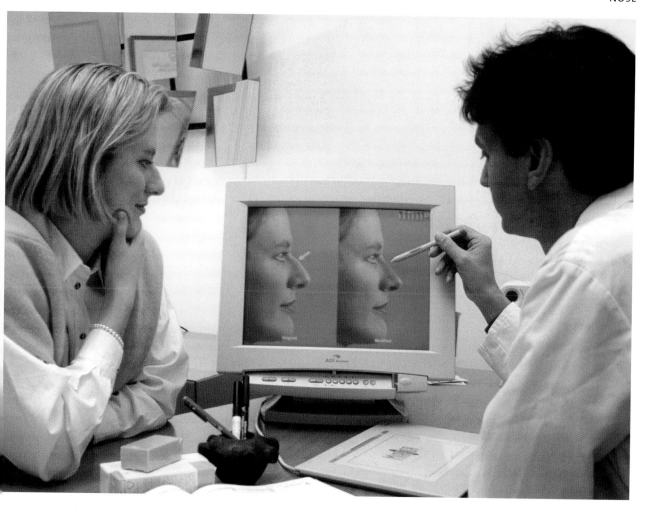

▲ *A nose job may resolve a deep psychological need in a person.*

Adenoids

The adenoids, two glandular swellings at the back of the nasal passage, are made up of lymph tissue. If the adenoids become infected, they block the nose, so that the sufferer is forced to breathe through the mouth. Chronic mouth-breathing in children is often caused by enlarged and enflamed adenoids.

Adenoidal children snore when they are asleep, because the enlarged adenoids partially obstruct the flow of air. These children may also have reduced hearing. Removal of the adenoids is a simple surgical procedure and will bring tremendous relief to the children.

Polyps

Although harmless, polyps are a nuisance and interfere with breathing. These soft growths of a jellylike texture, which are usually on a short stalk, are generally found in the middle concha. They rarely occur singly, and when one polyp is removed other polyps nearby may enlarge. Because of this, it may take several trips to the physician or ear, nose, and throat doctor to treat polyps. However, it is worth persevering with treatment. Some people worry that polyps are cancerous. This is not the case. They are benign tumors that respond well to surgery.

Foreign bodies

Small children have been known to push objects such as peanuts, buttons, lumps of foam from stuffed toys, wax crayons, peas, small stones, and other similar objects up their noses. The objects may cause no symptoms at first but will eventually result in swelling, discharge, headaches, and facial pain. Children who have pushed something up the nostril should be taken to a doctor.

Sinusitis

Bacterial infection is most common in the sinuses behind the eyebrows and those at the side of the nose. The condition usually develops as a complication of a viral infection, such as a cold. The symptoms of sinusitis include headaches, a discharge into the nose or throat, weakness, toothaches, and facial pain. Acute attacks may be precipitated by colds, hay fever, and damp weather. Acute sinusitis needs prompt treatment to avoid the risk that the infection will spread backward and cause meningitis to develop. Sinusitis may respond somewhat to menthol inhalation, but more severe cases may require the use of antibiotic treatment or an operation to wash out the sinuses.

See also: **Adenoids; Crying; Mucus; Smell; Sneezing; Stomach**

Optic nerve

Questions and Answers

What is my doctor looking for when he shines a light in my eye?

He is checking the pupillary light reflex. One sign of damage to the oculomotor nerve is a failure of the pupils to constrict when a light shines on them. Other problems in the brain or eye can also interfere with this reflex.

My sister has multiple sclerosis, and she lost all vision in her right eye for six weeks. Her vision then recovered. How can this be?

Multiple sclerosis commonly affects an optic nerve. The initial inflammation can block impulses from passing along the nerve, but it is unlikely to destroy more than a proportion of the million or so separate fibers that it contains. Vision usually returns, although special testing may show reduced resolution and color perception.

What is the blind spot, and what effect does it have on sight?

The blind spot is the point at which the optic nerve enters the eye. At this point there are no rods or cones—the cells that enable us to see movement and color. It doesn't affect our sight, because when we look at something the image falls on a sensitive part of the retina called the macula.

Since the optic nerves from the two eyes cross, if one of them is damaged, can the other one still channel the necessary information?

Not quite, because only half of the information from one eye passes across to the other side of the brain. However, all the information from the left side of each eye goes, via nerve fibers, to the brain, where the images are interpreted, so even if half of the visual field is lost, the patient can still read.

The optic nerve is a bundle of nerve fibers that carries impulses from the retina—the light-sensitive lining of the eye—to the back of the brain. Some of these nerve fibers cross over to the other side of the brain at the optic chiasma.

The back of the eye behind the lens is called the retina, and it is made up of a layer of light-sensitive cells. Each of these cells is connected by a nerve to the brain, where vital information about pattern, colors, and shapes is computed.

All the nerve fibers collect together at the back of the eye to form a single cable, which is known as the optic nerve. This runs from the eyeball, through a bony tunnel in the skull, to emerge inside the skull bone just beneath the brain, in the region of the pituitary gland; here it is joined by the optic nerve from the other eye.

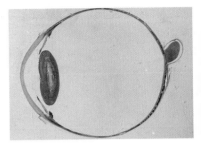

▲ *A vertical section through a human eye shows the optic nerve (the protrusion on the right).*

Half of the nerve fibers from each side then cross over so that some information from the left eye is passed to the right side of the brain and vice versa. Nerves from the temporal (side) part of each retina stay on the same side of the brain, whereas the fibers from the central part of the eye—the part that does most of the seeing—run to the opposite side of the brain. This crossover point is called the optic chiasma.

Structure of the optic nerve

Each optic nerve consists of a bundle of tiny nerve fibers that carry minute electrical impulses. Each fiber is insulated from the next by a fatty layer called myelin. At the center of the optic nerve, running its entire length, is a small artery called the central retinal artery. This emerges at the back of the eye, where vessels from it spread over the surface of the retina. A corresponding vein that drains the retina runs back down the optic nerve alongside the central retinal artery.

Nerve pathways

Nerves emerging from the retina are sensory. Unlike the nerves that supply muscle (motor neurons), which make only one connection on their way to the brain, optic neurons make more than one connection. The first connection or cell station is in the lateral geniculate body, which lies just behind the optic chiasma (where the sensory information from each eye swaps over). Here some information from the left and right eyes is swapped again across the midline. The function of this connection is linked with the reflexes of the pupils.

From the lateral geniculate body, the nerves fan out on each side around the temporal (side) area of the brain to form the optic radiation. They then turn slightly and meet to pass through the main exchange, the internal capsule, where all the motor and sensory information from the body is concentrated. From there the nerves pass to the visual cortex at the back of the brain.

Sensitive cells

There are two types of light-sensitive cells in the retina, the rods and the cones; and like photoelectric cells they convert light energy into electricity. The rods are used principally to detect objects in the dark—they are very sensitive to movement, so they register objects appearing from the extremes of the visual field in dim light. The cones are responsible for sharp color vision, and they are most plentiful at the fovea, the central point of the macula where the lens focuses light. However, there are no rods or cones at the point where the optic nerve enters the eye; this is known as the blind spot because light focused here is not perceived.

The diameter of the pupils is controlled in a way similar to the aperture in a camera. Light falling on the retina sends impulses up the optic nerve to the cell station behind the optic chiasma

THE MECHANICS OF SEEING

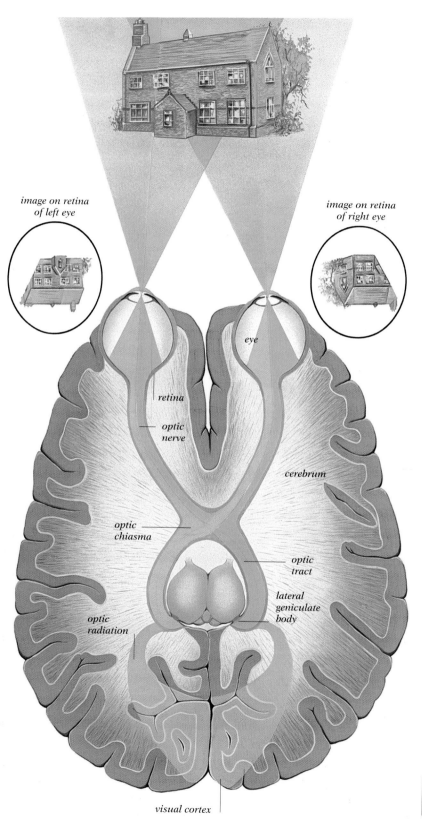

image on retina of left eye

image on retina of right eye

eye

retina

optic nerve

cerebrum

optic chiasma

optic tract

lateral geniculate body

optic radiation

visual cortex

◄ *The right and left eyes have slightly different fields of vision, each split into a right and left side. When light rays reach the retinas, they are transposed and inverted. They then travel down the optic nerves to the optic chiasma, where a crossover takes place. All the information from the left side of each eye travels down the optic tract through the lateral geniculate body and the optic radiation to the right visual cortex, and vice versa. Later these images are combined and interpreted by the brain.*

and then back to a motor nerve supplying the muscle of the pupil. The brighter the light, the more tightly the pupil constricts.

What can go wrong

Obstruction of the central retinal artery causes sudden total blindness. Although this is rare, it may occur if the optic nerve swells in its bony tunnel through the skull and presses on the artery. Vision can also be affected if the blood vessels running over the top of the retina rupture and bleed over the surface, thereby cutting out the light, as can occur in people with diabetes.

Some toxins can affect the retina and optic nerve; the most common is methanol (wood alcohol), which can cause blindness.

Multiple sclerosis can cause inflammation of the optic nerve, and visual loss occurs as a large central blind spot called a scotoma. If the optic nerve is sufficiently inflamed, the optic disc at the back of the eye becomes swollen and its edges are no longer distinct; this condition is called papilledema. Tumors of the pituitary gland may press on the optic chiasma, producing visual abnormalities depending on which fibers they constrict.

Strokes commonly interfere with the blood supply to the nerves that pass through the internal capsule, the main exchange center of the brain. Damage to this area results in a total loss of movement and sensation on one side of the body and an inability to see objects moving in from that side—a paralysis called hemiplegia. Sight, movement, and sensation are quite normal on the opposite side.

Damage to the visual cortex at the back of the brain, for example as a result of a gunshot wound, may cause total blindness.

See also: **Brain; Eyes; Nervous system; Reflexes**

Ovaries

No. Usually, when one ovary is removed, the other one grows slightly larger and takes over the work of both. The single ovary, as long as it is healthy, can still ovulate each month.

When do a girl's ovaries first begin to work?

The ovaries may begin to release their first hormones as early as a girl's seventh year. The effects of the hormones can be seen in a subtle change in her body shape, making it appear more womanly. The changes are followed by the start of breast development. The sooner these changes begin, the earlier a girl will have her first menstrual period.

Is it true that the Pill interferes with the way the ovaries work?

Yes. The Pill contains artificial sex hormones, similar to those made by the ovaries, and it prevents eggs from maturing. Its hormones alter the body's natural monthly rhythm of hormone production, and this prevents the release of eggs. In effect the body is tricked into thinking that it is pregnant, so the ovaries get the message that they need not release any eggs for the time being.

What happens to the ovaries and their eggs after menopause?

At menopause the ovaries stop making hormones and as a result also stop releasing mature eggs. The mature eggs that remain in the ovaries fail to develop any further. As time passes, the ovaries gradually shrink and become full of fibrous tissue. This tissue largely obliterates the remaining eggs.

The ovaries do more than produce and release eggs that are ready for fertilization. Their other vital role is to produce hormones that maintain a pregnancy and give a woman's body its feminine shape.

The ovaries are the parts of the female reproductive system that are designed to produce and release mature ova (egg cells). When an ovum meets and is fertilized by a male sperm, this event marks the start of a new human life. From the first menstrual period right up to menopause, usually one egg is released each month; this release can take place from either ovary. However, sometimes two or even more eggs may be released by an ovary in one month. The ovaries are also essential parts of a woman's hormonal system.

SITE, STRUCTURE, AND FUNCTION OF THE OVARIES

▶ *The ovaries are covered by a layer of cells. The cells that are destined to become eggs pass into the substance of the ovaries, where they are surrounded by a follicle membrane (egg sac). Each month a single follicle matures and bursts on the surface of one of the two ovaries, and an ovum is released. The corpus luteum then develops at the site of the egg's follicle. If the egg is fertilized, the corpus luteum grows and secretes the hormones that maintain pregnancy.*

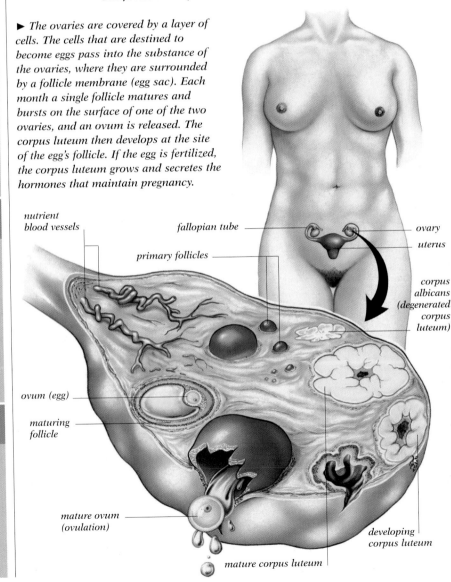

nutrient blood vessels

fallopian tube

ovary

uterus

primary follicles

corpus albicans (degenerated corpus luteum)

ovum (egg)

maturing follicle

mature ovum (ovulation)

developing corpus luteum

mature corpus luteum

Location and structure

The ovaries are two gray-pink almond-shaped structures each measuring about 1⅕ inches (3 cm) long and about ⅖ inch (1 cm) thick. They are found in the pelvis, the body cavity bounded by the hip or pelvic bones, and lie one on each side of the uterus. Each ovary is held in place by strong elastic ligaments. Just above each ovary is the feathery opening of the fallopian tube, which leads to the uterus. Although they are very close together, there is no direct connection between the ovary and the tube opening.

In a mature woman, the ovaries have a rather lumpy appearance. The reason for this can be seen by looking at an ovary's internal structure under a microscope. Covering the ovary is a layer of cells called the germinal epithelium. It is from these cells that the eggs or ova form; thousands of immature eggs, each in a round casing or follicle (the egg sac), can be seen clustered near the ovary edge.

More noticeable are the follicles that contain eggs in various stages of maturation. As these follicles enlarge, and after their eggs have been released, they produce the characteristic bumps on the ovary surface. The center of the ovary is filled with elastic fibrous tissue that supports the follicle-containing outer layer.

Ovulation

Under a microscope, the maturing follicles of the ovary can be seen as tiny balls enclosing a small mound of cells. In the center of the mound is the egg cell. A mature follicle forms a conspicuous swelling on the ovary about ¾ inch (2 cm) across. Exactly how the follicle ruptures to release the mature egg cell is not known. The ovum is then wafted by the feathery ends (fimbria) of the fallopian tubes into the tube openings.

In their role as egg producers, the ovaries also act as endocrine glands. The ovaries function under the control of the pituitary gland at the base of the brain. The pituitary makes a hormone called follicle-stimulating hormone (FSH), which travels in the bloodstream to the ovaries. FSH stimulates the follicles and causes the ova to mature; it also causes the secretion of the hormone estrogen. Under the influence of estrogen, the lining of the uterus thickens in preparation for receiving a fertilized egg. Estrogen also stimulates the buildup of body proteins and leads to fluid retention.

After a follicle has ripened and burst, another pituitary hormone, luteinizing hormone (LH), goes into action and brings about the development of the corpus luteum in the empty follicle. The corpus luteum helps to establish a pregnancy and also makes and releases its own hormone, progesterone. If the egg is not fertilized within 14 days, the corpus luteum shrinks, progesterone production is shut off, and the lining of the uterus is shed as the monthly menstrual period. FSH production then begins again and the whole cycle is repeated. If, however, the egg has been fertilized, the corpus luteum continues working until the placenta is established. There is no menstrual bleeding in this case.

Ovary development

Ovary development is largely complete by the time a female fetus is in the third month of life in the uterus, and few major changes will take place until puberty. By the time a baby girl is born, her two ovaries contain a total of between 40,000 and 300,000 primary follicles, each containing an immature egg. At most only 500 of these eggs will be released, and probably no more than six will go on to develop into new human beings.

▲ *Few of the eggs released by a woman's ovaries will be fertilized; fewer will go on to develop into a new human life.*

When the ovaries first start making estrogen, they are not yet capable of releasing mature eggs, but this early estrogen stimulates the physical changes of puberty, such as the growth of the breasts, the widening of the hips, and the growth of pubic hair. This happens at least a year before a girl has her first period, and it is a signal that the estrogens have stimulated the production of mature eggs.

What can go wrong

Aside from the normal failure of the ovaries at menopause, the most common problem is the formation of ovarian cysts. These growths, which are usually benign, can grow very large, making a woman's abdomen swell as if she were pregnant. Many small ovarian cysts disappear of their own accord, and cysts usually do not cause pain unless they are very large or if they bleed.

Ovarian cancer is a very dangerous condition in that diagnosis is often not made until the disease is advanced and thus the prognosis is poor.

Normal, healthy ovaries are not palpable on abdominal (external) examination. During an internal—or pelvic—examination, they may be palpable between the doctor's hands (one hand on the abdomen and the other inside the vaginal canal). For direct inspection and biopsy of the ovaries, a surgical procedure under general anesthesia is required.

See also: **Conception; Endocrine system; Estrogen; Hormones; Menopause; Menstruation; Pituitary gland; Puberty; Uterus**

Palate

As an integral part of the mouth, the palate helps in breaking up food. It also plays a crucial role in subtly changing the shape of the mouth to create an enormous variety in the sounds and character of speech.

Questions and Answers

Will a premature baby always have a cleft palate?

No. In a normal baby, the proper knitting together of the bones that form the palate occurs early in the fetus's life. Being born prematurely will make no difference to the structure of a baby's mouth. A baby destined to have a cleft palate will have one whether or not he or she is born before the due date.

I am 36 and pregnant with my first baby. Because of my age, is my baby more likely to be born with a cleft palate?

Surveys have found a relationship between the age of a mother and the likelihood of a baby's having a cleft palate, but the link is not very strong. About one in 750 babies is born with a cleft palate, and the number rises to one in 20 if someone in the mother's immediate family has the problem. If you have any concerns about this, talk to your doctor.

My daughter speaks with a lisp. Could it indicate that there is something wrong with her palate?

No. A lisp usually results from the faulty movement of the tongue inside the mouth. Although the palate, particularly the hard palate, plays a part in the formation of the sounds of speech, it has a passive role compared with the active influence of the tongue, mouth, and lips. A lisp can be corrected by teaching the correct movements to make. It may be useful for your daughter to see a speech therapist.

What is a "falling palate"?

"Falling palate" is a term used to describe an abnormal enlargement of the uvula, which hangs down from the back of the soft palate. This may cause constant coughing and is usually treated by surgery.

"Palate" is the technical word for the roof of the mouth. It is divided into two parts: the hard palate, toward the front of the mouth, and the soft palate at the back. The palate is involved in many of the functions of the mouth: eating, tasting, swallowing, breathing, and speaking. The palate can be injured, but the most common problem is faulty development, which can lead to the birth of a baby with a split or cleft palate.

Structure

The hard palate is created by the links between the maxillae, the bones of the upper jaw, and the palatine bones on either side of the face that connect with them. The soft palate has no such bony base. Instead it is underlaid with tough fibers and with muscles that allow it to move. At the back of the mouth, behind the tongue, the soft palate splits into two, and the gaps are occupied by the tonsils. Just in front of this divide hangs a fleshy projection called the uvula.

Covering the hard and soft palates, and forming the lining of the mouth, is a layer of mucous membrane that contains mucus-secreting glands. This membrane is subject to a great amount of

▲ *For a child, the feel of food is as important as its taste in determining whether or not it is enjoyable. The texture of food is picked up by nerves in the palate.*

Questions and Answers

My baby has suddenly developed white patches inside his mouth, with some on his palate. What should I do?

It sounds as if your baby has a yeast infection. This is caused by a fungus and needs treatment, so take him to a doctor as soon as possible. The usual treatment is either a topical or an ingestible antifungal medication.

Why is it that eating salty food makes my palate dry?

The sensation of a dry palate is part of the body's natural reaction to thirst. Salty food makes you feel thirsty because it temporarily upsets the body's internal water balance, drawing water out of the blood and into the tissues. The brain monitors this water level and also responds to the sensation of a dry palate. As a result, you are driven to search for a drink and therefore put the blood and body tissues back into their proper equilibrium.

Why is my palate so sensitive to the texture of certain foods? Sometimes the sensation makes me feel physically sick.

Like other parts of the body, the palate is endowed with a rich supply of nerves. When we eat, some of the nerves of the palate send signals to the brain about the nature of the food that is in the mouth. If these nerves send back the message "unpleasant," then the natural reaction of the brain is "reject." One of the most common rejection mechanisms is vomiting, which explains why food that is unpleasant to you produces such a strong physical sensation.

Does a cleft palate always have to be corrected?

Almost always. The only notable exception is if the split or cleft in the palate is very slight and affects only the area of the uvula at the back of the palate.

wear and tear, so it has to be tough and capable of renewing its surface cells constantly. The membrane on the hard palate is stuck tightly to the bony structure beneath to prevent it from becoming dislodged by the movements of the tongue. The ridge of bone which runs along the middle of the palate, and to which the membrane is attached, is called the raphe. Horizontal ridges of tissue called rugae extend from the raphe. These can be felt with the tongue, and are most prominent in childhood. The mucous membrane that covers the soft palate extends backward to join with the lining of the back of the nose. To aid lubrication of the throat during swallowing, it is more richly supplied with mucous glands than the lining of the hard palate.

What the palate does

The soft palate contains a few taste buds, which supplement the more numerous ones of the tongue. When people eat, the hard palate acts as the mortar onto which the pestle of the tongue pushes food to soften and mash it. When food is ready to be swallowed, the muscles of the soft palate contract and pull this part of the palate upward. This action not only helps push the food toward the esophagus but also blocks off the airway at the back of the nose. By helping to keep breathing separate from the eating process, the movements of the soft palate help to prevent choking.

The texture of food is sensed by nerves in the palate's mucous membrane. Any food that has an unpleasant texture is appropriately described as unpalatable. In extreme cases, unpalatable food may make a person feel sick because it causes "reject" messages to be sent to the brain, which trigger a vomiting reflex.

During normal breathing, the soft palate is held in a relaxed position to allow the free passage of air in and out of the lungs via the nose. When people speak, air is taken in through the mouth and molded by the tongue and lips as they exhale; movement of the soft palate creates subtle differences in the shape of the mouth's acoustic chamber; and the hard palate acts as a sounding board for the tongue, giving basic sounds the shape of speech.

Development

The palate begins to develop as early as the fifth week of life in the womb. At this stage the face is molded in gristly cartilage; later it hardens into

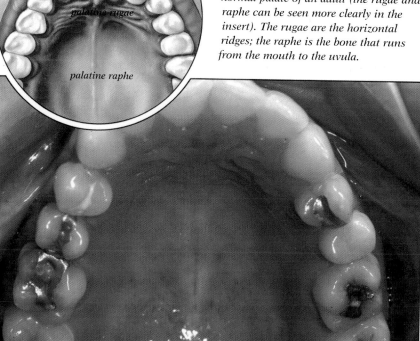

palatine rugae

palatine raphe

▼ *The photograph below shows the normal palate of an adult (the rugae and raphe can be seen more clearly in the insert). The rugae are the horizontal ridges; the raphe is the bone that runs from the mouth to the uvula.*

NORMAL AND ABNORMAL DEVELOPMENT OF THE PALATE

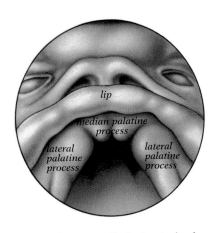

▲ The palate normally begins to develop when the fetus is about five weeks old; by eight weeks each part has formed.

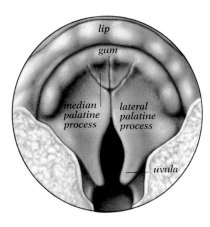

▲ By nine weeks the lateral processes have grown and fusion of the two sides has begun; a week later it will be complete.

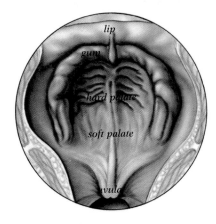

▲ The palate at birth, showing the prominent ridges of the hard palate; these become less pronounced with age.

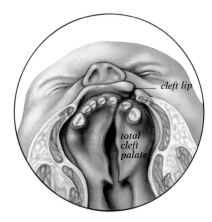

▲ Incomplete development of both the hard and the soft palate; this type of cleft may be accompanied by a cleft lip.

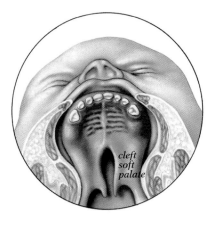

▲ A cleft affecting the whole of the soft palate occurs symmetrically along the midline from the hard palate to the uvula.

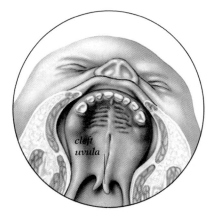

▲ A cleft uvula; unlike other types of cleft palate, this is not serious and does not usually require surgical treatment.

true bone. The palate does not develop from a single bone. Instead a pair of horizontal projections grow from the rudimentary upper jaw, under the eyes, to meet in the center of the skull. At the same time another projection grows down between the eyes to create the nose. The three pieces of cartilage finally fuse, and the fusion creates the palate. The process of growth and fusion demands precise timing and is synchronized with the formation of the outer tissues of the face, including the skin. For a perfect result, each part of the process must take place at the right time and at exactly the right rate.

Problems and treatment

The most serious and common defect of the palate comes from poor synchronization during the growth and fusion of the bones. The result is a split or cleft palate. A baby born with a cleft palate has a gap, usually Y-shaped, in the roof of the mouth. A cleft palate is often associated with a harelip. This second deformity is a puckering of the lip due to the faulty synchronization of cartilage growth with skin

growth. Both cleft palate and harelip are believed to result from defects that are hereditary.

Plastic surgery has made a cleft palate less serious than it used to be, but the necessary operations demand great skill and careful timing. A baby with a cleft palate needs special help to eat properly.

Sometimes the hard palate may be excessively concave in shape, impairing breathing. This problem is often associated with enlarged tonsils and tends to improve once the tonsils are removed surgically. Abnormalities of the soft palate can cause snoring and respiratory problems during sleep.

Any accident that involves a burn in the mouth, or an injury that leads to bleeding, should be treated by sucking on an ice cube. If there is bleeding from a palate injured in an accident, lay the victim on his or her side to prevent him or her from choking on blood.

See also: **Cartilage; Esophagus; Fetus; Mouth; Mucus; Snoring; Taste; Tongue; Tonsils**

Pancreas

The pancreas is one of the most important glands in the human body. It secretes some of the hormones, such as insulin, that are vital to life, as well as the enzymes that make digestion possible.

Questions and Answers

I have been told that the pancreas is so important that if it is removed or destroyed, the patient's life is in danger. Is this true?

It is true that the pancreas is one of the body's most important organs, but it can be removed by surgery and its function replaced in various ways. The first problem that has to be dealt with is diabetes, since the pancreas is the only source of insulin in the body. This problem is solved relatively easily with insulin injections.

The pancreas is also important in the digestion of food. When it is removed, the patient is usually treated with an extract of pancreatic digestive enzymes from animal sources, which is added to food. The drug omeprazole is used to stop the stomach from producing acids, thus preventing the breakdown of pancreatic enzymes by acid.

Can the pancreas be injured in an accident?

Yes; injuries to the pancreas most commonly occur as a result of car accidents, in which the upper part of the abdomen is struck with force. It can be difficult to tell if the pancreas is involved, and surgery may be needed if a pancreatic injury is suspected. Certain injuries, if left untreated, allow digestive juices of the pancreas to leak into the abdomen.

Does a person who has a diseased pancreas always get diabetes?

No; many pancreatic problems do not cause diabetes. However, if there is inflammation of the entire pancreas as there is in pancreatitis, the insulin-producing cells in the islets are almost bound to be involved, leading to a diabetic tendency that is often less marked than might be expected from the extent of the damage.

The pancreas, one of the largest glands in the body, is really two glands in one; almost all of it deals with secretion. It is an endocrine gland that secretes hormones, of which insulin is the most important. It is also an exocrine gland—one that secretes directly into the gut (or another body cavity) rather than into the blood.

The pancreas lies across the upper part of the abdomen in front of the spine and on top of the aorta and the vena cava (the body's main artery and vein). The duodenum is wrapped around the head of the pancreas; the rest consists of the body and tail stretched over the spine to the left. The basic structures in the pancreas are the acini, which are collections of

DUAL ROLE OF THE PANCREAS

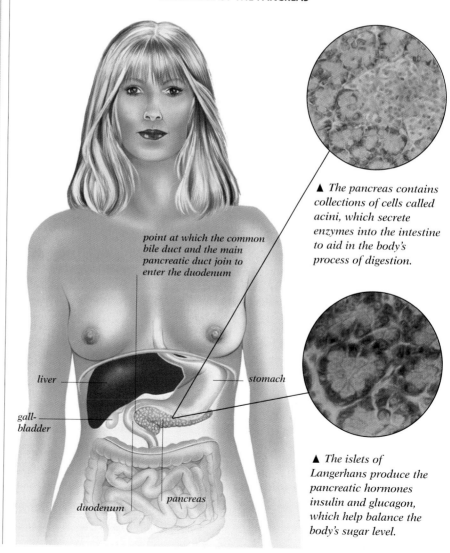

point at which the common bile duct and the main pancreatic duct join to enter the duodenum

liver

gall-bladder

stomach

duodenum

pancreas

▲ *The pancreas contains collections of cells called acini, which secrete enzymes into the intestine to aid in the body's process of digestion.*

▲ *The islets of Langerhans produce the pancreatic hormones insulin and glucagon, which help balance the body's sugar level.*

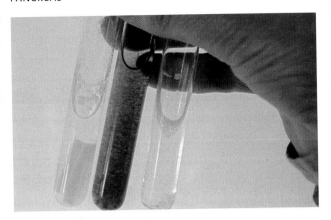

▲ *The enzyme amylase breaks down starch. From left to right: first, a vial filled with amylase; second, a vial that consists of a colored starch solution that has been added to the amylase; third, a vial of the colored starch solution after it has been broken down by the enzyme amylase, causing the colored starch solution to disappear.*

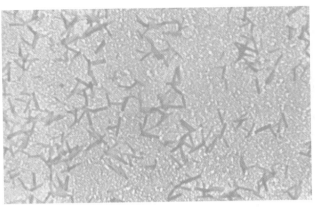

▲ *Crystals of insulin (an extremely important pancreatic hormone) control the level of blood sugar in the body by counteracting the effects of certain other hormones (such as adrenaline and cortisone) that raise the level of blood sugar. A deficiency of insulin results in the disease called diabetes, which can be treated with injections of insulin.*

secreting cells around the blind end of a small duct. Each duct joins with ducts from other acini until all of them eventually connect with the main duct that runs down the center of the pancreas.

Among the acini are small groups of cells of endocrine tissue, called the islets of Langerhans. They control glucose concentration in the blood and regulate other pancreatic hormones.

What the pancreas does

Exocrine pancreas: The pancreas produces essential alkali in the form of sodium bicarbonate to neutralize the heavily acidic contents of the stomach as they enter the duodenum. The pancreas also produces many important enzymes that help to break food down into basic chemical constituents; these are then absorbed by the intestinal wall.

Most of the main enzymes for the digestion of protein are produced by the pancreas; this could be a problem since the pancreas itself, like the rest of the body, is basically a protein-based structure. There would appear to be a risk that the pancreas might self-digest. This is avoided, however, because the main protein-digesting enzyme, trypsin, is secreted in an inactive form called trypsinogen, which changes to the active form once the pancreatic juices reach the duodenum. The pancreas also produces amylase and lipase—enzymes that break down starch and fats, respectively.

The digestive juices are powerful and cannot be released into the intestine safely unless food is present for them to act upon. Therefore, a sophisticated control system acts on pancreatic secretions. The vagus nerve—the main nerve of the parasympathetic system—stimulates the first small secretion as a result of the thought, taste, or smell of food.

Further secretion is stimulated by distension of the stomach, but most of the secretion takes place when the food finally reaches the duodenum. As this happens, cells in the wall of the duodenum release into the bloodstream two separate hormones —secretin and cholecystokinin—which travel in the blood to the pancreas and speed up secretion.

Endocrine pancreas: Insulin, a substance produced in the islets of Langerhans, lowers blood glucose levels. The islets also produce a hormone called glucagon, which has the effect of raising, rather than lowering, the level of sugar in the blood.

What can go wrong

Three diseases that can cause disorders of pancreatic digestive activity include acute pancreatitis (associated with sudden, severe abdominal pain), chronic pancreatitis (characterized by recurrent attacks of pain and failure of the pancreas to produce adequate amounts of digestive juice), and the inherited disease cystic fibrosis (in which many glands, including the pancreas, produce abnormally thick secretions). A complication of chronic pancreatitis is malabsorption from inadequate secretion of digestive juice. This condition is often associated with heavy alcohol consumption. Acute pancreatitis can also be associated with alcohol but may occur due to severely high triglyceride levels in the blood.

Another disease that can affect the pancreas is hemochromatosis, in which the gene that regulates the amount of iron absorbed from food is defective. This can cause an overload of iron in the body, with accumulation of iron in the vital organs, including the pancreas. This causes serious damage to the pancreas and can lead to diabetes.

Finally, cancer can affect the pancreas. Pancreatic cancer is very difficult to treat because it is usually advanced when diagnosed. Surgery as an attempt to cure pancreatic cancer is done only when the cancer has not spread to other organs. If the pancreas has to be removed, the body can still function. Insulin injections can be given, digestive enzymes can be sprinkled on food, and certain drugs can prevent the stomach from producing acids.

See also: **Abdomen; Autonomic nervous system; Digestive system; Duodenum; Endocrine system; Enzymes; Glands; Hormones; Insulin; Smell; Taste**

Parathyroid glands

The tiny parathyroids are among the most important glands in the body. They produce parathyroid hormone (PTH), which is vital for maintaining the delicately balanced quantities of calcium in the bloodstream.

Questions and Answers

I heard that years ago when the thyroid was removed during surgery, the parathyroids were often accidentally removed too. What happened in cases like these?

The patient would develop a very low level of calcium in the blood, which led first to tetany (an uncontrolled muscular spasm that occurs particularly in the hands and feet) and eventually to the loss of respiration, unless the problem was corrected. Once the condition was recognized, it was treated by giving the patient intravenous injections of calcium. The parathyroids may be removed if there is cancer in the thyroid region, but the patient would be treated with calcium supplements and possibly vitamin D, and would have regular blood tests to monitor the level of calcium.

My sister has an overactive thyroid gland. Is there any chance that her parathyroids will be affected?

No. Although the thyroid and parathyroid glands are close together, the disease processes that affect each are separate, and so her parathyroids should be normal. It is possible to develop a raised level of calcium in the blood simply as a result of a severely overactive thyroid gland.

My father recently underwent surgery on his parathyroids, and he was injected with blue dye. What was the reason for this?

The parathyroid glands are very small organs, so it is not surprising that they are difficult for a surgeon to find during surgery. For some unknown reason, the parathyroid glands are able to absorb a dye called Evan's Blue. This coloring allows the surgeon to see them more easily and to distinguish them from the rest of the tissues. Many surgeons use this technique to help with the operation.

The parathyroids are four tiny glands found behind the thyroid glands, which in turn are found just below the larynx in the throat. They play a major role in controlling the levels of calcium in the body. Calcium is a vital mineral, not only because it is a major structural element in the formation of bones and teeth, but also because it plays a central role in the workings of the muscles and nerve cells. The calcium levels in the body have to be kept within fairly constant boundaries, otherwise the muscles stop working and seizures (fits) may occur. The parathyroid glands keep the calcium levels in balance.

The absorption of dietary calcium into the bloodstream is controlled by vitamin D, which people get from sunlight and some foods, and by an important hormone produced by the

PARATHYROID GLANDS

▲ *This hip X ray shows a prominent bone cyst (center) in the head and neck of the femur, which was later operated upon. A parathyroid tumor was discovered and removed in the operation.*

thyroid cartilage

thyroid gland

superior parathyroid glands

trachea

inferior parathyroid glands

◄▲ *The tiny parathyroid glands are usually situated near the thyroid gland at the back of the larynx in the throat. The upper two, the superior parathyroids, are behind the thyroid. In this illustration, the inferior parathyroids are inside the thyroid.*

Questions and Answers

How are the parathyroids affected if the diet is deficient in calcium?

A low level of calcium in the diet will certainly tend to raise the output of parathyroid hormone from the parathyroids. In fact, it is more common for the diet to be deficient in vitamin D than in calcium. However, this, too, will result in a low blood calcium level, since vitamin D is essential for the absorption of calcium from the intestine into the bloodstream.

My brother had his parathyroids removed because of kidney trouble. Why was this done?

The kidney contains an enzyme that activates vitamin D. If there is a lack of the enzyme, low vitamin D levels cause blood calcium levels to fall, and the parathyroids may enlarge to compensate.

If you start having muscular spasms, do they necessarily mean that your parathyroids have failed?

There are many types of muscle spasms, and anything from an epileptic fit to abdominal cramps can be responsible. For example, in tetany—an uncontrollable contraction of the muscles, usually starting in the hands and feet—a lack of PTH may be responsible. However, the most common cause of this is hysterical overbreathing (hyperventilation). Loss of too much carbon dioxide changes the blood's acidity, and the calcium level drops. The treatment is to rebreathe for a time into a small paper bag.

Can PTH arise from any area aside from the parathyroid glands?

Yes. A number of different hormones can be manufactured by various cancers, and PTH is one of these. PTH can be produced by cancers of the lung and kidney. It used to be injected into patients to correct calcium levels, but this practice was discontinued because of its uncertain biological effects on the body.

What can go wrong

SYMPTOMS	CAUSES	TREATMENT
Overactive parathyroids (hyperparathyroidism): Thirst; increased urination. Pain in the stomach; loss of appetite; vomiting; kidney stones; fatigue; general feeling of ill health. Bone pain; spontaneous fractures.	Benign tumor of one or more glands. Hyperplasia (enlargement) of the glands, often due to kidney disease.	Surgical removal of the affected gland or glands.
Underactive parathyroids (hypoparathyroidism): Tetany; cramps; uncontrollable muscle spasms; seizures. Tiredness, irritability, depression, psychosis.	Idiopathic (this means "cause unknown"). There may be yeast infection of the nails. Surgical removal of the gland.	Vitamin D by mouth is very effective, but the level of calcium in the blood has to be monitored carefully.

parathyroids called parathyroid hormone, or PTH. If the level of calcium is too low, the parathyroids secrete an increased quantity of the hormone, which has the effect of releasing calcium from the bones to raise the level in the bloodstream. Conversely, if there is too much calcium, the parathyroids reduce or halt the production of PTH, thus bringing the level down.

The parathyroids are so small that they can be difficult to find. The upper two are situated behind the thyroid gland; the lower two, however, can actually be inside the thyroid or occasionally down inside the chest.

Like most endocrine hormone glands, the parathyroids can cause two main problems. They can be overactive, leading to a high level of calcium in the blood; or they can be underactive, leading to a dangerously low level.

INTERACTION BETWEEN BLOOD CALCIUM AND PTH

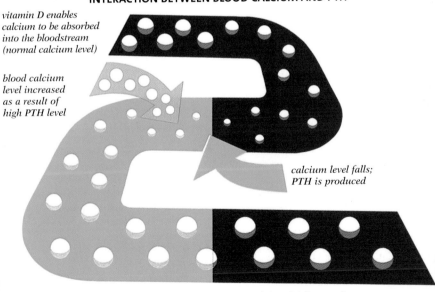

vitamin D enables calcium to be absorbed into the bloodstream (normal calcium level)

blood calcium level increased as a result of high PTH level

calcium level falls; PTH is produced

calcium level normal, PTH production is reduced

▲ *This diagram shows how, as the level of calcium in the blood drops, the parathyroids increase production of PTH; once the blood calcium level returns to normal, the production of PTH is reduced.*

▲ *Vitamin D, which is an essential vitamin for the absorption of calcium, is found in fish oils and is synthesized from sunlight.*

Overactive parathyroids

Hyperparathyroiditis, or overactive parathyroids, is a common problem. Doctors now measure the level of calcium in the blood as a routine part of the biochemical testing that is carried out on practically all hospital patients, and also on many patients by their primary care physicians. As a result, more instances of unexpectedly high blood calcium levels have been found in patients. It is now thought that as many as one person in a thousand may show some degree of parathyroid overactivity.

Symptoms

A raised blood calcium level may be caused by hyperparathyroidism. However, it is important to realize that there could be other causes.

For example, a common cause of a high level of blood calcium is a cancer that leads to the production of hormones that eat away bone. This causes an excess of calcium to be released into the bloodstream. An excessive intake of vitamin D can also cause a raised calcium level.

Symptoms of a raised calcium level include thirst and increased urination. There may also be fatigue, poor concentration, loss of appetite, and vomiting. When overactive parathyroids are the cause of a high blood calcium, many patients develop kidney stones. In this disease, the urine contains an excess of calcium, which tends to settle in the kidneys.

People with overactive parathyroids may suffer from indigestion. Also in about 10 percent of cases, the amount of calcium that is released from the bones as a result of the high level of PTH is so great that the bones themselves begin to show signs of strain—there may be bone pain, some loss of height, and even spontaneous fractures. X rays may show a characteristic picture of cysts in the bones, particularly the bones of the hands. The combination of bone problems, kidney stones, and indigestion has led to an old saying among doctors that the disease causes problems with "bones, stones, and abdominal groans."

Treatment

The only effective treatment for an overactive parathyroid is surgical removal of the overactive gland. In most cases all the glands are larger than normal (hypertrophied), and the standard surgical procedure is to identify all four glands and then to remove all but one half of one gland. The remaining half-gland provides enough PTH to keep the calcium level under control.

In the remainder of cases there is a tumor. Usually this affects only one gland, and only a tiny minority of patients will be found to have a tumor that is malignant.

It is not certain what the best treatment is for those people with a slightly higher blood calcium level but no symptoms. In general, "watch and wait" has been the usual approach, but recent studies now support surgery in some of these situations. The outlook following treatment is usually good.

Underactive parathyroids

In contrast, underactivity (hypoparathyroidism) is rare, unless, of course, the parathyroids are removed during thyroid surgery. People suffering from this disease are often tired. They may start having seizures and there may be signs of tetany—muscular spasm that initially affects the hands and feet.

There can also be marked psychological problems in cases of hypoparathyroidism. Many patients will have depression, anxiety, mood swings, or memory problems. People who have hypoparathyroidism may also have brittle nails.

This disease can be combated effectively by taking vitamin D by mouth. Even though a careful eye has to be kept on the patient's calcium level, the outlook after treatment is very good.

Another condition is called pseudohypoparathyroidism: the parathyroids are normal, but the body does not respond in a normal way to parathyroid hormone.

> *See also:* **Bones; Endocrine system; Glands; Hormones; Kidneys; Muscles; Thyroid**

Pelvis

Questions and Answers

Are women who have had babies more likely to get pelvic problems later in life?

Yes. The main reason is that some of the joints in the pelvis (especially those at the back called the sacroiliac joints) and their ligaments become loose during pregnancy to make the birth easier. In most women they never again become as firm as they were before the birth. The abnormal movement of bones at these joints causes backache; good posture and regular pelvic exercises, especially during pregnancy, should give at least a degree of relief.

My elderly mother will not believe me when I say that walking with a stick will help take the strain off her weak pelvis. Who is right?

You are correct in saying that a walking stick will help take the strain off her pelvis. Your mother may feel that using a stick is a sign that she has given in to the fact that she is getting a little unsteady on her legs. Instead of arguing with her, get a close family friend or your family doctor to talk with her. Or try tactfully to tell her that by helping herself she will in fact prolong her active life.

I have heard that the size of a woman's feet will give her an indication as to whether her pelvis is wide enough to allow the natural birth of her child. Is this true?

No. There is no rule relating foot length to pelvis width. Anyway, babies as well as their mothers vary in size. At a prenatal clinic all women have the size of the pelvis checked in relation to the size of their baby to see whether they will be able to give birth easily. In general, tall mothers usually have a large enough pelvis for this, but mothers who are less than 5 ft. (1.5 m) tall are more likely to have a problem giving birth.

Like a large bony hoop, the pelvis forms a complete ring around the lower part of the human body. It protects the organs within, forms a framework for muscles, and is the base to which the legs are hinged.

The pelvis is designed to bear the weight of the human body when it is running, walking, standing, or sitting. In women, the pelvis is relatively wide to help accommodate the presence of the growing fetus during pregnancy, and at the same time to partially protect it. The width of the pelvis gives a woman's hips a characteristic shape. Pelvic problems are most usually the result of damage or deformity of the bones and the muscles and ligaments connected to them. Many such problems manifest themselves as backache and associated pains.

The bones and joints

The pelvis is constructed from a group of immensely strong bones. The back of the pelvis is made up of the sacrum, a triangular structure that forms the base of the spine, and consists of five individual bones or vertebrae fused together to form a solid structure. No movement is possible between these bones. Attached to the base of the sacrum is a small projection of bone, the rudimentary human tail or coccyx, made from four fused vertebrae. The joint between the sacrum and the coccyx is padded with a disk of fiber-impregnated cartilage. In young people, some movement is possible at this joint, but it becomes rigid later in life. In young people, too, there are true joints between the bones of the coccyx, although this is more pronounced in girls.

Joined to each side of the sacrum is a massive hipbone or ilium; its curved top can easily be felt through the skin. The ilium is filled with marrow and is one of the major sites of blood cell

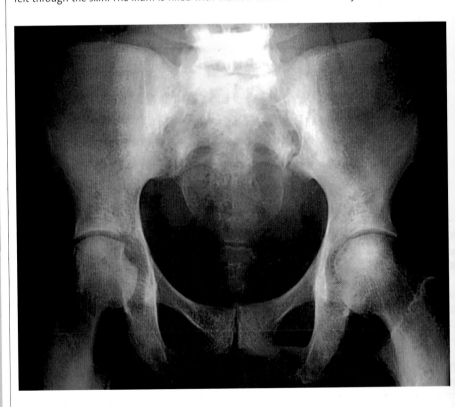

▲ *The pelvis is a large girdle of bone that protects many of the interior organs. Attached to the pelvis by ball-and-socket joints are the thighbones.*

PELVIC MUSCLES (FEMALE, FRONT VIEW)

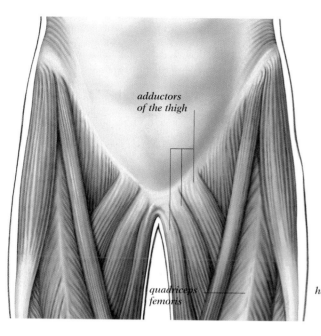

*adductors
of the thigh*

*quadriceps
femoris*

PELVIC MUSCLES (MALE, REAR VIEW)

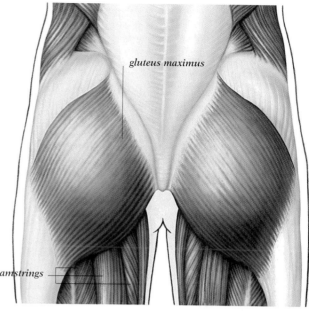

gluteus maximus

hamstrings

FEMALE PELVIC BONE

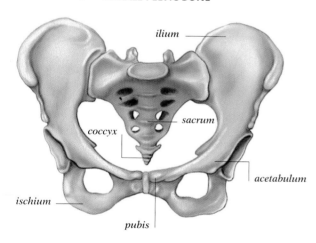

ilium

sacrum

coccyx

acetabulum

ischium

pubis

MALE PELVIC BONE

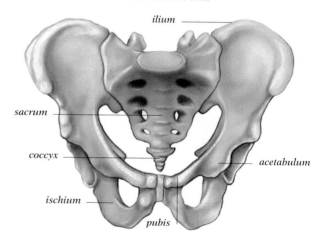

ilium

sacrum

coccyx

acetabulum

ischium

pubis

production. The vertical sacroiliac joints between the sacrum and ilium are toughened with fibers and bound with a crisscross series of ligaments. The surfaces of the bones are slightly notched, so they fit together like a loosely connected jigsaw, to give extra stability.

About two-thirds of the way down each ilium is a deep socket, the acetabulum, which is perfectly shaped to accommodate the ball at the end of the femur or thighbone. Below this socket, the hip bone curves around toward the front of the body. This part of the pelvis is the pubis and it is supplemented by a loop of bone known as the ischium, which forms the basis of the buttock. At the front of the body, the two pubic bones come together at a joint called the pubic symphysis. Padding the junction between the two bones is a disk of cartilage called the interpubic disk. More ligaments bind this joint together and also run from the top of it to the ilium to help keep the pelvis stable.

Sexual differences

Of all the bones in the body those of the pelvis show the most difference between male and female, for the simple reason that the female pelvis has to provide more space inside the body to allow for the development of the fetus. The pelvis of a man is relatively much longer and narrower than that of a woman, and because it has to bear a greater weight, consists of bone that is much less delicately molded. Thus the cavity created by a woman's pelvic bones is boat-shaped, and that of a man heart-shaped.

Because of the shape of her hip bones, and the shape and angle of placement of their sockets or acetabula, a woman stands with her feet relatively wider apart than a man, and with her legs at a different angle to her pelvis. The joints of a woman's pelvis also change during pregnancy to allow for expansion during the process of birth.

Why do women swing their hips more than men when they walk?

The reason for this lies in the shape and positioning of the pelvis. In women, the pelvis is, comparatively speaking, much wider than in men and more tilted, and the bones of the thighs join the female pelvis at a different angle. Therefore, a woman thrusts her legs out at a wider angle as she walks and this, in turn, makes her hips swing more than a man's.

I had a sneezing fit recently and felt weak in the pelvis afterward. Can sneezing strain your pelvis?

Yes, both sneezing and coughing can strain the internal muscles of the pelvic region because both actions lead to a large buildup of pressure inside the abdomen, and, as they expel air at high speed from the body, demand powerful contractions of the pelvic muscles. The reason the pelvic floor muscles contract strongly during a sneeze or cough is to ensure that bowel contents are retained.

I had my first baby normally, but I am much larger this time. Does this mean my pelvis may not be able to accommodate the birth?

Only your obstetrician can answer this question by doing an internal pelvic examination. However, often women seem much larger in their second pregnancies than in their first because their abdominal muscles are not as tight.

Can some exercises actually damage your pelvis?

If you are careful, you should have no problems. However, one exercise can damage the pelvic muscles: the one in which you lie on your back and lift both legs in the air, keeping the knees straight. If you have any hint of pelvic trouble, avoid this exercise. It can be particularly damaging for a mother who has recently given birth, and should never be attempted as a postnatal exercise.

Balance and movement

When a person is standing upright, his or her center of gravity lies near the middle of the pelvis and acts vertically downward. When the person moves his or her pelvis or feet, the center of gravity will readily move outside the supporting base of the feet. Unless the person makes appropriate correcting movements, he or she will fall over. Thus throughout each moment of a person's everyday life, except when he or she is lying down, the pelvis bears the weight of all the upper part of the body (the head, arms, and trunk).

The human skeleton is constructed in such a way that it is possible for the body to stay upright, on two legs, without falling over. As part of this design, the pelvis is not absolutely vertical but positioned at a tilt. This tilt, which is more pronounced in women than in men, makes it possible for a person to swing the hips and bear the weight on alternate legs as he

EXERCISES FOR THE PELVIS

▲ *This exercise, called pelvic rotation, helps loosen the back, pelvis, and hips. Lie on your back with your arms to your sides and your knees bent. Put your feet flat on the floor or as near to the buttocks as possible. Raise your hips off the ground and rotate your pelvis 10–20 times. Rest and repeat the rotation exercise.*

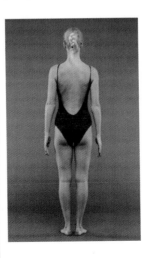
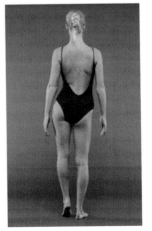
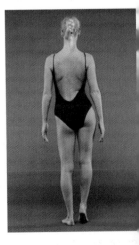

▲ *The pelvic side lift: standing, lift the left side of the pelvis toward the shoulder, but keep your shoulders in a straight line. Repeat with the right side.*

STRUCTURE OF THE MALE AND FEMALE PELVIS

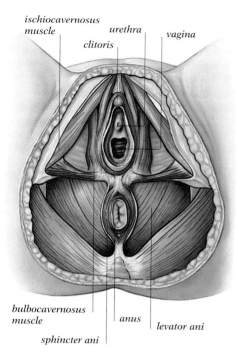

ischiocavernosus muscle
clitoris
urethra
vagina
bulbocavernosus muscle
anus
sphincter ani
levator ani

◄ *The boat-shaped female pelvis houses all the female reproductive organs and provides enough space inside the body to allow for the development of the fetus. The pelvic diaphragm protects the internal organs.*

► *The heart-shaped pelvis of a man is much longer and narrower than that of a woman's, and because a man's pelvis has to bear more weight, it consists of less delicately molded bone. The pelvis of a man also contains glands that act as lubricants for the external sex organs of the man.*

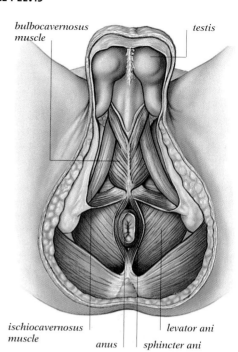

bulbocavernosus muscle
testis
ischiocavernosus muscle
anus
levator ani
sphincter ani

or she walks. Without the tilt, the hips would be too vertical and the person would fall flat on his or her face. At every footfall, the pelvis and particularly the hip joints act as living shock absorbers for the stress energy that passes up the legs, and, in the course of a lifetime, prevent the whole skeleton from crumbling under too much stress.

The pelvis muscles

The muscles of the pelvis do two very separate jobs. One is to make body movements possible; the other is to hold in the contents of the abdomen, and, quite literally, prevent them from falling out of the body.

The pelvic muscles used for movement are the piriformis, which run under the main muscles of the buttock; and the gluteus maximus and gluteus minimus, which join the top of the thighbone with the front of the sacrum. Contractions of the piriformis make it possible to move the thigh out sideways, as a person does when he or she is taking a step with the toes turned out to the sides.

The other main muscle used for movement in the pelvis is the obdurator internis. Fanned out so that it is attached at several places within the bony pelvic ring, the muscle forms two large triangular sheets that join up with a tough tendon. This, in turn, is connected to the femur (leg bone). The main job of the obdurator internis is not to move the body from place to place but to keep the body stable when it is standing still. This muscle makes the continual adjustments that are needed in order to keep the stationary body balanced.

The other category of pelvic muscles are grouped together to form an elastic sheet of tissue called the pelvic diaphragm or pelvic floor. The two main muscles in this diaphragm are the levator ani, which forms most of the lower margin of the pelvic cavity and can

be felt working if a person pulls in at the anus; and the coccygeus, which supports the coccyx, particularly during the act of defecation and while a baby is being born.

The layers of tissue, including ligaments and small muscles, that lie over the pelvic diaphragm, together form the perineum. In women, however, the word "perineum" is often used to describe only the tissues between the anus and the opening of the vagina.

Internal organs

The pelvic diaphragm does not form a complete seal over the base of the pelvis. Inevitably, there must be gaps to allow for the passage of urine and feces out of the body and, in women, to make both sexual intercourse and childbirth possible. These functions, affected by muscles, give a clue to the vital body organs that the pelvis protects—in both sexes, the bladder and the tube, the urethra, through which urine passes to the outside of the body; the lower part of the gastrointestinal tract; and the rectum and its exit, the anus. These exits are guarded by rings of muscle called sphincters which, in adults, can be relaxed by conscious control to allow urination and defecation.

In men, essential glands that act as lubricants for the externally placed sex organs are found within the pelvis, among them the prostate and seminal vesicles. In women, all the reproductive organs are housed within the pelvis—the ovaries, fallopian tubes, uterus (womb), and vagina. The vagina also has a sphincter muscle, which contracts powerfully during intercourse.

Like all parts of the body the pelvis has a supply of blood vessels and nerves, and the main ones lie near the bones. Passing in front of the pelvis are the femoral nerve and femoral blood vessel supplying the thigh. Beneath the sacroiliac joint, through a gap between the ilium and the sacrum, runs the sciatic nerve, which extends up the back.

303

Causes and treatment

Each of the separate organs within the pelvis can have specific things go wrong with it, and such problems may lead to pain either in the pelvis itself or in the back, legs, or abdomen. Some problems of the bony pelvis arise from disease of the bones that form the framework of the pelvis, or from conditions affecting the muscles that complete its base.

In both women and men, it is important that the pelvis is strong enough to support the body's weight and to take the strain

◄ *During pregnancy, there is additional strain on the muscles of the back, which can lead to backache if the correct posture is not maintained.*

of movement. Rickets, a disease that retards bone growth and may weaken bone, is caused by a lack of vitamin D and is a significant cause not only of poor pelvic development but also of pelvic weakness. It is, however, a rare disease in the United States.

The tilt of the pelvis that makes an upright, two-legged stance possible in the human body also leads to problems, the most common of which is backache. This arises from many causes, including, most commonly, strain of the muscles which are joined to the sacrum and whose contractions help move the pelvis, and problems with the sacroiliac joints. These joints often have such problems because they have little muscular support. The softening of the ligaments that bind the joint at the end of pregnancy results in a characteristic lower-back pain which may persist after the baby is born. A similar sort of pain arises from any trouble experienced at this joint.

Accidents that lead to pelvic injuries are uncommon, but they can happen. Falls from heights and crushing blows may break bones in the pelvis. If this happens, the great risk is that the broken ends of the bones may pierce one of the internal organs or major blood vessels. For this reason, and because this sort of accident may well involve spinal injuries, the victim should never be moved. The paramedics who are called to the scene will "log-roll" the patient onto the stretcher, minimizing any movement. Once at the hospital, appropriate imaging studies can be done to identify the extent of the pelvic bone injuries. With sufficient bed rest and prompt treatment a fractured pelvis usually heals in a couple of months.

Apart from tears that occur during labor, the most common problem affecting the muscles of the pelvic diaphragm is weakness leading to dropping or prolapse of the pelvic organs. This is especially the case following childbirth. One symptom of such weakness is so-called stress incontinence, that is, leakage of urine or feces when a person puts stress on the muscles.

Preventive measures

To avoid back pain resulting from strained pelvic muscles and ligaments, a person should attempt to adopt a good posture with the abdomen held well in and the back straight; lift things sensibly using the muscles of the arms and legs, and keeping the back straight so that excess strain is not put on the pelvic muscles and ligaments; and sit in a chair that provides support in the correct places.

The best preventive measure for the pelvic weakness that can lead to the prolapse of the pelvic organs following childbirth is a series of exercises that a woman can do to strengthen her pelvic floor muscles during her pregnancy and afterward. If necessary, a repair operation can be done to strengthen the pelvic diaphragm.

See also: **Anus; Back; Balance; Birth; Bones; Cartilage; Fetus; Hip; Joints; Ligaments; Movement; Muscles; Posture; Pregnancy; Prostate gland; Rectum; Skeleton; Spinal cord; Uterus; Vagina**

Penis

The penis performs two distinct and unrelated vital functions. It penetrates the vagina so that sperm can pass from the man to fertilize the women, and it is an outlet for urine to pass out of the body.

The penis consists of a central tube called the urethra through which urine passes when a man urinates. This is also the track through which semen passes during sexual intercourse.

The urethra connects the bladder, where urine is stored, to an opening at the tip of the penis (the meatus). Semen enters the urethra during intercourse through a pair of tubes called the seminal ducts, or vas deferens, which join it shortly after it leaves the bladder. A tight ring of muscle at the opening from the bladder into the urethra keeps the passage closed. Urine emerges only when this is intended.

The penis usually hangs down in front of the scrotum, which is a wrinkled bag containing the testes in a slack or flaccid state. Penis length varies from 2½ to 5 inches (6 to 12 cm). When the penis is sexually stimulated, it becomes stiff and erect, usually pointing slightly upward. It is then 4 to 8 inches (10 to 20 cm) long. The tip of the penis, called the glans, is the most sensitive area. The valley behind the glans is the coronal sulcus; the main length of the penis is the body or shaft; and the area of the penis where it joins the lower abdomen is called the root.

CROSS SECTION OF THE PENIS AND ASSOCIATED ORGANS

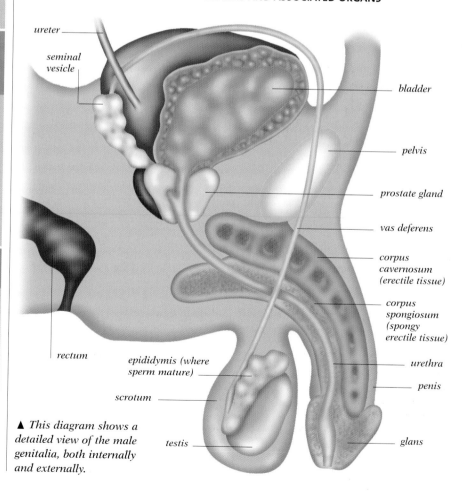

ureter

seminal vesicle

bladder

pelvis

prostate gland

vas deferens

corpus cavernosum (erectile tissue)

corpus spongiosum (spongy erectile tissue)

urethra

penis

glans

rectum

epididymis (where sperm mature)

scrotum

testis

▲ *This diagram shows a detailed view of the male genitalia, both internally and externally.*

ANATOMY OF THE PENIS

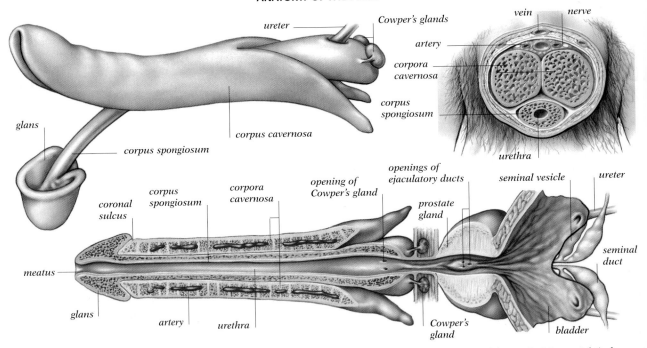

▲ *Above left is a detailed view of the penis, showing all of its parts. The section through the shaft of the penis (above right) shows the three groups of tissue responsible for erections. The longitudinal section of the penis (above) shows the path of the urethra.*

Erection

The largest part of the penis consists of three columns of spongy tissue that are responsible for erection. These areas are supplied with a rich network of blood vessels. When a man is sexually excited, the amount of blood that flows into these areas increases enormously. Engorgement with blood makes the penis longer, thicker, and rigid. It also rises as internal pressure increases. After ejaculation and after excitement subsides, blood flow diminishes and the penis returns to its flaccid state as the extra blood drains away.

The foreskin and the glans

The delicate glans is protected by a loose fold of skin called the foreskin or prepuce. As the penis becomes engorged with blood and enlarges during erection, the foreskin peels back to leave the glans exposed to the stimulation that eventually leads to orgasm.

Skin on the glans and foreskin produces a greasy substance called smegma that acts as a lubricant facilitating the movement of the foreskin over the glans. It is important to wash this away regularly. Failure to do so can result in soreness or inflammation of the foreskin and a condition called balanitis. Repeated or persistent balanitis is sometimes a medical reason for performing a circumcision, if it has not been performed at birth or for religious reasons.

Infections

The chief hazard to which the penis is exposed is infection, particularly sexually transmitted infections called sexually transmitted diseases (STDs). An inflammation of the urethra, when it discharges pus, is usually accompanied by discomfort or pain in passing urine. This condition is called urethritis. It can be caused by gonorrhea, when the discharge is copious and yellow; or by chlamydia, in which the discharge is likely to be less, as well as more mucuslike in appearance. These conditions are potentially dangerous, both to the patient and to his sexual partners, and they should be treated as soon as possible.

A more serious, but less common, disease that makes its initial attack on the penis is syphilis. This normally shows itself as a single ulcer near the head of the penis. This is a painless shallow, punched-out ulcer with a base like wet leather that is teeming with the spirochaetes of syphilis. The ulcer heals in a few days, but this is only the first stage. If it is left untreated, syphilis will continue to progress and may eventually become fatal.

Another condition that may affect the penis is phimosis, in which the foreskin is too tight to peel back during an erection or sticks to the glans. In paraphimosis, the foreskin forms a band around the coronal sulcus and causes the tip of the penis to swell up. Herpes genitalis causes small, painful ulcers similar to cold sores to appear on the penis. Except for herpes genitalis, these conditions respond well to treatment.

Other problems

Severe problems to do with impotence can be treated effectively with sildenafil (Viagra), which increases the blood flow to the penis or, if necessary, by surgically implantable devices.

See also: **Genitals; Prostate gland; Semen; Sperm; Testes; Urethra; Urinary tract**

Peritoneum

The peritoneum lines the abdominal cavity and covers all the organs inside the abdomen, allowing them to move freely. However, if it becomes inflamed, the patient can become very sick with peritonitis.

Questions and Answers

Will surgery for peritonitis leave me with a nasty scar?

Yes, usually. Since it oftentimes is not possible before surgery to determine what is causing the peritonitis, the incision must be placed in the middle of the abdomen, so that every part of it can be inspected. Infection is also a possibility, and this can lead to a scar.

What are my chances of getting peritonitis?

Quite low, especially if you have already had your appendix removed. As you get older and develop conditions that could lead to perforation of the bowel or other abdominal organs, your chance of getting peritonitis increases.

Why does appendicitis sometimes lead to peritonitis?

Peritonitis occurs when an infected appendix ruptures, releasing pus into the peritoneal cavity.

How would I know if my son had peritonitis?

He would have severe abdominal pain, which would be constant and made worse by movement. Your son would feel nauseated, would probably vomit, and would have a slight fever and a rigid abdomen. His breathing would probably be rapid and shallow.

Do cases of peritonitis always require surgery?

Almost always. The reason is to discover the cause and to do something to stop it. Some causes of peritonitis—inflammation of the pancreas, for instance—can be diagnosed by a blood test and may not need an operation.

The peritoneum is a thin membrane that lines the abdominal cavity. It also covers each of the organs contained within the abdomen.

The liver, stomach, and intestines are all covered with peritoneum, as are the spleen, gallbladder, pancreas, uterus (in women), and appendix. The peritoneum is so thin that if it was separated from the organs that it covers, it would be transparent. Despite this, it is also very strong. The way it is attached inside the abdominal cavity creates various spaces where fluid could collect in the event of leaking from one of the intra-abdominal organs.

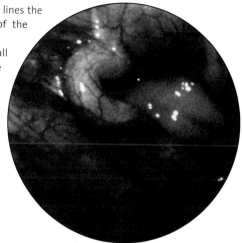

▲ *The peritoneum is the shiny membrane that covers this inflamed appendix; if the appendix ruptures, the peritoneum will become infected.*

The function of the peritoneum

The main function of the peritoneum in a healthy person is to allow the various bodily organs inside the abdomen to move freely. For example, when a person eats a meal, the stomach and the intestines become mobile and the muscles in the organ walls contract. This allows the food that has just been eaten to be mixed up and then propelled along on its journey through the digestive system (alimentary canal). During this process, both the stomach and the intestines are able to slide over one another largely because they are both covered with peritoneum; the two are also separated by a thin layer of lubricating fluid.

The peritoneum that covers the intra-abdominal organs, such as the stomach, pancreas, and so on, is called the visceral peritoneum. However, the peritoneum also lines the abdominal cavity, and where it does so, it is called the parietal peritoneum.

The parietal peritoneum has an extremely sensitive nerve supply, so that any injury or inflammation that occurs in this layer is felt by the patient as an acute localized pain. The visceral peritoneum, on the other hand, is not so sensitive and pain is experienced only if, for example, the intestine becomes stretched or distended. Even then, the pain is not very localized and is felt by the person as a dull ache, usually in the center of the abdomen. These differences in how pain is felt in the abdomen have an important bearing on the symptoms of various disorders of the intra-abdominal contents. In effect, differences in pain can often indicate the type of illness a person is suffering from.

The omentum

One structure that should be mentioned in connection with the peritoneum is an extension of it called the omentum. Shaped a little like an apron, the omentum consists of fat with a rich blood supply and is itself covered with peritoneum. The omentum hangs down from the stomach and the large intestine, and its lower part is free to move about in the space between the intra-abdominal organs and the abdominal wall. Hence, the omentum can be found between the visceral and parietal peritoneum, outside the intestines.

The role of the omentum is to act as a fat store and to help limit infections in the abdominal cavity by sticking to whatever area may be affected and so isolating the area to some degree.

Ascites and adhesions

Two of the ways in which the peritoneum can be affected by disease are by ascites and adhesions. In ascites, there is an excess amount of the lubricating fluid that is normally present between the parietal and visceral layers. Either it is caused by an imbalance between the production and absorption of the amount of fluid—such as occurs when a person is suffering from liver disease—or it happens when the peritoneum is irritated to a minor degree over a long period of

POSITIONS OF THE PERITONEUM AND OMENTUM

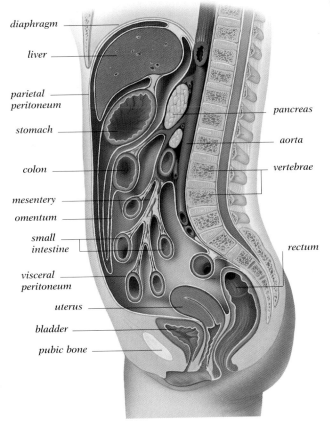

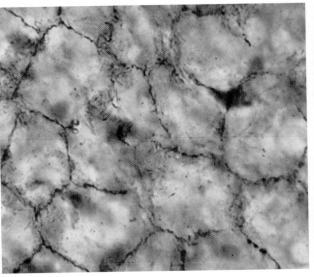

▲ *Mesothelium, the layer of flat cells that gives rise to the squamous cells of the peritoneum. The membranes that surround the heart and lungs are made up of similar cells.*

time, as can happen with a slow-growing tumor. A person with ascites usually has a very distended abdomen, although often the distention is not accompanied by any sort of pain.

Normally, the intra-abdominal organs, such as the stomach and intestines, are attached to, or suspended from, the peritoneal cavity by mesenteries, fused double layers of peritoneal membrane. The mesentery, which contains a series of branching arteries, veins, lymph vessels, and nerves, is the lifeline of the organ to which it is attached. The organs have a certain amount of movement other than the action of the mesenteries. However, adhesions occur where a part of one of these organs becomes stuck to the abdominal wall or to another organ. This can happen after a person has an abdominal operation or after he or she has had peritonitis.

The effect of adhesions is twofold. First, the mobility of the organ involved is impaired, and this in turn may lead to obstruction of the intestines. Second, the intestines may twist around an adhesion, cutting off its blood supply; this could eventually lead to gangrene of the bowels.

The symptoms of adhesions vary a great deal. They can range from recurrent attacks of abdominal pain to complete obstruction of the intestines, which causes pain, constipation, and abdominal distention. Bowel obstruction as a result of an adhesion sometimes corrects itself without surgery. However, if it continues for more than a few hours, surgery is needed to divide the adhesion and to check that the bowel has not become gangrenous. Adhesions that occur after an abdominal operation cannot be prevented. Some people are prone to recurrent adhesions.

Peritonitis

A third disease that can affect the peritoneum is peritonitis. In this the peritoneum becomes inflamed from infection, irritation by harmful substances, or injury. The main symptom is pain that is constant and may be poorly localized initially. A patient with peritonitis usually lies still, because any movement of the abdomen

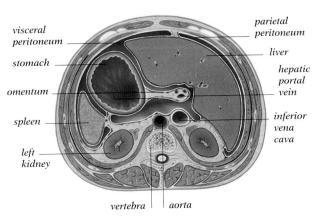

▲ *Cross section (above) and longitudinal section (top) of the abdomen, showing the parietal and visceral peritoneum.*

Questions and Answers

Is a hole in the peritoneum likely to be a serious problem?

A puncture of just the peritoneum is usually of little consequence. Much more worrying would be a puncture of the intestines, the stomach, or other organs, which could cause peritonitis. Also serious would be a puncture of a main blood vessel in the abdomen, which would cause hemorrhage into the peritoneal cavity.

Does peritonitis always lead to the formation of adhesions?

Yes. However, in some cases these adhesions disappear after a short time. The initial adhesions are made of a sticky substance secreted by the peritoneum. This may or may not eventually be converted into fibrous tissue.

If a person receives a knife wound to the abdomen, does peritonitis always result?

No. A knife can penetrate all the muscle layers and the peritoneum, enter the abdominal cavity some considerable distance, and still fail to puncture the intestine or a blood vessel. The intestine, which is covered with slippery peritoneum, may simply slide to one side of the knife blade.

I have a duodenal ulcer. How would I know if it had perforated?

You would experience severe pain all over the abdomen, quite unlike the indigestion-type pain you probably get now. It would be so severe that you would be unable to work or do anything else.

I have just had peritonitis from a burst appendix. How long will it be before I am back to normal?

If there are no complications, it usually takes about three months from the operation before you are completely back to normal. After this time, there shouldn't be any restrictions on what you can do.

is extremely painful. Even coughing and breathing may cause severe pain in the abdomen. However, patients who take narcotics or steroids, such as prednisone, may have peritonitis but because of the drugs would feel none of the usual pain.

With abdominal pain that is due to causes other than peritonitis, such as an obstruction with adhesions, the patient experiences waves of pain. When this pain reaches a peak, a person may roll around in agony, changing positions frequently. It is very unusual for someone with peritonitis to move around in this way.

When the peritonitis has been present for some hours, the peritoneum on the outside of the intestine becomes inflamed and the normal movements of the intestines (peristalsis) cease altogether. This state is known as paralytic ileus. Eventually, because nothing is passing through the alimentary canal, the stomach fills up with fluid, and this will cause the patient to vomit.

The spread of peritonitis can be prevented by the omentum, because it has the property of being able to stick to areas of inflammation, block infection, and so prevent the infection from spreading to the rest of the abdominal cavity.

When a doctor examines a patient for possible peritonitis, he or she will look for lack of movement of the abdominal wall, a feeling of rigidity when the abdomen is pressed, and an absence of intestinal sounds. The patient may show signs of shock—a fast pulse, low blood pressure, and pale and clammy skin.

Causes of peritonitis

Peritonitis can be caused by various diseases, including acute appendicitis. In this condition, the appendix becomes inflamed and ruptures, releasing pus into the peritoneal cavity. The initial symptom of appendicitis is sharp pain below the navel on the right side of the body, which is caused by the stretching of the appendix wall. However, when the pus is released, the parietal peritoneum becomes inflamed. As a result, the pain becomes localized in the area where the pus is—often the lower right side of the abdomen.

▲ *Boxers risk serious injuries, including blows to the abdomen. Such blows can in turn damage internal organs and lead to peritonitis.*

Although he had survived many death-defying stunts, the famous escapologist Harry Houdini died of peritonitis after being punched in the stomach.

If appendicitis is allowed to progress beyond this stage, it may become blocked off by the omentum and loops of small intestine, leading to the formation of a lump known as an appendix mass; alternatively, it may develop into widespread peritonitis. The latter situation is more common in young children, probably because the omentum is not yet fully developed. Peritonitis can have severe and sometimes fatal consequences.

Another cause of peritonitis is a perforated duodenal ulcer. In this case, a tiny hole is made by the ulcer through the wall of the duodenum, and this allows bile, pancreatic juice, and gastric juice to flood out into the space between the visceral and parietal peritoneum. These digestive juices have a corrosive effect, and if the resulting peritonitis is not treated at an early stage, a widespread infection will result and the patient will become extremely ill with bacterial peritonitis. Once again, this type of peritonitis can have fatal consequences.

Among the other causes of peritonitis is a condition called perforated diverticulitis. In this case a diverticulum—a blind-ended sac on the side of the large intestine—ruptures, with consequences similar to those of a ruptured appendix.

Peritonitis can also be caused by an injury to the stomach, such as a stabbing; as a result of a kick or heavy blow; or from an automobile accident. Peritonitis can be caused by infected fallopian tubes in women, and it can also be a complication of pancreatitis.

Symptoms of appendicitis

EARLY SYMPTOMS

Acute pain in the stomach that comes and goes.

Loss of appetite.

Constipation.

In children, a stomach flu may have symptoms that imitate appendicitis; these could delay diagnosis.

LATER SYMPTOMS (GET MEDICAL HELP AT ONCE)

More pain in the appendix area (right lower abdomen).

Pain may move up or down from umbilicus (navel).

Slightly raised temperature: for example, 99.5°F (37.5°C).

Slight increase in pulse rate.

In children, peritonitis can follow rapidly when the appendix ruptures, usually in a matter of hours after the first onset of pain. Peritonitis is particularly serious in a young child because the omentum—the abdominal "policeman"—is not well developed, so the infection can spread rapidly.

Treatment of peritonitis

The treatment of peritonitis obviously depends on the underlying cause. Most causes require an operation, but there is one cause—pancreatitis, which is diagnosed by a special blood test—in which surgery is considered unnecessary and even dangerous.

Because people with peritonitis will have constant vomiting, they will be given fluid intravenously. If infection is present, antibiotics will be given. A tube is usually passed down the esophagus and into the stomach to drain excess fluid.

The type of surgery (if it is to take place) will also depend on the cause of peritonitis. If it is caused by appendicitis, the appendix will be removed; if it is caused by a perforated ulcer, the perforation (hole) will be repaired. After the cause has been dealt with, the abdominal cavity is washed out with warm saline (salt) water.

Outlook

Most people make a complete recovery from peritonitis, and within a few months their health is generally back to normal. Occasionally a patient can be troubled by recurrent adhesions that may require further surgery. In cases of peritonitis that involve the peritoneum in the pelvis, a woman may be left with fertility problems, as the fallopian tubes sometimes become blocked.

See also: **Abdomen; Alimentary canal; Pancreas; Stomach**

Perspiration

Questions and Answers

I perspire heavily all year round, but I always feel more sticky and uncomfortable during the winter. What can I do?

In summer you probably wear light, loose-fitting clothes that allow air to circulate around your body, keeping you drier. In winter you wear heavier clothes to keep warm, but these trap perspiration, since little air can flow around your skin, making you feel sticky and uncomfortable. During the winter, try wearing looser clothes made of natural fibers, which allow freer evaporation of water vapor.

Whenever I take off my shoes and socks my feet are sticky and smelly. Is there anything I can do to stop my feet from sweating?

The short answer is no. The soles of your feet, like the palms of your hands, have many hundreds of sweat glands that are important for controlling your body temperature. As most people know, hands and feet also feel sticky when one is nervous or excited. You will feel less sticky if you wash your feet regularly with soap and water, dry them carefully, and then dust them with talcum powder. This is especially important if you wear nylon pantyhose or socks, and shoes made of plastic or another synthetic fiber. Cotton or wool socks and leather shoes are better for absorbing water vapor.

I find that I need to use the toilet more frequently during the winter, particularly when I am cold. Why does this happen?

Fluid loss from the body is a delicate balance between loss through perspiration, loss of water vapor from the lungs, and loss through feces and urine. When you are cold you lose very little fluid through perspiration and have to make up the balance through extra loss in urine.

Perspiration is one of the most underrated of all the vital functions of the body. Without the built-in thermostat of the sweat glands, people would overheat and eventually die.

Normal body temperature has been standardized at 98.6°F (37°C). Although there are variations and daily fluctuations from person to person, it is essential that the normal or core temperature is kept constant. If the outside temperature rises too much, the human body cleverly maintains its core temperature by losing heat through the process of perspiration.

How perspiration works

A small amount of body heat is lost each day directly through the lungs and the skin without involving the sweat glands at all. However, this is a fairly inefficient way of losing heat. It is not a very flexible method, because a person cannot increase his or her breathing, as a panting dog can, to lose excess heat if it gets too hot.

Most of the heat loss that occurs every day results from perspiration or sweat production from the sweat glands. However, the liquid sweat usually evaporates from the skin before it can be noticed, and for this reason it is called insensible perspiration. It is this evaporation that allows heat to be lost from the body.

Insensible perspiration works on the principle that liquid needs energy to help it evaporate, in the same way that boiling water transforms it into steam. In human beings, that energy comes

FUNCTIONING OF A SWEAT GLAND

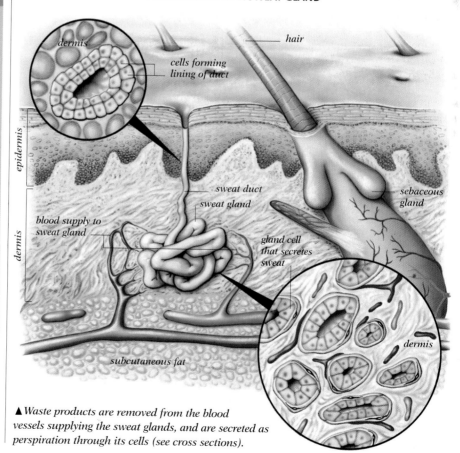

▲ Waste products are removed from the blood vessels supplying the sweat glands, and are secreted as perspiration through its cells (see cross sections).

Questions and Answers

My husband works in a steel mill, and a colleague was taken to the hospital with severe cramps. What caused them, and could my husband get them too?

If you work near a furnace, or even just live in a hot climate, when you are not used to it, you will perspire heavily. You lose more water than salt when you sweat, and too little of either can cause muscle cramps. It sounds as if your husband's colleague made the mistake of drinking lots of water without replacing the salt as well. If your husband gets very hot at work, make sure he takes extra salt as well as water.

I'm in my late forties and have just stopped having my menstrual periods. I also have hot flashes regularly, and I know this is normal during menopause. Sometimes I perspire a great deal as well. Is this normal?

Yes. During menopause, the hormonal and chemical balance in the body is disturbed. Hot flashes and sweats are caused when the body's thermostat overreacts to a garbled message. It's a bit like a furnace belting out heat because the thermostat registered that the room temperature had suddenly dropped to the freezing point. Cold sweats may also be a problem during menopause.

What's the best sort of deodorant or antiperspirant to use?

Body odor is caused by perspiration interacting with bacteria on the surface of the skin. Deodorants either mask the smell or inhibit the action of the bacteria, whereas the chemicals in antiperspirants stop the sweat glands from working by contracting the skin so that sweat cannot flow. Both types of protection can never be 100 percent effective and may cause irritation in people with sensitive skin. There is no substitute for regular washing with soap and water and for wearing clothes made from natural materials rather than from synthetic fibers.

▶ *During exercise, the human body can produce heat at a rate 10 to 20 times greater than when it is resting. To survive, the body must get rid of this excess heat, which is why people sweat so much when they are doing rigorous exercise.*

from the surface of the skin, and the effect of evaporating perspiration is to use up some of the heat and energy in the skin, leaving the person cooler. Once a person has become so hot that the perspiration is beginning to pour off his or her skin, the system has actually reached the stage where it can just barely cope. It takes only 1 calorie to raise 0.03 ounce (1 g) of water by 1.8°F (1°C), but it takes 539 calories to convert 0.03 ounce (1 g) of water from a liquid to a vapor state at the same temperature. This is the amount of heat that is taken from the body when each gram of sweat evaporates.

The sweat glands

The body is covered in sweat glands that produce liquid. Before puberty only one set of glands function, the eccrine glands. These are found all over the body except the lips and some parts of the sexual organs. There are many of these glands in thick-skinned areas such as the palms of the hands and the soles of the feet, and their activity is controlled both by the nervous system and by some hormones. This means that as well as responding to changes in temperature they also react under other conditions—hence the sweaty hands when someone is excited and, later in life, the unexpected hot flash of menopause.

The other glands, the apocrine glands, are more complicated than the eccrine glands. Under a microscope they look like worm casts—highly complex coils. They develop and start to function during adolescence and are found in the armpits, the groin, and the areola (nipple) of the breast. They are not associated with the nervous system, but the organic matter they produce does cause body odor if the body is not washed regularly. This is because organic matter reacts with bacteria in the skin, causing an offensive smell.

Keeping clean

Normal sweating is an important function and should not be prevented completely. Washing daily with soap and water should be sufficient, though many people like to minimize the possibility of body odor by using a deodorant. However, never use a deodorant without washing first, since it could irritate your skin.

People who perspire heavily often want to reduce the wetness by using an antiperspirant, but some antiperspirants, particularly the roll-on variety, are not ideal because they work by blocking the glands and preventing the escape of sweat.

Continued use of antiperspirants could lead to irritation, particularly for those who have sensitive skin.

▲ *The intense heat of the desert is made more bearable by the fact that the climate is very dry. Sweat can therefore evaporate more rapidly, cooling the skin.*

Overheating

Perspiration from the eccrine glands does not consist simply of water; it also contains a wide range of chemicals found in the body, the most important being salt. People who perspire very heavily because of their work, or because they live in a hot environment, may lose up to 1.75 gallons (6.5 l) of fluid a day. They have to replace not only the lost fluid but the lost salt as well. They can either eat salty food or, if they can't keep food down, take salt tablets. Failure to do so can result in dizziness and headaches, a condition known as heat exhaustion. It is possible, however, to adapt to living in a hot environment; the body itself adjusts and excretes less salt.

If the body does not adapt fully to very hot weather, a person can run a slight risk of suffering from heatstroke. This is a very serious condition in which the body stops sweating completely, and the core temperature rises dramatically. If the person is not cooled down quickly, the result may be brain damage or even death. A far more common condition in hot weather is prickly heat. When exposed to the sun, the skin develops a red itchy rash. This happens because the sweat glands become blocked so that perspiration does not escape and an irritation occurs in the skin surrounding the gland. The best treatment is to stay out of the sun.

Overheating can also occur when a person has a fever. Bacteria and viruses produce toxic substances that the body tries to kill by raising the thermostat. This raises the core temperature so that people with fevers sweat a lot.

Keeping cool

The body's cooling system works most efficiently in a drier atmosphere. If the atmosphere is humid as well as hot, perspiration cannot evaporate, and the film of perspiration that covers the skin stops the cooling process. This is why hot, humid climates are uncomfortable to live in, compared with hot, dry climates: it is impossible to remain cool. Likewise, tight-fitting clothes make people feel hot and sticky because their skin is bathed in a film of sweat, as if they were in a tropical rain forest. To stay cool in the heat, people should wear loose-fitting clothes so that air can circulate around their bodies. It is best to wear natural fibers, which will let the skin dry more easily.

See also: Body odor; Feet; Glands; Hand; Menopause; Nervous system; Temperature

Pharynx

Questions and Answers

My son complains constantly of a sore throat. Does he need to have his tonsils out?

This is a question that only your doctor can answer, after having examined your son's tonsils. Constant sore throats in children are often caused by infections of the tonsils and disappear once the tonsils have been removed. However, this is not the only cause. If you are a smoker, it could be that inhaling your cigarette smoke is causing the problem.

Why is it that so many illnesses seem to start with a sore throat?

First, some of the tissues found in the pharynx (throat) are part of the body's defense system against disease. These tissues become swollen and inflamed as they fight off bacteria and viruses. Second, if the body's defenses fail, disease-causing organisms attack the pharynx as well as other tissues.

When should I worry that a sore throat may be a strep throat?

Call your doctor if you cannot swallow liquids, if you have trouble breathing, if you have a fever over 101ºF (38.3ºC), if the nodes in your neck are swollen, if your tonsils are bright red or have white pus on them, or if your sore throat lasts longer than a week. Children with sore throats should get a throat culture if bacterial infection is suspected.

My daughter's sore throat has developed into a middle-ear infection. Why?

This is a common complication of a sore throat because the infection travels easily from the pharynx to the middle ear via the eustachian tube. The condition is easily treated. You should consult your doctor as soon as possible.

Commonly referred to as the throat, the pharynx provides a vital link between the nose, mouth, and voice box. In so doing, it plays a major role in the essential tasks of breathing, eating, and speaking.

The pharynx—usually called the throat—is the area at the back of the mouth that extends down inside the neck. Nearly everyone will have experienced a sore throat at some time—in most cases it is a symptom of the common cold or tonsillitis.

The pharynx is deep-lined with muscles and is shaped, very roughly, like an inverted cone. It extends for about 5 inches (12 cm) behind the arch at the back of the mouth to where it joins up with the gullet.

The upper, wider part of the pharynx is given rigidity by the bones of the skull, and at the lower, narrow end its muscles are joined to the elastic cartilages of the voice box. The outermost tissue layer of the pharynx, which is continuous with the lining of the mouth, contains many mucus-producing glands that help to keep the mouth and throat well lubricated during eating and speaking.

The parts of the pharynx

Anatomically, the pharynx is divided into three sections according to their positions and functions. The uppermost part, the nasopharynx, gets its name from the fact that it lies above the soft palate and forms the back of the nose.

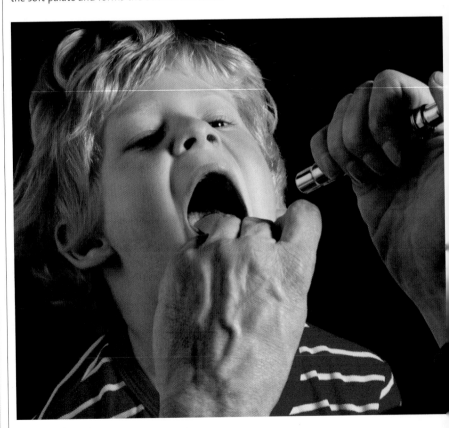

▲ *Using a small flashlight and a disposable tongue depressor, a doctor can look at a patient's throat to check for inflammation. In children, the tonsils are often implicated.*

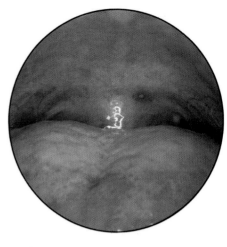

▲ *A form of pharyngitis called strep throat can be treated with antibiotics.*

◄ *Opera singers, such as the celebrated soprano Kiri Te Kanawa, must be careful to guard against infections of the pharynx.*

STRUCTURE OF THE PHARYNX

THE PHARYNX DURING SWALLOWING

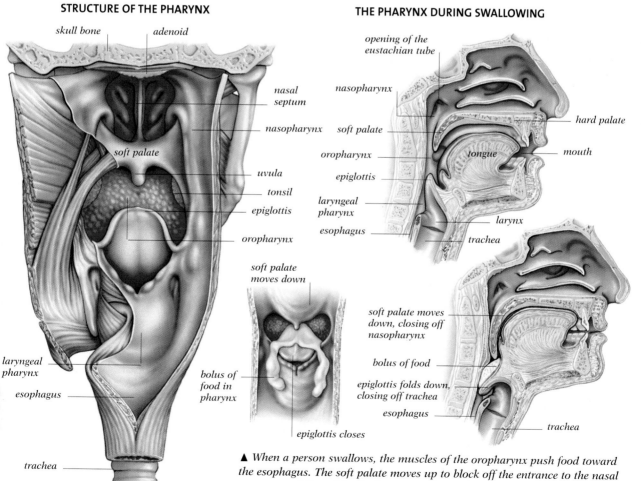

▲ *When a person swallows, the muscles of the oropharynx push food toward the esophagus. The soft palate moves up to block off the entrance to the nasal passage, and the epiglottis closes over the windpipe.*

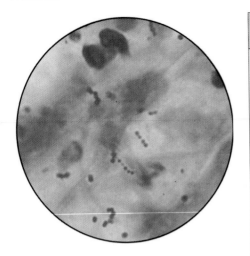

▲ *A magnified view of streptococci, the*
bacteria responsible for causing strep throat.

Throat obstruction

Small fish bones and other similar objects can get stuck in the throat very easily. If this happens, the best course of action is to eat one or two mouthfuls of bread and then take a drink of water. If the bread and water do not move the offending object, always seek medical help rather than try to get it out yourself. One reason for this is that the bone may not actually be stuck at all—instead it may have grazed the pharynx itself, leading to a sensation that mimics a stuck bone.

Below this, the nasopharynx is bordered by the soft palate itself. Upward movement of the soft palate closes off the nasopharynx when a person swallows, and this prevents food from being forced up and out of the nose. A failure in this coordination leads to the discomfort that is sometimes experienced when a person sneezes.

In the roof of the nasopharynx are the adenoids, two clumps of tissue that are most prominent in childhood. The nasopharynx also contains, on either side of the head, an entrance to the eustachian tube, the passage between the middle ear and the throat. Disease-causing microorganisms of the mouth, nose, and throat have easy access to the ears and can cause middle-ear infections.

The oropharynx, the area of the pharynx at the back of the mouth, is part of the airway between mouth and lungs. It is much more mobile than the nasopharynx. The squeezing actions of the muscles of the oropharynx help shape the sounds of speech as they come from the larynx. With the aid of the tongue, these muscles also help to push food down toward the entrance to the esophagus.

The most important organs of the oropharynx are the tonsils, two masses of tissue that are often implicated in the sore throats common in childhood. Like the adenoids, the tonsils are composed of lymphoid tissue characteristic of the body's defense system. This tissue produces specialized white blood cells that engulf invading bacteria and viruses.

The lowermost or laryngeal section of the pharynx is involved entirely with swallowing. This section lies directly behind the larynx, and its lining is joined to the thyroid and cricoid cartilages. The movements of these cartilages help in the production of sounds. Contraction of the muscles helps to propel mouthfuls of food through this part of the pharynx. Just above the laryngeal part of the pharynx is the epiglottis, a flap of tissue that closes down over the entrance to the airway as a person swallows and thereby prevents food from getting into the lungs and causing choking.

What can go wrong

By far the most common problem of the pharynx is inflammation, known medically as pharyngitis and experienced as a sore throat. Pharyngitis can appear suddenly (acute), or it can persist over several months or even years (chronic).

The most usual cause of acute pharyngitis is the common cold. A sore throat is a telltale sign of an impending infection, even before the first cough or sneeze. The common cold is caused by a virus. However, there are also bacteria—called streptococci—that can cause a form of pharyngitis known as strep throat.

As well as a direct infection, pharyngitis can be a subsidiary symptom of other diseases. Inflammation and infection of the parts of the body next to or within the pharynx, including the larynx, mouth, sinuses, and tonsils, can result in pharyngitis.

Other diseases that usually include pharyngitis as one of their many symptoms are glandular fever, measles, and rubella (German measles). Scarlet fever, once a lethal childhood disease but now easily controlled with antibiotics, is confined to the nose and throat; it is also associated with severe pharyngitis.

The chief culprits in the case of chronic pharyngitis are smoking and excessive drinking. Cutting down or—particularly in the case of smoking—stopping altogether is necessary for a cure. Another common cause of chronic pharyngitis is postnasal drip—a constant drip of fluid from the back of the nose. This results from persistent mouth-breathing due to a blocked nose. There are many causes of this, and a doctor will make a diagnosis.

Smokers may develop cancer of the pharynx. In the very obese, adipose tissue that has deposited in the soft tissues of the neck can partially obstruct the pharynx, causing snoring, sleep apnea, and other sleeping disorders.

Treatment

Treatment depends on the cause. Conditions caused by viruses, such as the common cold, do not respond to antibiotics. Bacterial cases, such as strep throat, can be cured with antibiotics.

People should use common sense in deciding whether or not they or their families need medical attention. When an acute sore throat first occurs, it can be soothed by sucking lozenges. If the discomfort persists, or if the patient has a very high fever, then medical advice should be sought.

> *See also:* Adenoids;
> Esophagus; Larynx; Palate;
> Throat; Tonsils

Pituitary gland

Questions and Answers

I heard of a woman who started to lactate when she had not had a baby. Could this have been due to a problem in her pituitary gland?

Yes. One fairly common pituitary problem is the secretion of excessive amounts of a hormone called prolactin, and this stimulates milk production. The cause is a slow-growing tumor in the gland called a prolactinoma. This can damage vision and may be fatal.

If the pituitary failed, would all the other glands also stop working?

No. The main glands affected would be the adrenal, thyroid, and sex glands. The pancreas would continue to produce insulin, and the parathyroid glands would continue to control the level of calcium. Important hormones from other glands, such as renin from the kidneys (one of the hormones concerned with retaining adequate amounts of salt and water in the body) would still be secreted.

Does your libido decline if your pituitary gland stops working?

Yes, if the condition is not treated. One of the main functions of the pituitary is to stimulate the production of sex hormones. These come from the ovaries and (to a lesser extent) from the adrenal glands in women and from the testes and adrenal glands in men. However, these hormones can be replaced if the pituitary fails.

Is the pituitary essential to life?

Yes. Pituitary failure is dangerous mainly because the adrenal glands depend on it to function. A lack of adrenal hormones, such as cortisone, is fatal. If the pituitary fails, the thyroid gland will also stop working; this too can be fatal, but replacement thyroid and cortisone can be given.

The pituitary is the main gland in the body's hormone system. Although it weighs only a few ounces, it is responsible for the control of many hormones, including those of the adrenal glands, thyroid glands, and sex glands.

The pituitary gland is found in the base of the brain. It is joined by a stalk of nervous tissue to the hypothalamus, and works closely with this part of the brain. Together, the pituitary and the hypothalamus control many aspects of the body's metabolism—the chemical processes that keep the body functioning.

Structure and function

The pituitary sits inside a protective bony saddle called the sella turcica—Latin for "Turkish saddle." The sella turcica, or sella, as doctors call it, can be seen clearly on an X ray of the skull; an enlarged sella is an indication that something is wrong with the pituitary gland.

The gland itself is divided into two totally separate halves. The rear half, or posterior pituitary, is linked to the hypothalamus through the pituitary stalk. It is concerned with the production of two major hormones that are produced in the hypothalamus. From there, the two hormones travel along specialized nerve cells to the posterior pituitary. They are released from the posterior pituitary when the hypothalamus receives appropriate messages about the state of the body.

The front half of the pituitary, the anterior pituitary, produces the hormones that activate other important glands in the body; it also produces two important hormones that act on the tissues directly. The anterior pituitary is not linked to the hypothalamus directly, but it is bound closely to this part of the brain in the way that it functions.

Since the anterior pituitary has no direct nerve paths to link it with the hypothalamus, it has to depend on a series of special releasing and inhibiting factors to control hormone release. Some of these factors are themselves specialized hormones, which are released by the hypothalamus and act on the pituitary gland. The specialized hormones are carried in a set of

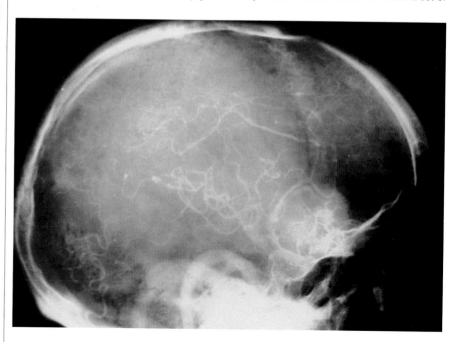

▲ *This X ray of the skull shows a tumor near the pituitary fossa, a hollow in the sella turcica. Such growths often prevent the pituitary from producing hormones normally.*

LOCATION AND STRUCTURE OF THE PITUITARY GLAND

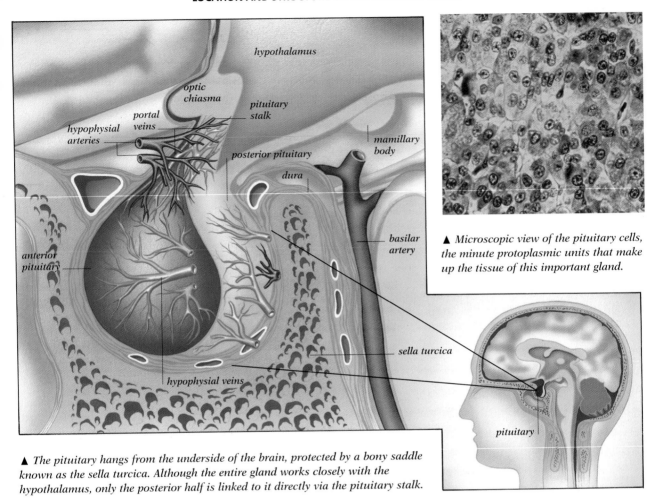

▲ *Microscopic view of the pituitary cells, the minute protoplasmic units that make up the tissue of this important gland.*

▲ *The pituitary hangs from the underside of the brain, protected by a bony saddle known as the sella turcica. Although the entire gland works closely with the hypothalamus, only the posterior half is linked to it directly via the pituitary stalk.*

blood vessels—known as the pituitary portal system—between the hypothalamus and the pituitary.

Although many of the instructions to release hormones come from the hypothalamus, the anterior pituitary also has independent control. In addition, the release of some of its secretions is inhibited by substances circulating in the bloodstream. One such secretion is the hormone TSH (thyroid-stimulating hormone), which stimulates the thyroid gland to produce its hormones. The release of TSH by the pituitary is inhibited when there are high levels of thyroid hormone already in the blood. Called negative feedback, this is an important principle in the control of many pituitary hormones.

This principle means that levels of hormones produced in glands that are far removed from the pituitary cannot rise too high; if hormone levels are already quite high, negative feedback acts on the pituitary and turns off the production of stimulating hormones.

Hormones of the pituitary

The posterior pituitary produces two hormones—antidiuretic hormone (ADH) and oxytocin. ADH is concerned with the control of water in the body. It acts on the tubules of the kidneys, so that the kidney tissue is able to withdraw more or less water (as necessary) out of the urine as it leaves the tubule. When ADH is

secreted into the blood, the kidneys tend to conserve water. When the hormone is not secreted, more water is lost in the urine. Alcohol stops the pituitary from secreting ADH; for this reason, drinking alcohol causes urination.

The role of oxytocin is less clear, but in women it is concerned with starting labor and causing the uterus to contract. It also plays an important part in starting the secretion of milk from the breasts during lactation. In males, oxytocin may be concerned with generating an orgasm.

The anterior pituitary produces six main hormones. Four of these are concerned with the control of other important glands in the body—the thyroid gland; the adrenal glands; and the gonads (the testes in the male; the ovaries in the female). The activity of the thyroid gland is triggered by TSH; the cortex (outer part) of the adrenal gland is affected by the hormone ACTH (adrenocorticotropic hormone). The overall levels of thyroid hormone and cortisone from the adrenal glands are maintained by a combination of negative feedback, which acts on the pituitary, and extra signals that come from the hypothalamus—in times of stress, for example.

The anterior pituitary also releases the hormones FSH (follicle-stimulating hormone), and LH (luteinizing hormone). These are known as gonadotropins, which are hormones that affect the sex

HORMONAL ACTIVITY OF THE PITUITARY

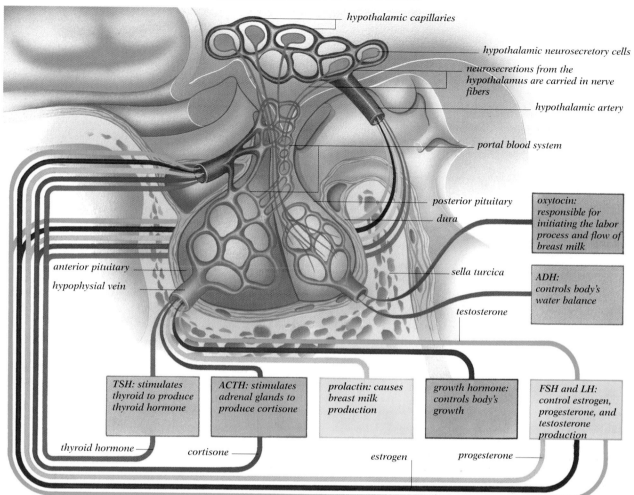

hypothalamic capillaries

hypothalamic neurosecretory cells

neurosecretions from the hypothalamus are carried in nerve fibers

hypothalamic artery

portal blood system

posterior pituitary

dura

oxytocin: responsible for initiating the labor process and flow of breast milk

anterior pituitary

hypophysial vein

sella turcica

ADH: controls body's water balance

testosterone

TSH: stimulates thyroid to produce thyroid hormone

ACTH: stimulates adrenal glands to produce cortisone

prolactin: causes breast milk production

growth hormone: controls body's growth

FSH and LH: control estrogen, progesterone, and testosterone production

thyroid hormone

cortisone

estrogen

progesterone

▲ *Four of the pituitary hormones activate an organ to produce another related hormone. Some of this hormone in the blood will then feed back into the pituitary, thus regulating its production; some of it will also pass through the hypothalamus, stimulating neurosecretions that travel to the portal blood vessels and back into the pituitary to control the release of various other hormones.*

glands. FSH and LH in turn stimulate the production of two major sex hormones: estrogen and progesterone. These control menstruation in the female and stimulate the production of hormones and sperm in the male.

The hormone prolactin is one of two hormones of the anterior pituitary that seem to act on the body's tissues without stimulating other glands. Like gonadotropins, prolactin is concerned with reproduction. Also in common with gonadotropins, prolactin has a more complicated role in the female than in the male. Its role in the male is unclear, although excessive amounts can have ill effects. In the female, prolactin stimulates the breasts to produce milk. In large amounts, it also inhibits ovulation and the menstrual cycle. This explains why women who are breast-feeding are unlikely to conceive.

The anterior pituitary also produces growth hormone; its role is to promote normal growth. This is of most importance during childhood and adolescence, but the hormone continues to be of some importance in later life, since it determines the way that body tissues handle carbohydrates.

What can go wrong
Since the pituitary plays such an important role, any malfunctions of the gland can be serious. However, such problems are quite unusual.

The pituitary may give rise to problems in three ways. Like any other gland, it may become either overactive or underactive; it may also be the site of a tumor. Since it lies in the base of the brain, tumors can cause problems by growing outward and pressing on important structures. For example, immediately above the sella is a structure called the optic chiasma, where the optic nerves that carry the information from the eyes cross over each other. Any slight pressure on the optic chiasma from an outward-growing pituitary tumor can lead to progressive loss of sight.

Tumors in the pituitary gland itself fall into two categories: those that produce excessive amounts of the various hormones, and those that do not. The most common type, prolactinomas, are very tiny;

however, they produce high levels of prolactin. In women, this overproduction will lead to amenorrhea (absence of periods), infertility, and sometimes galactorrhea (milk production from the breasts). In men, it can lead to impotence and sterility.

A second type of hormone-producing tumor leads to very high levels of growth hormone in the blood. If this starts before puberty, it leads to gigantism; this condition is rare, however. It is much more common for the condition to start in adulthood when the long bones of the arms and legs are no longer capable of growing. However, in such cases the affected person's hands and feet may grow thicker and his or her features may become gradually coarser as a result of new growth of facial bones. The nonbony parts of the

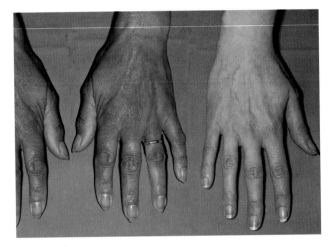

▲ *Darkening of the skin, shown in the two hands on the left, is a result of overactivity of the pituitary gland as it attempts to correct a deficiency of the adrenal glands.*

▲ *Two pituitary hormones are responsible for the production of breast milk: prolactin stimulates the breast to produce milk, and oxytocin triggers milk flow.*

body may also grow, leading to weight gain and thickening of the skin. When these problems occur, the condition is called acromegaly.

Another tumor makes ACTH, the hormone that stimulates the adrenals. This can lead to Cushing's syndrome, in which the adrenal glands produce too much cortisone. The condition results in obesity that is confined to the patient's abdomen and chest. The muscles in the arms and legs also become wasted; the skin becomes thin and bruises easily; and deep purple stretch marks may develop.

Tumors in or around the pituitary itself may affect the gland to the extent that it can no longer produce hormones. This can lead to hypopituitarism, which in adults may lead to a reduction in the gonadotropins secreted. Such a reduction usually results in a decline in sexual function in men, and amenorrhea in women; in both cases this can be distressing but it is not life-threatening.

If the thyroid gland or adrenal glands stop working properly, serious illness or even death can result. The main role of cortisone is to respond to stress, and it cannot do this if there is no ACTH to stimulate the adrenal glands when stress occurs. Doctors test the system by giving the patient an injection of ACTH and measuring the patient's response.

The posterior pituitary may also fail to produce hormones, although it is only the lack of ADH that causes difficulties. When ADH is absent, the body cannot retain water; this condition results in excessive thirst, and large amounts of urine are passed.

In children, an underactive pituitary can cause normal growth to be very slow, or to stop altogether. A lack of gonadotropins will also delay the onset of puberty. Tumors affecting the pituitary gland can occur in childhood, but are extremely rare. Children also occasionally suffer delayed growth as a result of a lack of growth hormone. In this condition, a tumor is not the cause.

Treatment

When problems with the pituitary cause disease, successful investigation depends on careful examination of the underlying hormonal problems. This can be exacting for both doctor and patient.

A small prolactinoma tumor may be treated by surgical removal. However, more often the resulting high prolactin levels are controlled with a drug called bromocriptine; this inhibits the release of prolactin. When there is an excess of growth hormone, surgery is more likely than it would be in a case of prolactinoma, because these tumors tend to be larger.

With Cushing's syndrome, tests must be performed to see if too much ACTH is produced by the pituitary, or if the problem is the adrenal glands. When the pituitary is at fault, the tiny pituitary tumor that is the cause may be removed by surgery.

For cortisone deficiency, cortisone tablets are given by mouth. Thyroid hormones, and sex hormones in women, are also given in this way. In men, sex hormone replacement is best given by applying testosterone gels to the skin or by long-lasting injections of testosterone (a male hormone). A deficiency of ADH is cured by giving the hormone in the form of a nasal spray.

Children who suffer from a lack of growth hormone can be treated successfully with injections of growth hormone.

See also: **Adrenal glands; Brain; Glands; Growth; Head; Hormones; Hypothalamus; Kidneys; Metabolism; Thyroid**

Placenta

During pregnancy the placenta provides the vital link between the mother and fetus. Both the feeding of the baby and the elimination of its waste products occur through this organ, which is expelled after the baby's birth.

Questions and Answers

Do twin babies share the same placenta?

When twins are identical it is because one fertilized egg has given rise to two babies. During their development they share a single placenta, but each baby grows in a separate sac.

Nonidentical twins grow from two separate eggs, so each baby has its own placenta.

My last baby was very small and I was told that this was because the placenta had not functioned well. What can I do next time to help my placenta function better?

Often there is no explanation for an unhealthy placenta, but as with all pregnancies it is important to eat well, get adequate rest, and attend prenatal clinics. Women with kidney disease, raised blood pressure, or heart disease are more likely to have small babies. If this applies to you, then you should talk to your doctor, who will be able to advise you about keeping as well as possible before you become pregnant. If you smoke, it is wise to stop before becoming pregnant, since smoking impairs placental function.

After I had my baby the doctor took a blood sample from the placenta. She said this was because my blood group is Rh-negative. Can you explain this?

Women whose blood group is Rh-negative can produce substances in their blood called Rh antibodies if the baby is Rh-positive. These antibodies cross the placenta, enter the baby's blood system, and can destroy some of the baby's blood cells, causing anemia. The blood in the large blood vessels of the placenta is the same as the baby's blood, and so it can be tested to check whether the baby is Rh-positive or Rh-negative.

ANATOMY OF THE PLACENTA

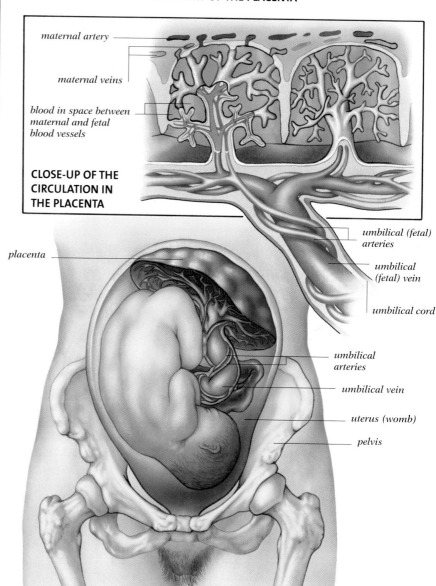

maternal artery

maternal veins

blood in space between maternal and fetal blood vessels

CLOSE-UP OF THE CIRCULATION IN THE PLACENTA

placenta

umbilical (fetal) arteries

umbilical (fetal) vein

umbilical cord

umbilical arteries

umbilical vein

uterus (womb)

pelvis

▲ The placenta consists of maternal blood vessels in the uterine wall and fetal blood vessels, which arise from the umbilical cord. The exchange of food, oxygen, and waste products takes place in the spaces between the maternal and fetal blood vessels, which are not connected. Deoxygenated (blue) blood leaves the fetus along the umbilical arteries, and oxygenated (red) blood reaches the fetus via the umbilical vein.

▲ *A placenta that supplied nourishment to twins. It was expelled shortly after the babies' birth.*

▶ *In the 10-week fetus the placenta is still immature and has a frilly, coral-like appearance.*

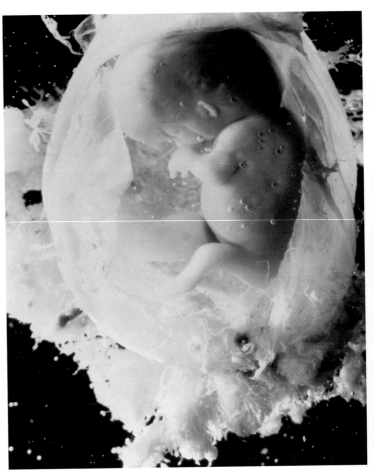

The placenta, or afterbirth, forms when a specialized part of the fertilized egg, the trophoblast, embeds in the wall of the mother's uterus. By the twelfth week of pregnancy, the placenta is an entirely separate organ. By the time the baby is born it will weigh a little over a pound (450 g) and will be dark red, spongy, and disk-shaped.

Two layers of cells keep the circulation of the fetal blood in the placenta separate from the mother's blood, but many substances can still pass from mother to baby.

The fetus receives all the food and oxygen it needs from its mother, and it is able to eliminate any waste products back into her circulation. This vital exchange function is carried out by the placenta, to which the fetus is attached by the umbilical cord. Oxygen, nutrients (simple carbohydrates, fats, and amino acids), and hormones pass from the mother to the fetus, and carbon dioxide, waste products, and hormones are transferred in the opposite direction.

The placenta also acts as a barrier to protect the fetus from potentially harmful substances, although many drugs can still cross the placenta and may damage the fetus. Some of the mother's antibodies also cross the placenta to protect the fetus.

Finally, the placenta produces hormones, some of which prevent the woman from releasing more eggs or menstruating while she is pregnant. These hormones also encourage breast development in preparation for breast-feeding, and the laying down of fat on the thighs, abdomen, and buttocks as a future energy store. Other placental hormones stimulate the growth of the uterus and probably prevent it from contracting before labor starts. There is also evidence to suggest that the amount of hormones released may be an important factor in determining when labor starts.

Monitoring the placenta

The absolute proof of a healthy placenta is the birth of a healthy baby. However, the efficiency of the placenta is often checked during pregnancy by measuring the amounts of hormones it releases into the mother's blood. Doctors assume that if the placenta is producing enough hormones, it is also working well in all other respects. If the hormone levels fall and it is suspected that the fetus is not receiving adequate nourishment, the baby may have to be delivered early.

Delivering the placenta

The placenta is normally delivered a few minutes after the baby's birth. The mother is given a shot of the hormone oxytocin, which causes the uterus to contract into a tight ball. The large maternal blood vessels to the placental site are then squeezed shut, and the placenta is sheared off the uterine wall. The obstetrician delivers the placenta through the vagina by pulling gently on the umbilical cord. He or she examines it afterward to ensure that it is complete.

Problems

Even a healthy placenta can cause occasional problems. In some rare cases it becomes partly or completely detached from the wall of the uterus before the baby is born, causing pain and bleeding. In such cases, urgent surgery is vital to save the lives of mother and baby.

In placenta previa, the placenta covers the neck of the womb completely, blocking the baby's passage through the birth canal. If this condition is present, the baby has to be delivered by cesarean section.

Sometimes a small piece of placenta does not come away in the normal manner at delivery and remains in the uterus. This can cause severe blood loss (secondary postpartum hemorrhage) several days after delivery. If an ultrasound shows a piece of retained placenta, an evacuation of the uterus must be performed.

See also: **Fetus; Pregnancy**

Plasma

Questions and Answers

Can a blood donor give just plasma without the other parts of the blood?

When people give blood, whole blood is normally taken, but it may then be separated into its component parts so that more patients benefit. In exceptional circumstances, blood can be taken with a cell-separating machine so that just the plasma is removed. For example, this technique might be used when antibodies to a disease such as hepatitis are needed to protect people who have been exposed to the virus.

Can someone run out of plasma or get sick by having too little?

Keeping an adequate amount of plasma in the circulation is such a central part of the body's activities that when this function breaks down, the whole body suffers. It is thus a matter of urgency when major bleeding or dehydration causes a loss of circulatory fluid. In emergency situations, doctors can replace much of the lost volume with saline rather than plasma, limiting the need for transfusions.

When my brother was in the hospital he was connected to a machine that took the blood from his body and then put it back again. Why was this done?

It is possible to take blood from the body, separate the cells from the plasma, and then return the cells in a citric-saline or similar solution—a process called plasmapheresis. The technique is used when the plasma contains a substance that is harmful to the patient. For example, the plasma may contain a harmful antibody, or an excessive amount of a certain protein that makes the blood thick and sticky so that it doesn't flow along the blood vessels properly.

Plasma is the fluid in which the red and white blood cells are suspended. It also contains many dissolved substances, including proteins and minerals, that are essential to life.

▲ *Blood bank technicians withdraw plasma from a machine that separates raw blood into its components. It is then frozen and stored until it is needed for transfusion.*

Plasma, the fluid component of blood, transports the red and white blood cells around the body. It consists mainly of water, but also contains water-soluble substances. Some of these substances are important to the body. They include body fuels such as glucose and basic fats; minerals such as iron, which is essential for the formation of the oxygen-carrying pigment hemoglobin; and other vital compounds, such as thyroid hormone. Others, such as carbon dioxide, are waste products that need to be expelled from the body.

Since plasma is a liquid, it can diffuse through the walls of the capillaries by osmosis, and mix with the extracellular fluid that bathes the surface of all the body's cells. As a result, soluble substances can be carried from cell to cell.

Plasma proteins

Protein is the most abundant of the soluble substances in the plasma: each quart (liter) of plasma contains about 2½ ounces (75 g) of protein. There are two main types: albumin and globulin.

Albumin, which is manufactured in the liver, is a source of food for the tissues and also provides the osmotic pressure that keeps the plasma inside the blood vessels and stops it from flooding out into the body tissues. Albumin can be thought of as a circulating liquid sponge that keeps necessary water in the bloodstream and so stops the whole body from degenerating into a damp mass of tissue. Of the globulins, possibly the most important are those that act as antibodies against infection. Others are active in the formation of blood clots.

Plasma transfusions

Blood from donors is often separated into its two main components, cells and plasma. Plasma transfusions may be given when people have lost a large amount of blood and lack the circulatory fluid to enable their heart to beat effectively, to counteract severe bleeding, and to boost immune function.

See also: Blood; Capillaries; Hormones; Immune system

323

Posture

Questions and Answers

Is the traditional military-style posture the best one to aim for?

In general, yes. The ideal is a straight spine with the back of the shoulders and the back of the buttocks in line, the shoulders drawn well back, and the chest pushed slightly forward. The chin should be tucked in and the eyes looking straight ahead.

What is the best posture to adopt during pregnancy?

The increasingly large and heavy forward bulge in the abdomen during pregnancy can become a source of back strain and great discomfort unless it is managed properly. Many women tend to bend backward so that the weight is carried by the spine rather than the abdominal muscles because they think that this is more likely to preserve their figure. Instead, it is likely to give them backache. The best course is to let both the spine and the abdominal muscles bear the load, by holding your posture as close to the normal straight-backed position as possible. If you have prepared your abdominal muscles by appropriate exercises, they will be strong enough to support the extra weight without sagging.

If you have bad posture habits when you are young, will this cause back trouble when you are older?

Yes. All the joints of your body suffer wear and tear over the years, but the joints in the lower part of the body—feet, ankles, knees, hips, and lumbar spine—are under even greater strain because they have to support the whole weight of your body. In later life, these joints are the most likely to become painfully arthritic. You can help prevent this by keeping your weight down and maintaining good posture so that the joints do not have to suffer unnecessary strain.

"Posture" is the term used to refer to the way people stand and move about. The sedentary nature of so many modern occupations has contributed to bad posture and many internal and back problems.

The normal upright stance that humans have adopted has taken millions of years to evolve, and the process is briefly repeated in the life of each new baby, who at first can move about only by crawling on hands and knees. The restricted positions adopted in the uterus are changed gradually by the baby's natural stretching and kicking reflexes, until at the age of 12 to 15 months he or she is strong and well-balanced enough to stand upright.

▲ *The military stance is everyone's idea of good posture—and indeed it makes a great deal of sense: a straight spine with the shoulders back, chest forward, and head erect.*

Posture and the force of gravity

The overwhelming physical force to which the human body is subject is the earth's natural pull, or gravity. The body's center of gravity—the point where its weight is balanced—is in the lower back and pelvis. Good posture is primarily a matter of efficiently balancing the body weight around this point. This must be done not only when a person is still but when he or she walks, sits, runs, works, plays games, and so on. Whatever movements are made, the weight on all sides of that point has to be equal, otherwise a person will tend to topple over.

Because this area of the back is the focus of so many of the body's movements, and because it bears the brunt of any lifting, it is particularly vulnerable to injury, damage, and disorders.

Muscle control

The body's erect posture is achieved and maintained only by a fine muscular adjustment that is concentrated around its center of gravity and supported by the ligaments of the spine.

Muscles can act only by shortening or contracting, not by stretching. Therefore, to reverse a movement, another muscle has to act in the opposite direction. Each muscle can also work together with other muscles to produce a balanced movement by pulling at the same time, but not with full power.

Well-balanced movements require not only muscular power but relaxation in those muscles that act in the opposite direction. Economy and precision of muscular effort, which are essential to good posture, improve when stress is removed. A person who is mentally relaxed will as a result be physically relaxed and therefore able to stand and move well.

The role of the nervous system

Posture is not simply a matter of muscle control. The muscles themselves have to be controlled, and this is done by the nervous system. In a child who is learning to stand and walk for the first time, this process of keeping balance is a conscious one. However, habit patterns soon develop and the action becomes automatic.

Messages indicating the position of various parts of the body in space, and in relation to each other, are transmitted from nerve endings in the skin and in the muscles along to the spinal cord and the brain. Signals are then sent to activate the appropriate muscles in the body for any necessary movement or change in position.

Standing correctly

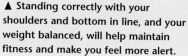

▲ Standing correctly with your shoulders and bottom in line, and your weight balanced, will help maintain fitness and make you feel more alert.

▲ Round shoulders and a drooping head and chest make for a sloppy stance and, over time, can give rise to respiratory problems.

Thus poor balance, abnormal posture, muscular weakness, and poor coordination of movements may each be due to disorders of the nervous system or disorders of the muscles. The connection is such that if, for example, a muscle loses its nerve supply, it can no longer contract sufficiently and so begins to waste away.

Good and bad posture

Apart from disorders of balance, people's posture may be either good or bad. Good posture depends largely on keeping all the postural

Simple exercises to improve your posture

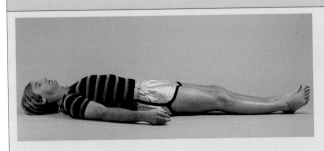

▲ Lie on your back and relax your whole body. Breathe deeply. Start from this position and return to it if you get tired.

▲ Turn onto your front and practice straight leg raising from this position, first lifting each leg alternately, as before.

▲ Tighten and relax your buttock muscles. Rest, then raise each leg in turn with the knee straight. Repeat four times.

▲ Now raise both legs. If you feel any muscle strain doing these exercises, adopt the rest position immediately.

▲ When you feel strong enough to do it without strain, try to raise both legs together, still keeping them straight.

▲ To strengthen the foot and ankle, move each foot in turn in a circular motion, first clockwise, then counterclockwise.

muscles of the body toned and balanced. Good posture requires the conscious effort of maintaining balance at all times, and of getting adequate exercise in order to keep all the joints and muscles healthy. To ensure correct posture and avoid strains and unnecessary accidents, posture training is needed. The appropriate exercises can be enjoyable and help people learn how the body works and how to take care of it. Maintaining good posture or correcting bad posture not only will transform the appearance but will substantially benefit health and make the person feel fresher at the end of the day.

Bad posture has specific ill effects. Bad stance and movement can affect the hip joints and the lowest joint of the back, which are more prone to developing arthritis in later life. Lower-back pain, slipped disk, and sciatica are also more likely to occur. Obesity increases the likelihood of bad posture because the extra weight increases the strain on the body's muscles.

Another disorder caused by poor posture is foot strain. This often occurs in people whose jobs involve long periods of standing and walking. The strain happens because they are standing and walking incorrectly and too much strain is placed on the ligaments and bones of the foot arches, causing pain. This may be avoided by doing simple exercises to strengthen the feet and ankles.

See also: Back; Balance; Bones; Coordination; Feet; Ligaments; Movement; Muscles; Nervous system; Pelvis; Pregnancy; Spinal cord

Pregnancy

In pregnancy, changes occur in the mother's body that are designed to meet the needs of the growing fetus. For most women it is an exciting and happy time; modern prenatal care has also made it much safer.

Pregnancy is the remarkable and highly complex process between conception and labor and lasts on average 38 weeks. Because the date of conception is often not known exactly, it is easier to date the pregnancy from the first day of the last menstrual period, which is usually about two weeks before conception, making a total of 40 weeks.

The first signs

The first sign of pregnancy is usually a missed period, although this can be caused by other conditions. However, if intercourse without contraception has taken place, pregnancy is the most likely cause.

Other early symptoms include a sense of fullness and tingling in the breasts and an urgent need to pass urine more frequently. Many women suffer from nausea, and even vomiting in early pregnancy. Although this is popularly called morning sickness, it can come at any time of day. It is often aggravated by preparing food. A cup of tea and a dry cracker first thing in the morning can sometimes help, and it is sensible to eat small amounts of nongreasy food at fairly frequent intervals through the day rather than large infrequent meals.

If a woman thinks she is pregnant

A woman should do a home pregnancy test as soon as she suspects that she might be pregnant. The pregnancy test will detect the presence or absence of a hormone called human chorionic gonadotropin, which is produced by the developing egg and excreted in the mother's urine. Simple home pregnancy tests are available from drugstores. If the woman carries out the test herself she should still consult her family doctor if the result is positive. If she does not wish to have her baby, she may need to find an adoption center, or talk to her doctor. If the fetus is not developing normally, this may result in a spontaneous abortion.

Once a woman's pregnancy is confirmed, her obstetrician will arrange for her care during pregnancy and plan for her to go to the hospital where the obstetrician has admitting rights. She will also need to make office visits, probably on a monthly basis throughout the first 28 weeks of

Questions and Answers

I think I might be pregnant. How soon would it be worthwhile to do a pregnancy test?

Pregnancy tests are now more sensitive. Most of the older tests were not reliable until five or six weeks after the last period, but now a test may be possible within a week after conception.

What can I do to prevent indigestion during pregnancy?

The hormones of pregnancy, and later pressure of the uterus on the stomach, tend to make the digestive acid pass back from the stomach into the esophagus, where it causes a burning sensation. This acid can be mopped up by drinking milk, eating small frequent meals, and taking antacids. Another way to prevent heartburn at night is to raise the head of the bed.

Can you prevent stretch marks from forming during pregnancy?

Some women, especially very fair women, seem more likely to get stretch marks, which form on the breasts, stomach, and thighs. Once the baby is born, the marks will lose their reddish color and eventually fade, but will never entirely disappear. Very little can be done to prevent them, although some women believe that a little olive oil rubbed into the skin every day helps.

My husband is worried about making love to me now that I am pregnant. Is he right to be?

No. You should continue to make love for as long as you feel comfortable enough to enjoy it. Occasionally, women who have had many miscarriages are advised to avoid making love in the first three months of pregnancy, but there is no absolute proof that this makes them any less likely to miscarry.

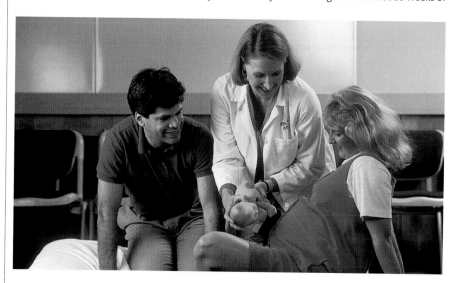

▲ *Fathers are welcome at prenatal classes, where they learn how to help their partners.*

Questions and Answers

I often feel faint when I lie on my back. Why?

This usually happens only late in pregnancy when the baby's weight in the uterus can press on a large blood vessel, the inferior vena cava, and decrease the blood supply to your heart and brain. If you feel faint, turn over onto your side and the faintness should soon pass.

Since I have become pregnant I need to urinate more frequently. Does this mean I have cystitis?

That is a possibility, and your urine will be tested at the doctor's office for any signs of infection. However, it's more likely that your uterus and bladder are competing for space in the pelvis and so your bladder feels full sooner. In the middle part of pregnancy, when the uterus has grown out of your pelvis, this will probably return to normal, but in the last weeks the baby's head often presses on the bladder, and you will find again that you need to urinate more frequently.

Is it true that a woman "blossoms" during pregnancy, and if so, why?

Pregnancy certainly suits some women very well. This is because the hormones of pregnancy often improve a woman's complexion and make her feel warmer, so that she has rosy cheeks. Even more important is the sense of well-being some women feel at this time, which is thought to be related to increased production of hormones called steroids.

My mother tells me that now that I'm pregnant I must eat for two, but I don't want to be fat after the birth. How much should I eat?

Your mother is wrong—you burn up your food more efficiently and use up less energy yourself as you become less active late in pregnancy. You simply need to eat a normal sensible diet with plenty of milk products, fish, meat, and fruit. Your obstetrician will keep an eye on your weight gain.

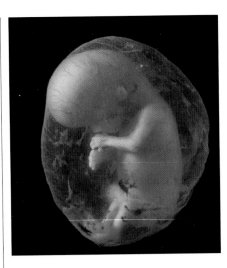

▲ *This 11-week-old fetus, although only 2 in. (5 cm) long, is recognizably human.*

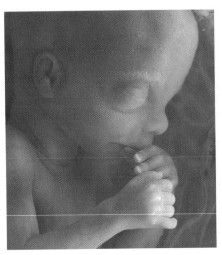

▲ *By four months the fetus has developed eyebrows, but the eyelids are still fused.*

pregnancy, then every two weeks until 36 weeks, and thereafter weekly until the baby is born. Obviously these arrangements have to be flexible to allow for any unusual circumstances.

Prenatal visits not only involve checks and tests to see if the mother and baby are both healthy, but may also include classes on baby care and preparation for labor and childbirth. Most courses will also include at least one session for prospective fathers to advise them on how to help their partners during pregnancy and labor and after the birth.

Minor discomforts

During pregnancy a woman may suffer from several minor discomforts which, although mostly trivial, can cause her some anxiety.

Nausea: Feelings of nausea noticed early in pregnancy usually continue until about 16 weeks. If they are severe enough to be incapacitating, the obstetrician may prescribe medicine that can help.

Vaginal discharge: Women have an increased vaginal discharge in pregnancy. Unless the discharge is offensive, irritating, or bloody, it does not require medical attention.

Backache and cramp: Muscular aches and pains may present a problem as the pregnancy advances. A pregnant woman instinctively tends to throw her shoulders back in an attempt to counteract the weight of the growing uterus, and this can put undue strain on the back. Improved posture and comfortable, low-heeled shoes may relieve it. Leg cramps are most common in late pregnancy and may be quite troublesome at night. They can sometimes be relieved by gently stretching the affected muscle.

Constipation and heartburn: Both are commonly experienced during a pregnancy. They are side effects of the hormone progesterone, which relaxes the smooth muscle fibers in the uterus. However, its action is not confined to the growing uterus; it has effects throughout the mother's body. It affects the intestines, making them sluggish, and relaxes the sphincter muscle at the opening from the esophagus into the stomach. This can allow the contents of the stomach to be regurgitated and cause heartburn. A mild antacid can help.

Hemorrhoids: These are fairly common in pregnancy because the blood flow to and from the woman's legs and pelvis is partially obstructed by pressure from the baby and the uterus. Straining through constipation can aggravate the problem.

Varicose veins: These may become worse during pregnancy, owing to the effect of progesterone on the blood vessels. They may also appear for the first time. Maternity support hose can help to prevent their formation, and this should be put on before the woman actually gets out of bed in the morning. Women with varicose veins should avoid standing still for long periods; it is better to walk around to keep the circulation going. When the woman is sitting, she should prop up her feet on a stool or a low chair.

Edema: This condition is caused by excess amounts of water in the pregnant mother's body. This fluid accumulates in certain areas, creating swelling, particularly around the ankles and the feet. Mild cases of edema are fairly common in later pregnancy; a low-salt diet and plenty of rest with the feet up should help. However, if edema is associated with an increase in blood pressure and protein in the urine, special treatment will be necessary, since these symptoms could indicate preeclamptic toxemia, which is a very serious condition.

Skin changes: As the pregnancy proceeds, stretch marks can appear across the abdomen, thighs, and breasts. Little can be done to prevent these from developing, but they will usually fade after the baby has been born.

Changes in the uterus and breasts

A pregnancy is divided into three trimesters of about 13 weeks each. In the first trimester, the uterus grows rapidly but remains within the pelvic cavity. It is during the second trimester, when the uterus moves up into the abdominal cavity, that a woman first becomes obviously pregnant.

By about 22 weeks the upper edge of the uterus, or fundus, reaches the navel, and at 36 weeks most of the abdominal cavity is occupied. The intestines are pushed upward and sideways so they press on the stomach and diaphragm. Because of pressure on the diaphragm, the lungs cannot expand fully, and many women find themselves short of breath quite regularly.

At about 22 weeks into a first pregnancy and 18 weeks in subsequent pregnancies the mother will start to feel her baby's movements. The fetal heartbeat is usually audible through a stethoscope by about 24 weeks.

The main change that a woman notices in the breasts during pregnancy is that they grow larger in preparation for feeding. The areola, the ring of darker skin around the nipple, becomes larger and darker in color and a secondary areola appears, which helps to improve the strength of the skin.

From about 12 weeks of a pregnancy the breasts produce a protein-rich substance called colostrum, and in the last few weeks this fluid may leak. Colostrum provides all the nutritional needs of the newborn baby until the milk appears on the third day after birth.

Prenatal care

At each office visit for a prenatal checkup, the pregnant mother will be weighed, her blood pressure measured, and a urine sample taken. On some visits, a blood sample will also be taken from the mother

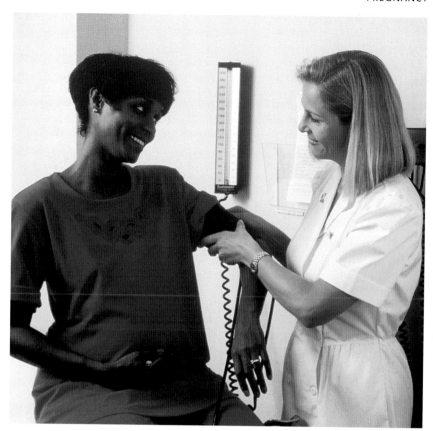

▲ *Expectant mothers are naturally concerned about the health of their babies, and modern techniques such as ultrasound help to reassure them that all is well. Other routine tests that will be needed during pregnancy include urine tests, blood samples, and regular checking of the mother's blood pressure.*

▼ *Women who are pregnant no longer feel that they must hide themselves away from the public gaze. In fact, their appearance is often enhanced by the rosy bloom of pregnancy and their own sense of well-being.*

6 WEEKS

▼ *At six weeks of pregnancy the embryo is still not recognizably human, and it is only 0.5 in. (about 1.3 cm) long. A pregnancy test now would be positive, and the mother may feel some symptoms such as breast sensitivity and nausea.*

12 WEEKS

▼ *At 12 weeks the uterus can just be felt above the pelvis. By now all the major fetal organs are formed; nails are appearing on the fingers and toes. The fetus is about 3 in. (7.5 cm) long and weighs about 0.5 oz. (14 g). The mother's breasts begin to produce colostrum.*

20 WEEKS

▼ *By 20 weeks of pregnancy the uterus has reached the level of the mother's navel, and she starts to become aware of some of the movements of her baby. The fetus now measures about 8 in. (21 cm) and is covered with fine, downy hair called lanugo.*

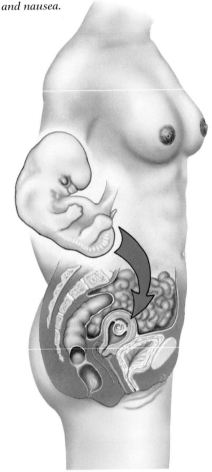
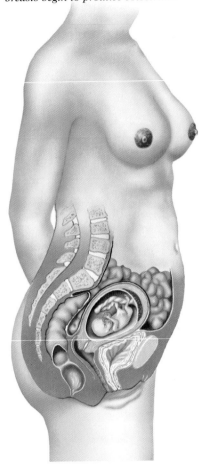
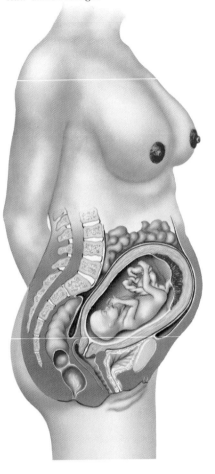

in order to establish her blood group and type, to check that she is not anemic, and to find out if there are specific disease-fighting antibodies present. This sample can also reveal if the placenta, through which the fetus is nourished in the uterus, is working efficiently. Some fetal abnormalities can also show up in tests of the mother's blood, although further tests are needed to confirm such problems in the fetus.

Every prenatal checkup includes an examination of the mother's abdomen to establish the baby's size and position. Internal examinations are usually done late in pregnancy and at the beginning of labor.

In addition to the routine checks, there are certain specialized tests which are carried out in certain cases to check the welfare of the fetus.

Amniocentesis: A sample of the amniotic fluid which surrounds the baby in the uterus may be tested if there is a risk that the baby is abnormal. For example, women over 40 who are pregnant for the

first time are more likely to give birth to a baby with Down syndrome. An alternative to amniocentesis is a biopsy of the chorionic membranes that surround the fetus. Amniocentesis can detect problems such as Down syndrome and spina bifida. If for any reason the baby might have to be delivered early—for example, if it is not growing well—amniocentesis can reveal whether the baby's lungs are mature enough for it to survive.

Ultrasound: This form of examination, using high-frequency, inaudible sound waves, involves passing a scanner over the woman's lubricated abdomen. The uterus and its contents can then be viewed on a video display. In this way, the size of the baby, the position of the placenta, and even the presence of twins or triplets can all be established.

X rays: These are seldom used and are carried out only in late pregnancy when they are safer for the baby. In rare cases they are used to show whether the mother's pelvis is wide enough to allow

28 WEEKS

▼ *At 28 weeks the uterus reaches about halfway between the navel and the breastbone. The fetal movements are more vigorous, and the mother may feel painless rhythmic contractions. The fetus is now viable, meaning that if it was born at this stage it could survive. Its skin is covered by a protective coating called vernix, and it can now open its eyes. It measures about 15 in. (38 cm).*

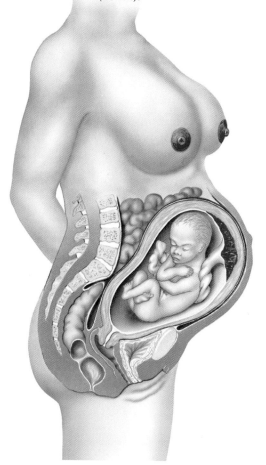

40 WEEKS

▼ *At 40 weeks the pregnancy is at full term and the mother is often impatient to get on with the delivery of her child. The upper edge of the uterus descends from its position high under the rib cage as the head of the baby moves down into the mother's pelvis. This is called engagement. The mother's breathing and digestion become easier, although pressure on her bladder increases.*

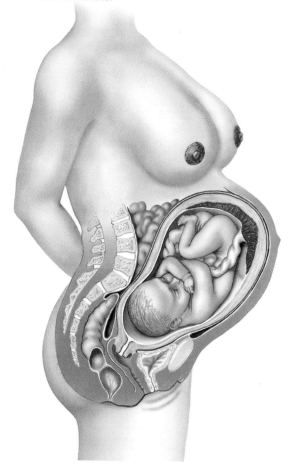

the baby's head to pass through easily, or, if not, whether delivery by cesarean section will be necessary.

During pregnancy

It is important for a pregnant woman to eat a well-balanced diet with plenty of protein (meat, fish, cheese, milk), fresh fruit, and vegetables. Too many cakes and cookies should be avoided, since they can lead to excessive weight gain, which will be difficult to shed after the birth.

The amount of weight gained should not be more than 26 pounds (12 kg), but a crash diets should never be embarked upon during pregnancy, because it will deprive the baby of nourishment. If a doctor has prescribed iron tablets, they should be taken regularly.

Moderate exercise is a good idea, although pregnancy is not the time to take up a new, strenuous sport. Women will be taught exercises at prenatal classes to strengthen their back and muscles.

Smoking cigarettes should be avoided at all costs. Smoking restricts the blood vessels in the placenta, decreasing blood flow. As a result the baby gets less nourishment and oxygen. Smoking even 10 cigarettes a day significantly reduces birth weight and increases the risk of mental and physical damage to the fetus. No drugs should be taken without the doctor's advice. While the occasional alcoholic drink will do no harm, any heavy drinking could damage the baby's brain and also slow its growth. Apart from these sensible precautions, the expectant mother should remember that pregnancy is a normal, healthy state for most women. With adequate rest, a good diet, and moderate exercise, most women pass happily through pregnancy without serious complications.

See also: **Birth; Breasts; Diaphragm; Fetus; Hormones; Menstruation; Pelvis; Placenta; Posture; Uterus**

Prostaglandins

Prostaglandins are chemical substances that are similar to hormones; they act as messengers in the body and are responsible for controlling many important physical functions.

Questions and Answers

My son has some sort of allergy to food which we think is caused by milk. Could prostaglandins give rise to his symptoms?

It is possible. Prostaglandins belong to a group of transmitting substances, some of which are involved in allergies. Recent research has shown that certain allergic reactions to food can be blocked by aspirin; this suggests that prostaglandins are involved in some cases.

Is it true that prostaglandins can cause heart attacks?

One type of prostaglandin is vital for making blood-clotting cells, or platelets, stick together to form a clot. This process may be a factor in forming blockages in the coronary arteries that lead to heart attacks. However, there is another prostaglandin that acts to reduce clotting. Prostaglandin-blocking drugs, such as aspirin, have real value in helping to prevent heart attacks by reducing the risk of thrombosis, but the major cause of heart attacks is the arterial disease atherosclerosis, which affects the coronary arteries and forms plaque sites on which blood is likely to clot.

Could my body work without prostaglandins?

Almost certainly not. They are essential in many ways, especially in dealing with injury and infection, and they may have other, as yet unknown, roles.

I have arthritis. Should I take prostaglandin-blocking drugs?

You may already be taking them. Many common painkillers have some prostaglandin-blocking activities. There are many new drugs available, but the older ones are still as effective.

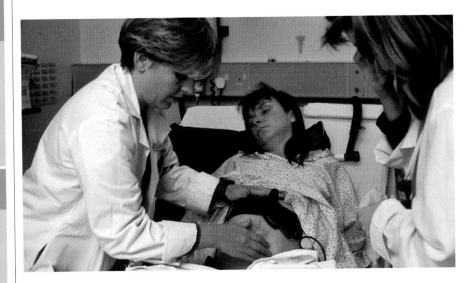

▲ *Prostaglandins E2 and F2 can stimulate muscle contractions during labor. They may be given in the form of pessaries to help speed up or induce labor.*

When prostaglandins arrive at a cell or tissue that is primed to respond to them, they trigger an important reaction. Unlike hormones, they are not produced just in specialized glands but may be produced in different tissues. Their effect is on cells and tissues nearby.

Discovery of prostaglandins

Prostaglandins were first found in the prostate gland, hence the name. Because they have a short lifetime (a few seconds between production and effect), they are difficult to study. They may not even enter the bloodstream, and so it is very difficult to measure the amounts of active prostaglandin that may be present. Most of the prostaglandins that have been found so far were discovered because they can produce a measurable effect on tissue samples in a laboratory.

How prostaglandins work

Many different body tissues respond to prostaglandins. They are central to the female reproductive system, and may help regulate menstrual bleeding. Menstrual pain may result from an imbalance in prostaglandins. Prostaglandins are relevant to men; a deficiency in seminal fluid is thought to reduce male fertility. Prostaglandins are associated with common aches and pains, as well as severe inflammation of arthritis. Aspirin, an effective painkiller, blocks prostaglandin production, and a range of more powerful drugs use the same principle. Prostaglandins produced by blood platelets are also closely involved with blood clotting and blood thinning.

There are many different prostaglandins; and each has a different effect, so creating drugs to manipulate them is a delicate process. For example, aspirin will reduce heart and circulatory problems by preventing blood clots, but it will block the production of prostaglandins that protect the stomach lining, and it may cause ulcers and stomach upsets. New drugs are being developed to block specific prostaglandins, and allow others to function normally.

See also: Hormones; Prostate gland

Prostate gland

What is the function of the prostate gland, and can it be removed safely?

It produces a special fluid that makes up part of the seminal fluid and allows the sperm to remain active, probably increasing the chances of fertilization. The prostate gland can be removed successfully without any adverse effect on health, although fertility and sexual performance may be affected. Because this operation is usually performed on elderly men, they may not care about infertility.

My father has been told he has prostate cancer, but his physicians are not going to operate. Does this mean that it has spread too far?

A high proportion of prostate cancers in older men are confined to the gland, progress very slowly, and do no harm. A "watchful waiting" policy is often adopted by urologists and is usually justified.

Do the hormones used to treat prostate cancer have side effects?

Yes, in some patients, but these are rarely serious enough to stop the treatment. They include increased retention of fluid in the tissues, enlargement of breast tissue, and loss of body hair and libido.

My husband is developing difficulty in urinating. He has to stand for ages before the urine starts to flow, and he is getting up several times a night. Does he need an operation on his prostate gland?

Possibly. There are other causes of the same symptoms, but he should see a specialist to have some tests carried out, including a blood test, a special X ray, and a urine flow test. These tests reveal whether or not the prostate gland is the cause of the trouble. If it is, surgery may be recommended.

The main function of the prostate gland is to aid male fertility. It is common for this gland to remain relatively trouble-free until late in life, and if problems do occur, they can usually be treated successfully.

The prostate gland is a walnut-shaped structure found only in males. It is situated at the base of the bladder and surrounds the urethra, the tube through which urine and seminal fluid pass out of the body. This gland produces the fluid that mixes with semen to make up part of the seminal fluid. Although the full function of the prostatic fluid is unknown, one of its roles is to help keep the sperm active so that fertilization can occur more easily.

Owing to its position in the body, problems associated with this gland can often affect the functioning of the bladder, though this condition is more common among elderly men.

Prostate problems

There are various things that can go wrong with the prostate gland during a man's lifetime. The gland can become inflamed as a result of bacterial infection. This condition, known as acute prostatitis, causes flulike symptoms and pain in the lower abdomen, groin, and perineum (tissue between the anus and external genitalia). More rarely, there may be a discharge from the penis.

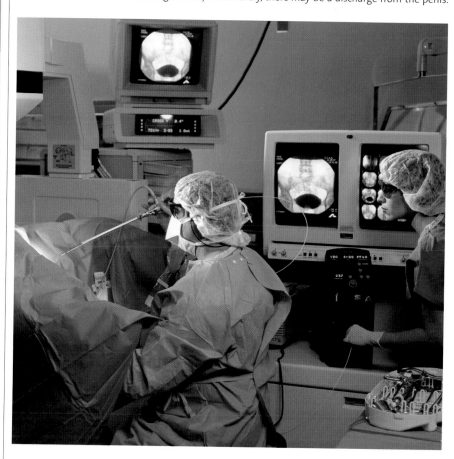

▲ *A surgeon performs minimally invasive interstitial (indigo) laser surgery on a patient with an enlarged prostate. The operation takes 30–60 minutes and uses a cystoscope to emit heat through a fiber-optic probe, thereby destroying the excess prostate tissue.*

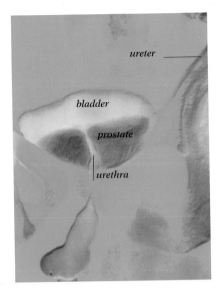

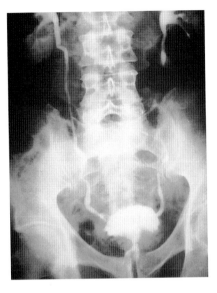

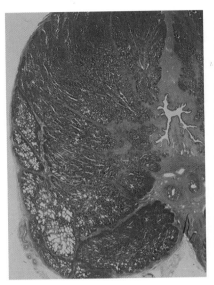

▲ *The state of the prostate gland, whether it is enlarged or not, can be determined by means of special X-ray techniques. A normal-size gland can be seen in the cystogram (above left) and an enlarged gland in the pyelogram (above right).*

▲ *As seen in this micrograph, the prostate is made up of glandular tissue and smooth muscle fibers.*

Occasionally, this infection can lead to an abscess in the gland, or to chronic prostatitis if there is a persistent low-grade infection.

As a result of chronic prostatitis, the gland may become calcified and gravel-like stones may be formed. Prostatitis is treated with antibiotics, although recurrent problems are common. Surgery may be required to treat the abscess. Another problem that can affect the prostate gland is the development of a tumor. The treatment of this condition varies, depending on whether the tumor is benign or malignant.

Signs of a benign enlargement

The most common disorder of the prostate gland is a benign increase in the size of the gland. This noncancerous condition affects many elderly men, although young men can suffer from it too.

Benign enlargement of the prostate gland (hyperplasia) is so common that many doctors believe that every man over the age of about 50 years has some degree of hyperplasia; it just has to be accepted as part of the aging process.

Because the prostate gland surrounds the urethra, and because it is so close to the base of the bladder, enlargement of the gland can seriously impair the normal mechanism of urination. A man with an enlarged prostate may notice the following symptoms: increased frequency of urination during the day; getting up at night to urinate; the development of a poor urinary stream; a tendency to stop and start, with the sensation that there is more to come; having to wait several seconds before urine starts to flow; dribbling of urine after the stream; and sometimes a sudden urge to urinate.

All these symptoms are due to distortion of the normal anatomy at the base of the bladder. The enlargement of the gland squeezes the urethra, causing it to become narrow. Sometimes the central part of

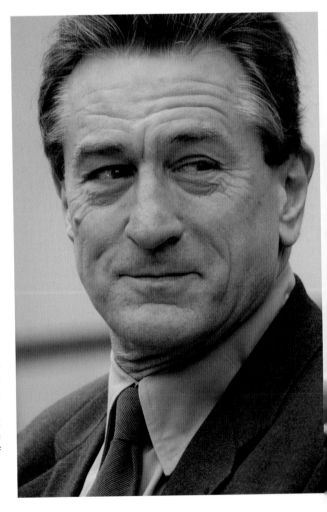

► *The actor Robert De Niro was diagnosed with prostate cancer during a routine medical test in October 2003.*

Why does the prostate gland become enlarged in older men?

No one really knows the answer to this. Finding an enlarged prostate gland is so common that it is difficult to equate it with factors earlier in life, such as sexual activity.

Is removal of the prostate gland using an open operation likely to be more permanent in its effect than removal with an instrument through the penis?

It can be, but doesn't have to be. Removal of the gland can be complete when done by the closed method, that is, by a transurethral resection (TURP), where a fine tube complete with a viewing piece and a cautery device is introduced into the penis and the prostate is actually chipped away. However, using the TURP method, some surgeons might only remove part of the gland in a very old person, just so that he can urinate easily.

Can anything be done for the pain in the bones that my father suffers as a result of his prostatic cancer spreading?

Yes. If hormone treatment has been tried and has not been successful, radiotherapy of the painful area is often extremely effective in relieving the pain of secondary tumors in the bones. Some experts also recommend removal of the testes, since this will decrease levels of testosterone and may cause the tumor to shrink.

I am due to have a prostatectomy soon. Is there a chance that I will be incontinent after I have my prostate gland removed?

There is a very small chance of this in the first few months after the operation because the muscles of the bladder that prevent leakage may be damaged during the operation, but it usually gets better in time and the incontinence may cease.

the gland becomes elongated and extends up into the bladder, where it can then block the entrance to the urethra and thereby inhibit the exit of urine. If a man continues to have these symptoms without seeking medical help, there can be a sudden and complete inability to urinate (acute retention), which is extremely painful and requires immediate treatment. A catheter (tube) has to be passed into the bladder to drain off the excess urine. The obstruction to outflow makes it impossible to empty the bladder, so filling up to the point of discomfort occurs much more rapidly than normal. Therefore, frequent visits to the toilet, night and day, are necessary. Eventually, he urinates in a dribble, and may even find that he is continually wetting his underwear and his bed. In the end, there is so much back pressure on the kidneys that they start failing, and waste products build up in the bloodstream, with extremely serious consequences, such as high blood pressure and anemia. This situation is called chronic retention of urine with renal failure.

Treatment

When a patient goes to his doctor complaining of difficulty in urinating or, perhaps, passing small quantities frequently, he will be sent for a series of special tests, including an X ray of the kidneys and bladder, and blood tests. Special equipment may also be used to measure his urine flow.

If the tests show that the prostate gland is enlarged, surgery may be recommended, because leaving the gland to enlarge further would complicate treatment later on. An operation would also be recommended for a patient who develops acute retention of urine, where urination is possible only with the aid of a catheter. Also, a permanent bladder catheter can be a social problem—as well as a medical one, because of the increased risk of infection in the urine.

If the prostate gland is causing symptoms because it is enlarged, there are drugs that can slow down the process and possibly reduce the size of the gland. While drugs may help the symptoms to a certain extent, they may only postpone inevitable surgery.

Prostate surgery

The prostate gland is situated in an extremely inaccessible part of the body, so surgery to remove it can be difficult. Also, because it is near the opening of the bladder, the delicate muscles that prevent urine from leaking can be damaged if great care is not taken. There are two types of surgery: the transurethral resection of prostate (TURP), also known as the closed method; and the retropubic prostatectomy, or open method.

The closed method, or TURP, is performed using a fine telescopelike instrument called a cystoscope, which is passed up the penis into the bladder. This instrument has a viewing piece and a special cautery (searing) device that has a wire loop on the end. Using this method, the prostate gland is chipped away from inside the urethra, with little pieces of tissue being cut away each time the wire loop, which is attached to an electric current, moves through the tissue.

The great advantage of the TURP method is that the patient is spared a surgical incision and there is less discomfort or pain after the surgery. A tube is left in the bladder, which is usually taken out after two or three days. Patients make a much quicker recovery from TURP because it is less traumatic to the body than the open method.

The open method is used for very large glands and also when surgery on the bladder has to be done at the same time. It is also often used to treat cancer of the prostate. The operation consists of making a cut across the lower abdomen and approaching the gland through the space between the back of the pubic bone and the bladder. The capsule of the prostate gland is opened and the gland is scooped out from inside the capsule. Fluid is then passed via a tube into the bladder to prevent the formation of blood clots. This tube is usually left in the bladder for about five days after the operation and then removed if there is no residual bleeding. A patient having this operation is usually in the hospital for seven to 10 days.

After a prostatectomy, the urinary stream is noticeably better, but there will be side effects. Because of the anatomy of the prostate gland, patients who have had either type of prostatectomy will find that they are unable to ejaculate semen normally. In this condition, known as retrograde ejaculation, the semen goes back into the bladder instead of traveling down and out the penis, in effect making the man infertile. This is because the muscle at the base of the bladder has to be cut when the prostate gland is removed. This muscle usually contracts during orgasm, preventing semen from going up into the bladder.

In addition to these two surgical options, minimally invasive surgery has also been used in recent years. This process can involve transurethral microwaves or transurethral removal using a needle to burn away tissue.

► *The size of the prostate gland can be approximated by a rectal examination. If it is markedly enlarged, surgery may be recommended.*

Since this surgery is usually performed on elderly men, they may not care about infertility (and sexual performance need not be affected), but a younger man suffering from an enlarged prostate may be worried by the prospect of retrograde ejaculation if he wants to have children. Sometimes the surgeon will be able to postpone treatment until the man has fathered a child, but this is not always possible if the symptoms are very severe. However, it is sometimes possible for the man's partner to be artificially inseminated if desired.

Cancer of the prostate

Cancer of the prostate is one of the most common cancers to occur in males. In fact, it has been found in routine postmortem examinations that nearly all elderly men have a tiny mass of cancer in the prostate gland. Most of these men would not have known that they had anything wrong. However, when a malignant tumor does manifest itself during a man's life, it does so in a number of ways.

First, it may be found on a routine examination by a doctor who notices a lump in the gland (the gland can be felt through the anterior wall of the rectum). Second, the patient may have difficulty passing urine because the tumor is so close to the urethra. Third, the patient may have no urinary symptoms, but develops symptoms from the spread of the tumor outside the gland. One of the most common sites for the secondary spread of the tumor is in the bones. The spread results when tiny clumps of cells break off the main tumor and circulate in the bloodstream.

Treatment and outlook

The treatment of cancer of the prostate depends on many factors, but one of the most important is the spreading of the tumor. Therefore, a patient who is suspected of having a malignant tumor of the prostate will have several tests, including X rays and radioactive scans of the bones, to determine the exact extent of the spread. If the tumor has been found by accident, and consists of a small nodule confined to the gland itself, most surgeons would treat this by removing the prostate.

If the tests show that the tumor has spread outside the prostate gland, then treatment in the form of hormone drugs will be given. Most cancers of the prostate have been found to be dependent on male hormones for their growth; so by counteracting their effect with estrogen, a female hormone, the cancer can be kept at bay. The drug is given in very small doses and often has dramatic effects on the primary and secondary tumors. Bones that have been riddled with secondary tumors become normal again, and the swelling in the gland becomes smaller.

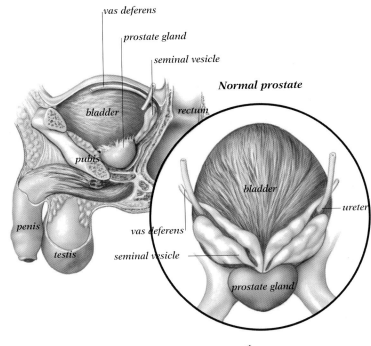

Normal prostate

vas deferens
prostate gland
seminal vesicle
bladder
rectum
pubis
penis
testis
vas deferens
seminal vesicle
bladder
ureter
prostate gland

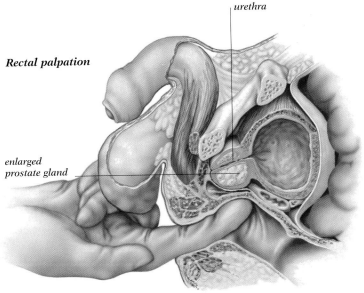

Rectal palpation

urethra
enlarged prostate gland

Estrogen does have side effects, however. It can decrease libido, promote the loss of hair, and increase the growth of breast tissue. It can also cause the body tissues to retain more fluid than usual, leading to swelling of the ankles and, in patients with heart problems, worsening heart failure.

A more recent treatment, now widely used, is the drug bicalutamide (Casodex). This drug blocks the male hormone receptors on the prostate cells and significantly reduces the tendency for the cancer to grow. It has largely replaced estrogen and has reduced the necessity to remove the testes—the source of the male hormones. Casodex is, however, likely to cause breast

TREATMENT OF PROSTATE PROBLEMS

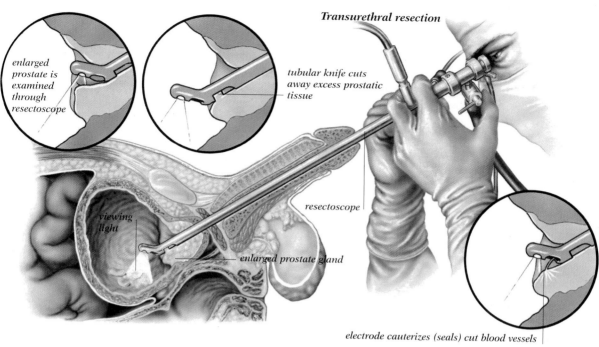

Transurethral resection

enlarged prostate is examined through resectoscope

tubular knife cuts away excess prostatic tissue

viewing light

resectoscope

enlarged prostate gland

electrode cauterizes (seals) cut blood vessels

Retropubic prostatectomy

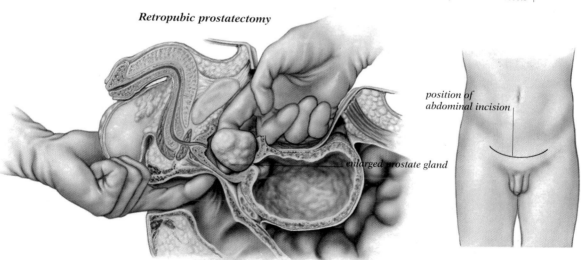

position of abdominal incision

enlarged prostate gland

enlargement (gynecomastia), tiredness, loss of libido, and other side effects.

Another treatment for prostate cancer involves the use of radioactive seeds, which are implanted in the prostate by needles that are guided by ultrasound.

No treatment may be recommended if the patient is elderly, if he has a limited life expectancy, and if the cancer is small and confined to the prostate gland. This is because many such patients will die from other unrelated causes without experiencing any symptoms as a result of their cancer, and surgery is an unnecessary ordeal.

There are a small number of patients whose tumors do not respond to drug treatment. Luckily, they are a very small proportion of patients with cancer of the prostate. The outlook for many patients with prostatic cancer is good. Some men live an active life for years, taking a small dose of the hormone drug daily.

See also: **Aging; Estrogen; Genitals; Hormones; Kidneys; Penis; Semen; Sperm; Testes; Testosterone; Urethra; Urinary tract**

Puberty

Puberty is the period at the beginning of adolescence when the physical changes that mark the end of childhood begin. Typically it is a time of emotional upheaval for both sexes.

Whereas the word "adolescence" refers to the whole period in human development between childhood and adulthood, the term "puberty" is more specifically a medical term describing the physical metamorphosis of a child into a sexually, if not emotionally, mature adult. Fired by the secretions of the hormonal system and involving both overt and less obvious changes, puberty is a period that demands understanding, patience, and frankness from parents.

The first female changes

The first signs that mark the end of a girl's childhood and the beginning of puberty are usually the budding development of the breasts and the enlargement of the nipples. These changes may start as early as age nine or 10, or be delayed as late as age 15.

After breast development has begun, the next obvious change is the growth of body hair under the arms and in the pubic region. There is a huge spurt in general growth at this time, accompanied by a gradual remolding of the body shape involving the development of the hips; the laying down of fat deposits over the thighs, hips, and buttocks; and a slimming of the waist.

Changes also take place in the skin; the oil-producing sebaceous glands of the skin become more active, as do the sweat-releasing glands, leading to increased greasiness of the hair and skin. There is also a likelihood of skin eruptions such as acne, and the development of body odor.

It is usually only after at least some of these alterations are under way that the most significant milestone in female physiology is reached, on average around age 13. This is the menarche or first menstrual period.

At first the periods are often scanty and irregular, but generally within about 12 months they settle down to a regular, approximately 28-day, pattern. Once the periods begin, and with them the release of ripe eggs from the ovaries at the rate of (usually) one each month, a girl is able to

▲ *Teenage girls often form close friendships that offer mutual support and the opportunity to discuss their lives with someone outside the family.*

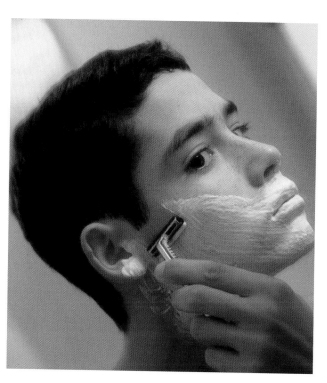

▲ *At puberty, the first downy growth of facial hair on the chin and upper lip begins to develop; boys usually then start to shave to remove this hair.*

conceive a child. Other developments take place to complete sexual maturity. The uterus enlarges so that it will be able to accommodate a developing fetus, vaginal secretions can now be released which are designed to facilitate sexual intercourse, and the ovaries enlarge to help sustain the monthly release of eggs.

The female hormones

A girl's endocrine system undergoes changes at puberty. The pituitary gland at the base of the brain begins, under the influence of the hypothalamus, to secrete a variety of stimulating (tropic) hormones. These travel to other hormone-secreting (endocrine) glands, which then release other hormones that directly affect sexual development.

Tropic hormones in the ovaries trigger the production of the hormones estrogen and progesterone, and these initiate and maintain the menstrual cycle. At first, the cycle is haphazard because the release of hormones does not conform to a regular pattern as it does in later years. Together with the hormones released by the adrenal glands (also stimulated by the pituitary), estrogen and progesterone effect changes in the shape of the body, and stimulate other secondary sexual alterations.

Emotional changes in females

In the years between about 11 and 16 a girl turns from a child into a physiological adult capable of childbearing. Emotionally, however, girls are far from mature when they begin their periods. Pubescent girls are likely to suffer from emotional ups and downs, manifested in tantrums toward the adult world and extremely close friendships

with other girls. Activities such as devotion to fashion and pop stars are an expression of their increasing awareness of themselves as sexual beings attractive to the opposite sex. At the other extreme, a girl who immerses herself in an activity such as horseback riding, or seems interested only in sports or schoolwork, may be trying to delay coming to terms with her own sexuality.

The first male changes

For boys, who seem to get left behind as their female peers sweep quickly into adulthood, the changes of puberty usually begin around age 13 although, as for girls, the normal upper age limit for the onset of puberty is about 15. Unlike girls, boys do not have a physiological equivalent to the start of menstruation to mark the end of childhood.

The first obvious alteration of male puberty is often a huge spurt in growth, usually much greater than the comparable growth spurt in girls. The accompanying changes in body shape also have a quite different pattern in that they involve the enlargement and strengthening of body muscles rather than the laying down of fat. Muscle development is most obvious in the limbs and torso and is accompanied by a broadening of the shoulders.

In boys, the voice box or larynx is enormously altered during puberty. The plates of gristle or cartilage of which the voice box is constructed enlarge and the internal elastic vocal cords thicken to give the voice a deep masculinity. A boy's voice may break suddenly, or he may go through an intermediate stage for months not knowing whether his speech will be mellow and low, or a high-pitched squeak.

The growth of body hair is much more extensive in pubescent boys than in girls and is more significant to the outward appearance of maleness. As well as pubic hair around the base of the penis and hair in the armpits, a downy growth begins to sprout on the upper lip and chin. This later becomes a coarser mustache and beard, and with the start of shaving, hair also begins to grow on the chest and

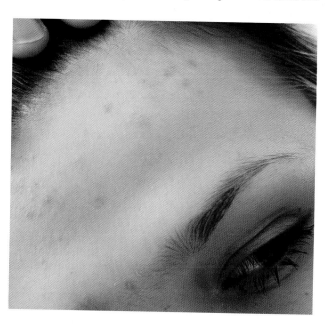

▲ *The onset of puberty leads to hormonal changes that often result in acne; this usually clears up after adolescence.*

sometimes on the back as well. The extent of this growth is largely a matter of genetics; a boy whose father has a smooth chest is also likely to be comparatively hairless.

The primary sexual changes of male puberty involve the testes, the penis, and the internal glands associated with the release of sperm. During puberty, the penis grows considerably in size and becomes more erectile, and the testes also enlarge. Within the testes, significant structural changes take place, which are much more extensive than the changes in a girl's ovaries. First, the testes start growing. As they grow, they begin to form sperm-producing tissue and continually make sperm until death. With the maturing of his sexual apparatus, a boy will begin to experience sexual dreams, often accompanied by the emission of semen (the sperm of which are fully formed and able to fertilize an egg released by a female ovary).

The male hormones

As in girls, it is the pituitary gland that is responsible for starting and continuing the changes in a boy's body during puberty. Sperm production is triggered by the action of the tropic hormones on the testes, stimulating the release of the hormone testosterone. In addition to helping sperm formation, testosterone is also responsible, along with other hormones from the adrenal glands, for completing the array of secondary sexual developments.

As with girls, growth hormone output increases, thus ensuring that by the time a boy is 17 or 18 he has reached full adult height, although growth may continue for one or two more years after that.

Emotional changes in males

Boys are just as prone to emotional conflicts during puberty as girls, but they often disguise these problems. However, under the surface they are also trying to come to terms with their sexuality. One aspect of this is masturbation, a subject that invariably embarrasses adolescents and parents.

Masturbation is of course just as common among girls, and modern thinking on this subject is that masturbation is not harmful to the physiology or the psyche and is an activity that falls into proper perspective as adolescence progresses. Similarly, boys may worry about their wet dreams and try to conceal them. Parents should take any opportunity offered to discuss such problems and not make an undue fuss about them.

Compared with their female peers, most pubescent boys may appear to be uninterested in the opposite sex, but this is often just a ploy. Unluckily for boys, they are affected more than girls by acne, which produces the traditional blemished face of adolescence: it can cause emotional and physical scars. While some boys may suddenly start to take an interest in girls, pop music, and the like, others play out their emotional turmoil by playing football, running track, or immersing themselves in automobiles.

Problems of puberty

The problems of puberty that usually concern families most—both parents and teenagers—are the rate at which the physical changes take place and their timing. It is important to remember that, except in rare cases, nature will take its predetermined course, either quickly or slowly, and not necessarily in any specific order. There are so many shades of normality that almost anything is acceptable.

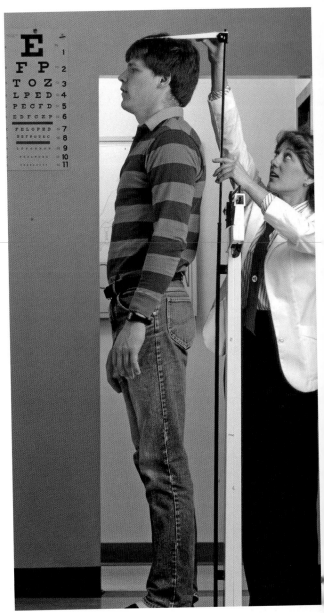

▲ *During puberty, dramatic changes can take place in the body, such as, for example, a growth spurt.*

Parents should be concerned if there is an obvious failure in physical and sexual development, in which case they should seek medical advice. A change of diet and, in extreme cases, hormone treatment may be needed to set puberty on its correct course. Puberty that begins before age 10 can cause problems, if only because it puts the child out of step with his or her contemporaries and parental care and tolerance are needed.

See also: **Endocrine system; Glands; Growing pains; Growth; Hormones; Larynx; Menstruation**

Pulse

The pulse is regarded as being the symbol of life itself, and it provides valuable information for doctors about the condition of the heart and major arteries of the body.

Questions and Answers

If you find someone collapsed, is it helpful to take his or her pulse?

Yes, it certainly could be. It is a common problem for doctors to be confronted with a patient who has collapsed, but has completely recovered on arrival at the hospital. In such cases it is very useful to know whether the pulse was very slow, fast, or totally absent. An abnormal pulse gives a strong clue that the heart is the cause of the trouble. The only time when it could be dangerous to take the pulse is if a bone might be broken. You should then try to take the pulse only if the wrist can be felt without moving the patient.

I feel very lethargic. Could my pulse be too slow?

There are various reasons for lethargy, from anemia to hormone problems to depression. It is true, however, that a very slow heart rate (due to a heart block, when the heart's electrical impulses are interrupted) can cause tiredness and lethargy. A simple test is to take your pulse when you have the symptoms. If it is under 55 beats per minute or irregular, see your doctor. If you are not sure, take your pulse again (when you no longer have symptoms) for comparison. Share these notes with your doctor.

An acupuncturist told me that diagnoses can be made by feeling both wrist pulses at the same time. Is this really true?

Most Western doctors don't believe it, and there is no real evidence to back it up.

However, many Western doctors are willing to try the effects of acupuncture when other methods have failed, and certainly it can sometimes be helpful.

When a doctor feels the pulse, with each beat he or she is feeling the action of the heart pumping blood through the arteries and around the body. Each beat of the heart transmits a force along the arterial walls just like a wave traveling across the surface of a lake. The walls of the arteries are elastic and expand to take the initial force of a heartbeat. Later in the course of the beat they contract and in this way push blood smoothly along the system.

The body's pulses

The pulse can be felt in the arteries that are near the surface of the body. The most common is the radial artery in the wrist, which can be felt on the inner surface of the wrist just below the thumb. It is customary to feel this pulse with one or two fingers rather than the thumb, which has its own pulse and can therefore cause some confusion. The brachial artery in the arm has a pulse which can be easily felt on the inside of the elbow joint almost in line with the little finger. A doctor may also examine the pulse of the carotid artery in the neck. This pulse can be felt with the fingertips on either side of the Adam's apple. Listening to a major artery like the carotid with a stethoscope can reveal a bruit—a regular whooshing noise with each heartbeat. This may indicate a partial blockage of the artery even though the pulse feels normal. There are also pulses in the groin, behind the knees, on the inside of the ankle, and on top of the foot.

The first thing that a doctor discovers from checking the pulses is the condition of the arteries themselves. This is easily discovered by checking if all the pulses are present and normal. It is particularly important to see if the pulses in the legs are present, since diminished pulses could be a sign of atherosclerosis (hardening of the arteries). If these pulses are weak or difficult to feel, atherosclerosis may have already developed.

A doctor also uses the pulse rate to evaluate the working of the heart. A regular rhythm suggests that the heart is beating steadily, since the pulse beat gives an exact indication of the

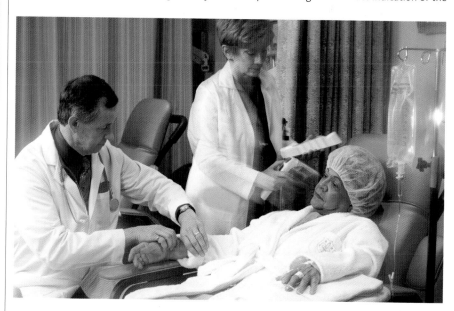

▲ *To take the pulse, place two or three fingers, not the thumb, on the inside of the wrist just below the thumb and count the number of beats felt in a minute.*

STRUCTURE AND FUNCTION OF THE ARTERIES

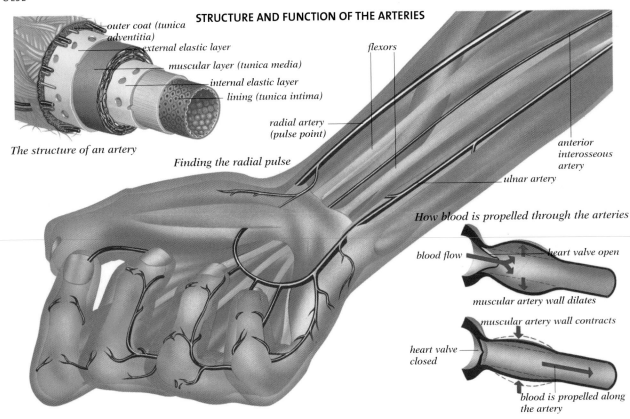

outer coat (tunica adventitia)

external elastic layer

muscular layer (tunica media)

internal elastic layer

lining (tunica intima)

The structure of an artery

flexors

radial artery (pulse point)

anterior interosseous artery

ulnar artery

Finding the radial pulse

How blood is propelled through the arteries

blood flow

heart valve open

muscular artery wall dilates

muscular artery wall contracts

heart valve closed

blood is propelled along the artery

▲ *The arterial walls are made up from several muscular layers of tissue that propel the blood smoothly along the arteries with each heartbeat. This action can be felt most easily at the radial pulse in the wrist.*

▶ *The pulse can be felt at a number of points in the body where a large artery passes close to the surface.*

heartbeat. Sometimes there can be a discrepancy between the heart rate and the pulse because some of the heartbeats are too weak to be transmitted along the arterial system. This may suggest heart trouble. A fast heartbeat (tachycardia), of more than 100 beats per minute (bpm), occurs as a normal response on exertion or as the result of anxiety or fever. It may also be caused by an abnormality in the heart's electrical conducting system. Slow heartbeats may be normal, particularly in athletes, but rates of less than 60 bpm are often abnormal in nonathletes, especially in elderly people. Heart block is caused by an obstruction to electrical impulses through the heart. Marked slowing of the heart

MAJOR PULSE POINTS

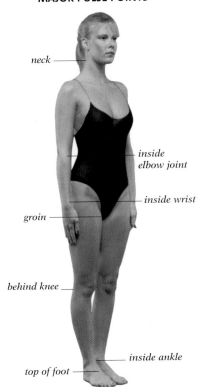

neck

inside elbow joint

inside wrist

groin

behind knee

inside ankle

top of foot

rate, to the extent that fainting results, can be caused by overstimulation of the parasympathetic system.

Diagnostic tool

Some rare diseases can be suggested by abnormal pulses. When the pulse in the legs is diminished in force and delayed in time compared with that in the arms, a rare congenital blockage of the body's main artery, the aorta, may be indicated. The pulses in the legs, and even in one or both of the arms, may disappear in a more common disease called dissection of the aorta, when the inner lining of the aorta is torn. This forces blood into the arterial wall and creates a blockage.

The shape and pattern of a pulse wave may help diagnosis. For example, a slowly rising pulse indicates obstruction of the aortic valve; and a pulse which rises fast and falls away again—a collapsing pulse—indicates a leaking aortic valve. Both conditions may require surgery.

See also: **Autonomic nervous system**

Rectum

I am about to undergo surgery to remove a growth in my rectum. Will I have to have a colostomy?

That depends on where in the rectum the growth is situated and on the type of growth it is. If it is in the lower part, and the doctor can easily feel it, then you will probably have to have a colostomy. If it is higher in the rectum, then the surgeon should be able to remove the growth and join the two ends of the intestine together using a special stapling device.

Is cancer of the rectum hereditary?

Yes, there is a link between cancer of the rectum and hereditary factors. If there is a family history of polyps in the colon or rectum, this can sometimes be a strong factor in the subsequent development of a cancer.

I passed blood in my feces and am going to have to see a doctor. Will a rectal examination be painful?

No. The doctor examines the lining of the lower rectum with a gloved finger, and, unless you have an anal fissure, it should not be painful. He or she may pass a small instrument into the rectum to look directly at the rectal lining, but this merely causes some discomfort. Do not put off seeing your doctor just because you are afraid of being examined.

My child has had a rectal prolapse. Is this serious? Is it permanent?

No. Rectal prolapse, or protrusion of the lining of the rectum out through the anus, is not serious in children and it usually gets better as the child gets older. It rarely requires surgery. Most cases are due to constipation, so altering the diet to provide more fluid and more bulk may help. Ensure that your child eats plenty of roughage.

The rectum is the relatively straight muscular tube that forms the lowermost part of the large intestine. It is involved in one of the most basic of the body's functions—defecation.

There is confusion among some people about the difference between the rectum and the anus. The anus is simply the short narrow tube surrounded by a ring of muscle that joins the rectum, the lowermost part of the large intestine, to the outside. The main function of the anus is to maintain continence of feces while the rectum acts as a reservoir for them. With a normally functioning anus and rectum a person can defecate when it is socially convenient and not just when feces happen to have passed through the whole of the large intestine.

The rectum itself, like the rest of the intestine, consists of a muscular tube lined with a special membrane called epithelium. In the rectum, this epithelium contains glands that produce mucus

POSITION AND STRUCTURE OF THE RECTUM

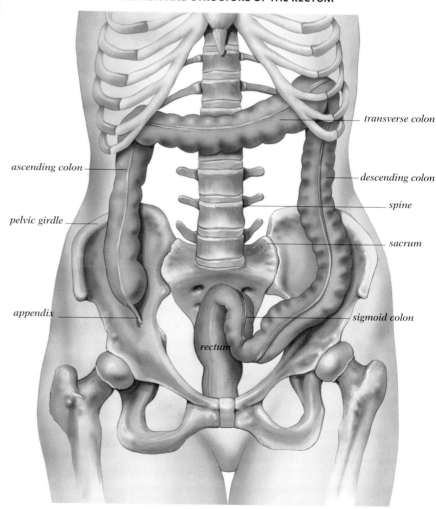

transverse colon

ascending colon

descending colon

spine

pelvic girdle

sacrum

appendix

sigmoid colon

rectum

▲ *Waste matter is passed through the loops and twists of the colon into the short, narrow rectum, which stretches to accommodate and retain the waste until it can be excreted.*

SMALL INTESTINE

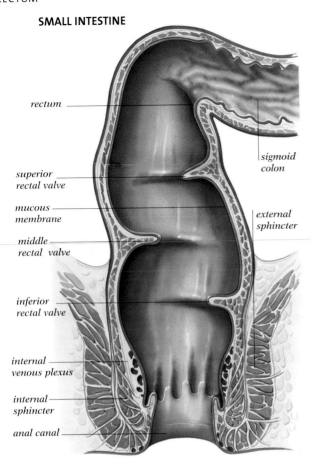

rectum

sigmoid colon

superior rectal valve

mucous membrane

middle rectal valve

external sphincter

inferior rectal valve

internal venous plexus

internal sphincter

anal canal

▲ *The rectum consists of a muscular tube lined with a mucous-producing membrane: the mucus acts as a lubricant to make the passage of feces easier. During defecation the rectal muscles contract to expel the feces through the anal canal.*

to lubricate the feces and make their passage easier. The muscular part of the rectum contracts during defecation to expel the feces, but at other times it is capable of stretching. It is this potential for increasing in size that enables the rectum to act as a reservoir.

What can go wrong

Several conditions can affect the rectum, including proctitis (inflammation), prolapse, and polyps. By far the most serious disease is carcinoma of the rectum.

Carcinoma of the rectum

This is a malignant tumor that arises from the epithelium. A patient with a carcinoma of the rectum is usually middle-aged or elderly, although on rare occasions the disease can occur in early adulthood. It shows itself in several ways. First, by bleeding: the blood is usually bright red and mixed in with the feces. Second, there can be an excess of mucus during defecation, probably caused by an irritation of the lining of the rectum. Third, the patient experiences a sensation of wanting to defecate, because there is a tumor in the rectum. However, there is no resulting intestinal action, so the patient is constantly going to the toilet, with no result. These

▶ *In a permanent colostomy, an opening from the intestine is made through the abdominal wall (shown here) so that the feces can pass through it. The operation can be lifesaving.*

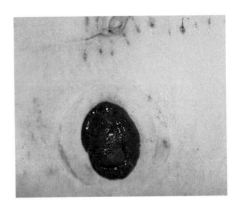

symptoms are described by the medical term "tenesmus." Finally, the tumor may have spread to other parts of the body, making the patient feel generally sick, without much in the way of symptoms in the rectum itself.

Treatment

The doctor will examine the rectum to detect any masses. Some growths can be felt easily with a gloved finger, but others can be detected only by passing a viewing tube into the rectum and inflating it with air. This examination, called a sigmoidoscopy, is painless, and it allows the doctor to see the growth and take a tiny piece with forceps so it can be examined in the laboratory.

The only successful treatment for a carcinoma of the rectum is surgery. The type of surgery involved depends on the growth in the rectum. Basically, if the growth is in the upper part, then it can be removed and the intestinal ends joined. However, if it is at the lower end, near the anal canal, then the lower rectum and anal canal have to be removed. That is, the patient has to have a permanent colostomy, or opening of the intestine, on the front of the abdomen.

It must be remembered that a colostomy can be a lifesaving procedure, and also that, with modern adhesives and plastics, a colostomy is not quite such an unpleasant device as it was in the days before colostomy bags, which stick onto the skin.

One of the new developments in the field of surgery for carcinoma of the rectum is the use of a special stapling device called a stapling gun to join the two ends of intestine together. As a result, much lower joins can be made and therefore some patients who would have had to have a colostomy are spared one.

Some patients have a temporary colostomy farther up the large intestine for a few weeks after the operation to remove the tumor. This acts as a safety valve until the join between the colon and the rectum has healed properly.

Outlook

After the surgery, the patient will have to visit the hospital for checkups for many years. The patient who had a growth in the upper rectum will have no noticeable side effects from the surgery and should be able to defecate normally. However, the patient with a colostomy will have to learn to look after it when he or she goes home. Most people manage to come to terms with this after a few weeks.

See also: Anus; Colon; Mucus

Reflexes

Do young babies have a sucking reflex or do they learn to suck?

Babies are born with a sucking reflex. If you gently stroke the cheek of a newborn baby near the mouth, his or her head will turn toward your finger. Once the finger is in his or her mouth, the sucking reflex is stimulated. The reflex is lost as the child grows.

Is blinking at strong light a reflex?

Yes. Strong light will cause a reflex blink, giving the pupil a chance to constrict (again by reflex). Blinking avoids overstimulation of the retina, which might otherwise be damaged by excessive light.

To what extent are reflexes learned behavior?

Simple reflexes, such as a knee jerk, are not learned, but are built into the nervous system. More complicated types of reflexes are partly learned, particularly the so-called conditioned reflexes. The stimulus producing the reflex is changed; we jump at the sound of a bell if the bell happens to have special significance because it is associated with another stimulus or reward.

Can I stop reflexes from occurring?

Simple tendon reflexes can be inhibited by tensing the muscles. Others, like the choking reflex, cannot be stopped at will.

Is treatment required for a lost reflex?

The absence of reflexes such as knee jerks may signify disease. If swallowing reflexes or those controlling breathing are absent, treatment is usually required, often as an emergency.

Why does a cat always land on its feet? What makes us blink? How do people continue to breathe even when they are asleep? These types of activities, which are governed by automatic responses outside conscious control, are known collectively as reflexes.

Reflexes vary in complexity from automatic withdrawal of a hand from a hot surface to more complex reflexes that maintain the body's position when balance is lost. Many primitive reflexes are inborn but are gradually overlaid by some of the many learned activities that make up a person's normal repertoire of behavior.

What a reflex is

A reflex is an automatic response to a specific stimulus that is not under conscious control. For a reflex to occur there must be a sense receptor, nerves to convey the report of the stimulus, an apparatus to convert this information into a response, and muscles or glands to provide the response, which is usually some type of movement. Any response that follows this pattern of automatic reaction to a stimulus is called a reflex.

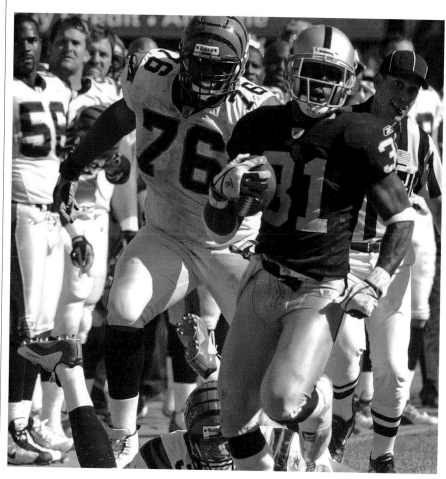

▲ *The system of reflexes is important for football players; if they are tripped, reflex action instructs the arms to move out in an effort to break their fall.*

▲ *A grasp reflex is shown by gently touching a baby's palm. The baby's fingers will curl around the adult finger and grip tightly. Pulling against the grip strengthens it.*

▶ *The baboon instinctively flees the leopard before turning at the last moment to face the aggressor. Reflex action in the form of the fight-or-flight response tunes the baboon's body to a high pitch in its desperate bid for survival.*

▲ *We don't have to think about moving our eyes to follow a bouncing basketball; they move automatically as the eye muscles contract and relax by reflex reaction.*

Conscious reactions

Conscious behavior is not a reflex response, because an analysis takes place between the stimulus and the response. The stimulus is analyzed and acted upon by a person according to previous experiences, mood and present desires.

Because of this method of deciphering a stimulus, an identical stimulus may produce different responses each time it is encountered. By contrast, a reflex response is exactly the same whenever the stimulus is presented.

A conscious reaction can overcome certain reflexes. For example, if we touch a hot stove, a reflex action would be to move away immediately. We can keep our hand on the hot stove, but we would have to apply conscious energy to do so. Reflexes therefore provide the body with protective and quick responses, especially to harmful or dangerous stimuli.

Some reflex movements, such as breathing, are so important that although they can be stopped by conscious effort for some time, they eventually break through the conscious control.

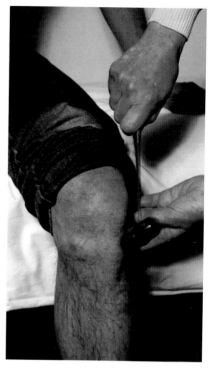

▲ *When the doctor tests a reflex by tapping a knee with a tendon hammer, he or she can establish the tone of the muscles or their readiness for action.*

Different types of reflexes

There are numerous different types of reflex. Some control muscle movements, basic bodily functions, and correct orientation in standing or sitting. Other, more complex reflexes are programmed responses to dangerous or frightening situations.

Muscle reflexes: These are more accurately called tendon reflexes because vibrations in the tendons set them off. When a doctor tests someone's reflexes by tapping his or her knee with a tendon hammer, the front thigh muscle is stretched; this stretching stimulates a motor neuron that conveys a signal to the spinal cord. Motor nerve fibers transmit a signal to the muscle, which then contracts and makes the leg jerk and kick out slightly. These reflexes are all part of a highly complex mechanism in the spinal cord that controls the tone of the muscles, that is, their readiness for action. The spinal cord mechanism is in turn controlled by the more advanced parts of the movement control hierarchy in the brain. Because of this control, the spinal reflexes can be brisker or more sluggish according to the setting imposed on them by the brain. The same spinal cord mechanism is linked to pain receptors in the skin and elsewhere in the body so that swift reflex responses can be made when harmful stimuli are received.

Orientation reflexes: If a cat is released from a few feet (meters) above the ground, the animal will always land squarely on its feet and come to no harm. This is an example of the operational speed of the reflexes that control posture and orientation. Similarly, if a person slips on an icy pavement, his or her body will twist in an attempt to right itself and usually a hand will stretch out in an attempt to prevent a fall. These complex responses are programmed by more advanced parts of the motor system. Sensitive receivers in a special part of the ears monitor a person's position in space. When someone starts to fall, messages from these receivers are quickly relayed to the cerebellum at the base of the brain, which selects the correct series of commands to the muscles of the arms and legs. This series of events occurs in a far shorter time than if there were a conscious decision about the appropriate set of movements. Primitive examples of this type of orientation reflex are seen in small babies. For example, if a baby's head is suddenly released, he or she will outstretch the arms in a sort of grasping movement. This is called the Moro reflex (Moro was the doctor who first described it); it disappears after some weeks if development is normal.

Bodily function reflexes: Young children who still wear diapers have no conscious control over the passage of urine from the bladder. When the bladder is full, the internal pressure sends a signal to the spinal cord to activate a reflex that will empty the bladder. As a child develops, he or she becomes able to suppress this reflex until the appropriate opportunity to pass urine arises.

However, even fully mature adults cannot suppress the reflex indefinitely; a point will be reached when the bladder empties not by conscious will but by reflex response, with the help of the spinal cord. Similar reflexes control several basic bodily functions, including breathing.

Some functions, such as bladder control, and to a lesser extent breathing, can be controlled consciously; others, like the control of the heartbeat, are purely automatic.

Behavioral reflexes: These are the most sophisticated reflexes in the body and are used to prepare the body's behavioral responses in a standardized way in extreme situations. The best example is the so-called fight-or-flight reaction—a pattern of reflex responses produced in answer to a threatening situation. If someone is suddenly confronted with a violent mugger, for example, he or she may either fight or run away; similarly, an animal faced with an aggressor has these two choices. The needs of the body are similar whether the brain decides to run or to stand and fight. Accordingly, this set of reflexes automatically produces the optimum conditions of heartbeat and breathing that will then be available for whichever course of action is chosen.

The same set of responses includes sweating (allowing heat loss during the fight or the flight) and pallor of the skin (because blood is being pumped away from the skin to more important muscles). This same reflex can be induced by the mere thought of any frightening or threatening situation when it becomes part of a so-called conditioned response.

Conditioned reflexes: A reflex is called conditioned when it is brought on by a stimulus other than that which first (or naturally) produced it. The reflex becomes attached to the second stimulus when it occurs repeatedly with the normal stimulus.

The famous Russian psychologist Pavlov first described this type of reflex when he noticed that if he rang a bell each time he gave a dog some food, after a time he could produce the response of salivation in the dog by ringing the bell alone: initially, only the food would prompt this reflex response. This type of conditioned reflex provides the basis for methods of training animals to perform.

What reflexes are for

Because reflexes are involuntary responses by the nervous system, they are very fast and more or less automatic; thus they save time and mental energy when prompt and often lifesaving action is required. Reflexes are a useful tool in medical diagnosis because they give a very precise way of testing various circuits in the nervous system, thus enabling the site of any trouble to be identified. For example, the tendon reflexes are organized in the spinal cord in segments. A neurologist can test each one in turn until he or she pinpoints the ones that are not functioning properly. The neurologist can then work out which part of the spinal cord is involved in the underlying disorder.

If reflexes are lost

Different reflexes vary in their relative importance in everyday life. Losing a knee jerk does not make much difference,

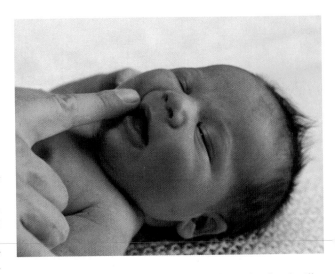

▲ *Gently stroke a new baby's cheek and his or her head will turn and the lips will move as he or she tries to get their mouth around your finger: this action is called a baby's sucking reflex.*

▲ *A reflex controls the action of blinking automatically. Blinking protects the retina from too much strong light by allowing the pupil to constrict, by reflex. In addition, potentially damaging particles are swept away from the surface of the eye; so if the reflex is lost, injury to the delicate cornea can result.*

although the underlying reason for the loss may cause other symptoms. Occasionally, people are born without some of the tendon reflexes, but they can live a normal life. Other reflexes, however, are more vital, including the orientation reflex.

The orientation reflex may be lost if a person is suffering from a disease of the cerebellum; the patient will then have considerable trouble maintaining balance unless he or she concentrates carefully.

Similarly if the blink reflex is lost, the eye can become seriously damaged by particles landing on the surface of the cornea. Because they are not swept away by blinking, they can scratch the eye, causing loss of vision.

People who lose the automatic reflex to breathe can have serious problems when they are sleeping, because the voluntary breathing system no longer works. Those who have lost the automatic breathing reflex must be protected by special treatment until the situation can be rectified. For example, it may be necessary for those who have lost the breathing reflex during sleep to be connected to a mechanical breathing device at night.

See also: Autonomic nervous system; Balance; Brain; Breathing; Cornea; Eyes; Homeostasis; Nervous system

Retina

The retina is by far the most complex part of the eye. Most eye problems affecting vision are not caused by retinal disorders, but when these do occur, the matter is likely to be serious.

My grandmother has age-related macular degeneration and is unable to read, watch TV, or even tell the time from her watch. Will she soon be totally blind?

No. Although she is badly disabled by her loss of central vision, the disease never affects any part of the retina other than the central macular region. Her peripheral vision will be preserved. Hold your fist against your nose and continue to look straight at it. This is what your grandmother's vision is like. She has what ophthalmologists call "navigational vision."

My father had a cataract operation years ago and has to wear thick, heavy glasses. His right eye is totally blind because he had a retinal detachment a month after the second operation and treatment failed. Now I am scheduled for cataract surgery. Is it likely I may suffer the same fate?

No. Cataract surgery has greatly improved in recent years. In your father's time, the whole internal lens was removed and there was about a 1 percent risk of retinal detachment following this kind of surgery. Cataract surgery today, with retention of the lens capsule and lens implant, causes very little risk of detachment. Also, retinal detachment surgery now has a much higher success rate than it had in those days.

I am a female with a family history of X-linked recessive retinitis pigmentosa. I know I have two X chromosomes, so am I going to go blind like my brother?

No, not unless you have inherited the affected gene from both your parents. You may be a carrier of the mutated gene on one of your X chromosomes, but the other one protects you. Males have only one X chromosome and in your brother's case it carries the gene.

The retina is the delicate but complex membrane that covers the whole of the rear inner surface of the eyeball. It is the screen on which the images formed by the lens system—the cornea and the crystalline lens—fall. It converts these images into electrical nerve impulses that are then passed back through the optic nerves and optic tracts to the rear part of the brain.

Structure and function of the retina

The retina, although very thin, has eight distinct layers, each with a different construction. Near the back is the layer of the rods and cones—specialized photoreceptor cells. In front of this layer are nerve cell and nerve fiber layers that act as a kind of computer to integrate changing visual images and connect the photocells to the fibers of the optic nerve. Each eye has over 100 million photoreceptors.

There are three kinds of cones, each responding maximally to one of the three primary colors. The cones are concentrated in the central retina, the macula, where they are tightly packed. It is only in this area that images are given sharpness (high acuity). The six muscles that move each eye are coordinated by the brain to ensure that objects looked at are focused on the macula. Visual acuity only 5 degrees away from the point of visual fixation is so low that the top letter on a sight-testing chart cannot be identified. The rods are more sensitive to light than the cones but are color-blind and, being peripheral and widely spaced, allow only low-resolution vision.

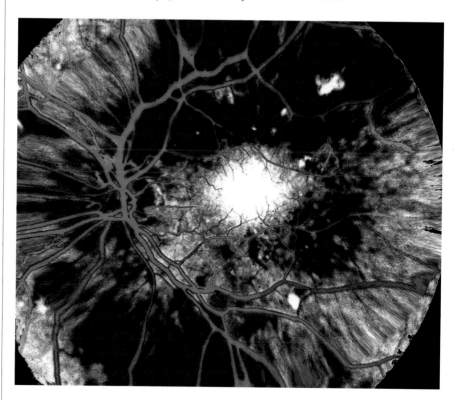

▲ *This image shows the destruction of the retina (white spot in center) and the degeneration of the blood capillaries (red) that are symptoms of diabetic retinopathy.*

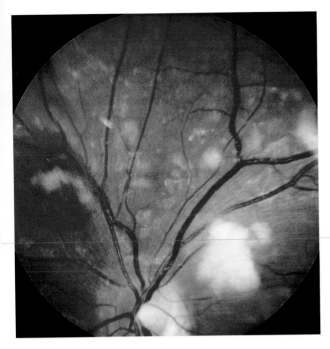

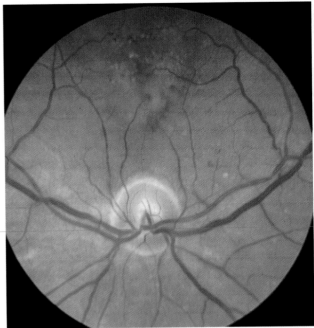

▲ *This image of a retina, taken through an ophthalmoscope, shows fluffy white lesions called cotton wool spots, caused by swelling of the surface layer of the retina. The condition can be found in patients with diabetes, high blood pressure, or AIDS.*

▲ *This image shows macular degeneration—a disorder that affects the macula (the central part of the retina), causing loss of visual acuity (ability to perceive detail) and central vision. The patient does retain peripheral vision, however.*

Retinopathies

The most common of these is diabetic retinopathy. It is a major cause of blindness, resulting from poor diabetic control, and affects about one person in 50. The risk increases with the duration of the diabetes. After 10 years, about 20 percent of diabetics have the disease; after 30 years, about 90 percent. The condition takes two forms: background retinopathy, which barely affects vision but which may progress to the more severe form, proliferative retinopathy, characterized by the production of new, fragile blood vessels and bleeding into the vitreous gel of the eye. Fibrous tissue can grow into the released blood and lead to retinal detachment. Proliferative diabetic retinopathy, caught in time, can be effectively treated by destroying the peripheral retina with hundreds of laser burns.

Hypertensive retinopathy is caused by severe high blood pressure; lowering of the blood pressure is urgently needed to avoid a potentially fatal stroke. Both types of retinopathy are diagnosed by examining the retinas with an ophthalmoscope.

Retinal detachment

This condition is quite rare, affecting about one person in 10,000. Contrary to popular belief, retinal detachment is seldom caused by blows to the eye or head. Most cases occur because of weakness in the retina that has led to tears or holes. These occur most often in very nearsighted people. Penetrating eye injuries or tumors behind the retina can also cause detachment. In most cases, the presence of a retinal hole leads to a progressive flow of fluid through the hole and behind the retina, stripping it off. The condition is painless but the field of vision opposite the site of the detachment is progressively obscured by a black area, like a curtain.

Treatment is by surgery. To achieve accuracy, silicone rubber sponges are stitched to the outside of the eyeball and indented over the holes. A freezing probe is applied to cause inflammatory adhesions.

Macular degeneration

This painless, age-related disorder affects about 30 percent of the population over age 75. It involves only the central, most sensitive part of the retina and may progressively destroy this area, causing visual distortion proceeding slowly or quickly to total loss of central vision. Although peripheral vision is preserved, direct gaze is impossible because of the large blind spot that obscures the point of interest. Usually one eye is affected first, but this may not be noticed until the second eye is involved. Macular degeneration can lead to severe visual disability and legal blindness. Its cause is unknown but it is more common in cigarette smokers than in nonsmokers. In some cases, the progression can be stopped by laser treatment.

Retinitis pigmentosa

This is a genetic disease affecting about one person in 4,000. It can be inherited in various ways. The mutated genes that cause it code for the protein called visual purple or rhodopsin, a substance needed for the proper functioning of the rods and cones. Retinitis pigmentosa causes night blindness and a progressive loss of peripheral vision so that form of tunnel vision gradually worsens. It varies in severity and may start at various ages, but usually begins in adult life. There is no effective treatment, but gene therapy is being tried.

See also: **Blood pressure; Cells and chromosomes; Cornea; Eyes; Optic nerve**

Rhesus factor

Questions and Answers

In blood transfusions, do Rh factors have to be matched as well as other groups like A and O?

Yes. A person with Rh-negative blood must be given a transfusion of the same group. Without this matching, the body may set up an immune reaction to the foreign substances in the transfused blood, even if not at once.

Long before I was married I had an abortion that my husband knows nothing about. Now I'm pregnant again and the doctor says my husband and I have different Rh groupings. Should I tell anyone about the abortion?

It is vital that you should reveal the information about your abortion to your doctor, even if you don't feel that you can tell your husband. The reason is that the abortion may have set up a potentially adverse reaction in your blood that could severely damage the baby. Tell your doctor about it as soon as possible so that proper tests can be carried out. Your doctor will respect your confidence completely.

Is it important for a husband's blood to be tested for Rh factors when his wife is pregnant?

This is usually necessary only when the wife is Rh-negative and has developed antibodies to Rh-positive blood. In this case, testing the husband's blood can give the doctor an indication of the chances that the fetus is Rh-positive.

If there is a risk of Rh incompatibility, can a baby be born normally?

Because of the possibility of complications, most doctors prefer that babies at risk be delivered by cesarean section or induced before the due date, usually about the 34th week of pregnancy.

Whether a person's blood is Rh-negative or Rh-positive is normally of little concern. However, if a mother who is Rh-negative conceives a baby who is Rh-positive, there can be serious consequences in subsequent pregnancies.

There are two main blood group systems: one is the ABO system; the other is the Rh (rhesus) system. In certain circumstances, incompatibility of two Rh factors between a mother and her baby can cause a disease that breaks down the red cells in the bloodstream of the fetus or newborn baby. Left untreated, Rh incompatibility in a baby can be fatal.

The Rh factor

The Rh system of blood groupings is based on the presence or absence in the blood of several different biochemical molecules or factors. Of these factors, the most important to the Rh system is one always designated as D when it is present and d when it is absent. All people inherit a pair of Rh genes: one from the mother, the other from the father. As a result, a person can be DD (both parents Rh-positive); Dd (one parent Rh-positive, the other Rh-negative); or dd (both parents with an Rh-negative gene). Because D is dominant over d, a person who is Dd will have the D factor in his or her blood, and so be Rh-positive. Only dd individuals make no D factor at all, and they are known as Rh-negative in this grouping.

The basis of incompatibility

The Rh factor causes problems when, for

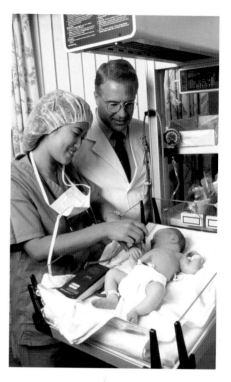

▲ *Babies at risk of Rh incompatibility can be given a transfusion of Rh-negative blood immediately after they are born.*

some reason, Rh-positive blood gets into the bloodstream of an Rh-negative person. The body then reacts as if it had been invaded by a disease-causing organism and begins to make antibodies ready to attack and eliminate that organism. When the blood types mix for the first time, only the antibody-manufacturing equipment is triggered. The Rh-negative blood is said to have become sensitized to the presence of Rh-positive blood.

If this sensitized blood subsequently becomes mixed with Rh-positive blood again, even years later, then the antibody-making machinery goes into action to make anti-D antibodies. The antibodies multiply and begin to break down the red cells of the Rh-positive blood so that it cannot do its job properly.

Pregnancy and problems

The likelihood of Rh incompatibility in pregnancy arises only if the ovum of an Rh-negative woman is fertilized by a sperm from an Rh-positive man. If the developing fetus has Rh-negative blood (which it may if the father is Dd, and the baby inherits his d gene), then there is no problem. However, if the fetus is Rh-positive, as is more likely because of the greater frequency of the D gene in the population, then trouble is likely, although not immediately.

When an Rh-negative mother is carrying an Rh-positive fetus, blood cells occasionally manage to escape from the fetal circulation, cross the placenta, and enter the mother's

RHESUS INCOMPATIBILITY AND ITS PREVENTION

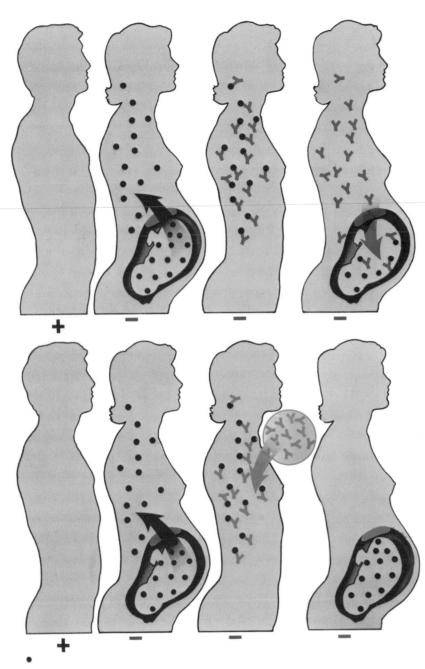

bloodstream. This transfer is thought to be more likely at birth or when a pregnancy is terminated.

The resulting sensitization in the mother's blood does not affect the first baby, but subsequent Rh-positive babies are definitely in danger. In response to the presence of Rh-positive blood cells from the fetus, the sensitized mother begins making large quantities of antibodies, which are small enough to cross the placental barrier. Once inside the baby's bloodstream, they break down the red cells in its Rh-positive blood.

If left untreated, this situation will damage the fetus. In the most severe cases, the fetus dies in the womb, usually because of extremely severe brain damage and anemia. This condition is called hydrops fetalis. Alternatively, the baby may be born alive but with a weak heart or with severe jaundice. The baby's yellow complexion is caused by the release of yellow bile pigments (bilirubin) from the destroyed red blood cells. This jaundice places the baby at risk because the bile pigments are toxic.

Treatment

One in roughly every 200 babies is at risk of Rh incompatibility. Treatment depends on the stage of pregnancy at which the problem is diagnosed and its severity. The aim is to transfuse the baby with Rh-negative blood or, if this is unnecessary, to watch the baby's progress carefully. If necessary (based on a specific blood test in the mother), amniocentesis is done to assess the concentration of bilirubin in the fetus. If a fetus is in great danger of dying in the womb, it is transfused with Rh-negative blood while it is still in the uterus. In less severe cases, an exchange transfusion after the birth replaces the baby's Rh-positive blood with Rh-negative blood.

When an Rh-negative mother gives birth to an Rh-positive baby for the first time, antibodies to Rh factor can be given to the mother to destroy the Rh-positive cells and prevent the mother from producing antibodies. This prevents problems from occurring during later pregnancies.

▲ *When an Rh-negative mother conceives an Rh-positive baby by an Rh-positive father, blood cells from the baby (shown in red) leak into the mother's circulation, usually during the birth. She begins to make antibodies (shown in blue) to the D factor. The next time she is pregnant with an Rh-positive baby, she produces large amounts of anti-D antibodies, which cross the placenta and attack the baby's red blood cells (causing Rh disease). To prevent this, an injection of anti-D antibodies (shown in green) can be given shortly after delivery. This destroys any of the baby's blood cells that have leaked into the mother's blood before she has time to make antibodies of her own. Therefore, a subsequent Rh-positive baby will be safe.*

See also: **Bile; Blood; Blood groups; Fetus; Immune system; Placenta; Pregnancy**

Ribs

The ribs are the curved bones in the chest wall that support and protect the heart, lungs, and liver, while allowing movement for normal breathing. Their unique structure enables them to perform these functions efficiently.

While strengthening the chest wall, the ribs also protect some of the body's vital organs. The heart, lungs, liver, and spleen all lie within the framework of the ribs, and become vulnerable to blunt injury only if the ribs are broken.

The structure of the ribs

The ribs are specifically designed to allow changes in the chest volume during breathing. If the chest was covered by a continuous sheet of bone there would be no freedom of movement; while without any ribs at all, breathing would be virtually impossible. The diaphragm alone could not adequately increase the chest volume.

The ribs have muscle in between them (intercostal muscles) and cover the top and sides of the chest. At the bottom of the chest lies the diaphragm (a thin, umbrellalike sheet of muscle), which is attached all around to the ribs.

The volume of air in the chest may be increased in two ways. The diaphragm can contract to make it flatter, and the muscles attached to the ribs can contract, causing the ribs to swing upward and outward. This movement can be felt if a person takes a deep breath at the same time as he or she rests the hands on the lower part of the chest.

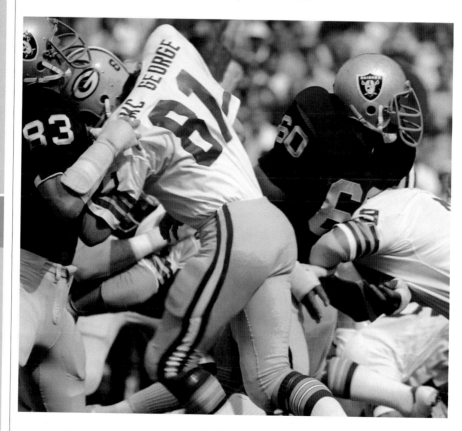

▲ *In rough contact sports like football, the players' ribs are particularly vulnerable to painful injuries.*

Questions and Answers

I've been to the emergency room with my broken ribs but was only given painkillers. Why wasn't my chest strapped?

Strapping the chest used to be a standard treatment for broken ribs. It relieves pain but does so by inhibiting movement of the chest wall during breathing. This increases the likelihood of chest infection, so the practice of strapping has been discontinued.

Why do surgeons operate on some broken ribs, whereas others are left to heal on their own?

Broken ribs are operated on only when a large segment of the chest wall has been pushed in. These patients normally have severe breathing problems because the collapsed segment moves independently of the rest of the chest wall. This affects the transfer of gases in and out of the lungs so surgery is needed to stabilize the injured chest wall. Less serious fractures do not require surgery; new bone will form and in time the ribs will be as good as new.

My left hand sometimes becomes numb and looks very pale. My doctor says I might have a cervical rib. What does this mean?

A cervical rib means an extra pair of ribs at the root of the neck. If you have a cervical rib, the artery that supplies your arm could be stretched up over the rib and thus compressed. After a special X ray of the artery you would probably need surgery in order to remove the extra pair and permanently eradicate the problem.

Are there really such things as floating ribs?

The floating ribs are the 11th and 12th ribs. They do not actually float in the body, but are embedded firmly in the muscles of the abdominal wall. They differ from the other ribs in that they are nonfunctional and not attached to the breastbone at the front.

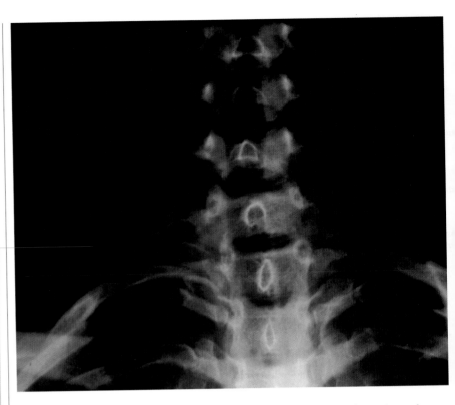

▲ *People are sometimes born with an extra pair of ribs, usually in the neck, as shown above. Because these extra ribs can compress the artery supplying the arms, they are removed by surgery to alleviate the problem.*

The ribs themselves are fixed to the spine by special joints that allow movement during breathing. At the front they are attached through a piece of cartilage to the sternum, or breastbone. The two lowest ribs (the 11th and 12th) are attached only at the back and are too short to be joined to the sternum. These are usually known as the floating ribs—they have little connection with breathing, and can be removed without affecting the patient.

The first and second ribs are closely connected with the collarbone or clavicle and form the root of the neck near which several large nerves and blood vessels pass on their way to the arms.

What can go wrong

A person can be born with some abnormality of the ribs or chest wall. Either the sternum may be too prominent or it may be collapsed inward so that there is a concavity in the middle of the chest. However, this usually does not impair normal chest function.

The presence of too many ribs is well known but causes serious defects in only a few cases. The extra ribs are usually in the neck and are called cervical ribs. They may take the form of a fibrous band or a complete bony rib; various combinations of the two are also possible. They can cause tingling or numbness in one hand by pressing on nerves and blood vessels leading to the arms. The radial pulse in the wrist may also be absent.

Such obstructions may not be easily diagnosed, since they do not always show up on an X ray and the symptoms are mainly in the patient's hand or arm, where other causes are also possible. Once the diagnosis has been made, however, surgery can be done to remove the extra rib, thus relieving the pressure on the nerves or the obstruction to the blood vessels and alleviating the symptoms.

Fractured ribs

Ribs may be broken in many ways: for example, during hard physical games such as football, in a fall, or in an automobile accident. Often the only significance of a rib fracture is that the underlying organs or tissue may be damaged, but if many ribs are broken there may be problems with breathing.

RIB CAGE: FRONT VIEW

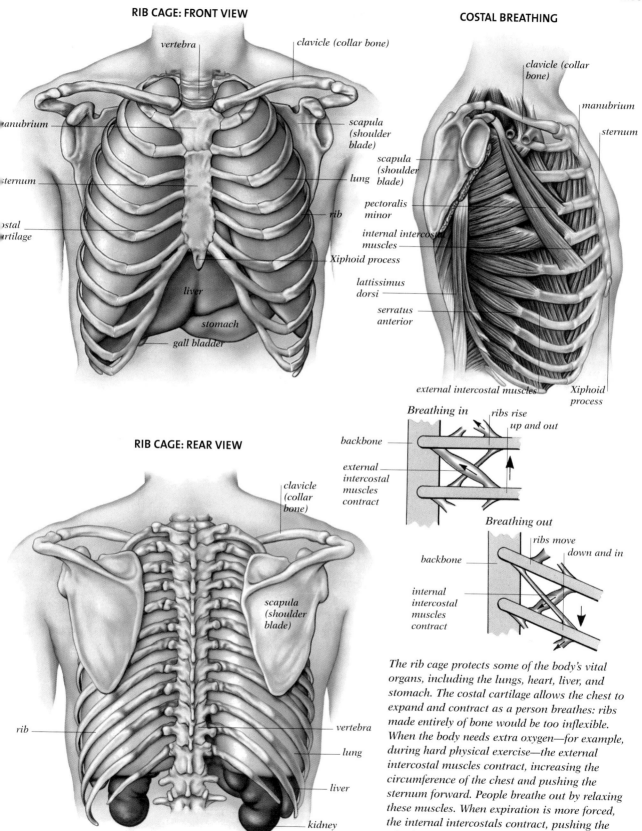

vertebra

clavicle (collar bone)

anubrium

sternum

ostal
rtilage

scapula (shoulder blade)

lung

rib

Xiphoid process

liver

stomach

gall bladder

COSTAL BREATHING

clavicle (collar bone)

manubrium

sternum

scapula (shoulder blade)

pectoralis minor

internal intercostal muscles

lattissimus dorsi

serratus anterior

external intercostal muscles

Xiphoid process

Breathing in

ribs rise up and out

backbone

external intercostal muscles contract

Breathing out

ribs move down and in

backbone

internal intercostal muscles contract

RIB CAGE: REAR VIEW

clavicle (collar bone)

scapula (shoulder blade)

rib

vertebra

lung

liver

kidney

The rib cage protects some of the body's vital organs, including the lungs, heart, liver, and stomach. The costal cartilage allows the chest to expand and contract as a person breathes: ribs made entirely of bone would be too inflexible. When the body needs extra oxygen—for example, during hard physical exercise—the external intercostal muscles contract, increasing the circumference of the chest and pushing the sternum forward. People breathe out by relaxing these muscles. When expiration is more forced, the internal intercostals contract, pushing the ribs downward and reducing the chest volume.

355

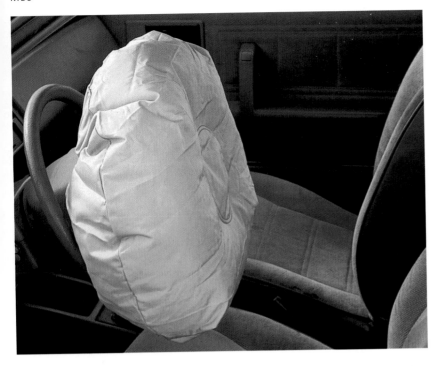

▲ *Inflatable airbags can prevent serious injury to the ribs should an automobile crash propel the driver toward the steering wheel.*

▲ *Cracked ribs, shown in this X ray, cause severe pain and can damage underlying tissue and organs.*

Fracture of one or two ribs

The main symptom is severe pain, which is made worse by breathing or if the chest is compressed. Sometimes breathing causes so much pain that the patient has to take shallow, frequent breaths.

A chest X ray will usually reveal any fractures and related damage. For example, sometimes a spike of rib pierces the lung and makes it collapse. There may also be associated bleeding from the end of the rib or the chest wall. However, sometimes a person with a broken rib may have a normal X ray; the sharp pain on breathing may be the only indication of a fracture.

There is no specific treatment for fractured ribs. Painkillers are usually given (by injection if necessary) and any complications must be dealt with as they arise. Some doctors give old people antibiotics to prevent infection, and breathing exercises are also helpful in this way. Strapping the chest is no longer considered a useful treatment, since it hinders chest movement and makes infection more likely. The pain often lessens within a week, but some pain is to be expected for several weeks, especially during exertion or when coughing. When the rib has healed, the bone should be as strong as it was before.

Fracture of many ribs

Several ribs may be fractured in at least two places in a severe crushing injury that pushes in a whole section of the chest wall. This is called a flail chest and can cause serious breathing problems. The normal movement of the chest wall during breathing is reversed; when the patient takes a deep breath the chest wall is sucked in instead of moving out.

If the area of injury is large, the patient may need mechanical ventilation to increase his or her oxygen supply. Sometimes the ribs need to be wired to prevent them from moving about. Again, the

outlook should be good, although any suspected tissue damage must always be investigated and its extent determined.

Tumors

It is rare to develop a primary tumor in a rib, but sometimes a secondary tumor may develop from a primary tumor that is located elsewhere in the body. This secondary development happens particularly with breast, lung, and prostate cancers. Symptoms include a dull pain that becomes progressively sharper and perhaps a fracture from a trivial injury. Secondary tumors show up on an X ray, but treatment depends on the nature and site of the primary tumor. Radiation treatment to the bone that has a secondary tumor may assist in pain relief.

Joint problems

In someone who has the disease ankylosing spondylitis, the ribs may become fixed at the joints with the spine so that the ribs cannot move normally during breathing. The patient can then breathe only by moving the diaphragm up and down, and the efficiency of the lungs is seriously reduced. These patients have difficulty meeting the increased demand for oxygen during exercise. However, this happens only in severe cases. Ankylosing spondylitis is thought to be hereditary because it is more frequently found in first-degree relatives of patients who have this disorder.

See also: Breathing; Cartilage; Chest; Circulatory system; Diaphragm; Joints; Lungs; Muscles; Neck; Nervous system; Skeleton; Spinal cord; Spleen

Saliva

Several important body functions depend on saliva. It is an important aid in chewing and digestion, it acts as a lubricant of the mouth, and it helps us to taste our food and drink.

Daily, the mouth produces about 3 pints (1.4 l) of saliva, a watery secretion consisting of mucus and fluid. Saliva contains an enzyme called amylase that aids digestion and a chemical called lysozyme that has an antibacterial function to protect the mouth from infection. Despite this, the mouth is always heavily contaminated by bacteria.

Saliva is produced by three principal pairs of salivary glands situated in the face and neck: the parotids, the submandibulars, and the sublinguals. Many smaller glands scattered around the mouth also contribute to the total production. Each salivary gland is composed of branching

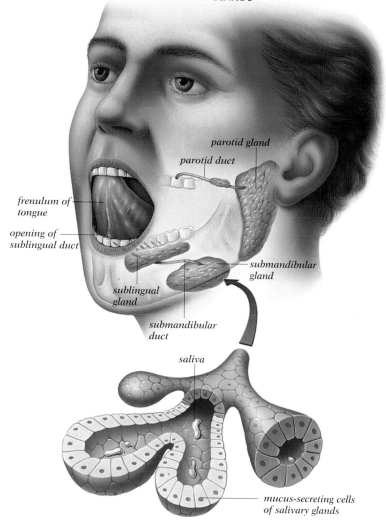

SALIVARY GLANDS

parotid gland

parotid duct

frenulum of tongue

opening of sublingual duct

sublingual gland

submandibular gland

submandibular duct

saliva

mucus-secreting cells of salivary glands

▲ *Saliva enters the mouth along ducts leading from the glands. These ducts drain into the upper part of the cheek and the floor of the mouth at a number of points.*

357

tubes that are packed together and lined with secretory cells. The function of the secretory cells varies between the glands and the fluids they produce.

Salivary glands

The parotids are the largest of the salivary glands, situated in the neck at the angle of the jaw and stretching up to the level of the cheekbone just in front of the ear. Saliva from the parotids empties into the cheeks from ducts running forward from the glands. Although the parotids are the largest glands, they produce only about a quarter of the total volume of saliva; their secretion is watery and contains an increased amount of the enzyme amylase, which is the enzyme that digests starch.

The submandibular glands, as their name suggests, lie under the jaw below the back teeth, and the sublinguals (also aptly named) lie under the tongue in the floor of the mouth. Both these pairs of glands discharge their contents at either side of the frenulum of the tongue (the small strip of tissue that sticks out from the base of the tongue and joins with the floor of the mouth).

The sublingual glands mostly secrete a very sticky, mucus-filled saliva. The submandibular glands produce a saliva that is an approximately equal mix of mucus and fluid containing amylase, and this makes up the bulk of the total volume of saliva secreted into the mouth.

The role of saliva

The major function of saliva is to help with digestion. Saliva keeps the mouth moist and comfortable when someone is eating and helps to moisten dry food so that it can be chewed and swallowed more easily. The mucus in saliva coats the ball of food, called a bolus, and acts as a lubricant to help swallowing.

The enzyme amylase that is secreted in saliva begins the first stage of digestion by breaking down starchy food into simpler sugars, until its action is stopped by acid in the stomach. However, if the bolus of food is large enough and well chewed, then the acid fails to penetrate to the center for some time and breakdown of starch continues.

Saliva also allows us to taste our food. The sensation of taste is created by the many thousands of taste buds that are mainly situated in the mucous membranes of the tongue. Taste buds are made up of clusters of receptor cells with tiny taste hairs that are exposed to saliva. Any food mixed with saliva interacts with the receptor sites. The receptors respond only to liquids; solid food in a dry mouth will produce no sensation of taste at all; that is why it is essential that foods are first mixed well with saliva to dissolve some of the food first. The fluid, containing food particles, can then wash over the taste buds which are chemically stimulated to transmit messages to the brain in the form of nerve impulses; the brain then interprets the flavor of the food. There are several types of receptor cells, which are constantly being replaced.

▲ *Eating citrus fruits or drinking fruit juice made from oranges, grapefruit, lemons, and limes stimulates the flow of alkaline saliva to counteract the acidity of the fruit.*

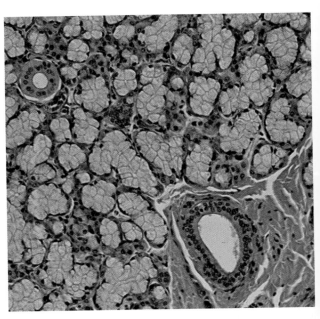

▲ *The sublingual gland (shown above in cross section) is composed primarily of cells that secrete mucus, unlike the watery solution of the parotids. These different secretions combine to form saliva.*

▲ *Just looking at food such as a creamy chocolate cake is enough to stimulate a reflex that causes a flow of saliva; the mouth waters.*

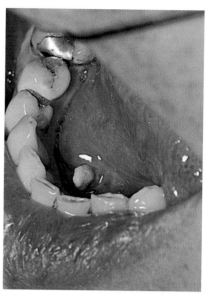

▲ *Occasionally a stone (above) forms in a salivary gland and blocks the duct. This can be painful but can be removed.*

When a person salivates

Saliva is produced continually day and night at a slow and steady rate. The amount generated is controlled by the autonomic nervous system that controls all of a person's unconscious activity. At times the rate of salivation is altered by nervous stimulation.

There are two divisions of the autonomic nervous system (ANS), the sympathetic and the parasympathetic. The ANS regulates all the functions of the body and helps to ensure homeostasis (maintenance of internal environment). The sympathetic division readies the body for stress, while the parasympathetic division helps to ensure there is consistent energy in the body. If the sympathetic system is stimulated, the flow of saliva is reduced; this is demonstrated when a person's mouth is dry because he or she is nervous. Speech can become difficult because the lips and tongue are not lubricated enough to move freely. By contrast, increased salivation is a reflex action mediated by the parasympathetic nervous system: the nerves that carry the sense of taste to the brain stimulate the flow of saliva when food is in the mouth. This is called an inborn reflex.

Because certain other stimuli are associated with the taste of food, saliva flow can increase at the mere thought of food. It is true that just looking at food can make the mouth water. This is called a conditioned reflex.

Salivary gland problems

The usual problems that arise in the salivary glands are from excess saliva or too little saliva and infections of the glands and the saliva itself. The medical term for an excess of saliva is "ptyalism," but it is not always easy to diagnose whether excess saliva is being secreted or the patient has difficulty in swallowing the normal volume that is produced. Sometimes both factors may play a part.

Some drugs, especially those containing mercury, bromide, iodine, arsenic, or copper salts, and even nicotine and tar from heavy smoking, can be causes of excessive salivation. Alternatively the condition may be caused by a general mouth infection that can have many origins, or by a reflex action brought on by irritation of one of the nerves supplying the mouth. This is usually as a result of dental or gum problems.

Lack of saliva can be caused by a variety of infections and can usually be treated by antibiotics and measures to ensure that the patient does not become dehydrated. It is also important for the mouth to be kept free from infection, since the natural protective effect from lysozyme in saliva is no longer present to act on any germs that enter the mouth.

Occasionally, stones form in the salivary glands and obstruct the ducts. After and during meals saliva flow is then blocked by a stone and the glands swell up, causing pain. X rays can reveal the exact site of a stone if it cannot be felt inside the mouth, and surgery may be needed to remove it.

The most common cause of swelling is mumps. However, this usually affects only the parotids and rarely affects the other glands.

The salivary glands, like most of the body's organs, are also subject to tumors, which are more likely to occur in the parotids than in the other glands. One interesting example is the so-called mixed tumor that forms a painless lump, usually behind the angle in the jaw. The tumor is made up of two kinds of tissues: one tissue is firm and the other is soft. Which one is predominant determines the consistency of the tumor. Mixed tumors usually affect women between 30 and 50, and men between 45 and 60. It is common for a tumor to grow intermittently over years.

See also: Autonomic nervous system; Digestive system; Gums; Mouth

Sebaceous glands

Questions and Answers

My skin is greasy; is something wrong with my sebaceous glands?

Sebaceous glands are built into the skin to produce a protective layer of grease over the skin. Many people think something is wrong if their skin looks oily, even though grease is normal. The amount of grease on people's skin varies throughout the population, but is largely controlled by hormones and genetic predisposition and is not an abnormality of the actual glands.

Does diet influence sebum production or greasiness?

For years researchers have tried to connect fat intake in the diet with sebum production but no connection has been shown. In other words there really is no connection between eating junk food and having greasy skin.

Does the Pill alter the activity of sebaceous glands?

Sebaceous gland secretions are controlled by hormones. Male hormones, such as testosterone, switch the glands on; the female hormone estrogen switches them off. For this reason, a pill containing estrogen has been used to reduce sebaceous gland activity and to treat acne. The progesterone in the Pill works the opposite way and may increase gland activity. The result depends on which hormone in the Pill the user responds to most. The Pill can make skin more greasy, or it can make skin drier.

What exactly is a sebaceous cyst?

When a sebaceous gland is blocked by dirt or hard wax, the production of sebum continues but sebum cannot escape, causing the gland to swell and to become a sebaceous cyst.

The human body is covered with a thin film of sebum, a greasy substance that is secreted by the sebaceous glands. The glands act as a protective coating for the skin and the hair.

Sebaceous glands are tiny glands in the skin that are found all over the body except for the palms of the hands and the soles of the feet. Although often associated with hair follicles, they also occur independently, over the nose, forehead, center of the back, and chest.

Sebaceous glands are so called because they produce a greasy substance called sebum, which is a mixture of water, fats, waxes, and cholesterol. It is produced as a result of cells dying within the gland and is secreted through hair follicles. Sebum covers the skin with a greasy film.

SEBUM PRODUCTION

The wall of each sebaceous gland is continually producing new cells. As they mature, they are pushed toward the center and then dissolve into sebum—a mixture of water, fats, waxes, and cholesterol.

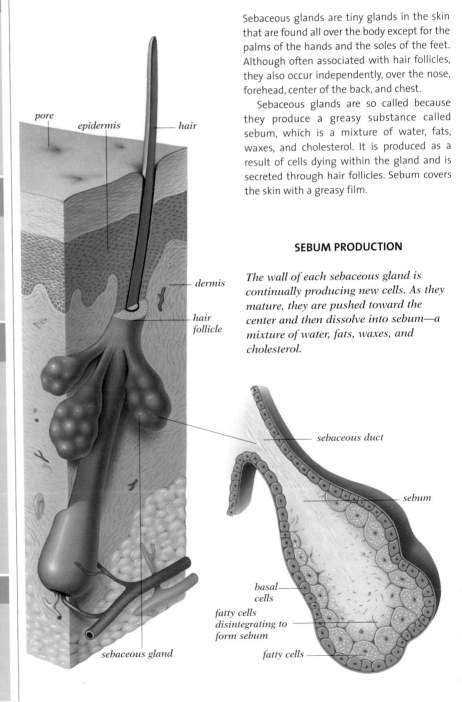

pore

epidermis

— hair

— *dermis*

— hair follicle

sebaceous gland

— sebaceous duct

— sebum

basal cells

fatty cells disintegrating to form sebum

fatty cells —

Functions

Sebaceous glands produce sebum, which has two important roles to play. First, it provides a waterproof layer to protect the skin. This works in two ways: in warm dry climates it helps to retain moisture and stops the skin from getting too dry; then, if the skin is submerged in water, sebum prevents overhydration and protects the skin by a waterproofing function.

Second, sebum appears to protect the skin against infection. Bacteria and fungi that might otherwise grow on the skin seem to be inhibited by substances in sebum. This is why people who are continually removing sebum—for example, when using detergents without protecting the hands with rubber gloves—seem to be particularly at risk of skin infections. Similarly, removing the sebum layer by repeatedly cleansing the skin with alcohol quickly leads to dry skin which can become infected.

How sebum is produced

Sebum production is controlled by hormones, in particular the sex hormones. Children, therefore, tend to produce relatively little sebum until they reach puberty, at which time production increases by as much as fivefold.

The male sex hormone, testosterone, seems to play an important part in sebaceous gland activity, and, as a result, men tend to produce slightly more sebum than do women. In women, sebum production is stimulated by male hormones made by the adrenal glands. The amount produced, however, is further controlled by the activity of the female sex hormones. Estrogen diminishes sebaceous gland activity, while progesterone encourages it. The fluctuation of these hormones during the female menstrual cycle accounts for changes in the greasiness of some women's skin. Other hormones are also important in promoting sebum production. Thyroid and pituitary hormones all influence secretion, and growth hormones seem to have a special controlling role.

▲ *Avoiding harsh soaps and applying moisturizing creams regularly is the best way to keep very dry skin supple.*

Everyone's skin has sebaceous glands, which all produce sebum. Some people, however, find that their skin is excessively greasy, while others have very dry skin.

Excessive greasiness is particularly common during adolescence, when sebaceous glands may be rather overactive. It is a condition that is often inherited and will usually clear in time. Meanwhile, people who have greasy skin will find that the best way to deal with the problem is to wash regularly with soap and warm water and to avoid greasy cosmetics. In some cases, a course of ultraviolet light treatment may also be recommended. Very dry skin is also a problem that can be inherited; it is caused by reduced sebum production. People with dry skin should avoid washing too much, and avoid harsh soaps. They should use moisturizing creams and emollients, since these seem to be the best remedies for the problem.

Comedones—or blackheads, as they are usually called—are plugs of dry grease and wax that get stuck in the pores of the skin. As a result sebum builds up and can cause the gland to swell into a sebaceous cyst. This can be prevented by squeezing out the blackhead using a special extractor. Blackheads should never be squeezed out with the fingernails, since this could burst any underlying cyst and spread infection. If the blackhead is not removed, the sebum behind it may become infected and produce a boil. When this happens extensively over the skin it is termed "acne." The heads of blackheads are black not from dirt but from the oxidation of the outer fatty material that is exposed to the air.

▲ *Sebum provides an effective, natural waterproof layer that prevents skin from becoming too dry.*

See also: **Hormones; Skin**

Semen

Semen, or seminal fluid, is the sticky material that forms the ejaculate produced during the male orgasm. It is a complex material that provides a transport medium, a nutritional supply, and an activating factor for the spermatozoa.

My boyfriend argues that the words semen and sperm mean the same thing. I say that the semen contains the sperm so they are different. Which of us is right?

You both are. If you use the word sperm as an abbreviation for spermatozoa—the microscopic swimming cells, one of which can fertilize the female egg and start a pregnancy—then "sperm" is different from "semen." The word "sperm" was used before much was known about spermatozoa and has, for centuries, described the generative substance of males, ejaculated during sexual intercourse; in that context the word means the same as semen.

How do doctors distinguish between "semen" and "sperm"?

When doctors talk about spermatozoa, or about a single spermatozoon, they use the full name or shorten the term to "sperm" or "a sperm." When they talk about seminal fluid, they either use the full phrase or shorten it to semen. They would never call semen "sperm" or confuse it with spermatozoa.

Is semen thick because of the number of sperm in it?

No. Although there are many millions of sperm in an ejaculate; the bulk of the sperm is only a few percent of the total volume.

My brother is worried about wet dreams and thinks there's something wrong with him. He is too embarrassed to talk to anyone but me about it. What can I do?

Explain that sperm are produced all the time. When the pressure gets too high semen is released spontaneously. Most adolescent boys have wet dreams; doctors call them nocturnal emissions.

There are numerous references to semen in ancient literature, including the Bible, in which it is commonly described as "seed." Semen is a sticky, opalescent fluid that, in addition to the spermatozoa, contains a large number of chemical substances that include sperm nutrients, chemical buffers to protect against vaginal acidity, prostaglandins, substances to increase the motile activity of the sperm, and mucus.

Sources and content

Semen is not fully constituted until just before ejaculation. The spermatozoa are formed in the testicles at the rate of about 30 million each day, and are carried up two tubes, one on each side, called the vasa deferentia. Each tube is called a vas deferens, and is filled with sperm, which form a kind of sludge. Contrary to popular belief, the sperm do not swim up the vas deferens but are pushed up from below by the accumulation of the new sperm that are being produced, and by peristaltic-like contractions of the smooth muscle in the walls of the vasa deferentia. The sperm remain nonmotile until stimulated by chemicals added to the fluid higher up.

The material in the vasa deferentia is not semen. Many constituents must be added before it is complete. The bulk of the fluid in semen comes from the prostate gland and from the adjacent seminal vesicles. The prostate provides most of the additional fluid bulk, but the chemicals that increase sperm motility come from the seminal vesicles. It was once believed that semen was

▲ *A graphic illustration of magnified swimming cells or spermatozoa. One of them will fertilize the female ovum (egg).*

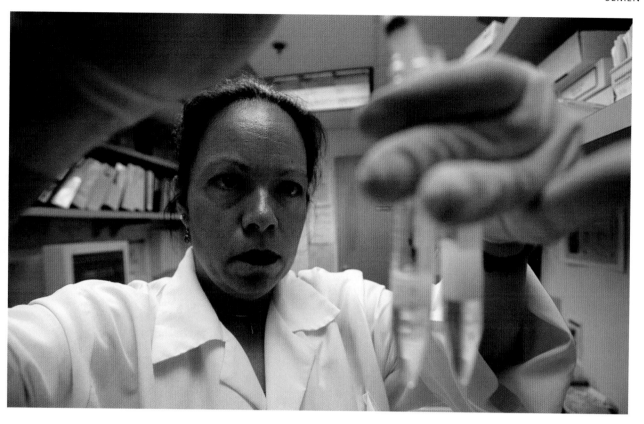

▲ *A lab manager at the Sperm Bank of California collects semen from donors that may later be used in artificial insemination.*

stored in the seminal vesicles, but this is not the case. The main storage site for sperm awaiting ejaculation are the vasa deferentia. The vesicles are filled with a solution of chemicals waiting to be ejected into the seminal fluid. Substances added to the fluid include choline, phosphocholine, citric acid, and acid phosphatase. The principal nutrient is the sugar fructose.

Formation and ejaculation of semen

The full formation of semen is not complete until shortly before ejaculation occurs; once the semen is fully formed by the addition of secretions from the prostate and seminal vesicles, ejaculation is virtually beyond control. This final stage is brought about by an almost simultaneous contraction of smooth muscle in the testicles, vasa deferentia, prostate gland, and seminal vesicles. This has the effect of mixing all the constituents and forcing the newly formed semen into the rear part of the urethra—the tube running though the prostate gland and penis that, at other times, also carries urine to the exterior.

The urethra has two rings of muscle (sphincters) capable of closing it off. The inner sphincter is immediately below the bladder and above the prostate gland. The outer sphincter lies immediately below the prostate. Expulsion of the semen is completed by a series of rapid, convulsive contractions of the smooth muscle of the urethral wall and the skeletal muscle at the base of the penis and a sudden relaxation of the outer urethral sphincter. The average volume of an ejaculate is about 0.1 fluid ounces (3 ml) and contains up to 300 million spermatozoa. During ejaculation, the sphincter immediately below the bladder is tightened strongly. This prevents urine from being passed during ejaculation and it also prevents semen from passing back and upward into the bladder. Men who have had surgery to remove excess prostate tissue, however, will usually no longer have control over the inner urethral sphincter, so during ejaculation, the semen passes back into the bladder and none is released from the end of the penis. This is called retrograde ejaculation. Although it may result in sterility, it does not affect the sensation of the orgasm.

Sperm quality

Semen looks much the same whether it is fertile or sterile. The difference is mainly in the number and quality of the spermatozoa. This can vary widely from man to man and is influenced by a number of factors. In different men, each 0.03 fluid ounces (1 ml) of semen may contain anything from about 20 million to about 90 million spermatozoa. Any man with a sperm count of less than 20 million per 0.03 fluid ounces (1 ml) is likely to be infertile.

Fertility

Factors other than sperm count can influence fertility; they include lack of adequately vigorous sperm movement, excessive numbers of abnormal forms, and various subtle defects that affect the ability of sperm to fuse with the egg. The investigation of infertility then starts with a detailed study of a sample of semen.

> *See also:* Conception; Genitals; Penis; Prostaglandins; Prostate gland; Sperm; Testes

Shivering

When it is cold I shiver long before others do. Why?

This is probably because your thermostat is set differently from theirs. Everyone has internal mechanisms that govern vital processes such as thirst, appetite, sleep, or temperature, and each person's setting is different. You may find your setting inconvenient, but you are not at any risk.

When my little girl had measles she shivered so violently that the whole bed shook. Why did this happen?

In many feverish illnesses germs or toxins push the "normal" setting of the brain thermostat to a higher level. The body immediately reacts as if it were too cold, and produces more heat. Its most efficient way of doing this is rapidly repeated muscle contractions—shivering.

How do you tell the difference between severe shivering and an epileptic attack?

If the shivering is only slight or the epileptic attack is severe, with the patient falling to the ground, jerking violently, and losing consciousness, there is little doubt. But if the shivering is severe or the epileptic attack mild, it may not be easy to distinguish the two just on appearance. If the circumstances are inappropriate for shivering—for example, if the attack occurs on a hot summer's day—then epilepsy is more likely. If there is any doubt, consult your doctor, who will arrange for neurologic tests.

Is shivering usually a sign of illness?

Shivering in cold weather is normal and one of the ways in which the body warms up. At other times it is a sign that something has upset the body's temperature control system, and is usually associated with a sharp rise in temperature.

Everyone knows what it's like to shiver in cold weather; this involuntary shaking movement of the skin muscles is one mechanism with which the body tries to keep warm by raising its temperature.

Because shivering is a familiar sensation and one that people tend to take for granted, they rarely stop to think about either the mechanisms involved or its purpose. Normal shivering consists of uncontrollable shaking that usually involves the whole body. It is uncontrollable in the sense that it cannot be stopped or started voluntarily by an act of will. There are several other body reactions that are similar, and shivering needs to be distinguished from them.

A response to cold

The most important feature of shivering is that it is a response to cold and that its purpose is to warm up the body. Trembling, on the other hand, is also an involuntary action that involves generalized shaking; however, the stimulus is not cold but fear, shock, or an emotional disorder such as hysteria.

Shaking or trembling that affects a particular part of the body is characteristic of some disorders of the brain or nerves—for example, Parkinson's disease or alcoholism. In medical parlance this is called tremor.

Shaking that is more severe or dramatic than is usually associated with shivering is called a convulsion or fit. This is characteristic of epilepsy and can usually be distinguished from true shivering. However, in slight attacks the distinction between the two movements may not be very great. The fact that the surrounding temperature is not low enough to account for shivering may be an indication that the shaking is really a minor convulsion and not the body's reaction to cold.

Temperature control

Shivering is an important mechanism in the body's complex and vital temperature control system. The temperature of the human body is normally in the region of 97°F to 99°F (36°C to 37°C) and

▲ *The best of both worlds: people having fun in a heated outdoor swimming pool in a spectacular setting. Swimming keeps them warm; the shivering starts later.*

remains remarkably constant, no matter how hot or cold the weather may be. This is not an accident, but a necessity, since this is the temperature at which the body's metabolism, the intricate complex of interrelated chemical reactions, functions best.

In fact, some of these metabolic reactions are so sensitive to temperature that if there are substantial departures from the normal temperature range for very long, they will be severely affected. Unless the situation is rectified and a return to normal body temperature is achieved, a progressive breakdown in the usual metabolic processes will occur until a point is reached at which life can no longer continue. This is what happens when people have very high temperatures, called hyperpyrexia, or, perhaps in circumstances of exposure, very low temperatures resulting in the condition called hypothermia.

Thus mechanisms for keeping the body's internal temperature at the right level, irrespective of how hot or cold the outside temperature is, are vital for our survival. These mechanisms, of which shivering is one of the most important, are under the control of the hypothalamus, one of the nonthinking, automatic centers in the brain. These kinds of temperature-regulating mechanisms are most efficient in healthy adults; they are much less effective in infants and elderly people, whose susceptibility to serious effects from temperature changes is therefore much greater. This is shown by the high incidence of hypothermia among elderly people.

The right temperature

The mechanisms by which our bodies maintain the correct temperature are ingenious. If the temperature-sensitive nerve endings, which are situated throughout the body, inform the thermostat in the hypothalamus that the temperature is becoming too high, a well-ordered chain of events takes place. Messages are sent to the arteries in the deeper parts of the body, and these become narrower so that less blood flows through them. At the same time, the blood vessels just below the surface of the skin are allowed to open wider; the result is that a higher proportion of the body's blood is flowing close to the surface of the skin. Here it can be cooled by the air and excess heat can be radiated away. If the body cannot be cooled sufficiently by radiation, the sweating mechanism comes into

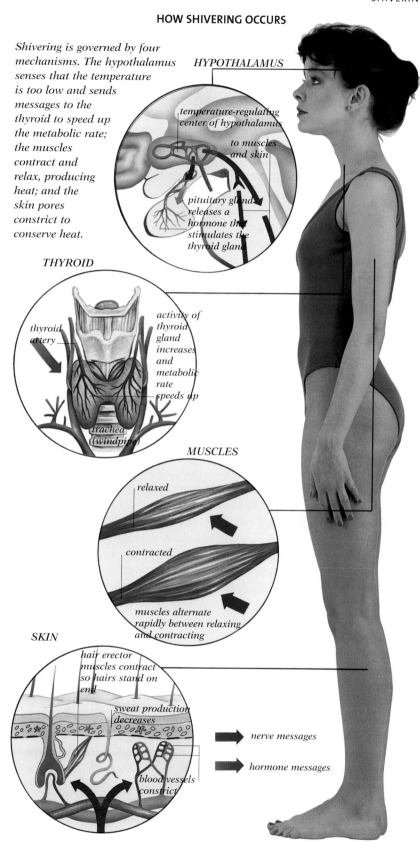

HOW SHIVERING OCCURS

Shivering is governed by four mechanisms. The hypothalamus senses that the temperature is too low and sends messages to the thyroid to speed up the metabolic rate; the muscles contract and relax, producing heat; and the skin pores constrict to conserve heat.

HYPOTHALAMUS

temperature-regulating center of hypothalamus

to muscles and skin

pituitary gland releases a hormone that stimulates the thyroid gland

THYROID

thyroid artery

activity of thyroid gland increases and metabolic rate speeds up

trachea (windpipe)

MUSCLES

relaxed

contracted

muscles alternate rapidly between relaxing and contracting

SKIN

hair erector muscles contract so hairs stand on end

sweat production decreases

blood vessels constrict

nerve messages

hormone messages

operation. Sweat is secreted onto the skin, and, as it evaporates, it cools both the skin and the blood flowing just beneath it, bringing the temperature down.

If, on the other hand, the outside temperature drops sufficiently for the body to react, then different mechanisms are brought into play to conserve or boost the body's internal temperature. The first of these, once again controlled by the hypothalamus, is a sequence to change the pattern of the blood circulation. Blood is directed away from the surface of the skin to the deeper regions of the body, so that heat loss through radiation is cut to a minimum. When, however, it is so cold that this mechanism is not enough, the additional and closely related measures of so-called goose pimples, a sudden puckering of the skin, and shivering come into play to try to increase the body's heat.

Goose pimples and shivering

The effectiveness of these reactions is based on the fact that the body's natural heat is produced as a side effect of metabolizing or burning up food materials. This happens to people most extensively when energy needs to be produced for muscular activity, such as after a long run. Thus any muscular activity will produce more heat and raise the temperature of the body; and this is what goose pimples and shivering achieve. The puckering of the skin that takes place when goose pimples are formed is the result of contraction of the muscle fibers in the skin. In a sense it is an involuntary isometric exercise. Shivering is much more energetic, and therefore a much more effective, heat-producing activity. It involves rapid alternating contraction and relaxation of the muscles of the skin.

Abnormal shivering

Shivering is normally a response to low external temperatures. When it occurs at other times it is said to be abnormal. A common example of this type of shivering is the bout of violent shaking that people sometimes experience as the body tries to raise its temperature rapidly. This happens with certain feverish illnesses, such as influenza, pneumonia, and tonsillitis; this type of shivering is called a rigor.

The explanation is that, in the course of the illness, the body suddenly produces an unusually large quantity of heat. This happens because the disease germs reset the thermostat in the hypothalamus to a higher level. The body systems read this as an indication that it is too cold and that heat is urgently required. The result is shivering, which, if the thermostat is set very high, may be violent. Later, however, the body's heat-losing mechanisms redress the balance by means of heavy sweating, another symptom of feverish illness.

◄ *If you insulate yourself from the cold, you won't shiver: the differences between internal and external temperatures will be stabilized. This little girl is warmly bundled up; her cats have built-in insulation thanks to their thick fur.*

See also: **Hypothalamus**; **Metabolism**; **Nervous system**; **Perspiration**; **Skin**; **Temperature**

Shoulder

The shoulders are the joints that connect the arms to the body and allow a vast range of arm and shoulder movements. However, the exceptional mobility of the shoulders makes them prone to a number of specific complaints.

The joint at the shoulder is the link between the upper arm bone, called the humerus, and the triangular-shaped scapula, or shoulder blade. Anatomically, the joint is classified as the ball-and-socket type because a ball-shaped projection at the top of the humerus articulates with a socket in the scapula. This socket, called the glenoid cavity or fossa, is actually rather shallow so that the head of the humerus perches on the cup, like a golf ball on a tee, rather than slotting deep into it. This allows the arm to rotate through a complete circle in several different planes.

The movement of the bones at the shoulder is assisted by lubricated synovial membranes between the surfaces of the two bones. Unlike other joints, the ligaments that are intended to hold the bone in place are not particularly strong; that is why dislocation of the shoulder is common. Much stability is provided by muscles that also control movement. The most important of these muscles are the series of fan-shaped ones that run from the scapula to the top of the humerus and to the end of the collarbone, or clavicle—the bone that runs from the breastbone, or sternum, to the top of the shoulder joint. These are the deltoid, pectoralis major, and trapezius muscles. The biceps and triceps muscles, located at the front and back of the upper arm, respectively, are also involved in shoulder movements.

Shoulder problems

The shoulder is prone to a number of injuries common to most other joints of the body, but there are one or two problems that are particularly associated with the shoulder.

Dislocation of the shoulder is a common problem with this joint because of its rather loose construction. A heavy fall may be sufficient initially to cause the injury. After a first dislocation, the shoulder often dislocates increasingly easily, even while a person is performing mundane tasks such as housework, digging the garden, or even decorating.

When the shoulder is dislocated, the ball-shaped projection at the end of the humerus can be seen as a bulge in the upper arm. The immediate treatment for a dislocated shoulder is manipulation of the joint to realign the bones correctly. In the long term, however, surgery will be needed to tighten the ligaments and provide a permanent cure.

A broken collarbone is the most common break associated with the shoulder and is an occupational hazard of many sports. The telltale symptom of a broken collarbone is an easing of

▲ *The butterfly action of this swimmer captures the power and mobility of the shoulder joint. Swimming is an ideal shoulder exercise, because it both builds up muscles and keeps the joint mobile.*

STRUCTURE OF THE SHOULDER

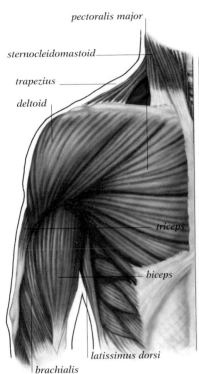

pectoralis major

sternocleidomastoid

trapezius

deltoid

triceps

biceps

latissimus dorsi

brachialis

clavicle

acromion

coracoid process

scapula

humerus

▶ *The synovial capsule protects and lubricates the joint, while the muscles and ligaments hold the two bones in place.*

pain when the head is held to the side of the injury. There is also a natural tendency for the injured person to support the elbow of the injured side of the body to take the strain off the shoulder and arm muscles. After any necessary manipulation the arm on the side of the injury is put into a sling for about three weeks. Most broken collarbones mend quickly and well, so that the patient is back to normal in two months at most.

Frozen shoulder is a common disorder of this part of the body and is caused by inflammation of some part of the joint. The shoulder gradually becomes increasingly difficult and painful to move. It is most common in middle-aged people, is often the result of an old injury, and is most likely to occur in sedentary and inactive individuals. Treatment consists of supporting the shoulder in a sling for a few days. The doctor may recommend ice packs to relieve the pain and may sometimes administer antirheumatic drugs to clear up the inflammation. After the inflammation has cleared up, physical therapy is used both to loosen the joint and to build up strength.

Stiffness in the shoulder is a similar problem and just as common. It affects mainly the elderly but sometimes occurs after injury. The cause is tearing or weakness of the muscles around the joint; but the stiffness can be eased or even cured by stringently following an exercise program drawn up by a physical therapist or your family physician.

Shoulder exercise

Exercising the shoulders regularly can help ward off problems with the shoulders in later years. Arm circling and circling of the shoulder joint are good for mobility. Swimming is an excellent exercise for both mobility and strength, and can also help soothe those muscles in the neck and shoulder that tend to tense up under stress.

See also: **Bones; Joints; Ligaments; Membranes; Movement; Muscles; Skeleton**

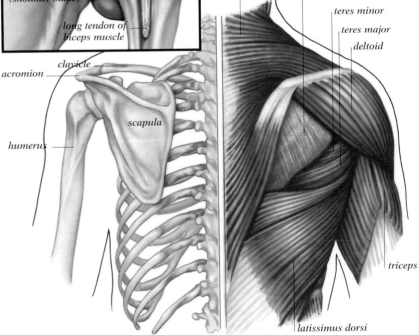

clavicle (collarbone)

ligaments

acromion

synovial capsule

humerus

scapula (shoulder blade)

long tendon of biceps muscle

acromion

clavicle

humerus

scapula

▲▼ *The shoulder joint is formed by the head of the humerus sitting in the socket of the scapula, seen here from the front (above) and the back (below).*

trapezius

infraspinatus

teres minor

teres major

deltoid

triceps

latissimus dorsi

Skeleton

Strong enough to protect the vital organs, intricately structured to keep the body upright, and flexible enough to allow great freedom of movement, the human skeleton is a marvel of mechanical and architectural design.

The skeleton of the average adult is made up of 206 bones, though there may be variation in the number of ribs, an additional pair occasionally occuring in the neck. The bones have a hard, thick, strong outer layer and a soft middle, or marrow. They are as strong and as tough as concrete and can support great weights without bending, breaking, or being crushed. Linked together by joints and moved by muscles that are attached at either end, they provide cages to protect the soft and delicate parts of the body while still allowing for great flexibility of movement. In addition, the skeleton is the framework or scaffolding on which the other parts of the body are hung and supported.

Questions and Answers

How much force is needed to break a bone?

It depends on the position, shape, and health of the bone. The long thin bones of the arms and legs are more prone to snapping fractures than are the plate bones of the shoulder and pelvis, which are more prone to crushing injuries. The undernourished and older people, whose bones have lost part of their protein framework and calcium mineralization, have brittle bones that break easily. Generally, it is amazing how much force a bone can withstand.

What is a greenstick fracture?

In this type of fracture, instead of the bone breaking into two or more separate fragments, only one side of it actually breaks. The other side is more bent than broken. The appearance and effect are similar to what happens if you try to break a stick of wood that is still green, hence the name. These fractures usually occur only in children whose bones have not yet fully ossified and are thus more supple.

Do bones ever just give way and break by themselves?

Yes, though a much more common situation is that the bone is exposed to some stress or injury, but to a lesser degree than would usually be required to break it. These are called spontaneous fractures and result from the bones' being thinner and weaker than normal, usually because of an inadequate diet or some hormonal or metabolic upset that interferes with the normal hardening or ossification process. Also, some diseases remove protein or calcium from the bones, weakening the bones and making them more prone to breakage, in which case a fracture is called pathological. Bones invaded by a tumor are an example of this.

▲ *The skeleton can adapt itself to a large range of sizes, as shown in this photo shoot for 1999's Guinness World Records' tallest man, Rhadouane Charbib (center).*

Men and women have the same number of bones, but in general the female skeleton is lighter and smaller. In order to accommodate the growing fetus during pregnancy, a woman's pelvis (right) is broader and more boat-shaped, giving her hips their characteristic shape. Her shoulders, however, are relatively narrow. In a man (left), the proportions are reversed: broad shoulders and slim hips.

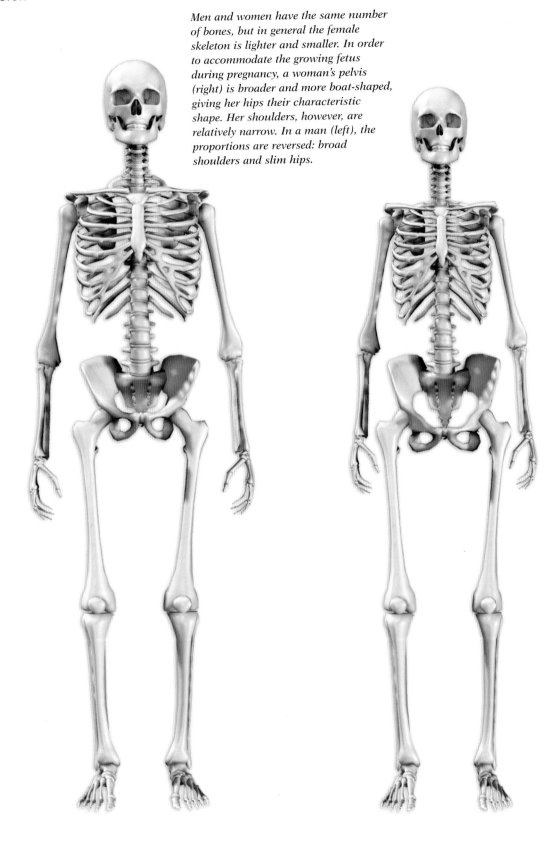

Structure of the skeleton

Each of the different parts of the skeleton is designed to do a particular job. The skull protects the brain, eyes, and ears. The lower jaw and teeth are attached to it, enabling us to eat. There are holes for the eyes, ears, nose, and mouth, and also one in the base of the skull where it joins the spinal column; the spinal cord passes through this, connecting the brain to the rest of the body.

The backbone, or spine, is made up of a chain of small bones, similar to spools of thread, called vertebrae, and forms the central axis of the skeleton. It has enormous strength, but, because it is a rod made up of small sections, instead of being one solid piece of bone, it is also very flexible. This enables people to bend down and touch their toes, and to hold themselves stiff and upright. The vertebrae also protect the delicate spinal cord. Between the vertebrae are shock-absorbing disks that act as cushions to help protect the vertebrae from rubbing on each other and becoming damaged as a person bends down and straightens up. The bottom end of the spinal column is called the coccyx. In some animals, such as the dog and cat, this forms a tail.

The rib cage is made up of the ribs at the sides, the spinal column at the back, and the breastbone, or sternum, in front. It is designed to protect the heart and lungs that lie inside it, since damage to these organs could prove fatal.

The arms are joined onto the central axis of the spinal column by the shoulder girdle, which is made up of the scapula (shoulder blade) and the clavicle (collarbone). The big bone of the

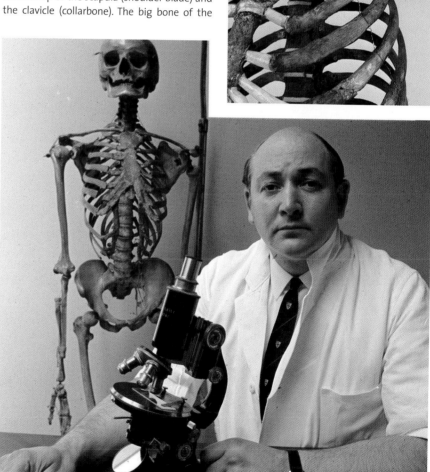

▲ *Forensic pathologists sometimes reconstruct skeletal remains to establish the cause of a violent death. The inset shows the rib cage of a woman who was buried for five years in a shallow grave: the fourth and fifth ribs are fractured—injuries due to a knife thrust.*

upper arm (the humerus) is joined at the elbow to the two bones of the forearm: the radius and ulna. The hand is made up of many small bones. This makes it possible for people to grip things and to make delicate, complicated movements in which each of the many parts of the hand moves in a different yet highly coordinated way.

The legs are attached to the spine by the pelvic girdle, which shields the reproductive organs and the bladder and gives protection to the developing baby which lies in this part of its mother's body. The femur (the thick bone of the thigh) is the longest bone in the body. Its round head fits into the socket in the pelvis to form the hip joint, which is designed to maximize freedom of movement of the leg. There are two bones in the lower leg: the shinbone, or tibia, and the much thinner fibula.

The foot, like the hand, is made up of a complicated arrangement of small bones. This enables people both to stand firmly and comfortably and to walk and run without falling over.

How the skeleton develops

Surprising as it may seem, a newborn baby has more bones in its body than an adult. At birth, around 350 bones make up the tiny frame; over the years some of these fuse together into larger units.

A baby's skull is a good example of this; during the birth process, it is squeezed through a narrow canal. Were the skull as inflexible as an adult's it would simply not be possible for the baby to pass through the mother's pelvic outlet. The fontanelles, or gaps between the sections of the skull, allow it to be molded sufficiently to accommodate itself to the birth canal. After birth, these fontanelles gradually close.

The skeleton of a child is made not only of bone, but also of cartilage, which is much more flexible. Gradually this hardens into bone—a process known as ossification, which continues well into adulthood. It is not until the age of about 20 that full skeletal maturity is reached.

The proportions of the human skeleton change dramatically as people mature. The head of a six-week-old embryo is as long as its body; at birth the head is still large in proportion to the body, but the midpoint has shifted from the baby's chin to the navel. Each person has his or her own timetable for skeletal growth, but it can be affected by such environmental factors as diet and disease. Certain glandular disorders can cause too much or too little growth; rickets is caused by a lack of vitamin D. Sometimes people are born with extra bones, such as an extra pair of ribs. In general, the frame of a woman is lighter and smaller than that of a man's. Her pelvis is proportionately wider to allow room for the growing fetus during pregnancy. The male shoulders are broader and the rib cage is longer but, contrary to popular superstition, men and women have the same number of ribs.

Evolution

About 25 million to 30 million years ago the primates (the subgroup of mammals to which humans belong) abandoned the trees which had been their habitat and took to terrestrial life. At this stage they probably walked on all four limbs. They relied on the sense of sight rather than smell and so the eye sockets were forward-looking and both hands and feet had the ability to grasp. Gradually, through evolution, the feet of these human ancestors became specialized: the ability to cling to branches was no longer needed. As humans began to walk upright, the foot had to be adapted to take the forces brought to bear on it: the fingers grew longer, the toes remained short and the big toe became nonopposable, that is, it could no longer be placed in opposition to the others, as the thumb can be to the other fingers. The heel has to come into contact with the ground at the beginning of the stride and the weight of the body is transferred over the arch to the big toe, which has to be in line with the direction of movement.

Some six million years ago our predecessor *Australopithecus*, though a biped, looked startlingly different from humans today. The bones of the skull were

▲ *The proportions of the human skeleton change dramatically as people mature. The midline of the adult frame would run through the symphysis pubis, just above the genitals; in a baby, the midpoint is the navel. If a newborn baby were as tall as an adult, as in the artist's impression above, the skull would be huge in comparison.*

EVOLUTION OF THE FOOT

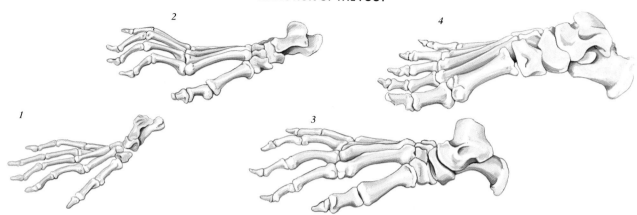

more massive; *Australopithecus* had a larger facial area, more prominent brow ridges, a flatter crown, and a smaller braincase. The gait was stooping, the legs were comparatively short, and the arms were comparatively long. Height was only about 4 feet (approximately 1.25 m). As these anthropoid primates evolved to become more like modern humans, and as they developed a more striding gait, the legs grew longer and the pelvis narrower. The fact that humans have developed great manual dexterity, however, is due less to skeletal sophistication (the hands are primitive adaptations from the fins of amphibious ancestors) than to the increasing capacity of a human's brain. In strictly anatomical terms, the human hand differs little from that of the apes, the relatives of humans.

The human skeleton of today is the result of millions of years of evolution, designed to equip it for the complex tasks it must perform. Questions such as whether the evolutionary process is complete, and whether the skeleton will develop further, and, if so, how, are the stuff of science fiction, not medicine. What modern medicine can ensure is that the injuries and disease to which our internal scaffolding is subject become better understood and more easily treated.

What can go wrong

While the bones of the skeleton are very resilient, if the weight or pressure on them becomes too great they will break. A bone may break from being crushed, from a sharp impact on its middle, or from force at one end causing it to snap. The amount of force required to break a bone varies with the individual: older people, whose bones are thinner and less resilient, are more prone to fractures. Treatment depends on the severity and position of the break, but involves ensuring that the broken bone is held together in the right position until the damaged ends grow together again.

Though the spine usually functions without trouble, it is hardly surprising that something in the long pile of vertebrae on their disks, rather precariously balanced one on top of the other, occasionally slips out of place. Sometimes the vertebrae become displaced; this is called spondylolisthesis and occurs most commonly in the lower back, or lumbar, region. At other times, one of the disks cushioning the vertebrae slips to become a prolapsed intervertebral disk. Rest and physical therapy are the best treatment for this condition. The spine is also prone to degenerative

▲ *Human evolution has involved many changes to the skeleton, including the foot. In humans' tree-living ancestors, the toes were long and the big toe could be opposed to the others (1, 2). Then an upright posture was adopted (3), and eventually humans developed a striding walk. The ability to grasp has been lost; the big toe is in line with the others and is large because it has to take much of the force of each step (4).*

disorders (which are usually brought about by aging) and can also become temporarily deformed by being held in abnormal positions when a patient is trying to avoid pain. The hunchback type of deformity is called kyphosis; forward curvature is called lordosis, and sideways curvature, scoliosis.

Osteoporosis, a bone disorder affecting many postmenopausal women (who have lost the anabolic effect of estrogen), results in bones that have become weakened by loss of both the protein (collagen) structure and the calcium and phosphate mineralization. Osteoporosis leads to pathological fractures from minor trauma, especially of the arm, spine, and neck of the femur, and is responsible for an immense amount of suffering, disfigurement, and disablement in older women.

Joints

The skeleton is a highly articulated piece of machinery. The many finger joints allow people to grasp objects; the elbow enables them to lift objects up; the knees ensure that they are able to walk, run, jump, and so on. The joints themselves are lined with lubricated cartilage, so the bone ends do not rub against one another; if the cartilage gets worn away, as in cases of osteoarthritis, pain and stiffness result. The bone may try to compensate for the subsequent damage by repairing itself. This bone growth is haphazard, however, and can lead to skeletal deformity: many people with arthritic finger joints, for example, develop extra knobs of bone known as Heberden's nodes. Overuse of certain joints can lead to tennis elbow or frozen shoulder. Heat, gentle exercise, local painkillers, and occasionally an injection of hydrocortisone can help.

> See also: Aging; Back; Bones; Cartilage; Elbow; Feet; Hand; Hip; Joints; Knee; Marrow; Muscles; Neck; Pelvis; Ribs; Shoulder; Spinal cord

Skin

Why are skin conditions so often inflamed by emotional upsets?

Most ill health is made worse by emotional upsets, and the skin is no exception. Emotion can also alter the state of the skin's irritability and sweating mechanisms. Conditions in which these factors are important, such as eczema, will be aggravated by anxiety, discontent, or depression.

I have developed an allergy to nickel because I wore cheap jewelry. Will I always be allergic?

Allergy to nickel is fairly common in those who either wear or handle it. There is always an interval between first contact with nickel and the development of the allergy. Traces of nickel are absorbed through the skin and the body reacts by forming antibodies so that any further contact results in an itchy skin rash. If no further contact occurs the allergy will gradually lessen. However, it usually remains to some degree throughout life.

I have bald patches on my scalp and beard area. Will they regrow?

Bald patches are a symptom of a condition called alopecia areata, which may run in families and frequently starts in childhood. Hair is lost in clear-cut round areas, often a few weeks after stress or shock. It usually calms down, and in most cases hair regrows in two to six months, although relapses are common.

My husband is 20 and is going bald. Is this normal?

Yes. Hair loss in men may begin at any time after puberty as a result of higher male hormone levels. Often there is a hereditary factor as well. The drug minoxidil (Rogaine), used locally under medical supervision, can aid hair regrowth.

The skin is the largest human organ, although it is not often thought of as such. It protects people from injury and infection, and it keeps the body's temperature and moisture content stable at all times.

The skin is much more than a simple wrapping for the body. It is an active and versatile organ that is waterproof so that people do not dry up in the heat or dissolve in the rain, and it protects them from the damaging radiation of sunlight. It is tough enough to act as a shield against injury, yet supple enough to permit movement. It conserves heat or cools the body as required, thereby keeping the internal temperature constant.

Skin diseases may be a nuisance and cause embarrassment, but they are seldom dangerous and are rarely fatal. However, they can cause a lot of ill health by their frequency and persistence.

Structure of the skin

The skin has two main parts. The outermost part, the epidermis, consists of several layers of cells, the lowest of which are called the basal cells, or mother cells. These cells are constantly dividing and the offspring cells are pushed up to the surface by subsequent offspring cells, where they

▲ *A person's skin type and his or her coloring are determined by heredity. Hair and nails are formed from skin cells, and these too are determined by genetic factors.*

374

Questions and Answers

Why does some people's skin age faster than others?

Inheritance is probably the most important factor in skin aging. Other influences involved are the environment, such as the amount of sun damage and hormonal changes throughout life. The loss of elasticity that causes wrinkles in old age is due to changes in the fibers of the supporting layer of skin. Skin also becomes drier and hair thinner with age.

Do commercial suntanning preparations really prevent burning and promote a suntan?

There are suntanning preparations that claim to prevent sunburn and enhance tanning. This can only happen with an efficient sunscreen, which enables you to sunbathe for longer periods, and builds up the protective pigmentation so that a gradual tan is produced. If you are very fair-skinned, you will find it virtually impossible to go brown, however much you sunbathe, since your skin produces little pigment, and you are especially vulnerable to sunburn. Many suntanning preparations contain dyes that artificially color the skin. Others contain a colorless substance that turns the outer skin layer brown. This is not a true tan, because the pigment-producing cells have not been activated to produce melanin.

My father gets a cold sore when he sunbathes. What causes this?

Cold sores are a common viral infection of the skin. Most people are first infected in childhood, and the infection may pass unnoticed, but the virus remains latent in the cells of the skin. It may be reactivated by a stimulus such as sunlight, giving rise to a cold sore.

Why do I get pimples before a period?

This is very common in young women and is probably due to the fall in the hormone estrogen before a period.

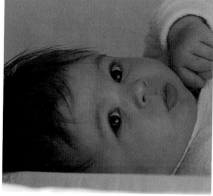
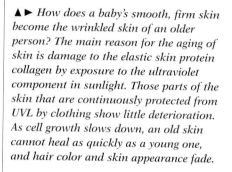

▲ ► *How does a baby's smooth, firm skin become the wrinkled skin of an older person? The main reason for the aging of skin is damage to the elastic skin protein collagen by exposure to the ultraviolet component in sunlight. Those parts of the skin that are continuously protected from UVL by clothing show little deterioration. As cell growth slows down, an old skin cannot heal as quickly as a young one, and hair color and skin appearance fade.*

progressively flatten as they approach the surface. The surface cells eventually die and are transformed into a material called keratin, which is finally shed as tiny, barely visible scales. It takes three to four weeks for a cell in the lowest layer to reach the surface.

This outer protective layer is firmly attached to an underlying layer called the dermis. Tiny, fingerlike bulges from the dermis fit into sockets in the epidermis, and this waviness at the junction of the two layers of skin gives rise to ridges, which are most obvious at the fingertips and give people their fingerprints. The dermis is made up from bundles of protein fibers called collagen and elastic fibers.

Glands and nerves

Embedded in the dermis are sweat, sebaceous, and apocrine glands; hair follicles; blood vessels; and nerves. The nerves penetrate the epidermis, but the blood vessels are confined to the dermis. The hairs and ducts from the glands pass through the epidermis to the surface of the skin.

Each sweat gland consists of a coiled tube of epidermal cells, which leads into the sweat duct to open out on the skin surface. The sweat glands are controlled by the nervous system and are stimulated to secrete by a release of emotion or by the body's need to lose heat.

The sebaceous glands open into the hair follicles and are made up of specialized epidermal cells that produce grease or sebum to lubricate the hair shaft and surrounding skin. They are controlled by sex hormones and are most numerous on the head, face, chest, and back.

The apocrine glands, a sexual characteristic, develop at puberty and are found in the armpits and breasts, and near the genitals. They are odor-producing and secrete a thick, milky substance.

There is a fine network of nerve endings in both the epidermis and the dermis, and they are particularly numerous at the fingertips. They transmit pleasurable sensations of warmth and touch, as well as cold, pressure, itching, and pain that may evoke protective reflexes.

Hair and nails

Hair and nails are specialized forms of the protein keratin. Nails are produced by living cells, but the nail itself is dead and will not hurt or bleed if it is damaged. The visible part of the nail is

STRUCTURE OF THE SKIN

The skin is made up from two layers of tissue: the epidermis and the dermis. Both layers contain nerve endings that transmit sensations. The sweat glands are vital in regulating the body's temperature; the sebaceous glands lubricate the skin and hair. The apocrine glands develop at puberty and are a sexual characteristic. The pigment-producing cells, called melanocytes, can cause freckles.

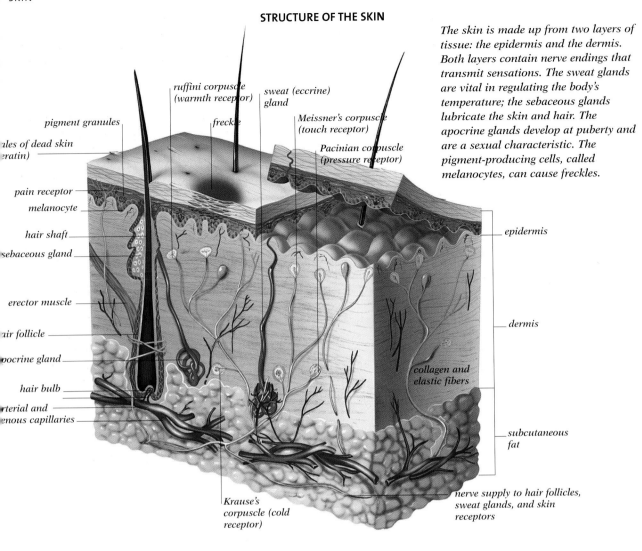

ruffini corpuscle (warmth receptor)

sweat (eccrine) gland

Meissner's corpuscle (touch receptor)

Pacinian corpuscle (pressure receptor)

pigment granules

freckle

...ales of dead skin ...eratin)

pain receptor

melanocyte

hair shaft

sebaceous gland

erector muscle

...air follicle

...pocrine gland

hair bulb

...terial and ...enous capillaries

epidermis

dermis

collagen and elastic fibers

subcutaneous fat

nerve supply to hair follicles, sweat glands, and skin receptors

Krause's corpuscle (cold receptor)

called the nail body, and its shape is partly determined by genetic factors. The base of the nail, the root, is implanted in a groove in the skin. The cuticle overlaps the root, which is the site of active growth. As the living cells divide and move upward they become thick and tough, and when they die they form part of the nail itself.

Hair is formed by cells in the hair follicles. There are two types: fine, downy hair, which is found over most of the body except the palms of the hands and soles of the feet; and thick, pigmented hair, which is present on the scalp, eyebrows, beard, and genital areas.

Hair grows in cycles, a long growing phase being followed by a short resting period. Hairs in the resting phase constitute up to 20 percent of the total 100,000 hairs on the scalp. The normal daily hair loss is between 20 and 100 hairs. Scalp hair grows about 0.3 inch (0.8 cm) per month. No single area is totally depleted at one time. The rapid growth of scalp hair makes it more susceptible to damage from disease, toxic drugs, and hormones.

The shape of our hair follicles is inherited, and this determines whether hair is straight or curly, together with the angle of the hair bulb in the shaft. If it lies straight, the hair will be straight; if bent, the hair will curl.

Skin color

Skin color is due to the black pigment melanin, which is produced by pigment cells in the lowest layer of the epidermis. There is the same number of pigment-producing cells in the skin of all races, but the amount of melanin produced varies. In dark-skinned people there is more melanin than in light-skinned people.

Other factors contributing to skin color are the blood in the blood vessels of the skin and the natural yellowish tinge of the skin tissue. The state of the blood within the blood vessels can greatly change skin color. Therefore, people become white with fear when small vessels close off, red with anger owing to an increased blood flow, and blue with cold because most of the oxygen in the blood moves out to the tissues when the blood flow slows down.

Wound healing

All wounds heal by scar formation unless they are very superficial, such as a graze. Children heal faster than adults, but they also produce a larger quantity of scar tissue. However, scars in young people tend to resolve in time. As a general rule, dark pigmented skin heals with an excessive amount of scar tissue compared with light pigmented skin.

The healing process involves many changes. First the wound bleeds and becomes filled with a blood clot, which dries to form a scab. Blood vessels and fibrous tissue grow in from the cut surfaces of the wound, and the result is a scar that gradually becomes paler in color with time.

Skin conditions in children

Birthmarks are marks that are present on a baby's skin at birth, or that appear soon afterward. They include strawberry marks, moles, and port-wine stains. Many birthmarks do not require treatment and disappear of their own accord. Strawberry marks, for example, appear a few weeks after birth and grow rapidly for a while, but the majority disappear completely by the time a child is old enough to attend school.

Moles are not usually present at birth but develop during childhood, gradually increasing in size in adult life and sometimes disappearing late in life. Moles are formed from collections of the pigment-producing cells in the skin, and they may occasionally become malignant.

Babies and children have their own particular skin complaints. These include infant cradle cap, diaper rash, and chilblains. Cradle cap is a common condition that appears as a collection of scales and

◄▼ The amount of melanin pigment in the skin is the major factor determining skin color, producing black and brown shades. Yellow tones are imparted by the pigment carotene.

grease which stick together and adhere to the scalp. This can be removed by gentle shampooing after the scales have been softened with olive oil the night before.

Diaper rash is a red rash in the diaper area, which can spread to include the thighs and lower abdomen. It results from irritation produced by the bacterial decomposition of urine and feces. Because diaper rash is caused by the friction of a wet or soiled diaper, it is essential to change diapers frequently, leaving them off whenever possible, and avoiding the use of plastic pants. The skin should be washed with emulsifying lotions rather than soap, and water-repellent ointments that act as a barrier should be applied.

Eczema (atopic dermatitis) is a very common childhood problem resulting from an inherited state called atopy. This is a form of allergy to various substances such as house dust mite droppings, pollens, cat and dog hair, and some food ingredients. The skin is locally inflamed, scaly, and very itchy. It is best to avoid the triggers of this problem, if known. Otherwise, topical hydrocortisone may be of benefit in many cases.

Chilblains are common in children who live in countries where the winters are cold. These sores occur on the toes, especially if tight-fitting shoes are worn, and on the fingers and ears. Sudden extreme changes in temperature should be avoided; although it is tempting to warm cold feet in front of the fire, this aggravates the condition. The affected area should be kept warm at all times.

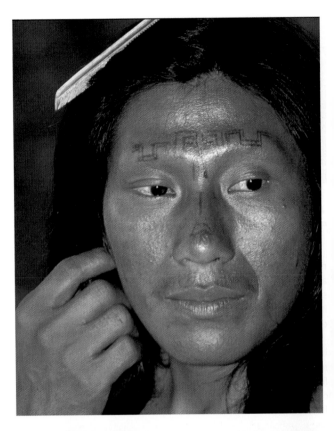

▼▶ *The pigment-producing cells are larger in dark-skinned people than fair-skinned races, but the number is constant. Dark-skinned people produce more melanin than fair people.*

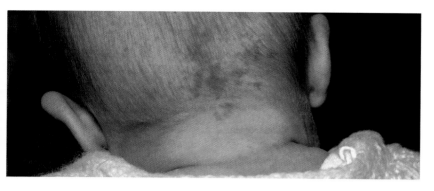

▲ *Many babies are born with superficial birthmarks. These do not need treatment and will disappear in time.*

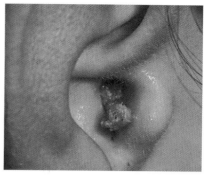

▲ *Warts are common in childhood, but simple treatments are usually effective.*

Infection of the skin often occurs in childhood because the skin's natural defenses have not yet been built up against bacteria, viruses, and fungi. Impetigo is a bacterial infection of the superficial layers of skin and is particularly likely to happen where skin is already damaged. It starts as a little red spot that enlarges and blisters to form a honey-colored crust. It is easily treated with antibiotics.

Ringworm is caused by several kinds of fungi and gets its name from the ringed appearance of the rash. Sites most commonly affected are the scalp, groin, and feet. This skin complaint is rare in adults because of their greater immunity, and it is thought that the grease glands on the scalp may have some protective effect. Treatment is with antifungal preparations.

Warts are the most common viral infection of the skin; most children catch warts as easily as they catch chicken pox. They tend to be found on the hands, knees, and soles of the feet, since the virus enters wherever the skin is broken. It lives in the outer layers of skin and causes very little damage, so a wart may pass unnoticed for months or even years. When the body becomes aware of the presence of the virus, it mobilizes its defenses and the wart may disappear as if by magic: hence the success of many wart charms and cures. Often the number of warts increases rapidly before they finally disappear.

Since warts heal without leaving a scar, no treatment destructive enough to cause scarring should be used. Only simple wart paints should be applied, and the dead skin regularly pared away or frozen off. Treatment is usually successful, but may take two to three months before having an effect.

Skin conditions in adolescence

The hormonal changes that occur at puberty affect the skin chiefly by activating the grease-producing glands.

Sweating and blushing can be annoying problems for teenagers. Excessive sweating of

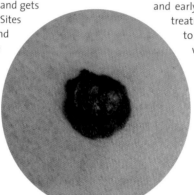

▲ *A malignant mole is characterized by itching, bleeding, or changes in color or size. Malignancies usually develop in adulthood, but if diagnosed early, a cure is possible.*

▼ *An allergic reaction is a common cause of dermatitis. For example, a person may find detergent irritating to the hands. Treatment would be to avoid contact by wearing rubber gloves.*

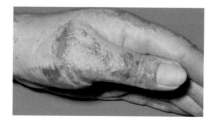

the armpits and hands, which is caused by emotion and heat, can be treated with aluminum chloride preparations applied locally. Blushing is not usually treated but tends to diminish as confidence increases.

Most teenagers develop a few acne spots that sometimes require medical attention. Acne appears sooner in girls than in boys because of the earlier onset of puberty. It usually resolves in the late teens and early twenties, but some people continue to need treatment up to the age of 30 or even 40. Acne is due to the excessive production of sebum or grease, which blocks the hair follicles. These then become infected with bacteria so that pimples and pustules form. Acne can and should be a successfully treated disease. Patients with mild acne require degreasing agents or keratolytics. Moderate or severe cases need oral antibiotics in addition to local treatment. Sunlight is usually beneficial because it dries up grease on the skin and helps to peel off the top layer.

Skin conditions in adults

As the skin ages there is a falling off in the production of natural emollients and it becomes dry. This drying process results in cracks in the skin, leaving it open to irritants at work or at home. Industrial dermatitis, where irritants are picked up in the workplace, is a more common cause of absence than any other industrial disease.

Dermatitis (skin inflammation) is thought to be mainly due to an external cause. Thus, those who use detergents at home may find their hands become irritated, while industrial dermatitis may be caused by many different substances, such as wet cement, chemicals, or fiberglass. Contact dermatitis is the result of an allergy to a substance, such as the nickel in jewelry. The cure is always avoidance of the cause.

Eczema is also common, giving rise to discomfort and disability at all ages. It presents different appearances at different stages, and may last from a few days to a lifetime. Initially there is reddening of the affected skin and itching, then pinhead swelling. Blisters then form, together with weeping and scaling. Eczema is often inherited, and is associated with asthma and hay fever. When all three occur in one person, the condition is known as atopic syndrome.

Caring for your skin

Skin care mainly consists of cleansing and moisturizing.

Cleansing creams are pure oil, and cleansing milks are oil in water. Both dissolve dirt and makeup without drying the skin too much. Cleansers that can be rinsed off are suited to all skin types.

Skin tonics refresh and stimulate the skin and should be applied after cleansing and before moisturizing. The simplest are distilled water and rosewater. Astringent lotions are more suited to oily skins because they help remove excess grease.

Moisturizers are creams with a low oil content. They slow down the rate of moisture evaporation from the skin.

Skin foods are heavy moisturizers with a high oil content; they are suitable for dry skins and nighttime use.

Cleanse oily skin with soap or a suitable cleanser and protect with a nonoily moisturizer.

Cleanse normal skin with mild soap or a liquid or cream cleanser. Use a moisturizer under makeup and at night.

Cleanse dry skin with a mild cleanser that can be rinsed away. Always use a moisturizer and night cream.

Eczema cannot be cured, but it can be controlled; it is not contagious and does not leave scars. The skin of an eczema sufferer is more sensitive than normal; it is usually itchy and has a tendency to dryness. Steroids are the mainstay of treatment, and the weakest effective steroid is used. Irritants such as soap should be avoided, since they exacerbate the condition.

Skin allergies are often manifested in the development of urticaria or hives. The rash consists of white welts surrounded by reddened skin, which itches but does not last longer than a few hours. New welts may appear, however, so that the condition persists for days or weeks. The cause is often something that has been eaten—fruit or shellfish, for example—and some people are allergic to preservatives and synthetic dyes.

The most common fungus infection of the skin is athlete's foot. The first signs are itching and peeling of the skin between the toes, which is often worse on one foot. Relapses are common in the summer, but the condition usually settles in cool weather and remedies are simple and effective.

Occasionally, a mole becomes malignant. Malignancy can occur in a mole that has been present for years, but if diagnosed early it can be removed and a cure is possible. Dubious-looking moles should always be removed. A malignant mole can arise at any age, though rarely in childhood. The main cause is excessive exposure to sunlight. Malignant melanoma has become much more common in recent years mostly because of increased sunbathing. Only one mole in a million becomes a malignant melanoma.

Hot flashes occur in women during menopause because of the hormonal changes at this time. Sometimes the flushing overstimulates the grease-producing glands to cause a condition called acne rosacea on the forehead, nose, cheeks, and chin. The sebaceous glands may enlarge to such an extent on the nose that it becomes bulbous and lumpy in appearance. The treatment of acne rosacea involves the avoidance of those things that aggravate flushing, particularly hot drinks. Prescribed oral antibiotics can also be very effective in severe cases.

Skin conditions in older people
As the skin starts to age it not only dries out but also loses its former elasticity and does not heal so quickly or easily. The smallest wounds leave slight scars. Many older people also develop skin tags, sun-induced keratoses—grayish patches of skin—and seborrhoeic warts. These are brownish-black warty lesions that commonly appear on the body after the age of about 50. They are easily treated by freezing or by scraping off the lesion with a special surgical instrument.

The most common type of skin cancer is basal cell cancer, commonly called the rodent ulcer. These ulcers commonly occur in the United States, but are more generally found among fair-skinned people in sunny climates. They usually occur on the face and neck and are slow-growing. These take the form of a hard, pearly-looking swelling, often with a dimple in the center and with a few fine red lines running over the edge. They do not spread remotely but can do much local deep damage. Such ulcers are easily treated by local excision if they are small enough, or by radiotherapy. They are the least dangerous of cancers.

See also: **Birthmarks; Blushing; Genitals; Glands; Hair; Melanin; Menopause; Nails; Perspiration; Sebaceous glands**

Smell

Of all the senses, smell is the most evocative; the slightest scent can start a rush of long-forgotten memories. Complex yet primitive, it retains its intimate associations with pleasure and warns of possible danger.

The sense of smell is probably the oldest and the least understood of the five senses. During evolution, smell retained connections with parts of the brain that interpret emotional responses and link smells to emotions. The sense of smell also plays an important role in sexual attraction, though this has become considerably muted during human evolutionary development. Its most important roles are those of a warning system and information gatherer, to warn of danger and give valuable information about the outside world.

Questions and Answers

Why is it that if one stays near a smell, the odor fades away until one is almost unaware of it?

This is called adaptation, and is characteristic of our sense of smell, though hearing and touch to some extent show the same phenomenon. If signals persist in coming from the smell receptors in the nose for some time, they are progressively ignored by the part of the brain that sorts them out. This fits in with the function of smell as a warning system.

Do children have a more acute sense of smell than adults?

Yes. The sensitivity of the smell receivers in the nose is at its maximum in childhood and gradually declines through adult life. This is compensated for by the progressively richer associations which smells can evoke as experiences increase.

When a friend quit smoking she remarked that her sense of smell improved. Does smoking really cause such a difference?

A common cause of a diminished sense of smell is heavy smoking, because smoke dries the delicate smelling apparatus and causes a mild inflammation in that part of the nose. The overall effect is to make sensitive-smell receivers much less efficient. When a person quits smoking, the smelling apparatus rapidly returns to normal, and it can be quite a dramatic change.

Do blind people have an enhanced sense of smell?

No. This point has been investigated by researchers who found that blind people's sense of smell is no more acute than sighted people's; neither are any of their other senses enhanced.

▲ *The fragrance of freshly baked bread makes it irresistible. No food aroma better illustrates how smell enhances taste.*

How we smell things

As with many organs in the body, the smelling apparatus is duplicated and each circuit acts independently.

The sensory receptors for smell are found in the roof of the nasal cavity, just beneath the frontal lobes of the brain. This is called the olfactory area and is tightly packed with millions of small cells (the olfactory cells). Each olfactory cell has about a dozen fine hairs, or cilia, which project into a layer of mucus. The mucus keeps the cilia moist and acts as a trap for odorous molecules, and the cilia effectively enlarge the area of each olfactory cell and thereby increase our sensitivity to smells.

It is not clearly understood how minute amounts of chemical substances trigger and stimulate the olfactory cells, but it is known that these substances dissolve in the mucus fluids, stick to the cilia, and then cause the cells to fire electrical signals. Each one of the enormous numbers of olfactory receptor cells carries one or at most a few different kinds of receptor sites. There are approximately 1,000 types of receptor sites, and each one will accept, lock onto, and respond to only one single type of chemically-related odorant molecule. The huge range of odorant molecules are differentiated by the specific chemical groups they carry. In this way, the brain receives information enabling it to discriminate between thousands of different odors.

Olfactory nerve fibers channel these signals across the bone of the skull to the two olfactory bulbs in the brain, where the information is gathered, processed, and then passed through a complicated circuitry of nerve endings to the cerebral cortex. Here the message is identified and the smell becomes a conscious fact.

What we smell

To have an odor, a substance must give off particles of the chemical of which it is made. This type of substance is generally chemically complex. Simple chemical substances, such as salt, do not have a smell, or have only a faint trace of odor.

The particles of a substance must remain in the air in a gaseous form in order to be swept into the nostrils and to the mucus surrounding the cilia. Once in the nostrils, they must be able to dissolve in the mucus for the olfactory apparatus to detect them. Substances that give off gas easily, such as gasoline, are usually very odorous because high concentrations of the chemicals are able to reach the cells.

Wetness also heightens smells. As the water evaporates from the substance, it carries particles of the substance into the air. Perfumes are structured in such a way that they are chemically complex and easily give off gas that can be breathed into the nose and hence into the mucus around the cilia.

▼ *An herb garden is not simply a source of culinary flavorings: its variety of fragrances—from the pungent to the sweet—can provide a relaxing environment.*

▲ *Animals are extremely sensitive to smells. For example, dogs can be trained to sniff out marijuana.*

▲ *Skunks defend themselves by emitting a stench sufficient to deter most enemies.*

Purpose of the sense of smell

In humans, the sense of smell appears to have four main purposes: it stimulates the salivary glands in preparation for eating; it has an important role in sexual attraction; it serves as a basic warning system; and it gathers information concerning the world about us.

In the animal world, greater dependence on the sense of smell for survival has given it more specific purposes; locating food, detecting an enemy, and choosing a mate are just a few examples.

The purposes of the human sense of smell have become less defined, because through the ages people have gradually become less dependent on it. Our main senses of vision, hearing, and touch

have developed further to provide people with the more accurate information that our higher intelligence requires from the world around them.

The continuing popularity of perfumes for both sexes illustrates that smell has a central but primitive role in the complexities of sexual attraction. This initial attraction between two people is dependent on a number of factors, and the individual smell of a person, known as pheromones, makes a persuasive contribution. Subconsciously, a personal scent helps someone to form an opinion about the other person. If the visual attraction between two people is very strong, but one of the two finds the smell of the other unpleasant, then this is very likely to be a deciding factor in the continuance of the relationship.

A sense of smell also serves as a warning system and gives a person a stream of information about his or her environment. It can be used to determine whether meat is good or bad, whether milk has soured, whether something is burning, or even whether the local wheat crop has been gathered. Although it is extremely sensitive to small quantities of odorous substances, the human sense of smell is not very efficient at detecting differences in the intensity of a smell. Intensity may have to change by a third before the difference is noticeable. In contrast, visual apparatus can detect a 1 percent change in the intensity of light. As a warning and information-gathering system, a sense of smell does not need to notice changes in intensity; it needs to notice only if a scent is actually there. When there is a need to experience as much of a smell as possible, people tend to sniff. This greatly increases the amount of air reaching the smell receivers at the top of the nose, giving them the best chance to sample the new smell.

Links between smell and taste

People are not always aware of the close link between the sense of taste and the sense of smell. However, when someone is suffering from a cold, it is usually noticed that not only has the sense of smell temporarily disappeared, but food has no discernible taste until the cold has cleared up.

◄ *Natural scents such as the fragrances in a springtime country garden, or the bracing intoxication of the sea air, can be appreciated as much by a sense of smell as by any of the other senses.*

Easy-to-grow fragrant herbs and flowers
Fragrant herbs:
Apothecary's rose (*Rosa gallica officinalis*)
Lemon balm (*Melissa officinalis*)
Lemon verbena (*Aloysia triphylla*)
Rosemary (*Rosmarinus officinalis*)
Lavender (*Lavandula angustifolia*)
Chamomile (*Chamaemelum nobile* "Flore Pleno")
Fragrant climbers:
Honeysuckle (*Lonicera* spp.)
Jasmine (*Jasminum* ssp.)
Sweet pea (*Lathyrus odoratus*)
Fragrant annuals:
Stock (*Matthiola* spp.)
Lilies (*Lilium* spp.)
Hyacinth (*Hyacinthus* spp.)
Daffodils (*Narcissus* spp.)
Roses (*Rosa* spp.)

▶ *Smell acts as a warning system: the acrid smell of smoke can alert us to a fire, the smell of burned toast signals to someone to turn off the broiler; the unmistakeable smell of decaying food tells someone that it would be unsafe to eat.*

Much of what people think of as taste is really smell. In fact, the sense of smell is 10,000 times more sensitive than that of taste. The taste buds in the tongue monitor relatively crude sensations of salt, sweet, sour, and bitter, while the more sophisticated taste sensations are manufactured by smell receivers in the nose. Faint vapors of whatever we are eating drift into the nasal cavity where the smell receptors add more detail to the information given by the taste buds. When the nose is blocked—for example, when someone has a cold—gas and vapors cannot flow over the receiver cells; he or she cannot smell anything and can taste only cruder tastes.

Smell, emotions, and memory

The part of the brain that analyzes messages coming from the receiver cells in the nose is closely connected with the limbic system, a part of the brain that deals with emotions, mood, and memory. It is called the primitive brain, sometimes even the smelling brain. The connection explains why smells are richly endowed with emotional significance. The smell of fresh rain on a summer's day makes people feel happy and invigorated; it may also evoke pleasant memories. The smell of fresh-baked bread will bring on instant pangs of hunger; the scent of perfume may bring anticipation of sexual pleasure. Conversely, unpleasant smells, such as rotten eggs, produce revulsion, and even nausea. Exceptions are the unpleasant smell of a ripe cheese, which actually attracts fervent fans.

Certain smells will bring memories of long-forgotten special occasions flooding back. This is because people tend to remember occasions that have special emotional significance, since the areas of the brain which process memories, and which are essential in their recall, are also closely linked to the limbic system, which, in turn, is linked to the centers in the brain governing the sense of smell.

Smelling disorders

There are two types of smelling disorder: anosmia (the loss of the sense of smell, either complete or partial, either temporary or permanent) and dysosmia (abnormal smell perception).

A decrease in the ability to smell is most commonly due to problems in the nose, as in the common cold, influenza, hay fever, or sinusitis. Heavy smokers also suffer loss of smelling ability, because the delicate smell tissue dries up.

Head injuries, even minor ones, can cause a loss of the sense of smell. This occurs when the delicate olfactory nerves are either bruised or even sheared off by the knock on the head. The loss of smell can be permanent and may affect only one nostril or both.

Diseases within the skull can also be associated with a loss of smell, though these are not very common. A tumor or aneurysm pressing on the olfactory nerves may impair the sense of smell, or temporarily interrupt the ability to smell. Meningitis and internal hemorrhaging may also affect the sense of smell.

Dysosmia sometimes occurs as a feature of severe depression or schizophrenia; sufferers can be plagued by illusory unpleasant odors. Similar delusions can also occur in some forms of epilepsy and during drying-out periods after severe alcohol addiction.

Smell and the visually impaired

For people who are blind or partially sighted, smell is an important sense. However, a blind person does not have an improved sense of smell; rather he or she tends to rely on a sense of smell, and get more information from it, than a sighted person. One of the ways that a blind or partially sighted person can enjoy the sense of smell is to walk or sit in an aromatic garden filled either with herbs or scented flowers. It can become an important and pleasurable part of daily life. The person may not be able to see the flowers but can touch the foliage and smell the various fragrances of the blooms. There are many sweetly fragrant plants that flourish in all seasons and exotic plants that are at their best in hot and tropical climates. Humidity and moist air release more fragrance from the scent glands of a flower, and blind and partially sighted people who live in regions which are both hot and humid will be able to appreciate a well-planned scented garden.

Creating an aromatic garden

When masses of plants are grouped in flower beds the full effect of their fragrance can be enjoyed. Just one or two plants do not create enough of an impact. However, there are some single plants that emit overpowering fragrances. The scented gardenia *Gardenia jasminoides* has been used in the creation of many famous perfumes and is relatively easy to grow in the right climate. A shrub called Daphne (*Daphne* spp.) has a sweet and strong smell. Frangipani are heavy with the scent of orange blossom and honey, particularly a variety called *Hymenosporum flavum*. A night-scented jasmine, *Cestrum nocturnum*, has tubular clusters of waxen green stars that give out an overpowering scent in the evening. Sweet peas are always fragrant, and make an attractive as well as fragrant addition to a garden. The climbers honeysuckle and jasmine and ginger plants have delightful fragrances that most people would find irresistible.

Roses are almost automatically associated with floral fragrance, but because there are many modern varieties that have no fragrance at all, it is important to check at the garden store when buying either a standard or a climbing rose for a scented garden.

There are many trees and shrubs with scented foliage and flowers that will enhance any garden for the visually impaired; for example, the lemonwood, *Pittosporum eugenifolium*, flowers freely, and after it has rained, the air is filled with the cloying scent of honey.

An herb garden bed planted with lavender, lemon thyme, marjoram, wild bergamot, golden sage, lady's mantle, and sage not only will bring pleasure to a blind person but is also practical. Many herbs have household uses and culinary uses as well as having an attractive appearance. People who have visual impairments and do not have a garden can grow scented herbs in containers. The beneficial effects of even a small herb garden can be immense.

SENSE OF SMELL

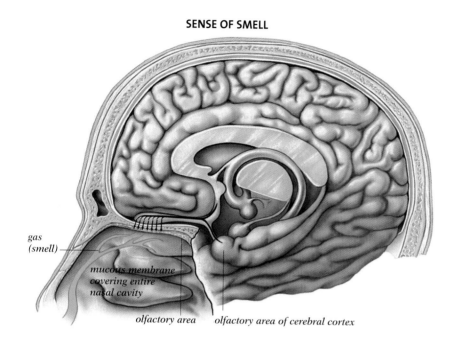

gas (smell)

mucous membrane covering entire nasal cavity

olfactory area   olfactory area of cerebral cortex

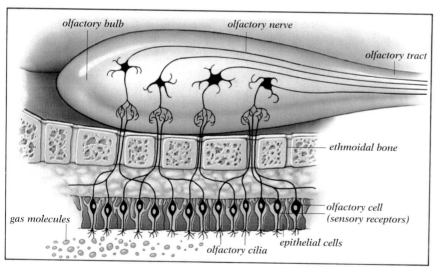

olfactory bulb   olfactory nerve

olfactory tract

ethmoidal bone

gas molecules

olfactory cell (sensory receptors)

epithelial cells

olfactory cilia

▲ *Gaseous substances are dissolved in the mucus surrounding the cilia. A biochemical process then takes place that stimulates the olfactory cells into electrical activity. These messages are passed across the ethmoidal bone via sensory nerve fibers, and into the olfactory bulb. Here the information is processed and then it is passed along the complex circuitry of the olfactory nerves to the cerebral cortex. At this point people become aware of the smell.*

See also: Head; Mucus; Nose; Taste

Sneezing

Like so many of our reflexes, sneezing is an important mechanism designed to protect the body. It may, however, sometimes be a sign of an abnormal state needing investigation and suitable treatment.

I used to sneeze a lot and have a permanently blocked nose, but I found that using a spray I bought from the drugstore helped. I now have to use more and more to get relief. Is it safe to continue?

See your doctor. Many people unwittingly cause permanent damage to the lining of the nose with these preparations, which tend gradually to lose their effectiveness. Your doctor can prescribe a medication that does not damage the nose, or send you to an ear, nose, and throat specialist.

I was told that my heart misses a beat when I sneeze. Is this true?

Not entirely. When you breathe in your heart speeds up and when you breathe out it slows down. The more pronounced the inspiration and expiration, the more your heart speeds up or slows down. Sneezing involves deep inspiration and expiration, so your heart rate may be affected by it. This will not affect your health; it is a normal phenomenon called sinus arrhythmia.

I heard that a girl who couldn't stop sneezing was sent to Switzerland. Why?

Mountain air is free from allergens, so patients with allergic rhinitis may improve in this environment. Sea air may have a similar effect.

I take antihistamines for hay fever. They relieve my symptoms but I hear it is dangerous to drive when taking them. Is this so?

No, unless they make you sleepy. If so, you should reduce the dose or ask your doctor to prescribe a nonsedative preparation. Alcohol also enhances the sedative effects of antihistamines.

The prime function of the nose is to clean up inspired air—the air we breathe in. The structure of the nose is designed for this function. A coarse sieve of hairs within the nostrils, called the nasal vibrissae, and a finer sieve, the nasal mucosa, line the nostrils with mucus. Both sieves filter out dust and other particles from the incoming air. The sense of smell monitors inspired air, and a reflex action, the sneeze, totally clears the nasal and mouth passages.

The sneezing mechanism

Sneezing is triggered by irritation of the nasal mucosa and results in a sudden, violent expulsion of expiratory gases through the nose and mouth. This outburst carries with it the irritant in the form of droplets, which may be projected up to 30 feet (9 m). During the explosive phase the eyes of the person sneezing close by reflex action.

▲ *Always use a handkerchief when you sneeze. A virus can travel from person to person in the droplets, which are often carried a long way by the violence of a sneeze.*

Causes

Sneezing can be a normal reaction to irritants of the nasal mucosa, such as smoke, aerosol sprays, or sudden changes in temperature or humidity. It may even follow a knock on the nose. The common cold is probably the most frequent cause of sneezing, because the virus and secondary bacterial infection make the nasal mucosa hypersensitive to minor stimuli.

There are, however, conditions in which the nose is irritated abnormally, or in which sneezing is induced by factors that do not affect the majority of people. These include the presence of foreign bodies, and an abnormally sensitive nasal lining.

Foreign bodies

Children are frequently admitted to the hospital when objects, such as beads or peanuts, have become lodged in the nasal passages. Within a few hours the child begins to sneeze as the foreign body irritates the nasal mucosa, secretions are trapped, and infection builds up. This results in a nasal discharge containing pus; it also results in bad breath and sometimes intermittent nosebleeds.

Removal of foreign bodies usually requires general anesthesia. The doctor will also check that no other foreign bodies are hidden in the ears, for instance.

Allergic rhinitis

Sneezing is also part of an abnormal reaction to an inhaled substance or allergen, so called because it is allergy-producing. The most common allergens are the grass and tree pollens that are responsible for hay fever, and the droppings of the house dust mite, which are found in most houses and may cause many allergic problems.

In a sensitive person antibodies are already attached to certain cells in the nasal mucosa. They are called mast cells. When an allergen enters, the antibodies fuse with them and in so doing tear the membranes of the mast cells. These then disintegrate and release histamine, which is highly irritant and produces an inflammatory response, in which patients are overcome by bouts of violent sneezing and have a profuse watery discharge and nasal obstruction. Treatment involves identification of the particular allergen so that it can be avoided or at least reduced. This is done by skin testing.

▲ *A close-up picture of willow catkins clearly shows pollen on the long clusters of flowers. Pollen is an allergen that affects people susceptible to hay fever and makes them sneeze.*

Treatment

There are three types of drugs used to control a patient's nasal symptoms. Antihistamines act by blocking the effect of histamine on surrounding cells. Disodium cromoglycate, taken as a spray or in drops, prevents the release of histamine from the mast cells, but is effective only if taken in advance of an attack. Topical steroids suppress the inflammatory reaction triggered by histamine release. Their activity is confined to the nose, and so they are deemed safe. When no medicine is effective, desensitizing injections under the care of an allergy specialist are an option.

Vasomotor rhinitis

A condition that produces symptoms similar to allergic rhinitis, but in which there is no obvious allergen, is called vasomotor rhinitis. Triggers may include irritants such as air pollution, dry air, or spicy foods.

Most cases respond to antihistamines and topical steroids, and some may be controlled by oral decongestant preparations, which shrink and soothe the nasal lining. Sneezing and nasal secretions are then reduced and airflow improves.

A few patients may need minor surgery to reduce the amount of nasal lining while at the same time preserving a sense of smell.

▲ *A greatly magnified computer-generated graphic image of pollen depicts the grains that cause pollen allergy or hay fever.*

See also: Breathing; Mucus; Nose

Snoring

Questions and Answers

Does drinking alcohol at night make people snore?

Alcohol does not actually trigger snoring, but the effect of alcohol makes it easier to fall asleep in a position in which the sleeper breathes through his or her mouth, making snoring more likely. Alcohol (and sedatives) will also cause very deep sleep: snoring may be continuous.

Is it true that snoring is common in pregnancy?

Snoring can occur in pregnancy. One theory is that in pregnancy fluid retention in the tissues affects the membranes in the breathing passages. They may become congested, forcing the woman to breathe through her mouth, making snoring likely.

When my husband snores I give him a push and he stops. Why?

Perhaps it wakes him up. More seriously, if he sleeps on his back or on his side with his head thrown back, his position may cause him to breathe through his mouth and snore. Pushing forces him into a position in which he breathes through his nose.

When I get hay fever I start snoring. Why is this?

Hay fever affects the mucous membranes of the nose and air passages, causing congestion and forcing you to breathe through your mouth. In sleep, the muscles of the soft palate and uvula relax, and as you inhale through your mouth they vibrate noisily. The way to stop snoring is to relieve the hay fever symptoms that are causing you to breathe through your mouth. Antihistamines will help. Sleep with your windows closed at the time you are most vulnerable to pollen.

Sleeping with the mouth open—whether it's because someone has a cold or has simply fallen asleep in an uncomfortable position—may cause snoring. It's a harmless phenomenon, but can be very irritating to others.

Snoring, breathing heavily through the mouth with a vibrating or snorting noise when asleep, is chronic in as many as one in eight sleepers. The noise can disturb partners and the snorers themselves and in some cases can cause great distress. Snoring can be a symptom of a much more serious condition called sleep apnea, in which breathing repeatedly stops during sleep. This condition can be confirmed by various tests and sleep studies.

The causes of snoring

Snoring is an involuntary act. The characteristic noise is created when, for some reason, the sleeper begins to breathe through his or her mouth, and the muscles of the soft palate and the uvula are allowed to relax. The passage through which air passes is narrowed, and, as the sleeper inhales, the air drawn into the lungs causes the soft palate and the uvula to vibrate. The quality and the intensity of the snoring will be governed by the shape of the mouth, the elasticity of the tissues, and the vigor with which the snorer inhales. Occasionally people snore so vigorously that they wake themselves up, but generally a snorer is oblivious of the noise he or she is making. Because snoring occurs when a person sleeps with the mouth open, a blocked nose or anything obstructing the nasal airways which forces breath out through the mouth will make snoring far more likely. A stuffy nose because of a cold, or enlarged tonsils or adenoids, may make someone

▲ *Falling asleep in a chair is likely to cause someone to snore. The head is thrown back and the mouth opens. This makes breathing through the mouth all too easy.*

likely to breathe through the mouth. If the muscles of the lower jaw and palate relax in sleep, snoring may start. A person who is sitting up and falls asleep loses control over these muscles and they relax. This is why so many people snore when they fall asleep sitting up on trains or in armchairs. Similarly, a person who lies on his or her back when asleep may also be prone to snoring because the lower jaw drops and the muscles of the palate relax.

Many studies have shown that obesity is linked to snoring. Reasons for this may involve the increased thickness of the neck or perhaps decreased muscle tone common in those who are obese. It has been shown, however, that if a snorer who is overweight loses a few pounds, there is likely to be a reduction of noise caused by snoring, if not a complete cessation of snoring.

The atmosphere where someone is asleep may also have an effect. A very dry, centrally heated room can lead to snoring in a susceptible individual, as can an exceptionally humid environment.

Stopping snoring

Since snoring involves breathing through the mouth in sleep, most forms of treatment to alleviate the condition aim at trying to reestablish breathing through the nose.

If someone who does not normally snore develops a cold and is told that he or she has started snoring, all that is needed to be done is to relieve the symptoms of the cold so that he or she can breathe through the nose again. In the same way, treating obstructions such as enlarged adenoids in the nasal airways will relieve, if not cure, the snoring.

Colds and nasal obstructions, however, contribute only to a temporary snoring problem, and other forms of treatment have to be tried in more indeterminate cases. Some measures are commonsense: for example, if people snore when they are sleeping on their back, they need to be persuaded to sleep on their side, or

▲ *If children have adenoid problems, they are quite likely to start snoring because the blocked nose, caused by the adenoids, forces them to breathe through the mouth.*

their stomach. There are also exercises designed to keep the mouth closed while the snorer is asleep, for example clenching the teeth for 10 minutes or so before retiring, and these have proved successful in some cases of snoring.

A more modern suggestion for helping persistent snorers involves using a cervical collar—the same type that is used for treating a sprained neck—when going to bed. The rationale is justifiable: snoring is often at its worst when the sufferer is lying on his or her back with the head sagging on the chest. The cervical collar will keep the chin and the lower jaw elevated, and so, it is hoped, prevent snoring from beginning.

Living with a snorer

The irritation that can be engendered by a persistent snorer, who is blissfully ignorant of the noise he or she is making, can be almost unbearable and very distressing. However, there is usually a physical reason for the condition, and it should be possible to relieve it. If snoring continues, earplugs for the other person may be the only solution.

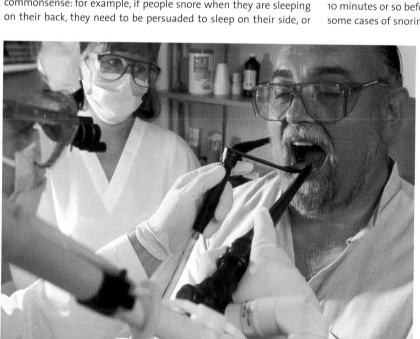

▲ *A doctor performs an assisted urula palatoplasty in his office. This surgical procedure is said to help reduce snoring.*

See also: Adenoids; Breathing; Mouth; Muscles; Nose; Palate; Tonsils

Sperm

The human male is an amazingly prolific producer of sperm: up to 350 million are released in one ejaculation. However, only one sperm may complete the journey to the female's egg and achieve fertilization.

"Sperm" is the name given to the male reproductive cells—the spermatozoa. It is also commonly used as another word for seminal fluid, but in this article it is used only to refer to both the single spermatozoon and the plural spermatozoa.

The sperm's only purpose is to achieve fertilization by union with the female cell, the ovum. Each sperm is about 0.002 inch (0.05 mm) in length and shaped like a tadpole. It has three main sections: a head, a midsection, and a tail. The front of the head—the acrosome—contains special enzymes

HOW SPERM MATURE

▼ *From puberty, sperm are constantly produced in the seminiferous tubules. To become sperm, the basic sperm cells go through three stages of cell division (bottom) before passing through the tubules and into the epididymis, where they are stored (below left). A mature normal sperm has a head, midsection, and tail (below right).*

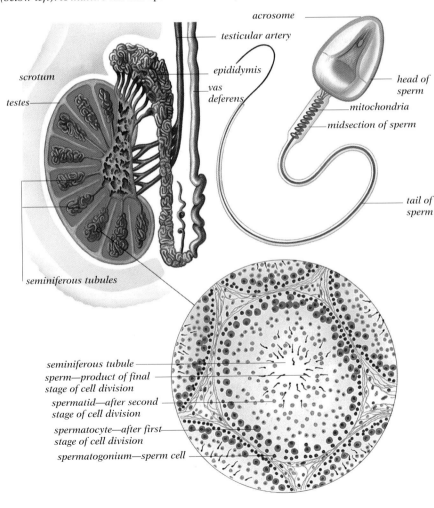

390

that enable the sperm to penetrate into the ovum and achieve fertilization. The midsection contains structures called mitochondria. These structures hold the vital source of energy needed by the sperm to fuel it on its journey to the ovum.

The tail's only function is to propel the sperm, which it does by moving in a whiplike fashion, generating a speed of about 0.12 to 0.14 inch (3 to 3.5 mm) per minute.

Sperm are made up of a number of essential chemicals and genetic material. These are the chromosomes that carry the genetic blueprint of the father and determine the paternally inherited characteristics of the child. A sperm carries either an X or a Y chromosome, producing females and males respectively.

The manufacture of sperm

The successful manufacture of sperm necessitates a temperature of about 1.8°F (1°C) lower than the rest of the body. Consequently, manufacture takes place outside the body, within the scrotum. Surrounding tissue helps to regulate the temperature of the testicles inside the scrotum by pulling them upward to the body in cold conditions, and by a rich supply of blood vessels which dissipate the heat when the temperature gets too high.

Sperm production—at the rate of 10 billion to 30 billion a month—takes place in the seminiferous tubules in the testicles. The newly formed sperm then pass through the seminiferous tubules into the epididymis, which is located behind the testicles. This serves as a storage and development area, the sperm taking between 60 to 72 hours to achieve full maturity. In fact, the epididymis can be emptied by three or four ejaculations in 12 hours; it takes about two days to be refilled. If ejaculation does not take place, the sperm disintegrate and are absorbed back into the bloodstream.

Ejaculation

Before ejaculation occurs, the sperm move along the vas deferens, two tubes connecting the testicles to the prostate gland, and into a further storage area, the ampulla. Here, the sperm receive a

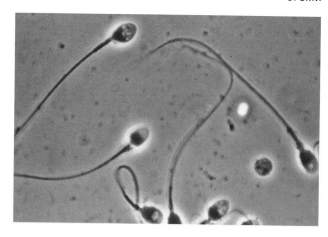

▲ *In order to reach and fertilize the female egg in the fallopian tube, the sperm must swim. Lashing their long tails is their sole method of propulsion.*

secretion from the seminal vesicles, two coiled tubes adjoining the ampulla. This secretion, called seminal fluid, stimulates the motility—the ability to move—of the sperm, and helps them survive in the vaginal secretion. The prostate gland, through which the sperm pass during ejaculation, produces a small amount of a similar fluid, giving the sperm full motility.

At the moment of ejaculation, the sperm and seminal fluid are forced out of the ampullae and epididymis into the urethra by a series of muscular contractions. If the sperm have been ejaculated into the vagina of a woman, they move as fast as they can through the cervix and into the uterus. They then make their way into the fallopian tubes, where fertilization may occur if an egg is present.

What can go wrong

Fertilization is unlikely to take place if the concentration of the sperm is too low, if the sperm are abnormal in form, or if the sperm are unable to move or stop moving too soon. The condition of the seminal fluid is another vital factor, since it both nourishes and protects the sperm. Blocked tubes, infection, stress, and ill health can also cause infertility.

The number of normal, healthy sperm in one ejaculate varies widely—anything from 20 million to 350 million in the semen (the seminal fluid and the sperm together). A sperm count lower than 20 million healthy sperm may well be responsible for infertility. When infertility is suspected, a sperm specimen will be tested in a pathology lab, and treatment and advice will depend on the cause. A man with a low sperm count may be advised to save up his sperm for a few days so as to produce the optimum number of sperm in his ejaculate. In other cases, artificial insemination is recommended. Several ejaculations are placed in a centrifuge, and the sperm concentrate is placed on the woman's cervix at her most fertile time.

Doctors are unlikely to think that investigation for infertility is needed until a couple have been trying to conceive for a year or more.

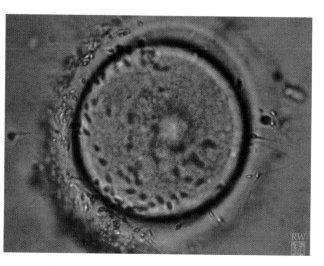

▲ *The moment of conception: chemicals in the tip of the sperm strip away the outer layer of the egg until one sperm can penetrate its smooth shell. Chemical changes in the outer layer of the egg then ensure that no further sperm can enter.*

See also: Cells and chromosomes; Cervix; Conception; Enzymes; Prostate gland; Puberty; Semen; Testes; Urethra; Vagina

Spinal cord

A vital link in the nervous system, the spinal cord gathers and analyzes information from the body and channels it to and from the brain. When this link is damaged or broken, however, permanent disability may result.

The spinal cord runs down most of the length of the bony part of the spine. It forms a vital link between the brain and the nerves connected to the rest of the body. However, the spinal cord is far more than simply a bundle of nerve fibers that go to and from the brain. It acts as an important initial analyzer for incoming sensations, and as a programming station for organizing some of the basic movements of the limbs.

SPINAL CORD

▼ *The spinal cord is protected by cerebrospinal fluid and membranes, and runs from the brain to the second lumbar vertebra before tapering into the filum terminale. A cross section of the cord (top right) shows sensory and motor pathways carrying messages to and from the brain. Reflex action occurs when messages cross the connector nerve.*

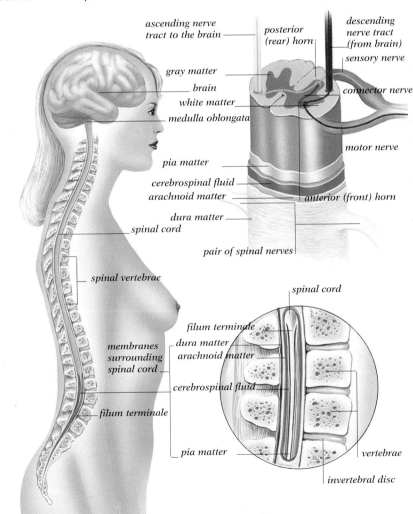

A number of conditions can affect the spinal cord, and injuries to this delicate structure can be devastating. The physical effects of any injury depend on which part of the cord is damaged, or which parts of it take the brunt of the injury.

Structure of the spinal cord

The spinal cord runs from the medulla oblongata in the brain stem down to the first or second lumbar vertebra.

The cord is well protected as it passes through the arches of the spinal vertebrae. Sensory and motor nerves of the peripheral nervous system leave the spinal cord separately just below the vertebrae and then join to form 31 pairs of spinal nerves (eight cervical, 12 thoracic, five lumbar, five sacral, and one coccygeal), each nerve corresponding to the vertebra that it leaves. These nerves branch out from the spinal cord, spreading to the surface of the body and to all the skeletal muscles.

The spinal cord is composed of collections of nerve cell bodies: neurons and bundles of nerve fibers. The gray matter, or the nerve cell collections, is H-shaped in cross section, with a posterior (rear) and anterior (front) horn (protuberance) in each half. The anterior is composed of motor neurons; the posterior horn contains cell bodies of connector neurons and sensory neurons.

The gray matter is surrounded by the white matter. The white matter is divided into columns and contains ascending and descending nerve tracts which connect the brain and the spinal cord in both directions. The descending tracts send motor impulses from the brain to the peripheral nervous system; the ascending tracts channel sensory impulses to the brain.

Surrounding these nerves and fibers is a series of membranes which are extensions of those membranes that surround the brain. Between the outer two of these three membrane layers is a small gap which contains cerebrospinal fluid. This circulates around the spinal cord and the brain, providing nutrients to the nerves and acting as a protective buffer.

Functions of the spinal cord

The spinal cord has two main functions: to act as a two-way conduction system between the brain and the peripheral nervous system, and to control simple reflex actions.

The spinal cord and the brain make up the central nervous system. Messages, in the form of electrical impulses created by the firing of interconnected neurons, from the surface of the body connect with the spinal cord via the sensory nerve fibers in the peripheral nervous system. The gray matter in the spinal cord rapidly processes the messages, and then relays some of them up the ascending tract of the spinal cord for more detailed analysis in the brain.

If some action is required, the brain sends messages of action, in the form of motor impulses, down the descending tract that result in coordinated muscular action involving many different muscles in the body.

For example, when an itch is felt in the hand, initial analysis takes place at the spinal cord. Further analysis then takes place in the brain, which may then send messages in response, instructing the appropriate muscles of the body to move accordingly.

In controlling the simple reflex action, the usual pattern of message transmission to the brain is drastically curtailed. If the skin touches something hot, streams of impulses are passed via the sensory neurons to the posterior horn in the gray matter of the spinal cord. Instead of then ascending to the brain, the messages are immediately processed, and then cross to the anterior horn of the gray matter via connector neurons. These allow messages to be transmitted from sensory neurons to motor neurons, giving an immediate physical response—the hand is rapidly and automatically withdrawn. This is known as the reflex arc. At the same time, information will be passed on to the brain, which will determine further action, if any.

Many of the body's important functions are controlled through reflex action and these occur at all levels of the spinal cord. Some movements involved in respiration, digestion, and especially excretion, for example, are reflex actions controlled in part by the spinal cord.

Spinal cord problems

A variety of conditions can seriously affect the spinal cord, from those that are acquired before birth, such as spina bifida, to others that can appear later on in life, like multiple sclerosis.

Spina bifida: In this fairly common birth defect, the spine and the spinal cord fail to develop normally in the womb. A flat plate of cells on the embryo's back normally folds itself into a tube, which then develops into the spine and spinal cord. In spina bifida, however, this

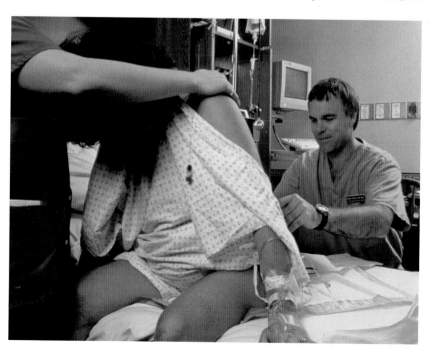

▲ *A woman receives an epidural—an injection of painkilling medication into the epidermal area between the dura mater and the more interior regions of the spinal cord—to ease her labor pains in the hospital delivery room during childbirth.*

Questions and Answers

In a spinal tap, is the fluid taken off through the needle from the spinal cord itself? If so, isn't there a risk of damage to the cord?

The spinal cord ends about three-quarters of the way down the spine. The spinal canal (the space enclosed by the bones of the spine), which is below that, is only partly filled by the nerves which go down to the legs. So there is room for a needle to remove fluid without damaging the nerves. The fluid is the same as that which circulates around the brain and through the center of the cord, so it is useful to examine in diagnosing conditions affecting the nervous system.

Are the cells of the spinal cord like brain cells or are they different?

The nerve cells, or neurons, of the spinal cord are the same as those of the brain. Although some are specialized for their particular job (like some brain cells) they are essentially the same.

I have heard that the spinal cord is affected by syphilis. Is this true?

This is now uncommon, but one of the delayed effects of syphilis is the attack of the sensory nerves just as they are entering the spinal cord. This causes the rear part of the cord, which is made up of the sensory fibers on their way to the brain, to wither. The main symptom from the loss of these nerves is a poor sense of where the joints are, making walking difficult.

Is the spinal cord always seriously deformed when a baby is born with spina bifida?

No, the spinal cord is not always deformed. There are degrees of severity. It is only in the most severe type that the spinal cord is involved and the bone fails to fold over as it should as the baby develops in the womb. Sometimes spina bifida involves only the bones of the spine, and there is seldom any spinal cord trouble.

AREAS OF THE BODY CONTROLLED BY THE SPINAL NERVES

▶ *The majority of signals that control the body's sensations and movement are fed to and from the brain via the 31 pairs of nerves joining the spinal column. These nerves control different areas of the body. If the cord is damaged, all areas fed by the nerves below the site of the injury will be affected.*

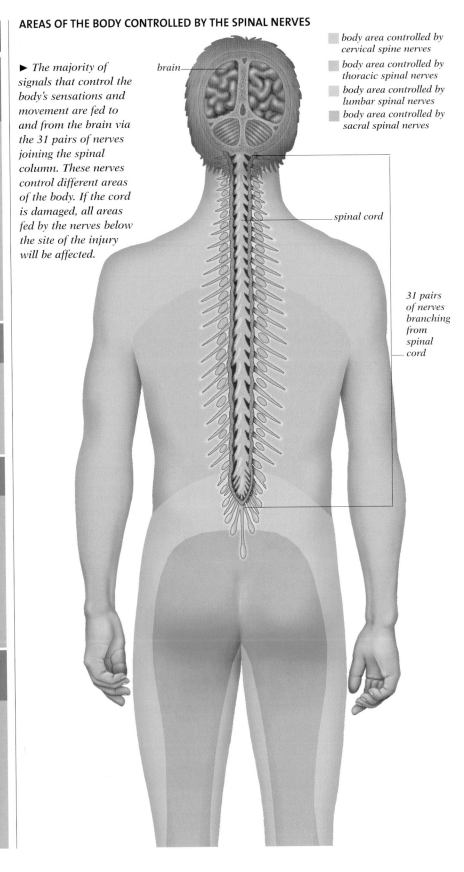

brain

spinal cord

31 pairs of nerves branching from spinal cord

body area controlled by cervical spine nerves

body area controlled by thoracic spinal nerves

body area controlled by lumbar spinal nerves

body area controlled by sacral spinal nerves

Tumors: These rarely occur inside the spinal cord itself, but the cord can be pressed on by tumors from the outside. The effects of this type of complaint on the cord, and what symptoms are produced, depend on where the tumor is located. Occasional pains around the trunk, or down an arm or leg, are common symptoms of a tumor pressing on the nerves emerging from the cord. Sensation may also be lost from either side of the body below the site of the trouble. As with spinal injuries, the person may also become incontinent.

Treatment depends on where the tumor is located, but it usually involves either surgical removal of the tumor or drainage of the abscess. In the case of a cancerous tumor, radiotherapy may be used.

Multiple sclerosis: The cause of this type of inflammation is still unknown. It attacks nerve tissue anywhere in the body, particularly the main nerves to the eye, the optic nerve, and nerves in the brain stem and the spinal cord, especially in the neck area. Spinal cord damage from multiple sclerosis can cause progressive loss of sensation and occasional tingling feelings in the hands and feet.

▲ *Competitive sports like basketball are actively pursued and enjoyed by many people with severe spinal injuries. Differing degrees of mobility and expertise can be achieved.*

process is disrupted, leaving the spinal cord malformed and exposed at the back.

The condition may leave the child with complete paralysis of the legs and no control over either urination or defecation. In some severe cases, brain damage may also occur. The type of treatment that may be possible will depend on how badly deformed the spinal cord is.

Injuries to the spinal cord: This is the most common cause of problems with the spinal cord, displaced vertebrae and whiplash injuries being the most frequent types of injury.

Exactly what functions are lost is determined by which part of the spinal cord is actually damaged. If the cord is damaged high in the neck, all the limbs will be paralyzed (this condition is known as quadriplegia), and there may even be difficulty in breathing. Immediately after the injury all the limbs become limp, with all feeling being lost below the level of the injury. In addition, bladder control is considerably affected. After a period of weeks or months, various changes appear in the paralyzed legs and arms, as the spinal cord below the injury recovers a little. The limbs become stiff and may respond briskly with reflex movements. In some cases, there may be more improvement, and the injured person may even achieve a stiff-legged walk. If the injury is in the middle of the back, then only the legs are affected (this condition is known as paraplegia). The bladder's function is usually affected in any spinal cord injury, since the nerves to the bladder leave the cord at its lower end. Sexual function is lost, too, because the nerves involved are also located at the lower end.

Myelitis: This term means inflammation of the spinal cord, and so includes multiple sclerosis. However, myelitis can be caused by some common viruses, when there is only one attack, unlike the repetitive attacks which characterize multiple sclerosis. The cord becomes acutely inflamed at a particular spot; below this, all function may be lost. This form of myelitis may follow an attack of flu, or some other trivial form of viral infection.

Vitamin deficiency: A lack of the vitamin B12 can produce a particular pattern of damage to the spinal cord (as well as to the peripheral nerves). The parts of the spinal cord affected include the muscle running down the side of the cord and the sensory nerves that convey sensations of touch and a sense of where the joints are. The affected person may suffer from a mild weakness in the limbs and have an odd, high-stepping walk because the person has difficulty in establishing where the feet are in relation to his or her body. Some recovery from this condition can be brought about when the vitamin deficiency is treated.

Many conditions affecting the spinal cord are long-term problems; recovery, if any, is slow and painstaking, and the patient will require devoted nursing. Recovery from a spinal cord injury may never be complete, especially if the damage is severe, but considerable movement and control can be regained through regular exercise.

See also: **Autonomic nervous system; Bladder; Bones; Brain; Membranes; Muscles; Neck; Nervous system; Reflexes; Skeleton**

Spleen

Questions and Answers

Situated in the top left-hand corner of the abdomen, the spleen plays a major role in blood formation and influences the development of immunity. In addition, it may signal disease elsewhere in the body.

I've heard that the spleen can burst during glandular fever. Is this true, and is it dangerous?

Reports of this are in the context of abdominal trauma. If the spleen bursts as a result of the glandular fever virus, then its contents will be released into the abdomen where they may inflame the membrane that lines the abdomen and give rise to peritonitis. A burst spleen can be life-threatening.

My brother insists that he has two spleens. Can this be true?

Yes. There is usually only one big spleen to be found in the abdomen, normally in the top left-hand corner, but occasionally there may be one or two accessory spleens in the same general area.

A friend of mine had to have his spleen removed after a traffic accident. Does this mean that his blood will be affected?

The spleen plays a major part in ridding the body of harmful bacteria, so its removal makes a person, particularly a child, more susceptible to serious bacterial infections. In some situations, prophylactic antibiotic treatment is recommended for postsplenectomy patients, especially children.

Does the spleen influence a person's emotional state?

No. It was once thought that the spleen was the seat of anger, and morose people were said to have a splenic personality. We still talk of venting our spleen, which means to get angry, but there is no scientific basis for this. No one knows why a blood-forming and blood-filtering organ should have acquired this reputation.

The spleen is an important organ of the body. Its main function is to filter the blood and to make antibodies. An enlarged spleen, which can be felt through the walls of the abdomen, is often an indication of disease somewhere in the body. The spleen is also an integral part of the lymphatic system—the basis of the body's defense against infection.

Location
The spleen lies just below the diaphragm at the top of the left-hand side of the abdomen. It is normally about 5 inches (13 cm) long, and it lies along the line of the tenth rib. The spleen usually weighs about ½ pound (about 200 g) in adults, but, in cases where it is enlarged, it can weigh up to 4½ pounds (2 kg) or more.

Appearance
If a spleen is examined with the naked eye, it will look like a fibrous capsule surrounding a mass of featureless red pulp. It may just be possible to make out little granulations called Malpighian corpuscles. The organ is supplied with blood via the splenic artery, which, like any other artery,

POSITION OF THE SPLEEN

The spleen is situated in the top left-hand corner of the abdomen, just below the diaphragm. It is in a relatively exposed position, and so it is frequently damaged in accidents and has to be removed, generally without any ill effects.

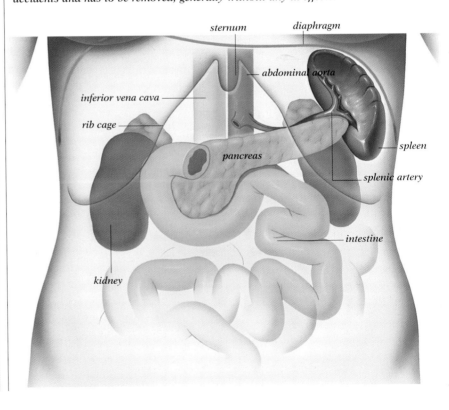

sternum · diaphragm · abdominal aorta · inferior vena cava · rib cage · pancreas · spleen · splenic artery · intestine · kidney

splits first into smaller arteries and then into tiny arterioles. However, the arterioles of the spleen are unusual in that they are wrapped in lymphatic tissue as they pass through the pulp of the spleen. The arterioles are unique in another way; instead of being connected to a network of capillaries, they appear to empty out into the main mass of the spleen.

The unusual way in which the spleen is supplied with blood is what enables it to perform two of its basic functions. First, the fact that the arterioles are wrapped with lymphatic tissue means that the lymphatic system comes into immediate contact with any abnormal protein in the blood and forms antibodies to it. Second, the way that the blood empties directly into the pulp of the spleen also allows the reticular cells of the organ to come into direct contact with the blood, filtering it of any old or worn-out cells.

▲ *The healthy spleen is an organ that usually weighs about ½ lb. (200 g) in adults and is 5 in. (13 cm) long.*

▲ *A diseased spleen: this spleen is greatly enlarged because of the presence of a lymphoma or lymphatic tumor.*

Functions of the spleen

The spleen is one of the main filters of the blood. Not only do the reticular cells remove old and worn-out blood cells, but they will also remove any abnormal cells. This applies, in particular, to red blood cells, but white cells and platelets are also filtered selectively when necessary by the spleen.

The spleen will also remove abnormal particles floating in the bloodstream. This means that it plays a major part in ridding the body of harmful bacteria. It is also instrumental in making antibodies—proteins circulating in the blood that will bind onto and immobilize a foreign protein so that white blood cells called phagocytes can destroy it. The Malpighian corpuscles, which are collections of lymphocytes, produce the antibody.

In some circumstances, the spleen has an important role in the manufacture of new blood cells. This does not happen in the normal adult, but in people who have a bone marrow disease the spleen and the liver are major sites of red blood cell production. In addition, the spleen makes a large proportion of the blood of an unborn baby.

Feeling the spleen

The spleen cannot be felt in normal healthy people, but there is a large range of diseases that cause enlargement of the spleen, which can then be felt through the walls of the abdomen. The procedure is that the patient lies on his or her back, and the doctor starts to feel (or palpate) the bottom of the abdomen, and then works up toward the top left-hand corner. The spleen moves as the patient breathes, so he or she is asked to take deep breaths so that this movement can be felt. Enlargement of the spleen can also be detected on some X rays or by using a radioactive isotope scan.

Enlargement of the spleen

The spleen may enlarge for many reasons. Since one of its main functions is to break down old and worn-out blood cells, those conditions where blood is broken down faster than normal are associated with an enlarged spleen. These diseases are called hemolytic anemias, and many of them, such as sickle-cell anemia or thalassemia, are inherited. Hemolytic problems can also have other causes; for example, some drugs such as methyldopa (used to control blood pressure) may cause hemolysis and thus a large spleen. Other blood diseases also cause the spleen to become enlarged. In some cases of leukemia, for example, the spleen

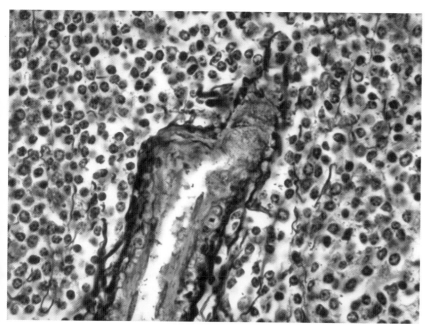

▲ *Spleen tissue, which is the site of phagocytosis and initiation of immune responses, is shown stained and magnified 160 times.*

GEOGRAPHICAL DISTRIBUTION OF KALA-AZAR

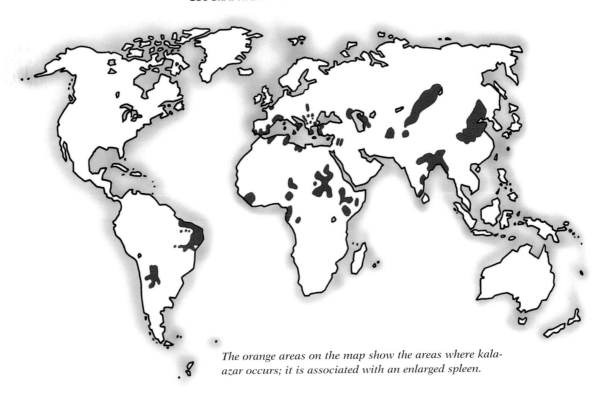

The orange areas on the map show the areas where kala-azar occurs; it is associated with an enlarged spleen.

grows so much that it stretches from the top left-hand corner of the abdomen to the bottom right-hand corner.

There are two other diseases that are associated with an enlarged spleen, malaria and the parasitic disease called kala-azar, in which the parasites actually inhabit the spleen. Because it is involved in the body's immune mechanisms against infection, many other infections can cause an enlarged spleen. A common disease associated with the enlargement of the spleen is glandular fever. Rarely, the large spleen found in this illness can rupture as a result of a comparatively minor injury to the abdomen, and an operation will be needed.

The veins of the spleen drain into the portal system of veins. These are the veins that drain blood that is rich with nutrients from the intestines into the liver. Thus, when liver disease is present, pressure on the system can rise, putting pressure on the spleen, which enlarges. Thus, an enlarged spleen can indicate that there is trouble in the liver. Also, it is a useful indicator of health of the body, since its enlargement may be a result of problems elsewhere. Little can go wrong with the spleen; cysts and benign tumors may form rarely.

Removal of the spleen

There are, however, a number of reasons why the spleen may have to be removed. Despite the fact that it is an important

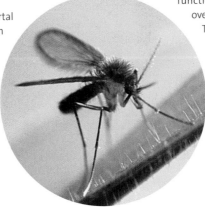

▲ *Kala-azar is caused by the bite of some sand flies; if the right treatment is not available, it can be fatal.*

filter of the blood, very few immediate effects seem to result from its removal. The susceptibility to severe infections is, however, increased.

If the spleen should rupture or become injured as the result of an accident, it tends to bleed profusely; the only option then is to surgically remove it. The spleen may also have to be removed during laparotomies (investigative operations during which the abdomen is opened) in order to investigate the extent of lymphomas or lymphatic tumors.

Occasionally the spleen becomes overactive in its function of breaking down blood cells, and this overactivity leads to excessive destruction of cells. This is likely to happen only when the spleen is already enlarged for some reason, such as a lymphoma or portal hypertension due to liver disease. In these cases, too, the spleen may be removed.

The spleen is therefore a unique organ; for whereas it has useful and straightforward functions, the body seems capable of functioning quite well without a spleen. Another anomaly is that although the spleen rarely malfunctions, it is an indicator of problems, and it is often involved in other defects elsewhere in the body.

See also: **Blood; Liver**

Stem cell

Questions and Answers

Is it true that stem cells can cure a whole range of human diseases?

A scientist would say that careful stem cell research on mice and other animals has shown that stem cells injected into damaged tissue of all kinds can lead to a regeneration of new tissue cells of the specific type that was damaged. The potential therapeutic value of such a discovery is considerable.

I hear that stem cells can be obtained from adults. If this is true, why is there so much argument about the use of stem cells from embryos or fetuses?

It is true that there are plenty of stem cells in adults. These, however, do not normally give rise to any kind of body cell, as is the case with embryonic stem cells. It seems that adult stem cells normally produce only cells of the type required in their own particular location. Liver stem cells differentiate into liver cells, nervous system stem cells produce nerve tissue, muscle stem cells produce muscle cells, and so on. Embryonic stem cells can, however, produce any kind of body cell.

Is it true that stem cells from one location in the adult body can produce different kinds of cells if transplanted elsewhere?

There have been many reports of adult bone marrow stem cells producing muscle, heart muscle, liver, brain, and artery lining cells after transplantation. Many of these reports, however, have been found to be unduly optimistic. For instance, the tagged transplanted cells often appeared to have fused with local cells so as to give the impression that they were forming new cells of the desired type. The experts are still discussing the issue.

The use of human stem cells in medicine is currently one of the most promising and potentially far-reaching areas of medical research and seems likely to lead to a revolution in medical science. There are, however, serious ethical problems associated with this work.

A human egg (ovum) that has been fertilized by a spermatozoon is, briefly, a single cell. From that single cell come all the billions of cells that make up a human body. Such a cell is said to be "totipotential." This means that it is capable of differentiating into any of many different kinds of cells in the body—muscle cells, nerve cells, heart cells, kidney cells, bone cells, skin cells, blood cells, and so on. Even after the fertilized egg has divided many times to form a mass of cells, each one of these is a stem cell and each remains totipotential. An embryo, however, is not the only place where stem cells are to be found. All the blood cells produced by the bone marrow—the oxygen-carrying red cells and the whole range of white cells—derive from a stem cell. Research has shown that adult stem cells are not limited to the bone marrow; rather, they exist in many parts of the body. Production of stem cells is controlled by various hormones.

How stem cells produce new tissues

Every living cell contains the whole DNA genome for the entire body of the person, animal, or plant, but at any particular location only some of the genes are required. Any cell will, typically, use only a fraction of all the genes in its DNA. If, for instance, muscle protein is needed, the genes

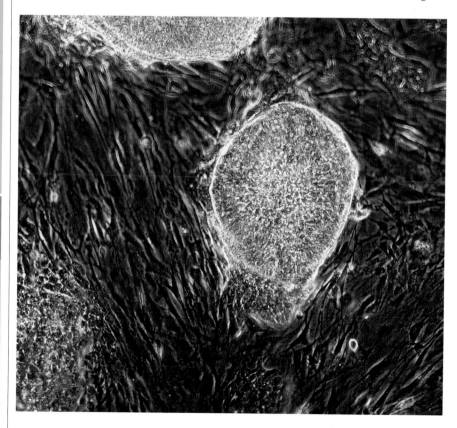

▲ *A greatly magnified stem cell seen through a microscope. All blood cells are derived from stem cells, the production of which is controlled by hormones.*

▲ *A laboratory technician is freezing a semitransparent sachet containing a blood sample taken from peripheral blood stem cell CD34.*

that code for this protein will be switched on and many other genes will be left permanently inactive. The term applied to switching on genes is "gene expression," and this occurs as a result of chemical signals from within the cell or from other cells. Gene expression can be regulated, and started or stopped, at various points in the pathway from DNA to protein.

There is now clear evidence that embryonic stem cells, if inserted into any tissue that is damaged or deficient, can continue to divide indefinitely and convert themselves into cells of the specific kind required. Thus, when a disease or disorder is caused by damage to, or shortage of, cells of a particular type, stem cell therapy should, in theory, be capable of bringing about a cure.

The list of such conditions is very long and includes heart failure, Alzheimer's disease, multiple sclerosis, liver disease, lung disease, Parkinson's disease, diabetic retinopathy, macular degeneration, and in combination with specific treatment, cancer. At present, such possibilities are, however, still in the experimental stage.

Ethical issue

Embryonic stem cells can be obtained from fetuses that have been aborted; or embryos formed in vitro and not required for implantation; or embryos created in the laboratory as a source of stem cells. Some people believe it is justifiable to use stem cells in scientific research; many others think the use of cells from fetuses or embryos is abhorrent and ethically unacceptable.

Legislation

In 2001, President George W. Bush authorized funding for stem cell research but only if that research used stem cell lines that existed on or before August 9 of that year. Eight years later, in 2009, President Barack Obama reversed those restrictions. However, the efforts to expand stem cell research were thwarted one year later, when in August 2010, a federal judge ruled that federal funding could not be used in such research because it involved the destruction of embryos.

See also: **Blood; Cells and chromosomes; Marrow**

Stomach

Questions and Answers

What happens when people have their stomach removed? Can they eat normally, and what happens to their digestion?

Usually only part of the stomach is removed. This can still affect them in many ways, however, and they usually find that they feel full after a small meal and have to eat little and often. They must see a doctor regularly as, in the long term, they may become anemic and develop other nutritional deficiencies. If the whole stomach is removed (this is not always necessary), they have more severe symptoms and will require regular injections of vitamin B12, since the stomach secretes a factor necessary for the natural absorption of this vitamin.

How big is the stomach and does its size vary?

It can vary considerably in size. It may be the size of a large pear or so large that it almost reaches the pelvis. Eating a lot of food may cause it to enlarge and a blockage to the outlet of the stomach could have the same effect. However, when someone is said to have a large stomach, this usually means that the abdomen is fat.

Is the stomach ever empty, or does it always contain some food?

Food entering the stomach takes an hour or so to pass into the intestines. After this, apart from a small amount of digestive juice, the stomach is empty.

Is drinking alcohol on an empty stomach harmful?

Alcohol would be more quickly absorbed into the bloodstream, so it is inadvisable to drink alcohol without food, since it can damage the lining of the stomach, causing inflammation and bleeding.

The stomach is the body's natural reservoir. It holds the food we eat and begins the digestive process. The products of digestion are then absorbed through the intestinal lining into the bloodstream and circulated.

The stomach is a muscular bag situated in the upper left part of the abdomen. It is connected at its upper end to the esophagus, and at its lower end to the duodenum. The wall of the stomach consists of a thick layer of muscle lined with a special membrane called epithelium. The stomach acts as a reservoir for food. The lining membrane produces a juice that contains acid and enzymes to break down the food and aid digestion. In the stomach the food is mixed with digestive juices until it forms a pulp, called chyme, which is then moved into the duodenum. At the junction between the stomach and the duodenum there is a ring of muscle, the pyloric sphincter or pylorus. The pylorus closes if large pieces of food leave the stomach; then anti-peristaltic waves in the stomach wall churn the food pieces back into the stomach for further enzyme and acid treatment. Chyme is then pushed along the intestines to be further digested and absorbed.

Common problems

Vomiting occurs when the stomach contracts forcibly. When this happens, the contents of the stomach may be ejected upward into the esophagus, an action known as vomiting. Vomiting can have several causes: any disturbance of the central nervous system, which controls the contraction of the stomach; an irritation of the lining of the stomach; or an obstruction of the

▲ *Stomach problems such as gastritis or ulcers can flare up for a variety of reasons, but perhaps most frequently as a result of eating spicy food, drinking alcohol, and taking certain drugs.*

outflow from the lower end of the stomach. Often, vomiting is due to a common condition known as gastroenteritis, in which the lining of the stomach becomes inflamed as a result of a viral infection, eating spicy foods, drinking alcohol, taking certain drugs, or stress. A mild attack of gastritis produces symptoms of nausea, vomiting, and occasionally some pain in the upper abdomen. Severe attacks can result in bleeding from the stomach lining.

The treatment of gastritis is to remove the cause and to drink bland fluids. Drugs which prevent vomiting may be prescribed by a doctor. Gastritis, however, is usually a self-limiting condition once the primary cause has been removed.

Gastric ulcers

Ulcers occur in the stomach when there is a local failure of the mucous layer that protects the lining and prevents it from the effects of the strong acid and enzymes naturally present in the gastric juices. The spiral germ *Helicobacter pylori* is a major cause of chronic gastritis and stomach ulcers. *H. pylori* infection can be detected by a urea breath test, by antibody tests, by stool tests, or by endoscopy with biopsy. The acid can erode the mucosal layer and then penetrate the muscle layer. This may lead to perforation of the stomach wall, causing an ulcer. Scarring and fibrous tissue are often found around the gastric ulcers, suggesting they have been present for many years prior to diagnosis.

People who develop gastric ulcers do not have a high level of acid in the stomach, unlike those people who develop duodenal ulcers. In fact, some patients with gastric ulcers have a low level of acid secretion.

Symptoms and dangers

Gastric ulcers are most commonly found on the upper aspect of the stomach (the lesser curve). They tend to appear, grow bigger, then disappear. Not all the causes of gastric ulcers are known, but spicy foods, alcohol, smoking, and stress are contributory factors.

The symptoms include burning pain, which comes shortly after eating; nausea; and sometimes weight loss. There may be episodes of these symptoms followed by long periods without any symptoms at all. If a gastric ulcer is left untreated for a long period of time, several things can happen. First, if the ulcer is situated over a main blood vessel supplying the wall of the stomach, it

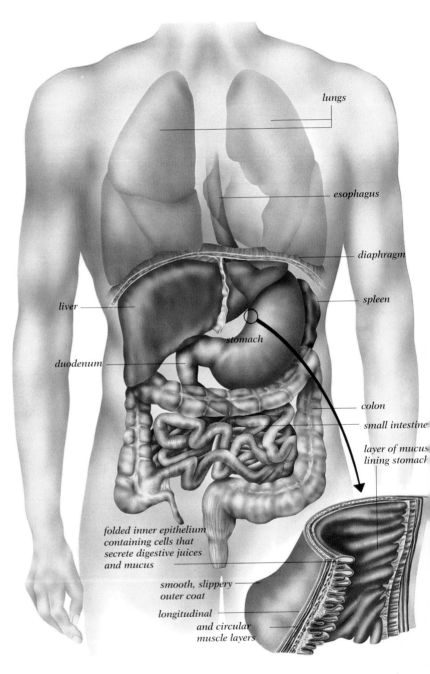

lungs

esophagus

diaphragm

spleen

liver

stomach

duodenum

colon

small intestine

layer of mucus lining stomach

folded inner epithelium containing cells that secrete digestive juices and mucus

smooth, slippery outer coat

longitudinal and circular muscle layers

Section through stomach wall

▲ *The stomach is situated higher up in the body than most people think—in fact, it is found just under the diaphragm. It is a muscular bag with a smooth, slippery outer coat and a corrugated inner lining that is protected from its own acidic digestive juices by a layer of mucus.*

Questions and Answers

My husband is 60 years old and he has started getting persistent indigestion pains. Antacids, such as milk of magnesia, do not really seem to be helping much. What should he do next?

It is important for someone of his age who has developed persistent indigestion to have an evaluation of this problem. Your doctor should arrange for him to have a number of tests, which will probably include a gastroscopy. The latter enables the doctor to see into the stomach by means of a flexible telescopic instrument (a gastroscope) passed through the mouth and esophagus.

Is it normal to lose weight with an ulcer?

It depends on what sort of ulcer it is. If the ulcer is in the duodenum, eating relieves the pain and the patient may put on weight. However, if the ulcer is in the stomach, eating causes pain so the patient tends to lose weight because he or she is afraid to eat.

I had an attack of vomiting and loss of appetite. My doctor said it was gastric flu. Is this possible?

Yes. You probably developed a viral infection of the stomach, causing its lining to become inflamed. However, there is no way of proving that this is what it was. The symptoms of viral gastritis, as it should be called, are usually short-lived, and respond to simple measures such as drinking bland liquids, and possibly taking tablets to prevent vomiting.

Does a stomach tumor always cause pain?

No. Sometimes there is no pain at all. The patient simply notices a loss of appetite, or loss of weight. In fact, pain can often confuse the issue, and sometimes leads to a misdiagnosis if the patient is simply thought to have attacks of indigestion.

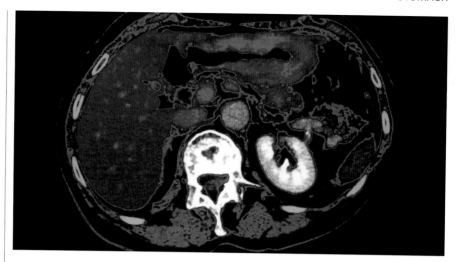

▲ *Computer-assisted imaging (CT) uses X-ray beams to produce a cross-sectional picture of a stomach. The tomogram produced shows that cancer is present.*

may erode through the blood vessel, leading to a massive hemorrhage. Second, the ulcer can perforate the wall of the stomach; the stomach contents can then leak out into the peritoneal cavity, causing peritonitis. Third, repeated attempts at healing the ulcer may lead to scarring and contraction of the tissues around the ulcer. This may eventually produce a narrowing in the middle of the stomach so that it assumes an hourglass appearance.

Treatment

Diagnosis of a gastric ulcer is usually made by means of a gastroscopy to ensure that the ulcer is benign. Eradication of *Helicobacter pylori* is the first step as this infection is often present in such cases. Recently developed drugs, such as proton pump inhibitors, are so effective in controlling gastric acid production that surgery for gastric ulceration is now required much less often than formerly.

Cancer of the stomach

Gastric cancer is one of the most common malignant tumors, and it affects more men than women. It can start as a small ulcer or as a small polyp and grows bigger to obstruct the passage of food through the stomach. Because the tumor involves the lining of the stomach, patients may lose blood into the stomach, and anemia may signal that there is a problem.

Other symptoms are constant nausea and loss of appetite, and weight loss. Pain may develop in the upper part of the abdomen. Because the pain can be similar to that caused by a benign gastric or duodenal ulcer, patients are often treated for a long time with antacids with no relief of symptoms.

It is for this reason that any patient over the age of about 50 years with indigestion should be treated with caution. If he or she does not respond quickly to conventional treatment for indigestion, further investigation of the case will be essential.

Like a benign ulcer, gastric cancer is usually diagnosed by a gastroscopy. Early diagnosis improves the overall outlook of the treatment. Whenever possible the treatment of gastric cancer involves surgical removal of the tumor, which often involves removing the whole stomach. Patients who have had their stomach, or a major portion of it, removed cannot eat large meals, but in other respects may live a normal life. They need yearly checkups, since they are more likely to develop anemia and nutritional disturbances, which can usually be corrected.

Sometimes the outlet of the stomach, the pyloric canal, becomes blocked, leading to a buildup of stomach contents followed by profuse vomiting. In infants this can occur as a result of overgrowth of the muscle in this region. In adults, it is caused either by a cancerous growth or by a long-standing duodenal ulcer, which has caused fibrous scarring.

See also: Digestive system

Subconscious

The subconscious is a part of the mind where our deepest fears, childhood experiences, and primitive memories are stored. The store of information is often outside conscious awareness, but it can be brought into consciousness.

Are dreams from the subconscious?

No one knows where dreams come from. They are the result of some kind of mental activity and occur when the brain is in a state of arousal that brings about bodily effects including rapid eye movements. Dreams are often related to anxiety. The general view is that they are an attempt by the mind to make sense of material that is not sufficiently organized into a unified whole to be effectively stored in memory.

Do artistic people have very active subconscious minds?

Yes. Artistic creativity is largely a matter of subconscious activity, although, to be effective, the artist must engage in hard conscious work. Literary artists are constantly surprised at what their subconscious minds produce. It seems that much of creativity consists of a subconscious synthesis of ideas and data.

I want to have psychotherapy, but worry about what is in my subconscious. Should I go ahead?

With a psychotherapist you respect and trust, you will find that he or she will support you through the journey of self-discovery. You may find that the more you understand about yourself, the stronger and less fearful you will feel.

If I do something my sister dislikes, she says I subconsciously resent her. Is she right?

Your sister may be right or wrong, but such an observation is not, by itself, helpful. It is used merely as a weapon in an argument. Explain that you can sort out your feelings only if you trust one another enough to talk about them without such accusations.

Human behavior is largely conditioned by past experience, beliefs, and feelings of which people are usually very much aware. Sometimes, however, people's actions may seem out of character, and although they felt that these actions were perfectly reasonable actions at the time, subsequently they find it difficult to explain why they acted as they did. This is just one piece of evidence for the existence of the subconscious mind.

There is other evidence: for example, people often hear their name mentioned in a conversation, even if they were not listening for it and could not hear anything else that was being said; or they will frequently notice a clock has stopped ticking, even though they were

▲ *Pyschotherapy can help to unbury thoughts or feelings that are destructive or are felt to be unacceptable. Once out in the open such feelings can be dealt with.*

unaware of the clock's existence before it stopped. The occasional ability some people seem to possess that enables them to solve a problem in an apparent flash of insight, rather than by conscious thought, also betrays the existence of the subconscious activity of the mind.

Development of a subconscious

From the moment of birth, feelings and images bombard the consciousness, and the mind begins the process of trying to make sense of them and to record them. Some of the images and feelings are almost certainly wrongly recorded and fail to fit later sensations and impressions; additionally, there is far too much information filtering into the brain for items to be recorded individually. Instead, general impressions and rules of behavior begin to be learned. These rules, however, are an individual's rules rather than the rules that someone teaches him or her.

Subconscious feelings

Later, however, when the experiences and rules of the individual come into conflict with what he or she decides are the wiser and stronger rules of others, his or her rules and particular experiences are filed away at the subconscious level. Yet forgetfulness does not prevent a later event from triggering the original feeling or the original reaction, and most people have by chance experienced a sound or a scent that vividly brings back a childhood feeling without being able to place the original event linked to the stimulus.

Problem solving

Ideas that occur are sorted by the subconscious into sections of the brain and filed as important or unimportant. When the important ideas are brought to conclusions, the subconscious breaks through to the conscious mind with the revelatory facts. This explains why when someone has been seeking the solution to an elusive problem, it can suddenly come unbidden into the mind.

Racial subconscious

Some of the experiences buried in the subconscious may be common to many people. This led the Swiss psychiatrist Carl Jung to postulate the existence of a racial subconscious that contained a set of memories common to those of a particular ethnic culture. Such a group memory would help to explain how myths, legends, and fairy tales that come from many countries often contain the same elements, even when the people of one country have no direct contact with the inhabitants of another.

Dealing with the subconscious

Most subconscious memories and feelings are harmless enough. However, strong feelings of anger, fear, or desire, which have become buried in a person's subconscious mind because they are regarded as

▲ *Fairy tales and folk stories are a possible expression of the most mysterious part of our common experience. They can also act as a focus for nameless childhood fears, and by making them fanciful, diminish them.*

unacceptable, will almost certainly emerge unexpectedly from time to time as behavior that is destructive to the possessor, or to another person. In such circumstances it must be helpful to draw out from the subconscious the memories and feelings that cause the unfortunate behavior in order that the person may recognize them for what they are and come to terms with them.

This process of examining the subconscious is the basis for many types of psychotherapy, and psychoanalytic therapy in particular. Because present thoughts can sometimes trigger subconscious memories under relaxed conditions, the examination of dreams (for which we seldom feel responsible) and the use of free association are practices that can be of value. In free association the patient is asked to say whatever comes into his or her head, wandering from image to image with the expectation that any pressure coming from the subconscious will emerge into consciousness by suddenly remembered associations.

The subconscious and psychotherapy

It must not be assumed, though, that the subconscious mind is the repository of the worst aspects of personality, or thoughts that ought to remain hidden. The subconscious mind is also likely to contain the unexpressed visions of the creative person, and joys and happiness of which we remember little.

Many people who have explored the subconscious through psychotherapy say that the journey can be life-transforming.

See also: **Brain; Head; Memory; Mind**

Taste

The word "taste" stands for discernment; yet in fact it is the most simple of the five senses, and we are capable of distinguishing only four basic taste sensations from our food and drink.

Of the five senses—sight, hearing, smell, taste, and touch—the sense of taste is the crudest. It is limited in both range and versatility and presents less information about the environment than any other sense. The exclusive role of the sense of taste is to select and appraise food and drink, a role that is helped by the more sensitive sense of smell, which adds information to the four basic tastes that the taste buds can recognize. Consequently, the loss of the sense of taste, for whatever reason, is less of a problem than the loss of the sense of smell.

How food and drink are tasted

Like smell, the taste mechanism is triggered by the chemical content of substances in food and drink. Chemical particles are registered in the mouth and converted into nerve impulses that are then transmitted to the brain and interpreted. The taste buds are at the heart of this system. Studding the surface of the tongue are many small projections called papillae. Inside the papillae are the taste buds. An adult has about 9,000 taste buds, mainly on the tongue's upper surface, but also on the palate, and even the throat.

Each taste bud consists of groups of taste receptor cells, and each receptor cell has fine hairlike projections called microvilli, which are exposed on the surface of the tongue through fine pores in the surface of the papilla. The taste receptor cells link up with a network of nerve fibers. The design of this network is complex, since there is a great deal of linking between nerve fibers and receptor cells. Two different nerve bundles, which make up the facial nerve and the glossopharyngeal nerve, carry nerve impulses to the brain.

The taste buds respond to only four basic flavors: sweet, sour, salt, and bitter. The receptor sites for these tastes are located on different parts of the tongue. The buds that respond to sweet are at the tip of the tongue, whereas those specializing in salty, sour, and bitter flavors are located progressively farther back. Odor and some other stimuli combine with the four crude flavors to produce more subtle taste sensations.

How the taste buds respond to the chemicals in food and initiate nerve impulses to the brain is not fully understood, but in order to be tasted, the chemicals must be in liquid form. Dry food gives little immediate sensation of taste, and acquires its taste only after being dissolved in saliva.

It is believed that chemicals in food can alter the electrical charge on the surface of the taste receptor cells, which in turn cause a nerve impulse to be generated in the nerve fibers.

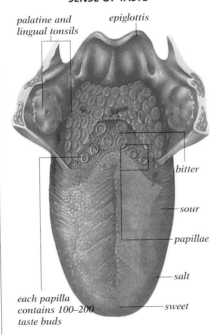

SENSE OF TASTE

palatine and lingual tonsils

epiglottis

bitter

sour

papillae

salt

sweet

each papilla contains 100–200 taste buds

◄ *The papillae on the tongue increase the area in contact with food and, except for those in the center, they contain numerous taste buds. The taste buds contain taste receptors that are distributed so that different parts of the tongue are sensitive to different tastes: sweet, salty, sour, or bitter.*

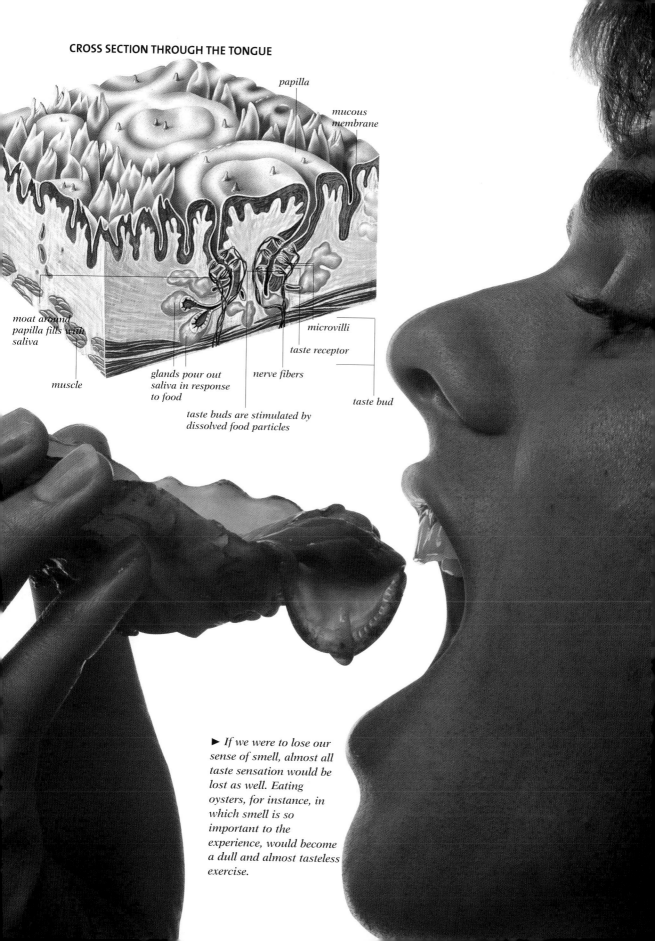

CROSS SECTION THROUGH THE TONGUE

papilla

mucous membrane

moat around papilla fills with saliva

microvilli

taste receptor

muscle

glands pour out saliva in response to food

nerve fibers

taste bud

taste buds are stimulated by dissolved food particles

► *If we were to lose our sense of smell, almost all taste sensation would be lost as well. Eating oysters, for instance, in which smell is so important to the experience, would become a dull and almost tasteless exercise.*

Analysis of taste by the brain

The two nerves carrying taste impulses from the tongue (the facial nerve and the glossopharyngeal nerve) first pass an impulse to specialized cells in the brain stem. This area of the brain stem also acts as the first stop for other sensations coming from the mouth. After initial processing in the brain stem center, taste impulses are transferred through a second set of fibers to the other side of the brain stem and ascend to the thalamus. Here there is another relay, where further analysis of the taste impulses is carried out before information is passed to the part of the cerebral cortex participating in the actual conscious perception of taste.

The cortex also deals with other sensations, such as texture and temperature, which are transmitted from the tongue. These sensations are probably mixed with the basic taste sensations from the tongue, and so produce the subtle sensations with which we are familiar when we eat.

The analysis of the food eaten, carried out in the lower part of the parietal lobe in the cortex, is further influenced by smell information being analyzed in the nearby temporal lobe. Much of the refinements of taste sensation are due to smell sensations.

Sensitivity of the taste buds

Compared with other senses (in particular smell), taste is not very sensitive. It has been estimated that a person needs 25,000 times as much of a substance in the mouth to taste it as is needed by the smell receptors to smell it. However, despite this insensitivity, the combination of the four types of taste buds responding to the basic tastes of salty, sour, bitter, or sweet enables a wide range of sensations to be determined as the brain analyzes the relative strength of the basic flavors. Some of the stronger tastes, such as the hot flavor of spicy food, are experienced through stimulation of pain-sensitive nerve endings in the tongue.

What can go wrong

Loss of taste usually comes about from a problem in the facial nerve. This nerve is connected to the muscles of the face, but a small branch carries the taste fibers from the front two-thirds of the tongue. For the part of the nerve involved in taste to become affected, the nerve must be damaged before the branch. This branch occurs just before the facial nerve passes near the eardrum. When frequent ear infections were common before the advent of antibiotics, operations had to be performed for mastoiditis and the facial nerve was frequently damaged.

However, even when the nerve on one side of the face is severely affected, the other side will continue to send taste information to the brain. If the nerve that connects to the back third of the tongue is also damaged, there may be considerable loss of taste.

Taste may also be affected in a condition called Bell's palsy, in which the facial nerve becomes inactive quite suddenly owing to a variety of conditions, such as infections, tumors, herpes simplex, and meningitis. It is very rare for all taste nerves to be affected at the same time; complete taste loss is very rare.

It is much more common for people who have lost their sense of smell on both sides (for example, as the result of a head injury) to complain of loss or reduction in their sense of taste. This is because without the sense of smell, subtler refinements of taste are lost.

Unpleasant alterations in taste

It is common for people suffering from depression to complain of unpleasant tastes in the mouth. The cause of this is not clear, but it may be related to the close relationship of taste and smell. Smell-analyzing centers of the brain have close connections with the emotional circuitry of the limbic system, and it has been suggested that certain moods can conjure up tastes and smells. Another type of unpleasant taste occurs in some people as an aura or warning sensation before an epileptic fit. This usually means that the abnormal electrical activity causing the fit is centered either low in the parietal lobe or in the neighboring temporal lobe.

▲ As well as enjoying the delicious taste, part of the pleasure of eating sweet corn is experiencing the crunchy texture and smell of the cooked corn. Eating the corn alfresco is also part of the fun.

▲ One of the cook's tasks is to ensure that the food tastes good and is appropriately seasoned. Tasting a dish during its cooking time can test these criteria and also test for doneness.

See also: **Brain; Nervous system; Palate; Saliva; Smell; Throat; Tongue**

Teeth

Everyone has two sets of teeth during his or her lifetime; the second set replaces the first throughout childhood. The function of teeth is to bite off pieces of food and grind them so they can be swallowed easily before digestion.

Questions and Answers

Why do some people have crooked teeth?

The development of the teeth and jaws is mainly controlled by inherited genetic factors. Each individual, however, has a unique assortment of genes and it is possible, for example, for a child to inherit large teeth from one parent and small jaws from the other, leading to overcrowding. However, teeth in irregular positions can often be aligned and made to bite together by using orthodontic appliances.

Does an impacted wisdom tooth always have to be removed?

An impacted wisdom tooth is unable to grow properly because its path is blocked, usually by the tooth in front. Some impacted wisdom teeth are highly prone to infection, especially those that are only partially through the gum. Such teeth are best removed. However, very deeply placed wisdom teeth may be best left alone if their removal requires the loss of an excessive amount of bone. Some impacted wisdom teeth, when developing, may overcrowd the incisors and should be removed. Your orthodontist will decide on the best treatment.

My niece was born with a tooth that had to be removed. Will she have a missing baby tooth when her teeth come through?

About one baby in 5,000 is born with one or two teeth already present in the mouth—Julius Caesar and Napoleon, for example, are both reputed to have had this distinction. It sometimes runs in families. These teeth are not fully formed and will fall out; occasionally, however, they present a danger of choking the baby, and so it is better to have them removed. The teeth will develop normally later on.

The teeth are hard, bonelike structures implanted in the sockets of the jaws. Two successive sets occur in a lifetime. Each tooth consists of two parts: the crown, which is visible in the mouth; and the root, which is embedded in the jawbone. The roots of the teeth are usually longer than the crowns. Front teeth have one root; those placed further back generally have two or three roots.

The major structural element of a tooth is composed of a calcified tissue known as dentine. Dentine is a hard, bonelike material that contains living cells. It is a sensitive tissue and gives the sensation of pain when stimulated either thermally or by chemical means. The dentine of the crown is covered by a protective layer of enamel, an extremely hard, cell-free, insensitive tissue. The root is covered with a layer of cementum, a substance similar to dentine, which helps anchor the tooth in its socket. The center of the tooth is a hollow chamber filled with a sensitive connective tissue known as dental pulp. This extends from within the crown right down to the end of the root, which is open at its deepest part. Through this opening, minute blood vessels and nerves run into the pulp chamber.

Each tooth is attached by its root to the jawbone; the part of the jaw that supports the teeth is known as the alveolar process. The mode of attachment is complex; teeth are attached to the jaw by the periodontal ligament. This consists of a series of tough collagen fibers that run from the cementum covering the root to the adjacent alveolar bone. These fibers are interspersed with connective tissue, which also contains blood vessels and nerve fibers. This method of attaching the teeth allows a very small degree of natural mobility. This serves as a buffer that may protect the teeth and bone from damage when biting. At the neck of the tooth where the crown and root merge, a cuff of gum bonded tight to the tooth protects underlying supporting tissues from infection and other harmful influences.

Types of teeth

There are two series of human teeth. Deciduous teeth are those present during childhood and are all usually shed. Deciduous teeth can be divided into three categories: incisors, canines, and molars. The permanent teeth are those that replace and extend the initial series. They are the same types as the deciduous teeth, but in addition there is a further category known as the premolars, which are intermediate, in form and position, between canines and molars. Incisors have a narrow, bladelike incised edge. The incisors in opposite jaws work by shearing past each other like scissor blades. Canines

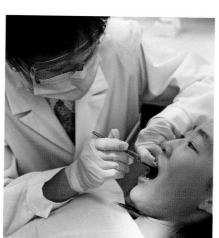

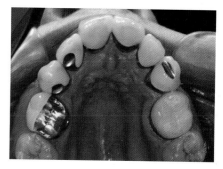

▲ *Dental decay can occur anywhere in a tooth and fillings will be the eventual result. To avoid this, it is important to take care of the teeth and visit a dentist regularly.*

Questions and Answers

I broke my front teeth in a car accident. Can they be repaired?

If only a small piece of enamel is broken off, then the sharp edges can be smoothed. If much of the tooth is missing, the tooth can be repaired by using a filling material bonded to the rest of the tooth, or a crown can be fitted. If the root of the tooth has been fractured, usually the tooth is extracted and replaced.

When milk teeth decay, why are they filled if they are going to fall out anyway?

It is usually preferable to fill them rather than extract them, since the early loss of milk teeth may cause the permanent teeth to drift into incorrect positions. Removal of teeth at the first signs of dental disease may make a child think that tooth loss is inevitable, and discourage the child from taking care of the teeth.

My son had a convulsion and then a high temperature. I thought it was due to teething, but my doctor sent him to the hospital. Was this necessary?

Teething does not cause high temperatures or any serious illnesses, although a child may be teething at the same time as he or she develops an illness. Your doctor was aware of this and wanted to ensure that there was no other cause for your son's fever and convulsion.

Should I give my baby fluoride tablets, which I've heard prevent tooth decay?

There is substantial evidence that the addition of fluoride to toothpaste and to water that contains a low level of natural fluoride helps prevent tooth decay, particularly in children. It may be a good idea to give your baby tablets if you live in an area where there is no extra fluoride in the water supply, but talk to your dentist and doctor before you go ahead with it.

DECIDUOUS (BABY) AND PERMANENT TEETH

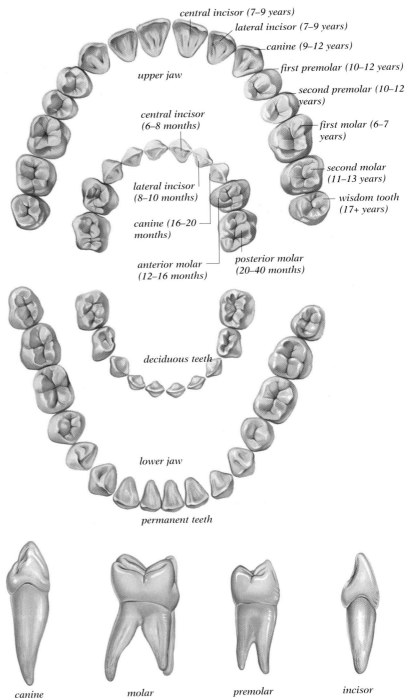

central incisor (7–9 years)
lateral incisor (7–9 years)
canine (9–12 years)
first premolar (10–12 years)
second premolar (10–12 years)
first molar (6–7 years)
second molar (11–13 years)
wisdom tooth (17+ years)

upper jaw

central incisor (6–8 months)
lateral incisor (8–10 months)
canine (16–20 months)
anterior molar (12–16 months)
posterior molar (20–40 months)

deciduous teeth

lower jaw

permanent teeth

canine   molar   premolar   incisor

In theory we all have 32 permanent teeth. The arrangement of these is exactly the same in the upper and lower jaws. In each jaw there are 4 incisors, 2 canines, 4 premolars, and 6 molars—16 in total. Babies and young children have only 20 deciduous teeth. Again, in each jaw there are 4 incisors, 2 canines, and 4 molars—10 in all. Incisors cut food; canines tear it; and molars and premolars grind it. As human beings have evolved, teeth have changed; canines have become far less pointed, and many people never develop any wisdom teeth.

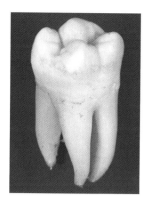

▲ *Each tooth consists of several layers: a stout shell of dentine (colored blue and yellow); cement covering the root (blue); and the crown's protective layer of enamel (yellow). The brown is plaque.*

tissue that the epithelium covers. These buds then become bell-shaped and gradually grow to map out the shape of the eventual junction between the enamel and dentine. Some cells form the dentine; others give rise to the enamel itself.

The edges of the bell continue to grow deeper and eventually map out the entire roots of the teeth, although this process is not complete until after about one year. At birth the only sign of the occlusion is provided by gum pads, which are thickened bands of gum tissue. Around the age of six months, the first of the lower incisors pushes through the gum, a process known as dental eruption. The age at which this occurs is variable: a few babies have teeth at birth; in others they may not emerge until age one.

and pointed teeth are used for a tearing action; molars and premolars are effective at grinding food rather than cutting it.

Teeth form an even, oval-shaped arch, with the incisors at the front and the canines, premolars, and molars progressively placed farther back. The dental arches normally fit together in such a way that, on biting, the teeth opposite interlock with each other.

Development of teeth

The first sign of the development of the teeth occurs when the fetus is only six weeks old. At this stage the epithelial (lining) cells of the primitive mouth increase in number and form a thick band that has the shape of the dental arch. At a series of points corresponding to individual teeth, this band produces budlike ingrowths into the

After the lower incisors have emerged, the upper incisors begin to erupt, and these are followed by the canines and molars, although the sequence may vary. Teething problems may be associated with any of the deciduous teeth. By age two and a half to three, the child will usually have a complete set of 20 milk teeth. Ideally, they should be spaced in a way that provides room for the larger permanent teeth. After age six, lower and then upper deciduous incisors become loose and are replaced by permanent teeth. The permanent molars develop behind the deciduous molars. The first permanent molars come through at age six, the second molars at age 12, and the third molars, or wisdom teeth, around age 18. There is, however, variation in the timing of the emergence of all the teeth. About 25 percent of people never develop one or more wisdom teeth. The reason may be

▼ *A dentist shows a woman an impression of her upper jaw; an impression is taken in preparation for making a crown.*

▼ *If teeth are crooked, spaced, or protruding, orthodontic appliances can be fitted.*

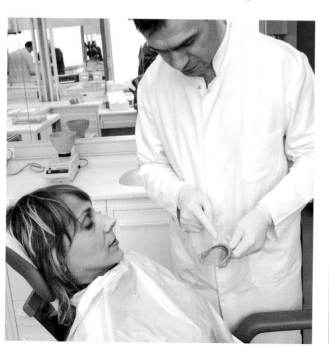

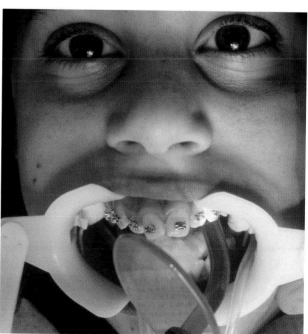

CROSS SECTION OF A MOLAR

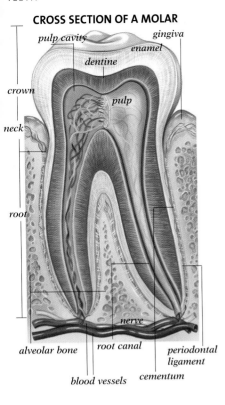

pulp cavity

gingiva

enamel

dentine

crown

pulp

neck

root

nerve

alveolar bone   root canal   periodontal
ligament

blood vessels   cementum

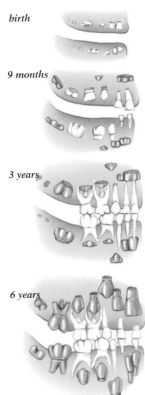

birth

9 months

3 years

6 years

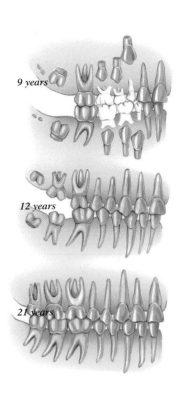

9 years

12 years

21 years

▲ *A cross section of a molar shows the soft pulp where nerves and blood vessels are located. Dentin is the sensitive tissue around the pulp. Enamel covers the tooth above the gum; below the gum, cementum is the outer layer of the tooth.*

▲ *The deciduous teeth erupt at about six months; these are the central incisors. The first permanent teeth are molars, at age six. By the early twenties, everyone normally has a full set of teeth.*

that the jaw is smaller, so the number of teeth has decreased. Some wisdom teeth never erupt through the gum and in 50 percent of people they become impacted and have to be removed.

Changes in teeth arrangement

The part of the jaw that supports the milk teeth increases very little in size from the age when all the milk teeth have erupted. It is only when the large permanent incisors have erupted that the final form of the dental arches become apparent. The upper permanent incisors often appear out of proportion to the child's face, but this naturally becomes less apparent as the face grows while the teeth remain the same size. Any tendency for the upper incisor teeth to protrude becomes obvious only when the milk teeth are replaced. Similarly, crowding often becomes clear only when the permanent teeth have erupted. During the six years that it takes for the milk teeth to be entirely replaced by permanent teeth, a gap may appear between the upper incisors. This gap usually closes when permanent canines push the incisors together, but the alignment or bite may need orthodontic treatment to bring the teeth into line.

Tooth eruption and teething

Many babies suffer discomfort as each tooth breaks through. Sometimes, a baby whose new tooth is hurting develops a red, inflamed patch on the cheek, and the gum may become red. There is excessive drooling, rubbing of the mouth, and crying; the cheeks

may appear pinker than usual. Symptoms such as fever or diarrhea are unlikely to be due to teething, and if they persist, medical advice should be sought. When teething causes the baby distress, a number of measures can be adopted, such as giving the infant a teething ring or a hard pacifier to suck or bite on. Candy must never be given; this will damage developing teeth. In some cases anesthetic cream can be applied on the child's sore gum, but it wears off quickly because a baby salivates so much.

Some babies may have problems. Sometimes a bluish swelling, called an eruption cyst, appears over a molar tooth before it erupts. This disappears once the tooth emerges and does not require treatment. Anodontia, a total absence of teeth, is very rare, although the absence of individual teeth is a phenomenon that commonly occurs in about 5 percent of people. Tooth development can be delayed in a number of diseases affecting growth; for example, rickets. Children with Down syndrome may also have delayed teeth, but this is rare. Some babies are as old as 18 months when the teeth erupt.

Tooth color varies from brown to deep cream. If a pregnant woman or a baby gets tetracycline antibiotics, the drug may cause a brown discoloration. Sometimes the incisors are twinned (joined together), but the second incisors usually develop normally.

See also: **Blood; Fetus; Gums; Mouth; Nervous system**

Temperature

The body's sophisticated temperature control mechanism is one of the major features that set humans aside from lower forms of life. The way that heat is produced and lost depends on a series of complex bodily functions.

Those who live in temperate climates often take very little notice of the temperature of their surroundings, yet the temperature of the environment is one of the most important factors in determining how people live.

The temperature of the outside world is very important in the way it affects our so-called internal environment. In a very advanced animal like a human, there are several sophisticated mechanisms that are designed to keep the internal environment balanced, despite any change in

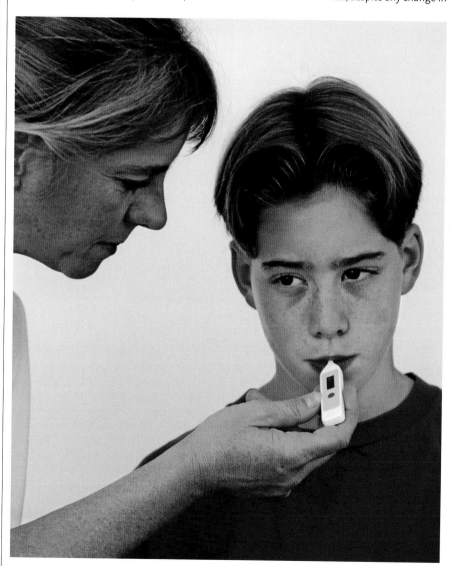

▲ *Temperature is measured by placing a thermometer in the mouth. In this case, a digital thermometer is used with a display that is easy to read.*

413

the external conditions. The maintenance of a constant temperature, constant levels of salt and water in the tissues, constant supplies of oxygen, and a constant balance of the amount of acid and alkali in the body is controlled by a process called homeostasis. Temperature control is perhaps the most important element of the homeostatic mechanism.

If, for instance, the mechanism that regulates temperature were to break down or to start to work inefficiently, then a number of potentially fatal disorders could result.

Temperature control

All forms of life depend on chemical reactions and the enzymes that regulate their reaction rate. It is the temperature at which these reactions take place that determines their speed and whether or not they happen quickly enough to be of use to the body. An internal temperature that is fixed within narrow limits ensures that chemical systems in the body work efficiently.

It is only the higher forms of life such as warm-blooded mammals or birds that can keep their internal temperatures constant within a very small range. This allows them an existence that is much less controlled by external circumstances than the lives of cold-blooded reptiles, whose less efficient chemical systems have to work over a much wider range of temperature.

How temperature control works

To enable the body to maintain a constant internal temperature during heat loss, the balance will be regained by the production of an equivalent amount of heat. If, on the other hand, a person is basking in hot sunshine, then heat is being absorbed by the body rather than being lost from it. In these circumstances, the body has to have an efficient way of losing heat.

The system of temperature control is one that determines that heat lost from the body must be made up, and heat gained must be reduced in order to maintain a stable temperature. The overall control of this temperature system is managed by the hypothalamus, the area at the base of the brain that governs so many of the body's vital functions.

Heat production

When the body temperature starts to fall, there are a number of processes that the body can set in motion to produce more heat. However, certain mechanisms in the body already constantly produce heat, since every one of the body's many chemical reactions liberates some heat as it takes place. This is called the heat of metabolism, and the general rate at which these reactions take place is called the metabolic rate. Although it cannot be controlled

▲ *This basketball player is sweating profusely to stay cool after a strenuous game.*

entirely by one organ, the metabolic rate is very closely related to the activity of the thyroid gland. An overactive thyroid will lead to a high metabolic rate, whereas an underactive thyroid causes a low metabolic rate. This close relationship between thyroid function and heat production explains why people who have an overactive thyroid are very intolerant of hot weather, and those who have an underactive thyroid feel the cold and so may be more prone to suffer from hypothermia.

If the heat of metabolism is insufficient to meet the demands of the homeostatic control system, then the hypothalamus will increase the activity of the muscles by working particularly on the main postural muscles up and down the spine.

What happens first is an increase in the tone of the postural spinal muscles, so more energy is used by each muscle fiber, and more heat is generated. We are not aware that the hypothalamus is making these subtle alterations in the tone of the main postural muscles, but they are the most important way of making extra heat.

▲ *Even Russia's frozen winters can't deter these swimmers from taking the plunge.*

same person is standing in the hot sun, then the opposite will happen; heat will be radiating into him or her.

If the amount of blood, and therefore the amount of heat, that flows through the skin is increased, then the amount of heat lost by both convection and radiation will also increase. This increase in the flow of blood through the skin is in fact the main way in which the hypothalamus controls heat loss from the body.

Following an increase in the flow of blood to the skin in order to lose heat, there will of course be an increase in the amount of sweating. The sweat glands work by liberating sweat onto the surface of the skin; the sweat is then evaporated with a consumption of heat in the same way that the evaporation of water from the lungs loses heat.

Most sweating takes place at a low level so that moisture is rarely noticed on the skin. However, when the system becomes overworked, and is therefore not working efficiently, sweat is clearly visible on the surface of the skin.

The efficiency of the function of sweating also depends on the humidity of the air. Humidity is the amount of water already present in the air in the form of water vapor. If someone sweats profusely when there is a lot of humidity, the sweat will not easily evaporate. This causes the body to be less efficient at losing heat than it should be.

Measuring temperature

The main reason people want to measure body temperature is to see if we have a fever. The standard way of measuring temperature is with a thermometer.

In all older children and adults the way of taking a temperature is to place the thermometer under the tongue for about 90 seconds, and then to remove it and read it. In very small children or babies, there is a possibility of the thermometer's being bitten off in the mouth, so the temperature may be taken using an aural thermometer, which is placed gently into the child's ear. Other methods are taking the temperature under the armpit or in the rectum. To take an axillary (armpit) temperature, the thermometer is put into the armpit and kept there, with the arm pressed against the side, for about two minutes. The temperature will read slightly lower than an oral temperature. In a baby, a rectal temperature is best taken with the baby face downward on your knee. The thermometer should be held gently between the fingers, with its tip in the rectum and the hand flat on the buttocks. The reading will be slightly higher than an oral reading.

Medical thermometers have an interior kink in them so that the column of mercury, which shows the temperature, will not immediately fall back to room temperature when taken from the mouth. Once the temperature has been taken, a thermometer should always be shaken so that the column falls back down. The thermometer should be rinsed in cold water or alcohol, since hot water

If the hypothalamic alterations in the amount of tone in the postural muscles fail to produce enough heat, another familiar and obvious mechanism comes into play: shivering. Shivering works by setting up alternate contraction and relaxation in many muscles in the body. This action uses up energy and so starts to heat the body.

Heat loss

Just as the main way of gaining heat is not really obvious to us, so the main mechanisms by which heat is lost function without our being aware of them.

The body has a certain amount of heat that it cannot avoid losing. Food and drink may have to be warmed by the body to its own temperature, and in cold conditions we have to warm the air we breathe. As we breathe there is also some unavoidable loss of water, in the form of vapor, from the lungs. Such vaporization or evaporation of water uses up heat the same way as heat is used up in steam from boiling water.

Apart from the heat that is lost in these various ways in the gut and the lungs, there are two main ways of heat loss through the skin: convection and radiation. Convection is a process in which heat is transferred by the circulation of currents, from a hotter region to a cooler region. In the case of the skin, it is usually cool air that moves across the surface of the skin and, as it does so, takes heat from the surface and carries it away from the body; the air becomes warmer in the process.

Radiation, on the other hand, is the way in which an electric fire produces heat, by pushing it out directly into the environment. In a cold climate a person will radiate quite a lot of heat; but when the

▼ *Cold-blooded reptiles like crocodiles have to live in a climate that remains sufficiently hot so that their essential chemical functions are not switched off by the cold.*

▲ *Warm-blooded animals like these emperor penguins are so well insulated by fat and feathers that they can live happily in the freezing conditions of Antarctica.*

causes the mercury to expand and may break the thermometer.

As well as checking for possible fever, there are other reasons for taking a temperature, such as wishing to know if and when ovulation has occurred.

In intensive care units, some patients require placement of a pulmonary artery catheter (commonly referred to as a Swan-Ganz catheter) to monitor closely their blood pressure, cardiac status, and respiratory status. This unit has a thermistor that can measure core temperature. Otherwise, core temperature is measured using an infrared tympanic membrane device.

See also: **Basal metabolic rate; Breathing; Enzymes; Homeostasis; Hypothalamus; Metabolism; Muscles; Perspiration; Skin; Thyroid**

Tendons

Although very simple and apparently minor parts of the body's structure, tendons play a very important role in transmitting muscular power and enabling movement.

Is it possible for tendons to be dislocated?

This is an unusual occurrence, but it can happen. Normally it is brought about by a sudden wrenching movement powerful enough to jolt a tendon out of the groove over a bone in which it runs. Treatment consists of manipulating the tendon back into its proper position.

My boyfriend accidentally cut through one of the tendons in his wrist. Will he be able to use his hand after the tendon has healed?

In most cases it is possible to sew the cut ends of the tendon together again. This is usually successful in restoring full movement to the affected muscle on recovery. But the tendon may not be as strong as it was and its use may give rise to discomfort.

Do severed tendons ever heal naturally?

If the tendon is nicked or only partially severed, scar tissue will probably form to fill in the gap. If the tendon is completely severed, the two ends are likely either to spring apart because they are under tension, or separate when the muscle moves. Unless they are identified and deliberately sewn firmly together, healing is unlikely.

What is likely to cause actual damage to the tendons?

Sudden sharp strains or twisting pressures on a tendon are likely to tear some of the tendon fibers away from the bone to which they were anchored. Alternatively, repetition of a movement over a long period of time may use up all the lubricating fluid in the tendon sheath and thus give rise to the friction and inflammation of tenosynovitis.

Tendons, or sinews, are vital to facilitating a wide variety of movements. A tendon is a very strong and tough band of fibrous connective tissue that joins the active section or body of a muscle to another part, usually a bone, which it is intended to move. The force of the contracting muscle fibers is concentrated in and transmitted through the tendon, achieving traction on the part concerned and thus making it move.

Tendons and muscles

Tendons are specialized extensions or prolongations of muscles. They are formed by the connective tissue, which binds the bundles of muscle fibers together, joining and extending beyond the muscles as a very tough, inelastic cord. They have very few nerve endings and, being essentially inactive tissues, little in the way of a blood supply. At one end they are formed from the belly of the muscle, and at the other they are very firmly tethered to the target bone, some of their fibers being actually embedded in the bone structure.

Location

Several tendons are located just beneath the surface of the skin, and so can be felt easily. For instance, the hamstring tendons, which control knee bending, are at the back of the knee.

▲ *Should this man miss with his mallet, he could damage a tendon. In addition, regular use of certain muscles can cause tenosynovitis.*

▲ *Tendons take fierce punishment during physical exertion.*

Because they take up much less space than muscles, tendons are also often found where there are a large number of joints to be moved in a relatively small space. Therefore both the backs and the fronts of the feet and the hands contain a whole array of different tendons. The muscles that make these tendons work are sited in the arms and legs.

An unusual tendon is found in connection with the muscle tissue that forms the wall of the heart and brings about its pumping action. Thickened, fibrous connective tissue forms tough strips within the heart muscle, which give it a firmer structure as well as forming firm supporting rings at the points where the large blood vessels join the heart.

TENDONS AND TENDON SHEATHS

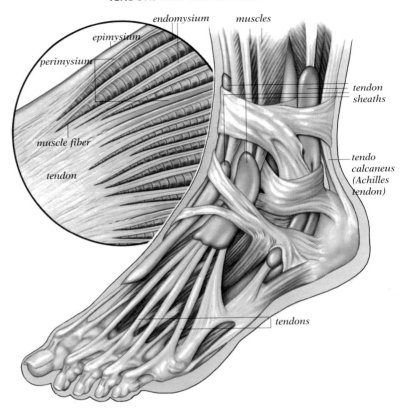

▲ *The membranes surrounding the muscles and muscle fibers (the endomysium, the perimysium, and the epimysium) join at the end of a muscle to form the tendon (inset above). Sheaths stop tendons from rubbing against other structures (above).*

Tenosynovitis

In order that they can move smoothly and without friction or the danger of abrasion, tendons are enclosed in sheaths at the point where they cross or are in close contact with other structures.

The tendon sheath is a double-walled sleeve designed to isolate, protect, and lubricate the tendon so that the possibility of damage from pressure or friction is reduced to a minimum. The space between the two layers of the tendon sheath contains fluid so that when the muscle is in action the two layers slide over each other like the parts of a well-oiled machine.

The human body, however, cannot sustain repeated movements in the same part of the body without sustaining damage in the form of inflammation, so rest periods are necessary for the lubricating fluid to be replenished. If this does not happen, and the system is run without adequate lubrication, the two layers of the tendon sheath begin to rub against each other and chafe. Continued movement will then be painful and cause a creaking sound called crepitus. This is the basis of the condition called tenosynovitis—inflammation of the tendon sheath. Any tendon sheath can be afflicted by this annoying and painful condition, which is common in keyboard operators, athletes, dancers, and others who use one particular set of muscles repeatedly. Sudden, unaccustomed use of a particular set of muscles is especially likely to cause tenosynovitis.

Injury

Virtually all the disorders of tendons are due to injury of one sort or another. A deep cut near the foot, ankle, hand, or wrist may sever one of the tendons that lie quite close to the surface.

It is usually possible to sew the two severed ends together, but there always remains the possibility of pain or some weakness of the muscle when it is used for a long time.

Extreme tension, overstretching, or sudden jerking on a tendon may damage it in a variety of ways. Some of the fibers of the tendon anchoring it to the bone become torn away from their moorings. The tendon itself is not really stretched, and only very rarely ruptured or snapped, since the force required to do this would already have pulled it away from the bone.

Injuries to tendons are usually treated with ice packs and rest, with the support of an elasticized bandage. Rather than exercise regularly, a gradual return to normal use is the best way to regain full use of the limb involved in the injury.

See also: Body structure; Bones; Feet; Hand; Muscles; Skeleton

Testes

Questions and Answers

If one testis has to be removed, will it affect a man's sex life?

No. The remaining testis can produce enough of the male hormone, and enough sperm, to ensure normal potency and fertility. If one testis is removed, it can be replaced with a prosthetic plastic one so that even the cosmetic defect is not apparent.

I have just noticed that there is a swelling in my scrotum. It is about the same size as a testis. What is this, and is it dangerous?

The most likely thing is that it is an epididymal cyst. This is a fluid-filled sac that develops for no particular reason and is not harmful. However, you should let your doctor examine you in case it is something more serious and requires treatment.

My 12-year-old son has been having pain in his testis. The pain is severe at first, but does not last very long. What is the cause?

There is a possibility that he is having recurrent attacks of torsion, or twisting, of the testis. Some people are born with a slight abnormality that allows this to happen. It may be possible for a specialist to confirm that the attacks are due to torsion, and a small operation can be done to prevent it from happening again.

Does vasectomy have any effect on the function of the testes?

There is still no evidence to show that a vasectomy has any effect on the natural hormone function of the testes.
  As far as sperm production is concerned, there may be changes that cannot be reversed in those cases when the vas deferens is joined up again following a vasectomy.

The testes are the twin organs that control male fertility and the production of the sex hormone testosterone. These functions are so important that prompt treatment of any problems is essential.

The normal human male has two testes that develop in the embryo from a ridge of tissue at the back of the abdomen. When the testes have formed, they gradually move down inside the abdomen so that, at the time of birth, each testis has arrived in its final position, usually within the scrotum.

Function and structure

The function of the testes is twofold. First, they provide the site where sperm is manufactured; each sperm contains all the genetic information for that particular male. Second, the testes contain cells that produce the male sex hormone testosterone, and consequently, masculine

STRUCTURE OF THE TESTES

The testes consist of seminiferous tubules, where sperm are made; and interstitial cells, which produce the male hormone testosterone. Sperm is stored in the epididymis before passing along the vas deferens to be ejaculated.

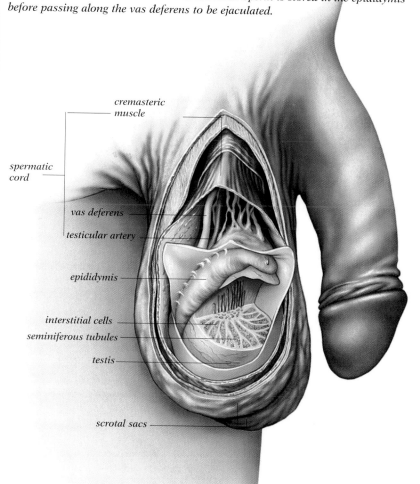

- cremasteric muscle
- spermatic cord
- vas deferens
- testicular artery
- epididymis
- interstitial cells
- seminiferous tubules
- testis
- scrotal sacs

Questions and Answers

Does damage to the testes cause changes in secondary sex characteristics, such as facial hair or voice?

If insufficient male hormone is produced by the testes at puberty, the secondary sex characteristics, such as facial hair and a deep voice, fail to develop. This results in what is known as eunuchism. Once the voice has deepened, then removal of, or damage to, the testes will have no effect, but there will be a loss of sexual drive if the testes are removed in adulthood. Male hormones can be administered by injection.

My two-year-old appears to have only one testis in his scrotum. Will the other one appear at puberty?

It depends. You should see your doctor. If the testis is retractile, that is, it has been pulled up into the groin by an overactive muscle, then it will come down at puberty. If, however, it never reached the scrotum, then he will need a small operation in a year or two to bring it down.

One of my husband's testes has become larger than the other. There is no pain. Should he have it checked?

Yes. It is very important that he is seen by a doctor right away. The enlargement could be caused by a collection of fluid around the testis, or by a small tumor inside the testis. It should be apparent to a doctor which of these it is. If it is a tumor, there is a good chance that it can be cured.

Is it possible to get an infection in the testis that makes it tender and swollen?

Yes, but this is quite unusual in young men, and tenderness and swelling would be more likely to be caused by a twisting of the testis on the spermatic cord, cutting off its blood supply. These symptoms may also arise as a result of an STD, usually with chlamydia.

CONNECTION OF TESTIS WITH PENIS

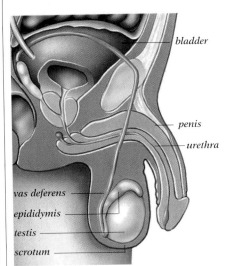

bladder

penis

urethra

vas deferens

epididymis

testis

scrotum

HYDROCELE

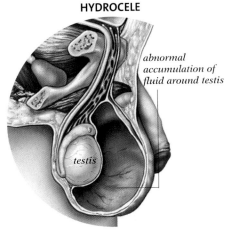

abnormal accumulation of fluid around testis

testis

▲ *An abnormal accumulation of fluid around the testes is a disorder called hydrocele.*

characteristics, such as a deep voice, male hair distribution, and typical distribution of fat. These two functions are carried out by separate sets of cells within each of the testes; one function can fail without the other one necessarily doing so.

The testes are ovoid structures. Attached to the posterior of each one is a smaller structure shaped like a long comma. This is called the epididymis. The epididymis consists of a series of microscopically tiny tubes that collect sperm from the testis. These tubes connect together to form one tube, called the vas deferens, which transfers the sperm toward the base of the bladder. All these structures, with the exception of the vas deferens, are microscopic in size.

Each testis is suspended in the scrotum by a spermatic cord, which consists of the vas deferens, the testicular artery, and the testicular vein. The spermatic cord is surrounded by a tube of muscle called the cremasteric muscle. The spermatic cord, therefore, serves two purposes: first, to provide a blood supply to the testis; and second, to conduct sperm away from the testis.

What can go wrong

For the vast majority of males, the testes carry out their complex functions without any problems. Sometimes, however, structural or functional difficulties do occur which can usually be successfully treated.

Undescended testes: In order for sperm to be manufactured, the testes have to be at a slightly lower temperature than the normal body temperature. For this reason the testes are suspended outside the abdominal cavity, in the scrotum.

At birth, however, it may be noticed that one testis has not descended into the scrotum. An undescended testis can have several effects: it will be unable to produce sperm; it may be damaged in the groin and be more prone to twisting, or torsion; and there is a slightly increased risk of developing a tumor. It is important to distinguish between an undescended testis and one that is retractile. The latter term means that the cremasteric muscle is overactive and has pulled the testis up into the groin. A retractile testis will probably descend into the scrotum at puberty; an undescended testis will not descend, and will require surgery. If the testis has not descended by the age of one, treatment is required. Sometimes a nasal spray is prescribed that stimulates the production of sex hormones which help to bring the testis into the scrotum, or a hormone may be injected into a muscle. However, this intervention is successful in very few cases.

If these measures fail, surgery is usually performed at the age of two or three. The operation for an undescended testis is called an orchiopexy. It consists of a small incision in the groin to find the testis, followed by stitching it into place in the scrotum. There is some evidence that if the operation is delayed beyond this age, irreversible damage to the testis may occur. However, even if a testis does not produce sperm because it has not descended, it still may be able to produce normal amounts of testosterone. One testis that is functioning normally should be perfectly adequate to produce sufficient sperm and testosterone.

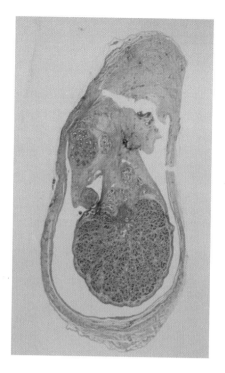

▲ *This cross section of the testes clearly shows the scrotum, the epididymis, the vas deferens, and the seminiferous tubules.*

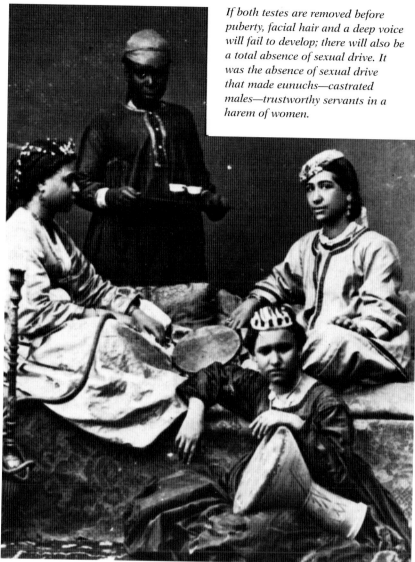

If both testes are removed before puberty, facial hair and a deep voice will fail to develop; there will also be a total absence of sexual drive. It was the absence of sexual drive that made eunuchs—castrated males—trustworthy servants in a harem of women.

Torsion: Some men have an abnormality in the way in which the testis hangs in the scrotum. This abnormality is present at birth, and usually affects both testes. The abnormality causes one of the testes to twist on its spermatic cord, leading to a sudden cutting off of the blood supply to the testis. The most common age for this to occur is the teens, but it can also happen in younger children or young adults.

The usual symptom of a torsion of the testis is extreme tenderness and the sudden onset of severe pain on one side of the scrotum. Untreated, the twisted testis swells and becomes inflamed, and eventually becomes gangrenous. Irreversible gangrene can occur within hours of the onset of the pain. Because of the swelling and inflammation, there is a danger that a case of torsion could be mistaken for infection. Infection in the testis in a teenager is rare, and all cases of inflammation of the testis in this age group should be treated as torsion until proved otherwise.

Treatment involves surgical exploration, untwisting of the testis, and fixing it with stitches so that it cannot twist again. If it is gangrenous, it will have to be removed. Because the abnormality is often present on both sides, a small operation is usually done on the other side to prevent torsion from occurring.

Infection: Sometimes, infection can pass from the bladder along the vas deferens, and then gain access to the epididymis. This infection, called epididymitis, is usually caused by bacterial infection. In men who are younger than 35 years of age, sexually acquired chlamydial or gonococcal infection is usually the cause. In older men, epididymitis is usually due to coliform bacteria (often in the setting of a urologic abnormality). The symptoms are pain and swelling in one side of the scrotum.

Inflammation: Inflammation of the testis itself is known as orchitis. Infection with the mumps virus can cause orchitis and, if both testes are affected, can lead to sterility. Fortunately, however, this type of complication is rare.

Cysts: There are several different types of cysts that can occur in the testes. They are usually associated with some congenital abnormality, but may not become apparent until the patient is older. The first type is an epididymal cyst and shows itself as a lump attached to the testis; the testis can still be felt as a separate entity. Shining a strong light from the scrotal surface through the lump confirms that it is not solid, but is full of clear fluid.

The second type of cystic swelling is known as a hydrocele. This may be larger and is different from an epididymal cyst in that the fluid-filled cavity surrounds the testis so that the testis cannot be

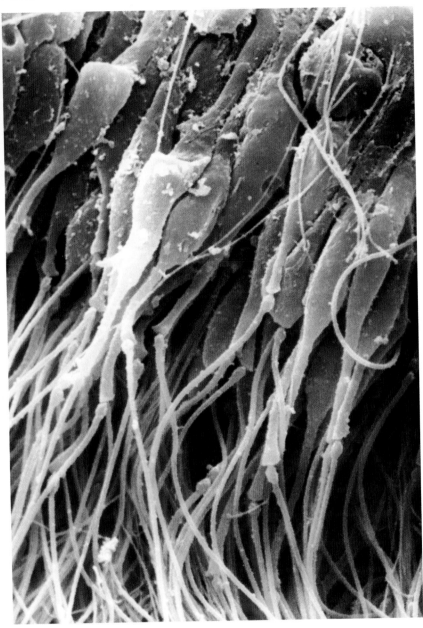

▲ *This image, taken using a scanning electron microscope and artificially colored, shows almost mature sperm embedded in the wall of a seminiferous tubule.*

significant because they can cause a rise in the temperature of the tissues in the scrotum. This can lead to impairment of sperm production and infertility.

Treatment involves surgical removal of the large veins, and this sometimes leads to a rise in the sperm count.

Tumors of the testis

Tumors of the testis are actually quite rare in comparison with the incidence of some other tumors. Many of them can be cured completely, even though they are, technically speaking, a form of cancer. Testicular tumors generally occur in young men, between the ages of 18 and 40 years, although they can occur in other age groups.

The most common way in which they become apparent is that the patient notices that one testis is bigger than the other or irregular. There may not be any pain, but occasionally there is some tenderness.

Treatment involves removing the testis, followed by further investigations to see whether the tumor has spread. Even if it has spread, the results of treatment with X rays and special drugs are good.

Infertility

Infertility can be caused by many factors, but the fault may lie in the failure of the man to produce adequate sperm in the semen. There may be many reasons for this. First, there may be a blockage in the vas deferens so that the normal number of sperm produced in the testis cannot pass into the urethra during intercourse. Second, the testes may be failing to produce sperm, a condition that is possibly associated with a lack of testosterone. The failure of sperm production may be inherent in the testes, or it may be caused by other factors, such as general illness, cigarette smoking, increased heat in the testis caused by wearing underpants that are too tight and push the testes up into the groin, or the presence of a varicocele, which raises the temperature of the testes.

felt separately. The former type of cyst is entirely harmless. The second type is usually harmless but occasionally may be associated with underlying inflammation in the testis.

The treatment of these cysts is to remove them surgically, but only if they are causing symptoms.

Varicocele: In this condition, the veins in the spermatic cord become enlarged and twisted, like varicose veins. The patient usually notices an aching sensation in the scrotum and may feel a lump just above the testis when he stands up. Because the lump is in fact the veins full of blood, it goes away when the patient lies down. Apart from causing a dragging, aching sensation, varicoceles are

Various measures can be taken to increase the sperm count. These measures include wearing loose pants and maintaining good health. Hormone tablets may also be prescribed. If there is a complete absence of sperm, and a blockage is suspected, surgery may be undertaken to bypass the blockage. This surgery will be performed only after a biopsy of the testes has shown normal production of sperm.

See also: **Genitals; Penis; Sperm; Testosterone; Urethra**

Testosterone

Testosterone is the most important of the male sex hormones, and it is responsible for producing typical male characteristics: a deep voice, beard growth, and increase in muscle bulk.

Questions and Answers

My son is 16 and shows no sign of puberty. Should he see a doctor?

Yes. In boys puberty normally starts around the age of 13, when they begin to grow more quickly. Their genitals enlarge; they develop facial, pubic, and general body hair; and their voices begin to deepen. If your son does not yet show any of these changes it could mean that he has a deficiency in the hormone that stimulates the production of testosterone. The doctor will arrange for his hormone level to be measured. Delayed puberty can be successfully treated with testosterone, and this treatment is often advised to prevent psychological problems.

My sister has an ovarian cyst. Her doctor told her that the cyst is producing testosterone, and this is why she has facial hair and her voice is getting deeper. Isn't testosterone a male hormone?

Testosterone is the most important male sex hormone, but it is also produced in small amounts in the ovaries. In women with healthy ovaries it is converted into estradiol. If for some reason the ovaries produce so much testosterone that it cannot all be converted it begins to have a masculinizing effect. Once the cyst is removed the male characteristics should disappear.

My baby has undescended testes. Does this mean he won't produce testosterone when he is older?

No. Your son's testes will probably descend during the next two or three years, and he will develop normally and reach puberty at the usual age. However, if the testes have not descended by the time he is about three, he can have an operation to lower them into a normal position. He should then develop normally.

The glands of the body produce many different hormones—substances that act as chemical messengers. They are made and released in one part of the body and then travel (mainly in the blood) to other body cells where they bring about their effects.

One important group of hormones is the sex hormones. These are different in men and women, and they are responsible for controlling fertility and reproduction, and for producing distinctive physical characteristics, such as deep voice, beard growth, and increased muscle bulk in men, and breast growth in women.

The male sex hormones are called androgens, and the most important of the androgens is testosterone. Like all the sex hormones it is a steroid hormone; it is known as an anabolic steroid.

How testosterone is produced

Testosterone is produced in specialized cells (called Leydig cells) in the testes. It is also produced in very small amounts in the ovaries of women.

Testosterone production begins when a boy reaches puberty. A hormone called luteinizing hormone (LH) is secreted by the pituitary. LH is then carried in the blood to the testes, where it stimulates the Leydig cells to secrete testosterone. During puberty, LH, and therefore testosterone, is secreted only at night; but later, as LH levels rise during the day, testosterone is secreted 24 hours a day.

In women, testosterone secreted by the ovaries is converted into a hormone called estradiol, which plays an important role in ovulation and in the menstrual cycle.

What testosterone does

The most noticeable effects of testosterone occur at puberty, when it is responsible for many of the physical changes of maturation that take place. During puberty, there are changes in both the external and the internal genitalia. The penis increases in length and width, and the scrotum becomes pigmented. Internally, the seminal vesicles, which lie just behind the bladder, and the prostate gland, which is tiny at birth, start to enlarge. The prostate increases to about the size of a walnut. They both begin to secrete seminal fluid.

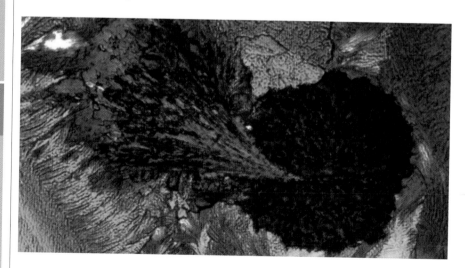

▲ *A colored and magnified (x260) photomicrograph of the male hormone testosterone.*

423

Testosterone causes the larynx to enlarge and the vocal cords to increase in length and thickness. These changes cause a boy's voice to break and become deeper. At around the same time a beard begins to appear, with a growth of chest, underarm, and pubic hair. The amount of body hair also increases over the whole body. All androgens have what is called an anabolic effect, that is, they raise the rate of protein synthesis and lower the rate at which protein is broken down. The effect is to increase muscle bulk, especially in the chest and shoulders, and to accelerate growth of the long bones (in the arms and legs), especially during early puberty. When puberty is finished, testosterone stops the growth of the long bones by ossifying the cartilage plates (epiphyses) at the end of the bones.

The skin changes that often occur in adolescence are also caused by androgens. Androgens stimulate the sebaceous glands in the skin to secrete sebum. If the glands secrete too much sebum, this can lead to acne.

Around this time the sex drive begins to increase, and boys start to show an interest in girls, and also to produce sperm. Ejaculation sometimes occurs during the night. Testosterone also promotes a more

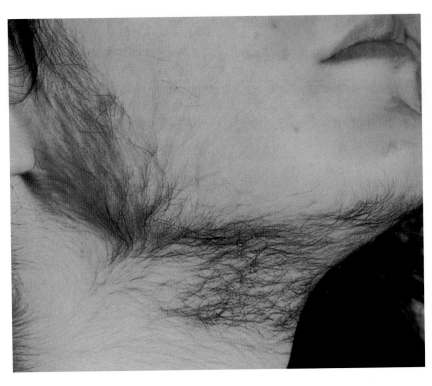

▲ *This young woman has developed excessive facial hair or hirsutism. It is thought to be associated with a hormonal disturbance producing excess testosterone.*

aggressive attitude, which is a characteristically male trait. The testes will produce testosterone throughout a man's life. After puberty, testosterone maintains the male sex characteristics, and with a hormone called follicle-stimulating hormone (FSH) it is involved in the production of sperm in the seminiferous tubules of the testes. Hereditary baldness in men is also linked to testosterone.

What can go wrong

Excess testosterone, either secreted by the body or taken as a drug, causes problems in both males and females. In women, overproduction of testosterone causes the development of masculine features such as an increase in facial hair and body hair, deepening of the voice, and some hair loss. Facial spots may develop and testosterone can also lead to an increase in body weight. It is very rare for a woman to produce an excess of testosterone, but occasionally it may be the result of an androgen-secreting ovarian tumor, or an ovarian cyst.

In adult males the presence of excess testosterone accentuates male physical characteristics. It can also cause a condition called priapism, which is a painful persistent erection. It may be the result of a testicular tumor. Excess testosterone given to stimulate puberty may interfere with normal growth or cause over-rapid sexual development. Initially the testosterone increases bone growth but adult height is reduced because the testosterone causes the long bones to stop growing too soon.

A deficiency of testosterone is far more common than an excess of the hormone. A deficiency may occur if the testes are diseased, or if the pituitary gland does not secrete LH. The effects vary, depending on whether the deficiency develops before or after

puberty. If it develops before puberty the boy's limbs continue to grow so that his final adult height will be increased, and he will have particularly long arms and legs. Other typical effects include decreased body hair and beard growth, smooth skin, a high-pitched voice, reduced sexual drive and performance, underdevelopment of the genitalia, and poor muscle development. Teenage boys with delayed puberty are usually treated with testosterone. This boosts puberty artificially, and so avoids psychological problems. Testosterone may also be given to men who have become infertile because of a deficiency of pituitary hormones.

Treatment

Testosterone can be given as a pellet implanted (under local anesthesia) under the skin of the abdominal wall, as an injection into the muscle, or as a gel applied to the skin. It can also be taken orally. All forms of treatment are effective and safe, although the resultant level of hormone in the patient's system may vary according to how the drug was delivered. Occasionally, if testosterone is given to an adult who has never been through puberty, testosterone can trigger aggression.

In the condition female transsexualism, the individuals have the appearance of femininity but believe that they belong to the male sex. After discussions with a doctor and psychiatrist, it may be decided that the transsexual should be given regular doses of testosterone to deepen the voice, promote hair growth on the face and chest, decrease feminine fat, and give an appearance of maleness.

See also: **Baldness; Hormones; Ovaries; Penis; Puberty; Testes**

Thirst

When it is very hot and I am thirsty, I get much more relief from having a cold drink than a tepid one. Is there a physical reason for this, or is it just psychological?

The amount of fluid your body receives is the same whatever the temperature of the drink you take. However, the relief of thirst depends to a large extent on the stimulation of your mouth and throat as you drink. Cold fluids tend to stimulate the linings here more than tepid fluids. Similarly, hot tea is thirst-quenching even on hot days, again because hotter drinks stimulate the throat more than tepid ones.

My husband has very bad bronchitis and has to breathe through his mouth most of the time. He says this makes him very thirsty. Why does this happen?

One of the things that make us thirsty is that the lining of the throat and mouth gets dry. People with bronchitis or asthmatic attacks who have to breathe through the mouth get very dry in the mouth through evaporation of the water on the lining. This stimulates the thirst centers in the brain, despite the fact that the actual water content of the blood is adequate.

Why do people who have diabetes get so thirsty before the disease is diagnosed and treatment begins?

In diabetics, the main change that occurs in the blood is that there is too much sugar. Excess glucose is dumped into the urine, and its presence there prevents the tubules of the kidney from controlling the amount of water that is lost. The most noticeable early indication of diabetes is excessive urine secretion (polyuria), causing the blood to become short of water and resulting in great thirst and an abnormally high fluid intake.

Taking a drink to quench the thirst is not only immensely pleasurable, it is also absolutely vital for people in order for them to maintain adequate amounts of water in the body fluids.

The delicate chemical processes that keep people alive demand that the amount of water in the body is kept very constant. The sense of thirst is one of the most vital human appetites because it ensures that the chemical equilibrium is maintained. Sensitive centers within the brain monitor the amount of water in the blood. These respond quickly to any significant change, producing the sensation of thirst that drives people to seek replenishment when they become short of water.

Remarkably, sensors in the throat seem to be able to assess accurately when a person has drunk enough, even before the centers in the brain signal that the amount of water in the body is adequate. As a symptom, excessive thirst is an important indicator that fluids have been lost. It can point to abnormalities in the kidneys, to the presence of excess sugar in the blood (which causes excess water loss), and to damage to those brain structures that participate in keeping the water balance in order.

The mechanics of thirst

The main control center for the sense of thirst is deep in the brain just below the thalamus and is known as the hypothalamus. Small groups of nerve cells in this gland are sensitive to the amount of water in the blood. If the amount of water in the blood compared with the amount of salts and other substances diminishes, these cells are stimulated and, in addition to producing the hormones that make the kidneys conserve water, produce the sensation of thirst. These cells are also stimulated by sudden changes in the volume of blood in the circulation, such as occur after a hemorrhage.

The other change that makes people thirsty is that the lining of the mouth and throat becomes dry. This stimulates nerve endings in the lining, which are also connected indirectly with the thirst centers in the hypothalamus. Simply moistening the mouth has a powerful effect in reducing the signals to the brain that fluid is in short supply.

▲ *Water fountains are a convenient way to satiate thirst when other forms of water may not be easily available. While people can remain for long periods without food, regular intake of liquids is vital. Thirst is a sign that a drink is absolutely essential.*

▲ *In hot climates, like those of the southern states, strenuous exercise can cause severe dehydration, so plenty of liquids must be drunk.*

▲ *To some extent, thirst can be stimulated by psychological factors; often the mere sight of a refreshing drink makes us thirsty.*

When people have been deprived of water and are then allowed to drink, they rapidly take enough water to replenish their stock; they stop before the water has had time to be absorbed and change the blood enough to reduce the stimulus to the water-sensitive cells in the hypothalamus. This illustrates the fact that there is some sort of metering mechanism in the body that assesses accurately when a person has drunk enough liquid.

The sensors that control this metering are in the mouth and throat. They coordinate with the thirst center in the brain through relays in the brain stem. In humans, this metering mechanism is not as accurate as it is in other animals, probably because people have superimposed on it certain psychological and habitual factors that are partly in control of our drinking behavior.

Thirst and illness

People are sometimes unable to compensate for dehydration by taking a drink—for example, when they are unconscious and cannot respond to messages from the thirst center in the brain. Thus, when people are very sick they may get dehydrated, especially if they lose fluid by vomiting and diarrhea.

Very occasionally, the thirst area in the brain is damaged by a stroke or tumor, and does not respond normally to the changes in the water content of the blood. More often, psychiatric disease, such as certain psychoses, disrupts a person's behavior so much that he or she ignores the signals from the hypothalamus and becomes either dehydrated or overhydrated. Thirst can be a symptom of various types of disease, and its presence in someone who is sick is often a useful clue to the doctor.

The common factor in people who suffer from abnormal thirst is that they are dehydrated. Dehydration may be brought about because of kidney damage, which may affect the kidney's ability to respond to signals from the brain to retain water. When there is an excess of sugar in the blood, as in diabetes mellitus, the excess spills over into the urine, taking with it an excessive volume of water. This is necessary to dissolve and dispose of the abnormally high levels of sugar in the blood. The result is that the blood becomes short of water and this causes great thirst and an abnormally high fluid intake. Once the blood sugar level is controlled by treatment the abnormal thirst will disappear.

In an uncommon disease called diabetes insipidus, the hormone normally produced by the pituitary gland to retain water is inadequately produced or, more rarely, the kidneys themselves may be unable to conserve water. Internal hemorrhage also produces thirst.

See also: **Appetite; Blood; Brain; Hormones; Hypothalamus; Kidneys; Mouth; Perspiration; Pituitary gland; Throat; Tongue; Urinary tract**

Throat

Questions and Answers

What should I do if a fish bone gets stuck in my throat?

First, try swallowing small amounts of bread. This may catch on the bone and dislodge it, carrying it down into your stomach. However, if this fails you should go immediately to a hospital emergency room, where the staff will remove it.

After I had difficulty swallowing, I had an X ray; then my doctor prescribed iron tablets. How could iron pills cure my throat trouble?

It sounds as if you had a condition known as pharyngeal web, which occurs in iron deficiency anemia. As well as difficulty in swallowing, patients complain of a sore tongue, cracks at the corner of the mouth, and brittle fingernails. In such cases, an X ray of the pharynx will reveal a fine web that appears partially to obstruct the food passage. This complex symptom is known as the Patterson Brown-Kelly syndrome. It is very important to recognize it and treat it by reversing the iron deficiency; if it is not treated the difficulty in swallowing can become much more serious.

Why is laryngitis more serious in children than it is in adults?

A child's airway is much smaller than that of an adult. When it becomes inflamed the lining swells and constricts the airway, and the smaller the airway the more potential there is for serious obstruction. In children, inflammation of the larynx makes it more sensitive to any agents that pass down the throat. Such agents can produce bouts of coughing and spasms of the vocal cords that make it impossible to breathe. In severe cases it is necessary to admit the child to the hospital to treat the inflammation intensively.

The air we breathe, as well as food on its way to the digestive tract, has to pass through the throat on its way to the lungs. Any obstruction to this vital passage can represent a serious threat to life.

The lay term "throat" describes the part of the neck between the chin and the collarbones and is the area that leads into the respiratory and digestive tracts. It extends from the oral and nasal cavities to the esophagus and the trachea, and is made up of two main parts: the pharynx and the larynx.

Structures of the throat

The pharynx is a muscular tube lined with mucous membrane. For practical purposes it is divided into three areas. The part behind the nasal cavity is the nasopharynx, the area behind the mouth the oropharynx, and the area behind the larynx the laryngopharynx. Clumps of lymphoid tissue

ANATOMY OF THE THROAT

The main component of the throat is the pharynx, a muscular tube about 5 in. (13 cm) long stretching from the base of the skull into the esophagus. It is the passage through which everything we eat, drink, and breathe has to pass, the junction point of all nasal and oral passages. It is also connected to the ears by drainage channels—the eustachian tubes—which help to equalize air pressure on each side of the eardrums.

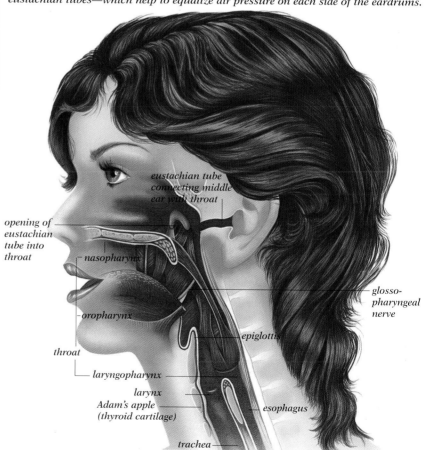

eustachian tube connecting middle ear with throat

opening of eustachian tube into throat

nasopharynx

glosso-pharyngeal nerve

oropharynx

epiglottis

throat

laryngopharynx

larynx

Adam's apple (thyroid cartilage)

esophagus

trachea

Questions and Answers

My husband has been told that he has cancer of the larynx. What are his chances of being cured?

Most forms of cancer of the larynx respond well to treatment. Cure rates for small laryngeal cancers that have not spread to the lymph nodes are 75 percent or better. However, the doctors will want to keep a close eye on your husband for the rest of his life.

What is the correct action to take when someone is choking on a bone or a peanut?

Take hold of the person from behind and use a fist to deliver an upward thrust to the abdomen. If the patient is a young child, place him or her face down on the lap and gently thump the child's middle back. Call for help at once.

Why are disorders of the throat often accompanied by pain in one or both ears?

Throat infections may cause pain in the ear because of the phenomenon of pain referral. This occurs when the same nerve supplies the two different but close-lying structures. The patient is unable to discern from which site the pain arises.

My son is always getting ear infections and we have been told that he should have an adenoidectomy. Will this definitely cure him?

It is never possible for a doctor to guarantee that any treatment will be totally successful. However, very enlarged adenoids are frequently implicated in recurrent otitis media (an infection in the middle ear) and it is only right to remove them in the hope that this is the cause of the problem. It is very important to try to minimize infectious attacks in the ear and prevent the deafness that can be associated with them, so if your ear surgeon recommends an adenoidectomy you should seriously consider it.

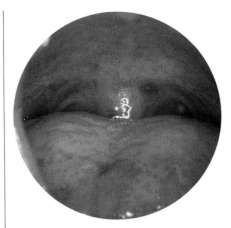

▲ *Caused by the streptococcus bacterium, strep throat is a common throat infection that can be very painful.*

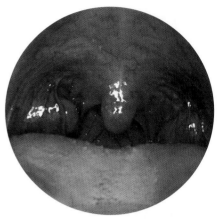

▲ *Infection can spread rapidly in the throat; in this case tonsillitis, usually caused by a streptococcus, has developed.*

lie in the lining of the pharynx; these are the adenoids in the nasopharynx and the tonsils in the oropharynx. They protect the entrance to the food and air passages. The other major part of the throat, the larynx, is in front of the laryngopharynx and is made up of a framework of cartilage swathed in muscles both internally and externally, and lined by a respiratory membrane.

The larynx is a specialized section of the windpipe. It has a flap valve—the epiglottis—hovering over the inlet to the airway, which acts as a type of umbrella against a shower of food and liquid when we eat or drink. The vocal cords are located in the larynx, and held in place by special cartilages. They are suspended across the airway and produce sound when vibrated by air movement.

Functions of the throat

Because the throat is an assembly of different components, it has a variety of functions. First is the pharynx: the channel through which food and liquid enter the digestive tract, and air enters the lungs. The movements of the pharynx must be coordinated to ensure that the respiratory gases end up in the lungs and food and liquid end up in the esophagus. This is achieved by a plexus, or network, of nerves called the pharyngeal plexus, the activity of which is controlled in the lower brain stem, where information is coordinated from the respiratory and swallowing centers higher in the brain.

When food is thrown into the oropharynx by the tongue, it is swiftly sent into the esophagus by a wave of muscular contractions that travel down the pharynx. At the same time mechanisms are triggered to prevent the food from entering the larynx.

No less important are the functions of the larynx, which are to produce sound and to protect the airway. Like the pharynx, the larynx achieves these functions through a complex coordinated nerve supply to its muscles. The nerves that supply the larynx are under the same central influence in the brain as the nerves that supply the pharynx.

Throat disorders

The pharynx and larynx are prone to a number of infections caused by viruses or bacteria, and also to damage by physical agents, such as excessive smoking or drinking. Either of these factors can lead to, for example, chronic laryngitis or pharyngitis. Other throat infections include nasopharyngitis, a viral infection of the mucous membranes lining the nasopharynx. This begins with a burning feeling in the throat that builds up in intensity and is aggravated by speaking and swallowing. The discomfort is accompanied by a general feeling of malaise, which lasts for a few days. Nasopharyngitis is often confused with tonsillitis. Because the tonsils and the nasopharynx are adjacent to each other, an infection in either area gives rise to similar symptoms. However, in tonsillitis the symptoms are much more severe.

The nasopharynx is also the site of the adenoids, and repeated infections in this area can cause the adenoids to enlarge. Patients with excessively large adenoids are unable to breathe

Home help for a sore throat

A sore throat that accompanies a common cold or other minor infection can be treated at home. Here are some proven remedies that can be tried.

Make up a gargling solution by dissolving two teaspoonfuls of household salt in a cupful of hot, but not boiling, water. Stir until the salt has dissolved, then use the solution to gargle. Make sure you spit the solution out of your mouth when you have finished.

Alternatively, you can use an aspirin gargle by dissolving two soluble aspirin tablets in a cup of hot water. If you swallow this after gargling, the aspirin will relieve some of the pain and reduce any fever you might have.

Drink plenty of hot liquids and try to eat only soft or liquid foods, such as soups, so that you do not take anything into your throat that might cause further damage or inflammation.

Lozenges can be effective in soothing a sore throat and can prevent the throat from becoming dry. Keep the lozenge as far back on your tongue as you can for maximum benefit.

▲ *The thyroid cartilage surrounding the larynx creates the projection called the Adam's apple. It is more prominent in men because they have larger vocal cords.*

through the nose and have a persistently gaping mouth. An increase in adenoid tissue may block the natural drainage channel, the eustachian tube, from the middle ear and cause an accumulation of fluid in this cavity. In severe cases deafness may result, and the condition also predisposes a person to recurrent attacks of otitis media. Children are most affected, and are frequently admitted to the hospital for drainage of the middle ear fluid and removal of the adenoids.

Papillomatosis is caused by a virus that affects the larynx; it occurs in children and adults. The virus is similar to that which produces warts on the genitals. As the papillomas increase in size they may restrict the passage of air. Like warts elsewhere, the growths can disappear spontaneously, but in severe cases laser surgery is required to remove them.

Pharyngeal cancer

Cancer of the pharynx is an unusual condition that occurs in middle and old age. Patients complain of a progressively painful difficulty in swallowing. The pain is experienced not only at the site of the disease but also radiating to the ear. In a large number of patients a lump appears on one side of the neck. Although cancer of the pharynx is an extremely serious condition, a high proportion of patients can be greatly helped by either radiotherapy or surgery.

Laryngeal cancer

Cancer of the larynx is a more common throat cancer. It is a form of cancer that can be cured if the disease is diagnosed and treated early. It occurs most frequently in late middle age and is much more common in men than in women. Most patients are, or have been, heavy smokers for much of their life. Progressive hoarseness is the most common symptom, and patients may also complain of an odd feeling in the throat, slight difficulty in breathing, pain in their ear, or pain radiating from the throat to the ear. Occasionally they may cough up a small amount of blood.

Most cases of early laryngeal cancer can be cured by radiotherapy. However, in a few cases the disease recurs or fails to respond to radiation treatment and surgery is necessary.

Foreign objects in the throat

A wide variety of foreign objects have been found in the pharynx, ranging from fish bones to coins and bottle tops. Any foreign object in the throat is potentially dangerous because it can perforate the pharyngeal wall and set up serious infection in the surrounding tissues of the throat.

Patients who have swallowed a foreign object that has become lodged in the throat are rarely in any doubt about its presence; every attempt to swallow even their own saliva is excruciatingly painful. If the object is lodged high in the pharynx it can usually be removed in the emergency room of a hospital. However, if it is lodged lower down—for example, in the laryngopharynx—it must be removed under general anesthesia.

Foreign bodies that are inhaled into the larynx can threaten a patient's life and little time should be lost in getting treatment. The younger the patient the more serious the condition, since the diameter of a child's airway is so narrow that even a small object is likely to obstruct it totally.

Removal of any object from the larynx is a matter of surgical urgency, and may require a temporary tracheostomy (an artificial opening in the trachea) to protect the lower airway and maintain adequate breathing. Recovery from this surgery is swift.

See also: Adenoids; Brain; Digestive system; Esophagus; Larynx; Mouth; Neck; Pharynx; Tongue; Tonsils; Vocal cords

Thymus

Questions and Answers

Is it possible to live without a thymus?

After the age of puberty, as we begin to grow older, the thymus starts to shrink. It may not be found in elderly people. This is because it has set up lifelong immunity. Normally, in the first few years of life, we will have come into contact with, and gained immunity to, most of the infections that the immune system is designed to repel. If a baby is born with an inadequate or absent thymus, he or she will not be able to fight infection. The thymus is essential early in life, but it can be removed later.

My husband said he had X-ray treatment for his thymus when he was a baby because it was too big. Is this common?

It is no longer the practice to irradiate large thymus glands in children. Before the importance of the thymus gland to immunity was realized, it was a common treatment.

My brother has myasthenia, and the doctors want to remove his thymus. I thought myasthenia was a nervous system disease, and I don't see the relevance of a gland in the chest. Please explain.

Myasthenia is a nervous system disorder and symptoms include weakness and tiredness of the muscles, which get worse during the day. The disease is caused by the formation of antibodies to the junctions between the nerves and muscles by the body's own immune system. These antibodies attack the junction, and the nerves cannot instruct the muscles to move. Since the thymus is very much involved with the control of the immune system, it has been found that removing the thymus can be effective in helping some myasthenia sufferers.

At one time, the role of the thymus in the body was a mystery. However research over the years has shown that it plays a vital role in the body's defenses against disease.

Over the last two decades it has become clear that the thymus sits at the center of a remarkable web of interconnected organs and tissues that make up the immune system, which defends the body from attack by infection.

There are still some questions about exactly how the thymus does its job, but it is now known to be essential for the proper running of the immune system, and its major function is carried out during the first few years of life.

Where the thymus is

The thymus is found in the upper part of the chest, where it lies just behind the breastbone (sternum). It consists of two lobes that join in front of the windpipe (trachea). In a young adult it measures a few centimeters in length and weighs about ½ ounce (15 g). However, unlike any other organ in the body, it is at its largest at around the time of puberty, when it may weigh as much as 1½ ounces (43 g).

In a baby, the thymus is very large compared with the rest of the body, and it may extend quite a long way down the chest behind the breastbone. It grows quickly until about the age of seven, after which it grows more slowly until the child reaches puberty.

After the age of puberty the thymus starts to shrink in size in a process called involution, until in an elderly person the only thymus tissue present may be merely a small amount of fat and connective tissue.

▲ *The thymus is at its most vigorous in the first years of youth, setting up immunity to disease; its function fulfilled, it shrinks into insignificance in old age.*

SIZE AND LOCATION OF THE THYMUS

left and right lobes of thymus

trachea

lungs

left and right lobes of thymus

heart

diaphragm

child

adult

▲ *The relative sizes of the thymus in an adult and a child demonstrate its importance in establishing the body's immune system early in life. In adulthood it actually shrinks.*

Structure and function

The thymus is made up of lymphoid tissue, epithelium, and fat. The lymphoid tissue consists of many small round cells called lymphocytes, which are the basic unit of the immune system. These cells are also found in the blood, the bone marrow, the lymph glands, and the spleen, and they travel into the tissues as part of the inflammatory reaction.

The outer layer of the thymus, which is called the cortex, has many lymphocytes. Inside this is an area called the medulla, which contains lymphocytes and also other sorts of thymus cells.

In the early years of life, the thymus is concerned with programming the way in which the immune system works. It is now known that the thymus is responsible for many of the most important aspects of the immune system. There are two main sorts of immune cells in the body and both are different types of lymphocytes. The first type are T or thymus cell lymphocytes that are controlled by the thymus and are responsible for the recognition of foreign substances and for many of the ways in which the body attacks them. The second type are B lymphocytes, which are responsible for making antibodies to foreign substances.

The exact ways in which the thymus goes about controlling the T lymphocytes are not known, but it certainly processes the T cells. One important mechanism has come to

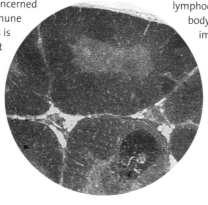

▲ *Low-power magnification of a section through a normal thymus shows large purple masses, which are the lobes of this vital organ.*

light. It seems that about 95 percent of the new types of lymphocyte that are made in the thymus are in fact destroyed there, before they ever have an opportunity to get out into the rest of the body. The probable reason for this is that they would have the potential for turning against the body itself, and the only cells that the thymus allows to develop are those that attack foreign substances.

The T cell processing in early life is affected by the environment to which the baby is exposed. This modulation of the T cells occurs in the thymus and its effect is that babies who are unduly protected against infection and other similar hazards may, later in life, have a less efficient immune system. Babies brought up on a farm, for instance, are less likely to develop asthma later.

As an isolated organ, the thymus rarely causes any trouble. However, it is important to remember that the T lymphocytes it controls are the most central part of the body's immune system, and therefore extremely important in almost all serious diseases.

In the thymus itself, as opposed to the wider aspects of its function, two main problems may occur. First, the thymus may fail to develop properly in babies. As might be expected, this leads to a failure of the immune system and a failure to resist infection, which may prove fatal. This, however, is not a common problem. Second, tumors can occur in the thymus. These are called thymomas, and they are treated by surgery followed by X-ray treatment.

See also: Immune system; Lymphocytes; Puberty

Thyroid

Questions and Answers

I am very nervous and anxious all the time and irritable with my children. Is it possible that I have an overactive thyroid?

Yes, although it may be difficult to differentiate between symptoms of the disease and those of pure anxiety. An overactive thyroid disorder is associated with weight loss, and the characteristic protruding eyes of Graves' disease, which is the main form of thyrotoxicosis or overactive thyroid. If you find that you are shaky and have a lot of difficulty tolerating heat, then your thyroid may be the reason. The tests for thyrotoxicosis are straightforward and if your doctor suspects this might be the trouble, he or she will arrange for you to have a blood test to confirm the diagnosis.

I have an overactive thyroid and am about to see a specialist. I'm scared of surgery; is there another treatment I could try?

You may be advised to have surgery, although your worries might lead the doctor to suggest alternative treatment. First, you could be given pills to take for about 18 months to suppress the activity of the gland. However, the condition might recur in the future. Second, you could be given treatment with radioiodine (radioactive iodine that is taken up by the thyroid), which reduces the level of thyroid activity. This has the advantage of being simple, and the condition is unlikely to recur. However, this type of treatment is not given to very young people or to women who might become pregnant, since there is a theoretical risk that it will cause cancer in the patient or in any children that may subsequently be born. There is also a definite risk of underactivity of the thyroid after treatment, but this, in turn, is very easy to treat.

Problems associated with the thyroid and the hormone it produces are fairly common. However, many of the disorders respond extremely well to treatment and can be completely cured once they have been identified.

Section through thyroid

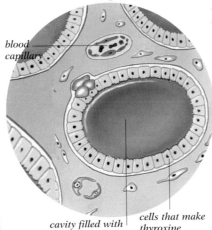

blood capillary

cavity filled with colloid in which thyroxine is stored

cells that make thyroxine

Problems associated with the thyroid gland are among the most common types of hormonal disorders, and affect large numbers of people. Many of these problems can be helped by administering a synthetic hormone, similar to that produced by the gland. Of the wide range of thyroid disorders, by far the most important are overactivity (hyperthyroidism) and underactivity (hypothyroidism). Both of

◄ *The insert is a section of the thyroid, which clearly shows the cells that produce and store the essential hormone thyroxine.*

▼ *The anatomic drawing shows the position of the thyroid gland in relation to the surrounding structures in the throat, which include the Adam's apple and the trachea.*

THYROID GLAND

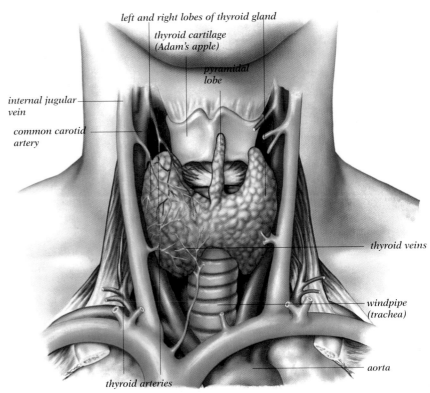

left and right lobes of thyroid gland

thyroid cartilage (Adam's apple)

pyramidal lobe

internal jugular vein

common carotid artery

thyroid veins

windpipe (trachea)

aorta

thyroid arteries

Questions and Answers

Are thyroid problems really so much more common in women than in men?

Yes. Underactivity of the thyroid happens in about 14 in every 1,000 women and only about one in every 1,000 men. Overactivity occurs in at least 20 in every 1,000 women, but in only about one or two in every 1,000 men.

Generally, thyroid disorders are quite common, and more than 3 percent of women are likely to have some type of thyroid problem. Most thryoid disorders can be treated effectively.

I had an overactive thyroid and was treated with pills. Although I have been better for the past two years, the clinic still insists on seeing me. Why is this necessary?

Thyrotoxicosis responds well to treatment with pills. However, the disease does tend to recur, although there may be several years between attacks. It is important to diagnose the disease in the early stages, when it is easier to treat; also, irreversible changes in processes such as the heart rhythm may occur if the disease is allowed to progress too far. For these reasons the doctors will want to keep an eye on you.

Is it true that your hair falls out if you have myxedema?

Myxedema is underactivity of the thyroid gland, and it leads to dry, coarse hair that is very difficult to manage. The disease is also associated with alopecia, in which the hair roots die and the hair falls out. However, this condition is not a direct result of the low thyroid levels.

Do thyroid disorders tend to run in families?

Yes they do, and this is quite common. It is also interesting to note that although one family member might have an overactive thyroid, another person might have an underactive thyroid.

▲ *In some areas of the world, such as the Matto Grosso in Brazil, the normal diet lacks iodine. This deficiency causes the thyroid to malfunction and swell, leading to endemic goiter, the disfiguring condition this woman is suffering from.*

these problems are much more common in women than in men, and up to 2 percent of the adult female population may suffer from an overactive thyroid at some time in life. An underactive thyroid is only slightly less common. Another disorder of the thyroid is thyroiditis, which is an inflammation of the thyroid as a result of a viral infection.

Position and function

The thyroid gland is located in the neck just below the level of the larynx, which can be seen or felt as the Adam's apple. There are two lobes to the gland, and these lie just in front and at either side of the windpipe, or trachea, as it passes down the front of the neck. The two lobes are connected by a small bridge of tissue, and there may be a smaller central lobe called the pyramidal lobe. In an adult the thyroid gland weighs about 0.7 ounce (20 g).

The function of the gland is to produce the thyroid hormone, thyroxine. When the gland is examined under a microscope many small follicles can be seen; these are islands of tissue containing collections of colloid, a protein substance to which the thyroid hormone is bound and from which it can be released under the influence of enzymes. Once thyroxine has been released from the gland into the bloodstream it is taken up by most of the cells of the body. A receptor on the surface of cell nuclei responds to the hormone. The overall effect is to increase the amount of energy that the cell uses; it also increases the amount of protein that the cell manufactures. Although the exact role of the hormone in the cell is not known, it is essential for life. The thyroid gland contains iodine, which is vital for its activity. The thyroid is the only part of the body that requires iodine and it is efficient at trapping available iodine from the blood. An absence of iodine in the diet results in malfunction of the thyroid and growth of the gland; this is endemic goiter.

Control of thyroid activity

The thyroid is one of the endocrine glands under the control of the pituitary gland. The pituitary produces thyroid-stimulating hormone (TSH), which acts on the thyroid to increase the amount of thyroid hormone released. The amount of TSH produced by the pituitary increases if the amount of thyroxine circulating in the system falls, and decreases if it rises. This system, called negative feedback, results in a relatively constant level of thyroid hormone in the blood.

The pituitary gland is controlled by the hypothalamus, and the amount of TSH it produces is increased when TRH (TSH-releasing hormone) is released from the hypothalamus. Most of the

Cancer of the thyroid

PROBLEM	INCIDENCE	TREATMENT
Papillary	Most common type; occurs in younger people	Surgery, then radioactive iodine if necessary. Outlook is good.
Follicular	Slightly less common; also occurs in young people	As for papillary carcinoma. Outlook is good.
Anaplastic	Uncommon; occurs in the elderly	Surgery is often impossible. X-ray treatment may be used. Outlook is poor.
Medullary	Uncommon. Often familial.	Surgery. Outlook is good if localized.

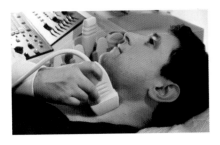

▲ A radiologist carries out an ultrasound scan of the thyroid.

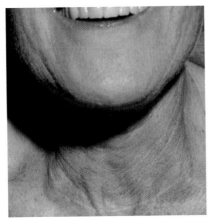

▲ A thyroidectomy—surgery to remove overactive tissue—may be necessary to combat hyperthyroidism. The scar left by the surgery is insignificant.

▲ This is a grossly enlarged thyroid gland that was removed. It measures over 10 in. (25 cm).

hormone released from the thyroid gland is in the form of tetraiodothyronine, which contains four iodine atoms and is known as T4. However, the active hormone at the cell level is triiodothyronine, which contains three iodine atoms and is known as T3. Although the thyroid releases some T3 into the blood, most of its output is T4, and this is converted into T3 in the tissues. Sometimes the tissues switch the way that they convert T4 to produce an ineffective compound called reverse T3. As a result, there will be less thyroid hormone activity in the tissues even though the amount of hormone in the bloodstream is adequate.

Any enlargement of the thyroid gland is called goiter. Small but visible goiters are found in about 15 percent of the population, and are about four times more common in women than in men. Usually these are of no significance.

In the past, iodine deficiency was the main cause of goiter, but now most goiters are caused by overactivity of the thyroid, or they are simple goiters that are not related to any abnormality of thyroid function. In a few cases, goiters are isolated lumps (nodules) in the substance of the thyroid, and these should be investigated using an ultrasound scan to see whether the lump is composed of functioning tissue. If it is not, and if an ultrasound scan shows that it is solid, then it could be malignant and may need a surgical exploration.

Overactive thyroid glands

Most cases of thyroid overactivity (thyrotoxicosis) are caused by Graves' disease. A goiter is usually present and the eyes become protruding and staring, a sign that many people associate with thyroid problems. It is the basic disease process, and not overactivity, that causes the symptoms.

Graves' disease is caused by the presence of antibodies in the blood. Although these antibodies do not destroy thyroid tissue, they stimulate the gland to produce thyroid hormone. It is not known why some people are more prone to making these antibodies than others, although there is certainly genetic susceptibility.

The effects of an overactive thyroid are: weight loss, increased appetite, anxiety and nervousness (sometimes with a tremor), palpitations of the heart, sweating, intolerance of heat, and irritability. In addition to eye problems, there may be weakness of the muscles, particularly at the shoulders and hips.

Once Graves' disease is suspected, the majority of cases can be diagnosed very simply by measuring the level of thyroid hormone in the blood. Often the T3 level is measured as well as, or instead of, the T4 level, since T3 is always raised in Graves' disease but it is possible to have the disease with a normal T4 level.

Treatment for Graves' disease is to suppress thyroid activity. This can be done with pills, which must be taken for a year or more. If the gland is very large then it may be appropriate to surgically remove some of it. The alternative is to give a dose of radioactive iodine. This is taken up by the thyroid so it presents no danger to other tissues. It will reduce the level of thyroid activity over the course of approximately six weeks. While the hormone levels are being brought under control, the symptoms can be alleviated by drugs

Thyroid problems

PROBLEM	CAUSE	EFFECTS	TREATMENT
Simple goiter	Unknown	Swelling of the thyroid, producing a swelling of the neck	Often unnecessary, but in many cases the problem responds to low doses of thyroid hormone pills
Endemic goiter	Lack of iodine in the diet	May lead to deficiency of thyroid hormone	Replacement of iodine in the diet
Myxedema or hypothyroidism	Inadequate levels of thyroid hormone in the blood. May be caused by Hashimoto's disease or autoimmune thyroid failure.	Problem develops slowly, leading to dry rough skin, tiredness, intolerance of cold, increase in weight, constipation, and hoarse voice	Replacement of thyroid hormone in pill form
Thyrotoxicosis or hyperthyroidism	Most commonly Graves' disease; others include nodules in the thyroid, either single or multiple	Increase in appetite (often with weight loss), sensitivity to heat, disorders of heart rhythm, nervousness, tiredness, and sweating. In some cases there is muscular weakness.	Various treatments, including use of pills, surgery to remove overactive thyroid tissue, and use of radioactive iodine administered by mouth
Graves' disease	Presence of antibodies in the blood, which stimulate the thyroid	Causes thyrotoxicosis; also affects the eyes, causing the protruding eyes associated with overactive thyroid	Troublesome eye problems can necessitate surgery to try to reduce the amount of eye protrusion
Thyroiditis	Inflammation of the thyroid gland resulting from a viral infection	Painful swollen gland that may come on suddenly. Mild thyrotoxicosis may result.	Usually unnecessary; painkillers may be given. In severe cases steroid drugs are sometimes used to reduce the inflammation.
Dyshormonogenesis	Inherited abnormality in the way the gland makes hormones. Of six different types the most common is Pendred's syndrome, which is associated with deafness.	A low level of hormone in the blood may cause the same effects as myxedema. There is often a large goiter.	Thyroid hormone
Congenital hypothyroidism (cretinism)	Failure of the fetal thyroid to develop	Mental and physical retardation is the major effect; it occurs in about one in every 4,000 births	Early diagnosis is vital; treatment will help normal development; retardism will otherwise occur

that block the effects of epinephrine, since high levels of thyroid seem to produce an increased response to epinephrine.

Underactive thyroid glands

Underactivity of the thyroid gland (myxedema) results in a lack of thyroid hormone, and a resultant slowing of the body's reactions. One of the most common reasons for hypothyroidism is Hashimoto's disease, in which antibodies in the body appear to damage the thyroid permanently.

In many cases weight gain results, together with a lack of energy, dry thick skin, thinning hair, intolerance of cold, a slow heartbeat, hoarseness, deafness, and a typical puffy face. The presence of hypothyroidism makes elderly people much more susceptible to the condition of hypothermia. Underactivity of the thyroid is readily diagnosed by blood tests. The level of T4 is reduced, but this is not conclusive, since a reduced T4 level can also occur when the thyroid

is functioning normally—in severe illness, for example. Much more significant is the high level of TSH that is found in the blood in myxedema, because the pituitary gland tries to stimulate the thyroid to produce enough hormones.

Once a diagnosis is made, the thyroid hormone T4 (thyroxine) can be given by mouth. The dose is built up fairly gradually, especially in those with heart disease, since there is a risk of making these patients worse; myxedema predisposes to coronary artery disease since it causes a very high level of cholesterol.

Patients with myxedema must continue to take medication for the rest of their lives. Although Graves' disease is not quite as easy to treat as myxedema, the results of treatment in both cases is usually satisfactory, and the outlook is very good in both conditions once the early difficulties are overcome.

See also: Glands; Hormones

Tongue

Questions and Answers

Is it possible to be tongue-tied, or is this just a figure of speech?

It was once thought that babies were tongue-tied if they could not put their tongues out, but it has now been found that as a child grows in the first year or two of life, so the strand of tissue (the frenulum) beneath the tongue, linking its underside to the floor of the mouth, elongates and makes the tongue properly mobile. Only in very exceptional circumstances is it necessary for the frenulum, to be cut to give the tongue greater mobility and improve a child's speech. In modern parlance, to be tongue-tied means to be incapable of speech for a short time, especially through shyness or surprise.

What should you do if someone accidentally swallows his or her tongue?

Unless you are trained in first aid it is important not to try to pull the tongue back to its normal position—you may do more harm than good and choke the patient. Lay the person down in the recovery position with the chin forward so that the airway is as clear as possible and call for help immediately.

Why does smoking make my tongue furred and yellow?

Smoking tends to dry the mouth and prevents a natural washing process from taking place. The cells of the tongue and mouth wear out and are replaced rapidly, and the remains of dead cells, plus dried mucus and the remains of food, form a furry deposit on the tongue if not washed off. Smoking inhibits the action of the salivary glands, and slows down cleaning. The yellow color is the staining action of the tars in tobacco. There is a simple answer to this problem—quit smoking.

The tongue is one of the body's muscles, but it is a muscle of an unusual kind. Not only are its movements vital for speaking and eating, but it also contains thousands of taste buds that give us our sense of taste.

The tongue is shaped like a triangle—wide at the base and tapering almost to a point at its tip. It is attached at its base or root to the lower jaw, or mandible, and to the hyoid bone of the skull. At its sides the root is joined to the walls of the pharynx, the cavity that forms the back of the mouth.

The middle part of the tongue has a curved upper surface, and its lower surface is connected to the floor of the mouth by a thin strip of tissue called the frenulum. The tongue's tip is free to move, but when a person is not eating or speaking it normally lies neatly in the mouth with its tip resting against the front teeth.

The sense of taste

The structures that give the tongue its characteristic rough texture are the ridged folds, or papillae, that cover the upper surface of the front two-thirds of the tongue. Largest of these are the eight to 12 V-shaped, or vallate, papillae that form the border between the ridged and unridged areas. Because the papillae are so numerous, they provide a huge surface area to accommodate the 9,000 or more taste buds with which the human tongue is equipped.

This large surface also creates a much increased area on which substances can be dissolved, tested, and tasted. Dissolving is a vital part of the tasting process because the taste buds will not work unless they are presented with molecules in solution. The taste buds nestle in the valleys of the papillae and consist of bundles of hairlike cells, which project into the valley. Nerve fibers

▲ *Licking a Popsicle is a study in contrasting sensations: flavor explodes onto the tongue, while the frostiness lends a biting effervescence to the taste.*

THE TONGUE: ITS POSITION AND STRUCTURE

frenulum attaches tongue to floor of mouth

lingual artery, vein, and nerve

salivary gland

pharynx

papillae giving upper surface its rough texture

palatoglossus and styloglossus muscles curl tongue up and back

hyoglossus muscle lowers tongue to resting position

hyoid bone

genioglossus muscle protrudes tongue

vallate papillae at base of tongue

filiform (leaf-shaped papillae

fungiform papillae

taste buds

mucous membrane

salivary gland

muscle

Section through tongue

▲ *These illustrations show the position of the tongue and surrounding musculature (above); the underside, with its blood and nerve supply (top right); and the upper surface with papillae (above right).*

connected to the hair cells run to the brain and to the salivary glands that are situated around the borders of the mouth.

When food is taken into the mouth, nerve cells in the tongue first feel it and test it for texture and temperature. As the taste buds become stimulated by chemical messages arriving from dissolved foodstuff molecules, they send messages to the salivary glands, which pour their secretions into the mouth, not just to help soften the food but also to assist in the dissolving process.

As a result of the combined action of the nerve cells in the tongue, messages are passed to the brain about the taste of food and about its texture and temperature, to give an overall impression of each mouthful. Because of the tongue's millions of minute undulations, chemicals may be trapped in the valleys for some time, allowing a taste to linger in the mouth long after a mouthful of food or drink has been swallowed.

Movements of the tongue

The actions of the tongue are determined by the muscles it is made up of and to which it is joined, and the way it is fixed in position.

The tongue contains muscle fibers running both longitudinally and from side to side, and these are capable of producing some movement. However, the actions of the tongue are given huge versatility by the contractions of a variety of muscles situated in the neck and at the sides of the jaws. The styloglossus muscle in the neck, for example, is responsible for bringing the tongue upward and backward; the hyoglossus muscle, also located in the neck, brings it back down again into the normal resting position.

In eating, one of the main functions of the tongue is to present the food to the teeth for chewing and to mold softened food into a ball, or bolus, ready for swallowing. These actions are performed by a range of curling and upward and downward movements. When the task has been completed the tongue pushes the bolus into the pharynx at the back of the mouth from where it enters the esophagus and is swallowed into the stomach.

The actions of the tongue are important to human communication through the enunciation of speech. The difference between a crisp, clearly spoken "s" sound and the fuzzy tone of a lisped one is, for example, all to do with tongue action. For someone to discover the range of positions the tongue must adopt during speech, he or she says the sounds of the alphabet out loud, then intones common characteristics such as "sh" and "th." There is a precise tongue position for each sound, and that is vital for enunciating consonants in particular.

Injuries and disease

It is easy for the tongue to be accidentally injured, and this often happens when a person bites his or her tongue.

Because it is so richly supplied with blood vessels, the tongue bleeds profusely when injured. If bleeding is severe, the best sort of first aid is to sit the person upright and, with a clean wad of tissue or a folded handkerchief, try to grip the bleeding area between your thumb and forefinger. If the injured person is an adult, and has not sustained severe injuries elsewhere on the body, he or she may be able to apply pressure to the tongue, but in any case a steady pressure should be maintained for about 10 minutes. If an injury is only minor, the injured person can suck an ice cube. Following accidents, the tongue can block the airway. If someone is unconscious after an accident, he or she should be placed in the recovery position, with the chin forward, so that the airway is left clear.

▲ *Sticking out the tongue can be an expression of revulsion or shock, or just an exaggerated reaction to something unpleasant.*

Action of saliva

The tongue is not often prone to disease, mainly because of the antiseptic action of the saliva that bathes it. The most common problems are ulcers and a yeast infection, *Candida albicans*.

Unless ulcers are very large or painful, or both, the best treatment is to use a proprietary mouthwash or gel, or to suck lozenges formulated to treat the problem. Cancer of the tongue is most often due to irritation from tobacco products. Cancers of the tongue usually start on the edge of the tongue as a local swelling or ulcer. They can spread widely, so it is vital that people see their doctor if they notice any abnormal swelling or pain in the tongue, or if they have persistent ulcers that do not clear up quickly.

The color, texture, and general state of the tongue can arouse a doctor's suspicions about disease, but the tongue rarely tells the whole story on its own. Normally, the ridged papillae are cleared of debris that accumulates on the tongue by the washing process, which is affected jointly by the saliva and the movements of the tongue.

If a coating of fur does build up on a person's tongue, the development could simply be due to a harmless disorder called hairy tongue. This problem goes away after treatment with antibiotics.

Early treatment

If a person has persistent discoloration or pain of the tongue or oral cavity, he or she should bring this condition to the attention of a doctor. As with so many other disorders, the sooner treatment begins, the better the chances of a total recovery.

▼ *To examine the throat, sometimes it is necessary to hold the tongue down with a implement called a tongue depressor.*

See also: **Blood; Brain; Digestive system; Mouth; Muscles; Nervous system; Saliva; Taste**

Tonsils

How soon after having an attack of tonsillitis is it safe to have surgery to remove the tonsils?

Two to three weeks is the accepted period, after which time the risk of abnormal bleeding after surgery returns to normal. If a child is having attacks of tonsillitis every 10 days and is due for a tonsillectomy, the doctor will put him or her on a course of antibiotics for three weeks before surgery. This eliminates the possibility of the child's being sent home when he or she is taken to the hospital for surgery.

Should I suck or swallow aspirin when I have a sore throat?

Sucking aspirin can be harmful, because it can cause a chemical burn on the mucous membranes lining the mouth and the throat.

Will I be more prone to infections after I have had my tonsils out?

No. Your tonsils probably have a significant function only during the first few years of life. That is one of the reasons why doctors prefer not to remove the tonsils of a very young child.

My three-year-old son has very frequent attacks of tonsillitis, but the surgeon is reluctant to remove his tonsils. Why is this?

At three years of age your son will probably weigh between 26 and 40 lb. (12 to 18 kg) and therefore his total blood volume will be between 2 and 3 pt. (950 and 1400 ml). A loss of 0.3 pt. (120 ml), which is the average loss for the operation, would deprive him of between 8 and 12 percent of his blood and cause significant anemia. If he should bleed heavily after the operation his condition would become very serious.

The tonsils have an important role as part of the body's defense system. Viruses and bacteria, however, can cause the tonsils to become inflamed and painful. Recurrent infections may make removal of the tonsils necessary.

The tonsils are part of a ring of lymphoid tissue (Waldeyer's ring) that encircles the entrance to the food and air passages in the throat. Although the tonsils are present at birth they are relatively small. They grow rapidly during the first few years of life, only to regress after puberty. However, they do not disappear completely.

◄▼ *The tonsils encircle the entrance to the food and air passages, in their role of defense against infection. But when the tonsils are infected—and the most common infection is tonsillitis—they become swollen and inflamed, making swallowing, and sometimes breathing, difficult and painful.*

POSITION OF TONSILS

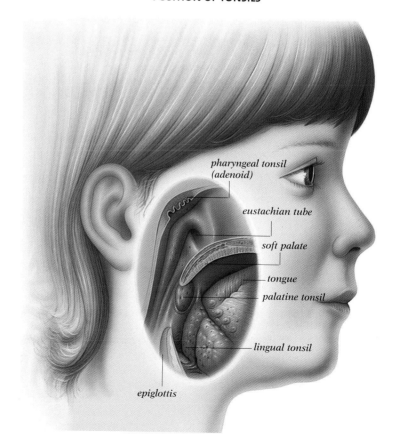

- pharyngeal tonsil (adenoid)
- eustachian tube
- soft palate
- tongue
- palatine tonsil
- lingual tonsil
- epiglottis

Questions and Answers

My husband is going into the hospital for a tonsillectomy. How long will he have to stay off work and how should I look after him?

Most adult patients take about two weeks to recover completely. These operations are, many times, done in same-day surgery centers. It is important that you encourage your husband to eat foods that are comfortable; he should avoid spicy foods and extremely hot foods. If it was your child who was having surgery, the recommendation would be the same, with the additional precaution that he or she should not come into contact with other children for a week.

If a person's tonsils are larger than normal, should they be removed?

The size of the tonsils alone rarely necessitates their removal. However, in a few cases the tonsils are so enlarged that they obstruct the passage of air and a tonsillectomy must be performed. The obstruction occurs mainly at night. While the person is asleep he or she stops breathing intermittently and therefore wakes up. The consequence of this condition is called sleep apnea. Subjects may become lethargic, undergo personality changes, and develop high blood pressure. Very rarely, the chronic deprivation of oxygen can lead to heart failure and irregularities of the heart rate.

Is it true that some children have to have their tonsils removed to stop them from going deaf?

Yes. In children, attacks of tonsillitis may precipitate attacks of otitis media (an infection of the middle ear) or prevent a complete recovery from serous otitis media (known as glue ear). In these cases the surgeon may advise tonsillectomy to prevent possible damage to the middle ear and to avoid the decreased hearing that, although temporary, is associated with these conditions.

▲ *Although children recover very quickly from a tonsillectomy, being in the hospital can be a lonely experience. The reassuring presence of a parent, especially just before and after surgery, helps speed recovery.*

Function

In childhood, the tonsils are an important part of the immune system and play a significant role in the body's defense mechanisms. They are ideally situated to monitor ingested material and to react to those materials that could pose a threat to the well-being of the body. The tonsils provide immunity against upper respiratory tract infections by producing lymphocytes. In addition, the tonsils produce antibodies that deal with infections locally.

Tonsillitis

Tonsillitis is an acute infection of the tonsils, pharynx, or both. The organism causing the infection is sometimes a potentially dangerous bacterium called streptococcus.

Because there is a considerable overlap between the two illnesses, it is not always easy to distinguish streptococcal tonsillitis from a simple sore throat that is caused by a virus. However, streptococcal tonsillitis usually lasts longer—approximately one week. Symptoms of streptococcal tonsillitis include marked discomfort in the pharynx, making swallowing painful. Pain from the throat may also be felt in the ears. Some patients also experience discomfort on turning their head, because of swelling of the glands in this region. A fever almost always accompanies streptococcal tonsillitis, but it varies in degree. Children tend to develop higher temperatures and consequently more symptoms (such as malaise and vomiting) than adults. Some children may have no symptoms in the throat but will complain of abdominal pain instead.

When the tonsils are infected they become enlarged and inflamed. Specks of pus may exude from the surfaces of the tonsils with streptococcal infection. This infection responds well to antibiotics, and improvement can be expected within 36–48 hours. Symptoms can be alleviated by eating soft foods and drinking plenty of liquids. Painkillers such as aspirin both relieve the pain and reduce the temperature. Aspirin is not recommended for children under the age of 16, however, because there is a very slight risk of a disorder called Reye's syndrome.

Tonsillitis tends to occur most frequently between the ages of four and six years, and then again around puberty. The more often the tonsils are infected, the more prone they are to persistent and recurrent infection. A stage is reached when removal of the tonsils is the only sensible way of controlling the illness.

In some cases an infection is so severe that an abscess forms in the tissue around the tonsils. This is called peritonsillar abscess, or quinsy. Quinsy usually affects one side of the tonsils and is very rare in children. The affected tonsil swells to a considerable extent, and may prevent swallowing altogether. Local inflammation contributes to this by limiting the opening

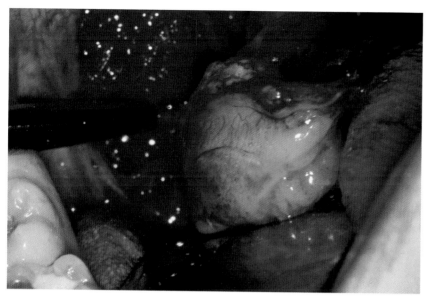

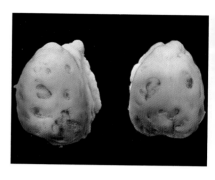

▲ *These tonsils that have been removed show patches of scar tissue from repeated bouts of tonsillitis.*

◀ *Only as a last resort will a person's tonsils be removed. They are dissected from the pharyngeal wall while the patient is under general anesthesia.*

of the jaw. Oral antibiotics are not only difficult to swallow but also rarely effective. Higher doses of antibiotics are given intravenously for 24–36 hours, followed by oral antibiotics. If the quinsy is ripe—that is, the abscess is pointing—recovery may be accelerated by lancing the abscess and allowing the pus to drain.

In exceptional cases, an infection is not limited to the tonsils, but spreads both down the neck to the chest and up toward the base of the skull; this is a parapharyngeal abscess. This is a life-threatening condition and requires urgent admission to the hospital, where the abscess can be drained and treated with massive doses of powerful antibiotics. Patients who have had quinsy are thought to be more susceptible to this complication and are therefore advised to have their tonsils out even if they have not been troubled previously by recurrent tonsillitis.

Viral infections of the tonsils

Tonsillar tissue can be affected by viral infections, which commonly lead to a sore throat. Symptoms are similar to those of tonsillitis but are usually milder and last for only a few days.

The tonsils are also affected in infectious mononucleosis, in which a sore throat is accompanied by severe lassitude, joint pains, and generalized swelling of all the lymph nodes. In this condition, the tonsils may be covered by a white membrane, and the adjacent palate is dotted with splinterlike hemorrhages. There is no specific treatment available for this viral illness. Supportive measures such as rest, fluids, and anti-inflammatory pain medicines can be of benefit.

Tonsillectomy

A tonsillectomy is performed under general anesthesia. The tonsils are dissected away from the pharyngeal wall and the resulting bleeding is controlled by ligatures. On average about 1/3 pint (120 ml) of blood is lost during surgery, irrespective of the age of a patient. Surgeons are therefore very reluctant to operate on children who are very small or below the age of four, since this amount of blood loss is a significant proportion of their total blood volume. If surgery is necessary, a child will be given an intravenous infusion of fluids after the tonsillectomy and until the child can handle oral intake.

The only serious complication that may occur after a tonsillectomy is further hemorrhage. When this occurs it is usually within the first 24 hours after surgery and requires a return to the operating room for further ligation.

Bleeding may also occur six to ten days after surgery if the tonsil bed becomes infected. Patients most commonly affected are those who eat poorly postoperatively or who have had an attack of tonsillitis immediately before admission to the hospital. Delayed bleeding is treated with antibiotics, but if the patient has lost a lot of blood, a transfusion may be necessary to replace the lost blood. Children tend to recover from surgery more quickly than adults and require only one or two days in the hospital. Adults, however, may need to be in the hospital for four to five days until they are fit to be discharged.

Tumors of the tonsils

Tumors of the tonsils are uncommon but may occur at any age. Lymphomas may cause a sudden tonsillar enlargement and are usually associated with swollen glands in other parts of the body—for example, under the arms and in the groin. They generally respond well to treatment; many patients can be cured. Another type of tumor is known as squamous cell carcinoma. It occurs in the older age group and men are more frequently affected than women. The condition results in an enlarged and painful tonsil, which may be ulcerated. Variable degrees of difficulty and pain on swallowing are experienced, and in advanced cases it may be impossible for the patient to swallow his or her saliva. Similarly, as the disease progresses, the tumor spreads to the glands in the neck. With early treatment a recovery rate of at least 50 percent can be expected.

See also: Immune system; Lymphatic system; Lymphocytes; Pharynx; Temperature

Touch

Questions and Answers

Why do babies touch as well as look at everything around them?

Babies train their brain to match the sight of an object with its feel. When they are older, these earlier experiences enable them to predict what the texture of an object or surface is without touching it.

When someone is paralyzed down one side from a stroke, does he or she lose the sense of touch on that side?

Not necessarily. Some people who have been paralyzed by a stroke retain their sense of touch on that side, if the damage has been confined to the parts of the brain that control movement. If the area of damage is sufficiently large to have involved the touch analyzers in the brain or their connections, then the sense of touch will be damaged.

On very cold days my sense of touch is poor. Why?

In cold weather, two things are working against the touch receptors just below the skin in the fingers. The cold itself reduces their efficiency, and blood is diverted away from the skin in order to minimize heat loss; this poor blood supply further impairs the ability of these nerve endings to send concise messages to your brain.

Do blind people have a better sense of touch than the sighted?

Blind people have the same equipment in their nervous system for touch perception. What makes them able to use it more effectively than sighted people is the practice that this sense has had in the absence of sight. The brain has to rely much more on touch, so the analysis of touch has become more efficient, enabling, for example, the rapid reading of braille.

Touch is so fundamental to life that most people never think of how the many sensations they feel are produced: how, for instance, people can tell silk from sandpaper, or recognize an object simply by the way it feels on the skin.

Touch is one of the first ways in which young babies explore their world, and it remains people's most intimate way of relating to their environment. People have a wide range of receptors in their skin that are sensitive to different types of pressure. Through these they are constantly able to monitor their immediate surroundings and keep the brain in touch with the surfaces on which they sit, the objects they grasp, and so on. However, the sensation of touch is complex and is therefore sensitive to disturbances in many parts of the nervous system.

The sensory receptors

Just below the surface of the skin there are many nerve endings with varying degrees of sensitivity. These allow the nervous system to be supplied with different types of touch sensations.

Wrapped around the base of the fine hairs of the skin are the free nerve endings, which respond to any stimulation of the hair. These touch receptors are the least sophisticated in structure and rapidly stop firing if the hair continues to be stimulated. Receptors found in greater numbers in the hairless part of the skin—for example, on the fingertips and lips—are formed into tiny disks. Because the nerve fibers are embedded within these disks they respond more slowly to pressure and continue to fire when the pressure is maintained. Other more structurally complicated receptors are formed by many membranes wrapped around a nerve ending like an onion skin, and give responses to more constant pressure. In addition, the information that all receptors send into the nervous system tends to be influenced by the temperature at which they are operating. This explains why people's sense of touch can be impaired in cold weather.

▲ *The sense of touch is an important method of communication in relationships, not only between people, but also between people and animals.*

TOUCH PATHWAYS

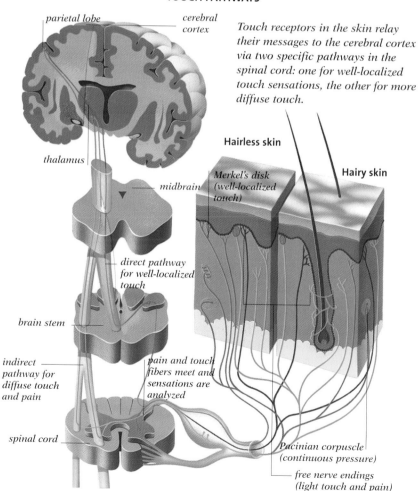

parietal lobe

cerebral cortex

Touch receptors in the skin relay their messages to the cerebral cortex via two specific pathways in the spinal cord: one for well-localized touch sensations, the other for more diffuse touch.

thalamus

midbrain

Hairless skin

Merkel's disk (well-localized touch)

Hairy skin

direct pathway for well-localized touch

brain stem

indirect pathway for diffuse touch and pain

pain and touch fibers meet and sensations are analyzed

spinal cord

Pacinian corpuscle (continuous pressure)

free nerve endings (light touch and pain)

▲ *The touch receptors in our skin are so sensitive that they respond to the gentlest stroking of a blade of grass.*

The distribution of the different types of touch receptors reflects their particular job. The receptors around the base of body hair send messages from large areas of the skin about the pressure stimulating them. They rapidly stop their flow of information once a person has been warned of the presence of an object—for example, an insect on the skin. On the hairless skin, more sophisticated receptors give continuous information, allowing objects to be felt as the brain gathers this information into a coherent picture.

Analysis in the spinal cord

Some of the fibers that convey touch information pass into the spinal cord and, without stopping, go straight up to the brain stem. These fibers deal mainly with sensations of pressure, particularly a specific point of pressure. Therefore, they need to send their messages directly to the higher centers of the brain so that this well-localized sensation can be assessed without confusion from any analysis in the spinal cord.

Other nerve fibers bringing information of more diffuse touch enter the gray matter of the spinal cord and there meet a network of cells that perform an initial analysis of their information. This is the same area that receives messages from the pain receptors in the skin and elsewhere. The meeting in the

▲ *Touch can physically benefit the body. A massage after strenuous exercise can help to ease muscle aches and pains.*

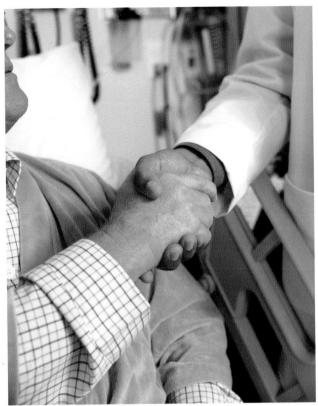

▲ *Touch is often a way of showing friendliness, as in the handshake—a common gesture used to greet people.*

spinal cord of messages dealing with both touch and pain allows for the mixture of these two sensations, and explains such responses as the relief of painful stimuli by rubbing.

This spinal cord analysis filters the sensations, which are then sent upward to the brain. The gray matter of the spinal cord here acts as an electronic gate so that pain information can be suppressed by the advent into the cord of certain types of touch impulse, limiting the amount of trivial information that needs to be transmitted to the higher centers.

This division of the touch pathways to the brain into two streams—one of which goes fairly directly up to the brain stem, and the other that is first analyzed by the cells of the spinal cord—enables the fine discriminating aspects of touch to be preserved. As a result, a person can estimate accurately the amount of pressure in a touch and its position, but if the pressure is too great, or too sharp, the pain analyzers become involved through the connections in the spinal cord and tell the person that the touch is painful as well.

The sensory sorting house

Touch sensations from the skin can come by the more direct route or via analysis in the spinal cord. Either way, they eventually end in the thalamus, a compact knot of gray matter deep in the center of the brain.

The direct touch fibers will have already relayed once in the brain stem and then will have crossed over to the other side, streaming to

the thalamus in a compact bundle. The other fibers will have crossed over to the opposite side of the spinal cord after their relay in the gray matter; so all the touch sensations from one side of the body are analyzed by the opposite side of the brain.

In the thalamus, these pieces of information from various different types of receptor in the skin are assembled and coordinated. This enables the brain's highest centers in the cerebral cortex to put together a picture of the sensations of touch of which a person becomes conscious.

The final analysis

The area of the brain that enables the complex array of touch sensations entering the nervous system to be consciously perceived is the middle section of the cerebral cortex. As with all other sensory information, touch is analyzed by the cortex in a series of steps, each increasing the complexity of the sensory perception. From the thalamus, the raw data are projected to a narrow strip in the front of the parietal lobes.

This primary sensory area of the cortex processes the information before relaying it to the secondary and tertiary sensory areas. In these latter areas the full picture of the site, type, and significance of the touch sensations a person feels is produced and correlated, along with memories of previous sensations and sensory stimuli coming via the ears and eyes. The latter coordination is achieved easily because the areas for vision and hearing back up to the areas for touch.

Touch sensations are also coordinated in the secondary and tertiary sensory areas with the sensations of what position the limbs, joints, and digits are in. This is vitally important, as it enables people to determine an object's size and shape, and helps them to distinguish one object from another.

Problems

Damage to the nervous system at many different levels can alter a person's ability to feel and notice things that touch his or her skin. How this affects the person depends largely on the exact place in the nervous system that the damage occurs.

Damage to the peripheral nerves, which may happen in alcoholism or in diabetes, for example, can affect the sense of touch. However, it takes extensive damage for the sense of touch to be lost completely or severely diminished.

Often people with such disorders feel pins and needles in their hands and feet for some time before any alteration in their sense of touch. The ability of the fingers to make fine touch discrimination may be involved, and sufferers may report feeling as if they have gloves on all the time. Instead of being lost or diminished, the sense of touch can also become distorted as a result of damage to the peripheral nerves, so that a sufferer may say that smooth surfaces feel like sandpaper or warm surfaces feel hot.

Much greater distortion of the sense of touch, however, arises from disease in the spinal cord, for example in multiple sclerosis. If the spinal cord is diseased—or even compressed from the outside—the cross connections that arise here produce distortions of touch that can be disabling and unpleasant. Apart from noticing a feeling

▲ *People tend to show their affection for each other through touch. Hugging, for example, is felt as a pleasant sensation by babies and children as well as adults.*

of numbness, the hands may have lost their ability to make properly coordinated touch perception, for example, in picking the correct coin from a pocket, or the feet may feel as if they are walking on cotton balls instead of firm ground.

Similar types of symptoms can arise from damage to the same touch pathways through the brain stem all the way to the thalamus. Thalamic damage, which happens after strokes, for example, can produce bizarre alterations of touch so that a simple pinprick produces unpleasant, spreading, electric shocklike sensations, or the gentle stroking of a finger may be felt as an unpleasant burning spreading over the skin.

Damage to the parietal lobes of the cerebral cortex, common in strokes and tumors of the brain, may disrupt touch sensations in other ways. If the thalamus is still intact (it is often involved in the disease as well), then the touch will be felt, but the localization of the touch will be inaccurate—it may, for example, be felt on the other side of the body. If the parietal lobe is not functioning, the correlation of different types of sensation will not occur. For instance, usually, when the hand or skin is drawn upon, a person will be able to distinguish letters and numbers, but someone with parietal lobe damage will not recognize shapes, although he or she will be aware that a touch has occurred.

▼ *Touch can help us to discern shapes, so it is obviously important to someone who is blind, since the ability to define shapes is fundamental to reading braille.*

See also: Brain; Hand; Nervous system; Skin; Spinal cord

Transplants

Questions and Answers

Can someone who has had a heart transplant be given another one if this fails or is rejected?

This has been attempted, but there are many problems involved. The rejection of a transplanted heart is often sudden and a further donor heart has to be available at very short notice, which is usually not the case. In addition, a second operation is much more difficult because there is always the problem that rejection can occur more quickly with the second operation. In the case of kidney replacements, however, repeated transplantation happens fairly often, since the patient can be maintained on a kidney machine until an organ becomes available.

My father, who is past retirement age, has bad heart trouble and his medication doesn't work well anymore. Would he be a candidate for a heart transplant?

Very few people are suitable for heart transplantation. The reason for this is that after a long period of heart disease, the lungs and other organs suffer damage; therefore, replacing the heart will not be enough to cure the problem. In general, heart transplants are performed on those whose heart failure is recent but so severe that it is clear that survival is not otherwise possible.

Is it possible to perform a brain transplant, and has it been tried?

To transplant the brain really means to perform a body transplant, since it is the brain that makes each of us an individual. This is not feasible now, nor is it likely to be, since all the nerves would stop functioning irreversibly once cut. The whole spinal cord would have to be transplanted as well, making this an inconceivable technical feat.

Not so long ago the idea of successfully transplanting an organ would have been considered pure science fiction. Today the number of people who have been given a new lease on life by a transplant is rapidly increasing.

The idea of replacing diseased parts of the body with spares is quite old, but it has become a reality only in the last few decades. There are considerable problems that must be overcome before transplants can be done, including locating suitable replacements and resolving important ethical questions that might arise. Despite this, the field of transplant surgery is expanding, since it offers the hope of treating illnesses that must otherwise be disabling or fatal.

Important and valuable work also takes place in the transplanting of other body tissues. Corneal grafting, and bone marrow transplants can be counted among some of the most successful procedures in this field of medicine.

In theory, any organ of the body except the brain can be transplanted in surgery, but a consistent level of success has been achieved only in operations involving the kidneys, heart, and liver. Transplantation of a lobe of a liver from a living donor can be successful and, recently, the injection of donated liver cells has been shown to be valuable in children with liver disease. Heart and lung transplantation has become fairly frequent; and much experimental work on the transplantation of other organs, and even hands, has been undertaken, with variable results.

▲ *Before any transplant surgery can be performed, vital matching of both blood group and tissue type is carried out. Tissue types are found by testing white blood cells.*

Questions and Answers

Although heart transplants were considered experimental and very risky only two decades or so ago, 26,704 heart transplants have now been performed. Will they soon be considered a routine operation?

Transplants will become more common as problems such as rejection and organ preservation for a pending transplantation are overcome. However, they may be just a stopgap in the history of spare part surgery, since it is likely that synthetic organs will eventually be designed. There would then be no problem of finding donors, and rejection of donor organs would not be an issue.

If I want to donate my body to medical research, can my relatives refuse to give their permission after my death?

No. If a person has expressed a positive wish to donate his or her organs in a will, by carrying a donor card, or by stating the intention in the presence of two witnesses during the last illness, then the relatives have no legal right to be consulted after the death. However, when death takes place in the hospital and the medical examiner wants to perform an autopsy, no part of the body can be removed without the medical examiner's permission.

I have heard that some people have had pancreas transplants to treat their diabetes. Is this true, and were the operations usually successful?

Pancreas transplants for diabetics have been mainly limited to patients who have kidney failure due to diabetes, because a person with a transplanted pancreas must remain on antirejection drugs indefinitely. If, however, this is required because the patient has to have a kidney transplant, the additional benefit of a fully functional pancreas and the cure of the diabetes can both be achieved at no extra disadvantage to the patient.

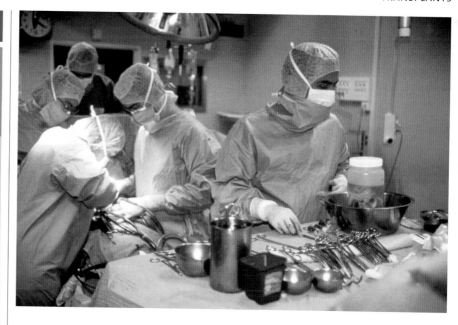

▲ *Surgeons perform a heart transplant on a patient. This type of surgery requires more than one surgeon; it is intricate and can take many hours to accomplish.*

Physical and ethical problems

Finding donor organs to transplant into patients who depend on them for continued life is one of the most difficult of the immediate problems that a surgeon has to face.

The organ should come from a person who was fit and preferably young at the time of his or her death. The tragedy of modern living is that the most likely way for this to occur is through a traffic accident.

The organ to be used in a transplant will deteriorate quite quickly if it is no longer supplied with blood after death. Therefore it must be removed as soon as possible. It is this dilemma that has provoked the most controversy in recent years because the exact point of death was sometimes difficult to define.

The solution to this problem centered on whether the brain was alive in the sense of being capable of recovering independent life support. After careful study of the survival of many victims of brain damage there have now emerged clear-cut ways in which doctors can determine the point at which brain death has occurred.

Once brain death has been confirmed, the organ for transplant can be removed immediately after the respirator has been switched off. Although this can seem macabre, the reality is no more unpleasant than the death of a victim in other circumstances.

There is a crucial shortage of available organs, particularly where kidney transplants are concerned. This is because of the difficulty of coordinating the teams of surgeons doing the transplant, and the doctors caring for the trauma victim.

When an organ has been taken from a dead donor it must be preserved until it can be placed in the recipient. The organ must be placed on ice and special fluid pumped through its blood vessel system to keep the system open and free of blood clots.

The technology required during this crucial period is being constantly updated and in the case of kidneys is quite advanced. The kidney is simply flushed through to remove the blood. It is then put into a polyethylene bag that is surrounded by ice. It can be kept in this state for up to 36 hours before being used.

Living donor transplants

Transplants from living donors are common in kidney transplantation, since a normal person can easily survive in good health with only one kidney. If a relative is prepared to give one of his or her kidneys, this is often the best match, since it lessens the risk of rejection. A close relative such as a brother, sister, or parent will share many genetic characteristics of the recipient; there is thus

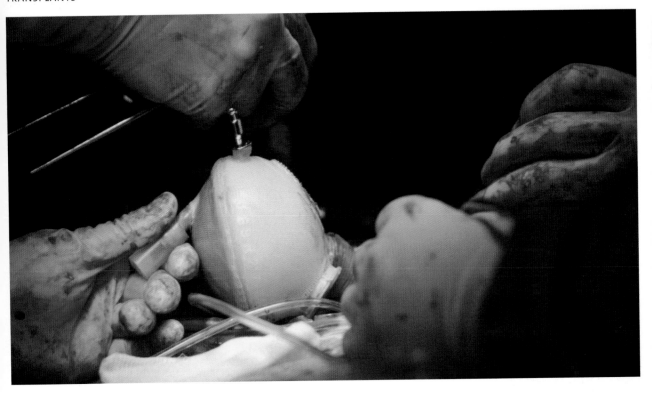

a chance that the recipient's immune system will not recognize the graft as foreign and reject it. The kidney for transplantation can be removed calmly without the urgency that is called for in a recently dead donor, allowing more time to plan the operation.

Rejection of the transplant

Rejection is the main problem of transplant surgery, and directly or indirectly leads to most of the failures that occur. The body has a powerful defense system designed to repel the invasion of its domains by bacteria and viruses. The invaders are recognized because their chemicals are subtly different from those of our body. Unfortunately when a transplanted organ comes into contact with the body cells that act as soldiers in this defense force, the chemical makeup of the transplant is recognized as foreign and a reaction is mounted against it as though it were a host of bacteria.

For this reason attempts are made before transplant surgery to match the type of tissue of the donor organ and the recipient. This is similar to blood grouping, since everyone has a tissue type as well as a blood group. Tissue types are discovered by testing the white blood cells; blood groups reflect the chemical makeup of the more numerous red blood cells.

Apart from trying to match the blood group and tissue type of the donor organ with that of the patient receiving it, there are various things that can be done to stop rejection. For example, it has been discovered that the more transfusions a patient has prior to surgery, the less likely he or she is to reject the transplant. This preventive effect occurs because the body's immune system is confronted many times with foreign tissues of donated blood. It seems to become more tolerant of other invasions and less likely to reject a transplanted organ. More deliberate prevention

▲ *At the University of Utah in Salt Lake City, surgeons transplant an artificial heart into a calf.*

takes the form of using powerful drugs to control the immune system so it cannot mount its attack on the new organ. This is a double-edged sword, since the body needs some immunity to fight against infecting bacteria, viruses, or fungi. The development of the drug cyclosporin was a major breakthrough. This drug selectively discourages rejection of tissue without unduly interfering with the other protective functions of the immune system.

Other complications

Apart from the complications caused by the drugs used in preventing rejection, other problems can arise. These include surgical problems that are related to the considerable task of sewing the organ in place. The blood vessels into the transplant must be carefully joined to prevent bleeding once the circulation is reestablished. Similarly, in the case of the kidney, the tubes carrying urine from the transplant must be delicately sewn into the recipient's bladder; and in a liver transplant, bile ducts must be implanted into the intestine.

The surgery

Usually, with the kidneys or the liver, the whole of the organ is transplanted into the patient's body, although segmental liver grafts can be performed from a live donor. The liver is placed similarly to its normal position, though the arteries used to supply it with blood are changed by grafting them to nearby intestinal arteries.

It is difficult to insert a new kidney in the position of the diseased one, which is usually left where it is. The new kidney is

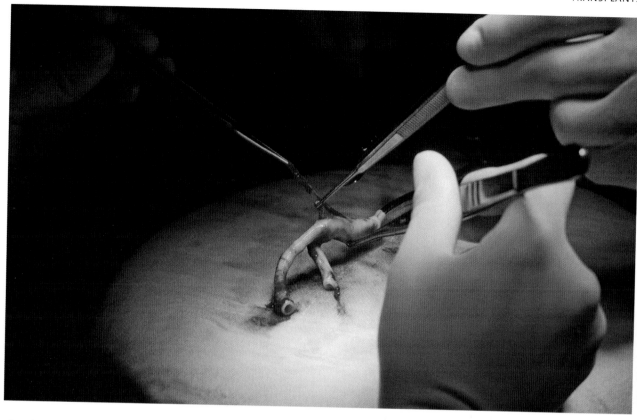

placed in the pelvis, which is conveniently spacious; this allows the new kidney's blood vessels to be joined to the large vessels to and from the legs. Another consideration is that in this position the joining of the ureter, the tube that carries urine to the bladder of the recipient, is made far easier. When a heart is transplanted the whole of the diseased heart is not removed. This is both

▼ *Tissue expanders, traditionally used in plastic surgery, are to be used under the skin in pioneering transplant surgery.*

▲ *Surgeons prepare a section of artery for use in a liver transplant operation. The artery being used comes from the dead victim of an automobile accident.*

unnecessary and unnecessarily complicating. Two large veins enter the right upper chamber and four large veins enter the left upper chamber of the heart. Because of this, the greater part of the two upper chambers of the patient's own heart is commonly left in place so that six fewer major blood vessels have to be connected. It is usually the lower chambers, the ventricles, that are severely diseased and it these that are replaced by the donor heart, leaving only two large arteries to be reconnected.

Postoperative care

After any surgery there is a risk of infection, but the recipient of a transplant is at particular risk. Measures are taken to protect the patients from germs, by isolating them in a special room, or by ensuring that people attending them and visiting them wear special clothing and masks—or both. Acute rejection is a risk a few days after the operation, and even after this there may be a risk of chronic rejection, which is slower in causing the loss of the organ but equally dangerous. Once the drugs suppressing the rejection process start to work, the person may gradually return to a fairly normal life. In the case of many kidney transplants, the recipients can return to work, and many women who have had a kidney transplant have given birth to healthy children.

See also: **Blood; Blood groups; Brain; Immune system; Kidneys; Marrow**

Urethra

Questions and Answers

Is urethritis always caused by a sexually transmitted disease?

No, but a sexually transmitted infection is the most common cause. This is because the penis in men and the area around the vagina that includes the opening of the urethra in women are usually in close contact during intercourse. Women may also get urethritis through contamination from the anus; to avoid this, they should always wipe themselves after defecating, or dry themselves after bathing, from the front toward the back, never the opposite.

Can urethritis always be cured?

Yes, almost always, though this depends on doing tests to identify the precise cause so that the appropriate treatment, usually an antibiotic, can be given. However, just taking the antibiotic may not be enough to produce a cure, since like any other infected area of the body, the urethra needs a period of rest for full recovery. New lining tissue can then become established and replace that which has been destroyed by the inflammation. It is usually necessary to refrain from intercourse for about two weeks or the mechanical stress and friction on the urethra will damage the new lining tissue before it has had a chance to settle down, causing a further attack of urethritis.

A friend told me that her little boy had hypospadias. What is that?

Hypospadias is an abnormality in the development of the penis. This occurs before birth and as a result the opening, or meatus, is on the underside of the penis rather than at the tip. In most cases there is no difficulty in passing urine, having intercourse, or fathering children, and no treatment is necessary. In the few cases where there is a problem, it can be corrected with minor surgery.

The urethra is a channel from the bladder to an opening on the outside of the body along which urine is discharged. It is a common site of urethritis—an infection that can be very painful and always needs medical attention.

CROSS SECTION OF THE URETHRA

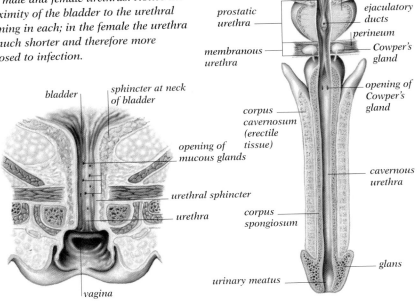

The male and female urethras. Notice the proximity of the bladder to the urethral opening in each; in the female the urethra is much shorter and therefore more exposed to infection.

The mature male urethra averages 8 inches (about 20 cm) in length and consists of three sections. The first, or prostatic, section is about 1 inch (2.5 cm) long and passes from the sphincter, or valve, at the outlet of the bladder through the middle of the prostate gland. The middle part of the urethra is only about 0.5 inch (12 mm) long and is often called the membranous urethra. The final—and, at over 6 inches (15 cm), the longest—section is called the spongy or cavernous urethra. This section is within the penis and opens at the slit in the tip, the urethral meatus. In men, the urethra is also the channel through which semen is ejaculated.

In women the only function of the urethra is as a channel for urine disposal. Surrounded by mucous glands, it is much shorter—about 1.5 inches (4 cm) in length. Because of its shortness, and because it opens into a relatively exposed, contaminated area, women often get urinary infections.

Urethritis

Inflammation of the urethra, called urethritis, is the most common urethral disorder. It can have many causes, the most common of which is a sexually transmitted disease. The symptoms in the male are a discharge that leaks out from the urethral meatus, increasing pain on urinating, and a desire to urinate frequently. In women, usually only the pain (dysuria) and the frequency of urination are present. These symptoms are often attributed to cystitis or inflammation of the bladder, but it is more commonly the urethra that is involved. If urethritis is not fully treated, usually with antibiotics, permanent damage can be done both to the urethra itself and to the reproductive organs.

See also: Bladder; Genitals; Prostate gland

Urinary tract

The urinary tract is one of the major systems of the body. It has the important function of disposing of waste products that would, if retained, eventually cause serious illness or death. The urinary tract system is prone to a wide range of disorders.

Questions and Answers

My son, who is seven, has urinary tract problems. A urologist arranged for him to have an X-ray while he was urinating. Is this a proper kind of test?

Yes. It is used extensively in children. It is called micturating cystourethrography (MCU), which just means an X-ray of the bladder and the outlet tube (the urethra) while the patient is urinating. It is useful in investigating abnormal bladder emptying and abnormalities in the urethra.

Is it true that women are more prone to cystitis than men? And if so, why is this?

Urinary tract infections are more common in women than in men. This is generally thought to be because the urethra is much shorter in women than in men and its external opening is more easily contaminated than a man's. The idea is encouraged by the fact that most urinary infections in women are due to coliform organisms. This is not the whole story; the symptoms of cystitis, in the absence of any infection, are also more common in women than in men.

My mother is 70 and often has urinary infections. She needs to go to the toilet several times during the night but produces only a small amount of urine. Last week I persuaded her to see the doctor and he prescribed an estrogen vaginal cream. Why?

Many cases of postmenopausal urinary infection are the direct result of an atrophic condition of the vagina from estrogen deficiency. This leads to a change in normal healthy vaginal bacteria and frequent bladder infections. There's a good chance that this prescription will clear up your mother's problems.

The urinary tract consists of two kidneys; two tubes called ureters that run down from the kidneys to enter the back of the base of the bladder; the expansible muscular bladder itself with its control sphincter; and a single outlet tube, the urethra, that carries urine from the bladder to the exterior. In women, the reproductive system is entirely separate from the urinary tract, but in men, these functions are partly combined and the urethra is a conduit for seminal fluid as well as urine. Another important difference between the sexes is that in males the part of the urethra immediately below the bladder is surrounded by a gland, the prostate gland, that tends, in elderly men, to enlarge and obstruct the outflow of urine.

Function of the urinary tract

The kidneys have several vital functions. They regulate the total volume of fluid in the body and the chemical composition of that fluid. They remove the waste material produced by the body's metabolic processes, mainly urea from protein breakdown, uric acid from DNA and RNA, creatinine from muscle, and bilirubin from red blood cells. The kidneys excrete drugs. They can synthesize glucose during starvation; they also secrete hormones that promote new blood cell formation, control blood pressure, and influence calcium balance.

Urine produced by the kidneys, at an average rate of about 1 ml per minute, passes down the ureters to be stored in the bladder. This is an unconscious process for a time because the bladder expands easily as it fills. The adult bladder has a capacity of 12–15 fluid ounces (350–450 ml), but can expand to about 24 fluid ounces (700 ml) if urination is deliberately delayed. At that point of expansion, however, voluntary control is lost.

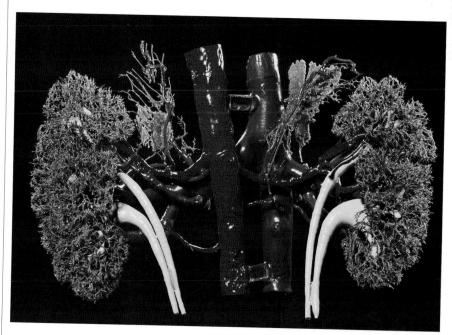

▲ *A resin corrosion cast shows the blood vessels of the kidneys. The arteries are red, the veins are blue, and the ureters and renal pelvis are yellow.*

Nephritis

The most common form of kidney inflammation affecting the urine-producing parts of the organ is called glomerulonephritis. This is not the result of an infection but is caused by abnormal functioning of the immune system (autoimmune disease) following the production of antibodies to a streptococcal infection. Most cases cause no serious long-term effects, but a small proportion proceed to a long-term or chronic state that may end in total loss of kidney function, that is, kidney failure. This will require dialysis or kidney transplantation. Nephrosis is a chronic degenerative disease of the kidney substance, and there is severe loss of protein in the urine. It may follow glomerulonephritis, other diseases, or drug reactions.

Urinary tract infections

Like every other system in the body, the urinary tract is subject to a wide range of diseases and disorders. The most common of these are infections. Urethritis is inflammation of the lining of the urethra, usually from sexually acquired infection. The common causes are chlamydia and gonococci.

Infection of the urinary bladder is very common, especially in women; 80 percent of the germs are coliform species from the lower bowel. Many other germs can, however, infect the bladder. Infection can spread upward from the bladder to the kidneys to cause pyelonephritis, an infection of the kidneys and the tubes that lead away from the kidneys. Abscesses can also occur in the kidney tissue, and primary infections of the kidneys with germs such as the tubercle bacillus may occur.

Urinary tract cancer

Cancer of the penis is rare, especially if personal hygiene is good, with daily washing. Apart from prostate cancer, which is one of the most common of all kinds of cancer, bladder cancer is the most frequent form of urinary tract cancer. It accounts for 7 percent of new cancers in men and 3 percent in women. It is mainly due to

▲ *A scanning electron micrograph of nephrons, the filtration tubules in the kidney, magnified 140 times. Nephrons are the functional units of the kidney; they filter out metabolic waste.*

cigarette smoking and exposure to industrial carcinogens. Bladder cancer can be diagnosed by direct visual examination through a cystoscope and the taking of biopsy samples. The outlook depends on the stage which the disease has reached at the time of diagnosis.

Cancer of the kidney accounts for about 3 percent of adult cancers. The only consistently linked risk factor is cigarette smoking. Again, the outlook depends on the stage at the time of detection. Wilms' tumor (nephroblastoma) is a rare congenital kidney cancer that shows itself in the first three years of life. Treatment of both bladder and kidney cancers is mainly surgical.

Other urinary tract disorders

Congenital defects of the urinary tract are fairly common. They include absent kidneys, underdeveloped kidneys, joined kidneys (horseshoe kidney), and multicystic kidneys, a condition in which cysts replace normal tissue until the kidneys fail completely. Polycystic kidneys in adults are due to the late effects of congenital defects in the tubular system.

Stone formation is common in the urinary tract and this may occur in the kidneys or in the bladder. Kidney stones in the ureters cause severe pain called renal colic. Stones can sometimes be broken up by focused sound waves (extracorporeal shock wave lithotripsy).

> *See also:* Bladder; Immune system; Kidneys

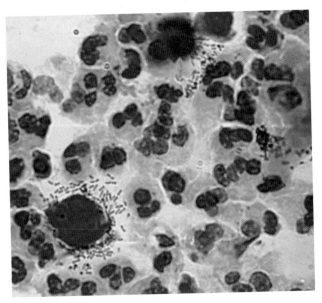

▲ *A gram-stained micrograph of a urethral discharge shows the presence of* Trichomonas vaginalis.

Uterus

The workings of the uterus are something of a medical mystery. What is undeniable, however, is that it is perfectly adapted for the protection and nurturing of an unborn child.

Questions and Answers

Does the Pill affect the uterus?

Yes. There are several types of contraceptive pill, all containing a synthetic form of progesterone at levels that make the endometrium, or lining of the uterus, unsuitable for the implantation and growth of a fertilized egg. The Pill affects the secretions from the cervix to prevent sperm from swimming through the cervix to fertilize the egg. Many contraceptive pills also contain a synthetic estrogen. This helps to prevent pregnancy by inhibiting the release of eggs from the ovary.

Is it true that some women may go into premature labor because the cervix is too weak to remain closed until the baby is mature?

Doctors are often unable to tell a woman why she had her baby prematurely, and one preterm birth is a source of anxiety for any subsequent pregnancies. Occasionally a woman has an abnormal cervix that can be stretched open too easily; it can usually be treated by putting a purse-string type of suture into the neck of the womb to keep it closed. This is inserted under anesthesia, and is removed when the woman is in the 38th week of pregnancy.

Why do women stop menstruating at menopause?

When a woman menstruates she sheds the lining to the womb, called the endometrium. The endometrium is stimulated to grow by two ovarian hormones, estrogen and progesterone. At menopause the ovaries no longer secrete the hormones, the endometrium does not regrow, and the woman has no more periods. The onset of menopause is around 45 to 55. If periods are heavier or more frequent during this time, a doctor should be consulted.

In the past, the uterus has been blamed for almost every mental and physical ailment women suffer. Today we have a more rational, though still incomplete, understanding of this vital organ.

The uterus is composed of two main parts; the corpus or body of the organ, and its cervix or neck. It is capable of undergoing major changes during a woman's reproductive life.

From puberty to menopause, the lining of the womb (endometrium) develops each month under the influence of hormones to provide nutrition for a fertilized egg. If the egg is not fertilized, the endometrium is shed during menstruation and is slowly replaced in the course of the next menstrual cycle.

During pregnancy the uterus expands, allowing the fetus to grow and providing it with protection and nutrition. Simultaneously, contraction of the large muscle fibers is prevented. When the fetus is mature, the uterus suddenly changes its role and begins to contract in order to open the cervix and allow both the baby and the placenta to pass through. It then contracts tightly to close off the large blood vessels that have been supplying the placenta. After birth the uterus rapidly returns to its prepregnant state, ready to accept another fertilized egg. Reportedly, this is known to have happened as early as 36 days after a delivery. The uterus seems to have almost no function prior to puberty and after menopause, when it would obviously be unsuitable, both mentally and physically, for a woman to have a baby. All these changes in the functioning of the uterus are caused by hormones released from the pituitary gland, and from the ovaries, and by similar substances called prostaglandins, released by the uterine tissue. How these substances interact is still not fully understood.

Position

In an adult woman, the uterus is a hollow organ approximately the size and shape of a small pear. It lies inside the girdle of pelvic bones. The narrow end of the pear is equivalent to the cervix, which protrudes into the vagina; the remainder forms the body of the uterus. This organ is connected to two fallopian tubes that carry the monthly egg from the ovaries. The uterus forms part of a channel between the abdominal cavity and the exterior of the body. Special mechanisms exist to prevent the spread of infection by this route; the lining of the uterus is shed when a woman menstruates; the cervix secretes antibodies; and the acidity of the vagina inhibits the growth of bacteria.

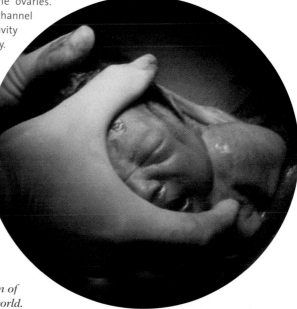

▶ *Pain, joy, and wonder combine in this unique moment, as a new baby safely completes the journey from the protection of the uterus to the outside world.*

Problems of the uterus

PART OF THE UTERUS AFFECTED	NAME OF CONDITION	POSSIBLE SYMPTOMS	TREATMENT
Entire uterus	Absent	No periods, infertile	None
	Congenital malformation (a double uterus or an abnormal division in the cavity of the uterus)	Often no symptoms (when pregnant, a woman may go into premature labor)	Very rarely it may be necessary to do surgery to make the uterus a normal shape
	Prolapse	No symptoms or the sensation of "a lump coming down into the vagina"	Special pelvic floor exercises, avoidance of constipation, weight loss if necessary. Sometimes surgery.
Endometrium (lining of the cavity of the uterus)	Endometrial polyps	Bleeding from the vagina between periods or after menopause	Dilatation and curettage (D & C)
	Endometrial hyperplasia (overgrowth of the lining of the uterus)	Heavy irregular periods, usually as a woman approaches menopause	Diagnosis by D & C followed by a course of hormone pills
	Endometritis (inflammation of the lining of the uterus)	Lower abdominal pain and heavy periods	Diagnosis by D & C followed by a course of antibiotics
	Endometrial carcinoma	Bleeding from the vagina after menopause or between periods	Hysterectomy and possible radiotherapy or hormones, or both
	Trophoblastic disease (formation of placenta in the uterus with no fetus present)	A feeling of pregnancy and irregular bleeding from the vagina	D & C, avoidance of pregnancy until the condition is cured, drugs if the placental tissue spreads outside the uterus
	Dysfunctional (abnormal) uterine bleeding	Heavy or frequent periods, or both, with no obvious physical abnormality of the uterus	Diagnosis by D & C, hormone treatment followed by hysterectomy if hormones are unsuccessful
Myometrium (muscular wall of the uterus)	Fibroids	Often no symptoms—possibly heavy periods and enlarged uterus	If they cause a problem, fibroids may be surgically removed (myomectomy) or a hysterectomy may be performed
	Sarcoma (cancerous form of fibroid)	Often no symptoms—sometimes the uterus becomes enlarged	Hysterectomy and radiotherapy
	Adenomyosis (endometrial tissue deposited inside the muscular layers of the uterus)	Painful, heavy periods	May be treated by hormonal therapy but often only diagnosed by a hysterectomy
Cervix	Polyp	Often no symptoms—sometimes vaginal bleeding after sexual intercourse or between periods	Removal of polyp
	Erosion or ectropion (cells which normally line the cervix begin to grow on the outside)	Often no symptoms—sometimes a watery discharge	No treatment needed, but the area may be burned so that new cells form (cautery)
	Cervical dysplasia (abnormal cells which may revert to normal or become cancerous)	Abnormal cervical smear	Repeated cervical smears, colposcopy, and occasionally removal of the area of abnormal cells
	Cancer of the cervix	Bleeding from the vagina after sexual intercourse and between periods	Radiotherapy and sometimes hysterectomy

Questions and Answers

I have a retroverted uterus. Can I still get pregnant?

A retroverted uterus tilts backward from its attachment at the top of the vagina instead of forward (an anteverted uterus). This is quite normal and occurs in about 20 percent of women with no ill effect. There is certainly no evidence that you will be less fertile. Rarely, a disease such as endometriosis or severe pelvic infection will cause a normal anteverted uterus to become retroverted. These diseases can decrease a woman's fertility, and since they are associated with a retroverted uterus, it is sometimes said that retroversion of the uterus causes infertility.
A problem like endometriosis can be treated by surgery or hormone therapy, which does not always cure the retroversion but will improve the woman's chances of conceiving.

Is it true that some women have two uteri?

During the development of the uterus in the fetus, two ducts, called the Müllerian ducts, fuse. This happens about the 65th day of pregnancy, and the fused portion forms the uterus. Occasionally the fusion of the ducts is incomplete, and the woman may have an abnormality such as a dimple at the top of the uterus or two separate uteri, which is rare. Women with this problem seldom have difficulty in becoming pregnant, but they are more likely to have miscarriages or go into premature labor.

Is it possible for a baby girl to have a period?

Estrogen and progesterone can cross the placenta from the mother to stimulate the growth of the lining of the uterus in the developing fetus. Once the baby is born the levels of hormones in the baby's blood rapidly fall, and she sheds the lining, which shows as a pink stain in the diaper a few days after birth. This does not occur again until puberty.

NORMAL CHANGES IN THE UTERUS

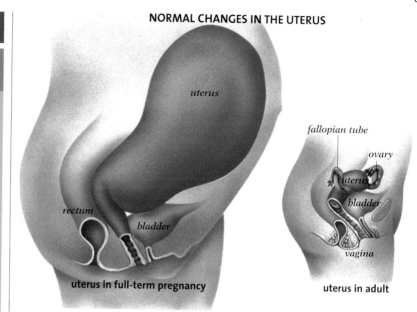

uterus in full-term pregnancy

uterus in adult

▲ *The adult nonpregnant uterus is usually tilted forward at an angle of about 90° in relation to the vagina; its muscular walls are thick and its cavity is a mere slit. During pregnancy, the walls expand dramatically to accommodate the fetus and the amniotic sac.*

The anterior (front) of the uterus sits on the bladder and the posterior (back) lies near the rectum. The uterus is normally supported inside the pelvis by the pelvic floor muscles, and by bands of connective tissue and blood vessels from the side wall of the pelvis, which are attached to the cervix.

During pregnancy the uterus enlarges so that by the 12th week it can just be felt inside the abdominal cavity above the pubic bone. At about 38 weeks it usually reaches the lower end of the rib cage, and about two weeks after the baby is born the uterus can normally no longer be felt in the abdomen. After menopause, the uterus shrinks, but not to its former size. These size variations are controlled by sex hormones, which also control the nature of the glandular tissue lining the uterus (the endometrium).

▼ *It is astonishing that a baby of 8 lb. (3.5 kg) or more could grow inside this organ, which is normally about 3 in. (7.5 cm) long.*

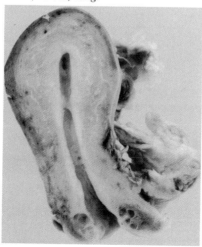

During the first half of the menstrual cycle, the endometrium increases in thickness until an egg is released, then stops growing and begins to secrete substances rich in nutrients to allow growth of the egg should it be fertilized; if it is not fertilized, the endometrium is shed during menstruation.

Congenital variations

During the development of the female reproductive organs in the fetus, two tubes of tissue, called Müllerian ducts, grow from the side wall of the abdominal cavity and meet centrally. These tubes continue to grow downward until they fuse with tissue that will later form the lower vagina. The upper portions of these tubes become the fallopian tubes, and the lower central portions fuse to form the uterus and upper vagina. Very rarely, both Müllerian ducts fail to form, with the result that an adult woman will have a short vagina and no uterus or fallopian tubes. Nothing can be done to cure this condition, although

NORMAL CHANGES IN THE UTERUS

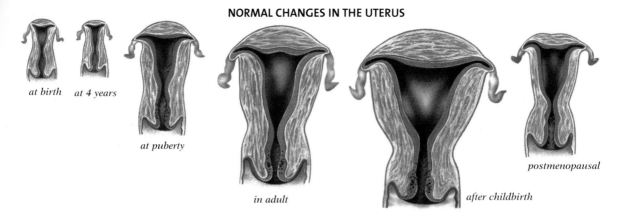

at birth   *at 4 years*

at puberty

in adult

after childbirth

postmenopausal

▲ *In a female fetus, the growth of the uterus accelerates during the last two months before birth, probably owing to the high level of maternal hormones present. Within a few days of birth the uterus shrinks, and it remains static until a year or two before the menarche, when the ovaries start to produce hormones. These stimulate the uterus to grow, so that by the time a girl is about 15, it has reached adult size. Pregnancy enlarges the uterus but it shrinks again after menopause. The adult nonpregnant uterus is usually tilted forward at an angle of about 90 degrees to the vagina; its muscular walls are thick and its cavity is a mere slit. In pregnancy, the walls expand to accommodate the fetus and the amniotic sac.*

sometimes surgery is performed to lengthen the vagina. Such women are infertile and do not menstruate.

If only one of the Müllerian ducts develops, the woman will have a uterus and vagina but only one fallopian tube. This does not cause any major problems.

Another rare occurrence during fetal development is incomplete fusion of the Müllerian ducts. This may result in any abnormality

from a double uterus to a small dimple at the top of the uterus. If such abnormalities create difficulties for the woman in carrying a pregnancy to term, plastic surgery can be performed to re-form the uterus into a single cavity.

Considering the complex mechanisms that control the normal functioning of the uterus, it is remarkable how few women have any problems. Several different conditions can give rise to the same symptoms as menstruation. For example, bleeding from the vagina between periods or after intercourse is often due to a minor condition such as a polyp, which can be easily treated. However, such symptoms may be caused by uterine cancer, which can be completely cured if it is detected early enough. For this reason, it is very important for any woman who has these symptoms to promptly seek her gynecologist's advice.

Treatment

Abnormalities in the uterine cavity are difficult to diagnose because a direct examination is not possible; a woman may need to have a dilatation and curettage (D & C) to diagnose the cause of a menstrual problem or vaginal bleeding after menopause. Once a correct diagnosis has been made, a doctor can then prescribe appropriate antibiotic drugs or hormones.

▼ *During the menstrual cycle, the lining of the uterus thickens and becomes rich in nutrients (left). It disintegrates and sloughs off during menstruation (right).*

Doctors, however, are not always able to control all abnormal menstrual symptoms in this way. Sometimes, the patient and the doctor may agree that the only cure is a hysterectomy. This will stop a woman from having any more periods, and will also make her infertile, but should not otherwise alter her life. However, modern medical therapies are improving and this operation is being performed less frequently. It is hoped that in the future most problems of the uterus will be treated simply with medication.

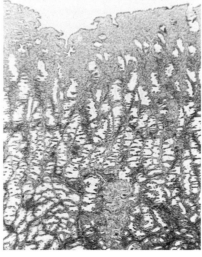

See also: **Cervix; Hormones; Menopause; Menstruation; Ovaries; Pelvis; Pituitary gland; Placenta; Pregnancy; Puberty; Vagina**

Vagina

Questions and Answers

Part of a woman's genitalia, the vagina is the passage for the creation of life, for giving birth, and for sexual pleasure. Although it is prone to some minor disorders, most can be treated easily if medical advice is sought promptly.

The vagina is the channel that leads from the vulva to the uterus. During a woman's life the vagina undergoes several changes. A child's vagina is smaller than that of a mature woman. The lining of the wall of the vagina is also thinner in a child or postmenopausal woman than in a woman in the reproductive years of her life. These changes are influenced largely by a group of hormones called estrogens, which are released by the ovaries.

The vagina plays an important role during intercourse and childbirth. The role during childbirth is relatively passive: the vagina forms the lower portion of the birth canal and is capable of opening sufficiently to allow the birth of the baby. It is only relatively recently that the vaginal changes during sexual intercourse have been described fully.

The vagina is a canal 2¾ inches (7 cm) to 3½ inches (9 cm) long. It is surrounded by fibrous and muscular tissue, and is lined with a layer of cells called squamous epithelium. The walls of

Can using a vaginal douche after intercourse prevent pregnancy?

No. Scientists have shown that sperm can swim from the vagina into the uterus (womb) within 90 seconds. This is far too rapid to make a vaginal douche effective.

Is a vaginal discharge normal, or does it show something is wrong?

All women have some vaginal discharge during their fertile years; the amount varies from woman to woman. The discharge also increases at certain times— such as when a woman ovulates or feels sexually aroused, or during pregnancy. However, certain types of discharge do indicate an infection. Women should seek medical advice if the discharge is thin, foamy, and foul-smelling; if it is thick and white; or if it causes soreness or irritation.

During the recent birth of my daughter, my vagina tore and needed stitches. How long should I wait before I have intercourse?

These tears usually take between three and six weeks to heal. You could examine yourself to see if there is any tender area, and if the tear feels healed, then you could attempt sexual intercourse. However, remember to use contraception unless you want another baby soon; some women have become pregnant again as early as 36 days after giving birth.

Could a cyst at the entrance of my vagina be a sign of cancer?

This is very unlikely. It is probable that a duct from the Bartholin's gland has become blocked. Although you should see your gynecologist, the cyst does not necessarily need treatment. However, if it is uncomfortable, it can be cured by minor surgery.

STRUCTURE OF THE VAGINA

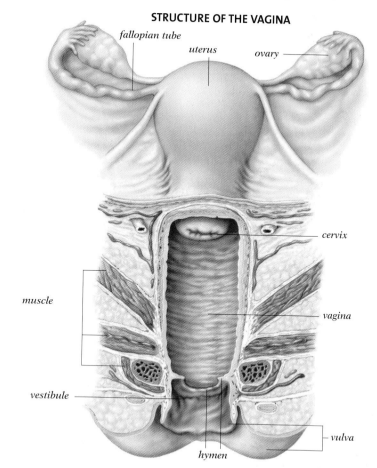

fallopian tube

uterus

ovary

cervix

muscle

vagina

vestibule

vulva

hymen

▲ *The vagina is a tough muscular canal situated between the uterus and the vulva. Its corrugated structure is designed specifically to give it the elasticity necessary to allow the passage of the baby during childbirth.*

Questions and Answers

Does a woman's vagina become smaller after menopause, and can this reduce sexual enjoyment?

After menopause the vagina ceases to have a high estrogen level—the hormone that seems to be responsible for its elastic properties and the thickness of the lining wall. This may cause the vagina to become narrower and more rigid, especially if the woman stops having regular sexual intercourse. However, it is seldom a problem if she continues to have regular coitus.

How long after giving birth can you have a diaphragm fitted?

It takes about six weeks after the birth of a baby for the birth canal, including the vagina, to return to its normal prepregnancy state. A diaphragm cannot be fitted until after this time.

Are vaginal deodorants necessary, or can they be harmful?

Simple practices such as regular showers and clean underwear are adequate hygiene measures. Vaginal deodorants can cause a reaction in some women, resulting in a heavier, more unpleasant vaginal discharge than normal. If a woman suffers from an unpleasant, foul-smelling vaginal discharge she should consult her gynecologist rather than use a deodorant.

Can an abnormal vaginal discharge be a symptom of gonorrhea or syphilis in a woman?

Discharge is a common symptom in the early stages of gonorrhea. It should not be ignored, since a reduction in the discharge often means the infection is spreading to other organs to produce pelvic inflammatory disease. Syphilis does not cause a discharge and may go undetected, since the initial symptom is painless ulcers on the genitalia. Most women learn they have syphilis only when their partners are diagnosed.

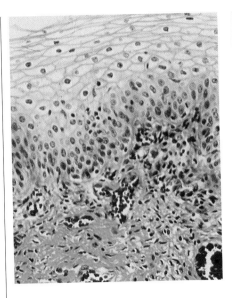

▲ *The vagina is lined with a layer of cells called squamous epithelium. During sexual intercourse, it is lubricated by secretions that seep through these cells.*

Maintaining good health

Take a bath or shower every day
Change your underwear daily
Wear only cotton underwear
Never wear nylon panty hose directly against your skin
Avoid tight pants and jeans
After urinating, always wipe from front to back
Never use douches or vaginal deodorants
See your gynecologist if you notice an unpleasant vaginal discharge

the canal are collapsed normally and folded onto one another. These properties make it easy for the vagina to be stretched during intercourse or childbirth. The urethra lies on the front wall of the vagina and the rectum lies on the upper third of the back of the vagina. The anus is separated from the vagina by a fibromuscular tissue called the perineal body. The ducts from two glands called Bartholin's glands enter on either side of the outer end of the vagina; the cervix protrudes into the top of the vagina.

Function of secretions

During a woman's reproductive years the vaginal secretions are slightly acidic, and this acidity tends to inhibit the growth of harmful bacteria in the vagina. However, during the prepubertal and postmenopausal years the vagina becomes mildly alkaline. In these years bacteria can thrive and occasionally make the vagina sore and uncomfortable, a condition called atrophic vaginitis.

The walls of the vagina are well lubricated with secretions from the cervical canal and Bartholin's glands. During intercourse, secretions also seep through the vaginal epithelium into the vaginal canal. A certain amount of discharge from the vagina is normal in all women. The amount increases during ovulation and sexual arousal.

During sexual arousal, a woman's genital organs, especially the labia minora and lower vagina, become engorged with blood, and the amount of vaginal secretion increases. During an orgasm the muscles of the pelvis, including those surrounding the vagina, contract involuntarily.

If a woman is particularly tense or anxious during intercourse, the muscles surrounding the vagina will go into spasm. This makes the vagina narrower and makes sexual intercourse painful. This condition is called vaginismus. It can be cured by help from a psychosexual counselor, but it may take many months before the woman can enjoy sex fully.

Disorders

Some women find it embarrassing to consult their gynecologist about problems having to do with the genitalia. However, because most of the problems are relatively minor and respond to simple forms of treatment women should seek medical advice early before complications occur.

One of the most common problems is an irritating vaginal discharge called yeast, which is caused by the *Candida albicans* fungus. This can be treated easily with vaginal pessaries. However, because it is easy to become reinfected, the woman's sexual partner should seek treatment at the same time. She should also be careful that towels and underwear are laundered thoroughly, since the spores of the fungus can lodge in these articles.

Vaginal disorders and treatment

AGE	SYMPTOM	CAUSE	TREATMENT
Infancy and childhood	Vaginal discharge	Foreign body	Hygiene and antibiotics
Puberty	Failure to menstruate and have regular monthly periods	Imperforate hymen (membrane at entrance to vagina that prevents menstrual loss, so blood is trapped in the vagina)	Minor surgery to incise the membrane
		Hormonal problem	Evaluation of hypothalamus and pituitary gland
	Failure to menstruate in an otherwise normal female due to absent vagina	Failure of vagina to form in the embryo; usually uterus is also absent	Surgery to make artificial vagina. Patient very unlikely to be fertile or menstruate.
Reproductive years	Vaginismus	Psychological or physical causes	Psychosexual counseling
	Vaginal discharge	Physiologically normal	No treatment
		Foreign body in the vagina such as forgotten tampon	Removal of foreign body
		Vaginal infection with *Candida albicans* or *Trichomonas vaginalis*	Treatment by uterine creams or pills; also treat patient's partner
	Cyst to one or other side of entrance to the vagina	Bartholin's cyst (drainage duct from Bartholin's gland is blocked)	Minor surgery to allow it to drain (marsupialization)
	Abscess (tender red swelling at entrance to vagina)	Bartholin's abscess	Minor surgery to open and drain abscess. Antibiotics are sometimes prescribed at the same time.
	"Lump coming down in the vagina" or incontinence of urine when laughing or talking	Vaginal wall prolapse	Weight loss and pelvic floor exercises. Fit patients, surgical operation; unfit patients, vaginal ring insert.
	Bloodstained discharge from vaginal wall	A very rare cause is carcinoma of the vagina	Surgical removal of tumor
Menopause	"Lump coming down in the vagina"	Vaginal wall prolapse or prolapsed uterus	Weight loss and pelvic floor exercises. Surgery or vaginal ring insert.
	Sore vagina, pink-stained vaginal discharge	Atrophic vaginitis (infection due to nonspecific bacteria)	Mildly acidic jelly or estrogen creams inserted into the vagina to inhibit bacteria
	Pain during intercourse due to dry vagina	Lack of estrogen in the woman's circulation	Use of lubricating jelly or estrogen cream in the vagina, or hormone replacement
	Pain during coitus due to narrow vagina	See above	May respond to application of estrogen cream to the vagina

Another common problem is atrophic vaginitis. This condition often affects women in their mid-sixties and it tends to make the vagina sore and uncomfortable. This occurs because women of this age no longer have high enough levels of estrogen in their circulation to stimulate the growth of the vaginal epithelium. The vagina then loses its acidity, so that the growth of harmful bacteria is encouraged. The condition can be treated easily with estrogen creams or by the woman's inserting a slightly acidic jelly into the vagina to inhibit growth of the bacteria.

See also: Cells and chromosomes; Estrogen; Genitals; Hormones; Menopause

Valves

Although valve problems are no longer the most common kind of heart disease, many people still suffer from them. Because of advances in investigation and surgery, much can be done for patients who are affected.

When my father had his aortic valve replaced, he had surgery for his coronary arteries. Is this usual?

It's not uncommon. If there is significant coronary disease both operations can be performed at the same time. Disease of the coronary arteries is very common in the general population, and most people who are investigated for heart valve abnormalities will also have their coronary arteries checked. If abnormalities are discovered, then grafts to bypass the blockages in the coronary arteries are made in addition to putting in a new valve; this results in a better recovery rate.

What is a prolapsing mitral valve and is it a serious condition?

The mitral valve is located between the atrium and the ventricle on the left side of the heart. When this valve prolapses (becomes displaced), blood leaks back into the atrium while the ventricle is pumping. The condition affects 5 percent of women and 0.5 to 1 percent of men. It is not serious and is detected mainly by ultrasound.

I have aortic valve problems but I don't want to have surgery. However, my cardiologist seems anxious that I should. Why is this?

No doctor is going to force surgery on an unwilling person. However, replacement of the aortic valve is a safe procedure in all but the sickest of patients. The reason your cardiologist is so anxious is probably that your aortic valve is obstructed—a disease called aortic stenosis. Not only will the symptoms of the disease be relieved effectively by surgery but, more important, your life may be saved. Without surgery most patients with symptoms from aortic stenosis die within three years.

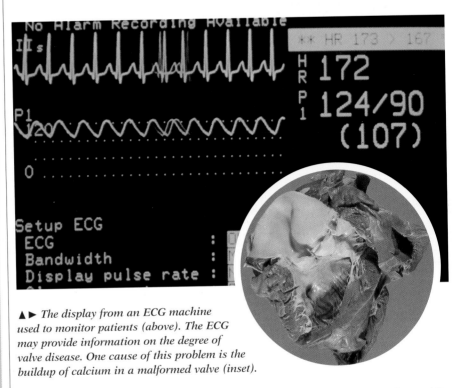

▲▶ *The display from an ECG machine used to monitor patients (above). The ECG may provide information on the degree of valve disease. One cause of this problem is the buildup of calcium in a malformed valve (inset).*

The heart is a muscular pump and its function is to maintain the circulation of blood around the body. The mechanism is similar to many pumps in that the heart depends on a series of valves to work properly. On the right-hand side are the pulmonary and tricuspid valves; on the left-hand side are the aortic and mitral valves. Changes in blood pressure on either side of the valves cause them to open and close. A closed valve prevents the blood from flowing in the wrong direction.

The pulmonary and aortic valves are similar in structure. They have three leaf-like cusps, or leaflets, and are made of thin fibrous tissue that is nonetheless very tough. The mitral and tricuspid valves are more complicated, though they are similar in structure. The mitral valve has two leaflets, while the tricuspid valve has three.

Each of these valves sits in a ring between the atrium and the ventricle. The bases of the leaflets are attached to the ring, and the free edges touch each other and close the passage between the ventricle and atrium when the valve is closed. These free edges are also attached to a series of fine strings called the chordae tendineae; these strings pass into the ventricle and stop the valve from springing back into the atrium when under pressure.

Problems

Only two things can go wrong with a valve. Either it can become partially blocked so that blood cannot pass easily through (stenosis), or it can allow blood to leak backward in the direction opposite to the normal circulation (incompetence or regurgitation).

In the past, rheumatic fever was the most common cause of valve disease. In cases of this disease, the inflammation affected the valves on the left side (aortic and mitral) almost exclusively, and subsequently could lead to stenosis or incompetence.

VALVES—VIEWED FROM ABOVE

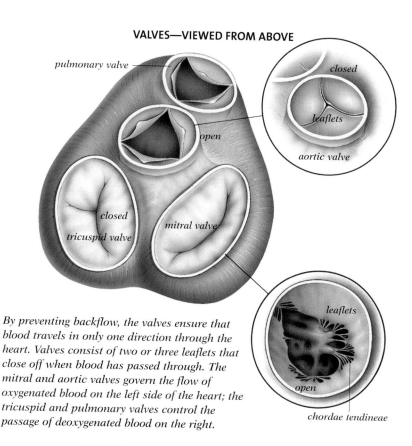

pulmonary valve

closed

leaflets

aortic valve

open

closed

tricuspid valve

mitral valve

leaflets

open

chordae tendineae

By preventing backflow, the valves ensure that blood travels in only one direction through the heart. Valves consist of two or three leaflets that close off when blood has passed through. The mitral and aortic valves govern the flow of oxygenated blood on the left side of the heart; the tricuspid and pulmonary valves control the passage of deoxygenated blood on the right.

POSITION OF VALVES—VIEWED FROM THE FRONT

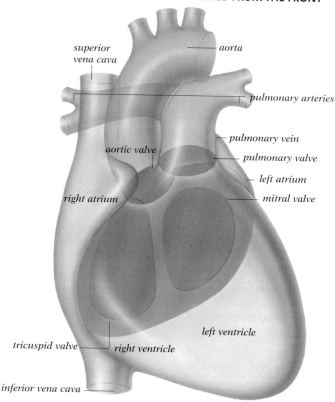

superior vena cava

aorta

pulmonary arteries

pulmonary vein

aortic valve

pulmonary valve

left atrium

mitral valve

right atrium

tricuspid valve

right ventricle

left ventricle

inferior vena cava

Congenital abnormalities are probably the most common forms of modern valve disease. Stenosis or incompetence can occur as a result of minor structural abnormalities of any of the valves; however, the aortic and pulmonary valves are most likely to be affected. Problems may not emerge until late in life when wear and tear starts to put strain on the valve. An abnormal valve may thicken and get deposits of calcium in it over the years, causing reduced efficiency. Abnormal valves also appear to be at risk of picking up germs from the bloodstream, which grow on the valve and start to destroy its substance. This gives rise to a potentially serious disease called infective endocarditis.

There may also be congenital problems that affect many parts of the heart and give rise to problems immediately after birth. One of the most common of these is Fallot's tetralogy: this involves a pulmonary valve obstruction, together with a hole between the two ventricles (ventricular septal defect).

Another abnormality that can cause minor problems is mitral valve prolapse. This happens because there is some slack in the valve and its chordae tendineae, which allows it to balloon back into the atrium, resulting in a leakage of blood.

Symptoms and treatment

Murmurs are usually picked up by a doctor using a stethoscope; aortic and mitral problems may come to light because of breathlessness caused by a buildup of fluid in the lungs. Thickening of the muscular wall of the heart, a result of the extra strain that a blocked or leaky valve puts on it, may lead to heart pain (angina). A final symptom that may lead to diagnosis of valve problems is a disturbance in the heart's rhythm as the orderly contraction of muscle breaks down.

Often the symptoms of valve problems can be controlled by pills, but in some cases surgery may be needed. To decide if this is necessary, a cardiologist will arrange for an electrocardiogram (ECG) and a chest X ray; these give a good idea of the level of strain the heart is under. An echocardiogram, which uses ultrasound to look at the heart, may be performed to picture the valves. Finally, a cardiac catheterization procedure may be carried out to measure the pressure in the chambers of the heart.

See also: **Blood pressure; Heart**

Veins

After giving up oxygen and nourishment to the tissues, blood is carried back to the heart by the veins. These specially designed channels play as important a part as the arteries in the efficient working of the circulation.

The vein from the leg is used to form a bypass for the blocked coronary arteries that cause angina. Veins are designed to carry blood at much lower pressure than arteries, but they can cope very well with the extra strain usually imposed on the arteries.

Yes, it is possible. If you are sitting still for a long time, then the rate of blood flow in the deep veins of the legs slows down. This is largely because the muscles in your legs, which are active when you are moving about, also help pump the blood back toward the heart; this pumping is lost when you sit down. There is also pressure from the seat on your thighs as you sit, which will tend to reduce the rate of blood flow. Once the rate of flow falls, the blood clots more easily. The danger of blood clots in the deep veins of the legs is that they can break off and pass through the heart to get stuck in the lungs. You should keep your feet moving when you are on a long flight, and make sure that you get out of your seat every hour or so if possible.

It is not a needle that is left in the arm; it is a thin plastic tube called a catheter. This catheter can remain in the arm for a few days. In intensive care units, longer and sometimes wider catheters are used, and these can be put into the big veins in the chest. These central lines are designed to last longer than the small catheters in the arm.

Veins are similar to arteries in their distribution; the arteries and veins associated with a particular organ or tissue often run together, but there are major differences. Many veins have valves, which the arteries do not, and the walls of an artery are always thicker than those of a vein of the same size, whereas the central channel, or lumen, will be much bigger in the vein than the artery.

Structure and function

Veins are tubes of muscular and fibrous tissue. Their walls have an outer layer, the tunica adventitia; a middle layer of muscle fiber, the tunica intermedia; and an inner lining, the tunica intima. They contain only a thin layer of muscle. After passing through the capillaries from the arteries, blood enters the venous system. It first passes into very small vessels called venules, which are the venous equivalent of arterioles. It then makes its way into small veins and back toward the heart along the veins that are large enough to be seen under the skin. Veins of this size contain valves that prevent blood from flowing back toward the tissues. The valves have little half-moon-shaped cups that project into the lumen of the vein, and these make the blood flow in only one direction.

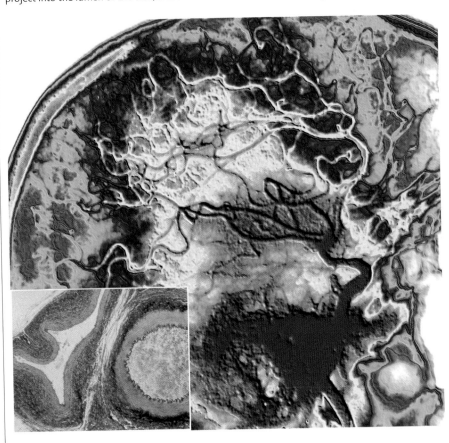

▲ *The carotid angiogram (above)—a type of X ray of blood vessels—graphically illustrates the network of veins carrying blood away from the head. The micrograph (inset) shows transverse sections of a vein and an artery; the vein is on the left.*

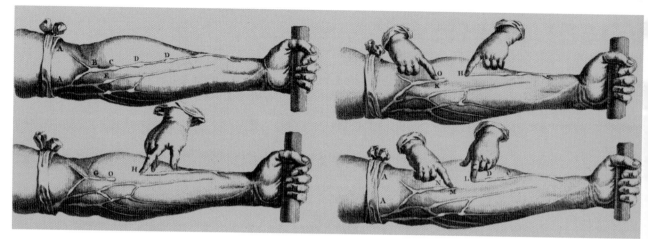

▲ *The English physician William Harvey, who proved that the blood circulates (1628), illustrated how blood in the arm flows continuously in one direction, controlled by valves in the veins.*

Eventually, blood flowing back to the heart enters one of two large veins; the inferior vena cava receives blood from the lower half of the body, and the superior vena cava receives blood from the head and arms. These vessels are about 1 inch (2–3 cm) wide, and they enter the right atrium of the heart. Blood passes from here into the right ventricle, and then into the lungs via the pulmonary arteries. It leaves the lungs by means of the pulmonary veins, which enter the left atrium of the heart.

Special types of vein

There is one area where the veins are arranged in a very different way from the arteries, and this is in the intestines. Here, instead of draining into a vein that passes straight into the heart, blood from the intestines is drained into what is called the hepatic portal system of veins. This allows the blood, which may be rich in digested food, to be carried directly to the liver.

Once blood from the intestines reaches the liver, it passes in among the liver cells, in special capillaries that are called sinusoids, and then enters the system of veins called the hepatic veins. These eventually lead on to the inferior vena cava, and thus into the heart. This system ensures that food passed into the venous system from the intestines is brought to the liver for chemical processing in the most efficient way.

Other areas where there are special kinds of venous structure are in the extremities—the hands, feet, ears, and nose. Here it is possible to find direct communications between the small arteries and veins,

VEIN NETWORKS

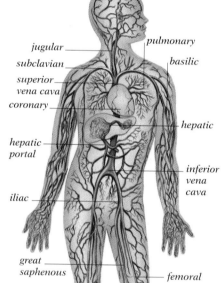

jugular
subclavian
superior vena cava
coronary
hepatic portal
iliac
great saphenous

pulmonary
basilic
hepatic
inferior vena cava
femoral
popliteal
tibial

◄ *The body's complex system of veins returns blood in the direction of the heart (as opposed to arteries, which carry it away) through the conduits called the venae cavae.*

where blood may flow through from one to the other without having to go through a system of capillaries in the tissues. The main function of these arteriovenous connections is related to the control of body temperature. When they are open, heat loss increases and the body cools down.

There are similar connections between the arteries and veins in the genital areas. These allow for the engorgement of blood that occurs in the genitals as a result of sexual excitement.

What can go wrong

One serious problem that affects veins is the tendency of blood clots to form in them. This is likely to happen because of the slowness with which blood flows along the veins, in contrast with the rapid flow maintained in the arteries. Smoking and the Pill increase the risk of clotting, although the low dose of estrogen in modern contraceptive pills has much reduced this danger.

Another major trouble is caused by the upright stance of human beings—this leads to pressure in the veins of the legs, since they are supporting a high column of blood. This can result in twisted, engorged veins in the legs called varicose veins. However, problems affecting the veins are minor compared with those arising from arterial disease.

See also: Arteries; Blood; Capillaries; Circulatory system; Genitals; Heart; Liver; Lungs; Temperature; Valves

Vocal cords

The vocal cords are two strong bands of tissue inside the larynx that vibrate as air exhaled from the lungs passes between them, producing the sound that is a basis for speech.

POSITION AND STRUCTURE OF THE VOCAL CORDS

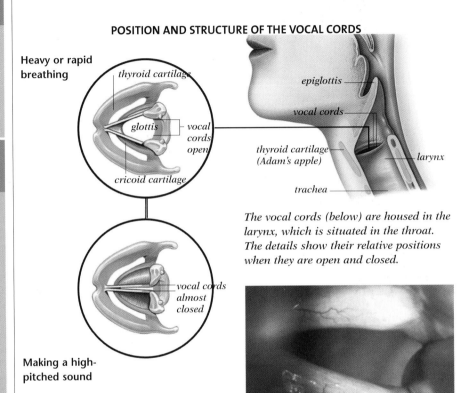

Heavy or rapid breathing — thyroid cartilage, glottis, vocal cords open, cricoid cartilage

Making a high-pitched sound — vocal cords almost closed

epiglottis, vocal cords, thyroid cartilage (Adam's apple), larynx, trachea

The vocal cords (below) are housed in the larynx, which is situated in the throat. The details show their relative positions when they are open and closed.

The vocal cords are similar to the reed in a wind instrument such as a bassoon. When a musician blows air between the reeds, the thin wood strips vibrate, producing the basic sound that is modified by the pipe length and holes of the instrument. Similarly, the vocal cords vibrate when someone vocalizes, and the sounds produced are modified by the throat, nose, and mouth.

The vocal cords are in the larynx. They consist of two delicate ligaments, shaped like lips, which open and close as air passes through them. One end is attached to a pair of movable cartilages called the arytenoids; the other is anchored to the thyroid cartilage, which is part of the Adam's apple. The arytenoid cartilages alter position so that the space between the cords (the rima) varies in shape from a narrow V during speech to a closed slit during swallowing.

The vibration of the vocal cords during speech occurs when the rima narrows and air from the lungs is expelled between the cords and through the larynx. The loudness of the voice is controlled by the force with which air is expelled, and the pitch by the length and tension of the cords. The depth and timbre of the voice are due to the shape and size of the throat, nose, and mouth. Men usually have a large larynx, longer vocal cords, and deeper voices than women.

Disorders of the vocal cords cause hoarseness due to inflammation of the larynx. It can occur as a result of cheering at a football game or a viral infection such as a cold. Laryngeal tumors may cause laryngitis: they are usually benign polyps, cysts, or growths known as singers' nodes because of constant rubbing of the vocal cords, and have to be surgically removed.

See also: Larynx; Neck; Throat

Vulva

Questions and Answers

Is soreness of the vulva a sign of a sexually transmitted disease?

No. There are many reasons why the vulva becomes inflamed and sore; but since it is the part of the body most intimately involved in sexual intercourse it is inevitable that sexually transmitted infection is a common cause. If you may have been in contact with infection, it is important that you go to your doctor or a clinic to make sure. Meanwhile, you should avoid further sexual intercourse.

I have developed small growths on my vulva. What could they be?

They are probably vulval or genital warts. This condition is a result of a viral infection and both men and women can be infected. It is often passed by sexual contact, though sometimes only one partner is affected. Because there are as yet no effective antiviral drugs for warts, treatment consists of regularly treating the warts with a caustic liquid such as podophyllin. These preparations are very powerful and have to be applied carefully and accurately, so the treatment has to be given by a doctor or nurse and cannot safely be self-administered.

Can I get cancer of the vulva?

Yes, cancer can develop on the vulva, but it is rare and usually confined to women between 50 and 70. It is often preceded by an area of thickening and itching on the labia, called leukoplakia. If you get any spots, lumps, or sores on your vulva, seek medical advice.

My doctor says I have pruritis vulvae. What does this mean?

It means irritation of the vulva. It is associated with vulvitis, but you should consult your doctor to have the basic cause diagnosed.

The word "vulva" describes the sexually sensitive outer region of the female reproductive system. This area is susceptible to a variety of infections that are collectively known as vulvitis.

Most prominent among the parts of the vulva are the two pairs of lips or labia. The outer and larger lips—labia majora—consist of thick folds of skin that cover and protect most of the other parts. They become thinner at the base and merge with the perineum (the skin over the area between the vulva and the anus). At the top the outer lips merge with the skin and hair on the pad of fatty tissue that covers the pubic bone, the mons pubis or mons veneris, which is often referred to as the "mound of Venus."

Within the labia majora are the labia minora or lesser lips. They join at the top to form a protective hood over the sensitive clitoris, dividing into folds that surround it. They also protect the opening to the urethra. The area between the labia minora is largely taken up by a space called the vestibule. Before a woman is sexually active, the space is mostly covered by the hymen. This varies in shape, size, and toughness, and although it is usually either torn or stretched during the first sexual intercourse, it may either be strong enough to make intercourse difficult or have been previously ruptured by strenuous exercise, masturbation, or tampons. The tags of skin that many women have around the vestibule are the remains of the

STRUCTURE OF THE VULVA

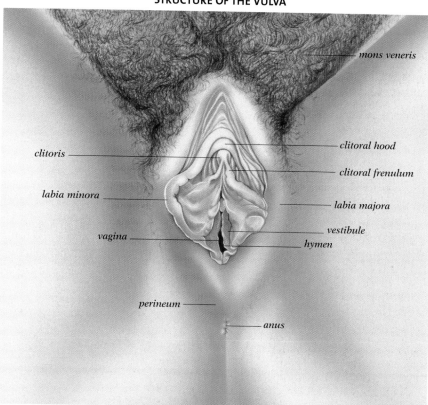

▲ *Situated at the entrance to the vagina, the vulva consists mainly of outer and inner lips called the labia. These folds of skin cover and protect the sensitive interior, including the main organ of sexual excitement—the clitoris.*

465

hymen, and are called the carunculae myrtiformes. At the back the labia minora join to form the fourchette, which is often ruptured during the first childbirth.

The clitoris and glands

The clitoris is actually similar in structure to the penis, even to the extent of having a hood of labia, the equivalent of the foreskin, and a small connecting band of tissue called the frenulum. It is primarily an organ of sexual excitement. It is extremely sensitive, and when stimulated its spongy tissue fills with blood and becomes erect. Friction on the erect clitoris—either by movement of the penis during intercourse or by some other means—will usually lead to orgasm. Other parts of the vulva also respond to sexual stimulation—the labia contain erectile tissue and often become enlarged during lovemaking; and the Bartholin's glands become active. Two pairs of glands are associated with the vulva. The first are Skene's glands, which lie just below the clitoris and secrete an alkaline fluid that reduces the natural acidity of the vagina. The other, larger pair lie in the bottom of the vestibule. These are Bartholin's glands and they secrete clear mucus when a woman is sexually aroused so that the entrance to the vagina becomes lubricated and can more easily accept the penis. These glands are normally about the size of a pea and are not prominent. They are prone, however, to venereal and other infections, becoming swollen, red, and tender. This condition, called Bartholinitis, requires treatment with antibiotics. In some cases, an abscess forms in one of the glands—a Bartholin's abscess—and may need to be incised to release the pus.

Vulvitis

Vulvitis is an inflammation of the vulva or of a part of it, the labia being the structures most often involved. Although vulvitis is mostly due to an infection, such as a yeast infection (monilia) or

▲ Wearing tight-fitting jeans occasionally is fine, but if they are worn every day they can trigger vulvitis.

trichomoniasis, it can also result from the friction of tight underwear or jeans, excessive rubbing or scratching, damage from stale urine or sweat, the chemical effects of vaginal deodorants, or allergy to some material or cosmetic preparation with which it comes in contact.

Vulvitis can be a complication of diabetes and obesity. Senile vulvitis among the elderly is a result of decreased hormone levels. Currently on the increase, and a cause of painful vulvar ulcers, is infection from the herpes simplex virus.

Symptoms and treatment

Irrespective of cause, the symptoms of vulvitis are basically the same. The skin becomes red, sore, and itchy, and there may be some swelling. If there is a great deal of irritation, known as pruritus vulvae, scratching may make the problem worse, causing the labia to become more sore and inflamed.

In herpes, small blisters develop; these burst, leaving sore, tender ulcers that allow bacteria to penetrate the skin and cause further infection.

Treatment of vulvitis depends on the cause. It is usually necessary to consult a doctor or go to a public health clinic for diagnosis and appropriate treatment. The symptoms can be relieved by wearing loose underwear or none at all, and scrupulously avoiding scratching. Also, fragranced talcum powder should be avoided, since this can exacerbate the condition.

See also: **Birth; Genitals; Glands; Hormones; Hymen**

Wrinkles

Questions and Answers

Why do some people get wrinkles earlier than others?

Prolonged exposure to the sun is one of the causes of wrinkling. People who work outdoors tend to get wrinkles earlier than those who work indoors. Skin pigmentation gives protection; people with darker skins are less prone to wrinkling than fair-skinned people. Other factors that influence the early development of wrinkles are a poor diet, ill health, smoking, and poor skin care.

Will doing facial exercises delay the onset of wrinkles?

Almost certainly not. Some people believe that facial exercises stimulate muscle growth, tighten the skin, and prevent wrinkles. Others claim that the overuse of facial expressions accelerates wrinkling, since a pattern of facial lines will be fixed. In fact there are no muscles in the skin, so the effect of strengthening the facial muscles lying under the skin would be to stretch it a little.

Does skin cream containing collagen prevent wrinkles?

Collagen is a component of healthy skin, but wrinkling is believed to be caused by changes in skin structure. There is little scientific basis for the claim that collagen can rejuvenate skin. Creams with collagen may delay the appearance of wrinkles, but the effect may be due to the moisturizing properties of the cream rather than its collagen.

Can heavy bags under the eyes be surgically removed?

Yes. This operation is called a blepharoplasty and is like a face lift. Excess skin and fat are removed, and the remaining skin stretched and restitched.

Like graying hair, the appearance of a few wrinkles is a classic sign of aging. How do they come about, and can anything be done to avoid or put off this universal, and seemingly inevitable, phenomenon?

Skin wrinkling is usually apparent in anyone over 40 years of age, but it often occurs much earlier. Although wrinkles can give the face character, many people feel they are something to be avoided or postponed.

Causes

The most important underlying change that brings about wrinkling involves the connective tissues just under the outer layer of the skin, which are made up of two types of protein fiber: collagen and elastin. The collagen provides the matrix for the tissue, and the smaller number of elastin fibers give elasticity and suppleness. With time, however, the amount of elastin diminishes and the collagen fibers become disorganized, cross-linking and enmeshing with each other. As a result, the tissue gradually loses elasticity. A general thinning and drying out of the skin is part of the process of aging, and this also predisposes to wrinkling.

▲ *This Algerian woman is deeply wrinkled, not only from age but also because she has spent her life in a country where the sun is intensely strong for most of the year.*

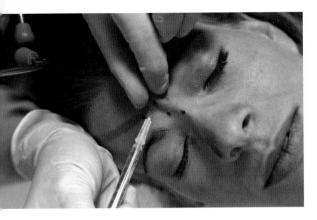

▲ *Botox injections temporarily plump out wrinkled skin for three to eight months. To achieve a smooth appearance, the injection paralyzes the muscles under the skin.*

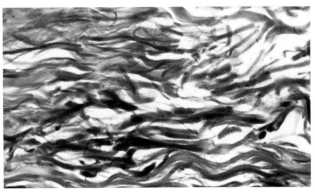

▲ *The breakdown that occurs in the skin's supportive tissue is a basic cause of wrinkling. With time, the elastin fibers (blue) diminish and the collagen fibers (pink) become loose and disorganized, resulting in loss of firmness and elasticity.*

Sunlight affects the skin, and accounts for wrinkling on exposed areas of the body, particularly the face, neck, and backs of the hands. The ultraviolet component of sunlight accelerates the chemical changes in skin that cause wrinkling, and its effects are pronounced in fair-skinned people who have little natural protection against the sun's rays.

In the same environment, the difference between skin exposed to solar radiation and skin protected from sun is strikingly illustrated in cloistered nuns living in tropical countries. Doctors have been struck by the absence of wrinkling and other effects of sunlight on the skin of women whose dress and way of life have provided lifelong avoidance of direct sunlight.

Ultraviolet radiation has a direct effect on the DNA in the skin cells. The damage to DNA is normally repaired by natural processes of DNA repair and replication, but these are not capable of repairing all the damage, and some DNA remains abnormal. This effect is cumulative, and eventually abnormal skin is the result. This abnormality is mainly manifested in wrinkles, but it can also appear as several forms of skin cancer, including rodent ulcers.

Facial expression can influence wrinkles. When someone smiles, grimaces, or frowns, furrows are formed in the skin according to which facial muscles are being contracted. Over the years these facial lines become ingrained so that they are visible all the time.

A further influence on the pattern of wrinkling is the amount and distribution of subcutaneous fat deposits in and around the face. Excess fat deposits tend to be drawn downward by gravity to form bags under the eyes, double chins, and heavy jowls. In older people, jowl creases may extend up to the cheeks, and tend to run perpendicular to the lines of facial expression, producing a crosshatch pattern of wrinkles. Finally, smoking speeds up the process of wrinkling, perhaps because smokers screw up their faces to prevent smoke from irritating their eyes and this grimacing accentuates the lines of expression. Also, cigarette smoke contains at least 3,000 different chemical substances, many of which are absorbed into the bloodstream and carried to every part of the body, including the skin, where they may damage the proteins of the skin.

The effects of aging cannot be put off forever, but the onset and development of wrinkles can be delayed. Perhaps the most important protective measure is to avoid prolonged exposure to the sun and to use sunscreens with a high sun protection factor, which help to prevent the type of skin damage that promotes wrinkling. Dry skin has a greater tendency to wrinkle than oily skin, so using a moisturizing cream is a worthwhile preventive measure. Avoiding or cutting down on smoking may also be beneficial.

For many women, wrinkles become a particular problem around menopause, and this may be due to hormonal changes. Hormone replacement therapy may slow down wrinkling, but this is not yet proved and this therapy has other disadvantages to health.

There are a number of antiwrinkle creams available that vary in effectiveness. They cannot remove wrinkles but can give temporary camouflage. Some work by moisturizing and plumping up the skin; others fill in the wrinkles. The implication that collagen can be restored by skin creams should be viewed with skepticism.

Surgical treatment

Once wrinkles are established, surgical treatment is probably the only effective way of removing or reducing them, either by a face-lift or by chemical abrasion. Face-lifts involve making incisions at the borders of the face, stretching the skin upward and outward, and restitching. The initial results are good, but wrinkling may recommence a year or two after surgery.

Cosmetic procedures to remove redundant stretched skin are not without risk. There have been cases in which too much redundant eyelid skin has been removed, with the result that the eyes have been unable to close. This quickly leads to severe exposure damage to the peripheral parts of the corneas.

In the method of chemical abrasion, a caustic gel is spread over the face and neck to break down the outer layers of skin, which are then removed with the gel. The new skin that grows in its place is usually considerably less wrinkled, but the procedure can be rather painful. Healing may take several weeks, and sometimes the skin is left looking patchily discolored.

Skin that has been treated in this way is extremely sensitive to sunlight for several weeks, and it is important to protect it and avoid ambient sunlight during this time.

See also: Aging; Melanin; Skin

Wrist

Composed of eight bones and surrounded by tendons, the wrist is very flexible and surprisingly trouble-free. Like all bones, the wrist can break, but the most painful problems occur when the tendons become inflamed.

The pulse at your wrist (the radial pulse) belongs to the radial artery. The artery may be in a slightly different position in all wrists. Your pulse may be deeper than usual or it may lie slightly to the right or left. In either case, it does not signify anything of importance, although your doctor might appreciate a warning that your pulse is difficult to find.

My wrists have always been weak. How can I strengthen them?

Why not take up a sport such as tennis or squash? Any activity in which you have to grip a racket or a club could strengthen the wrists and make them more flexible. It will have the added advantage of improving your general fitness and health at the same time.

My daughter is an avid gymnast and is always doing handstands. Could she damage her wrists?

Wrist injuries are always a risk in sports such as gymnastics, but your daughter's wrists are more likely to be strengthened than hurt. Rarely, deformities of the wrist occur as a result of exercise, but generally this happens only if the bones are already diseased or weakened.

I broke my wrist and have a cast on my arm. Can I still drive?

No. If you drive during the first few weeks you may stop the bones from mending properly. More important, you should also remember that with a cast on your arm you are a potential hazard on the road. The flexibility of your wrist will be reduced and you could be a danger both to yourself and to others.

Each wrist is actually a complex of numerous joints between lots of little bones. This gives the joint great flexibility but makes it a potential weak spot. It is strengthened by a web of ligaments and tendons that link the bones and make lifting possible.

The structure of the wrist

The wrist is made up of eight separate bones called carpals. They are like small pebbles arranged in two rows and bound together by about twenty ligaments and tendons. The carpals sit between the metacarpals of the hand and the long bones of the arm.

The bones in the row nearest the arm, which run from the thumb to the little finger, are the scaphoid, lunate, triquetral, and pisiform. The second row consists of the trapezium, trapezoid, capitate, and hamate. The only one of these bones that is visible on the skin surface is the pisiform, which can be seen as the bumpy wrist bone.

The ligaments covering the wrist bones form a tunnel, the carpal tunnel, which prevents the long muscle tendons from springing away from the bones when the wrist is bent.

▲ *To be able to perform this action, the ligaments in the wrist must be strong enough to support the bones in the wrist.*

Movement of the wrist

The carpal joints are relatively immobile, although as a unit the wrist is very flexible indeed. The exception is the joint between the trapezium and the thumb bone. This type of joint makes it possible to grasp an object between the finger and thumb. An opposable thumb makes humans particularly adept at using tools.

Anatomically, the wrist joint is described as ellipsoid. This means that, although it enables up and down actions, side to side actions, and some circular movement, it cannot rotate like the hip and shoulder joints. This limitation helps to ensure the stability of the wrist joint. However, it is thought that the wrist joint is properly stable only when the tendons, ligaments, and muscles are acting to keep all the components of the joint in the right place. This tension is necessary, even when the body is completely at rest.

Such a fragile joint is clearly easy to damage. Lots of people will have experienced a slight sprain or strain and noticed how much it affects manual manipulation—every tiny action hurts the damaged joints.

Fractures and dislocations

Of all the injuries that involve the wrist, the most common is a break at the lower end of the radius—one of the two long bones in the forearm. This is called a Colles fracture, and is

treated by manipulation to reset the bones and immobilization in a cast.

The small carpal bones can also suffer hairline cracks. Sometimes there is a swelling on the back of the hand, just below the thumb, but often there are no external signs, only pain and stiffness in the joint. Often even X rays do not show these small cracks, but they can still be a problem. A hairline crack can separate a small portion of bone from its blood supply and this can result in bone death. If there is any suspicion that this is happening, the doctor will immediately immobilize the wrist joint to prevent any more damage.

The bones of the wrist can also become dislocated if banged or moved awkwardly, especially the lunate bone in the center of the wrist and the triquetral below the little finger. The dislocation shows up as a bulge on the outside of the wrist and should be manipulated into position by a doctor as soon as possible.

Problems with the tendons

The most common problem to afflict the wrist is called carpal tunnel syndrome. The fibrous carpal tunnel encloses all the wrist bending tendons and one of the main nerves supplying the hand, the median nerve. If the fibers in the tunnel become swollen or compressed they press the nerve against the wrist bones, causing pain.

The syndrome can be caused by simple overuse of the thumb and fingers. It is also common in late pregnancy when edema (swelling) can put pressure on the median nerve. It is generally more common in women than in men.

Whatever the cause, carpal tunnel syndrome begins with a sensation of pins and needles or numbness, especially in the thumb and next two fingers. The wrist may swell up near the thumb, and the forearm and thumb are often very painful. Usually the symptoms will gradually ease as the swelling or pressure is reduced. Occasionally, however, the syndrome may be persistent or recurrent, in which case surgery may be required to effect a permanent cure.

Another problem is tenosynovitis, in which the lubricated tendon sheaths become inflamed as a result of a bacterial infection or rheumatoid arthritis. It becomes difficult and painful to uncurl the fingers, and movement may result in audible grating noises. The fingers and thumb may also feel numb as if they have permanent pins and needles. A doctor may prescribe antibiotics for bacterial infection and aspirin to relieve pain.

▲ ▼ The eight wrist bones are strengthened by ligaments and are at their most vulnerable when having to bear the brunt of the body's weight.

STRUCTURE OF THE WRIST

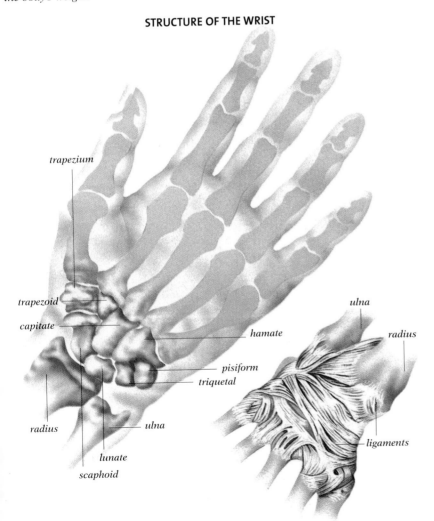

trapezium

trapezoid

capitate

hamate

pisiform

triquetal

radius

ulna

lunate

scaphoid

ulna

radius

ligaments

See also: **Hand; Joints; Ligaments; Nervous system; Pregnancy; Tendons**

FURTHER RESEARCH

BIBLIOGRAPHY

Clark, William R. *In Defense of Self: How the Immune System Really Works.* New York: Oxford University Press, 2008.

Cohen, Barbara Janson, and Jason James Taylor. *Memmler's Structure and Function of the Human Body,* 9th edition. Baltimore, MD: Lippincott, 2009.

Davies, Juanita. *Essentials of Medical Terminology,* 3rd edition. Clifton Park, NY: Delmar, 2008.

Heffner, Linda J., and Danny J. Schust. *The Reproductive System at a Glance,* 3rd edition. Hoboken, NJ: Blackwell, 2010.

Marieb, Elaine N., and Katja Hoehn. *Human Anatomy and Physiology,* 8th edition. San Francisco: Benjamin Cummings, 2010.

Smith, Margaret E., and Dion G. Morton. *The Digestive System: Basic Science and Clinical,* 2nd edition. New York: Elsevier, 2010.

Standring, Susan, ed. *Gray's Anatomy,* 40th edition. New York: Elsevier, 2008.

Watkins, James. *Structure and Function of the Musculoskeletal System,* 2nd edition. Champaign, IL: Human Kinetics, 2010.

West, John B. *Respiratory Physiology: The Essentials,* 8th edition. Philadelphia, PA: Lippincott, 2008.

Whittemore, Susan. *The Circulatory System.* New York: Chelsea House, 2009.

Wolsey, Thomas A. A., Joseph Hanaway, and Mokhtar H. Gado. *Brain Atlas: A Visual Guide to the Human Central Nervous System.* Indianapolis, IN: Wiley, 2002.

Zelman, Mark, et al. *Human Diseases: A Systemic Approach.* Upper Saddle River, NJ: Prentice-Hall, 2010.

INTERNET RESOURCES

Accident Prevention Corporation
www.safetyman.com

Action on Smoking and Health
www.ash.org

Addiction Search
www.addictionsearch.com

Administration on Aging
www.aoa.gov

Aerobics and Fitness Association of America
www.afaa.com

Alexander Graham Bell Association for the Deaf and Hard of Hearing
http://nc.agbell.org

Alternative Medicine Foundation
www.amfoundation.org

Alzheimer's Association
www.alz.org

American Academy of Allergy, Asthma, and Immunology
www.aaaai.org

American Academy of Dermatology
www.aad.org

American Academy of Facial Plastic and Reconstructive Surgery
www.aafprs.org

American Academy of Family Physicians
http://familydoctor.org

American Academy of Pediatrics
www.aap.org

American Alternative Medicine Association
www.joinaama.com

American Association of Colleges of Osteopathic Medicine
www.aacom.org

American Cancer Society
www.cancer.org

American Chiropractic Association
www.amerchiro.org

American College of Rheumatology
www.rheumatology.org

American College of Sports Medicine
www.acsm.org

American Council on Science and Health
www.acsh.org

American Dietetic Association
www.eatright.org

American Foundation for the Blind
www.afb.org

American Geriatrics Society
www.americangeriatrics.org

American Health Care Association
www.ahcancal.org

American Heart Association
www.heart.org

American Liver Foundation
www.liverfoundation.org

American Lung Association
www.lungusa.org

American Osteopathic Association
www.osteopathic.org

American Pain Society
www.ampainsoc.org

American Podiatric Medical Association
www.apma.org

American Prostate Society
www.americanprostatesociety.com

American Psychiatric Association
www.psych.org

American Psychological Association
www.apa.org

American Red Cross
www.redcross.org

American Thyroid Association
www.thyroid.org

American Urological Association
www.auanet.org

Arthritis Foundation
www.arthritis.org

Birth Defect Research for Children
www.birthdefects.org

Body Positive
www.bodypositive.com

Brain Injury Association of America
www.biausa.org

Centers for Disease Control and Prevention
www.cdc.gov

Child Development Institute
www.childdevelopmentinstitute.org

Council for Responsible Nutrition
www.crnusa.org

Deafness Research Foundation
www.drf.org

Eating Disorder Referral and Information
Center
www.edreferral.com

Genetic Alliance
www.geneticalliance.org

Harvard School of Public Health
www.hsph.harvard.edu

HealthCentral
www.healthcentral.com

International Chiropractors Association
www.chiropractic.org

KidsHealth
http://kidshealth.org

MayoClinic
www.mayoclinic.com

MedHelp
www.medhelp.org

MedlinePlus
www.nlm.nih.gov/medlineplus

National Association of the Deaf
www.nad.org

National Campaign to Prevent Teen and
Unplanned Pregnancy
www.thenationalcampaign.org

National Council on the Aging
www.ncoa.org

National Institute of Allergy and Infectious
Diseases
www.niaid.nih.gov

National Institute of Arthritis and
Musculoskeletal and Skin Diseases
www.niams.nih.gov

National Institute on Aging
www.nia.nih.gov

National Institutes of Health
www.nih.gov

National Kidney Foundation
www.kidney.org

National Library of Medicine
www.nlm.nih.gov

National Spinal Cord Injury Association
www.spinalcord.org

National Stroke Association
www.stroke.org

National Women's Health Information
Center
www.womenshealth.gov

Nutrition Source
www.hsph.harvard.edu/nutritionsource

Planned Parenthood
www.plannedparenthood.org

Prevent Blindness America
www.preventblindness.org

Sports Medicine on the Web
www.sportsmedicine.com

TeenGrowth
www.teengrowth.com

Women's Health Resource
www.imaginis.com

HOTLINES

Alzheimer's Association
800-621-0379

American Academy of Family Physicians
800-274-2237

American Pregnancy Hotline
888-672-2296

American Trauma Society
800-556-7890

Mental Health InfoSource
800-447-4474

National Cancer Institute
800-4-CANCER (800-422-6237)

National Drug Abuse Hotline
800-662-HELP (800-662-4357)

National Eating Disorders Association
800-931-2237

National Health Information Center
800-336-4797

National Sexually Transmitted
Diseases Hotline
800-227-8922

National Youth Crisis Hotline
800-448-4663

Smokers' Helpline
800-NO-BUTTS (800-662-8887)

TalkZone (Peer Counselors)
800-475-TALK (800-475-2855)

INDEX

PICTURE CREDITS